T0299178

THE LIVES AND DEATH-THROES
OF MASSIVE STARS

IAU SYMPOSIUM 329

COVER ILLUSTRATION: R136 in the 30 Doradus Nebula in the LMC.

By NASA, ESA, F. Paresce (INAF-IASF, Bologna, Italy), R. O'Connell (University of Virginia, Charlottesville), and the Wide Field Camera 3 Science Oversight Committee,
http://hubblesite.org/newscenter/archive/releases/2009/32/image/a/
Public Domain:
https://commons.wikimedia.org/w/index.php?curid=8819141.

IAU SYMPOSIUM PROCEEDINGS SERIES

Chief Editor
PIERO BENVENUTI, IAU General Secretary
IAU-UAI Secretariat
98-bis Blvd Arago
F-75014 Paris
France
iau-general.secretary@iap.fr

Editor
MARIA TERESA LAGO, IAU Assistant General Secretary
Universidade do Porto
Centro de Astrofísica
Rua das Estrelas
4150-762 Porto
Portugal
mtlago@astro.up.pt

INTERNATIONAL ASTRONOMICAL UNION

UNION ASTRONOMIQUE INTERNATIONALE

THE LIVES AND DEATH-THROES OF MASSIVE STARS

PROCEEDINGS OF THE 329th SYMPOSIUM OF THE INTERNATIONAL ASTRONOMICAL UNION HELD IN AUCKLAND, NEW ZEALAND NOVEMBER 28–DECEMBER 2, 2016

Edited by

J.J. ELDRIDGE
Department of Physics, University of Auckland

JOHN C. BRAY
Department of Physics, University of Auckland

LIAM A.S. McCLELLAND
Department of Physics, University of Auckland

and

LIN XIAO
Department of Physics, University of Auckland

Shaftesbury Road, Cambridge CB2 8EA, United Kingdom

One Liberty Plaza, 20th Floor, New York, NY 10006, USA

477 Williamstown Road, Port Melbourne, VIC 3207, Australia

314–321, 3rd Floor, Plot 3, Splendor Forum, Jasola District Centre, New Delhi – 110025, India

103 Penang Road, #05–06/07, Visioncrest Commercial, Singapore 238467

Cambridge University Press is part of Cambridge University Press & Assessment, a department of the University of Cambridge.

We share the University's mission to contribute to society through the pursuit of education, learning and research at the highest international levels of excellence.

www.cambridge.org
Information on this title: www.cambridge.org/9781107170063

First published 2017

A catalogue record for this publication is available from the British Library

ISBN 978-1-107-17006-3 Hardback
ISSN 1743-9213

Table of Contents

Session 3: Observations and surveys of massive stars: hot stars, cool stars, transition objects and binaries

Session 4: Theory of stellar evolution & atmospheres: beyond standard physics, rotation, duplicity, mass loss and magnetic fields and instabilities

Session 5: Massive stars and their supernovae as galactic building blocks and engines: Milky Way, nearby galaxies and the early Universe

Session 6: Conference Summary

Splinter Session: X-ray observations of massive stars

Posters

Preface

Research on Massive Stars is undergoing a period of rapid progress. While these stars are relatively few in number they are the main driver of chemical and dynamical evolution in galaxies via their stellar winds and explosive deaths in core-collapse supernovae. Our understanding of massive stars is going through a remarkable time of change with long held convictions being shown to be incomplete. This evidence arises from new research concerning the formation and evolution of massive stars and linking this to their deaths in core-collapse supernovae. Now is a fortuitous time to make significant advances in massive star research. This meeting was proposed with the central rationale to bring together the two communities that study massive stars and their supernovae.

The impact of massive stars is widely recognized. They are used as tools to interpret the conditions and processes arising in different environments (studies of Galactic structure, chemical and dynamical feedback, population synthesis, Starbursts, high-z galaxies and cosmic reionization). In parallel, the development of new instrumentation, analysis techniques and dedicated surveys across all possible wavelengths have delivered large amounts of exquisite new data. This data is now providing a harsh test for the current state-of-the-art theoretical calculations of massive star birth, evolution and death.

We are beginning to gain some measure of success understanding how complex phenomena such as magnetic fields, pulsations, rotation, mergers and multiplicity act within massive stars. This enables us to revolutionize our understanding of short-lived and enigmatic phases such as seen in Wolf-Rayet stars, Red Supergiants, the Luminous Blue Variables and B-Supergiants. But at the same time, mysteries persist surrounding these phases and the supernovae produced by these stars. For example there is growing evidence that all these stars, except the Wolf-Rayet stars, give rise to supernovae.

Finally, while we know individual stars are important, the impact of massive star populations via their evolution and death, including the influence of X-ray and gamma-ray binaries, is of high interest to those studying the high-z Universe. Locating the source of photons needed to reionize the early Universe remains unsolved. Uncertainties in our understanding of massive star populations impacts our interpretation of galaxies at the edge of the observable Universe and how the Universe became transparent.

In view of recent developments and the significant impact massive stars have in the broader community, a new IAU Symposium in late 2016 was proposed. The meeting summarized recent progress and established stronger links between the massive star community and closely-linked fields, particularly those studying end stages of massive star evolution and massive star cosmic implications.

The massive star community has traditionally held IAU Symposia with a frequency of 4-5 years (Argentina, 1971; Canada, 1978; Mexico, 1981; Greece, 1985; Indonesia, 1990; Italy, 1994; Mexico, 1998; Spain, 2002; USA, 2007). More recently, a last meeting was held in Greece in June 2013, without IAU sponsorship, but with great success (225 participants from 27 countries). Seeking for a long-term geographical balance, the IAU MSWG selected New Zealand as location for this meeting among a total of seven proposals. We hoped that the selection of this venue will allow greater participation by countries in the Asia and Pacific area.

The meeting covered 4 broad areas related to massive stars and included one splinter session. The topics were:

(*a*) Massive stars deaths in core-collapse supernova and other events.
(*b*) Observations of massive stars.
(*c*) Theoretical modelling of massive stars.

(*d*) Massive stars as building blocks of galaxies through the history of the Universe.
(*e*) Splinter session: X-ray observations of massive stars

During the week the linkage between these sessions was strong with several talks being able to be placed in any session. This reflects the growing collaborative nature of the study of massive stars. Formal discussion during the week was kept to a minimum. However long breaks were worked into the schedule to encourage informal discussion during the week. This was also further encouraged by setting the room up in a cabaret style with round tables rather than the normal lecture room format. Thus allowing people to form round table discussion groups. There was an extended discussion on Thursday run by Nathan Smith and Jose Groh concerning the best observations to test our theoretical models. In some ways bringing together all the topics of the week so far.

In addition to IAU sponsorship for travel grants it was also sponsored by the Department of Physics from the University of Auckland as well as CAASTRO (ARC Centre of Excellence for all-sky astrophysics). We also acknowledge the work of the LOC in organizing all the numerous details required to make such a symposium a success. In particular we thank Aimee Crawshaw for her tireless effort and dedication to making sure the meeting was a success.

There were two conference excursions, one to the Auckland Domain museum in Auckland the other a trip to Rangitoto island. A volcano in the bay of Auckland that erupted some 600 to 800 years ago. Surprisingly (to the tour organizer) 150 of the attendees went on the latter trip. Many enjoyed the race to the summit, exploring the lava caves afterwards and investigating a Kiwi bach (holiday home). The tour operator who organized the trips also made themselves available during the week to answer questions from attendees to plan their future travel in New Zealand after the conference. On the day after the meeting several conference attendees contributed to a public symposium. Where the highlights of the meeting were discussed along with varied subjects from ancient astronomy, the structure of the Galaxy and science-fiction comic books.

At the end of the meeting and afterwards we received great positive feedback on our organization of the meeting with the worst points being around trouble with the AV systems. Students and early-career researcher seems to be those who enjoyed the meeting the most and we hope that many productive collaborations will arise due to the meeting.

J.J. Eldridge, Margaret Hanson and Artemio Herrero, co-chairs SOC.

Equity Report

We are also particularly proud that our meeting from the onset worked hard to achieve an equitable meeting. Key to this was setting up an balanced SOC with 7 men, 7 women and 1 person of diverse gender. Also in selecting invited and contributed abstracts members were reminded to think about achieving an equitable balance at the meeting our efforts appeared to pay off. We also had the impression that the population of junior scientists, students and post-docs, was higher than usual at an international conference. 21% of the talks were given by PhD students with two-thirds of these being women. This was another indication of our positive approach.

For all talks 68% were by men and 32% by women. This was representative of the number of abstracts submitted and similar to the balance of attendees (Men 68.5%, Women 30.9%, Diverse 0.6%). The first day we were able to achieve equity with the ratio at 48% to 52%. We list the relevant equity statistics below in two tables.

The meeting also had a code of conduct to make the meeting spaces safe, inclusive and free of harassment. In the shortest form it said,

> *NZstars2016 is dedicated to providing an equitable and harassment-free conference experience for everyone regardless of gender, gender identity and expression, sexual orientation, disability, physical appearance, body size, race, age or religion. We do not tolerate harassment of conference participants in any form. Conference participants violating these rules may be sanctioned at the discretion of the conference organizers*

A reporting proceeding was explained on the first day as well as identifying those on the "equity committee" who were responsible for enforcing the code. The experiment worked well but in future we recommend that the process to deal with breaches of the code be outlined before the meeting.

We also recorded further equity information such as the gender of those who asked questions. We are still analyzing this data but of the 160 questions asked 74% were asked by men, 17% were asked by women and 9% by asked by people of diverse gender. We should note that in all the above statistics we are using the presumed gender. We only know of one gender diverse person who attended and asked those questions.

Our treatment of equity at the meeting also went beyond the talks and discussion. During the conference dinner for example we worked hard to provide entertainment that did not appropriate local culture and continue the safe inclusive environment. The result of this was to use a conference quiz (based on the "pub quiz" format common in the UK, Australia and New Zealand).

Finally we note that the aim of future meetings should build upon the work of the IAUS329 organizing committees to make future meetings more inclusive and equitable. We summarize our list of recommendations for future meetings as follows:

(*a*) Have an equity committee which at least is balanced in gender with representatives from all career stages.

(*b*) Publicize consequences of breaking the code of conduct in advance and have a clear process for dealing with issues.

(*c*) From the outset select and task the SOC from the onset with creating a speaker list that is diverse in terms of gender, race, nationality, and career stage.

(*d*) Ask participants for their gender and other demographic information at registration so that these can be analyzed as much as possible without assumptions.

(*e*) Take steps to encourage more people to ask questions. For example, highlight that asking questions is a skill to practice. Or have prizes for first student question.

Group	Number by assumed gender (F/M/D)	Fraction in % by assumed gender (F/M/D)
SOC	7:7:1	47%:47%:6%
LOC	3:2:1	50%:33%:17%
EC	2:2:1	40%:40%:20%
Invited speakers	7:21:0	25%:75%:0%
Contributed speakers	16:27:0	37%:63$:0%
Splinter session speakers	2:6:0	25%:75%:0%
Student prizes	4:0:0	100%:0%:0%
Chairs	6:2:1	67%:22%:11%
Question askers	27:119:14	17%:74%:9%
Participants	55:122:1	30.9%:68.5%:0.6%
Submitted abstracts	81:168:1	32%:67%:0.4%

Country	Number of Attendees
Argentina	1
Australia	4
Austria	3
Belgium	5
Brazil	3
Bulgaria	1
Canada	8
Chile	13
China	2
Czech Republic	4
Finland	1
France	5
Georgia	2
Germany	15
Ireland	1
Israel	1
Japan	11
Malaysia	1
Mexico	4
New Zealand	10
Russia	1
South Korea	2
Spain	13
Sweden	1
Switzerland	2
Taiwan (R.O.C.)	1
The Netherlands	3
UK	17
U.S.A.	43

THE ORGANIZING COMMITTEE

Scientific

J. Anderson (Chile)
B. Davies (UK)
J.J. Eldridge (co-chair, New Zealand)
A. Herrero (co-chair, Spain)
C. Neiner (France)
L. Oskinova (Germany)
N. St-Louis (Canada)
S.-C. Yoon (Korea, Rep of)
M. Cantiello (USA)
S. Ekström (Switzerland)
M. Hanson (co-chair,USA)
D.J. Hillier (USA)
M.-F. Nieva (Austria)
A. Soderberg (USA)
J.S. Vink (UK)

Local

A. Crawshaw
J.J. Eldridge (chair)
N. Rodrigues
J.C. Bray
L.A.S. McClelland
L. Xiao

Equity

J.J. Eldridge
J.L. Hoffman
L.A.S. McClelland
M. Hanson
P. Massey
L. Xiao

Acknowledgements

The symposium was sponsored and supported by IAU Divisions C (Education, outreach and Heritage), D (High Energy Phenomena andFundamental Physics), G (Stars and Stellar Physics), H (Interstellar Matter and Local Universe) and J (Galaxies and Cosmology), as well as Commissions C26 (Double and Multiple Stars), C29 (Stellar Spectra), C33 (Structure and Dynamics of the Galactic-System), C35 (Stellar Constitution), C36(Theory of Stellar Atmospheres) and C47(Cosmology).

The Local Organizing Committee operated under the auspices of the
Department of Physics, University of Auckland.

Funding by the
International Astronomical Union,
Department of Physics, University of Auckland,
ARC Centre of Excellence for All-Sky Astrophysics (CAASTRO),
is gratefully acknowledged.

CONFERENCE PHOTOGRAPH

Photograph by Gantcho Gantchev

Participants

Michael **Abdul-Masih**, Institute Of Astrophysics, Ku Leuven, Belgium mabdul02@villanova.edu
Claudia **Agliozzo**, Departamento De Ciencias Fisicas, Universidad Andres Bello, Chile c.agliozzo@gmail.com
Robert **Andrassy**, Physics and Astronomy, University of Victoria, British Columbia, Canada andrassy@uvic.ca
Jennifer **Andrews**, Astronomy, University Of Arizona, United States jandrews@as.arizona.edu
Ignacio **Araya**, Instituto De Física Y Astronomía, Universidad De Valparaíso, Valparaíso, Chile ignacio.araya@uv.cl
Catalina **Arcos**, Instituto De Física Y Astronomía, Universidad De Valparaíso, Chile catalina.arcos@uv.cl
Kyle **Augustson**, CEA-Irfu Service D'astrophysique, France kyle.augustson@cea.fr
Dietrich **Baade**, European Southern Observatory, Germany dbaade@eso.org
Rodolfo **Barba**, Departamento De Fisica Y Astronomia, Universidad De La Serena, Chile rbarba@userena.cl
Christopher **Bard**, NASA, Maryland, United States christopher.bard@nasa.gov
Emma **Beasor**, Astrophysics Research Institute, Liverpool John Moores University, United Kingdom e.beasor@2010.ljmu.ac.uk
Chris **Benton**, Auckland Astronomical Society, New Zealand cbenton@xtra.co.nz
Sopia **Beradze**, Abastumani Astrophisical Observatory, Illia State University, Georgia sopia.beradze.1@iliauni.edu.ge
Melina Cecilia **Bersten**, Astronomy, Ialp-fcaglp, Argentina, Argentina mbersten@fcaglp.unlp.edu.ar
Federica **Bianco**, New York University, New York, United States fbianco@nyu.edu
John **Bray**, Physics, University Of Auckland, New Zealand john.bray@auckland.ac.nz
Fabio **Bresolin**, University Of Hawaii, Hawaii, United States bresolin@ifa.hawaii.edu
Aaron **Brocklebank**, Jeremiah Horrocks Institute, University Of Central Lancashire, United Kingdom ajbrocklebank1@uclan.ac.uk
Bram **Buysschaert**, LESIA - Paris Observatory, France bram.buysschaert@obspm.fr
Saida **Caballero-Nieves**, Physics & Astronomy Department, University Of Sheffield, United Kingdon saida.caballero@gmail.com
Inés **Camacho**, IAC, Tenerife, Spain icamacho@iac.es
Matteo **Cantiello**, Kavli Institute For Theoretical Physics, Uc Santa Barbara, California, United States matteo@kitp.ucsb.edu
Alex **Carciofi**, IAG, Universidade de São Paulo, Sp, Brazil carciofi@usp.br
Luiz Paulo **Carneiro Gama**, Universität-Sternwarte München / LMU, Germany luuizpaulo@gmail.com
Heon-young **Chang**, Department of Atmospheric Sciences And Astronomy, Kyungpook National University, South Korea hyc@knu.ac.kr
Emmanouil **Chatzopoulos**, Physics & Astronomy, Lousiana State University, Louisiana, United States chatzopoulos@phys.lsu.edu
Andre-nicolas **Chene**, Aura Inc. - Gemini Observatory, Hawaii, United States achene@gemini.edu
Grant **Christie**, Auckland Observatory, New Zealand grant@christie.org.nz
Sang-Hyun **Chun**, Department Of Physics And Astronomy, Seoul National University, Seoul, South Korea shyunc.m@gmail.com
Wonseok **Chun**, Seoul National University, South Korea wonseok@astro.snu.ac.kr
Benoit **Cote**, University Of Victoria (Canada) / Michigan State University (USA), Quebec, Canada bcote@uvic.ca
Andrea **Cristini**, Faculty Of Natural Sciences, Astrophysics, Keele University, Staffordshire, United Kingdom a.j.cristini@keele.ac.uk
Paul **Crowther**, Dept Physics & Astronomy, University of Sheffield, S. Yorkshire, United Kingdom Paul.crowther@sheffield.ac.uk
Michel **Cure**, Instituto de Fisica y Astronomia, Universidad de Valparaiso, Chile, Chile michel.cure@uv.cl
Simon **Daley-yates**, Astrophysics & Space Research Group, University Of Birmingham, West Midlands, United Kingdom sdaley@star.sr.bham.ac.uk
Augusto **Damineli**, Astronomy Dept, IAG, University of Sao Paulo, SP, Brazil augusto.damineli@gmail.com
Alexandre **David-uraz**, Department of Physics & Space Sciences, Florida Institute Of Technology, Florida, United States adaviduraz@fit.edu
Ben **Davies**, Astrophysics Research Institute, Liverpool John Moores University, United Kingdom b.davies@ljmu.ac.uk
Antonio **De Ugarte Postigo**, Iaa-csic, Granada, Spain adeugartepostigo@gmail.com
Trevor **Dorn-Wallenstein**, Astronomy Department, University Of Washington, Washington, United States tzdw@uw.edu
Maria **Drout**, Carnegie Observatories, California, United States mdrout@carnegiescience.edu
JJ **Eldridge**, Physics, University Of Auckland, Auckland, New Zealand j.eldridge@auckland.ac.nz
Christiana **Erba**, Department of Physics & Astronomy, University Of Delaware, Delaware, United States cerba@udel.edu
Thomas **Ertl**, Max Planck Institute For Astrophysics, Germany tertl@mpa-garching.mpg.de
Celia Rosa **Fierro**, Laboratory of Applied Mathematics & Computing of High Perfomance Abacus, Cinvestav, Mexico celia.fierro.estrellas@gmail.com
Corinne **Fletcher**, Florida Institute Of Technology, Florida, United States cfletcher2013@my.fit.edu
Morgan **Fraser**, School of Physics, University College Dublin, Dublin, Ireland mf@ast.cam.ac.uk
Kotaro **Fujisawa**, Advanced Research Institute For Science And Engineering, Waseda University, Japan fujisawa@heap.phys.waseda.ac.jp
Jim **Fuller**, Astronomy, Caltech, California, United States jfuller@caltech.edu
Gantcho **Gantchev**, Department of Astronomy, Physics Faculty, University of Sofia, Bulgaria pgantcho@yahoo.fr
Kristen **Garofali**, University of Washington, Washington, United States kgarofali@gmail.com
Cyril **Georgy**, Department of Astronomy, Genenva University, Switzerland cyril.georgy@unige.ch
Douglas **Gies**, Center for High Angular Resolution Astronomy, Georgia State University, Georgia, United States gies@chara.gsu.edu
Avishai **Gilkis**, Physics, Technion, Israel agilkis@technion.ac.il
Ylva **Goetberg**, Anton Pannekoek Institute for Astronomy, University of Amsterdam, North Holland, Netherlands y.l.l.gotberg@uva.nl
Víctor **Gómez-González**, Astrophysics, Instituto Nacional De Astrofísica, Óptica Y Electrónica, Puebla, Mexico mau.gglez@gmail.com
Michael **Gordon**, Minnesota Institute for Astrophysics, Minnesota, United States gordon@astro.umn.edu
Alex **Gormaz-matamala**, Instituto De Física y Astronomía, Universidad De Valparaíso, Chile alex.gormaz@postgrado.uv.cl

Eric **Gosset**, Institut D'astrophysique - STAR, Université De Liège, Belgium, Belgium gosset@astro.ulg.ac.be
Götz **Gräfener**, University of Bonn, Germany ggraefener@gmail.com
Jose **Groh**, Trinity College Dublin, Ireland jose.groh@tcd.ie
Sergei **Gulyaev**, Institute for Radio Astronomy & Space Research, New Zealand sergei.gulyaev@aut.ac.nz
Diah Y.A. Setia **Gunawan**, Universidad de Valparaíso, Valparaíso, Chile diah.gunawan@uv.cl
Rainer **Hainich**, Institute of Physics & Astronomy, University of Potsdam, Germany rhainich@astro.physik.uni-potsdam.de
Kenji **Hamaguchi**, NASA/GSFC & UMBC, Maryland, United States Kenji.Hamaguchi@nasa.gov
Wolf-Rainer **Hamann**, Institut Für Physik Und Astronomie, Universität Potsdam, Germany wrh@astro.physik.uni-potsdam.de
Margaret **Hanson**, Department of Physics, University Of Cincinnati, Ohio, United States margaret.hanson@uc.edu
Xavier **Haubois**, European Southern Observatory, Chile xhaubois@eso.org
Gerhard **Hensler**, Dept. of Astrophysics, University of Vienna, Austria gerhard.hensler@univie.ac.at
Artemio **Herrero**, Instituto De Astrofisica De Canarias, Spain ahd@iac.es
Anthony **Hervé**, Academy of Science of Czech Republic, Czech Republic herve@asu.cas.cz
Erin **Higgins**, Armagh Observatory, United Kingdom eh@arm.ac.uk
D. John **Hillier**, Department of Physics & Astronomy, University Of Pittsburgh, Pennsylvania, United States hillier@pitt.edu
Jennifer L. **Hoffman**, Physics & Astronomy, University of Denver, Colorado, United States jennifer.hoffman@du.edu
Gonzalo **Holgado Alijo**, Instituto De Astrofisica De Canarias, Instituto De Astrofisica De Canarias, Spain gholgado@iac.es
Leah **Huk**, Physics & Astronomy, University Of Denver, Tennessee, United States Leah.Huk@gmail.com
Delphine **Hypolite**, CEA, France delphine.hypolite@cea.fr
Natalia **Ivanova**, Physics, University of Alberta, Alberta, Canada nata.ivanova@ualberta.ca
Hans-Thomas **Janka**, Max Planck Institute for Astrophysics, Germany thj@mpa-garching.mpg.de
Yan-Fei **Jiang**, Kavli Institute for Theoretical Physics, University Of California, Santa Barbara, Ca - California, United States yanfei@kitp.ucsb.edu
Stephen **Justham**, University of the Chinese Academy of Sciences & NAOC, China sjustham@ucas.ac.cn
Lex **Kaper**, Astronomical Institute Anton Pannekoek, University of Amsterdam, Netherlands L.Kaper@uva.nl
Chinami **Kato**, Waseda University, Japan chinami@heap.phys.waseda.ac.jp
Dylan **Kee**, Institute of Astronomy & Astrophysics, University of Tübingen, Germany nathaniel-dylan.kee@uni-tuebingen.de
Zsolt **Keszthelyi**, Royal Military College Of Canada, Ontario, Canada zsolt.keszthelyi@rmc.ca
Megan **Kiminki**, Dept. of Astronomy / Steward Observatory, University of Arizona, Arizona, United States mbagley@email.arizona.edu
Robert **Klement**, Astronomical Insitute, Charles University, Czech Republic robertklement@gmail.com
Nino **Kochiashvili**, Abastumani Astrophysical Observatory, Ilia State University, Georgia nino.kochiashvili@iliauni.edu.ge
Erik **Kool**, Physics And Astronomy, Macquarie University / AAO, NSW, Australia erik.kool@students.mq.edu.au
Peter **Kretschmar**, European Space Agency, Madrid, Spain peter.kretschmar@esa.int
Jiri **Krticka**, Masaryk University, Czech Republic krticka@physics.muni.cz
Jonathan **Labadie-bartz**, Lehigh University, Pennsylvania, United States jml612@lehigh.edu
Ryan **Lau**, Caltech/JPL, California, United States ryanlau@caltech.edu
Claus **Leitherer**, Space Telescope Science Institute, Maryland, United States leitherer@stsci.edu
Maurice **Leutenegger**, X-ray Astrophysics, NASA/GSFC/CRESST/UMBC, Maryland, United States maurice.a.leutenegger@nasa.gov
Emily **Levesque**, University of Washington, Washington, United States emsque@uw.edu
Jamie **Lomax**, Department of Astronomy, University of Washington, Washington, United States jrlomax@uw.edu
Thomas **Madura**, Department of Physics & Astronomy, San Jose State University, California, United States thomas.madura@sjsu.edu
Jesús **Maíz Apellániz**, Centro de Astrobiología, CSIC-INTA, Madrid, Spain jmaiz@cab.inta-csic.es
Grigoris **Maravelias**, Astronomical Institute, Czech Academy Of Sciences, Greece maravelias@asu.cas.cz
Anthony **Marston**, ESA, STScI, Maryland, United States tmarston@sciops.esa.int
Philip **Massey**, Lowell Observatory, Arizona, United States phil.massey@lowell.edu
Seppo **Mattila**, Department of Physics & Astronomy, University of Turku, Finland sepmat@utu.fi
Liam **McClelland**, Department of Physics, University Of Auckland, New Zealand liam.mcclelland@auckland.ac.nz
Andrea **Mehner**, ESO Chile, Chile amehner@eso.org
Tobias **Melson**, Max-Planck-Institute for Astrophysics, Germany melson@mpa-garching.mpg.de
Athira **Menon**, Monash University, VIC, Australia athira.menon@monash.edu
Maria **Messineo**, Dep. Astronomy, Univ. Science and Technology Of China, China, China messineo@ustc.edu.cn
Georges **Meynet**, Astronomy, Geneva University, Geneva, Switzerland georges.meynet@unige.ch
Takashi **Moriya**, Division of Theoretical Astronomy, National Astronomical Observatory of Japan, Japan takashi.moriya@nao.ac.jp
Bernhard **Mueller**, Astrophysics Research Centre, Queen's University Belfast, United Kingdom b.mueller@qub.ac.uk
Melissa **Munoz**, Queen's University, Ontario, Canada munoz@astro.umontreal.ca
Jeremiah **Murphy**, Florida State University, Florida, United States jeremiah@physics.fsu.edu
Francisco **Najarro**, Cab Csic-inta, Spain najarro@cab.inta-csic.es
Ko **Nakamura**, Faculty of Science and Engineering, Waseda University, Japan nakamura.ko@heap.phys.waseda.ac.jp
Makoto **Narita**, National Institute of Technology, Okinawa College, Japan narita@okinawa-ct.ac.jp
Tim **Natusch**, Institute for Radio Astronomy & Space Research, Aut University, New Zealand tim.natusch@aut.ac.nz
Yael **Naze**, University of Liege, Belgium naze@astro.ulg.ac.be
Ignacio **Negueruela**, Universidad De Alicante, Alicante, Spain ignacio.negueruela@ua.es
Kathryn **Neugent**, Lowell Observatory, Arizona, United States kathrynneugent@gmail.com
Maria-Fernanda **Nieva**, Institute for Astro- & Particle Physics, University Of Innsbruck, Austria maria-fernanda.nieva@uibk.ac.at
Kenichi **Nomoto**, Kavli Institute for the Physics & Mathematics of the Universe (Kavli IPMU), The University of Tokyo, Japan nomoto@astron.s.u-tokyo.ac.jp

Takaya **Nozawa**, National Astronomical Observatory of Japan, Tokyo, Japan takaya.nozawa@nao.ac.jp
Keiichi **Ohnaka**, Universidad Catolica Del Norte, Instituto de Astronomia, Chile k1.ohnaka@gmail.com
Atsuo **Okazaki**, Hokkai-Gakuen University, Japan okazaki@lst.hokkai-s-u.ac.jp
Mary **Oksala**, California Lutheran University, California, United States meo@udel.edu
Lidia **Oskinova**, Institute for Physics & Astronomy, University Of Potsdam, Germany lida@astro.physik.uni-potsdam.de
Stanley **Owocki**, Department of Physics & Astronomy, University Of Delaware, Delaware, United States owocki@udel.edu
Herbert **Pablo**, University of Montreal, Quebec, Canada hpablo@astro.umontreal.ca
Manfred **Pakull**, Strasbourg University, Observatoire Astronomique de Strasbourg, France manfred.pakull@astro.unistra.fr
Veronique **Petit**, Department of Physics & Space Sciences, Florida Institute Of Technology, Florida, United States vpetit@fit.edu
Philipp **Podsiadlowski**, Physics, University Of Oxford, UK, United Kingdom podsi@astro.ox.ac.uk
Konstantin **Postnov**, Sternberg Astronomical Institute, Russian Federation kpostnov@gmail.com
Vincent **Prat**, DRF/IRFU/SAp, CEA, France vincent.prat@cea.fr
Joachim **Puls**, University Observatory, Ludwig-Maximilians-Universitaet Munich, Germany uh101aw@usm.uni-muenchen.de
Alan **Rainot**, Institute of Astrophysics, Ku Leuven, Belgium alan.rainot@kuleuven.be
Sebastián **Ramírez Alegría**, Universidad de Valparaíso / MAS, Chile sebastian.ramirez@uv.cl
Oscar Hernan **Ramirez-Agudelo**, UK Astronomy Technology Centre, STFC, United Kingdom oscar.ramirez@stfc.ac.uk
María C. **Ramírez-Tannus**, Anton Pannekoek Institute, North Holland, Netherlands m.c.ramireztannus@uva.nl
Thomas **Rivinius**, European Southern Observatory Chile, Chile triviniu@eso.org
Nicole **Rodrigues**, Department of Physics, University Of Auckland, New Zealand, New Zealand nrdo653@aucklanduni.ac.nz
Sara **Rodriguez Berlanas**, Instituto de Astrofisica De Canarias, Tenerife, Spain srberlan@iac.es
Eliceth Yaneire **Rojas Montes**, Armagh Observatory, Armagh, United Kingdom erm@arm.ac.uk
Margarita **Rosado**, Instituto de Astronomia, UNAM, margarit@astro.unam.mx
Marcelo **Rubinho**, IAG / Departamento de Astronomia, Universidade de São Paulo, São Paulo, Brazil esteemeuemail@gmail.com
Maria Del Mar **Rubio Diez**, Centro De Astrobiología (csic-inta), Spain mmrd@cab.inta-csic.es
Christopher **Russell**, X-ray Astrophysics Laboratory, NASA Goddard Space Flight Center, Maryland, United States crussell@udel.edu
Stuart **Ryder**, International Telescopes Support Office, Australian Astronomical Observatory, NSW, Australia sdr@aao.gov.au
Carolina **Sabin-Sanjulian**, Universidad De La Serena, Chile cssj@dfuls.cl
Hugues **Sana**, Institute of Astrophysics, KU Leuven, Belgium hugues.sana@kuleuven.be
Andreas **Sander**, Institute for Physics & Astronomy, University Of Potsdam, Germany ansander@astro.physik.uni-potsdam.de
Norbert S. **Schulz**, Kavli Institute for Astrophysics & Space Research, Massachusetts Institute Of Technology, Massachusetts, United States nss@space.mit.edu
Peter **Scicluna**, Institute of Astronomy & Astrophysics, Academia Sinica, Taiwan peterscicluna@asiaa.sinica.edu.tw
Tomer **Shenar**, Astrophysik I, University of Potsdam, Germany shtomer@astro.physik.uni-potsdam.de
Isaac **Shivvers**, Astronomy Department, University Of California, Berkeley, California, United States ishivvers@berkeley.edu
Susan **Shoebridge**, Auckland Astronomical Society, New Zealand susan@polarconsult.co.nz
Matthew **Shultz**, Astronomy, Uppsala University, Ontario, Canada matt.shultz@gmail.com
Sergio **Simon-diaz**, Instituto De Astrofisica De Canarias, La Laguna, Tenerife, Spain ssimon@iac.es
Linda **Smith**, Space Telescope Science Institute, European Space Agency, Maryland, United States lsmith@stsci.edu
Nathan **Smith**, Astronomy, University of Arizona, Arizona, United States nathans@as.arizona.edu
Elizabeth **Stanway**, Department of Physics, Universirty Of Warwick, United Kingdom e.r.stanway@warwick.ac.uk
Heloise **Stevance**, Physics & Astronomy, University Of Sheffield, United Kingdom fstevance1@sheffield.ac.uk
Ian **Stevens**, School of Physics & Astronomy, University Of Birmingham, United Kingdom irs@star.sr.bham.ac.uk
Nicole **St-louis**, Département de Physique, Université De Montréal, Canada stlouis@astro.umontreal.ca
Yasuharu **Sugawara**, ISAS/JAXA, Kanagawa, Japan sugawara.yasuharu@jaxa.jp
Alexander **Summa**, Max Planck Institute For Astrophysics, Bavaria, Germany asumma@mpa-garching.mpg.de
Katie **Tehrani**, Physics & Astronomy, University of Sheffield, South Yorkshire, United Kingdom k.tehrani@sheffield.ac.uk
Christina **Thöne**, HETH, IAA-CSIC, Spain cthoene@iaa.es
Alexey **Tolstov**, Kavli IPMU, The University of Tokyo, Japan alexey.tolstov@ipmu.jp
Frank **Tramper**, European Space Astronomy Centre, European Space Agency, Madrid, Spain ftramper@sciops.esa.int
Asif **ud-Doula**, Penn State Worthington Scranton, Pennsylvania, United States auu4@psu.edu
Miguel A. **Urbaneja**, Institute for Astro & Particle Physics, University Of Innsbruck, Austria Miguel.Urbaneja-Perez@uibk.ac.at
Jorick **Vink**, Armagh Obsevatory & Planetarium, Armagh Obsevatory And Planetarium, Northern Ireland, United Kingdom jsv@arm.ac.uk
Gregg **Wade**, Department of Physics, Royal Military College Of Canada, Ontario, Canada marmottecam@yahoo.ca
Michael **Wegner**, University Observatory Munich, Germany wegner@usm.lmu.de
Lin **Xiao**, University of Auckland, New Zealand lin.xiao@auckland.ac.nz
Yu **Yamamoto**, Applied Physics, Waseda University, Shinjuku-ku, Japan yamamoto@heap.phys.waseda.ac.jp
Sung Chul **Yoon**, Physics & Astronomy, Seoul National University, South Korea yoon@astro.snu.ac.kr
Norhasliza **Yusof**, Department of Physics, Faculty Of Science, University Of Malaya, Malaysia norhaslizay@um.edu.my
Janos **Zsargo**, Department of Physics & Mathematics, ESFM-IPN, Mexico City, Mexico jzsargo@esfm.ipn.mx

Session 1: Introduction

The Lives and Death-Throes of Massive Stars
Proceedings IAU Symposium No. 329, 2016
J.J. Eldridge, J.C. Bray, L.A.S. McClelland & L. Xiao, eds.

doi:10.1017/S1743921317003283

Massive stars, successes and challenges

Georges Meynet[1], **André Maeder**[1], **Cyril Georgy**[1], **Sylvia Ekström**[1], **Patrick Eggenberger**[1], **Fabio Barblan**[1] **and Han Feng Song**[2,3]

[1]Department of Astronomy, University of Geneva, Switzerland
[2]College of Science, Guizhou University, Guiyang, 550025, Guizhou Province, PR China
[3]Key Laboratory for the Structure and Evolution of Celestial Objects, Chinese Academy of Sciences, 650011, Kunming, PR China
email: georges.meynet@unige.ch

Abstract. We give a brief overview of where we stand with respect to some old and new questions bearing on how massive stars evolve and end their lifetime. We focus on the following key points that are further discussed by other contributions during this conference: convection, mass losses, rotation, magnetic field and multiplicity. For purpose of clarity, each of these processes are discussed on its own but we have to keep in mind that they are all interacting between them offering a large variety of outputs, some of them still to be discovered.

Keywords. Massive stars, convection, mass loss, rotation, magnetic field, multiplicity

1. Massive stars in the Universe

Stars are at the crossroad of many topical questions in astrophysics, cosmology and physics. In astrophysics, the way they form, evolve and end their nuclear lifetimes has a large impact on the evolution of the matter as a whole in the Universe, on the evolution of galaxies. Also stars are bridges for connecting processes occurring at different space and time scales. For instance, stars are important producers of dust grains in the Universe (e.g. Todini & Ferrara 2001) and at the same time are the sources of light and of new elements in galaxies (e.g. Chiappini *et al.* 2006). They also provide links between evolution of stars at various redshifts and metallicities covering the whole cosmic history, from the origin of the very peculiar composition of the most iron poor stars (Frebel & Norris 2015) to the origin of the short lived radionuclides in the nascent solar system (Gounelle 2015). At a cosmological level, stars are sources of ionizing photons. This is especially true at low metallicities and for the first stellar generations that may have contributed significantly to the reionization of the Universe (see e.g. Amorìn *et al.* 2017). Stars offer powerful physics laboratories (e.g. in the case of the solar neutrinos) to explore some questions at the frontiers of physics as the possible variations with time of the fundamental constants (Ekström *et al.* 2010), or properties of new particles as Weakly Interactive Particles or axions (see e.g. Taoso *et al.* 2008). The ranges of temperature and density conditions they span vastly outrange the domains that may be explored in the laboratory. Of course, to address all these questions, a first prerequisite is to sufficiently well understand their physics. In the following, we discuss a few selected questions giving the opportunity to present some new ideas.

2. Convection

Convection, and more generally turbulence, is a long standing problem in stellar evolution. Actually these processes are difficult to describe already on Earth and thus it is no surprise that indeed they are difficult to be well described in stars. These processes involve many different space and time scales, they are fundamentally 3 dimensional processes and thus their incorporation in 1D models is done at the expense of some simplifications.

Then why not to address this question through 3D hydrodynamical modelling? Actually 3D modelling is indeed used to probe the physics of convection, but this approach is possible only for relatively short time intervals. If one wants to explore the evolution of stars along their whole lifetime, exploring different initial masses, chemical compositions, rotation, magnetic field,... then 1D models are still the only tool that allows exploring large areas of the parameter space. This is the reason why some authors (Meakin & Arnett 2007ab; Arnett *et al.* 2015; Arnett & Meakin 2016) follow the approach consisting in deducing from multi D hydrodynamical simulations more physical recipes for incorporation into 1 D models. This is by far not an easy task. Actually it consists as written by Arnett & Meakin (2016) to extrapolate the weather to determine the climate! Said in other words, it is not straightforward to deduce from short time simulations, prescriptions for convection that can be used on long term evolution.

Stellar models usually use either the Schwarzschild or the Ledoux criterion for determining the size of the convective core. In addition some overshooting can be applied extending the size of the convective cores. During the Main-Sequence phase of massive stars, there is no changes expected whether the Schwarzschild or the Ledoux criterion is used. Indeed, in massive stars, during the Main-Sequence phase, the convective core decreases in mass, thus at the border of this receding core there is no μ-gradient (μ being the mean molecular weight). But during the core He-burning phase where the core increases in mass, depending which criterion is used lead to different outputs.

Gabriel *et al.* (2014) discuss various numerical methods to determine the size of the convective core in 1D models. During the iterative process for computing the structure of the star at a new time step, some numerical method has to be chosen in order to determine the size of the convective core at the new iteration. The boundary of the core is physically given by the condition $L_{\rm rad} = L_{\rm total}$, where $L_{\rm rad}$ is the luminosity due to the radiative transport and $L_{\rm total}$, the total luminosity also accounting for the thermal and kinetic energy transported by convection. In the local Mixing Length Theory and without overshooting, this condition is equivalent to the equality of the temperature gradients at the border of the convective core, *i.e.* $\nabla_{\rm rad} = \nabla_{\rm ad}$ *on the convective side of the boundary* (Biermann 1932). This condition on the gradients may not be reached on the radiative side of the boundary if a μ-gradient is present in the radiative layers above the convective core.

At the end of a given iteration, we shall have values for the adiabatic and radiative gradients at every mesh points. The quantity $\Delta\nabla = \nabla_{\rm rad}$-$\nabla_{\rm ad}$ changes sign at the border of the core (passes from a positive value inside the convective core to a negative one in the radiative zone in case no semiconvective zone is present). To determine the position where $\nabla_{\rm rad} = \nabla_{\rm ad}$, there are three possibilities: 1) take the two last positive values of $\Delta\nabla$ at the border of the convective core and extrapolate to find the place where $\Delta\nabla$ becomes zero; 2) take the last positive value inside the convective core and the first negative one in the radiative one and interpolate to find the place where $\Delta\nabla$ becomes zero; 3) to extrapolate (inwards) the first two negative values of $\Delta\nabla$ in the radiative region.

Gabriel *et al.* (2014) first shows that depending on that choice, various values for the convective core can be obtained. As a numerical example, they show that the mass of the convective core in a 16 $M_\odot$ during the MS phase (when the mass fraction of hydrogen in the core is about 0.15) is 22% of the total mass when the rule 1 is applied and only 17% when the rule 3 is applied (here adopting the Ledoux criterion). This produces a difference of more than 20%! A second important result that they have obtained is that the only physically justified method is the method number 1, that means interpolating from inside the convective core. Any other choice can lead to unphysical results (1) in case a convective core is expanding and/or (2) in case the Ledoux criterion is adopted.

In case of a shell convective zone, at the moment no definite conclusion can be reached about the proper numerical procedure since in case of a convective shell forming in a mu-gradient region, it is impossible to satisfy the condition $L_{\rm rad} = L_{\rm total}$ at both boundaries. Actually, depending on which criterion for convection is chosen (although in the case of a pre-existing mu-gradient region, the Ledoux criterion should be adopted), significantly different results are obtained (see *e.g.* Georgy *et al.* 2014).

3. Mass losses

Mass loss by stellar winds is a key process in massive star evolution. It changes the evolutionary tracks in the HR diagram, the surface abundances and rotation velocities. The wind contributes to the chemical enrichment in new elements of the interstellar medium, to the rate of injection of momentum and energy in the interstellar medium. It has also an impact on the flux of ionising photons. Mass loss has also an impact on the nature of the supernova event (if any) and on the properties of the stellar remnant (if any) (see *e.g.* the review by Smith 2014).

For hot stars, the line driven winds theory provides a sophisticated theoretical frame that can account for the amplitude of the mass losses by stellar winds (with uncertainties limited to a factor of a few) and their dependence with the metallicity (see e.g. the review by Puls *et al.* 2008). Note however that already a factor 2 difference during the Main-Sequence phase may produce significant changes at solar or higher metallicities for the stars with masses larger than about 30-40 $M_\odot$ (Meynet *et al.* 1994). Greater uncertainties are present when the stars evolve away from the MS phase and enter into the regime of Luminous Blue Variable or of the red supergiant stage. In both cases, observations tell us that these stars may show outbursts. The LBV iconic case is η Car that in the middle of the eighteen century lost during one or two decades mass at a rate of about 1 $M_\odot$ per year (see e.g. Humphreys & Martin 2012). Less extreme, but still showing some sporadic mass ejection events are for instance the star VY CMa, a bright and extended evolved cool stars (See Humphreys 2016 and the references therein).

Outbursts are difficult to be accounted for in stellar evolution computations mainly because their physics is still to be unravelled. These outbursts may however have an important impact removing in short timescales large amounts of mass (see e.g. Smith & Owocki 2006). When occurring just before the supernova, they might produce superluminous supernovae by converting part of the mechanical energy of the ejecta in radiations when the ejecta crash into the stellar wind.

These outbursts make typically the modelling of the LBV and also of the red supergiant phase still uncertain. To illustrate this point, it is interesting to compare the post red supergiant evolution obtained with different prescriptions for the red supergiant mass loss rate (Salasnich *et al.* 1999; Vanbeveren *et al.* 2007, Georgy 2012, Meynet *et al.* 2015). As is well known, removing mass during the red supergiant stage may make the star to evolve back to the blue side of the HR diagram. In general the bluewards evolution for stars with masses above about 15 $M_\odot$ begins when the mass of the core becomes a fraction higher than 60-70% the total mass of the star (Gianonne 1967). Increasing the mass loss has thus for effect to shorten the red supergiant lifetime of those stars that eventually evolve bluewards. It produces core collapse supernova progenitors that are yellow, blue supergiants, even sometimes LBV's or Wolf-Rayet stars. An interesting result of this kind of evolution is that it might affect, if frequent enough, the distribution of stars as a function of the luminosity along the red supergiant branch. If we look at this argument the other way around, it means that the observed luminosity distribution of the red supergiants provides some hints about the time averaged mass loss rate during that phase. Larger are the mass losses, larger will be the decrease as a function of luminosity of the number of red supergiants (see for instance the left panel of Fig. 5 in Meynet *et al.* 2015). This might be interesting to measure from complete red supergiant samples the slope of the luminosity distribution function and to compare with predictions of population synthesis models based on various mass loss rates during the red supergiant phase.

4. Rotation

Axial rotation is a very interesting feature of stars for many reasons:

• The angular momentum content of a star on the ZAMS results from the star formation process. As is well known, during the formation process large amounts of angular momentum have to be removed from the collapsing cloud otherwise stars cannot be formed. How this happens depends on processes like for instance disk locking, disruption of the collapsing cloud in multiple systems. Haemmerlé *et al.* (in press) computed pre-MS evolutionary models with accretion and rotation. They concluded that during the phase between the formation of the small mass hydrostatic core of 0.5 $M_\odot$ and the arrival on the ZAMS of the star with its final mass, in order to avoid the star to reach the critical velocity a braking mechanism is needed. This mechanism has to be efficient enough to remove more than 2/3 of the angular momentum from the inner accretion disc. They also conclude that due to the weak efficiency of angular momentum transport by shear instability and meridional circulation during the accretion phase, the internal rotation profiles of accreting stars reflect essentially the angular momentum accretion history.

• During its nuclear lifetime, rotation can induce many changes in the observed properties of stars (see e.g. the review by Maeder & Meynet 2012 and references therein). It changes the surface abundances as a results of the mixing processes induced by the same instabilities indicated above that transport angular momentum. It modifies the evolutionary tracks in the HR diagram making a star of a given mass more luminous than its non-rotating sibling. Rotation may activate a dynamo in convective regions and it might also activate one in differentially rotating ones. When the star is rotating sufficiently fast, typically with surface angular velocity larger than about 70-80% the critical angular velocity, the star becomes significantly oblate, it will show anisotropic stellar polar winds (more intense winds in the polar rather in the equatorial direction). At very high rotation, the rotational mixing may be so strong that the star will follow a homogeneous evolution. The surface velocity, together with other observed properties represents thus an important pieces of information of stellar physics, either for single stars or stars in close binaries. It varies as a results of the internal redistribution of the angular momentum by convection, shear, meridional currents, magnetic fields instabilities, also by processes like stellar mass losses, tidal interactions with a companion, or mass accretion from a companion.

• At the end of its stellar lifetimes, rotation may change the consequences of the core collapse. In case the core rotates sufficiently fast, it may favor a luminous supernova explosion. It may also have an impact on the rotational properties of the stellar remnant if any. It is however not obvious how the state of rotation in the presupernova structure is linked to the rotation rate of the neutron stars and black hole. Indeed the explosion mechanism itself and/or some braking mechanism operating in the early phases of the evolution of the new born neutron star may have a significant impact and thus hide the pre-explosion conditions. In case the angular momentum would be conserved, present

day stellar models predict in general too fast rotating neutron stars when compared to the rotation rate of young observed pulsars (Heger *et al.* 2005). If true it would indicate that massive star models with rotation still miss some angular momentum transport mechanism, a process that appears well confirmed for small mass stars (see Beck *et al.* 2012, Eggenberger *et al.* 2017 and references therein).
The physics of rotation is complex and involves turbulence, a feature that, as recalled above cannot be described easily in numerical simulation. In 1D models, the unknown aspects of turbulence are accounted for in the choice of a few parameters. As an example, in the Geneva stellar evolution code, we use the theory proposed by Zahn (1992) that is based on the hypothesis that the star settles into a state of shellular rotation due to a strong horizontal turbulence. In those models two parameters related to the shear turbulence have to be fixed. One is the value taken for the critical Richardson number. This value governs the efficiency of the mixing by the shear instability along the radial direction. A second one intervenes in the expression of the horizontal turbulence.

The Richardson criterion comes from the fact that for the mixing to occur, the excess energy from differential rotation has to be larger than the energy needed to overcome the gradient of density (see the textbook by Maeder 2009). The excess of energy in the shear can be expressed by $1/4\rho(\delta V)^2$, where ρ is the density, and δV the differential of velocity over a given distance in the radial direction. The energy needed to overcome the vertical density gradient can be written $g\delta\rho\delta z$, where g is the gravity, $\delta\rho$ the difference of density between the interior of the blob and the exterior, δz the length scale over which the blob moves. The Richardson criterion tells that one has mixing when $g\delta\rho\delta z < \frac{1}{4}\rho(\delta V)^2$, or $R_{\rm i} = g\delta\rho\delta z/\rho(\delta V)^2 < R_{\rm i,crit} = \frac{1}{4}$. In this approach the critical Richardson number is 1/4. Some numerical simulations indicate that turbulence begins to appear already for a value of $R_{\rm i,crit} = 1$ (Brüggen & Hillebrandt 2001). The choice of $R_{\rm i,crit}$ is thus confined between 1/4 and 1 and this is one of the important parameter that has to chosen.

The Richardson criterion is actually the key expression to find the value of the diffusion coefficient due to shear in the radial direction. A diffusion coefficient can be expressed as $(1/3)vl$ where v is a typical velocity and l a typical size of the moving blobs. In a turbulent medium, the transport is dominated by the largest eddies, *i.e.* the largest l values. The size of the eddies enter into the expression of $\delta\rho$ through the way the transport changes the temperature and molecular weight gradients (see Maeder 2009 and references therein). Thus looking for the size of the eddies that satisfies the Richardson criterion, it is possible to deduce the vertical diffusion coefficient. In case of secular shear, *i.e.* shear occurring on timescales that are long compared to the thermal diffusion timescale, it is in general always possible to find a size of the eddies that satisfies the Richardson criterion. Note that it is also important to check that the largest eddies that satisfy the Richardson criterion are not too small to be damped by the viscous forces.

The second parameter intervenes into the expression for the horizontal turbulence, *i.e.* the turbulence along an isobar. In the last Geneva grid at solar metallicity (Ekström *et al.* 2012), the choice of these two quantities was mainly driven by requiring that stars with initial masses between 9 and 15 $M_\odot$ at solar metallicity, presenting a surface velocity during the Main-Sequence phase compatible with the observed averaged velocities present nitrogen surface enrichments in agreement with the mean observed values. Note that this way of doing is shared by other groups (Brott *et al.* 2001, Chieffi & Limongi 2013) and thus the stellar models will predict similar values in this mass and velocity range at the solar metallicity. However, outside these ranges, different models may actually predict significantly different behaviors depending on the way rotation is accounted for.

In shellular rotating models, when no magnetic field is present, the mixing of the chemical elements is mainly due to shear instabilities, while the transport of the angular

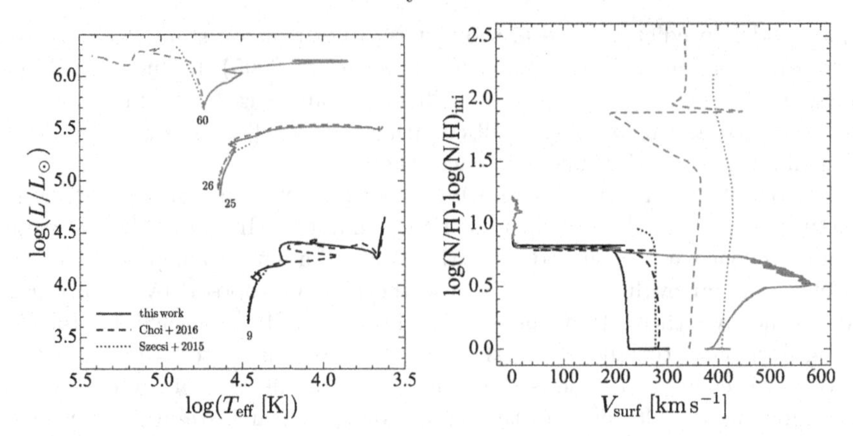

Figure 1. *Left panel:* Comparisons of the evolutionary tracks for different initial mass models in the theoretical HR diagram for a metallicity, Z around 0.00004, *i.e.* corresponding to the metallicity of the galaxy I Zw 18. The code MESA without an internal dynamo has been used by Choi *et al.* (2016) and a [Fe/H]=-1.5. The code STERN with an internal dynamo has been used by Szecsi *et al.* (2015, note that they computed a 26 $M_\odot$ model). *Right panel:* comparison in the plane surface nitrogen abundance versus surface rotation. Only the 9 and 60 $M_\odot$ models are shown for purpose of clarity. Figure taken from Groh *et al.* (in preparation).

momentum is mainly driven by the meridional currents. In models where a strong internal magnetic field is considered, as for instance due to the Tayler-Spruit mechanism (Spruit 1999, 2002), the mixing of the chemical elements is mainly driven by meridional currents.

The stellar models predict that the surface enrichments in products of the CNO burning, observable for example by the N/H, or N/C ratios, increase with stellar mass M and rotation velocities v, because mixing gets stronger. The enrichments also increase with the age t, since more and more new nuclear products reach the stellar surface. The enrichments are also stronger at lower Z, shear mixing being favored in more compact stars. Tidal interactions, as will be seen below, may influence the mixing. Thus, the chemical enrichments are a multivariate function (Maeder, 2009), e.g. (N/H)=f(M, Ω, age, Z, B, multiplicity,....). This fact has for consequence that it is difficult from observed stars to isolate the specific correlation between the surface enrichments and rotation. When sufficient care has been taken for selecting the stars presenting properties allowing to isolate the effect of rotation (isolated stars of about the same initial mass, age and metallicity) then a good agreement is found between rotating stellar models and the observations (Maeder *et al.* 2009, see also Przybilla *et al.* 2010, Martins *et al.* 2015).

5. Magnetic fields

Magnetic fields can intervene in various parts of the star: starting from the core region, a magnetic field can be attached to the convective core. Then in the radiative envelope, a magnetic field can be amplified by the Tayler-Spruit dynamo (Spruit 1999, 2002). Finally at the surface, a magnetic field can be present (Wade *et al.* 2006; Grunhut *et al.* 2017), either produced by some dynamo attached to the small convective region that is present even in massive stars in the outer layers (Cantiello *et al.* 2009), or more probably being a fossil field, *i.e.* having its origin in a previous phase of the evolution of the massive star. In that last case, it is reasonable to think that this magnetic field will actually be present in the whole star taking much larger values in the interior.

A main impact of a surface magnetic field is to exert a torque at the surface of the star†. This happens in case the energy density associated to the magnetic field is larger than the density of kinetic energy in the wind (ud-Doula & Owocki 2002), so typically when $B^2/(8\pi)$ is larger than $1/2\dot{M}v_\infty^2$, where B is the equatorial magnetic field, $\dot{M}$ the mass loss rate, and v_∞ the terminal wind velocity, *i.e.* the velocity of the wind when the process of acceleration is terminated. The impact of this wind magnetic braking is to slow down the star (ud-Doula *et al.* 2008, 2009). This slowing down may be accompanied or not by strong surface enrichments depending whether the magnetic field is due to a surface dynamo attached to the outer convective zone or if it results from a fossil magnetic field present in the whole star (Meynet *et al.* 2011).

The impact of a strong magnetic field in the interior of the star is mainly to impose a nearly rigid rotation at least during the MS phase. This is the case for instance when the Tayler-Spruit mechanism is used to compute stellar models. Although this process will be active only in differentially rotating layers, it will actually make the whole star to rotate as a solid body because convective zones are assumed to rotate as solid bodies. As already indicated above, in those models, the mixing of the elements is not long driven by shear instabilities since there is no strong differential rotation, but by meridional currents. A question that might be asked at that point is whether meridional currents will not be prevented to be active by the magnetic field itself! Likely this depends on the geometry and strength of the field (Zahn 2011).

In Fig. 1, the evolutionary tracks for different initial mass models having similar initial rotations are plotted for purpose of comparison. The metallicity of these models corresponds to that of the Galaxy I Zw 18 and is about 1/50 the solar metallicity. One of the main differences between the Geneva tracks and the tracks computed with MESA and STERN is the efficiency of the angular momentum transport, more efficient in MESA and in STERN than in the Geneva models. The most striking difference occurs for the 60 $M_\odot$ models. While the Geneva track, for an initial velocity of about 430 km s^{-1} evolves as usual to the red part of the HR diagram, the MESA and STERN models, starting with an initial rotation of respectively 410 and 340 km s^{-1} evolve homogeneously. This is an effect mainly due to the different physics considered. Actually the Geneva code with the account of the Tayler-Spruit dynamo would also produce a similar behavior. Thus we see that the inclusion of the Tayler-Spruit dynamo favors the homogeneous evolution. This comparison also shows that it is important to be aware that behind the terms *rotating models*, very different physics may be considered leading to significantly different outputs.

If we compare the changes of the surface abundances predicted by these two types of models (see the right panel of Fig. 1), we see that indeed the MESA and STERN tracks are much more enriched than the Geneva track, quite consistently with the behavior in the HR diagram. We can note another difference in this right panel. The MESA and STERN tracks evolves from the beginning vertically, while the Geneva track evolves first horizontally (the surface velocity is decreasing) and then evolves vertically. The initial decrease of the velocity occurs on a very short timescale and is due to an initial redistribution of angular momentum by the meridional currents inside the star. This redistribution is triggered by meridional currents that transport angular momentum from the outer layers to more central region, hence the decrease of the surface velocity.

† A strong magnetic field may reduce the mass loss by stellar winds, having interesting consequences for the formation of massive black holes even at solar metallicity (Petit *et al.* 2017) or for Pair Instability Supernova to appear at solar metallicities (Georgy *et al.* 2017).

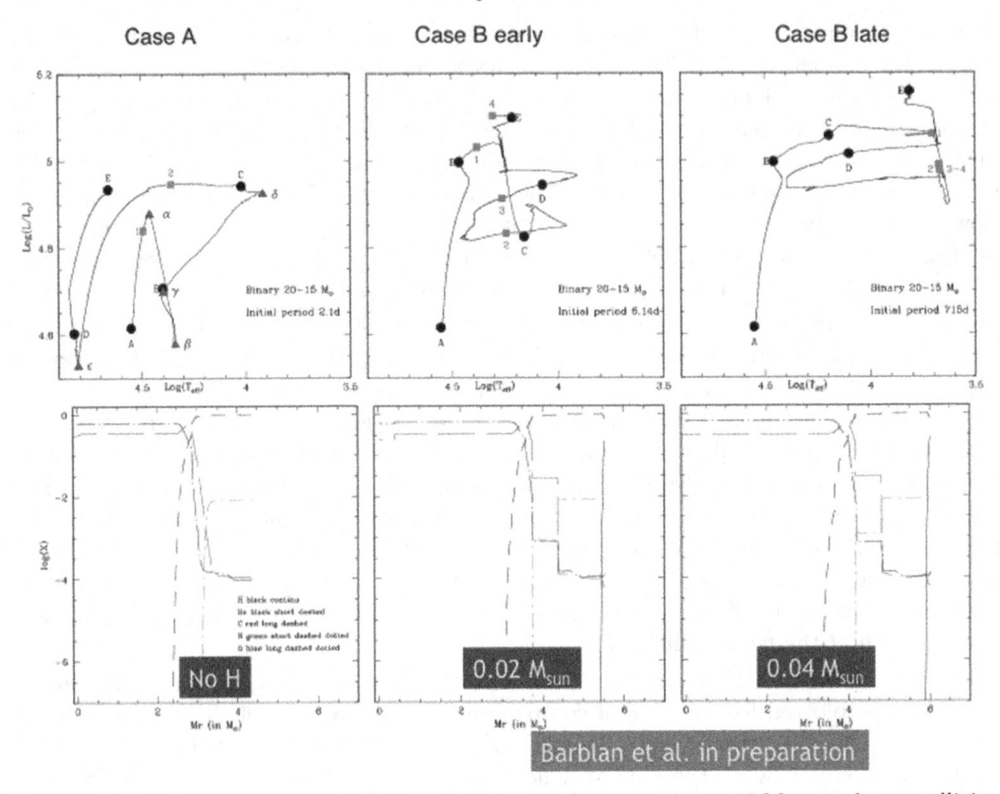

Figure 2. The three upper panels show the evolution of a non-rotating 20 M$_\odot$ at solar metallicity in a binary system with a 15 M$_\odot$. The initial orbital period is respectively from left to right equal to 2.10, 6.14 and 715 days. In each panel, point A corresponds to the ZAMS, B to the end of the core H-burning phase, C to the ignition of helium in the core, D to the end of the core He-burning phase, and E to the end of the core carbon burning phase. A first mass transfer episode occurs between point 1 and 2. A second mass transfer occurs between points 3 and 4. The lower panel shows the chemical structure of the stars at point E.

6. Multiplicity

Many massive stars are in multiple systems and part of those may follow a different evolution because of interactions with their close companions (Sana *et al.* 2012, 2013). This has triggered many recent works exploring close binary evolution and their consequences for explaining the origin of various stellar populations (see e.g. Eldridge & Stanway 2009, 2016; Stanway *et al.* 2016; Yoon *et al.* 2010, 2017). Examples of the evolution of a primary star of 20 M$_\odot$ in a short period binary systems with a 15 M$_\odot$ are shown in Fig. 2 (non-rotating stellar models). Three different cases are shown, corresponding to different initial orbital periods and thus different times for the first mass transfer episode. These models were computed in order to see whether the primary could evolve into a low luminous WC star at the end of the evolution. Such low luminous WC stars are observed (Sander *et al.* 2012) and thus the question is how they are formed. Do they result from the single star channel? In that case it would required very high mass loss rates. It might be the case if some outbursts occurred during the evolution of the progenitors as can be reflected by the fact that many WR stars present ring nebulae (see e.g. Esteban *et al.* 2016). Do they results from close binary evolution? To provide at least a partial answer to that last question we computed the models shown in Fig. 2. We chose for the primary a 20 M$_\odot$ model because its luminosity, in case it would evolve into the WC stage would

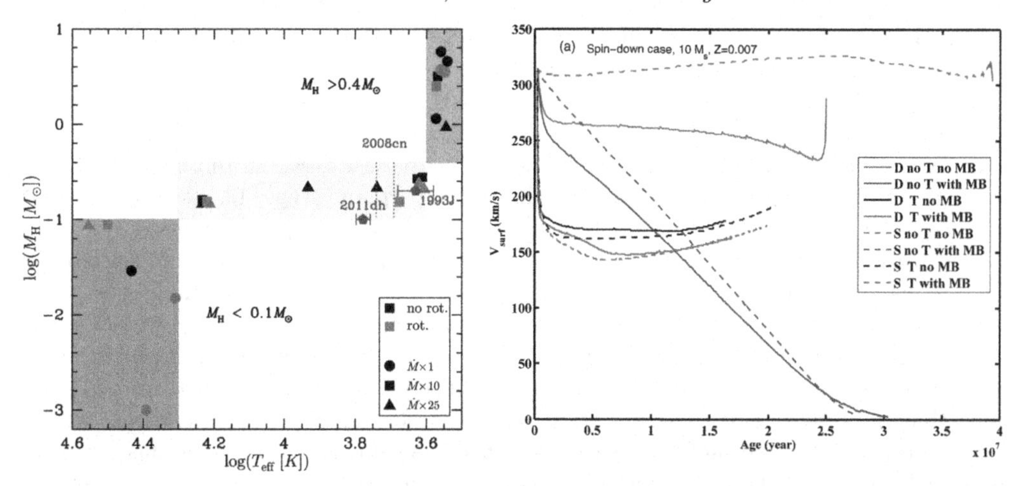

Figure 3. *Left panel:* Mass of hydrogen in solar masses at the pre-supernova stage for the various models with initial masses between 9 and 25 $M_\odot$ at solar metallicity, for various initial rotation between 0 and 40% of the critical velocity on the ZAMS and for different prescriptions of the mass loss rates during the red supergiant phase (see more detail in Meynet *et al.* 2014). Positions in this diagram of some supernovae are indicated by pentagons with error bars. The big red dots correspond to the positions in that diagram of the three 20 $M_\odot$ binary models shown in Fig. 2. *right panel*: Evolution of the surface equatorial velocity as a function of time for different 10 $M_\odot$ stellar models at a metallicity Z=0.007 with $v_{\rm ini}$ = 310 km s^{-1}. The symbols D means (radially) differentially rotating models, S, solid body rotating models, T, models with tides in a close binary system (the companion is a 7 $M_\odot$ star and the initial orbital period is 1.2 days), MB is for wind magnetic braking (the equatorial surface magnetic field is 1 kG).

more or less match the luminosities observed for the low luminous WC stars. Actually, the primary will evolve into WNE stars but it will never reach the WC phase. Thus this channel does not provide a solution for the origin of these stars. It has then to be checked whether the secondary might evolve into that stage. Another question is why, in case such an evolutionary channel would be frequent enough, the WNE stars that it produces are not observed. A possibility is that these stars may be difficult to detect hidden in the light of their more massive companion having accreted the mass. This point needs also to be confirmed by more detailed investigations.

In the following, we shall use the models of Fig. 2 to illustrate another effect of binarity. Looking at the bottom panels of Fig. 2, we can see the chemical structures of the different models at the core carbon-ignition. The outer layers at that stage have already reached their final structures, unless very strong mass loss episode would still occur in the very last moments of the evolution just before the supernova explosion. A feature that does appear as a kind of special features that is seen in these close binary models is the tiny amounts of hydrogen that is left in the cases B mass transfer (early and late). Actually single star evolution may have difficulties in producing such structures. This is illustrated in the left panel Fig. 3 where the mass of hydrogen in the envelope of pre-supernovae models is given for various models as a function of the effective temperature. All the models, except the large red starry dots are from single rotating and non rotating stellar models computed with various mass loss rates during the red supergiant phase. The models with a mass of hydrogen larger than 0.4 $M_\odot$ explode as red supergiants, and all the models with masses of hydrogen below about 0.1 $M_\odot$ are Wolf-Rayet stars just before the core collapse. Only those models that have a mass of hydrogen intermediate between these two values are yellow-blue supergiants. If we plot on that same diagram the models obtained by close binary evolution, two of them fall in the blue region as other models obtained by single star evolution. However there is the interesting case of the late case B mass transfer that falls below the yellow shaded region, indicating that this channel is able to produce a core collapse supernova progenitor with a lower hydrogen content at a given effective temperature. Interestingly this late case B would more or less mimic an increase of the mass loss rate during the red supergiant stage, however it does not give a similar structure as the models with an enhanced red supergiant mass loss. This is due to the fact that the single and the close binary models have different mechanism for putting an end to the strong mass loss episode. In the case of the single star, what makes the star to evolve into a lower mass loss rate regime is simply the evolution out of the red supergiant stage when a critical amount of mass has been lost. In the case of the close binary, it is the the decrease of the primary radius below the critical Roche limit that puts an end to the mass transfer. Thus increasing the mass loss rate at the red supergiant stage will not necessarily produce the same structure as a mass transfer episode occurring at the red supergiant phase. One can wonder, whether more generally a low hydrogen content (at least in this mass domain between 15 and 25 $M_{\odot}$) might be an indication favoring a core collapse supernova progenitors having gone through a mass transfer episode. Much more computations need to be performed. If true it might be an interesting signature since the mass of hydrogen in the progenitor can be sometimes trace back from properties of the supernova light curve.

The physics of rotation is important to model close binary stars. The reason is that in a binary there is a huge reservoir of angular momentum in the orbital movement. Through tidal interactions, exchanges between the orbital and the axial orbital reservoirs happens, sometimes spinning up the stars, sometimes spinning them down and of course changing the parameters of the orbit. The impact of these changes of angular momentum in stars and the way this angular momentum is redistributed inside the star by various instabilities is important for questions regarding many aspects of the evolution of such systems like synchronization, circularization, induced tidal mixing etc... Thus rotation and binarity might be tightly intertwined.

Also rotation, multiplicity and magnetic winds can interact. Recently Song *et al.* (in preparation) have investigated the case of massive stars with a strong surface magnetic field in a close binary system. The question these authors want to address is how tidal and magnetic torques interact and what are the consequences for the axial rotation of the two stars and for the orbital evolution. The right panel of Figure 3 shows what happens in a system composed of a 10 and a 7 $M_{\odot}$ star orbiting around their center of mass with an orbital period of 1.2 days. Cases with and without a surface magnetic field for the 10 $M_{\odot}$ has been considered (the equatorial surface magnetic field considered is taken equal to 1kG) The plot shows the evolution of the surface velocity with time of the primary. Two different angular momentum distribution has been considered, solid body rotation driven by the Tayler-Spruit dynamo and the case of differential internal rotation driven by shear and meridional currents. The cases of single stars with and without magnetic braking are also shown. We see that the single stars when braked reach very low surface velocities at the end of the Main-sequence phase . The same star with a companion will actually be maintained at a much higher rotation by the tidal torques. Tidal torques tap angular momentum from the orbit to transfer it to the star. This tends to reduce the distance between the two stars and thus to increase the tidal torque. This is why the velocity of the primary increases after 5-10 Myr. We see that the difference between the solid body and differentially rotating case is the most important in the case of the single non-magnetic stars. As soon as either the wind magnetic braking or the tidal torque (or both) occur, the two cases show very similar behaviors in the surface velocity versus time plane. For what concerns the evolution of the tracks in the HR diagram and the surface abundances, the differences are much larger.

7. Conclusion

The few questions addressed above do not pay credit to many other interesting questions that massive star evolution still trigger. One sees that the picture becomes more complicated adding to the impact of the initial mass and metallicities, other quantities as the initial rotation, the magnetic field and the multiplicity. We saw that these effects can interact strongly: for instance rotation has an impact on the lifetimes and the evolutionary tracks, this in turn changes the quantities of mass and angular momentum lost by the stellar winds, changing the rotation of the star. In a close binary system, rotation may change under the impact of mass loss and by tidal torques. These processes have an impact on the orbital evolution and thus on the tidal torques.

Likely the challenges for the future will be on one side to obtain the correct physics to account in a proper way of all these effects in stellar models. But even if that stage will be reached, then will remain the challenge of exploring the consequences of many different initial conditions and to find a reasonable way to compare with the observations. While these challenges are severe, there is some hope that they will be at least in part overcome in the future thanks to the ever increasing observational channels that provide new constraints about the way stars are evolving. The recent detection by LIGO of gravitational wave well illustrates this point. Large surveys collecting data on huge number of stars and thus offering unbiased samples of observed stars are also essential for making progress. Even short lived phases may have very important consequences. Thus we are living in a very exciting time for exploring the physics of massive stars that are so important for driving short timescales processes in our Universe.

Acknowledgement

GM thanks Prof Arlette Noels for her help and the hospitality of the Kavli Institute in Santa Barbara.

References

Amorín, R., Fontana, A., Pérez-Montero, E., *et al.* 2017, *Nature Astronomy*, 1, 0052
Arnett, W. D., Meakin, C., Viallet, M., *et al.* 2015, *ApJ*, 809, 30
Arnett, W. D. & Meakin, C. 2016, *Reports on Progress in Physics*, 79, 102901
Beck, P. G., Montalban, J., Kallinger, T., *et al.* 2012, *Nature*, 481, 55
Biermann, L. 1932, *Zeitschrift für Astrophysik*, 5, 117
Brott, I., de Mink, S. E., Cantiello, M., *et al.* 2011, *Astron. & Astrophys.*, 530, A115
Brüggen, M. & Hillebrandt, W. 2001, *MNRAS*, 320, 73
Cantiello, M., Langer, N., Brott, I., *et al.* 2009, *Astron. & Astrophys.*, 499, 279
Chiappini, C., Hirschi, R., Meynet, G., *et al.* 2006, *Astron. & Astrophys.*, 449, L27
Chieffi, A. & Limongi, M. 2013, *ApJ*, 764, 21
Eggenberger, P., Lagarde, N., Miglio, A., *et al.* 2017, *Astron. & Astrophys.*, 599, A18
Ekström, S., Coc, A., Descouvemont, P., *et al.* 2010, *Astron. & Astrophys.*, 514, A62
Ekström, S., Georgy, C., Eggenberger, P., *et al.* 2012, *Astron. & Astrophys.*, 537, A146
Eldridge, J. J. & Stanway, E. R. 2016, *MNRAS*, 462, 3302
Eldridge, J. J. & Stanway, E. R. 2009, *MNRAS*, 400, 1019
Esteban, C., Mesa-Delgado, A., Morisset, C., & García-Rojas, J. 2016, *MNRAS*, 460, 4038
Frebel, A. & Norris, J. E. 2015, *Annual Review of Astronomy and Astrophysics*, 53, 631
Gabriel, M., Noels, A., Montalbán, J., & Miglio, A. 2014, *Astron. & Astrophys.*, 569, A63

Georgy, C. 2012, *Astron. & Astrophys.*, 538, L8
Georgy, C., Saio, H., & Meynet, G. 2014, *MNRAS*, 439, L6
Georgy, C., Meynet, G., Ekström, S., *et al.* 2017, arXiv:1702.02340
Giannone, P. 1967, *Zeitschrift für Astrophysik*, 65, 226
Gounelle, M. 2015, *Astron. & Astrophys.*, 582, A26
Grunhut, J. H., Wade, G. A., Neiner, C., *et al.* 2017, *MNRAS*, 465, 2432
Heger, A., Woosley, S. E., & Spruit, H. C. 2005, *ApJ*, 626, 350
Humphreys, R. M. 2016, Journal of Physics Conference Series, 728, 022007
Humphreys, R. M. & Martin, J. C. 2012, *Eta Carinae and the Supernova Impostors*, 384, 1
de Jager, C., Nieuwenhuijzen, H., & van der Hucht, K. A. 1988, *Astron. & Astrophys. Suppl.*, 72, 259
de Mink, S. E., Cantiello, M., Langer, N., *et al.* 2009, *Astron. & Astrophys.*, 497, 243
Maeder, A. 2009, Physics, Formation and Evolution of Rotating Stars: *Astronomy and Astrophysics Library.* ISBN 978-3-540-76948-4. Springer Berlin Heidelberg
Maeder, A., Meynet, G., Ekström, S., & Georgy, C. 2009, *Communications in Asteroseismology*, 158, 72
Maeder, A. & Meynet, G. 2012, *Reviews of Modern Physics*, 84, 25
Martins, F., Hervé, A., Bouret, J.-C., *et al.* 2015, *Astron. & Astrophys.*, 575, A34
Meakin, C. A. & Arnett, D. 2007, *ApJ*, 667, 448
Meakin, C. A. & Arnett, D. 2007, *ApJ*, 665, 690
Meynet, G., Maeder, A., Schaller, G., Schaerer, D., & Charbonnel, C. 1994,*Astron. & Astrophys. Suppl.*, 103, 97
Meynet, G., Eggenberger, P., & Maeder, A. 2011, *Astron. & Astrophys.*, 525, L11
Meynet, G., Chomienne, V., Ekström, S., *et al.* 2015, *Astron. & Astrophys.*, 575, A60
Petit, V., Keszthelyi, Z., MacInnis, R., *et al.* 2017, *MNRAS*, 466, 1052
Przybilla, N., Firnstein, M., Nieva, M. F., Meynet, G., & Maeder, A. 2010, *Astron. & Astrophys.*, 517, A38
Puls, J., Vink, J. S., & Najarro, F. 2008, *The Astronomy and Astrophysics Review*, 16, 209
Salasnich, B., Bressan, A., & Chiosi, C. 1999, *Astron. & Astrophys.*, 342, 131
Sana, H., de Mink, S. E., de Koter, A., *et al.* 2012, *Science*, 337, 444
Sana, H., de Koter, A., de Mink, S. E., *et al.* 2013, *Astron. & Astrophys.*, 550, A107
Sander, A., Hamann, W.-R., & Todt, H. 2012, *Astron. & Astrophys.*, 540, A144
Smith, N. 2014, *Annual Review of Astronomy and Astrophysics*, 52, 487
Smith, N. & Owocki, S. P. 2006, *ApJL*, 645, L45
Song, H. F., Meynet, G., Maeder, A., Ekström, S., & Eggenberger, P. 2016, *Astron. & Astrophys.*, 585, A120
Song, H. F., Maeder, A., Meynet, G., *et al.* 2013, *Astron. & Astrophys.*, 556, A100
Spruit, H. C. 2002, *Astron. & Astrophys.*, 381, 923
Spruit, H. C. 1999, *Astron. & Astrophys.*, 349, 189
Stanway, E. R., Eldridge, J. J., & Becker, G. D. 2016, *MNRAS*, 456, 485
Taoso, M., Bertone, G., Meynet, G., & Ekström, S. 2008, *Physical Review D*, 78, 123510
Todini, P. & Ferrara, A. 2001, *MNRAS*, 325, 726
Ud-Doula, A., Owocki, S. P., & Townsend, R. H. D. 2009, *MNRAS*, 392, 1022
Ud-Doula, A., Owocki, S. P., & Townsend, R. H. D. 2008, *MNRAS*, 385, 97
ud-Doula, A. & Owocki, S. P. 2002, *ApJL*, 576, 413
Vanbeveren, D., Van Bever, J., & Belkus, H. 2007, *ApJL*, 662, L107
Wade, G. A., Neiner, C., Alecian, E., *et al.* 2016, *MNRAS*, 456, 2
Yoon, S.-C., Dessart, L., & Clocchiatti, A. 2017, arXiv:1701.02089
Yoon, S.-C., Woosley, S. E., & Langer, N. 2010, *ApJ*, 725, 940
Zahn, J.-P. 1992, *Astron. & Astrophys.*, 265, 115
Zahn, J.-P. 2011, *Active OB Stars: Structure, Evolution, Mass Loss, and Critical Limits*, 272, 14

Session 2: Death throes: supernovae, stellar deaths and progenitors

The Lives and Death-Throes of Massive Stars
Proceedings IAU Symposium No. 329, 2016
J.J. Eldridge, J.C. Bray, L.A.S. McClelland & L. Xiao, eds.

doi:10.1017/S1743921317002575

The Core-Collapse Supernova Explosion Mechanism

Bernhard Müller[1,2]

[1]Astrophysics Research Centre, Queen's University Belfast
BT7 1NN, Belfast, Northern Ireland
email: b.mueller@qub.ac.uk

[2]Monash Centre for Astrophysics, Monash University
Clayton, VIC 3800, Australia

Abstract. The explosion mechanism of core-collapse supernovae is a long-standing problem in stellar astrophysics. We briefly outline the main contenders for a solution and review recent efforts to model core-collapse supernova explosions by means of multi-dimensional simulations. Focusing on the neutrino-driven mechanism, we summarize currents efforts to predict supernova explosion and remnant properties.

Keywords. Supernovae: general, stars: evolution, stars: interiors, hydrodynamics, instabilities, convection, neutrinos, turbulence, methods: numerical

---

1. Introduction

Many open questions still surround the terminal gravitational core collapse of massive stars. Theory has by now established a clear picture of the first two acts of their death throes. The collapse of the iron (or in some cases O-Ne-Mg) core is triggered by deleptonization and/or photo-disintegration of heavy nuclei. The collapse is then halted once the core reaches supranuclear densities, and its elastic rebound launches a shock wave that quickly stalls due to energy losses by nuclear dissociation and neutrino losses. Theory has yet to fully explain the subsequent acts in the drama – the revival of the shock and the development of a supernova explosion – let alone the full systematics of supernova explosion and remnant properties. Among the proposed explosion scenarios (see Janka 2012 for an overview), the neutrino-driven mechanism and various flavours of a magnetorotational mechanism have been most thoroughly explored.

Here we shall mostly focus on the neutrino-driven mechanism, which has the virtue of not requiring special evolutionary channels for producing progenitors that spin more rapidly than generically predicted by current stellar evolution models (Heger *et al.* 2005; Cantiello *et al.* 2014) for the bulk of massive stars. In the neutrino-driven scenario (sketch in Figure 1), shock revival is accomplished thanks to the increase of the post-shock pressure due to partial reabsorption of neutrinos that stream out from the young proto-neutron star (PNS) and the cooling layer of accreted material on its surface. Except for the low-mass end of the progenitor spectrum (e.g. Kitaura *et al.* 2006; Melson *et al.* 2015b), this mechanism has been found to depend critically on multi-dimensional (multi-D) instabilities such as buoyancy-driven convection (Herant *et al.* 1994) and the standing accretion shock instability (SASI, Blondin *et al.* 2003) to foster neutrino-driven runaway shock expansion by a combination of effects including the mixing of the post-shock region and the provision of turbulent stresses (Murphy *et al.* 2013). Demonstrating quantitatively that shock revival can be achieved in this manner is a notoriously hard problem and requires sophisticated multi-physics simulations that incorporate neutrino transport,

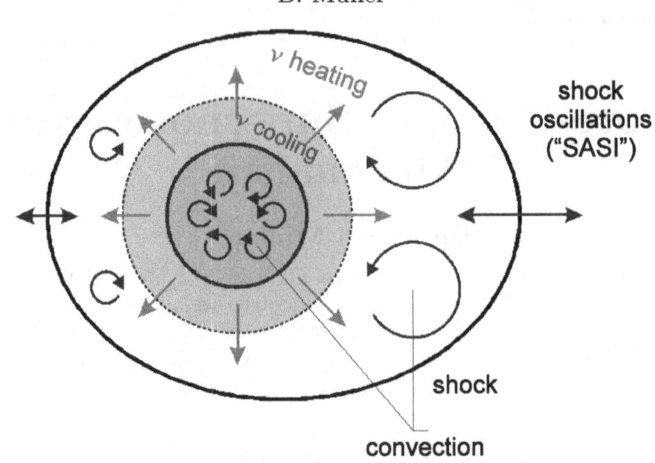

Figure 1. Sketch of the crucial elements of the neutrino-driven explosion mechanism. Neutrinos are emitted from the accretion layer (grey) around the PNS and from its core. A fraction of the electron-flavour neutrinos are reabsorbed in the "gain layer" behind the shock. In the gain region, non-spherical flow can develop because neutrino heating drives convective overturn and because of an advective-acoustic instability (SASI) that manifests itself in the form of large-scale oscillatory motions of the shock. Energy and lepton number loss also drives convection inside the PNS (cyan), which does not play a prominent role for shock revival, however.

multi-D fluid flow, general relativity, and the pertinent microphysics (neutrino interaction rates, nuclear equation of state) at an appropriate level of accuracy.

2. Progress and Challenges in Modelling Neutrino-Driven Explosions

As they progressed in sophistication over several decades, simulations of neutrino-driven explosion followed a tortuous path between success and failure. The most recent step to three-dimensional (3D) multi-group neutrino hydrodynamics simulations has been no exception. After experiments with simple "light-bulb" models in 3D (Nordhaus *et al.* 2010; Hanke *et al.* 2012; Couch 2013), fully-fledged 3D three -flavour neutrino hydrodynamics simulations soon became available (Hanke *et al.* 2013), and by now there is already a modest number of successful explosion models for progenitors from $9.6M_\odot$ to $20M_\odot$ obtained with rigorous (Melson *et al.* 2015b,a; Lentz *et al.* 2015) or simplified (Takiwaki *et al.* 2014; Müller 2015) energy-dependent neutrino transport. By and large, the 3D models have exhibited a trend towards *missing* (Hanke *et al.* 2012; Tamborra *et al.* 2014) or *delayed* explosions, i.e. they show less optimistic conditions for shock revival compared to the massive corpus of successful axisymmetric (2D) simulations that has been amassed by different groups (Janka *et al.* 2012; Bruenn *et al.* 2013; Nakamura *et al.* 2014; Summa *et al.* 2016; O'Connor & Couch 2015). The adverse effect of the third dimension has mainly been attributed to the different behaviour of turbulence in 3D and 2D (though this is only the most important element of a more nuanced view as pointed out by Müller 2016 and Janka *et al.* 2016). In 2D the inverse turbulent cascade proves conducive to the emergence of large-scale modes and allows convection and the SASI to become more vigorous (Hanke *et al.* 2012). The heating conditions are not *substantially* more pessimistic in 3D, however. A quantitative analysis reveals that even non-exploding models, such as the $27M_\odot$ (Hanke *et al.* 2012), $11.2M_\odot$, and $20M_\odot$ (Tamborra *et al.* 2014) runs of the MPA group, come close to the critical conditions for neutrino-driven runaway shock expansion.

Nonetheless, the results of recent simulations pose a problem. On the one hand, 3D models have not even been able to clearly establish the efficacy of the neutrino-driven mechanism for progenitors in the mass range of 10–15$M_\odot$, where there is clear observational evidence for "explodability" (Smartt 2015). This may merely be a result of poor sampling as only one case in the middle of this range has been simulated with state-of-the-art transport so far (Tamborra *et al.* 2014). On a more serious note, the delayed onset of explosions in 3D models makes it more difficult to reach significant explosion energies, as the decline of the neutrino luminosities and the mass in the gain region reduces the amount of energy that can be pumped into the ejecta by neutrino heating.

The more reluctant development of explosions in 3D has therefore prompted efforts to identify missing physical ingredients for earlier and more robust shock revival. The gist behind these ideas can be understood if the problem of shock revival is phrased in terms of a critical limiting luminosity $L_{\nu,\mathrm{crit}}$ as a function of accretion rate $\dot{M}$ for stationary accretion flow in spherical symmetry (Burrows & Goshy 1993). This concept has been generalized to account for other parameters of the accretion flow and multi-D effects (Janka 2012; Müller & Janka 2015; Janka *et al.* 2016) in a phenomenological manner and captures the transition to shock revival remarkably well even in multi-D (Summa *et al.* 2016). In terms of the averaged root-mean-square energy E_ν of ν_e and $\bar{\nu}_e$, the accretion rate $\dot{M}$, the PNS mass M, the gain radius R_g, the average net binding energy $|e_{\mathrm{tot,g}}|$ in the gain region, and the average $\langle \mathrm{Ma}^2 \rangle$ of the (squared) turbulent Mach number in the gain region, Summa *et al.* (2016) find

$$(L_\nu E_\nu^2)_\mathrm{crit} \propto \left(M\dot{M}\right)^{3/5} |e_{\mathrm{tot,g}}|^{3/5} R_\mathrm{g}^{-2/5} \left(1+\frac{4}{3}\langle \mathrm{Ma}^2\rangle\right)^{-3/5}. \tag{2.1}$$

The key to more robust explosions consists in identifying effects that change the terms in Equation (2.1) in a favourable direction.

2.1. *Uncertainties in the Microphysics*

The potential impact of uncertainties in the microphysics on the supernova explosion mechanism has long been recognized. We cannot cover the full spectrum of ideas, which even includes less defensible scenarios for achieving shock revival in spherical symmetry (Fischer *et al.* 2011). Instead we merely highlight some developments that illustrate that modest (and therefore credible) variations of the input physics could tip the balance in favour of shock revival in multi-D models that are already close to the explosion threshold by increasing L_ν and E_ν. This can be achieved, e.g., by nuclear equations of state that are "softer" (Janka 2012; Suwa *et al.* 2013) in the sense that they result in a faster contraction of the warm PNS and hence in hotter neutrinospheres. Contrary to superficial expectations, the increase of the neutrino luminosities and mean energies outweighs the adverse effect (increase of $|e_{\mathrm{tot,g}}|$) of the contraction of the gain radius.

A similar beneficial effect on the heating conditions could result from faster contraction of the PNS because of enhanced heavy flavour neutrino losses. This was recently explored in 3D simulations by Melson *et al.* (2015a), who found that a reduction of neutral-current scattering opacities of the order of ~20% in the neutrinospheric region was sufficient to tip the scale in favour of an explosion in a 20$M_\odot$ progenitor. Melson *et al.* (2015a) achieved this opacity reduction by assuming a stronger strangeness contribution $g_\mathrm{a}^\mathrm{s} = -0.2$ to the axial vector coupling of the nucleon than measured in current experiments. This is merely a parameterization of nuclear physics uncertainties that can lead to similar effects such as nucleon correlations; here the accuracy of extant rate calculations (Burrows & Sawyer 1998; Reddy *et al.* 1999) based on the random phase approximation (RPA) in the relevant density regime may be questioned. More reliable calculations of

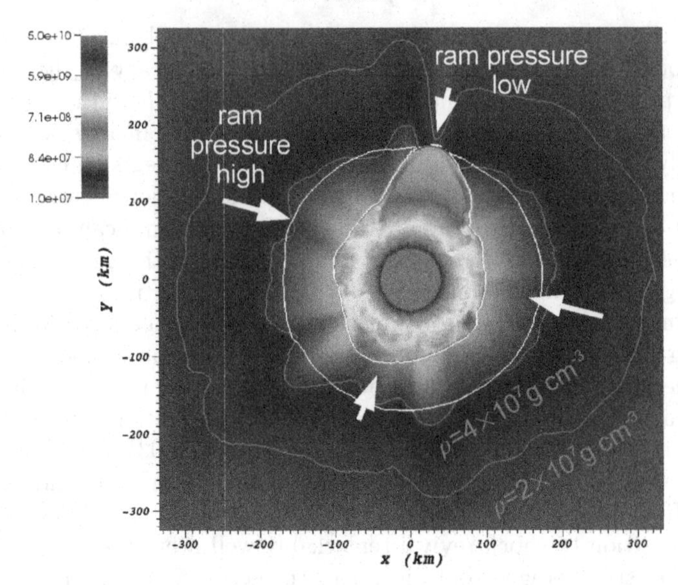

Figure 2. Illustration of the perturbation-aided version of the neutrino-driven mechanism, showing a 2D slice of the density in a simulation of an $18M_{\odot}$ star using 3D initial conditions from Müller (2016) at a time of 293 ms after bounce. The convective velocity perturbations in the oxygen shell are translated into density perturbations during the collapse as can be seen by the distortion of density isocontours (red), whereas variations in the pre-shock radial velocity v_r are small (outer white curve: isocontour for $v_r = -4.5 \times 10^9\,\mathrm{cm\,s^{-1}}$. The variations in ram pressure significantly distort the shock (inner white curve) and facilitate runaway shock expansion by creating large and stable buoyant bubbles.

neutral-current scattering opacities at neutrinospheric densities due to correlation effects are now emerging (Horowitz *et al.* 2016) and being tested in supernova simulations (Burrows *et al.* 2016), but whether there are sizable changes to extant models based on RPA opacities remains to be seen.

2.2. *Perturbation-Aided Explosions and Uncertainties in the Progenitor Structure*

Another idea for more robust neutrino-driven explosion models targets the term $\langle \mathrm{Ma}^2 \rangle$ in Equation (2.1) and seeks to identify effects that increase the violence of convection and the SASI in order to reduce the critical luminosity. This requires some additional forcing for the instabilities in the gain region, which could arise naturally due to the infall of shells with sizable seed asymmetries from the late convective burning stages (Couch & Ott 2013; Müller & Janka 2015). During the collapse, the initial velocity perturbations in such shells give rise to density and pressure perturbations by advective-acoustic coupling (Lai & Goldreich 2000). As illustrated in Figure 2, the anisotropic pre-shock ram pressure then leads to increased shock deformation, the development of fast lateral flows downstream of the shock as a result of shock obliquity (Müller & Janka 2015), and post-shock density perturbations that are translated into turbulent kinetic energy by buoyancy (Müller *et al.* 2016b).

Various studies employing parameterised initial perturbations (Couch & Ott 2013, 2015; Müller & Janka 2015; Burrows *et al.* 2016) have been conducted and established the conditions under which this "perturbation-aided" mechanism can be effective. Initial convective Mach numbers $\gtrsim 0.05$ and large-scale flow with dominant wavenumbers $\ell \lesssim 4$ are required for a significant reduction of the critical luminosity (Müller & Janka 2015). More definitive statements require supernova simulations starting from more self-consistent 3D initial models. First attempts to evolve massive stars through the last few minutes of convective shell burning before collapse in 3D have been made by Couch *et al.* (2015) for silicon burning in a $15M_{\odot}$ progenitor (assuming octant symmetry and a somewhat problematic acceleration of the deleptonization of the iron core) and by Müller *et al.* (2016b) for oxygen shell burning in an $18M_{\odot}$ progenitor using a contracting inner boundary condition. Numerical modelling of the final stages of massive stars is still in its infancy, however. Silicon burning in particular presents a major technical obstacle because it needs to be treated with larger nuclear reaction networks than used in current 3D progenitor simulations. Moreover, the 3D progenitors of Couch *et al.* (2015) and Müller *et al.* (2016b) have been computed under the proviso that the evolution up to the last minutes before collapse is captured correctly by spherically symmetric stellar evolution models (see below).

The first multi-group neutrino hydrodynamics simulation (Müller 2016) based on the 3D progenitor model of Müller *et al.* (2016b) showed a strong dynamical effect of the initial perturbations in the oxygen shell. Thanks to the forced shock deformation, a neutrino-driven explosion develops 250 ms after bounce, whereas shock revival does not take place in a control run with spherically symmetric initial conditions at least for another 300 ms. By contrast, the effect of perturbations was modest in the simpler leakage-based simulations conducted by Couch *et al.* (2015) using their $15M_{\odot}$ model. Despite the vastly different simulation methodology employed by these two groups, this may already indicate that large initial perturbations are not a panacea for the neutrino-driven mechanism: 1D stellar evolution models in fact show a huge spread of the parameters determining the efficiency of the perturbation-aided mechanism. The convective Mach numbers in the relevant shells range between $\sim$0.01 and $\sim$0.15 at the onset of collapse, and the shell thickness (which determines the dominant angular wavenumber ℓ of convective eddies) also varies tremendously. Pre-collapse asphericities may therefore significantly help shock revival in certain progenitors with violent shell burning, but play a subdominant role in the explosion mechanism in others.

While the idea of a perturbation-aided mechanism does not require any *fundamental* break with 1D stellar evolution models based on mixing length theory (and merely needs an additional initialization step before the onset of collapse), it has also been speculated that the secular evolution of supernova progenitors may be more seriously affected by multi-D processes such as turbulent entrainment at convective boundaries (Meakin & Arnett 2007), which might change M and $\dot{M}$ in Equation (2.1). At this stage, it still appears premature to draw conclusions based on 3D simulations of convective boundary mixing (Meakin & Arnett 2007; Müller *et al.* 2016b; Cristini *et al.* 2016; Jones *et al.* 2017) during brief intervals of advanced burning stages. More work is needed to translate the findings from such multi-D models into suitable recipes for 1D stellar evolution.

2.3. *Strong SASI, Rapid Rotation, and Magnetic Fields*

Several other mechanisms for facilitating neutrino-driven shock revival have also been explored recently, although it is less clear whether they could operate generically. Fernández (2015) found a *reduction* of the critical luminosity in the strongly SASI-dominated regime in 3D compared to 2D in his light-bulb models, which he traced to the ability of the non-axisymmetric SASI spiral mode to store more non-radial kinetic energy than the axisymmetric sloshing mode in the non-linear regime. Whether and when this regime is realized remains to be further investigated; the effect has not yet been replicated by self-consistent simulations that probe the SASI-dominated regime (e.g. the $27M_{\odot}$ and $20M_{\odot}$ models of Hanke *et al.* 2013 and Melson *et al.* 2015a), and may be restricted to massive progenitors with sustained high accretion rates.

Janka *et al.* (2016) and Takiwaki *et al.* (2016) showed that rapid rotation is conducive to shock revival even without the help of magnetic fields. For increasing rotation rate, this beneficial effect initially comes about because the contribution of rotational energy reduces the average net binding energy $|e_{\rm tot,g}|$ in the gain region, and because angular momentum support lowers the pre-shock velocity (Janka *et al.* 2016). Above a critical rotation rate corresponding to initial iron core spin periods of $\sim$1 s, the effect of rotation becomes more dramatic as a strong corotation instability develops (Takiwaki *et al.* 2016).

There may also be a regime where magnetic fields play a subsidiary role in the neutrino-driven mechanism instead of acting as the primary driver of the explosion if the initial fields can be sufficiently amplified by a small-scale turbulent dynamo, which could be provided by convection (Thompson & Duncan 1993) or by the SASI (Endeve *et al.* 2010, 2012) even in the absence of rapid rotation. If the fields come close to equipartition strength, they may help organize the flow into stable large-scale bubbles and thereby prove conducive to shock revival (Obergaulinger *et al.* 2014). However, strong initial fields of $\sim 10^{12}$ G are still required for this scenario in Obergaulinger *et al.* (2014)

If either rotation or magnetic fields are to play at least a subsidiary role in neutrino-driven explosions, it thus appears that special evolutionary channels are already required considering that current stellar evolution models predict pre-collapse spin periods of $\gtrsim 30$ s and magnetic fields of $\lesssim 10^{10}$ G (Heger *et al.* 2005) for solar-metallicity supernova progenitors.

2.4. *The Need for Alternative Mechanisms for Hypernovae*

That extremely fast rotation rates are encountered at least in a small fraction of supernova progenitors is suggested by observations of broad-lined Ic supernovae or "hypernova", whose explosion energies reach up to $\sim 10^{52}$ erg (Drout *et al.* 2011). These energies are likely out of reach for the neutrino-driven mechanism and thus require a different explosion mechanism altogether; even optimistic parameterised models suggest an upper limit of $\sim 2 \times 10^{51}$ erg for neutrino-driven explosions. Various magnetohydrodynamic mechanisms (Usov 1992; MacFadyen & Woosley 1999; Akiyama *et al.* 2003) that tap the rotational energy of a rapidly spinning and highly magnetized "millisecond magnetar", black hole, or accretion disk remain the most promising explanations for the most energetic supernovae and the gamma-ray-bursts associated with some of them, but we must refer to the review of Janka (2012) for a more detailed discussion.

3. Outlook: From Neutrino-Driven Explosion Models to Observables

It is conceivable that 3D simulations of core-collapse supernovae may need no more than a combination of relatively minor changes in the input physics and slight improvements in numerical accuracy (for a discussion see Janka *et al.* 2016 and Müller 2016) to produce neutrino-driven explosions over a wide range of progenitor masses. This, however, would only provide a solution to the problem of shock revival. The bigger challenge lies in accounting for the observed explosion (e.g. explosion energy and nickel mass) and remnant properties (neutron star mass, spin, and kick).

In neutrino-driven supernovae, the key explosion properties only reach their asymptotic values after a phase of simultaneous accretion and mass ejection that lasts for $\gtrsim$1 s after shock revival (or even later in the case of the neutron star kicks, see Wongwathanarat *et al.* 2013). Such time-scales are only marginally within reach even for 3D hydrodynamics simulations with simplified multi-group neutrino transport (Müller 2015, 2016). Axisymmetric models do not offer a serious alternative not only because of their higher proclivity to explosion. 2D effects are even more problematic during the explosion phase than prior to shock revival (Müller 2015) and may be responsible for the low explosion energies (compared to typical observed values of $5 \ldots 9 \times 10^{50}$ erg, see Kasen & Woosley 2009) and high neutron star masses found in typical 2D simulations (Janka *et al.* 2012; Nakamura *et al.* 2015; O'Connor & Couch 2015, but see also the more energetic models of Bruenn *et al.* 2016). Because of these obstacles, it yet remains to be demonstrated by self-consistent models that the neutrino-driven mechanism can account for the explosion properties of the majority of core-collapse supernovae (and this statement holds *a fortiori* for the magnetohydrodynamic mechanism).

At present, the only alternative for understanding the systematics of the explosion properties is to rely on parameterised models of neutrino-driven supernovae (e.g. O'Connor & Ott 2010; Ugliano *et al.* 2012; Pejcha & Thompson 2015; Ertl *et al.* 2016; Sukhbold *et al.* 2016; Müller *et al.* 2016a) that are based on 1D simulations and/or analytic theory. Although objections may be raised against the predictiveness of such an approach, it has undoubtedly proved valuable and led to significant results. Although the available parameterised models differ greatly in detail, some findings such as trends in "explodability" (with explosions up to $\sim 15 M_\odot$ and several islands of explodability alternating with black hole formation up to $\sim 30 M_\odot$), or the upper limit of $\sim 2 \times 10^{51}$ erg (Ugliano *et al.* 2012; Ertl *et al.* 2016; Sukhbold *et al.* 2016; Müller *et al.* 2016a) have proved quite robust, which suggests that they indeed reflect the inherent physics of the neutrino-driven mechanism. Furthermore, first-principle simulations are increasingly starting to inform parameterised models (Müller *et al.* 2016a), and these in turn are proving useful for selecting appropriate targets for full-scale multi-D simulations. Given the prohibitive costs of state-of-the-art models, such a two-pronged approach likely remains the best strategy for explaining the diversity of core-collapse supernovae in the foreseeable future.

Acknowledgements

The author acknowledges long-term assistance by his collaborators, especially A. Heger, H.-Th. Janka, and T. Melson, and support by the STFC DiRAC HPC Facility (DiRAC Data Centric system, ICC Durham), the National Computational Infrastructure (Australia), the Pawsey Supercomputing Centre (University of Western Australia), and the Minnesota Supercomputing Institute.

References

Akiyama, S., Wheeler, J. C., Meier, D. L., Lichtenstadt, I. Meier, D. L., & Lichtenstadt. 2003, *ApJ*, 584, 954

Blondin, J. M., Mezzacappa, A., & DeMarino, C. 2003, *ApJ*, 584, 971

Bruenn, S. W. *et al.* 2016, *ApJ*, 818, 123

——. 2013, *ApJL*, 767, L6

Burrows, A. & Goshy, J. 1993, *ApJL*, 416, L75+

Burrows, A. & Sawyer, R. F. 1998, *Phys. Rev. C*, 58, 554

Burrows, A., Vartanyan, D., Dolence, J. C., Skinner, M. A., & Radice, D. 2016, ArXiv e-prints, 1611.05859

Cantiello, M., Mankovich, C., Bildsten, L., Christensen-Dalsgaard, J., & Paxton, B. 2014, *ApJ*, 788, 93

Couch, S. M. 2013, *ApJ*, 775, 35

Couch, S. M., Chatzopoulos, E., Arnett, W. D., & Timmes, F. X. 2015, *ApJL*, 808, L21

Couch, S. M. & Ott, C. D. 2013, *ApJL*, 778, L7

——. 2015, *ApJ*, 799, 5

Cristini, A., Meakin, C., Hirschi, R., Arnett, D., Georgy, C., & Viallet, M. 2016, *Phys. Scr.*, 91, 034006

Drout, M. R. *et al.* 2011, *ApJ*, 741, 97

Endeve, E., Cardall, C. Y., Budiardja, & Mezzacappa, A. 2010, *ApJ*, 713, 1219
Endeve, E., Cardall, C. Y., Budiardja, R. D., Beck, S. W., Bejnood, A., Toedte, R. J., Mezzacappa, A., & Blondin, J. M. 2012, *ApJ*, 751, 26
Ertl, T., Janka, H.-T., Woosley, S. E., Sukhbold, T., & Ugliano, M. 2016, *ApJ*, 818, 124
Fernández, R. 2015, *MNRAS*, 452, 2071
Fischer, T. *et al.* 2011, *ApJS*, 194, 39
Hanke, F., Marek, A., Müller, B., & Janka, H.-T. 2012, *ApJ*, 755, 138
Hanke, F., Müller, B., Wongwathanarat, A., Marek, A., & Janka, H.-T. 2013, *ApJ*, 770, 66
Heger, A., Woosley, S. E., & Spruit, H. C. 2005, *ApJ*, 626, 350
Herant, M., Benz, W., Hix, W. R., Fryer, C. L., & Colgate, S. A. 1994, *ApJ*, 435, 339
Horowitz, C. J., Caballero, O. L., Lin, Z., O'Connor, E., & Schwenk, A. 2016, ArXiv e-prints, 1611.05140
Janka, H.-T. 2012, Annual Review of Nuclear and Particle Science, 62, 407
Janka, H.-T., Hanke, F., Hüdepohl, L., Marek, A., Müller, B., & Obergaulinger, M. 2012, Progress of Theoretical and Experimental Physics, 2012, 010000
Janka, H.-T., Melson, T., & Summa, A. 2016, *Annual Review of Nuclear and Particle Science*, 66, 341, 1602.05576
Jones, S., Andrassy, R., Sandalski, S., Davis, A., Woodward, P., & Herwig, F. 2017, *MNRAS*, 465, 2991
Kasen, D. & Woosley, S. E. 2009, *ApJ*, 703, 2205
Kitaura, F. S., Janka, H.-T., & Hillebrandt, W. 2006, *A&A*, 450, 345
Lai, D. & Goldreich, P. 2000, *ApJ*, 535, 402
Lentz, E. J. *et al.* 2015, *ApJL*, 807, L31
MacFadyen, A. I. & Woosley, S. E. 1999, *ApJ*, 524, 262
Meakin, C. A. & Arnett, D. 2007, *ApJ*, 667, 448
Melson, T., Janka, H.-T., Bollig, R., Hanke, F., Marek, A., & Müller, B. 2015a, *ApJL*, 808, L42
Melson, T., Janka, H.-T., & Marek, A. 2015b, *ApJL*, 801, L24
Müller, B. 2015, *MNRAS*, 453, 287
——. 2016, *PASA*, 33, e048
Müller, B., Heger, A., Liptai, D., & Cameron, J. B. 2016a, *MNRAS*, 460, 742
Müller, B. & Janka, H.-T. 2015, *MNRAS*, 448, 2141
Müller, B., Viallet, M., Heger, A., & Janka, H.-T. 2016b, *ApJ*, 833, 124
Murphy, J. W., Dolence, J. C., & Burrows, A. 2013, *ApJ*, 771, 52
Nakamura, K., Kuroda, T., Takiwaki, T., & Kotake, K. 2014, *ApJ*, 793, 45
Nakamura, K., Takiwaki, T., Kuroda, T., & Kotake, K. 2015, *PASJ*, 67, 107
Nordhaus, J., Burrows, A., Almgren, A., & Bell, J. 2010, *ApJ*, 720, 694
Obergaulinger, M., Janka, H.-T., & Aloy, M. A. 2014, *MNRAS*, 445, 3169
O'Connor, E. & Couch, S. 2015, ArXiv e-prints, 1511.07443
O'Connor, E. & Ott, C. D. 2010, Classical and Quantum Gravity, 27, 114103
Pejcha, O. & Thompson, T. A. 2015, *ApJ*, 801, 90
Reddy, S., Prakash, M., Lattimer, J. M., & Pons, J. A. 1999, *Phys. Rev. C*, 59, 2888
Smartt, S. J. 2015, *PASA*, 32, 16
Sukhbold, T., Ertl, T., Woosley, S. E., Brown, J. M., & Janka, H.-T. 2016, *ApJ*, 821, 38
Summa, A., Hanke, F., Janka, H.-T., Melson, T., Marek, A., & Müller, B. 2016, *ApJ*, 825, 6, 1511.07871
Suwa, Y., Takiwaki, T., Kotake, K., Fischer, T., Liebendörfer, M., & Sato, K. 2013, *ApJ*, 764, 99
Takiwaki, T., Kotake, K., & Suwa, Y. 2014, *ApJ*, 786, 83
——. 2016, *MNRAS*, 461, L112
Tamborra, I., Hanke, F., Janka, H.-T., Müller, B., Raffelt, G. G., & Marek, A. 2014, *ApJ*, 792, 96
Thompson, C. & Duncan, R. C. 1993, *ApJ*, 408, 194
Ugliano, M., Janka, H.-T., Marek, A., & Arcones, A. 2012, *ApJ*, 757, 69
Usov, V. V. 1992, *Nature*, 357, 472
Wongwathanarat, A., Janka, H.-T., & Müller, E. 2013, *A&A*, 552, A126

The Lives and Death-Throes of Massive Stars
Proceedings IAU Symposium No. 329, 2016
J.J. Eldridge, J.C. Bray, L.A.S. McClelland & L. Xiao, eds.

doi:10.1017/S1743921317003271

Core-Collape Supernova Progenitors: Light-Curve and Stellar-Evolution Models

Melina C. Bersten

Instituto de Astrofísica de La Plata (IALP), CCT-CONICET-UNLP. Paseo del Bosque S/N (B1900FWA), La Plata, and Facultad de Ciencias Astronómicas y Geofísicas, Universidad Nacional de La Plata, Argentina, Kavli Institute for the Physics and Mathematics of the Universe (WPI), The University of Tokyo, 5-1-5 Kashiwanoha, Kashiwa, Chiba 277-8583, Japan
email: mbersten@fcaglp.unlp.edu.ar

Abstract. A very active area of research in the field of core-collapse supernovae (SNe) is the study of their progenitors and the links with different subtypes. Direct identification using pre- and post-SN images is a powerful method but it can only be applied to the most nearby events. An alternative method is the hydrodynamical modeling of SN light curves and expansion velocities, which can serve to characterize the progenitor (e.g. mass and radius) and the explosion itself (e.g. explosion energy and radioactive yields). This latter methodology is particularly powerful when combined with stellar evolution calculations. We review our current understanding of the properties of normal core-collapse SNe based chiefly on these two methods.

Keywords. binaries: general, supernovae: general

1. Introduction

Most of massive stars ($M_{\rm ZAMS} \gtrsim 8M_{\odot}$) end their lives as core-collapse supernovae (CCSNe), leaving behind either neutron stars or black holes, and ejecting heavy elements into space with large explosion energies. Thus, these events are important in our understanding of the chemical and dynamical evolution of the Universe. Nonetheless, it is not yet clear what is the mechanism that transforms the core collapse into a SN explosion.

It has long been known that there are different types of CCSNe. The classification is mainly based on spectral properties. The main division comes from the presence (Type II) or absence (Type I) of hydrogen (H) lines in the spectra (Filippenko(1997)). Type II SNe (or H-rich) are the most common type of the explosion in the Universe, more than 50% of CCSNe belong to this Type (Smith *et al.* (2011)). In section 3.1 we provide an overview of our knowledge of these objects. Type I (or H-deficient) SNe are usually also called “Stripped-envelope SNe” (SESNe). Depending on the degree of envelope stripping, they are classified as Type IIb, Ib, or Ic. We review these objects in section 3.2.

An important remaining problem in astrophysics is finding the links between SN Types and their progenitor stars. In particular, it is important to know what kind of CCSNe come from single stars, as opposed to interacting binary systems. Recent studies of open clusters have indicated that the incidence of interacting binaries among massive stars is particularly large (about 70%; Sana *et al.* (2012)). This clearly shows the relevance of close binary evolution in connection with SN progenitors. There are also others arguments that support the idea that the majority of SESNe originate from binary systems. For example, the fractions of different SN Types (Smith *et al.* (2011)) and the ejecta mass estimates in SESNe (Drout *et al.* (2011), Lyman *et al.* (2014)). In particular for the case of SN IIb, there is further evidence of binarity with the detection of the companion star in late-time observations, as in the case of SN 1993J (Maund *et al.* (2004), Fox *et al.* (2014)) and possibly in SN 2011dh (Folatelli *et al.* (2014)).

2. Progenitor Identification Methods

There are several methods proposed in the literature to analyze physical properties of the SN progenitors, namely:

- Archival pre-explosion image searches (combined with post-explosion observations)
- Environmental and metallicity studies
- SN rates
- Mass-loss rates from radio and X-rays
- Spectropolarimetry
- Very early ("flash") spectroscopy
- Light-curve and spectrum modeling

The search for progenitor stars in deep pre-explosion images is a powerful, direct approach to understand the origin of SNe. It provides a critical test for stellar evolution models. Important results have been achieved using this technique, which we describe in section 2.1. However, in most cases, either because the SN is too distant or simply due to lack of pre-supernova images, other methods are required to infer progenitor properties. One such method is the hydrodynamical modeling of SN observations which we describe in section 2.2.

2.1. *Archival imaging method*

This technique is based on the identification of a possible progenitor star using archival pre-explosion images of the SN location. The association then needs to be confirmed using post-explosion imaging to confirm the disappearance of the progenitor candidate. This method has been largely favored with the use of the HST archive. Important results have been found in recent years. Currently, there are 20 progenitor detections and $\approx$30 upper limits (see Smartt (2015) for a review). In general terms, there are two main groups leading efforts in this area, one led by S. Smartt and the other by S. Van Dyk. Using pre-explosion photometry an estimate of the main-sequence mass of the progenitor can be derived by assuming some stellar evolution model. Below we list some of the main conclusions achieved with this method:

- Confirmation that Type II-P SNe arise from the explosion of red supergiant (RSG) stars, as previously predicted by stellar evolution calculations and by hydrodynamical analysis of SN light curves.
- SN II-P progenitors have $M_{ZAMS} \lesssim 16 - 18\ M_{\odot}$
- One blue supergigant (BSG) progenitor detected in association with the famous SN 1987A, a peculiar Type II SN.
- One Luminous Blue Variable (LBV) progenitor associated with a Type IIn SN. †
- Some yellow supergiant (YSG) progenitors detected, mostly associated with Type IIb SNe.
- No detection of Type Ib or Ic SN progenitors with the possible exception of SN iPTF13bvn (see more details in section 3.2)

† Type IIn SNe are objects that show narrow lines in their spectra. The narrow lines are indicative of interaction between the SN ejecta and a previously existing circunmstellar medium (CSM).

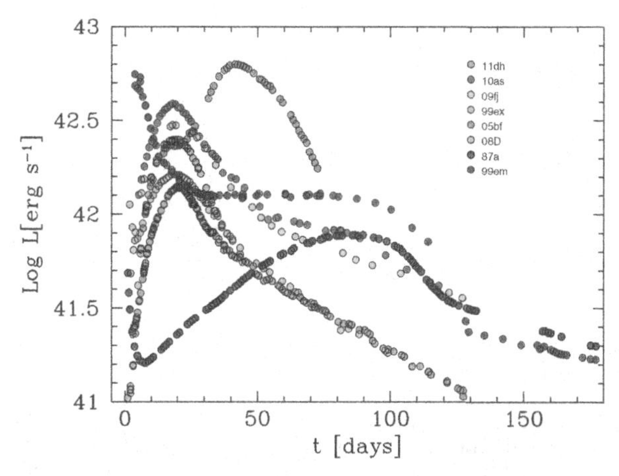

Figure 1. Bolometric light curves of CCSNe showing the diversity among these Types of SNe.

We emphasize that with the current technology, this method cannot be applied to events further than ~30 Mpc. Therefore, more indirect methods are required to analyze the progenitors of the majority of SNe.

2.2. *Hydrodynamical modeling*

CCSNe light curves (LC) are very heterogeneous (see Figure 1), as opposed to the standard-candle Type Ia SN. This heterogeneity is associated to varying progenitor properties. It is a very well-known fact that the LC morphology is sensitive to the physical characteristics of the SN progenitor. Therefore, LC modeling, ideally combined with modeling of spectra or photospheric velocities, provides a useful way to constrain progenitor properties, such as mass and radius, as well as explosion parameters (explosion energy and production of radioactive material). This methodology is particularly powerful when combined with stellar evolution calculations.

It must be noted that there is currently no self-consistent model for the origin of the SN explosion. Nevertheless, SN problem is usually decoupled into two independent processes: the explosion trigger, and the ejection of the envelope. The propagation of the explosion through the envelope can be simulated independently of how it is triggered. In this way, it is possible to study the observational outcome of the explosion, such as LCs and spectra. Bersten *et al.* (2011) developed a one-dimensional Lagrangian code with flux-limited radiation diffusion and gray transfer for gamma-rays to artificially explode the hydrostatic structures and analyze the outcome after the shock propagation. Figure 2 shows how sensitive the LC is on the initial progenitor structure assumed. A typical SN II-P LC is produced when one explodes a RSG structure. For BSG progenitors a typical 87a-like SN is produced. If one adopts a compact progenitor, such as a helium (He) star, the typical ^{56}Ni-powered LC is produced.

Different phases of the LC evolution can be distinguished, each one with different dependence on physical parameters (see Figure 3). For the case for RSG, with thick H envelopes, we can distinguish a cooling phase (mainly dependent on the stellar radius), a plateau phase (dependent on H mass, explosion energy and radius), and a tail phase (mainly determined by the radioactive material). For more compact structures, most of the SN evolution happens during the radioactive-power phase. However, the early emission provides unique information about the structure of the star previous to the explosion, as well as the mixing processes. The small progenitor radius is responsible for the rapid degradation of the shock energy, leading to a fast initial peak that is usually

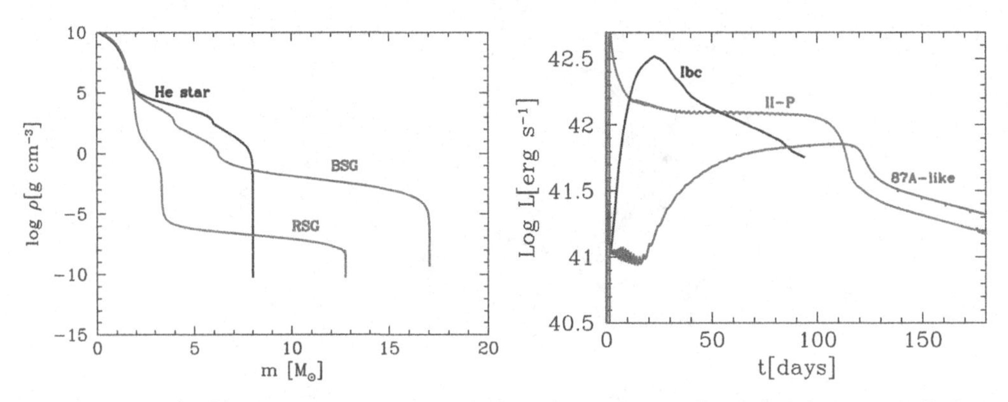

Figure 2. (Left) Progenitor structures before the explosion. (Right) Bolometric LCs for different progenitor structures assumed.

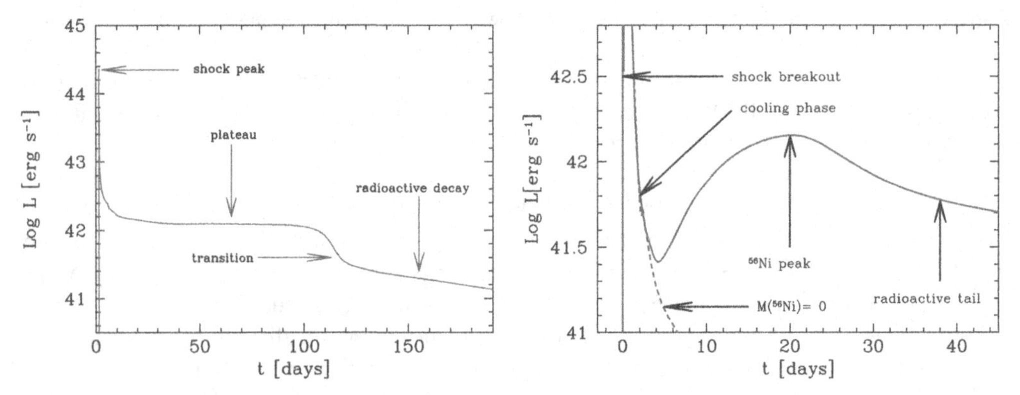

Figure 3. (Left) Bolometric LC for a RSG structure. (Right) Bolometric LC for SESNe. Different phases of LC evolution are indicated in the figures.

unobserved. Bersten *et al.* (2012) have shown the dependence of the early emission on the progenitor radius for Type IIb SNe (see their Figure 10), and for Type Ib SNe in Bersten *et al.* (2014) (see their Figure 2). Until recently, only a handful of SESNe were observed during the cooling phase. However, given the important information provided during this phase, large efforts are being made by current surveys to catch the SNe as early as possible. Therefore, it is expected to have an interesting progress in this area in coming years.

3. SN Progenitors

In the following section we briefly summarize some recent results on CCSNe progenitors, separated into H-rich and H-poor (or SESNe) events.

3.1. *Hydrogen-rich*

H-rich SNe are the most common type of explosion in nature, amounting to ~50% of all CCSNe. Additional interest on SNe II-P has arisen from the fact that they have been proposed as good distance indicators, independent of Type Ia SNe, for example via the standard-candle method (SCM) (Hamuy & Pinto (2002)). Regarding the range of masses of their progenitors, pre-explosion imaging connected with stellar evolution calculations suggest a range of masses of $M_{\rm ZAMS}$ $8-16$ $M_{\odot}$. This is in contradiction with hydrodynamical LC modeling (see for example Utrobin (2007)), which predicts a larger

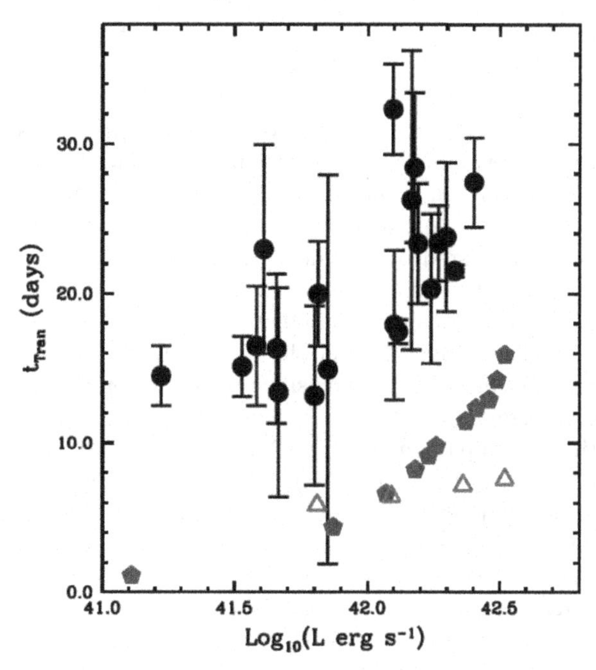

Figure 4. Bolometric luminosity at the mid-point of the plateau as a function of t_{trans} for CSP data (black points) and for models with different radii (blue points) and different energies (red triangles)

range of masses, as noted by Smartt (2009). Until now, the reason of this discrepancy is not clear.

There are some recent works focused on the properties of the early LC. For example González-Gaitán *et al.* (2015) have shown that the typical radii of SNe II progenitors should be $\approx 500\ R_{\odot}$, or alternatively, some CSM needs to be present nearby. This conclusion was based on rise-time studies of a sample of early LC. Recent hydrodynamical modeling has also suggested the presence of CSM in order to explain the LC of some SNe II (Morozova *et al.* (2016)). Figure shows the relation between t_{trans} (defined as the epoch of the beginning of the plateau phase) and the plateau luminosity (black points) for a large sample of very well-observed SNe II observed by the *Carnegie Supernova Project* (Anderson *et al.* (2014)). A set of hydro models is also shown as a comparison (blue points for different progenitor radii and red triangles for different explosion energies). Although the models can reproduce the dependency of the data there is a clear offset between models and observations. This can be solve by assuming some CSM close to the progenitor star (Anderson *et al.*, prep). Our results indicate that most of the H-rich progenitors need to loose some material previous to the explosion, in concordance with previous works based on fewer objects.

Finally, it is important to mention that SNe II have recently been proposed as good metallicity indicators (Anderson *et al.* (2016)).

3.2. *Hydrogen-poor*

The two most appealing mechanisms to remove the H envelope necessary to produce SESNe progenitors are strong stellar winds in very massive stars ($M \gtrsim 25 M_{\odot}$) and mass transfer in close binary systems (see Langer (2012) for a recent review). Pre-explosion image analysis as well as LC modeling suggest low mass progenitors (see e.g. Drout *et al.* (2011), Lyman *et al.* (2014)), which can only be reconciled with the binary scenario.

In addition, there are three confirmed YSG progenitors of type IIb, namely SN 1993J (Maund and Smartt (2009)), SN 2008ax (Folatelli *et al.* (2015)) and SN 2011dh (Van Dyk *et al.* (2013)). And one candidate SN 2013df (Van Dyk *et al.* (2014)). The YSG progenitor is difficult to explain by single stellar evolution models, unless mass-loss rates due to winds are assumed to be several times higher than the standard values (Georgy (2012)). On the contrary, a YSG star can be naturally produced in an interacting binary system (see e.g. Benvenuto *et al.* (2013)). Moreover, the possible detection of the companion of SN 1993J have been suggested by Maund *et al.* (2004) and more recent data appear to confirm this suggestion (Fox *et al.* (2014)). For the case of SN 2011dh, a blue point source with compatible properties as the predicted companion star has been detected in deep UV HST post-explosion images (Folatelli *et al.* (2014)). However, the optical post-explosion data of SN 2011dh are not conclusive about the companion detection (Maund *et al.* (2015)). Further observations need to be carried out in the coming years to confirm the association with the companion star.

Until very recently no firm progenitor identification was reported for H-deficient SNe (Eldridge *et al.* (2013)). The exception is SN iPTF13bvn, whose progenitor candidate was identified by Cao *et al.* (2013) and recently confirmed by Folatelli *et al.* (2016) and Eldridge & Maund(2016) using HST post-explosion observations. Therefore, iPTF13bvn represents the first and until now only H-deficient progenitor detected. At the moment no detection has been done of He-deficient SN Ic progenitors.

4. Final Remarks

Several pieces of evidence suggest that a large fraction of massive stars belong to interacting binary systems. This suggests that most SESNe arise from binaries. There is only one firm companion detection and possibly a second case.

Regarding progenitor masses, there is consistency between pre-explosion imaging and hydrodynamical modeling for H-poor objects. On the contrary, the situation is problematic among H-rich events. According to the pre-explosion imaging method for SNe II, initial masses of most "ordinary" CCSN progenitors seem to be $\lesssim 20\ M_{\odot}$. This rises the question about what is the outcome of more massive stars. The current situation seems to indicate that efforts must be made in improving our understanding of massive star evolution.

References

Anderson J. P., Bersten, M. C., *et al.*, in preparation
Anderson, J. P., Gutiérrez, C. P., Dessart, L., *et al.* 2016, *A&A*, 590, C2
Anderson J. P., *et al.*, 2014, *ApJ*, 786, 67
Benvenuto, O. G., Bersten, M. C., & Nomoto, K. 2013, *ApJ*, 762, 74
Bersten, M. C., Benvenuto, O. G., Folatelli, G., *et al.* 2014, *AJ*, 148, 68
Bersten, M. C., Benvenuto, O. G., Nomoto, K., *et al.* 2012, *ApJ*, 757, 31
Bersten, M. C., Benvenuto, O., & Hamuy, M. 2011, *ApJ*, 729, 61
Cao, Y., Kasliwal, M. M., Arcavi, I., *et al.* 2013, *ApJl*, 775, L7
Drout, M. R., Soderberg, A. M., Gal-Yam, A., *et al.* 2011, *ApJ*, 741, 97
Eldridge, J. J. & Maund, J. R. 2016, *MNRAS*, 461, L117
Eldridge, J. J., Fraser, M., Smartt, S. J., Maund, J. R., & Crockett, R. M. 2013, *MNRAS*, 436, 774
Filippenko, A. V. 1997, *ARA&A*, 35, 309
Folatelli, G., Van Dyk, S. D., Kuncarayakti, H., *et al.* 2016, *ApJl*, 825, L22
Folatelli, G., Bersten, M. C., *et al.* 2015, *ApJ*

Folatelli, G., Bersten, M. C., Benvenuto, O. G., *et al.* 2014, *ApJl*, 793, L22
Fox, O. D., Azalee Bostroem, K., Van Dyk, S. D., *et al.* 2014, *ApJ*, 790, 17
Georgy, C. 2012, *A&A*, 538, L8
González-Gaitán, S., Tominaga, N., Molina, J., *et al.* 2015, *MNRAS*, 451, 2212
Hamuy, M. & Pinto, P. A. 2002, *ApJl*, 566, L63
Langer, N. 2012, *ARA&A*, 50, 107
Lyman, J. D., Bersier, D., & James, P. A. 2014, *MNRAS*, 437, 3848
Maund, J. R., Arcavi, I., Ergon, M., *et al.* 2015, *MNRAS*, 454, 2580
Maund, J. R. & Smartt, S. J. 2009, *Science*, 324, 486
Maund, J. R., Smartt, S. J., Kudritzki, R. P., Podsiadlowski, P., & Gilmore, G. F. 2004, *Nature*, 427, 129
Morozova, V., Piro, A. L., & Valenti, S. 2016, arXiv:1610.08054
Sana, H., de Mink, S. E., de Koter, A., *et al.* 2012, *Science*, 337, 444
Smartt, S. J. 2009, *ARA&A*, 47, 63
Smartt, S. J. 2015, *PASA*, 32, e016
Smith, N., Li, W., Silverman, J. M., Ganeshalingam, M., & Filippenko, A. V. 2011, *MNRAS*, 415, 773
Utrobin, V. P. 2007, *A&A*, 461, 233
Van Dyk, S. D., Zheng, W., Fox, O. D., *et al.* 2014, *AJ*, 147, 37
Van Dyk, S. D., Zheng, W., Clubb, K. I., *et al.* 2013, *ApJ*, 772, L32

The Lives and Death-Throes of Massive Stars
Proceedings IAU Symposium No. 329, 2016
J.J. Eldridge, J.C. Bray, L.A.S. McClelland
& L. Xiao, eds.

doi:10.1017/S1743921317002885

The progenitors of core-collapse supernovae

Morgan Fraser†,

School of Physics, O'Brien Centre for Science North, University College Dublin, Belfield, Dublin 4, Ireland
email: morgan.fraser@ucd.ie

Abstract. Linking core-collapse SNe to their stellar progenitors is a major ongoing challenge. To date, H rich Type IIP SNe have been shown to come from red supergiants, while there is increasing evidence that the majority of stripped envelope SNe come from binary systems. The first candidates for failed SNe, where a massive star collapses to form a black hole without a bright optical display have been identified, while the range of outbursts and eruptions from pre-SN stars are just beginning to be revealed.

Keywords. supernovae: general, supergiants

1. Introduction

The tremendous diversity observed in core-collapse supernovae (SNe) is believed to arise from the wide range of possible pre-SN progenitor configurations. The chemical composition of the ejecta, bolometric luminosity, ejected ^{56}Ni mass, explosion energy and timescales, are hence expected to be a function of the progenitor mass, radius, composition and multiplicity. While it is relatively easy to measure SN properties, it is much harder to connect these to the pre-SN progenitor characteristics. In this review, I discuss the use of archival imaging, hydrodynamic and spectroscopic modeling, and mass-loss diagnostics to probe the progenitors of core-collapse SNe.

2. Progenitor diagnostics

Progenitor detections Since the first identification of the progenitor of SN 1987A in archival data (West *et al.* 1987), around fifty core-collapse SNe have either detected progenitors, or had restrictive limits placed upon them (e.g. Van Dyk *et al.* 2003, Smartt *et al.* 2009). For each of these SNe, a pre-explosion image was available with sufficient depth and spatial resolution to potentially detect an individual massive star. In most cases, these images came from the Hubble Space Telescope (HST), which has now observed over 40% of nearby, massive, low inclination, star-forming galaxies.

To identify a progenitor candidate in pre-explosion images, an image of the SN with comparable depth and resolution is required. While some SNe have been observed with HST to this end, more often 8-m class telescopes equipped with adaptive-optics (AO) are used. Ground-based AO can match the $\sim$0.1″ resolution of HST, albeit over a smaller field of view and at near-infrared (NIR) wavelengths. Using differential astrometry between the pre- and post-explosion images, the position of the SN can typically be localised to a few tens of milliarcseconds on a pre-explosion image. Once the SN position is determined, it is simply a matter measuring the flux from an identified, co-incident progenitor candidate, or setting an upper limit to its flux if no progenitor candidate is detected.

† Royal Society - Science Foundation Ireland University Research Fellow

To infer a progenitor mass from measured photometry requires the use of either empirical relations between observed colours and effective temperature, or model atmospheres and synthetic photometry (e.g. the MARCS models; Gustafsson *et al.* 2008). In either case, the inferred temperature allows the bolometric correction to be estimated, and a progenitor luminosity calculated. Once the luminosity and temperature of a SN progenitor are known (or at least constrained), a zero-age main sequence (ZAMS) progenitor mass can be determined through comparison to stellar evolutionary tracks.

While in principle this is a relatively straightforward process, in practice several complications arise. If a SN progenitor is only observed in a single band, then a guess must be made as to its temperature. For Type IIP SNe, which require a progenitor with an extended H envelope to give their characteristic plateau, the progenitor is assumed to be a red supergiant (RSG). The precise choice of temperature will also affect the calculated luminosity, for example the bolometric correction to V-band changes by 1 mag when going from a 3400 K RSG, to a star that is only 200 K hotter. In such cases, it is prudent to be conservative, and set a restrictive *upper limit* to the luminosity and mass based on the largest plausible bolometric correction.

A second complication comes from the (often unknown) circumstellar extinction. While in many cases, attempts have been made to estimate the extinction local to a SN from the strength of the NaD lines (e.g. Turatto, Benetti & Cappellaro 1997), these relations have considerable scatter for individual objects (Poznanski *et al.* 2011). Furthermore, dust local to the progenitor can be destroyed by the shock breakout of the SN, as was seen for SN 2012aw (Fraser *et al.* 2012). Measured extinction towards the SN can hence only be taken as a lower limit to the extinction towards the progenitor, and the safest approach is to fit the reddening simultaneously with the progenitor temperature (if colours are known for the progenitor). It is also important to remember that circumstellar dust around a progenitor should not be treated with standard interstellar extinction laws (such as Cardelli, Clayton & Mathis 1989), as the dust is a spherical shell rather than a foreground screen (see Kochanek, Khan & Dai 2012 for a comprehensive discussion of this point).

Finally, when determining the ZAMS mass of the SN progenitor from its luminosity, it is essential to make a physically meaningful comparison. For a core-collapse SN to explode the progenitor must have a massive Fe core, and so it is necessary to compare the observed luminosity of the progenitor to the *pre-SN* luminosity of a model as close as possible to Si-burning, regardless of its effective temperature. As was seen for SN 2011dh (Maund *et al.* 2011, Van Dyk *et al.* 2011), it is possible for stars to explode at a position in the HR diagram where stellar tracks do not end, as mass loss due to rotation or binary mass transfer can easily change the effective temperature of the progenitor. However, the final luminosity of the core should still be a good indicator of the progenitor ZAMS mass. It is also possible to calculate a conservative upper limit to the progenitor ZAMS mass by comparing to the luminosity of a model at the end of core He-burning.

The ultimate test of whether a claimed SN progenitor was indeed the star that exploded, is to observe the disappearance of that star at late times, after the SN has faded. Only a handful of progenitor candidates have been thus confirmed (Maund & Smartt 2009; Fraser 2016), and in fact, in several cases progenitor candidates have been demonstrated to have been mis-identified (Maund *et al.* 2015). Late time observations also offer the possibility of detecting a surviving binary companion (even in the absence of a progenitor detection; Van Dyk, de Mink & Zapartas 2016).

Nebular spectroscopy Once the ejecta of a SN becomes optically thin, its spectrum is dominated by a mixture of permitted and forbidden lines such as Ca, O, Na and Mg. These lines arise from a combination of the primordial metal content of the progenitor envelope, the products of hydrostatic nuclear burning in the progenitor, and the products of explosive SN nucleosynthesis. It is hence possible to use line strengths during the nebular phase to estimate the SN progenitor mass (e.g. Fransson & Chevalier 1989). In particular, the strength of the [O I] $\lambda\lambda$6300,6364 lines depend strongly on progenitor mass.

This approach has been applied to both stripped envelope SNe (in particular for broad-lined Type Ic SNe which are associated with GRBs; Mazzali *et al.* 2001) and for Type IIP SNe (Jerkstrand *et al.* 2014). Encouragingly, the results of nebular modeling have been found to be consistent with direct imaging of progenitors (Jerkstrand *et al.*). Similarly to direct imaging, nebular spectra for a population of Type IIP SNe again tend to favour progenitors in the mass range 10-15 $M_{\odot}$(Jerkstrand *et al.* 2015; Silverman *et al.* 2017).

Nebular spectroscopy is an attractive tool to probe core-collapse progenitors as it does not depend on pre-explosion imaging, but only requires that a SN is sufficiently bright that spectra can be taken at late times. This has proven especially useful for the class of superluminous SNe, the closest of which have been found at z$\sim$0.1 (Gal-Yam 2012). Using late time spectra, Jerkstrand, Smartt & Heger 2016 demonstrated that the progenitors of these events are inconsistent with very massive stars exploding as pair instability supernovae (as had been proposed previously).

Lightcurve modeling Attempts to infer SN progenitor masses from SN lightcurves have met with mixed success. For H rich SNe, there has been a tendency for hydrodynamic lightcurve modeling to favour higher progenitor masses, when compared to masses found from direct detections. For example, Utrobin & Chugai (2009) found a progenitor mass of between 25 and 29 $M_{\odot}$ for SN 2004et from lightcurve modeling, while Crockett *et al.* (2011) estimated the progenitor mass to be 7 to 13 $M_{\odot}$ from pre-explosion imaging. Morozova *et al.*(2015) compared SNEC and CMFGEN models of the lightcurve resulting from the explosion of a 15 $M_{\odot}$ RSG, and find plateau durations that differ by around two weeks. The cause of such differences is not yet clear, and as noted by Dall'Ora *et al.* (2014) and others, further work is needed to explore the differences between various hydrodynamic codes.

Efforts to measure the ejecta mass of stripped envelope SNe have met with more success. Hydrodynamic models have been applied to Type IIb SNe such as SN 2011dh (Bersten *et al.* 2012), and in general these have yielded progenitor masses consistent with those found from direct imaging. Along with comprehensive hydrodynamic modeling, attempts have also been made to apply simple semi-analytic models (Arnett 1982) to large samples of stripped envelope SNe (Drout *et al.* 2011; Lyman *et al.* 2016). In general, these authors have found ejecta masses $\lesssim$2 $M_{\odot}$ for Ibc SNe, considerably lower than would be expected for single Wolf-Rayet stars, but consistent with the ejecta mass from stars <15 $M_{\odot}$ in binaries.

Mass loss diagnostics The recent advent of wide field imaging surveys, coupled with the capability to rapidly identify and follow up transients, has opened up a new window on SNe. "Flash spectroscopy" exploits spectra taken of SNe discovered less than one day after explosion (Gal-Yam *et al.* 2014), when the extended progenitor wind has not yet been over-taken by the SN ejecta, and is illuminated by the shock-breakout of the SN. By modeling these spectra, which are dominated by narrow, high-ionization emission lines, it is possible to determine mass loss rates, and the abundances in the wind (Groh 2014). While this has only been done for a handful of SNe to date, this will likely increase as high cadence surveys cover progressively larger areas of the sky.

X-rays and radio observations can also be used to constrain the pre-SN mass loss experienced by a progenitor. Dwarkadas (2014) noted that there were no x-ray bright Type IIP SNe consistent with mass loss rates above 10^{-5} $M_{\odot}$ yr^{-1}. As mass loss scales with luminosity, and hence with mass, they argued that this implied that most Type IIP progenitors had ZAMS masses below 19 $M_\odot$. However, as pointed out by Beasor & Davies (2016), the significant scatter in x-ray mass loss diagnostics and in the relation between mass loss and luminosity, means that there is a large uncertainty on this limit. Similarly, radio observations can be used to probe the circumstellar environment of a SN, and hence its mass loss rate. For Type IIb SNe, Chevalier & Soderberg (2010) suggested that radio lightcurves could be split into two classes, those coming from either extended or compact progenitors. However, in the case of SN 2011dh, the radius of the progenitor from pre-explosion observations is inconsistent with that inferred from the radio (Soderberg *et al.* 2012).

3. Implications for SN progenitors

To date, the majority of detected SN progenitors have been RSGs associated with Type IIP SNe. This is unsurprising as this is the most common subtype of SNe, and RSGs are bright in the *I*-band filter which is frequently used for HST observations of nearby galaxies. What is more surprising is the apparent deficit of Type IIP SN progenitors above $\sim$ 16 $M_\odot$, given that RSGs up to about 30 $M_\odot$ are seen in the Milky Way. Unless these stars evolve back across the HR diagram before they explode (Groh *et al.* 2013a), then will presumably explode as Type IIP SNe. While the initial mass function favours lower mass progenitors, if we consider the 26 observed detections and upper limits to progenitors from Smartt (2015), then four of these should have been above 20 $M_\odot$. Naïvely, the most massive progenitors will also be the easiest to find, as they will be the most luminous. The absence of detected high-mass RSG progenitors has been termed the "Red Supergiant Problem" by Smartt *et al.* (2009).

A simple, if unexciting, explanation for the RSG problem is that the most massive progenitors are missed due to extinction. Beasor & Davies (2016) used IR photometry of RSGs in a young cluster in the Large Magellanic Cloud to study mass-loss rates, and found that the most evolved stars in the cluster were the most reddened. From this, they suggested that reddening may explain the apparent lack of high mass SN progenitors. While this is an appealing explanation, it is not clear that it will explain all the missing RSGs. Walmswell & Eldridge (2012) found that when they took circumstellar dust into account, the maximum mass for a Type IIP progenitor increased to only $\sim$22 $M_\odot$. Furthermore, even if the most massive progenitors of Type IIP SNe were dust enshrouded in pre-explosion imaging, their nebular spectra should still match a higher mass star.

The finding by Smartt *et al.* (2009) that there is a paucity of Type II SN progenitors with ZAMS mass >16 $M_\odot$ raises the exciting possibility that these stars are collapsing to form black holes without a bright SN. Kochanek *et al.*(2008) proposed an ambitious strategy to find these failed SNe - monitor 10^6 RSGs for a number of years to see if any disappear. A search of the HST archive was conducted by Reynolds, Fraser & Gilmore (2015) for disappearing RSGs in nearby galaxies, while Gerke *et al.* (2015) attempted to find disappearing massive stars using regular monitoring of a sample of galaxies with 8-m telescopes. In both cases, candidates for failed SNe were found, and in one case, shown to have remained well below its original flux level some years later (Adams *et al.* 2015).

For stripped envelope SNe (Types IIb, Ib and Ic), it appears that binary interaction is a necessary requirement in a large fraction (or perhaps the majority) of cases. This is chiefly evidenced by the small ejecta masses seen for stripped envelope SNe (Lyman *et al.* 2016), along with the clear detections of IIb progenitors and their binary companions in archival imaging. The fact that only a single Ibc progenitor has been observed to date does not provide strong constraints on their progenitors, as most Wolf Rayet stars would be below the sensitivity of the available pre-explosion images (Eldridge *et al.* 2013).

While several Type IIb SNe have been linked with yellow supergiants (Maund *et al.* 2011, Van Dyk *et al.* 2011), only a single Type Ibc SN has an identified progenitor (Cao *et al.* 2013). The pre-explosion source found to be coincident with iPTF13bvn was originally suggested to be a Wolf-Rayet star (Groh, Georgy & Ekström 2013), but later shown to be a more convincing match to a lower mass He star progenitor (Eldridge *et al.* 2015). A low mass progenitor was also favoured by hydrodynamic modeling of the SN lightcurve (Bersten *et al.* 2014; Fremling *et al.* 2014). Late time imaging reveals that the progenitor candidate identified for iPTF13bvn has faded, although whether the binary companion can now be seen remains unclear (Eldridge & Maund 2016; Folatelli *et al.* 2016).

Aside from Type II and stripped envelope SNe, a handful of interacting Type IIn SNe have associated progenitor detections (e.g. Gal-Yam & Leonard 2009). At least some of these SNe appear to come from Luminous Blue Variables (LBVs), stars which were classically not expected to explode. However, the connection between Type IIn SNe, so-called supernova impostors, and LBVs remains unclear.

4. Future avenues for progenitor studies

While it is tempting to expect that JWST and future 30-m class telescopes will dramatically increase our sample of SN with progenitor detections, this will likely take some time, as new facilities must first build up an archive of deep imaging of nearby galaxies. For Type II SNe, JWST and E-ELT will allow for fainter progenitors to be detected, in principle allowing the lower mass limit for Fe-core collapse (or indeed electron capture SNe) to be better constrained. Furthermore, NIR imaging (*IJHK*) is less affected by extinction, and can provide a sensitive test of RSG temperatures (Davies *et al.* 2013). While possible Wolf-Rayet progenitors of stripped envelope SNe are expected to be luminous, as they are hot they emit most of their flux in the UV (as do lower mass He-giant progenitors; Kim, Yoon & Koo 2015). Hence JWST, or 30-m telescopes with next-generation AO working in the NIR are unlikely to lead to a progenitor detection for these classes of SNe.

While most observed SN progenitors have been identified in a single image, in a few cases we are fortunate enough to have repeated observations of the progenitor over a number of years (Fraser *et al.* 2014, Kochanek *et al.* 2017). This has enabled us to be sensitive to pre-explosion variability in some progenitors. In particular, monitoring on the timescale of years to decades is well matched to the timescales for shell C-burning and core-O burning, and can potentially probe pre-SN instabilities (Smith & Arnett 2014). In the case of ASASSN-16fq, pre-explosion imaging limited the change in luminosity in the years prior to core-collapse to around 300 $L_\odot$ yr^{-1} or less (Kochanek *et al.* 2017). To date only six SNe have repeated progenitor detections, but as the number of galaxies with multiple epochs of pre-explosion imaging increases, this will improve. The Large Synoptic Survey Telescope (LSST) will play a significant role in constraining progenitor variability for host galaxies $<$10 Mpc. If we consider SN 2012aw, which was at only 10 Mpc, the progenitor had an magnitude of $I = 23.4$ (Fraser *et al.*, 2012). This progenitor would have been detectable in a single epoch of LSST imaging. In combination with high spatial resolution HST or JWST imaging to disentangle the progenitor from the surrounding stellar population, LSST can in principle supply a pre-explosion lightcurve for a significant sample of nearby SNe. As an aside, it is worth noting that most SNe that are close enough to have a potential detection will likely be saturated in LSST data (at least at early times). Regular pre-explosion monitoring is also of interest for Type IIn SNe, which may eject material shortly before exploding.

Alongside observational effort, advances in stellar evolutionary modeling can help better understand SN progenitors. The observational evidence for binaries producing many stripped envelope SNe is consistent with the finding of Sana *et al.* (2012) that over 70% of massive stars will undergo some mass transfer during their lives. If we consider that half of these stars will undergo mass loss, then we obtain a fraction of stars which is comparable to the overall relative rate of stripped envelope SNe (37 %; Smith *et al.* 2011). However, the implication of this is that a significant fraction of the SNe where the progenitor has retained a relatively massive H envelope must have been the mass-*gainers* in a binary system. In some cases, the progenitor may even be the product of a merger (as was proposed for SN 1987A by Podsiadlowski, Joss & Rappaport 1990). While in most cases, the limited observational data on the progenitor does not justify the additional free parameters that a binary model would entail, as more SNe are discovered with multi-band progenitor detections, binary progenitor models will become testable.

Finally, the first detection of gravitational waves by advanced LIGO (Abbott *et al.* 2016) has opened up a new window on the Universe, that can potentially shed light on the mechanism of core-collapse. Unfortunately, even for an exceptionally nearby SNe such as SN 2008bk (4 Mpc), the associated gravitational wave signal is likely not to be detectable with Advanced LIGO (Ott 2009; Gossan *et al.* 2016). Similarly, neutrino detectors such as IceCube will be unable to detect neutrinos associated with a core-collapse at this distance (IceCube Collaboration, 2011). For now, and pending the next Galactic supernova, visible and NIR observations of nearby SNe and their host galaxies will continue to shed light on their progenitors.

Acknowledgements

I thank the organisers of IAUS 329 for hosting a stimulating and thought provoking meeting. I would also like to thank my collaborators with whom I have had the pleasure of hunting for supernova progenitors, especially Stephen Smartt, JJ Eldridge, Seppo Mattila, Nancy Elias de la Rosa, Justyn Maund, Rubina Kotak, Andrea Pastorello, Tom Reynolds, Anders Jerkstrand and Cosimo Inserra. This work has been supported by a Royal Society - Science Foundation Ireland University Research Fellowship.

References

B. P. Abbott *et al.* (LIGO Scientific Collaboration and Virgo Collaboration) 2016, *Phys. Rev. Lett.*, 116, 061102
Adams, S. M., Kochanek, C. S., Gerke, J. R., Stanek, K. Z., & Dai, X. 2016, arXiv:1609.01283
Arnett, W. D. 1982, *ApJ*, 253, 785
Beasor, E. R. & Davies, B. 2016, *MNRAS*, 463, 1269
Bersten, M. C., Benvenuto, O. G., Nomoto, K., *et al.* 2012, *ApJ*, 757, 31
Bersten, M. C., Benvenuto, O. G., Folatelli, G., *et al.* 2014, *AJ*, 148, 68
Cao, Y., Kasliwal, M. M., Arcavi, I., *et al.* 2013, *ApJL*, 775, L7
Cardelli, J. A., Clayton, G. C., & Mathis, J. S. 1989, *ApJ*, 345, 245
Chevalier, R. A. & Soderberg, A. M. 2010, *ApJL*, 711, L40
Crockett, R. M., Smartt, S. J., Pastorello, A., *et al.* 2011, *MNRAS*, 410, 2767
Dall'Ora, M., Botticella, M. T., Pumo, M. L., *et al.* 2014, *ApJ*, 787, 139
Davies, B., Kudritzki, R.-P., Plez, B., *et al.* 2013, *ApJ*, 767, 3
Dwarkadas, V. V. 2014, *MNRAS*, 440, 1917
Eldridge, J. J., Fraser, M., Smartt, S. J., Maund, J. R., & Crockett, R. M. 2013, *MNRAS*, 436, 774

Eldridge, J. J., Fraser, M., Maund, J. R., & Smartt, S. J. 2015, *MNRAS*, 446, 2689
Eldridge, J. J. & Maund, J. R. 2016, *MNRAS*, 461, L117
Folatelli, G., Van Dyk, S. D., Kuncarayakti, H., *et al.* 2016, *ApJL*, 825, L22
Fransson, C. & Chevalier, R. A. 1989, *ApJ*, 343, 323
Fraser, M., Maund, J. R., Smartt, S. J., *et al.* 2012, *ApJL*, 759, L13
Fraser, M., Maund, J. R., Smartt, S. J., *et al.* 2014, *MNRAS*, 439, L56
Fraser, M. 2016, *MNRAS*, 456, L16
Fremling, C., Sollerman, J., Taddia, F., *et al.* 2014, *A&A*, 565, A114
Gal-Yam, A. 2012, *Science*, 337, 927
Gal-Yam, A., Arcavi, I., Ofek, E. O., *et al.* 2014, *Nature*, 509, 471
Gerke, J. R., Kochanek, C. S., & Stanek, K. Z. 2015, *MNRAS*, 450, 3289
Gossan, S. E., Sutton, P., Stuver, A., *et al.* 2016, *Phys. Rev. D*, 93, 042002
Groh, J. H., Meynet, G., Georgy, C., & Ekström, S. 2013a, *A&A*, 558, A131
Groh, J. H., Georgy, C., & Ekström, S. 2013b, *A&A*, 558, L1
Groh, J. H. 2014, *A&A*, 572, L11
Gustafsson, B., Edvardsson, B., Eriksson, K., *et al.* 2008, *A&A*, 486, 951
IceCube Collaboration, Abbasi, R., Abdou, Y., *et al.* 2011, arXiv:1108.0171
Jerkstrand, A., Smartt, S. J., Fraser, M., *et al.* 2014, *MNRAS*, 439, 3694
Jerkstrand, A., Smartt, S. J., Sollerman, J., *et al.* 2015, *MNRAS*, 448, 2482
Jerkstrand, A., Smartt, S. J., & Heger, A. 2016, *MNRAS*, 455, 3207
Kim, H.-J., Yoon, S.-C., & Koo, B.-C. 2015, *ApJ*, 809, 131
Kochanek, C. S., Beacom, J. F., Kistler, M. D., *et al.* 2008, *ApJ*, 684, 1336-1342
Kochanek, C. S., Khan, R., & Dai, X. 2012, *ApJ*, 759, 20
Kochanek, C. S., Fraser, M., Adams, S. M., *et al.* 2016, arXiv:1609.00022
Lyman, J. D., Bersier, D., James, P. A., *et al.* 2016, *MNRAS*, 457, 328
Maund, J. R. & Smartt, S. J. 2009, *Science*, 324, 486
Maund, J. R., Fraser, M., Ergon, M., *et al.* 2011, *ApJL*, 739, L37
Maund, J. R., Fraser, M., Reilly, E., Ergon, M., & Mattila, S. 2015, *MNRAS*, 447, 3207
Mazzali, P. A., Nomoto, K., Patat, F., & Maeda, K. 2001, *ApJ*, 559, 1047
Morozova, V., Piro, A. L., Renzo, M., *et al.* 2015, *ApJ*, 814, 63
Ott, C. D 2009, *Classical and Quantum Gravity*, 26, 063001
Podsiadlowski, P., Joss, P. C., & Rappaport, S. 1990, *A&A*, 227, L9
Poznanski, D., Ganeshalingam, M., Silverman, J. M., & Filippenko, A. V. 2011, *MNRAS*, 415, L81
Reynolds, T. M., Fraser, M., & Gilmore, G. 2015, *MNRAS*, 453, 2885
Sana, H., de Mink, S. E., de Koter, A., *et al.* 2012, *Science*, 337, 444
Silverman, J. M., Pickett, S., Craig Wheeler, J., *et al.* 2017, *MNRAS* in press
Smartt, S., J. J., E., Crockett, R., & Maund, J. 2009, *MNRAS*, 395, 1409
Smartt, S. J. 2015, *PASA*, 32, e016
Smith, N., Li, W., Filippenko, A. V., & Chornock, R. 2011, *MNRAS*, 412, 1522
Smith, N. & Arnett, W. D. 2014, *ApJ*, 785, 82
Soderberg, A. M., Margutti, R., Zauderer, B. A., *et al.* 2012, *ApJ*, 752, 78
Turatto, M., Benetti, S., & Cappellaro, E. 2003, From Twilight to Highlight: The Physics of Supernovae, 200
Utrobin, V. P. & Chugai, N. N. 2009, *A&A*, 506, 829
Van Dyk, S. D., Li, W., & Filippenko, A. V. 2003, *PASP*, 115, 1
Van Dyk, S. D., Li, W., Cenko, S. B., *et al.* 2011, *ApJL*, 741, L28
Van Dyk, S. D., de Mink, S. E., & Zapartas, E. 2016, *ApJ*, 818, 75
Walmswell, J. J. & Eldridge, J. J. 2012, *MNRAS*, 419, 2054
West, R. M., Lauberts, A., Schuster, H.-E., & Jorgensen, H. E. 1987, *A&A*, 177, L1

The Lives and Death-Throes of Massive Stars
Proceedings IAU Symposium No. 329, 2016
J.J. Eldridge, J.C. Bray, L.A.S. McClelland & L. Xiao, eds.

doi:10.1017/S174392131700343X

Radiation Hydrodynamical Models for Type I Superluminous Supernovae: Constraints on Progenitors and Explosion Mechanisms

Ken'ichi Nomoto[1,6], Alexey Tolstov[1], Elena Sorokina[2], Sergei Blinnikov[3,1], Melina Bersten[4,1] and Tomoharu Suzuki[5]

[1]Kavli Institute for the Physics and Mathematics of the Universe (WPI), The University of Tokyo, Kashiwa, Chiba 277-8583, Japan
[2]Sternberg Astronomical Institute, M.V.Lomonosov Moscow State U., 119991 Moscow, Russia
[3]Institute for Theoretical and Experimental Physics, 117218 Moscow, Russia
[4]Instituto de Astrofisica La Plata, 1900 La Plata, Argentina
[5]College of Engineering, Chubu University, Kasugai, Aichi 487-8501, Japan
[6]Hamamatsu Professor
email: nomoto@astron.s.u-tokyo.ac.jp

Abstract. The physical origin of Type-I (hydrogen-less) superluminous supernovae (SLSNe-I), whose luminosities are 10 to 500 times higher than normal core-collapse supernovae, remains still unknown. Thanks to their brightness, SLSNe-I would be useful probes of distant Universe. For the power source of the light curves of SLSNe-I, radioactive-decays, magnetars, and circumstellar interactions have been proposed, although no definitive conclusions have been reached yet. Since most of light curve studies have been based on simplified semi-analytic models, we have constructed multi-color light curve models by means of detailed radiation hydrodynamical calculations for various mass of stars including very massive ones and large amount of mass loss. We compare the rising time, peak luminosity, width, and decline rate of the model light curves with observations of SLSNe-I and obtain constraints on their progenitors and explosion mechanisms. We particularly pay attention to the recently reported double peaks of the light curves. We discuss how to discriminate three models, relevant models parameters, their evolutionary origins, and implications for the early evolution of the Universe.

Keywords. magnetar, stellar mass loss, supernovae, superluminous supernovae

1. Superluminous Supernovae

Superluminous supernovae (SLSNe) are brighter than −21 magnitude in any optical band at the maximum brightness, which is ≈ 30 times brighter than the average of normal supernovae (Quimby *et al.* 2012, Gal-Yam *et al.* 2012). They are divided into Type I (SLSN-I) and Type II (SLSN-II) according to the absence and presence of hydrogen feature in the spectra. It is evident that there is significant dispersion in both rise and decay time scales. These differences could indicate some diversity in the progenitors of SLSNe.

SLSNe-II, like SN 2006gy, may be powered by interactions of the SN ejecta with CSM in the form of steady wind or a shell (Moriya *et al.* 2013). The origin of SLSNe-I is still a matter of debate (Gal-Yam *et al.* 2012). Currently three possibilities are under consideration (a) pair instability supernovae, (b) supernovae powered by ejecta/wind interaction, and (c) supernovae powered by a spin down of a rapidly rotating young magnetar. In the present paper, we focus on the magnetar and circumstellar interaction models for SLSNe-I and discuss possible engines powering the observed super-luminosities.

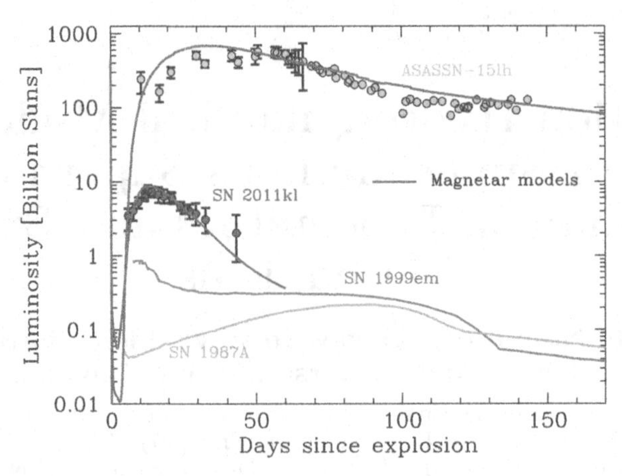

Figure 1. The light curves of ASASSN-15lh and SN 2011kl compared with SN 1999em and SN 1987A, and with the magnetar models (Bersten *et al.* 2016).

The existence of SLSNe opens the possibility of studying extremely luminous supernovae in the very early Universe. Understanding their nature will open new paths to better comprehend the origin and evolution of stars in the Universe.

2. Magnetar Driven Supernovae

A rapidly rotating neutron stars with large magnetic fields, $\approx 10^{14} - 10^{15}$ Gauss, or as usually called "magnetar" could be an extra source to power the light curve of SNe (see, e.g., Bersten *et al.* 2016 for references). Important question is if the rotation energy of the magnetar is efficiently deposited into the supernova ejecta.

SN 2011kl represents the first SLSN associated with an ultra long GRB (Greiner *et al.* 2015). This gamma-ray burst lasted several hours whereas the typical long-duration GRBs (LGRBs) fade in a matter of minutes. A handful of SNe have been observed coincident with LGRB but SN 2011kl was more than three times more luminous than normal supernovae. The peculiar properties of both the LGRB and the supernova suggest a common extreme nature.

ASASSN-15lh is possibly the most luminous and powerful explosion ever seen (Dong *et al.* 2016). It was 200 times more powerful than a typical supernova. And it outshone all of the other SLSNe by at least a factor of two. During over one month its luminosity was twenty times larger than that of the whole Milky Way galaxy with its billions of stars. Such an enormous release of energy has challenged the current models that attempt to explain the SLSN phenomenon.

The spectra of both SN 2011kl and ASASSN-15lh showed no evidence of hydrogen, i.e., these are SLSNe-I. Bersten *et al.* (2016) have analyzed two mysterious cases among SLSNe. Despite the differences between them, both events could be explained as explosions boosted by a highly magnetized, rapidly spinning neutron star (magnetar).

Bersten *et al.* (2016) have performed numerical hydrodynamical calculations to explore the magnetar hypothesis for these two intriguing objects. They found that both SLSNe could be understood in the framework of magnetar-powered supernovae (Fig. 1). In particular for the extreme ASASSN-15lh, they were able to find a magnetar source with physically allowed properties of magnetic field strength and rotation period. The solution avoided the prohibited realm of neutron-star spins that would cause the object to breakup due to centrifugal forces. To confirm the proposed model, it is critical to obtain further observations when the material ejected by the supernova is expected to become thin.

The magnetar models above were performed to reproduce observations of ASASSN-15lh until ~150 days. However, recent publications shows that ASASSN-15lh has undergone an UV re-brightening (Brown, *et al.* 2016, Godoy-Rivera *et al.* 2016). The long and bright UV plateau of ~120 days duration (Godoy-Rivera *et al.* 2016) could be in contradiction with a purely magnetar source. An interesting possibility could be to invoke a combination of multiple power sources (including circumstellar interaction) (Chatzopoulos *et al.* 2016, Tolstov *et al.* 2017).

3. Circumstellar Interaction Models

SN light curves for circumstellar interacting models have been constructed analytically for both optically thin and optically thick cases in a number of papers (Chevalier *et al.* 2011, Moriya *et al.* 2013, Chatzopoulos *et al.* 2013). However, radiation hydrodynamical models are noticeably different from the analytic models (Sorokina *et al.* 2016).

Sorokina *et al.* (2016) calculated radiation-hydrodynamical models of circumstellar interaction to reproduce LCs of both SN 2010gx and PTF09cnd for optically thick CSM. PTF09cnd and SN 2010gx reach peak luminosities too high to be explained by the ^{56}Ni production allowed by their late time photometric limits (Pastorello *et al.* 2010). For the LC shape, SN 2010gx has one of the narrowest LC while PTF09cnd has one of the broadest LC among SLSNe-I.

In the models by Sorokina *et al.* (2016), when the shock wave structure starts to form due to collision between the ejecta and the CSM, the envelope is cool and transparent. Then the emission from the shock front heats the gas in the envelope, thus making it opaque, and the photosphere moves to the outermost layers rather quickly. When the photospheric radius reaches its maximum, one can observe maximal emission from the supernova.

The speed of the growth of the photospheric radius depends on the mass of the envelope, since more photons must be emitted from the shock to heat larger mass envelopes. This is why the photosphere moves to the outermost layers in model N0 faster than in model B0. At the beginning of this post-maximum stage all gas in the envelope is already heated by the photons which came from the shock region and diffused through the envelope to the outer edge, and the whole system (ejecta and envelope) becomes almost isothermal. Model N0 seems to provide the best match to SN 2010gx. The maximmu emission and the slopes of the light curves after the maximum are in excellent agreement.

Model B0 is the best for PTF09cnd. For the models with a massive envelope the rising time is determined by diffusion time of the photons from the shock wave and by the movement of the photosphere to the outermost edge of the envelope. The rising time which more or less fits the observations was obtained for the model B0 with 49 $M_\odot$ in the envelope and 5 $M_\odot$ in the ejecta of the explosion with the energy of $4 \cdot 10^{51}$ ergs.

For these SNe, moderate explosion energy $E_{51} \sim 2-4$ is enough to explain both LCs. Main difference between the two SLSNe are the masses of CSM. Only $\sim 5\,M_\odot$ of non-H (He, C, O) is needed to reproduce SN 2010gx light curve (Fig. 2), while almost $55\,M_\odot$ CSM to reproduce the broad LC of PTF09cnd (Fig. 3).

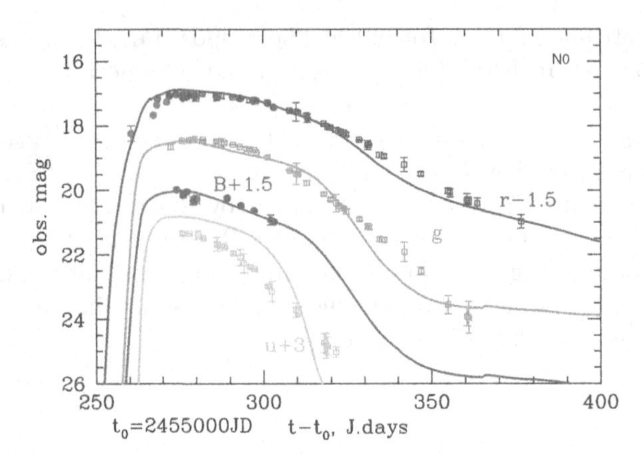

Figure 2. Synthetic light curves for SN 2010gx, in r, g, B, and u filters compared with Pan-STARRS and PTF observations (Sorokina *et al.* 2016). Pan-STARRS points are designated with open squares (u, g, and R bands), PTF points, with filled circles (B and r bands). Four pink points in the beginning of the r band shows PTF observations in the Mould R-band which is similar to the SDSS r band.

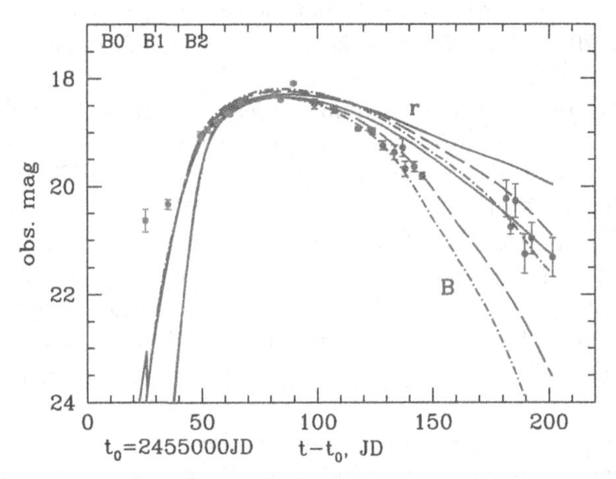

Figure 3. Observed (*dots*) and synthetic (*lines*) light curves for PTF09cnd in r and B filters . *The solid, dashed,* and *dash-dotted* lines correspond to the models with the static outer envelope, expanding envelope, and faster expanding envelope, respectively (Sorokina *et al.* 2016).

4. Origin of Circumstellar Matter

SN ejecta-CSM interaction is a possible energy source for SLSNe-I, if the CSM is depleted of its hydrogen. Evolutionary calculations of very massive stars ($\gtrsim 100 M_\odot$) of different metallicities show that most of H burns into He during main sequence before the intensive mass loss (Yusof *et al.* 2013). Such H-burning occurs in a very extended convective core that reaches outermost layers of the star. In addition, the rotation of the star forms a meridional circulation which also helps H (and, possibly, some part of He) to be burned even near the surface. In later phase, the star undergoes a loss of a large amount of envelope which is no longer H-rich, but He with enhanced C+O. This could be a plausible way to produce CSM of $\sim 50 M_\odot$ He with enhanced C+O.

Materials cast off by instabilities in the cores of very massive stars in their final years, e.g., pulsational pair-instability (Woosley *et al.* 2007, Ohkubo *et al.* 2009), are possible sources of C+O CSM.

In close binary systems, a merger (or even multiple mergers) of the C+O core from an evolved WR star with another hydrogen-deficient star, like a neutron star, a white dwarf, or WR star could occur (Taam *et al.* 2000, Glebbeek *et al.* 2009, Chevalier *et al.* 2012). Such processes could form a dense C+O CSM or common envelope. In a dense stellar cluster, the (multiple) collision of massive stars (in particular, Wolf-Rayet stars) can make a massive C+O star surrounded by a massive non-Hydrogen CSM (Portegies Zwart *et al.* 2007, Suzuki *et al.* 2007).

The result of those events could be an envelope or an expanding wind with the size large enough to provide the long living shock wave. It may break out through the envelope in a few months. The explosion within such an envelope could give rise to an SLSN-I.

The work has been supported by the World Premier International Research Center Initiative (WPI), MEXT, Japan, and by the Grants-in-Aid for Scientific Research of the JSPS (JP26400222, JP16H02168, JP17K05382).

References

Bersten, M., Benvenuto, O., Orellana, M., & Nomoto, K. 2016, *ApJ* **817**, L8
Brown, P. J., Yang, Y., Cooke, J., *et al.* 2016, *ApJ* **828**, 3
Chatzopoulos, E., *et al.* 2013, *ApJ* **773**, 76
Chatzopoulos, E., Wheeler, J. C., Vinko, J., *et al.* 2016, *ApJ* **828**, 94
Chevalier, R. A. & Irwin, C. M. 2011, *ApJ (Letters)* **729**, L6
Chevalier, R. A. 2012, *ApJ (Letters)* **752**, L2
Dong, S., Shappee, B. J., Prieto, J. L., *et al.* 2016, *Science* **351**, 257
Gal-Yam, A. 2012, *Science* **337**, 927
Godoy-Rivera, D., Stanek, K. Z., Kochanek, C. S., *et al.* 2016, *MNRAS* **466**, 1428
Glebbeek, E., Gaburov, E., de Mink, S. E., *et al.* 2009, *A&A* **497**, 255
Greiner, J., Mazzali, P. A., Kann, D. A., *et al.* 2015, *Nature* **523**, 189
Moriya, T., *et al.* 2013, *MNRAS* **428**, 1020
Ohkubo, T., Nomoto, K., Umeda, H., Yoshida, N., & Tsuruta, S. 2009, *ApJ* **706**, 1184
Pastorello, A., Smartt, S., Botticella, M., *et al.* 2010, *ApJ (Letters)* **724**, L16
Portegies Zwart S. F., van den Heuvel E. P. J. 2007, *Nature* **450**, 388
Quimby, R. 2012, IAU Symp. 279, Death of Massive Stars: Supernovae and Gamma-Ray Bursts, ed. P. Roming, *et al.*, 22
Sorokina, E., Blinnikov, S., Nomoto, K., Quimby, R., & Tolstov, A. 2016, *ApJ* **829**, 17
Suzuki, T. K., Nakasato, N., Baumgardt, H., *et al.* 2007, *ApJ* **668**, L19
Taam, R. E. & Sandquist, E. L. 2000, *ARAA* **38**, 113
Tolstov, A., Nomoto, K., Blinnikov, S., Sorokina, E., Quimby, R., & Baklanov, P. 2017, *ApJ* **835**, 266
Woosley, S. E., Blinnikov. S. & Heger, A. 2007, *Nature* **450**, 390
Yusof, N., Hirschi, R., Meynet, G., *et al.* 2013, *MNRAS* **433**, 1114

The Lives and Death-Throes of Massive Stars
Proceedings IAU Symposium No. 329, 2016
J.J. Eldridge, J.C. Bray, L.A.S. McClelland & L. Xiao, eds.

doi:10.1017/S174392131700312X

SN 2015bh: an LBV becomes NGC 2770s fourth SN... or not?

Christina C. Thöne[1], Antonio de Ugarte Postigo[1,2] and Giorgos Leloudas[3]

[1]IAA-CSIC, Glorieta de la Astronomía s/n, 18008 Granada, Spain
email: cthoene@iaa.es

[2]Dark Cosmology Centre, Niels-Bohr-Institute, Univ. of Copenhagen, Juliane Maries Vej 30, 2100 Copenhagen, Denmark

[3]Department of Particle Physics & Astrophysics, Weizmann Institute of Science, Rehovot 76100, Israel

Abstract. Massive stars in the final phases of their lives frequently expel large amounts of material. An interesting example is SN 2009ip that varied in brightness years before its possible core-collapse. Here we present SN 2015bh in NGC 2770 that shows striking similarities to SN 2009ip. It experienced frequent variabilities for 21 years before a smaller precursor and the "main event" in May 2015. Its spectra are consistent with an LBV during the outburst phase and show a complex P-Cygni profile during the main event. Both SN 2009ip and 2015bh were always situated red-wards of LBVs in outburst in the HR diagram. Their final fate is currently still uncertain, SN 2009ip, however, is now fainter than in pre-explosion observations. If the star survives this event it is undoubtedly altered, and we suggest that these "zombie stars" could be LBVs evolving into a Wolf-Rayet (WR) star over a very short timescale.

Keywords. supernovae: individual (SN 2015bh, SN 2009ip), stars: evolution, stars: mass loss

1. Introduction

The distinction between massive stellar outbursts, so-called SN "impostors", and real core-collapse supernovae (SNe) is not always straightforward. In particular SN Type IIn-like events seem to span a large range of luminosities and also include super-luminous SNe. The progenitors of those SNe are typically massive, probably LBV stars (e.g. Smith *et al.* 2011). Massive stars often show outbursts in their final stages of evolution ranging from small variations like S-Dor type outbursts to massive mass ejections such as observed for Eta Carinae (see e.g. Kiminki *et al.* 2016). Ofek *et al.* (2014) find that almost all Type IIn SNe seem to show outbursts in the years prior to explosion. A famous event in this respect was SN 2009ip which experienced several outbursts of ~ 2 mag difference in luminosity for several years prior to its possible explosion as a Type IIn SN (Pastorello *et al.* 2013, Mauerhan *et al.* 2013).

Here we present observations of the peculiar event SN 2015bh in NGC 2770, a MW-like spiral galaxy at a distance of $\sim$27 Mpc which has hosted 3 other SNe previous to SN 2015bh. Our observations span the entire event including the pre-cursor, the main event and up to 200 days using the telescopes at Observatorio de Sierra Nevada (OSN), the Gran Telescopio Canarias (GTC) and *Swift*. The observations are supplemented by archival data of NGC 2770 spanning 21 years and a serendipitous spectrum of the outburst phase in November 2013.

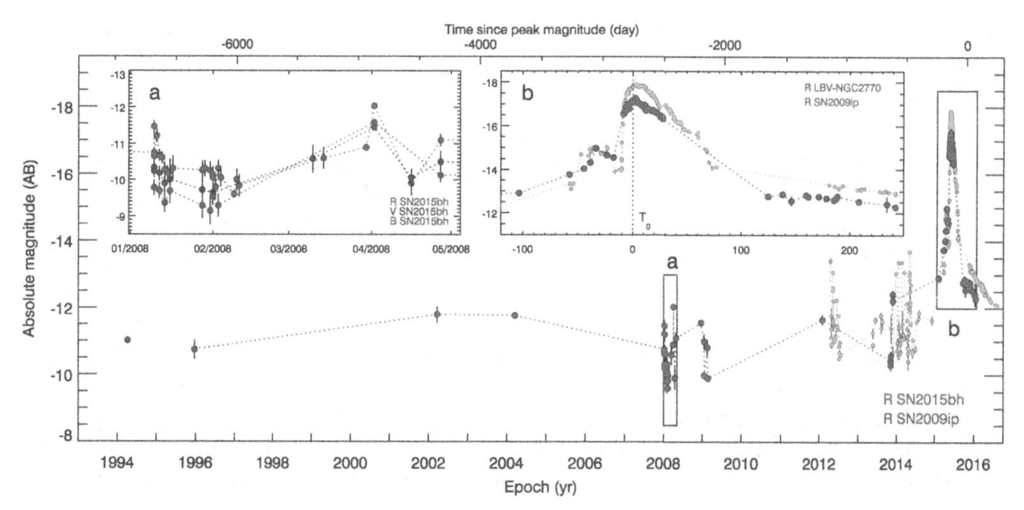

Figure 1. Lightcurve of SN 2015bh and the short term variations of its progenitor over 21 years.

2. The timeline of SN 2015bh and its progenitor

Our search for the progenitor in archival data of NGC 2770 revealed that the star had been in outburst and varying by ∼2 magnitudes since at least 1994 (Thöne *et al.* 2017). iPTF discovered the progenitor in outburst phase in December 2013 and named it PTF13efv, but the detection was not reported until much later (Duggan *et al.* 2015). On Feb. 7, 2015, SNhunt reported a new transient which was then characterized as SN impostor and/or possible IIn (Elias-Rosa *et al.* 2015), which turned out to be the precursor of SN 2015bh with a maximum ∼40 days before the main event and a peak magnitude of R=–13.4 mag. We initiated a follow-up campaign with the 0.9 and 1.5m OSN telescope. In the observations from May 16, 2015, a massive rebrightening by 2 mag marked the onset of the main event reaching an absolute magnitude of –17.5 mag on May 24 (which we define as T_0). We started a dense monitoring campaign with OSIRIS at the GTC in spectroscopy as well as *Swift* in UV and X-rays. Neither X-ray nor radio emission was detected from the SN and/or its progenitor . After an observational gap due to sun constraints, the SN light curve had dropped to ∼ –12.5 mag and shows only a very shallow decay which could imply that the material is not any more accelerated at this time. During all the phases, the light curve resembles very closely that of SN 2009ip. The light curve is shown in Fig. 1.

The spectra from the LBV outburst in 2013 shows a typical LBV spectrum with a single P-Cygni profile and prominent Balmer and Fe-lines as well as weak He lines. During the precursor, a second absorption component emerges with a velocity of ∼ –2000 $\mathrm{km\,s^{-1}}$ in addition to the still present absorption at –700 $\mathrm{km\,s^{-1}}$, suggesting a connection of the ejected material with the event that caused the precursor. During maximum of the main event, the absorption components nearly disappear showing only the typical narrow IIn-type profiles. After maximum the double P-Cygni profile clearly emerges in all lines while the He-lines have largely disappeared. After the sun gap, the spectra had changed to broad nebular lines and interestingly, the second, faster absorption now shows up as emission line accompanied by a small bump at the same velocity, but redshifted, suggesting some asymmetry in the ejecta. Our toy model for the explosion assumes a dense shell expelled some years before the main event which we see as the absorption component at –700 $\mathrm{km\,s^{-1}}$. The faster material is expelled during the precursor and interacts with the dense wind of the progenitor giving rise to the luminosity of the main event. Some time around 100 days post explosion it catches up with the first larger shell and shows up in emission.

The main event had a total energy release of 1.2 – 1.8 $\times 10^{48}$ erg depending on the evolution of the light curve during the sun gap. We model the SED during the main event with a simple BB model and derive the temperature and radius evolution. Until about 20 days past maximum the BB has an expansion velocity of $\sim$ 2200 km s^{-1} and a cooling rate of 660 K/day after which it slows down to 880 km s^{-1} and 150 K/day. The Hα EW behaves very similar to those of other IIn SNe with a drop around maximum below 100 Å and a subsequent rise to very high values of $>$ 3000Å which could be indicative of a true core-collapse (Smith *et al.* 2014). The Balmer decrement (Hα / H β) drops below the value for case B recombination around maximum which could be explained by a very dense plasma thermalizing the emission. During the "nebular phase" it rises to $>$10, which is also seen in the nebular phase of other types of events. An explanation by increasing extinction from the ejecta is excluded as the SED shows no signs of reddening and dust production.

3. The host and environment

The serendipitous spectra during the LBV phase were part of a drift-scan spectroscopy campaign to observe the satellite galaxy of NGC 2770 and happened to also cover the centre of the galaxy and the site of the progenitor. All spectra are contaminated by the emission of the progenitor and we hence can only study the neighboring HII regions. The LBV/SN is located in one of the outer spiral arms of NGC 2770, different from SN 2009ip which was found at a rather large distance from the host. The regions near the SN site have a metallicity of around half solar and a stellar population age of 7-10 Myr, consistent with single star LBV models.

Despite its 3 or possible 4 SNe, NGC 2770 does not show a particularly large star-formation rate, the only remarkable fact is a large HI mass (Thöne *et al.* 2009). Likewise, there are little indications for a major merger event triggering large star formation (Thöne *et al.* in prep.). The HI mass could be indicative of an inflow of gas onto the galaxy, giving rise to an enhancement in star-formation.

4. The evolution of the progenitor star and its possible future fate

The progenitor of SN 2015bh as well as SN 2009ip are suggested to be LBV stars due to its spectral properties. Despite this, the progenitors were always located red of typical LBVs in S-Dor outbursts at similar temperatures as Eta Carinae during its great eruption but at lower luminosities (see Fig. 2). During the pre-explosion event, the temperature stayed at similar values while the luminosity increased. The colors during the SN were typical of other IIn SNe. At late times, both SN 2009ip and SN 2015 show a turn towards higher temperatures whose origin is still unclear. SN 2009ip now seems to be below the level of the proposed progenitor as observed in 1999 (Foley *et al.* 2011).

At this point we cannot determine whether any of the two events actually did explode or whether the observed main event was only yet another giant outburst of the star. It it survived this outburst, the star might have lost most of its outer envelope and might settle at a lower luminosity some time in the next few years, while it seems it also turns towards higher temperatures similar to WR stars. Originally, LBVs have been suggested to be only a transitionary period on the evolution to a WR star (Maeder & Meynet 2000), but lately LBVs have also been observed to directly explode as CC-SNe (Gal-Yam

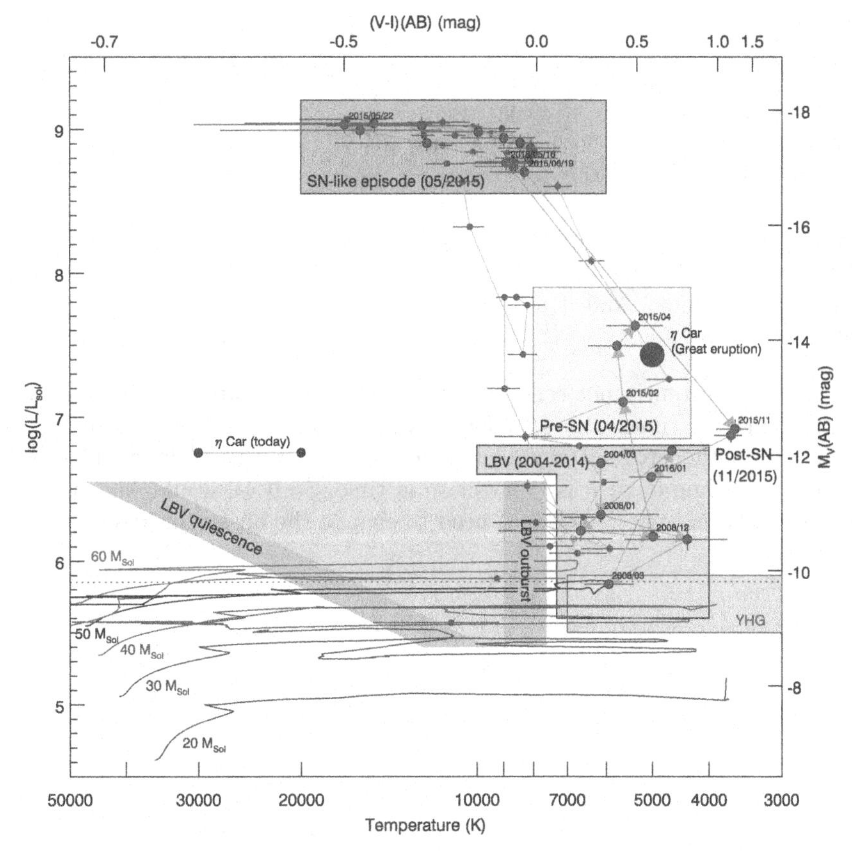

Figure 2. The evolution of SN 2015 (red thick dots) and SN 2009ip (blue thin dots) in the HR diagram.

& Leonard 2009). So far, it has been unclear how the star can loose enough mass in short time to evolve from a LBV to a WR star (Smith 2014a). This kind of event might be the solution to this problem, expelling enough mass to transform a LBV in very short time into a WR star. Infact, many WR stars have been found to be embedded in SNR like shells. Another possibility is that these events happen in a binary system of which one star experiences core-collapse and the other, blue, companion emerges at late times. Very late time observations of the SN could prove or reject the possibilities where the star 1) became a WR star 2) underwent core-collapse and/or 3) has a binary companion, although for tight binary systems, this might be difficult.

References

Smith, N., Li, W., Miller, A. A., *et al.* 2011, *ApJ*, 732, 63

Kiminki, M. M., Reiter, M., & Smith, N. 2016, *MNRAS*, 463, 845

Ofek, E. O., Sullivan, M., Shaviv, N. J., *et al.* 2014, *ApJ*, 789, 104

Pastorello, A., Cappellaro, E., Inserra, C., *et al.* 2013, *ApJ*, 767, 1

Mauerhan, J. C., Smith, N., Filippenko, A., Blanchard, K. B. *et al.* 2013, *MNRAS*, 430, 1801

Thöne, C. C., de Ugarte Postigo, A., Leloudas, G., *et al.*, 2017, *A&A* in press

Duggan, G., Bellm, E., Leloudas, G., Kasliwal, M. M., & Kulkarni, S. R. 2015, *The Astronomers Telegram* 7515

Elias-Rosa, N., Benetti, S., Tomasella, L., *et al.* 2015, *The Astronomers Telegram*, 7042

Smith, N., Mauerhan, J. C., & Prieto, J. L. 2014, *MNRAS*, 438, 1191
Thöne, C. C., Michałowski, M. J., Leloudas, G., *et al.* 2009, *ApJ*, 698, 1307
Foley, R. J., Berger, E., Fox, O., *et al.* 2011, *ApJ*, 732, 32
Maeder, A. & Meynet, G. 2000, *A&A*, 361, 159
Gal-Yam, A. & Leonard, D. C. 2009, *Nature*, 458, 865
Smith, N. 2014, *ARA&A*, 52, 487

Discussion

BIANCO: Would LSST be suited to study the variability phase of those objects using multi-band observations? And if not what would have to be changed in the observing strategy?

THÖNE: LSST is currently not scheduled to observe in all filters during one night. As these events vary on timescales of a few days to weeks as we have seen in the light curve and also in the progenitor modeling in Elias-Rosa *et al.*, we would need nearly simultaneous modeling or at least very close in time. So if we would want to use LSST to characterize those events, we would need to change the observing strategy.

The Lives and Death-Throes of Massive Stars
Proceedings IAU Symposium No. 329, 2016
J.J. Eldridge, J.C. Bray, L.A.S. McClelland
& L. Xiao, eds.

doi:10.1017/S1743921317002915

Emission-line Diagnostics of Nearby HII Regions Including Supernova Hosts

Lin Xiao[1], J. J. Eldridge[2], Elizabeth Stanway[3] and L. Galbany[4]

[1,2]Department of Physics, University of Auckland, NZ
email: lin.xiao@auckland.ac.nz, j.eldridge@auckland.ac.nz

[3]Department of Physics, University of Warwick, Gibbet Hill Road, Coventry, CV4 7AL, UK
email: e.r.stanway@warwick.ac.uk

[4]PITT PACC, Department of Physics and Astronomy, University of Pittsburgh, Pittsburgh, PA 15260, USA
email: lgalbany@das.uchile.cl

Abstract. We present a new model of the optical nebular emission from HII regions by combining the results of the Binary Population and Spectral Synthesis (BPASS) code with the photoionization code CLOUDY (Ferland *et al.* 1998). We explore a variety of emission-line diagnostics of these star-forming HII regions and examine the effects of metallicity and interacting binary evolution on the nebula emission-line production. We compare the line emission properties of HII regions with model stellar populations, and provide new constraints on their stellar populations and supernova progenitors. We find that models including massive binary stars can successfully match all the observational constraints and provide reasonable age and mass estimation of the HII regions and supernova progenitors.

Keywords. Binaries: general, HII regions, Supernovae: general.

The spectra of star-forming HII regions carry a wealth of information about properties of gas cloud and stellar populations. A number of emission line features in a spectrum allow us to derive element abundances and set valuable constraints on the age and mass of stellar population (e.g., Byler *et al.* 2016; Levesque *et al.* 2010) and supernova (SN) progenitors (Galbany *et al.* 2016). These studies all assume single star models, the inclusion of interacting binaries in stellar populations is a major step forward in modelling populations accurately (e.g. Stanway *et al.* 2016). Here we discuss the importance of including interacting binaries in modelling HII regions.

1. HII Regions Sample

We consider two surveys of HII regions - one isolating such regions in nearby galaxies, and a second identifying the emission line properties at the location of known core-collapse supernovae (CCSNe). We study these separately to place constraints on massive star populations and CCSN progenitors.

First we use the van Zee sample of nearby individual HII regions. There are a total of 254 HII regions with 188 from 13 spiral galaxies (van Zee *et al.*, 1998) and 66 from 21 dwarf galaxies (van Zee & Haynes 2006). The one-dimensional spectra of these HII regions were corrected for atmospheric extinction, flux-calibrated and optical emission line fluxes were measured relative to Hβ intensity. In the work of van Zee *et al.* (1998), abundances for several elements (oxygen, nitrogen, neon, sulfur, and argon) were determined and radial abundance gradients were derived for 11 spiral galaxies.

The second sample comes from the PMAS/PPAK Integral-field Supernova hosts COmpilation (PISCO, Galbany *et al.* in prep). It is composed of observations of SN host galaxies selected from the CALIFA survey and other programs using the PMAS/PPAK Integral Field Unit with large FoV and a spatial resolution of 1 arcsec. Galbany *et al.*(2014) have compiled 128 SNe and, for the first time, spectroscopically probed the association of different SN types to different formation environments. Moreover, the differences in the local environmental oxygen abundance for different SNe have been also investigated and they find the location of SNe Ic and II show higher metallicity than those of SNe Ib, IIb, and Ic-BL (Galbany *et al.* 2016).

2. Nebular Emission Model

To reproduce the observed nebular emission lines we combine BPASS v2.0 stellar population models described in detail in Stanway *et al.*, (2016) and reference therein with the photoionization code CLOUDY (Ferland *et al.*, 1998). In brief, the BPASS models are a set of publicly available stellar population synthesis models which are constructed by combining stellar evolution models with the latest stellar synthetic spectral models. Our aim is to investigate how binary evolution, which dominates in star-forming regions, can differ from single evolution in terms of nebular emission production.

Binary interactions lead to a variety of evolutionary pathways, such as preventing stars from becoming red giants by removing their hydrogen envelopes. These stars then become helium-rich dwarfs and evolve at much higher surface temperatures. Also a companion star can accrete material or two stars can merge, creating a star that is more massive than it was initially. This can upset the simple relationship between the most massive or luminous star and the age of the population of which it is a member.

Interacting binaries thus can lead to very different input ionizing spectra for CLOUDY models. In our work, we assume the gas nebula without any dust to be spherical, ionization-bound and to have a constant, non-evolving hydrogen density n(H) spanning logarithmically from 0 to 3 with 0.5 dex intervals. Moreover, we consider 13 values of Z between 0.05 per cent of and 2 times solar (Z=0.020), corresponding to metallicities at which stellar population models are available. We compute models for 21 values of U logarithmically spaced in the range from -3.5 to -1.5, in steps of 0.1 dex at 21 stellar population ages between 1 and 100 Myrs in 0.1 dex bin. In addition, we specify the 9 essential elements' abundance of the nebula in terms of metallicity, Z, with Hydrogen $= 0.75 - 2.5 \times Z$; Helium $= 0.25 + 1.5 \times Z$; Carbon $= 0.173 \times Z$; Oxygen $= 0.482 \times Z$; Nitrogen $= 0.053 \times Z$; Neon $= 0.099 \times Z$; Magnesium $= 0.038 \times Z$; Silicon $= 0.083 \times Z$; Iron $= 0.072 \times Z$.

As a consequence, we obtain thousands of photoionization models and find the spectral difference between single and binary model varies with stellar age. At stellar ages of less than 5 Myrs, both single-star and binary-star models produce almost same UV radiation and emit the same strong emission lines. However, at ages up to 10 Myrs, there is more UV radiation produced by binary-star models than single-star models. Therefore the population still emits emission lines that are lost from single-star models. This is due to production of hot helium stars or WR stars in the binary models which are not made in the single-star population due to the limited strength of stellar wind mass loss. More detailed discussion can be found in Stanway *et al.* (2016) which investigates the relationship between stellar population and ionizing flux, and is summarised elsewhere in this proceedings (Stanway 2017). Therefore, with these binary-star population to heat the nebulae, the HII regions can survive longer than 10 Myrs, assuming the interstellar medium remains sufficiently dense.

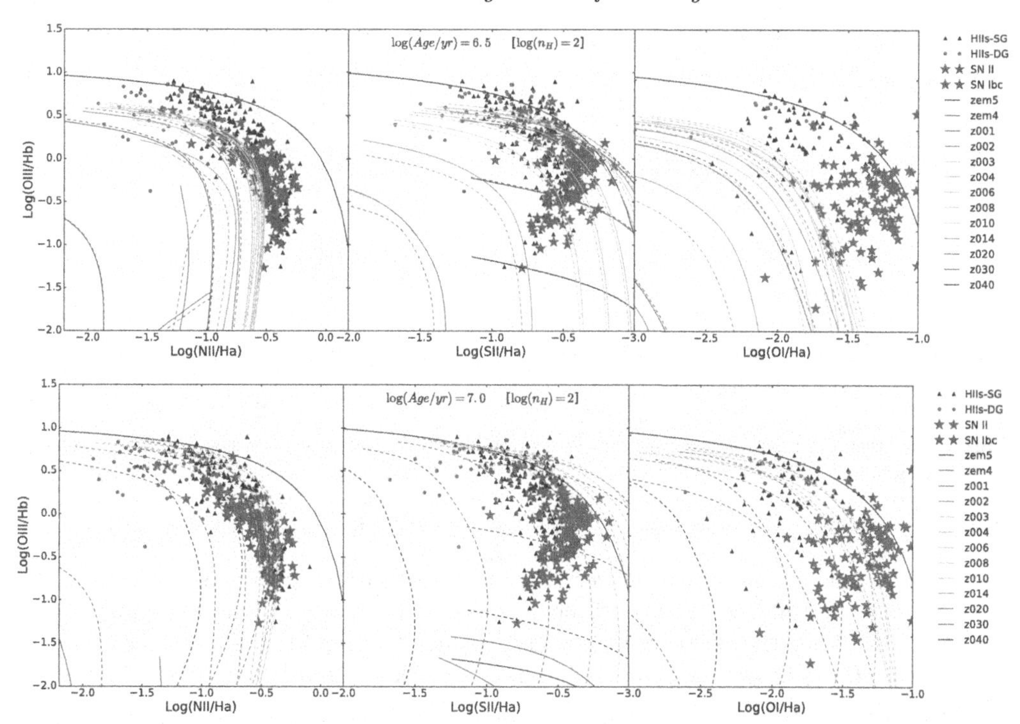

Figure 1. The BPT diagrams at 2 different timescales separately from top panel to bottom panel, 3 Myrs, 7 Myrs and 10 Myrs. The tracks are in solid lines are single models and in dashed lines are binary models. Colour indicates metallicity, ranging from Z=0.0001 (blue) to 0.040 (red). The values of ionization parameter of the models is reduced following the track from upper left ($log(U) = -1.5$) to lower right ($log(U) = -3.5$).

3. Comparison to Observations

The BPT diagrams use strong optical emission line ratios, such as [OIII]λ5007/Hβ, [NII]λ6584/Hα, [SII]λ6724/Hα, [OI]λ6300/Hα, to probe the hardness of the ionizing radiation field in nebular emission regions (Baldwin *et al.*1981). In Figure 1 we show the BPT diagrams with $\log(n_H) = 2$ to discuss the effect of binary evolution, metallicity and age on emission line ratios.

We find the 13 different metallicity models produce their separated tracks in the BPT diagrams, extending from high log([OIII]/Hβ) and low log([NII]/Hα), log([SII]/Hα) and log([OI]/Hα) as the ionization parameter decreases. Both the highest and lowest metallicity models fail to match the distribution of observed HII regions. Models with metallicities ranging from Z = 0.001 to 0.020 (roughly a twentieth Solar to Solar) pass through the observational data at ages of around 5 Myrs for both single and binary Models. In the single models this emission drops away quickly, while binary models continue to match the data through to ages of 10 Myrs.

The location of a model HII region in the BPT diagram depends on numerous factors. We find the best-fitting models to match these individual HII regions respectively by varying all of the input nebular parameters including age, hydrogen density, metallicity and ionization parameter. This allows us to derive the [O/H] and age of these observed HII regions. In the left panel of Figure 2, we compare our derived [O/H] metallicities with those determined by van Zee survey based primarily on strong line ratios. These are calibrated from photoionization models much like our own, but assuming a different relation between ionizing spectrum and stellar metallicity. We find that both our single

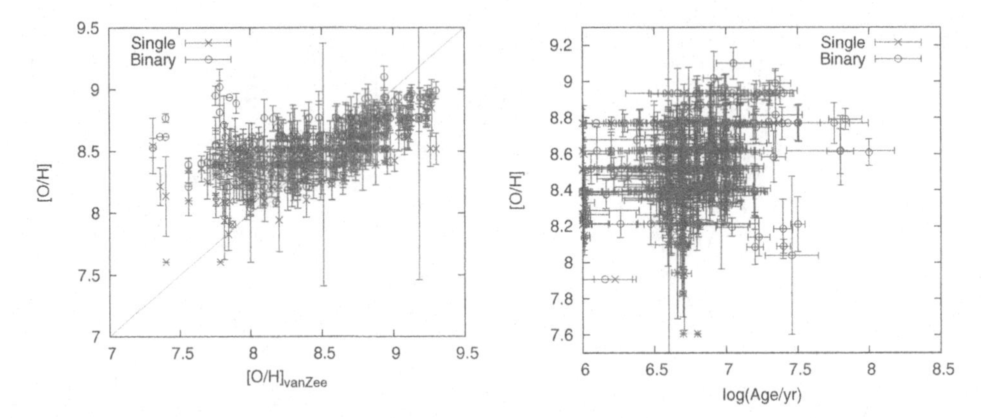

Figure 2. Best-fitting models from single model, the red crosses with error bars, and binary model, the blue circles with error bars, for van Zee sample are presented here. The left panel shows [O/H] ($12 + \log$ [O/H]) estimation from single and binary models compared to the [O/H] from van Zee survey. The right panel presents age from these best-fitting models.

and binary models suggest a narrower range of metallicities for these sources than the van Zee survey estimates. With both ends of the distribution moving back towards to sample mean. When the effects of binaries are included, our moderate metallicity models ($12 + \log[O/H]{\sim}8.5$) produce similar line ratios to those used by van Zee survey with $12 + \log[O/H] \sim 7.5$.

Perhaps the most important property we derive for these HII regions is age. It provides an insight into the evolution history of massive stars and may also provide constraints on core-collapse supernova progenitors. In the right panel of Figure 2 we present the estimation on age of these HII regions from the van Zee sample and a clear difference between single and binary model stands out. Firstly for these HII regions, binary model can extend to 30 Myrs, while most regions have ages around 10 Myrs. In comparison single-star models are younger within 10 Myrs and mostly around 4 Myrs. These results are consistent with the binary evolution pathway of massive stars which become cool supergiant stars due to mass loss, rather than hot helium stars as in single star evolutionary models.

4. Constraining the Progenitor Ages of Core-collapse Supernovae

We can perform a similar analysis of the emission spectra found from stellar populations at the sites of core-collapse supernovae. Using the single-star models we find that the age of all core-collapse supernova progenitors would be less than 3 Myrs. In Figure 3 we show the estimated ages from binary models for SNe II and SNe Ibc explosion sites. We find little difference in age for these two types of SNe. The important difference is that most of the ages for the supernova progenitors is around 10 Myrs or greater. This implies that the majority of progenitor stars for most supernovae have masses less than 20 $M_{\odot}$ for both type II and type Ibc SNe. This is approximately the mass at which it is expected that core-collapse produces black-holes that may not lead to visible supernova (Eldridge & Maund 2016 and references therein). We note that there is also a strong metallicity preference for type Ib/c to arise from metallicities close to solar with most low metallicity events being of type II.

We caution these findings are still preliminary, but indicate that modelling the nebular emission from stellar populations including interacting binaries will provide a new method to understand core-collapse supernovae.

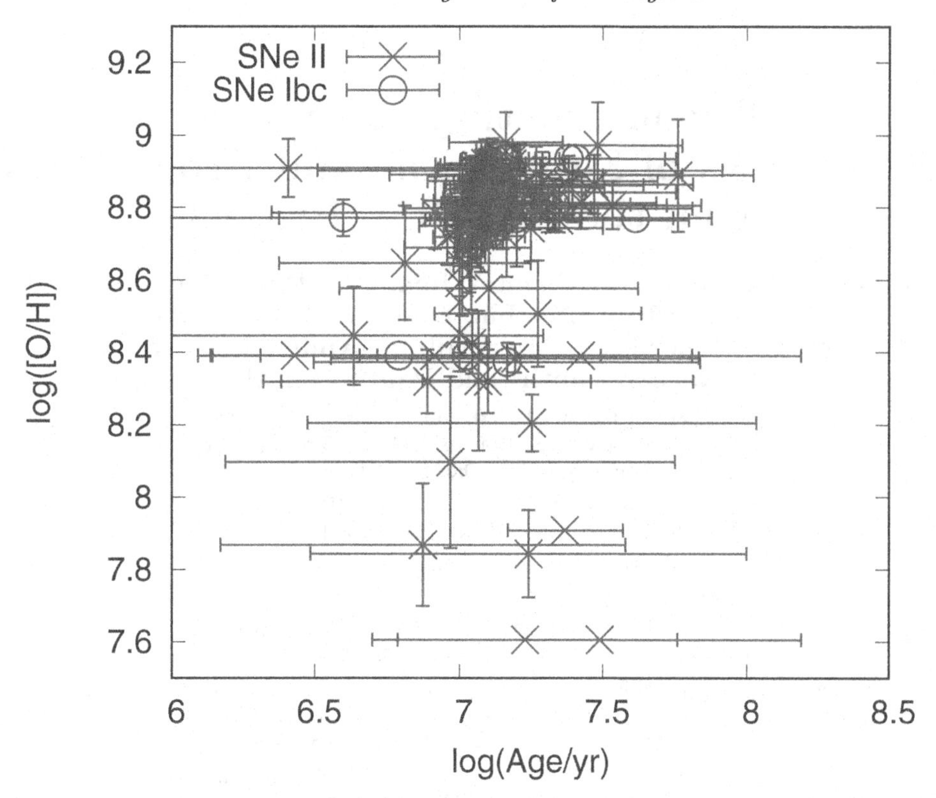

Figure 3. The relation between [O/H] and age for SNe II and SNe Ibc from binary models. The red crosses with errorbars correspond to SNe II and blue circles with errorbars stands for SNe Ibc.

References

Byler, N., Dalcanton, J. J., Conroy, C., & Johnson, B. D. 2016, arXiv:1611.08305

Eldridge, J. J. & Maund, J. R. 2016, *MNRAS*, 461, L117

Ferland, G. J., Korista, K. T., Verner, D. A., *et al.* 1998, *PASP* 110, 761

Galbany, L., Stanishev, V., Mourão, A. M., *et al.* 2014, *A&A*, 572, A38

Galbany, L., Stanishev, V., Mourão, A. M., *et al.* 2016, *A&A*, 591, A48

Levesque, E. M., Kewley, L. J., & Larson, K. L. 2010, *AJ*, 139, 712

Stanway, E. R., Eldridge, J. J., & Becker, G. D. 2016, *MNRAS*, 456, 485

van Zee, L. & Haynes, M. P. 2006, *AJ*, 636, 214

van Zee, L., Salzer, J. J., Haynes, M. P., O'Donoghue, A. A., & Balonek, T. J. 1998, *AJ*, 116, 2805

The Lives and Death-Throes of Massive Stars
Proceedings IAU Symposium No. 329, 2016
J.J. Eldridge, J.C. Bray, L.A.S. McClelland & L. Xiao, eds.

doi:10.1017/S1743921317003052

Reconstructing the Scene: New Views of Supernovae and Progenitors from the SNSPOL Project

Jennifer L. Hoffman[1], G. Grant Williams[2,3], Douglas C. Leonard[4], Christopher Bilinski[3], Luc Dessart[5], Leah N. Huk[1], Jon C. Mauerhan[6], Peter Milne[3], Amber L. Porter[7], Nathan Smith[3] and Paul S. Smith[3]

[1]Department of Physics & Astronomy, University of Denver,
2112 E. Wesley Ave., Denver, CO 80210-6900, USA
email: jennifer.hoffman@du.edu

[2]MMT Observatory, P.O. Box 210065,
University of Arizona, Tucson, AZ 85721-0065, USA

[3]Steward Observatory, University of Arizona,
933 N. Cherry Avenue, Tucson, AZ 85721, USA

[4]Department of Astronomy, San Diego State University,
5500 Campanile Drive, San Diego, CA 92182-1221, USA

[5]Unidad Mixta Internacional Franco-Chilena de Astronomía (CNRS UMI 3386),
Departamento de Astronomía, Universidad de Chile,
Camino El Observatorio 1515, Las Condes, Santiago, Chile

[6]Department of Astronomy, University of California,
Berkeley, CA 94720-3411, USA

[7]Department of Physics and Astronomy, Clemson University,
118 Kinard Laboratory, Clemson, SC 29634

Abstract. Because polarization encodes geometrical information about unresolved scattering regions, it provides a unique tool for analyzing the 3-D structures of supernovae (SNe) and their surroundings. SNe of all types exhibit time-dependent spectropolarimetric signatures produced primarily by electron scattering. These signatures reveal physical phenomena such as complex velocity structures, changing illumination patterns, and asymmetric morphologies within the ejecta and surrounding material. Interpreting changes in polarization over time yields unprecedentedly detailed information about supernovae, their progenitors, and their evolution.

Begun in 2012, the SNSPOL Project continues to amass the largest database of time-dependent spectropolarimetric data on SNe. I present an overview of the project and its recent results. In the future, combining such data with interpretive radiative transfer models will further constrain explosion mechanisms and processes that shape SN ejecta, uncover new relationships among SN types, and probe the properties of progenitor winds and circumstellar material.

Keywords. polarization, astronomical databases:miscellaneous, supernovae: general

---

1. Supernovae and polarization

Because supernovae (SNe) can be seen at great distances, they provide clues about stellar evolution in faraway galaxies and in the cosmological past. SNe are broadly classified into "thermonuclear" and "core-collapse" categories. Thermonuclear (Type Ia) supernovae likely arise from binary stars in which a white dwarf accretes material from its companion and becomes unstable (although white-dwarf mergers may be a viable mechanism for some SNe Ia; see, e.g., Ruiz-Lapuente 2014). Core-collapse supernovae (CCSNe;

Types Ib/c and II) are thought to arise from massive stars that exhaust the fuel in their cores and undergo catastrophic collapse and subsequent neutrino-driven explosion. The wide range of different subtypes of CCSNe is not well understood, but seems to be related to mass loss in late stages of evolution, which affects the amount of circumstellar material (CSM) surrounding the doomed progenitor star (e.g., Smart 2009).

All types of SNe display intrinsic polarization signatures that vary over time, but the behavior of this polarization differs among subtypes (Wang & Wheeler 2008). SN polarization is caused primarily by light scattering from free electrons in ionized ejecta material. An unresolved spherical scattering region yields no net polarization, but an elongated or otherwise aspherical electron-scattering region produces a net wavelength-independent polarization signal. Thus, the simplest case of SN polarization is a constant continuum signal that reveals the presence of aspherical ejecta (Höflich 1991). Tracing this continuum polarization over time as the SN evolves can reveal the complex nature of the surrounding material. For example, SN 2004dj and related objects have very low continuum polarization at early times but show a polarization spike at the end of the plateau phase, revealing an inner, aspherical core as the the spherical outer envelope becomes transparent (Leonard *et al.* 2006; Chornock *et al.* 2010; Dessart & Hillier 2011).

More complex wavelength-dependent scenarios occur when clumps or inhomogeneities within the ejecta obscure portions of the scattering region (Kasen *et al.* 2003), when circumstellar material surrounds the ejecta and contributes its own scattering signal (Hoffman *et al.* 2008), or when line photons interact differently with the ejecta than do continuum photons. Combinations of these effects are common. The time domain adds an additional dimension of complication, as the relative contributions of each of these effects may change as the SN evolves. (However, the constant contribution of interstellar polarization [ISP] can be constrained with time-series observations, e.g., Leonard *et al.* 2000; Hoffman *et al.* 2008.) The vast majority of SNe studied with more than one epoch of polarimetry have displayed variability over time in the continuum, lines, or both.

Time-dependent spectropolarimetry thus provides us with an unprecedentedly detailed view into the complex and variable structures surrounding SNe of all types. This allows us to probe not only the SN mechanism itself, but also to constrain progenitor properties by measuring the characteristics (density, composition, velocity, geometrical structure) of the stellar winds and eruptions that preceded the final explosion.

2. The SNSPOL Project

Begun in 2012 and funded by the National Science Foundation, the SNSPOL Project (PI: G.G. Williams) continues to amass the largest database of time-dependent spectropolarimetric data on supernovae of all types. The SNSPOL Project obtains observations of bright SNe roughly once per month using the CCD Imaging/Spectropolarimeter (SPOL; Schmidt, Elston, & Lupie 1992) at either the 61" Kuiper telescope on Mt. Bigelow, the 2.3-m Bok on Kitt Peak, or the 6.5-m MMT on Mt. Hopkins. Supporting observations include optical and near-IR photometry from the Mount Laguna Observatory (Khandrika *et al.* 2014) and ISP probe star observations from the HPOL spectropolarimeter at the University of Toledo's Ritter Observatory (Davidson *et al.* 2014). To date, the SNSPOL project has obtained spectropolarimetric data for over 80 SNe, of which 59 have been observed at more than one observational epoch.

SNSPOL's monthly observation cadence allows us to monitor the spectropolarimetric evolution of our targets over their optically visible lifetimes in more detail than has previously been possible. In the sections below, we review some of the most significant results from the SNSPOL Project to date.

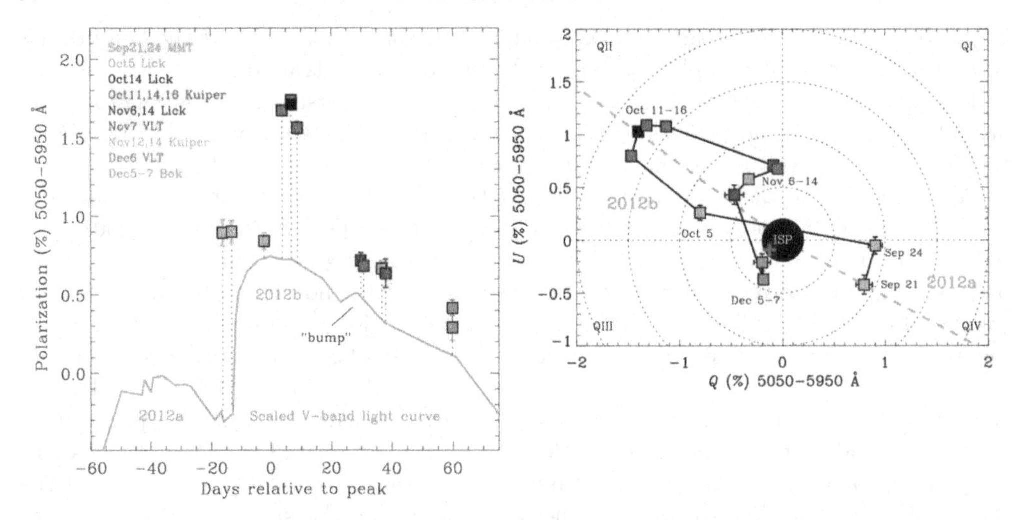

Figure 1. *Left:* Temporal evolution of the total V-band polarization for SN 2009ip in 2012. Filled squares represent polarization measurements; the gray line represents the V-band light curve. *Right:* Filled squares represent the same polarization measurements as at left, now plotted in the Q–U plane. ISP is constrained to be less than 0.2%, illustrated by the black dot. Dashed lines represent the position angles 166° and 72°, as measured at the first epoch on Sep. 21 and peak polarization on Oct. 14, respectively. The approximate point-reflection symmetry between 2012a and the peak of 2012b suggests two separate and roughly orthogonal components of polarization on the sky. Figure from Mauerhan *et al.* (2014), courtesy Oxford University Press and the Royal Astronomical Society. For the full-color figure, see that paper.

2.1. *SN 2009ip*

SNSPOL observations of the peculiar transient SN 2009ip provided new insights into the nature of its two 2012 brightness peaks (Fraser *et al.* 2013; Mauerhan *et al.* 2013; Levesque *et al.* 2014; Margutti *et al.* 2014). SNSPOL data showed that these two peaks possessed orthogonal polarization position angles (Fig. 1; Mauerhan *et al.* 2014). This suggests that two distinct scattering regions produced the polarization at these two epochs. We concluded that the 2012a peak arose from the maximum brightness of a true core-collapse SN explosion and the corresponding polarization was due to asphericity of the SN ejecta (Mauerhan *et al.* 2014). The 2012b peak was likely due to CSM interaction, and its polarization angle suggests this CSM was equatorially concentrated orthogonal to the ejecta axis. This geometry is similar to the one suggested by Levesque *et al.* (2014) based on spectral modeling. This CSM likely arose from the progenitor's mass loss in late evolution, and its flattened nature could indicate a binary origin.

2.2. *SN 2011dh*

In the case of the Type IIb supernova SN 2011dh, we combined four SNSPOL observations with additional spectropolarimetry from the Lick and Palomar Observatories. The commonalities in position angle between the early continuum polarization and that of the Hα and He I lines led us to favor a picture in which fast-rising plumes of radioactive ^{56}Ni produced clumpy excitation of the SN envelope (Mauerhan *et al.* 2015; Fig. 2). This scenario may link SN 2011dh with other SN types and with SNRs such as Cas A.

2.3. *SN 2014J*

Seven SNSPOL observations of the nearby Type Ia SN 2014J revealed that this object displayed typical polarization behavior for a SN Ia, with nearly spherical ejecta combined with a clumpy distribution of silicon; circumstellar dust may also contribute to the

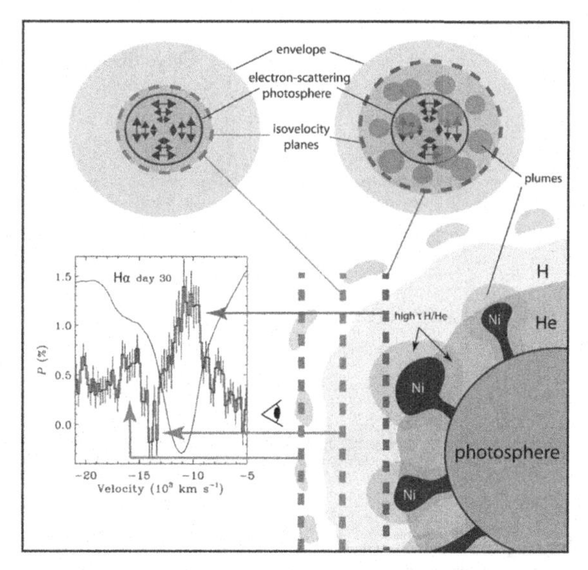

Figure 2. Illustration of the formation of the polarized Hα line in SN 2011dh. The upper diagrams are face-on views; the lower diagram illustrates the transverse view. Several isovelocity planes (*dashed lines*) are depicted along with their predicted polarization characteristics and locations within the polarized line profile. Figure from Mauerhan *et al.* (2015), courtesy Oxford University Press and the Royal Astronomical Society. For the full-color figure, see that paper.

continuum polarization in this object (Porter *et al.* 2016). Our spectropolarimetric data also allowed us to investigate the nature of the ISP in the host galaxy, M82. The dust grains along this sightline in M82 are smaller than those making up typical Milky Way dust.

2.4. *SN 2012au*

The polarization of SN 2012au evolved from a large magnitude of $\sim 2\%$ at early times to nearly zero within 3 months, while its well-defined polarization angle gradually disappeared (Hoffman *et al.* 2014; Fig. 3). Rather than a dramatic change in large-scale ejecta geometry, we attribute this behavior to a receding of the photosphere through an outer region with greater asymmetry to an inner region with greater symmetry (the opposite behavior from SN 2004dj; Leonard *et al.* 2000). The He I λ5876 emission line displayed complex intrinsic polarization behavior with a position angle nearly 90° different from that of the continuum, a possible disk-jet signature. Analysis of this SN is still underway.

3. The future of SNSPOL

As the SNSPOL Project continues to collect data on a broad variety of SNe, our focus will shift from observing as many objects as possible to improving wavelength and time coverage for particular targets of interest. We aim to improve the statistical sample size of the various SN types studied with spectropolarimetry; this will allow us to move from analyzing individual SNe to considering the relationships among SN subtypes and among different events within a given subtype. These larger-scale studies will allow us to draw connections between SNe and their progenitors more clearly than ever before.

Our collaboration is also developing radiative transfer modeling capabilities that include complex polarization effects (Dessart & Hillier 2011; Huk *et al.* 2017, this volume). These will allow us to better interpret the spectropolarimetric data obtained by the

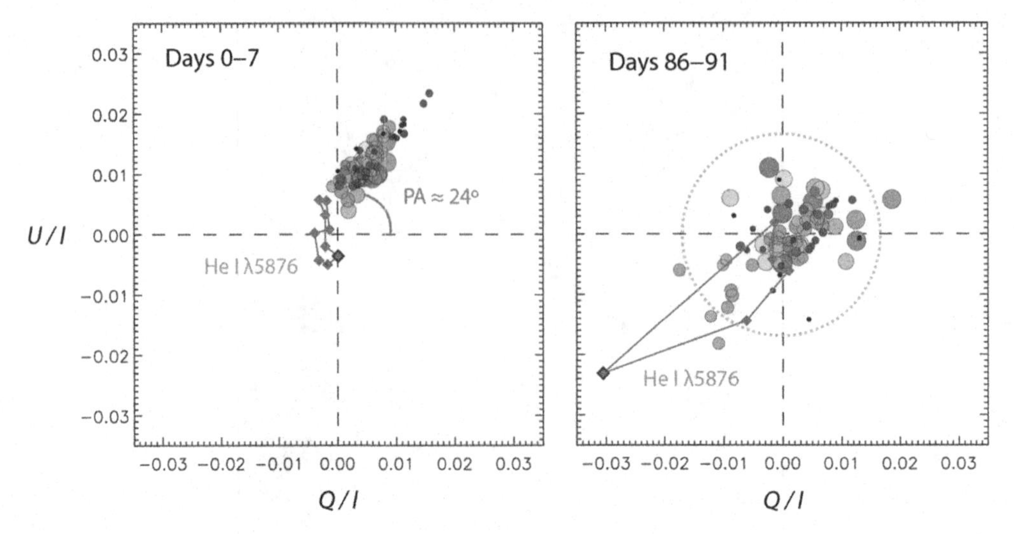

Figure 3. Two epochs of polarization of the Type Ib SN 2012au from SNSPOL. Wavelength points (binned to 50 Å) run from small to large circles; the connected diamond points represent the He I λ5876 emission line, binned to 10–20 Å. The continuum data show a clear polarization axis at $\sim 24°$ at early times, but scatter around 0 at later times (suggesting a low ISP contribution; dotted circle). The helium line is intrinsically polarized differently from the continuum at all times. Figure from Hoffman *et al.* (2014).

SNSPOL Project. The confrontation of observations with models will ramp up in the near future as the project shifts its focus from data collection to larger-scale analysis.

Acknowledgments. JLH and LNH thank the IAU for generous travel support. The SNSPOL Project is supported by NSF awards AST-1210599, AST-1210311, and AST-1210372; the SN 2009ip research was also partly funded by NSF award AST-1211916.

References

Chornock, R., Filippenko, A. V., Li, W., & Silverman, J. M. 2010, *ApJ*, 713, 1363
Davidson Jr., J. W., Bjorkman, K. S. Hoffman, J. L., *et al.* 2014, *JAI*, 03, 1450009
Dessart, L. & Hillier, D. J. 2011, *MNRAS*, 415, 3497
Fraser, M., Magee, M., Kotak, R., *et al.* 2013, *ApJ*, 779, 8
Hoffman, J. L., Leonard, D. C., Chornock, R., *et al.* 2008, *ApJ*, 688, 1186
Hoffman, J. L., Smith, N., Bilinski, C., *et al.* 2014, *AAS*, 223, 35421 (http://goo.gl/hLOeYq)
Höflich, P. 1991, *A&A*, 246, 481
Kasen, D., Nugent, P., Wang, L., *et al.* 2003, *ApJ*, 593, 788
Khandrika, H. G., Leonard, D. C., Horst, C., *et al.* 2014, *AAS*, 224, 121.16
Leonard, D. C., Filippenko, A. V., Barth, A. J., & Matheson, T. 2000, *ApJ*, 536, 239
Leonard, D. C., Filippenko, A. V., Ganeshalingam, M., *et al.* 2006, *Nature*, 440, 505
Levesque, E. M., Stringfellow, G. S., Ginsburg, A. G., *et al.* 2014, *AJ*, 147, 23
Margutti, R., Milisavljevic, D., Soderberg, A. M., *et al.* 2014, *ApJ*, 780, 21
Mauerhan, J. C., Smith, N., Filippenko, A. V., *et al.* 2013, *MNRAS*, 430, 1801
Mauerhan, J., Williams, G. G., Smith, N., *et al.* 2014, *MNRAS*, 442, 1166
Mauerhan, J., Williams, G., Leonard, D. C., *et al.* 2015, *MNRAS*, 453, 4467
Porter, A. L., Leising, M. D., Williams, G. G., *et al.* 2016, *ApJ*, 828, 24
Ruiz-Lapuente, P. 2014, *NewAR*, 62, 15
Schmidt, G. D., Elston, R., & Lupie, O. L. 1992, *AJ*, 104, 1563
Smartt, S. J. 2009, *ARA&A*, 47, 63
Wang, L. & Wheeler, J. C. 2008, *ARAA*, 46, 433

The Lives and Death-Throes of Massive Stars
Proceedings IAU Symposium No. 329, 2016
J.J. Eldridge, J.C. Bray, L.A.S. McClelland & L. Xiao, eds.

doi:10.1017/S1743921317003386

The evolution of red supergiants to supernovae

Emma R. Beasor and Ben Davies

Astrophysics Research Institute, Liverpool John Moores University, Liverpool, L3 5RF, UK
email: e.beasor@2010.ljmu.ac.uk

Abstract. With red supergiants (RSGs) predicted to end their lives as Type IIP core collapse supernova (CCSN), their behaviour before explosion needs to be fully understood. Mass loss rates govern RSG evolution towards SN and have strong implications on the appearance of the resulting explosion. To study how the mass-loss rates change with the evolution of the star, we have measured the amount of circumstellar material around 19 RSGs in a coeval cluster. Our study has shown that mass loss rates ramp up throughout the lifetime of an RSG, with more evolved stars having mass loss rates a factor of 40 higher than early stage RSGs. Interestingly, we have also found evidence for an increase in circumstellar extinction throughout the RSG lifetime, meaning the most evolved stars are most severely affected. We find that, were the most evolved RSGs in NGC2100 to go SN, this extra extinction would cause the progenitor's initial mass to be underestimated by up to 9M$_\odot$.

Keywords. circumstellar matter, stars: supergiants, stars: evolution, stars: mass-loss

---

1. Introduction

Archival imaging now provides a vital tool in identifying the progenitors to supernovae (SNe). Red supergiants (RSGs) end their lives as Type IIP core collapse SNe, of which there have been 7 progenitors confirmed with pre-explosion imaging, most recently the 12.5 ± 1.2 M$_\odot$ progenitor to SN 2012aw (Fraser *et al.* 2016). Theory predicts that these RSG progenitors should be exploding between masses of 8 to 25M$_\odot$ (Ekström *et al.* 2012) but so far it seems that the exploding stars are of a relatively low mass with no progenitors appearing in the upper end of the mass range (17 to 25M$_\odot$; Smartt *et al.* 2009, 2015).

Since Smartt *et al.* (2009), various scenarios have been proposed to solve the RSG problem. From a supernova perspective, it was considered whether these high mass RSGs were ending their lives as other types of core collapse supernovae (CCSNe). Smartt *et al.* (2009) stated that the fractions for type IIn and IIb still did not make up for the lack of high mass RSG progenitors. However, Smith *et al.* (2011) disagreed, and suggested that high mass progenitors could indeed be exploding as other CCSNe.

It is possible that extreme levels of mass loss cause stars to evolve away from the RSG phase and explode as a different class of star. Currently, stellar evolution models rely on observational or theoretical mass loss rate prescriptions, often based on large studies of field stars (e.g. de Jager *et al.* 1988) or stars that are known to be heavily dust enshrouded (e.g. Van Loon *et al.* 2005). A potential weakness of using field stars for these studies is that the parameters of initial mass ($M_{\rm initial}$) and metallicity (Z) are left unconstrained, possibly causing the large dispersions in the observed trends, while studies targeting heavily dust enshrouded stars are biased towards high-$\dot{M}$ objects.

Georgy (2012) and Meynet *et al.* (2015) discussed the implications of increasing these standard $\dot{M}$ prescriptions by factors of 3, 5 and 10 times. These studies found that increasing $\dot{M}$ caused stars to evolve away from the RSG phase at lower masses than predicted by models with standard $\dot{M}$, matching the upper mass limit found from progenitor studies.

There is also the potential that the heaviest RSGs end their lives with no explosion at all. It has been suggested that RSGs with masses higher than 17$M_\odot$ may collapse immediately to black hole with little or no explosion (Kochanek *et al.*, 2015). Large observational searches are currently being conducted to find these disappearing stars, with so far only a yellow supergiant as a potential candidate (Reynolds *et al.*, 2015).

However, there could be a simpler solution to the lack of high mass progenitors in the form of circumstellar dust. It has been long established that RSGs form dust in their winds (e.g. de Wit *et al.* 2008) and it is possible that if a large enough mass of dust built up around the RSG it would appear less luminous, and hence a lower mass would be inferred (as discussed by Walmswell & Eldridge, 2012).

To investigate to what degree dust accumulates around the star, and how it might cause the observer to underestimate its initial mass, we have measured the amount of circumstellar material surrounding 19 coeval RSGs, allowing us to investigate whether this is correlated with evolution.

2. Application to NGC 2100

NGC 2100 is a young massive cluster in the LMC rich in RSGs. We assume the cluster is coeval, as any spread in the age of the stars will be small compared to the age of the cluster. This also means the spread in mass between the stars currently in the RSG phase is small, within a few tenths of a solar mass. Using mass tracks and isochrones we find the initial mass for the stars within NGC 2100 to be ~14-17 $M_\odot$ with an age of 15 Myrs. Any difference in luminosity for the RSGs can be considered an evolutionary effect, as the slightly more massive RSGs will evolve at a slightly faster rate, however all the RSGs will follow the same path across the HR diagram. Luminosity can therefore be used as a proxy for evolution.

2.1. *Modelling results and discussion*

We ran our fitting procedure for 19 RSGs located in NGC 2100. Figure 1 shows the model fit for the most luminous star in our sample with observed photometry. The plot shows our best fit model spectrum (green line), the models within our error range (blue dotted lines) and various other contributions to the total output flux, including scattered flux, dust emission and attenuated flux. It also shows the photometric data (red crosses) and model photometry (green crosses).

Our model fits allowed us to derive mass loss rates and luminosities for all 19 RSGs in our sample. Figure 2 shows a positive correlation between luminosity and mass loss rate. Since we are using luminosity as a proxy for evolution, the stars with the lowest luminosity can be considered to be the early stage RSGs and the stars with the highest luminosities closest to supernova. Our results suggest $\dot{M}$ increases by a factor of 40 throughout the lifetime of an RSG, approximately 10^6 years for a 15$M_\odot$ star (Georgy *et al.* 2013). Overplotted are commonly used mass loss rate prescriptions. Our correlation is well matched by the prescription of de Jager (de Jager *et al.* 1988) as it provides the best fit for the more evolved stars (where the mass loss mechanism is stronger). We also find a tight correlation between $\dot{M}$ and luminosity, which we conclude is due to keeping M_{initial} and Z constrained.

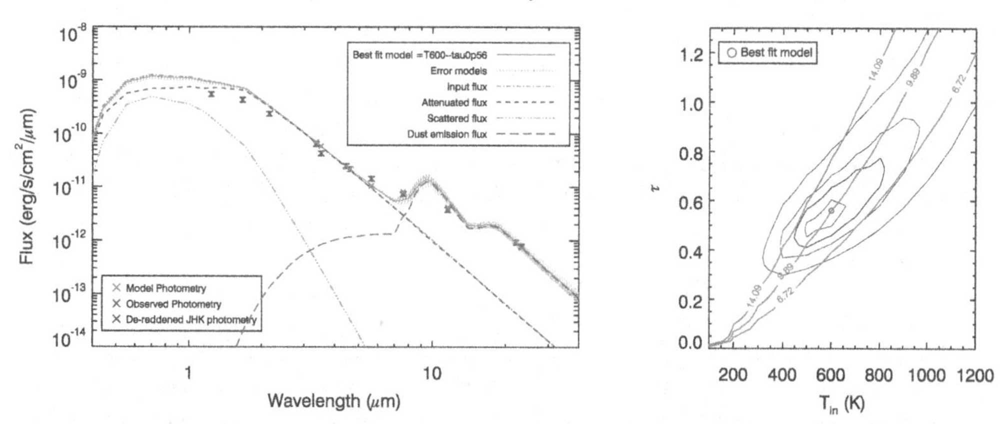

Figure 1. *Left panel:* Model plot for the star with the highest $\dot{M}$ value in NGC 2100 including all contributions to spectrum. *Right panel:* Contour plot showing the degeneracy between χ^2 values and best fitting $\dot{M}$ values in units of 10^{-6} M$_\odot$ yr^{-1}. The green lines show the best fit $\dot{M}$ and upper and lower $\dot{M}$ isocontours. It can be seen that while there is some degeneracy between inner dust temperature and optical depth the value of $\dot{M}$ is independent of this. Figures taken from Beasor & Davies (2016).

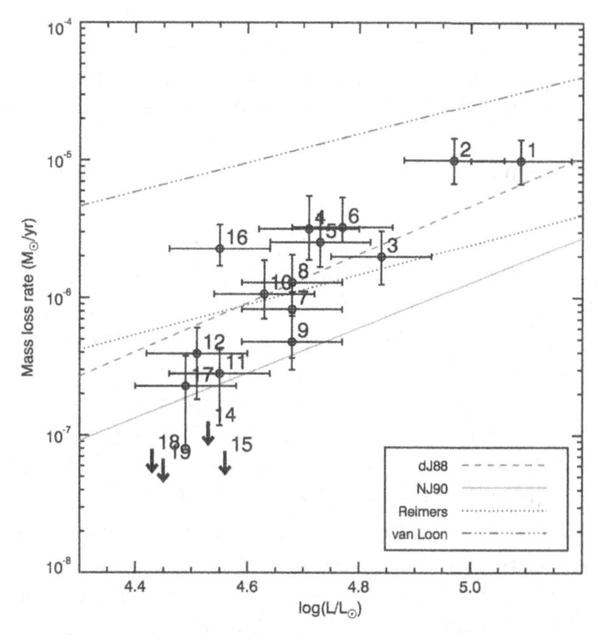

Figure 2. Plot showing $\dot{M}$ versus $L_{\rm bol}$. A positive correlation can be seen suggesting $\dot{M}$ increases with evolution. This is compared to some mass loss rate prescriptions. The downward arrows show for which stars we only have upper limits on $\dot{M}$. Figure taken from Beasor & Davies (2016).

We were also able to work out what level of extinction would result from the warm inner dust shell of each star, finding very low levels of extinction that would only have minor effects on mass calculations. However, we did observe an increase in reddening for the two most evolved stars in the cluster. As a check to our model fits, we overplotted JHK photometry (de-reddened for foreground extinction) for all of the stars (which we had not included in our fitting procedure). For the majority of RSGs, the JHK photometry fit the best fit model with no tweaking required. The best fit models for stars #1 and

#2, the two most evolved stars, instead over predicted the flux at these wavelengths (see Fig. 1 where JHK photometry is shown by the blue crosses). We considered many different scenarios that could be causing this increase in reddening, including assuming an effective temperature that was too high or extreme variability, none of which provided satisfactory solutions (see Beasor & Davies 2016 for full discussion).

We next considered the possibility that this increase in reddening was due to cold, clumpy dust at large radii from the stars which would not be detectable with mid-IR photometry. It is known that RSGs have extended, asymmetrical dust shells, a famous example being μ Cep (de Wit *et al.* 2008). If we were to move μ Cep to the distance of the LMC, the cold dust from the extended nebula would be too faint to be observable, at a level of around 0.2 Jy (even before we account for a factor of 2 lower dust to gas ratio in the LMC). It is therefore plausible that the enhanced extinction we observe for stars #1 and #2 is caused by the stars being surrounded by a similar amount of dust that is too faint to detect at the distance of the LMC.

3. Implications

3.1. *Stellar evolution*

Our results show a clear increase in $\dot{M}$ with RSG evolution, by a factor of ~ 40 throughout the lifetime of the star. For this metallicity and initial mass, we see this $\dot{M}$ is well described by current mass loss rate prescriptions, in particular de Jager's, suggesting there is no need for evolutionary models to increase $\dot{M}$ by significant amounts during the RSG phase. For this $M_{\rm initial}$ ($\sim$14-17 $M_\odot$) and at LMC metallicity altering the $\dot{M}$ prescriptions by factors of 10 or more seems unjustified (Georgy 2012, Meynet *et al.* 2015).

3.2. *Application to SNe progenitors*

We also found evidence for increased reddening to the two most evolved stars in our sample. We now ask the question, if star #1 were to go SN tomorrow, what would we infer about its initial mass from limited photometric information? Progenitor studies often rely on single-band photometry or upper limits from non-detections, requiring assumptions to be made about spectral type and level of circumstellar extinction. If we apply similar assumptions to those of Smartt *et al.* 2009 to #1, without considering the extra reddening we have observed, we find a mass of 8 $M_\odot$. From mass tracks, we have determined the initial mass of the NGC 2100 stars to be 14 - 17 $M_\odot$. Hence the mass of the most evolved star in the cluster from single band photometry is clearly underestimated when applying the same assumptions as used by Smartt *et al.* When we take into account the additional reddening, the mass increases to $\sim$17$\pm$5 $M_\odot$, in good agreement with the mass inferred from mass tracks. This is shown in Fig. 3, where the green star represents the mass estimate with no additional extinction considered, the red star is the mass estimate when the additional extinction is considered and the orange line shows the best fit mass track for this cluster.

With this in mind, we went on to see what effect this level of extinction could have on previously determined progenitor masses. We considered three case studies, the progenitors to SN 1999gi, 2001du and 2012ec (of which SN 1999gi and 2001du are based on upper limits, with SN 2012ec having a detection in one band). The inferred masses of these progenitors increased by 10$M_\odot$, 7$M_\odot$ and $\sim$6$M_\odot$ (see Beasor & Davies 2016 for full discussion). We have shown that by including similar levels of reddening that we find in the most evolved stars in NGC 2100, the initial mass estimated for Type IIP SNe progenitors increase substantially. If we were to apply this to all objects in the Smartt

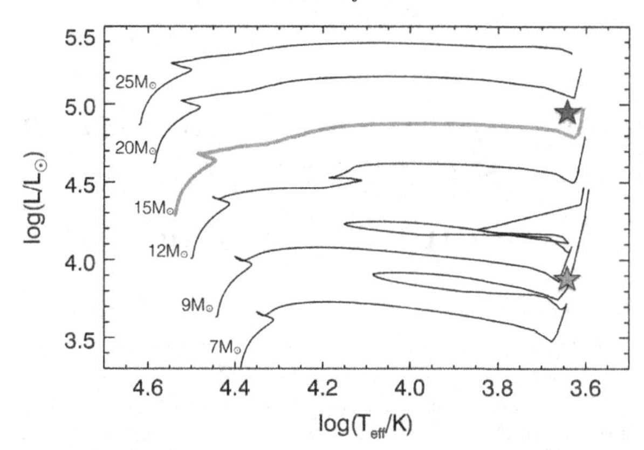

Figure 3. Plot showing the effect of including the additional reddening we observe on progenitor mass estimates for star #1. The green star represents the mass estimate when no circumstellar extinction is considered, the red star shows the estimated progenitor mass when we take into account additional extinction and the orange line shows the best fit mass track for this cluster. The mass tracks are from Georgy *et al.* (2013) and are at $Z = 0.002$.

et al. (2009) sample this may resolve the inconsistency between theory and observations and hence solve the red supergiant problem.

References

Beasor, E., R. & Davies, B. 2016, *MNRAS*, 463, 1269

de Jager, C., Nieuwenhuijzen, H., & der Hucht, KA 1988, *Astron. & Astrophys. Suppl. Series* , 72, 259

Fraser, M. *et al.* 2016, *MNRAS*, 456, L16

Ekström, S., Georgy, C., Meynet, G. *et al.* 2012, *Astron. & Astrophys.*, 537, A146

Georgy, C. 2012, *Astron. and Astrophys.*, 538, L8

Georgy, C., Ekström, S., Eggenberger, P. *et al.* 2013, *Astron. and Astrophys.*, 558, AI03

Kochanek, C. S. , 2015, *MNRAS*, 446, 1213

Meynet, G. & Maeder, A. 2003, *Astron. and Astrophys.*, 404, 975

Meynet, G. *et al.* 2015, *Astron. and Astrophys.*, 575, A60

Reynolds, T., Fraser, M., & Gilmore, G 2015, *MNRAS*, 453, 2885

Smartt, S., Eldridge, J., Crockett, R., *et al.* 2009, *MNRAS*, 395, 1409

Smartt, S., *et al.* 2015, *Publ. of the Astron. Soc. of Australia*, 32, e016

Smith, N., Li, W., Filippenko, A. *et al.* 2011, *MNRAS*, 412, 1522

Walmswell & Eldridge 2012, *MNRAS*, 419, 2054

de Wit, W. *et al.* 2008, *The Astrophys. J. Lett.*, 419, 2054

The Lives and Death-Throes of Massive Stars
Proceedings IAU Symposium No. 329, 2016
J.J. Eldridge, J.C. Bray, L.A.S. McClelland & L. Xiao, eds.

doi:10.1017/S1743921317003003

Blue supergiant progenitors from binary mergers for SN 1987A and other Type II-peculiar supernovae

Athira Menon[1] and Alexander Heger[1,2,3]

[1]Monash Centre for Astrophysics (MoCA) and School of Physics and Astronomy, Monash University, Clayton, VIC 3800, Australia
email: athira.menon@monash.edu
[2]School of Physics and Astronomy, University of Minnesota, Minneapolis, MN 55455, U.S.A.
[3]Centre for Nuclear Astrophysics, Shanghai Jiao Tong University, Shanghai, China
email: alexander.heger@monash.edu

Abstract. We present results of a systematic and detailed stellar evolution study of binary mergers for blue supergiant (BSG) progenitors of Type II supernovae, particularly for SN 1987A. We are able to reproduce nearly all observational aspects of the progenitor of SN 1987A, Sk –69 °202, such as its position in the HR diagram, the enrichment of helium and nitrogen in the triple-ring nebula and its lifetime before its explosion. We build our evolutionary model based on the merger model of Podsiadlowski *et al.* (1992), Podsiadlowski *et al.* (2007) and empirically explore an initial parameter consisting of primary masses, secondary masses and different depths up to which the secondary penetrates the He core during the merger. The evolution of the post-merger star is continued until just before iron-core collapse. Of the 84 pre-supernova models ($16\,M_\odot - 23\,M_\odot$) computed, the majority of the pre-supernova models are compact, hot BSGs with effective temperature > 12 kK and $30\,R_\odot - 70\,R_\odot$ of which six match nearly all the observational properties of Sk –69 °202.

Keywords. (stars:) supernovae: individual (SN 1987A), stars: evolution, (stars:) binaries (including multiple): close

1. Introduction

The remarkable supernova SN 1987A that exploded in the Large Magellanic Cloud (LMC) (West *et al.* 1987) was an unusual Type II SNe– it had a dome-shaped light curve (Catchpole *et al.* 1988, Hamuy *et al.* 1988), unlike typical Type II-P SNe. The compact BSG progenitor that exploded as SN 1987A, was a B3 Ia supergiant in the LMC, called Sk –69 °202 (Walborn *et al.* 1987, Blanco *et al.* 1987). Another surprising aspect of the progenitor was the presence of a triple-ring nebula of circumstellar material, enriched in helium and nitrogen (Lundqvist & Fransson 1996, France *et al.* 2011), ejected by the progenitor about 15-20 kyr before the explosion (Wampler *et al.* 1990, Burrows *et al.* 1995, Crotts *et al.* 2000). The angular momentum required for the expulsion of the nebular material and the formation of its complex geometry, could only have been possible from a binary merger (Podsiadlowski 1992, Morris & Podsiadlowski 2007, Morris & Podsiadlowski 2009).

There has not been so far, an evolutionary model that can consistently reproduce all the key signatures of the progenitor of SN 1987A (Smartt 2009), despite the many attempts from single star models. Podsiadlowski *et al.* 1992, and Podsiadlowski *et al.* 2007, described a binary merger scenario for the progenitor, that could explain all the observations of Sk –69 °202. In these works, they demonstrated that the merger of a $15\,M_\odot$ primary red supergiant (RSG) with a main-sequence secondary of $5\,M_\odot$ could

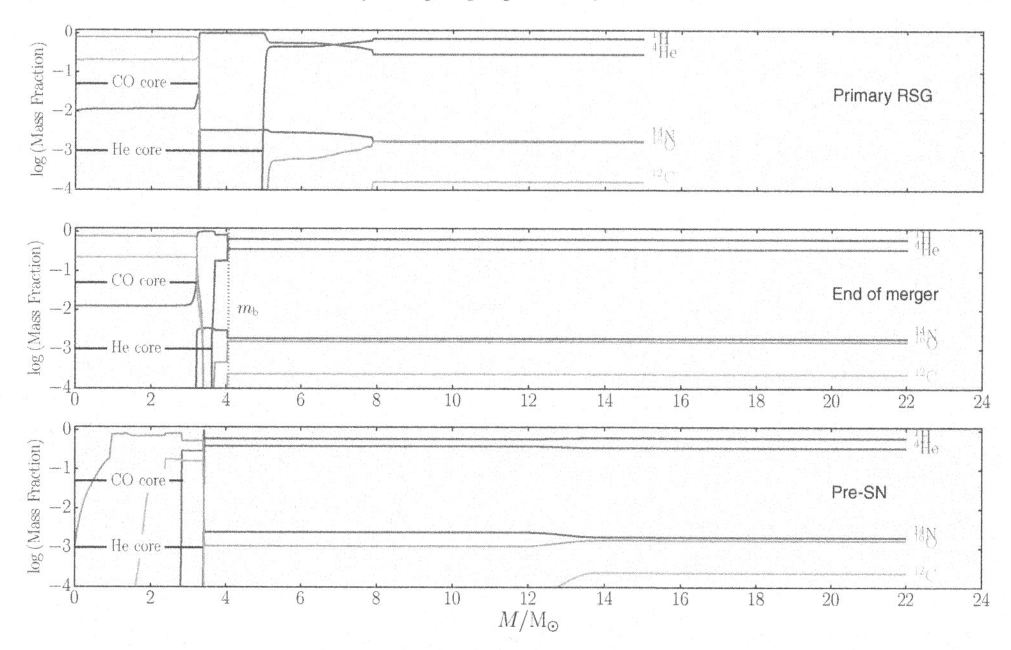

Figure 1. Top panel: composition of the primary RSG model of $M_1 = 16\,M_\odot$ just prior to the merger. Middle Panel: composition at the end of the merger with $M_2 = 7\,M_\odot$, for mixing boundary m_b (dotted vertical line) set for $f_{sh} = 50\,\%$. Bottom panel: Composition of pre-SN model.

lead to a BSG, whose explosion would resemble SN 1987A. The scenario begins with a wide binary system of a $15\,M_\odot - 20\,M_\odot$ primary and a $1\,M_\odot - 5\,M_\odot$ secondary, with an initial orbital period greater than 10 yr. When the primary evolves to an RSG with a CO core, it transfers mass on a dynamically unstable timescale to the secondary main-sequence star. A common envelope (CE) phase is ensued and the envelope is partially ejected. The secondary star is engulfed by the CE and spirals in towards the core of the primary. Eventually the secondary undergoes a Roche Lobe Overflow and is completely dissovled in the envelope. Hydrodynamic simulations of the merger showed the He core being penetrated by the secondary, shrinking it in the process and resulting in the dredge-up of He in the envelope (Ivanova *et al.* 2002). The merger lasts for an order of ~ 100 yr and the post-merger star could evolve to a BSG (Ivanova & Podsiadlowski 2002).

2. Methodology

We construct an evolutionary model based on the above merger scenario, over an initial parameter consisting of the main-sequence primary mass, $M_1 = 15, 16, 17\,M_\odot$ with an initial rotation velocity of $v/v_{cri} = 0.30$, main-sequence secondary mass, $M_2 = 2, 3, ..., 8\,M_\odot$ and fraction of the He shell from the He core dredged up, $f_{sh} = 10, 50, 90, 100\,\%$. We evolve 84 such initial models until just prior to the core-collapse stage (the pre-supernova (pre-SN) model), using KEPLER, an implicit one-dimensional hydrodynamics code that can compute stellar evolution models with rotation and nucleosynthesis (Heger *et al.* 2000, Heger *et al.* 2005, Woosley *et al.* 2002). The initial (solar-scaled) composition is that of the LMC– $X_H = 0.739$, $X_{He} = 0.255$ and $Z = 0.0055$ (Brott *et al.* 2011).

We compare the pre-SN models computed with the three confirmed signatures of Sk –69°202:

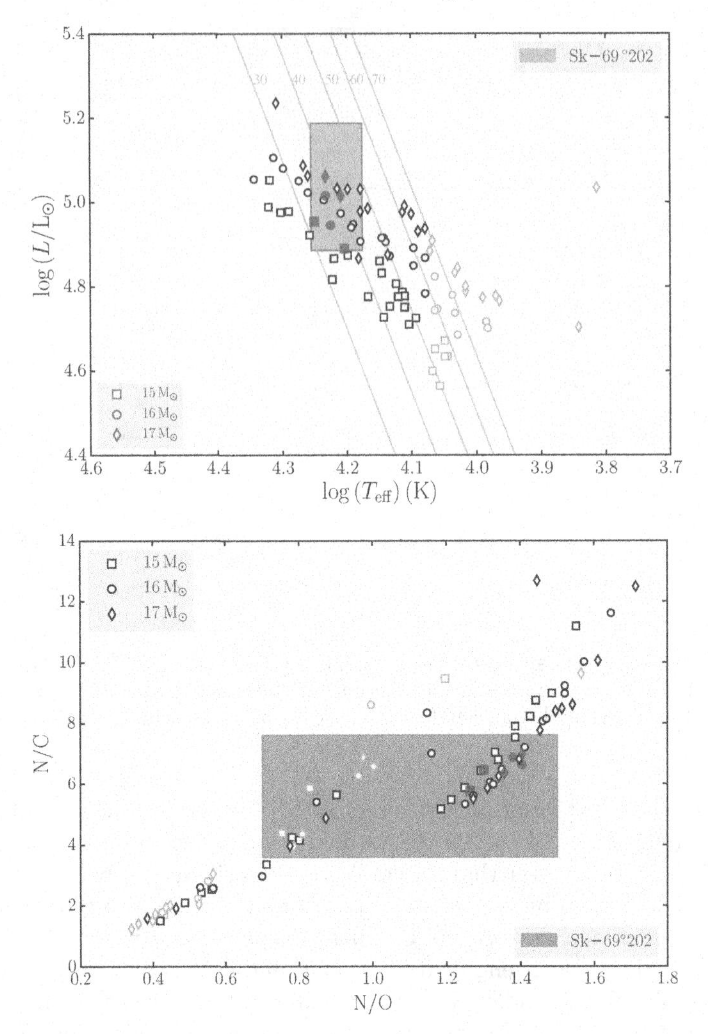

Figure 2. Distribution of all computed pre-SN models in the HR diagram (top) and their relative number ratios of N/C and N/O. Shade regions denote the observational limits for Sk −69 °202. Yellow symbols are YSGs, blue symbols are BSGs and filled blue symbols are progenitor models for SN 1987A.

(*a*) Lies in the HR diagram where Sk −69 °202 was observed before exploding– $\log(L/\mathrm{L}_\odot) = 5.15 - 5.45$, $T_{\mathrm{eff}} = 15\,\mathrm{kK} - 18\,\mathrm{kK}$ and $R/\mathrm{R}_\odot = 28 - 58$ (Woosley 1988).

(*b*) Relative elemental ratios by number in the surface from the BSG model match those of the triple-ring nebula; N/C $\sim 5 \pm 2$, N/O $\sim 1.1 \pm 0.4$ (Lundqvist & Fransson 1996) and He/H $= 0.14 \pm 0.06$ (France *et al.* 2011).

(*c*) Lifetime of the post-merger BSG phase before explosion is at least 15 kyr.

In this first study, the merger is a simplified process that does not include heating or angular momentum loss from the CE phase. We first choose a primary mass, M_1 and evolve it to an RSG with a CO core. Since the primary RSG model is rotating, its envelope becomes enhanced in nitrogen and helium. The secondary main-sequence star, M_2, is evolved to the same age as the primary RSG. M_2 gets accreted on the primary over 100 yr and is simultaneously mixed throughout its envelope using a numerical prescription. Mixing is restricted to a specified boundary inside the He core, depending on the value set for f_{sh}. The mixing causes the He core to shrink; larger values of f_{sh} lead to smaller

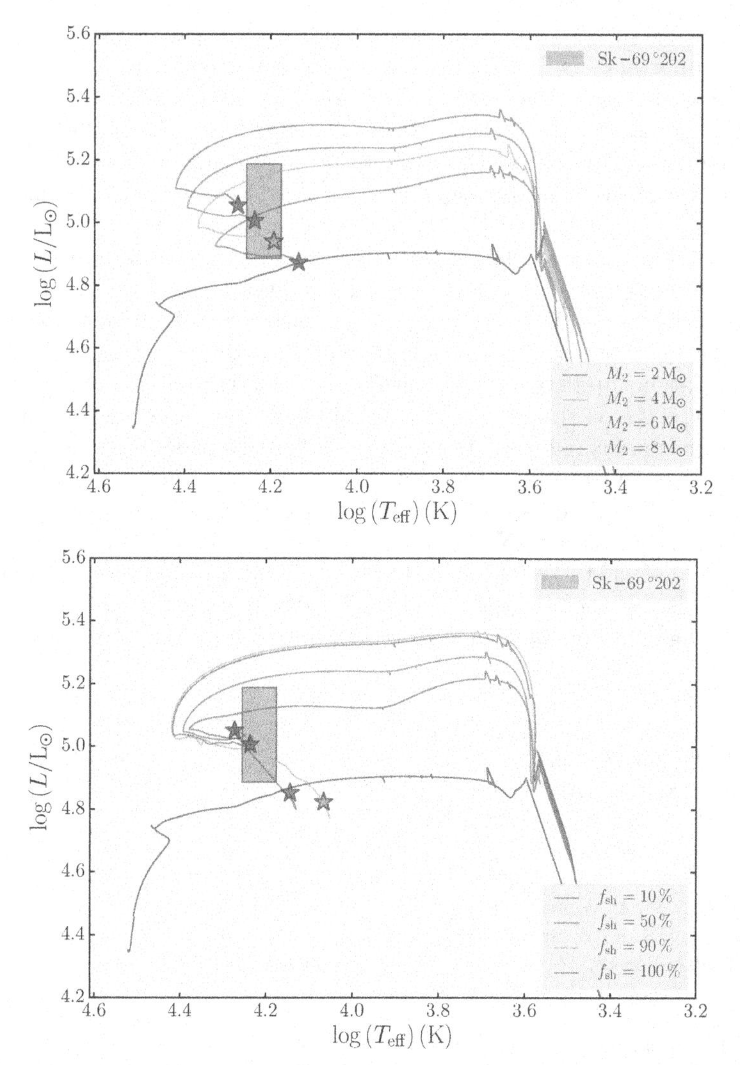

Figure 3. Top: Evolutionary tracks from the merger of $M_1 = 16\,M_\odot$ and $M_2 = 6\,M_\odot$, with varying values of f_{sh}. Bottom: Tracks from the merger of $M_1 = 16\,M_\odot$ and various values of M_2 keeping $f_{sh} = 50\,\%$.

He cores. The envelope is further enriched in He and also mixed with carbon and oxygen dredged up from the He core. At the end of the merger, the post-merger star has a small He core and a massive envelope with relative enhancements of N/C, N/O and He/H (Fig. 1).

3. Results

The majority of pre-SN models amongst the 84 we computed, are BSGs (i.e., have $T_{\rm eff} \geqslant 12\,{\rm kK}$), while the rest are yellow supergiants (YSGs) ($12\,{\rm kK} < T_{\rm eff} \leqslant 4\,{\rm kK}$). Of these, six pre-SN models match the three observational signatures of Sk −69 °202 (Fig. 2).

The most influential initial parameter that determines how hot the surface of the pre-SN model is, is f_{sh}. Across the range of M_1 and M_2 studied, pre-SN models are blue when the He core does not become smaller than $f_{sh} = 10\,\% - 50\,\%$. These values of f_{sh}

also result in values of N/C and N/O in the surface identical to Sk –69°202. The next parameter of importance is M_2. Increasing M_2 for a particular He core mass, increases the $T_{\rm eff}$ of the pre-SN star (Fig. 3) but decreases N/C and N/O in the surface. Finally, the parameter that affects the lifetime of the BSG star before its explosion, is the age of the primary model at the time of the merger. Younger primary models on the red giant branch lead to longer-lived post-merger BSG models, for a given combination of initial parameters.

We were unable to draw a relation between the three initial parameters, $M_1, M_2, f_{\rm sh}$ and the position of the pre-SN model in the HR diagram. The results of Podsiadlowski *et al.* 1992 indicated that the accretion of M_2, which led to high envelope to core mass ratios, was sufficient for a star to become blue at the time of explosion. We do not find such a linear relationship between core-envelope mass ratio and $T_{\rm eff}$ of the pre-SN model (Fig. 3). In fact, if M_2 is simply accreted in our model without He core dredge-up, the post-merger star remains red until its explosion. There also does not exist any correlation between N/C and N/O ratios in the surface and the position in the HR diagram of a pre-SN model, implying they are independent constraints (Fig. 2).

The progenitor models for SN 1987A are more massive than their single star counterparts (16 $M_\odot$ – 23 $M_\odot$ as against 14 $M_\odot$ – 20 $M_\odot$) and have smaller He cores (2.4 $M_\odot$ – 4.5 $M_\odot$ as against 4 $M_\odot$ – 7 $M_\odot$, (Woosley 1988)). In a future study we shall compare the light curves from their explosions to that of SN 1987A.

References

Brott, I. *et al.*, 2011 *AAP*, 530

Burrows, C. J. *et al.* 1995, *ApJ*, 680, 452

Catchpole, R. M. *et al.* 1988, *MNRAS*, 75, 231

Crotts, A. P. S. & Heathcote, S. R., 2000 *ApJ*, 426, 528

Hamuy, M., Suntzeff, N. B., Gonzalez, R., & Martin, G., 1988 *AJ*, 63, 95

Heger, A., Langer, N., & Woosley, S. E., 2000 *ApJ*, 368, 528

Heger, A., Woosley, S. E., & Spruit, H. C., 2005 *ApJ*, 350, 626

Ivanova, N., Podsiadlowski, P., & Spruit, H., 2002 *MNRAS*, 819, 334

Ivanova, N. & Podsiadlowski, P., 2002 *Astronomical Society of the Pacific Conference Series*, 245, 279

Morris, T. & Podsiadlowski, P., 2007 *Science*, 1103, 315

Morris, T. & Podsiadlowski, P., 2009 *MNRAS*, 515, 399

Podsiadlowski, P., Joss, P. C., & Hsu, J. J. L. 1992, *ApJ*, 246, 391

Podsiadlowski, P. 1992, *PASP*, 717, 104

Podsiadlowski, P., Morris, T. S., & Ivanova, N. 2007, *AIPCS*, 125, 937

Smartt, S. J., 2009 *ARAA*, 63, 47

Walborn, N. R., Prevot, M. L., Prevot, L., Wamsteker, W., Gonzalez, R., Gilmozzi, R., & Fitzpatrick, E. L. 1989, *AAP*, 229, 219

Woosley, S. E., 1988 *ApJ*, 218, 330

Woosley, S. E., Heger, A., & Weaver, T. A., 2002 *Reviews of Modern Physics*, 1015, 74

The Lives and Death-Throes of Massive Stars
Proceedings IAU Symposium No. 329, 2016
J.J. Eldridge, J.C. Bray, L.A.S. McClelland & L. Xiao, eds.

doi:10.1017/S1743921317003301

The mass-loss before the end: two luminous blue variables with a collimated stellar wind

C. Agliozzo[1], C. Trigilio[2], C. Buemi[2], P. Leto[2], G, Umana[2], G. Pignata[1], N. M. Phillips[3], R. Nikutta[4] and J. L. Prieto[5]

[1]Departamento Ciencias Fisicas, Universidad Andres Bello, Av. Republica 252, Santiago, Chile
email: c.agliozzo@gmail.com

[2]INAF-Osservatorio Astrofisico di Catania, Via S. Sofia 78, I-95123, Catania, Italy
[3]European Southern Observatory, Alonso de Córdova 3107, Vitacura, Santiago, Chile
[4]National Optical Astronomy Observatory, 950 N Cherry Ave, Tucson, AZ 85719, USA
[5]Núcleo Astronomía Facultad Ingeniería, Universidad Diego Portales, Av. Ejército 441, Santiago, Chile

Abstract. We gathered a multiwavelength dataset of two well-known LBVs. We found a complex mass-loss, with evidence of variability, such as has been seen previously. In addition, our data reveal signatures of collimated stellar winds. We propose a new scenario for these two stars where the nebula shaping is influenced by the presence of a companion star and/or fast rotation.

Keywords. stars: individual (RMC127, HR Car), stars: mass loss, stars: winds, outflows

---

1. Introduction

Recently the classical view of luminous blue variables (LBVs) as a phase of the post-main sequence (MS) evolution of a single massive star has been challenged by Smith & Tombleson (2015), who proposed that LBVs are mass-gainers in binary systems with Wolf Rayet (WR) stars, which are the mass-donors. Humphreys *et al.* (2016) defended the accepted description of LBVs as evolved massive stars that have to quickly lose their H envelope through severe mass-loss, before to evolve as WRs. There is a growing evidence that H-poor Core Collapse-Supernovae (SNe) progenitors consist of both binary and single stars. The mechanism that induces some LBVs to explode as Type IIn SNe is not known. The debate is very active and the mass-loss mechanism that triggers the LBV Giant Eruptions (Humphreys & Davidson 1994) is still not understood. Our understanding is made difficult by the rarity of this phenomenon in our Galaxy. On the other hand, we know that at least 70% of massive stars have to exchange masses with a partner at least once in their life (Sana *et al.* 2012). Is the asymmetry of several LBV nebulae due to intrinsic asymmetry of the stellar mass-loss or due to the influence of a companion star? What can we learn from resolved imaging by the latest instrumentation? In this talk I will discuss two interesting cases: the LBVs HR Carinae (HR Car) and RMC127.

2. The Galactic LBV HR Car

The study I am going to discuss has been published in Buemi *et al.* 2017. The nebula associated with HR Car has a bipolar filamentary structure seen in Hα by Nota *et al.* 1997. These authors suggested a similarity with the Homunculus nebula of η Car. White 2000 observed an asymmetric nebula emitting in the radio, co-spatial with the optical emission. During the quiescence, HR Car is a B2 supergiant and is not hot enough to ionize the circumstellar nebula, therefore White 2000 suggested that HR Car may have a

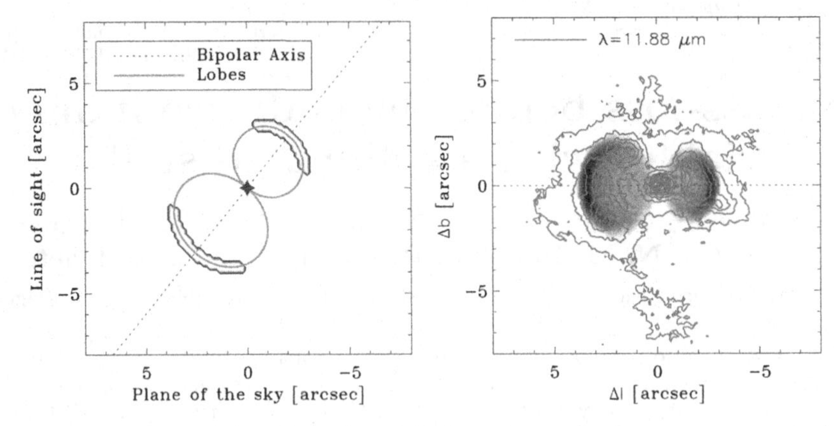

Figure 1. Figure readapted from Buemi *et al.* 2017. Left panel: schematic representation of the bipolar model for HR Car's nebula. Right panel: simulated emission from the two polar lobes (grey) and VISIR mid-IR emission (green contours) as seen at 11.88 μm.

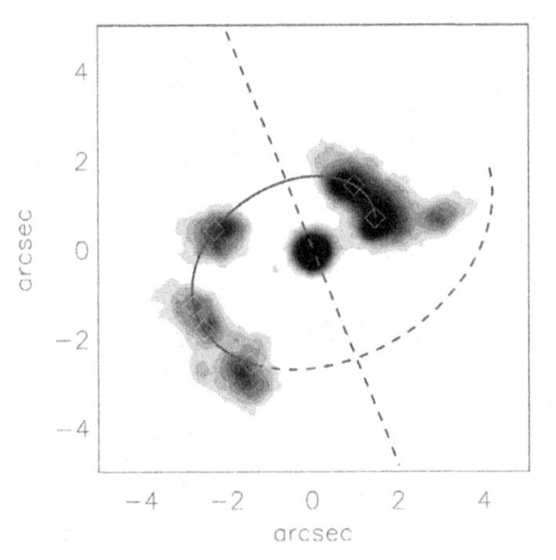

Figure 2. VISIR mid-IR image of the inner dust envelope and spiral model (line). Figure readapted from Buemi *et al.* 2017.

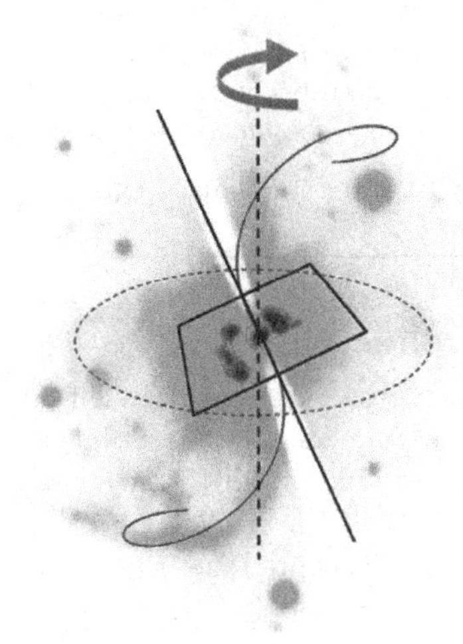

Figure 3. Hα image (green/yellow), VISIR mid-IR image (red), jet-precession model (lines). Figure readapted from Buemi *et al.* 2017.

companion star responsible for the ionization of HR Car's nebula. Recently, Boffin *et al.* 2016 obtained a direct detection of at least one companion star for HR Car, very likely a red supergiant (RSG). RSGs do not have sufficiently ionizing photon flux. Therefore, the ionization of the circumstellar material around this system is still unexplained.

In Buemi *et al.* 2017 we presented a new study of the circumstellar environment of HR Car, based on a multiwavelength dataset. This comprises radio ATCA observations at two different epochs, far-IR *Herschel*/PACS images and mid-IR VLT/VISIR images. The *Herschel* images indicate that HR Car's optical and radio nebula is surrounded by an extended and asymmetric outer nebula of optically thin dust.

From the radio data we found that the mass-loss rate increased by a factor of 2 between 1994 and 2014. During the observations at both epochs the star was in the quiescent phase. This variation does not seem correlated with the star S Doradus cycle, so we proposed that it could have been induced by the periastron passage of the companion star, constrained by Boffin *et al.* 2016 between 2013.2 and 2015.5.

Our VISIR images show IR emission from an inner dust envelope with a two arc shape (see the green contours in Fig. 1, right panel, and the grey image in Fig. 2). These structures can be fitted with an expanding bipolar-lobe model (Fig. 1), as shown in Buemi *et al.* 2017, consistently with the model by Nota *et al.* 1997. In Buemi *et al.* 2017, we also considered an alternative scenario, where these two arcs close to the star form an Archimedean spiral (Fig. 2). In this case the spiral nebula would be generated by equatorial mass-loss events. The direction of this mass-loss depends on the position of the companion star at the moment of the ejection. In this scenario the ionized gas, that seems confined within this arc-shaped material, is channeled along an axis perpendicular to the plane of the spiral (Fig. 3). Since HR Car is in a binary or even multiple system, the star may suffer a torque that causes an axial precession and then a helical outflow. The stellar wind would be collimated along the rotational axis of HR Car, and gives origin to the Hα and free-free emission. The ionization of the ejecta has to occur thanks to a third companion, a spectral type BO V star, as suggested by White 2000.

3. The Magellanic LBV RMC127

RMC127 is another classical LBV in the Large Magellanic Cloud. First optical observations of RMC127's nebula revealed a diamond-shaped nebula (Clampin *et al.* 1993). Schulte-Ladbeck *et al.* 1993, and later Davies *et al.* 2005, found a high-degree of polarization associated with the central star, suggesting an aspherical mass-loss, possibly a disk in the equatorial plane of the star. Later the *HST* provided the highest resolution image of the Hα nebula (Weis 2003). Weis 2003 found that the nebula deviates from a spherical symmetry, with a bipolarity in the North-South direction, forming two *Caps*, perpendicular to two *Rims* in the East-West direction. This configuration resembles the nebular features reproduced by Nota *et al.* 1995 by simulating a fast outflow in a preexisting dense medium.

We gathered a unique dataset at long wavelengths, consisting of multi-frequency and multi-resolution radio maps acquired with ATCA, and of a sub-millimeter map obtained with ALMA. We also observed RMC127 with VISIR at VLT and with the high-dispersion spectrometer MIKE on the Clay Telescope at Las Campanas Observatory.

In the optical spectrum, acquired in January 2013, we found a broad and shallow photospheric absorption line Si IV λ4088, as shown in Fig. 4, which suggests that RMC127 is a fast rotator. From P Cyg profiles of He I lines with the absorption component saturated we could also determine the terminal velocity of the wind ($148 \pm 14\,\mathrm{km\,s^{-1}}$), an important parameter to estimate the mass-loss rate.

Very likely the IR emission detected from the space telescopes arises from extended optically thin dust, rather than a compact shell close to the star. This would explain why we did not obtain any detection with VISIR.

We detected RMC127 in the radio and in the submillimeter. The interferometric maps offer an inside look at the core of the nebula, where the *HST* image is limited by artifacts due to the bright star. The nebula was detected with ATCA at 1.3 cm (17 GHz), 3.3 cm (9 GHz) and 6 cm (5 GHz), as shown in Fig. 5. In the maps we recognize the two Northern and Southern *Caps* and the Eastern and Western *Rims*. In addition, there are diagonal *Arms* crossing the nebula center. The brightest emission has a "Z" shape and is not reproduced in the simulation by Nota *et al.* 1995.

At 9 GHz the central object emission emerges from the nebula and becomes brighter at higher frequencies. At 23 GHz and at the ALMA frequency (349 GHz) this object is the only detected component of emission and it appears as a point source. By fitting the central object flux density distribution from 9 to 349 GHz, we found that the spectral

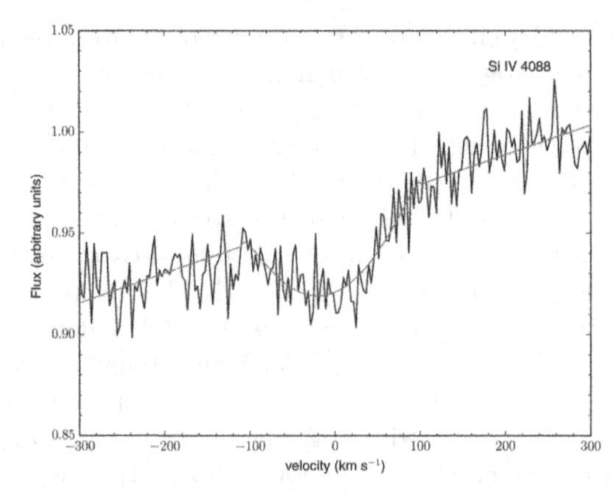

Figure 4. Black line: observed Si IV absorption line in our MIKE spectrum of RMC127 acquired in 2013. Red line: best-fit obtained with a rotational broadening profile. The fit gives a projected rotational velocity of $\sim 105\,\mathrm{km\,s^{-1}}$ and was obtained with a routine in the package PyAstronomy.

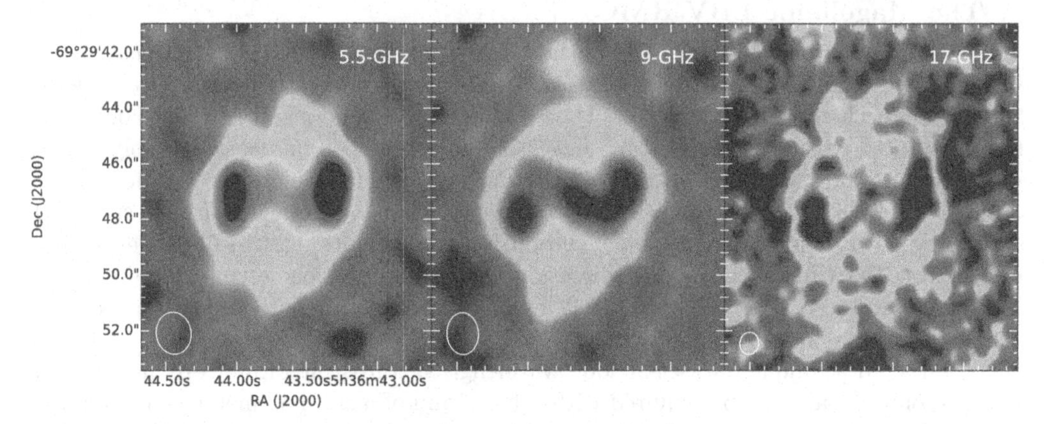

Figure 5. ATCA radio maps at different frequencies and resolutions. The synthesized beam (resolution) in each map is indicated with a white ellipse. N is up and E is left.

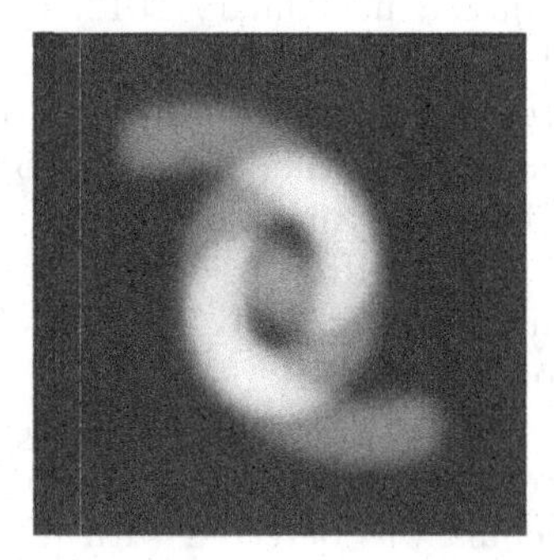

Figure 6. A frame extracted from the simulation of a flight around a 3-D double helix nebula. The simulation was performed by assuming that the axial precession of the star has completed one period. An animated version can be found at http://vimeo.com/151528747.

index α ($S_\nu \sim \nu^\alpha$) is positive (0.78 ± 0.05), suggesting thermal emission from an ionized stellar wind. However, the high value indicates an excess of emission, departing from the classical case of an ionized spherical wind ($\alpha = 0.6$). We examined different mechanisms that could give origin to such high spectral index. The asymmetries found in RMC127's stellar wind and optical nebula by the previously mentioned authors make all the models based on spherical symmetry unsuitable for RMC127. Therefore, we modeled the central object emission with the collimated stellar wind model formulated by Reynolds 1986. We found that, in this scenario, the mass-loss rate of the outflow would be $\sim 8 - 9 \times 10^{-6}\,M\odot\,yr^{-1}$, a factor of 3 less than the spherical case.

Collimated stellar winds, jets and bipolar outflows are usually associated with fast rotation and/or dense disks. Such disks can form as a result of mass-transfer in binary systems, magnetic fields or fast stellar rotation. We did not find evidence of a dense disk in our multiwavelength data, but we know that RMC127 is indeed a fast rotator.

Finally, we attempt to explain the Z-shape morphology of the outer nebula with a new geometry. We explored the possibility of a highly inclinated bipolar nebula, as suggested first by Schulte-Ladbeck *et al.* 1993. Fig. 6 shows a frame extracted from the simulation of a flight around a 3-D double conical helix nebula that we created by using our newly public code RHOCUBE (that allows modelling 3D density distributions and fitting the observed maps; Agliozzo *et al.* 2017, Nikutta & Agliozzo 2016). We found remarkable similarities with the radio maps (Fig. 5). This geometry would imply that the nebula was formed by the precession of a bipolar-outflow. The precession needs a torque from a companion star to occur, similarly to HR Car. However, the binarity nature of RMC127 has not been yet demonstrated.

4. Final comments and outlook

The circumstellar material around HR Car and its mass-loss seem strongly influenced by the presence of at least one companion star. To explain the nebular morphology, we suggest that RMC127 may be an analogous case to HR Car. The mechanism discussed in these studies seems to us an interesting case to further investigate, to understand the role of binarity in the post-MS evolution of massive stars and the physical phenomenon that triggers the Giant Eruptions in LBVs and in SN impostors. The next step will be the verification of the jet precession scenario with integral field unit observations.

References

Agliozzo, C., Nikutta, R., Pignata, G., *et al.* 2017, *MNRAS*, 466, 213

Boffin, H. M. J., Rivinius, T., Mérand, A., *et al.* 2016, *A&A*, 593, A90

Buemi, C. S., Trigilio, C., Leto, P., *et al.* 2017, *MNRAS*, 465, 4147

Clampin, M., Nota, A., Golimowski, D. A., Leitherer, C., & Durrance, S. T. 1993, *ApJ* (Letters), 410, L35

Davies, B., Oudmaijer, R. D., & Vink, J. S. 2005, *A&A*, 439, 1107

Humphreys, R. M. & Davidson, K. 1994, *PASP*, 106, 1025

Humphreys, R. M., Weis, K., Davidson, K., & Gordon, M. S. 2016, *ApJ*, 825, 64

Nikutta, R. & Agliozzo, C. 2016, Astrophysics Source Code Library, ascl:1611.009

Nota, A., Livio, M., Clampin, M., & Schulte-Ladbeck, R. 1995, *ApJ*, 448, 788

Nota, A., Smith, L., Pasquali, A., Clampin, M., & Stroud, M. 1997, *ApJ*, 486, 338

Reynolds, S. P. 1986, *ApJ*, 304, 713

Sana, H., de Mink, S. E., de Koter, A., *et al.* 2012, *Science*, 337, 444

Schulte-Ladbeck, R. E., Leitherer, C., Clayton, G. C., *et al.* 1993, *ApJ*, 407, 723

Smith, N. & Tombleson, R. 2015, *MNRAS*, 447, 598

Weis, K. 2003, *A&A*, 408, 205

White, S. M. 2000, *ApJ*, 539, 851

The Lives and Death-Throes of Massive Stars
Proceedings IAU Symposium No. 329, 2016
J.J. Eldridge, J.C. Bray, L.A.S. McClelland & L. Xiao, eds.

doi:10.1017/S1743921317001843

The Progenitor-Remnant Connection of Neutrino-Driven Supernovae Across the Stellar Mass Range

Thomas Ertl

Max-Planck-Institut für Astrophysik,
Postfach 1317, D-85741 Garching, Germany
email: tertl@mpa-garching.mpg.de

Abstract. We perform hydrodynamic supernova (SN) simulations in spherical symmetry for progenitor models with solar metallicity across the stellar mass range from 9.0 to 120 $M_\odot$ to explore the progenitor-explosion and progenitor-remnant connections based on the neutrino-driven mechanism. We use an approximative treatment of neutrino transport and replace the high-density interior of the neutron star (NS) by an inner boundary condition based on an analytic proto-NS core-cooling model, whose free parameters are chosen to reproduce the observables of SN 1987A and the Crab SN for theoretical models of their progenitor stars.

Judging the fate of a massive star, either a neutron star (NS) or a black hole (BH), solely by its structure prior to collapse has been ambiguous. Our work and previous attempts find a non-monotonic variation of successful and failed supernovae with zero-age main-sequence mass. We identify two parameters based on the "critical luminosity" concept for neutrino-driven explosions, which in combination allows for a clear separation of exploding and non-exploding cases.

Continuing our simulations beyond shock break-out, we are able to determine nucleosynthesis, light curves, explosion energies, and remnant masses. The resulting NS initial mass function has a mean gravitational mass near 1.4 $M_\odot$. The average BH mass is about 9 $M_\odot$ if only the helium core implodes, and 14 $M_\odot$ if the entire pre-SN star collapses. Only ~10% of SNe come from stars over 20 $M_\odot$, and some of these are Type Ib or Ic.

Keywords. nuclear reactions, nucleosynthesis, abundances, stars: black holes, stars: massive, stars: neutron, supernovae: general, hydrodynamics, neutrinos

1. Introduction

Recent observations in the past years have provided a picture of the population of core-collapse supernovae (CCSNe) in various aspects: Progenitor stars of Type-IIP SNe (Smartt, 2015) have been identified and suggested a lack of high-mass stars exploding as SNe; mass measurements of black holes (BHs) and neutron stars (NSs) in binary systems have provided a sketch of the mass distributions (e.g., Özel & Freire, 2016, and references therein); the elemental abundance patterns (for the sun see Lodders, 2003) pose a mixed imprint of all the sources of heavy elements including the contribution by CCSNe. These observational signatures connect progenitor stars with remnant and explosion properties.

Theoretical studies of CCSNe, however, have barely tapped these observational resources and a converged theoretical picture has not yet emerged, although considerable progress has been made in recent years with the first successful self-consistent explosions in full geometry (3D; Melson *et al.*, 2015b;a, Lentz *et al.*, 2015). However, computational power restricts evolving these simulations for sufficiently long time to gain converged observable signatures and a large number of models is also computationally unaffordable. Nevertheless, any successful explosion mechanism needs to reproduce the observed

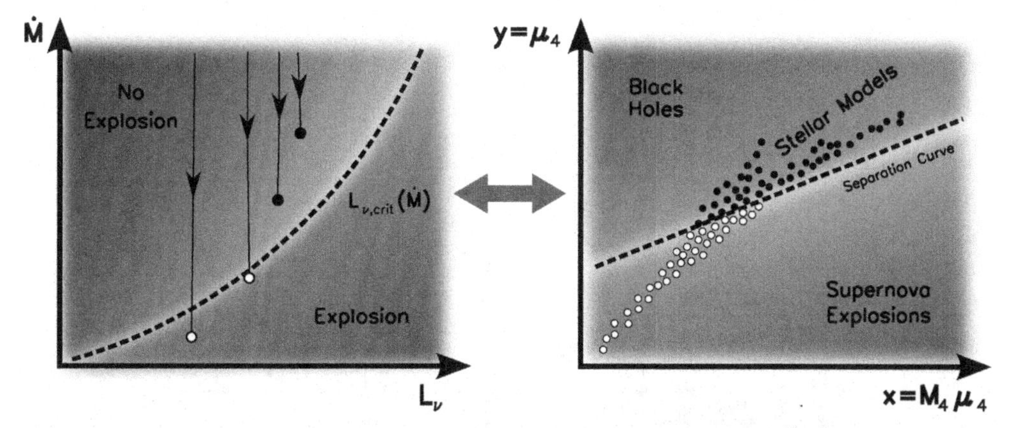

Figure 1. (Taken from Ertl *et al.*, 2016, Fig. 6) Correspondence of L_ν-$\dot{M}$ plane with critical neutrino luminosity $L_{\nu,\text{crit}}(\dot{M})$ (*left*) and x-y plane with separation curve (*right*). In the left plot post-bounce evolution paths of successfully exploding models (white circles) and non-exploding models (black circles) are schematically indicated, corresponding to white and black circles for pre-collapse models in the right plot. Evolution paths of successful models cross the critical line at some point and the accretion ends after the explosion has taken off. In contrast, the tracks of failing cases never reach the critical conditions for launching the runaway expansion of the shock. The symbols in the left plot mark the "optimal point" relative to the critical curve that can be reached, corresponding to the stellar conditions described by the parameters ($M_4\mu_4$ and μ_4) at the $s = 4$ location, which seems decisive for the success or failure of the explosion of a progenitor, because the accretion rate drops strongly outside.

population of SNe and their remnants. The delayed neutrino-driven mechanism is the best candidate for a mechanism and certainly the most elaborate.

We investigated the properties of the progenitor stars and their SN explosions by a *systematic parameter approach.* The novel aspect of the approach is that the explosions are not triggered artificially by a piston or a thermal bomb, but are based on the current understanding of the neutrino-driven mechanism. We use an approximative neutrino transport solver and excise the inner core of the proto-NS and replace it by an analytic one-zone core-cooling model (Ugliano *et al.*, 2012), whose free parameters are tuned to reproduce SN 1987A and the Crab SN for theoretical models of their progenitors. Both SNe act as our observational anchors of the survey. The obtained parameter choice is then applied to progenitor models of different zero-age main-sequence (ZAMS) masses and evaluated for the explosion properties.

2. A Two-Parameter Criterion for the Explodability of Massive Stars

In Ertl *et al.* (2016), we published a criterion for the "explodability" of massive stars, which is solely based on the pre-collapse structure of the star. Two parameters were identified based on the "critical luminosity concept" (Burrows & Goshy, 1993). One is the normalized mass inside a dimensionless entropy per nucleon of $s = 4$,

$$M_4 \equiv m(s=4)/M_\odot\,, \tag{2.1}$$

and the mass derivative at this location,

$$\mu_4 \equiv \left.\frac{\mathrm{d}m/M_\odot}{\mathrm{d}r/1000\,\mathrm{km}}\right|_{s=4}, \tag{2.2}$$

which allow for a nearly perfect prediction of exploding and non-exploding cases in our study ($\gtrsim 97\%$ of all models are correctly predicted). Fig. 1 depicts the correspondence.

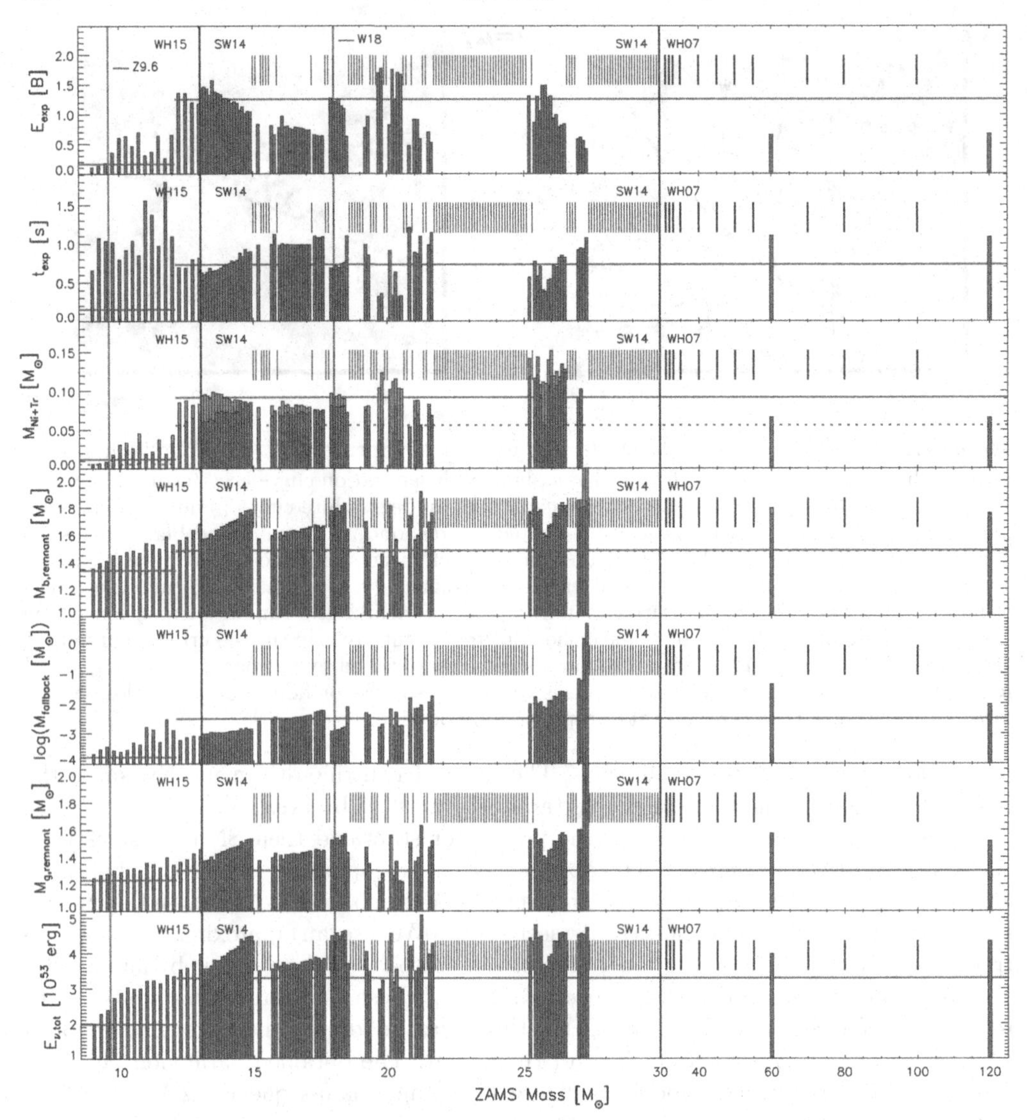

Figure 2. (Taken from Sukhbold *et al.*, 2016, Fig. 8) Explosion properties for all models exploded with the Z9.6 and W18 calibrations. Black vertical lines mark the boundaries between the different progenitor sets our model sample is composed of. The panels, *from top to bottom*, show the final explosion energy, E, in units of 1 B = 1 Bethe = 10^{51} erg, the time of the onset of the explosion, t_{exp}, the mass of finally ejected, explosively produced ^{56}Ni (red bars) and tracer element environment (orange bars), the baryonic mass of the compact remnant with the fallback mass indicated by orange sections on the bars, the fallback mass, the gravitational mass of the compact remnant, and the total energy radiated in neutrinos, $E_{\nu,\mathrm{tot}}$. The masses of the calibration models are indicated by vertical blue lines, and the corresponding results by horizontal solid or dashed blue lines, which extend over the mass ranges that are considered to have Crab-like or SN1987A-like progenitor properties, respectively. Non-exploding cases are marked by short vertical black bars in the upper half of each panel.

3. Nucleosynthesis, light curves, explosion energies, and remnant masses

Continuing the simulations beyond the point of shock break-out, we were able to determine explosion energies and remnant masses as well as nucleosynthesis yields and light curves in a post-processing step. The results for a dense set of progenitor models with solar metallicity are reported in Sukhbold *et al.* (2016). Fig. 2 shows the results of our study for one core-model parameter choice as an example. The parameters were calibrated employing the blue supergiant model "W18" (see Sukhbold *et al.*, 2016, for details).

This work was supported by Deutsche Forschungsgemeinschaft through grant EXC 153 "Excellence Cluster Universe" and the European Research Council through grant ERC-AdG No. 341157-COCO2CASA.

References

Burrows, A. & Goshy, J. 1993, *ApJ* (Letters), 416, L75
Ertl, T., Janka, H.-T., Woosley, S. E., Sukhbold, T., & Ugliano, M. 2016, *ApJ*, 818, 124
Özel, F. & Freire, P. 2016, *ARA&A*, 54, 401
Lentz, E. J., Bruenn, S. W., Hix, W. R., *et al.* 2015, *ApJ* (Letters), 807, L31
Lodders, K. 2003, *ApJ*, 591, 1220
Melson, T., Janka, H.-T., & Marek, A. 2015, *ApJ* (Letters), 801, L24
Melson, T., Janka, H.-T., Bollig, R., *et al.* 2015, *ApJ* (Letters), 808, L42
Smartt, S. J. 2015, *PASA*, 32, e016
Sukhbold, T., Ertl, T., Woosley, S. E., Brown, J. M., & Janka, H.-T. 2016, *ApJ*, 821, 38
Ugliano, M., Janka, H.-T., Marek, A., & Arcones, A. 2012, *ApJ*, 757, 69

Session 3: Observations and surveys of massive stars: hot stars, cool stars, transition objects and binaries

The Lives and Death-Throes of Massive Stars
Proceedings IAU Symposium No. 329, 2016
J.J. Eldridge, J.C. Bray, L.A.S. McClelland & L. Xiao, eds.

doi:10.1017/S1743921317003167

Highly accurate quantitative spectroscopy of massive stars in the Galaxy

María-Fernanda Nieva and Norbert Przybilla

Institut für Astro- und Teilchenphysik, Universität Innsbruck,
Technikerstrasse 25/8, 6020 Innsbruck, Austria
email: maria-fernanda.nieva@uibk.ac.at

Abstract. Achieving high accuracy and precision in stellar parameter and chemical composition determinations is challenging in massive star spectroscopy. On one hand, the target selection for an unbiased sample build-up is complicated by several types of peculiarities that can occur in individual objects. On the other hand, composite spectra are often not recognized as such even at medium-high spectral resolution and typical signal-to-noise ratios, despite multiplicity among massive stars is widespread. In particular, surveys that produce large amounts of automatically reduced data are prone to oversight of details that turn hazardous for the analysis with techniques that have been developed for a set of standard assumptions applicable to a spectrum of a single star. Much larger systematic errors than anticipated may therefore result because of the unrecognized true nature of the investigated objects, or much smaller sample sizes of objects for the analysis than initially planned, if recognized. More factors to be taken care of are the multiple steps from the choice of instrument over the details of the data reduction chain to the choice of modelling code, input data, analysis technique and the selection of the spectral lines to be analyzed. Only when avoiding all the possible pitfalls, a precise and accurate characterization of the stars in terms of fundamental parameters and chemical fingerprints can be achieved that form the basis for further investigations regarding e.g. stellar structure and evolution or the chemical evolution of the Galaxy. The scope of the present work is to provide the massive star and also other astrophysical communities with criteria to evaluate the quality of spectroscopic investigations of massive stars before interpreting them in a broader context. The discussion is guided by our experiences made in the course of over a decade of studies of massive star spectroscopy ranging from the *simplest* single objects to multiple systems.

Keywords. stars: parameters, stars: abundances

---

1. Introduction

We focus our discussion on stars with masses between ~20 and ~6 solar masses. From the spectral characteristics, this corresponds to the latest O stars to early B stars on the Main Sequence, the most numerous but also the *simplest* massive stars to be studied. In contrast to earlier spectral types, supergiants or Be stars, they are not significantly affected by mass outflow or complex geometries. And, in comparison to many mid/late B-type stars, they do not present signatures of atomic diffusion in their atmospheres. Many model atmospheres and analysis methods for massive star spectra have therefore been optimized to study these simple objects. The typical assumptions are that they are single, they have stationary, hydrostatic and chemically homogeneous atmospheres, are described well by plane-parallel geometry and (non-)local thermodynamic equilibrium.

However, in order to address the topic of massive stars in its entirety, one has to analyze earlier O stars as well, OB supergiants, Be and late B-type stars, chemically peculiar, magnetic, pulsating, fast-rotating stars, and also pre-main-sequence stars. The large variety of massive stars challenges the basic assumptions of standard atmospheric models and spectral analysis techniques that have been often used to estimate their fundamental parameters and chemical composition. One has to consider e.g. (inhomogeneous) hydrodynamic outflows, oblateness of the stars because of high rotational velocities and the presence of an accretion/decretion disk surrounding them (i.e. non-spherical geometries), the effects of spots on stellar surfaces due strong magnetic fields, elemental abundance stratification due to diffusion, deformation of spectral lines due to stellar pulsations, etc. Practically all these topics have been addressed by dedicated studies. However, a unique tool to (automatically) analyze spectra formed under such a variety of conditions is not implemented currently. The limiting factor is the large number of physical assumptions to be made for the general case, preventing an efficient calculation of models covering the entire parameter space.

A more feasible approach to reliably study each kind of star is via tailored model atmospheres and analysis techniques that incorporates the most realistic physical background for each particular case. Dedicated spectral grids can then be computed for each kind of object when the range in all relevant parameters is known. The final scope is to derive the stars' atmospheric parameters to reliably place them in a Kiel diagram (surface gravity $\log g$ vs. effective temperature $T_{\rm eff}$). In cases where fundamental parameters can also be derived (with additional information, like e.g. parallaxes), the objects can be placed in a Hertzsprung-Russell diagram, and basic relationships between fundamental stellar parameters can be checked. Related to the atmospheric and fundamental parameters is the determination of the stars' spectroscopic distances, that – when compared with other independent distance indicators – allow us to constrain the accuracy of the stellar parameters. Spectral energy distributions from the UV to the IR allows us to derive the extinction and reddening for a particular star, and to cross-check for it's global energy output.

A subsequent step after the stellar parameter determination is the chemical analysis of trace elements, which also differs among the various kind of stars because of e.g., temperature constraints, where lines of different species/ionization stages vary in their strength, different blends appear in different parts of the spectrum, or some lines can turn from absorption to emission. The projected rotational velocity will also determine which lines can be analyzed in isolation and which blends can be consistently taken into account. A detailed chemical analysis can yield on accurate elemental abundances when four basic conditions are met: i) the stellar atmospheric models are applicable to the objects, ii) the spectra have a good quality (spectral lines should be resolved and measurable at good S/N), iii) the analysis methodology takes into account a good selection of spectral lines that can be well reproduced by the model spectra and iv) the atmospheric (and when possible fundamental) parameters have been consistently derived.

This manuscript is biased towards our experience on the spectral modeling and analysis of normal and peculiar massive single and recently also multiple stars. It discusses some successes but also challenges to current modeling capabilities. It is intended as a guideline for colleagues working in other areas to *assess* the accuracy and precision of published fundamental parameters and chemical abundances for massive stars, before they are interpreted in various astrophysical contexts. A special emphasis is put on the single OB stars on the Main Sequence, because they allow us to derive most parameters and chemical abundances at highest accuracy and precision, and therefore, we can consider them as reference objects. Shorter descriptions are dedicated to particular classes of more complex objects for which we have recently extended our spectral analyses.

2. Late-O and early B-type on the Main Sequence: the *simplest* stars

For over a decade, we have improved the spectral modeling and analysis technique for stars with spectral classes from O9 to B2 and luminosity classes V to III – the so-called OB subgroup because of their similar spectral characteristics. Given that standard stellar model atmospheres like e.g. ATLAS9 (Kurucz 1993) meet most requirements to reproduce their atmospheric structures well (Nieva & Przybilla 2007), our efforts were invested into realistic level population and line-formation computations in non-LTE by building and testing different configurations of new input atomic data to provide with robust model atoms for different elements (see Przybilla *et al.* 2016 for updated references). The codes used for the non-LTE level population and line-formation calculations are DETAIL (Giddings 1981) and SURFACE (Butler & Giddings 1985)†. Our new spectral modeling in combination with a self-consistent spectral analysis that accounts for multiple ionization equilibria, applied to high-quality observations, resulted in a drastic minimization of stellar parameter and chemical abundance uncertainties, particularly reducing several systematic effects previously unaccounted for. A comprehensive study and first applications of our work on single and normal early B-type dwarfs and giants were discussed by Nieva & Przybilla (2012, 2014, hereafter NP12 and NP14) and Nieva (2013). These studies allowed us to put constraints not only on their atmospheric parameters and chemical composition at high precision and accuracy, but also to derive spectroscopically other stellar parameters like radius, luminosity, mass and to explore whether theoretical relations between them hold using our observationally derived parameters. For such stars, uncertainties in effective temperature, surface gravities and chemical abundances as low as 1-2%, 15% and 25%, respectively (NP12) and in evolutionary masses, radii and luminosities better than 5%, 10%, 20%, respectively and in absolute visual and bolometric magnitudes lower than 0.20 mag (NP14) are achieved. Moreover, practically a perfect match between the observed and the computed spectrum per star for one set of parameters confirm the robustness of models and analysis. Figure 1 shows a precise location of a sample of single early B-type stars (NP12) in the Kiel diagram in comparison to detached eclipsing binaries (DEBs). Figure 2 shows their mass-radius relation and Fig. 3 their mass-luminosity relation (NP14), indicating that the derived stellar parameters lie within theoretically expected values and agree with more accurate results from DEBs. The empirical relation between absolute magnitudes and spectral types in NP14 show an offset with respect to classical older calibrations from the literature (Cox 2000). The latter are often adopted, affecting the computation of e.g. the stellar luminosities. The level of accuracy reached for this kind of objects can hardly be reproduced for more complex stars or systems. We therefore use these results as references for further studies.

3. BA and OB supergiants

Supergiants of spectral types late-B and early-A are descendants of OB stars on the Main Sequence. Their atmospheres can be well modeled with ATLAS9, DETAIL and SURFACE, as described in Przybilla *et al.* (2006). These objects also have the advantage that they outshine practically any companion in the visual, facilitating to treat them as single stars. The spectral analysis of early-O stars and OB supergiants is beyond the modeling capability of this hydrostatic approach, however other robust hydrodynamic non-LTE codes that treat simultaneously the spherically symmetric stellar atmospheric structure

† For this type of stars, LTE atmospheric structures computed with ATLAS9 and stellar fluxes computed with ATLAS9 and DETAIL are practically identical to those calculated with TLUSTY Hubeny & Lanz (1995) in non-LTE, see Nieva & Przybilla (2007) for a detailed discussion.

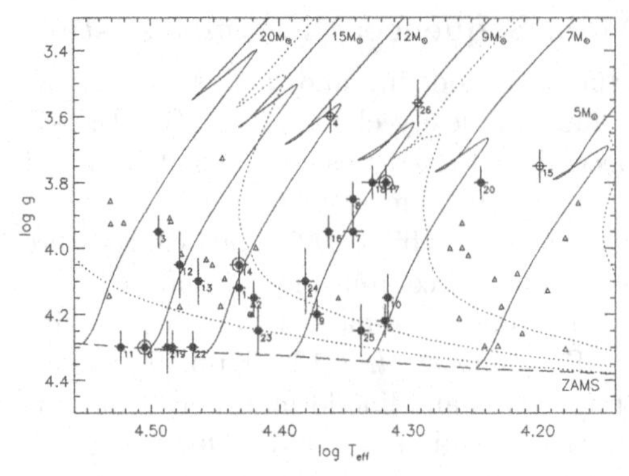

Figure 1. $T_{\rm eff}$ and $\log g$ of a sample of early B-type stars on the Main Sequence (black dots). Open thick circles are objects beyond core H-exhaustion. Wide circles surrounding the dots mark magnetic stars. Data from double-lined detached eclipsing binaries are shown as small triangles. Evolutionary tracks and isochrones from Ekström *et al.* (2012) are shown. See NP12 and NP14 for details.

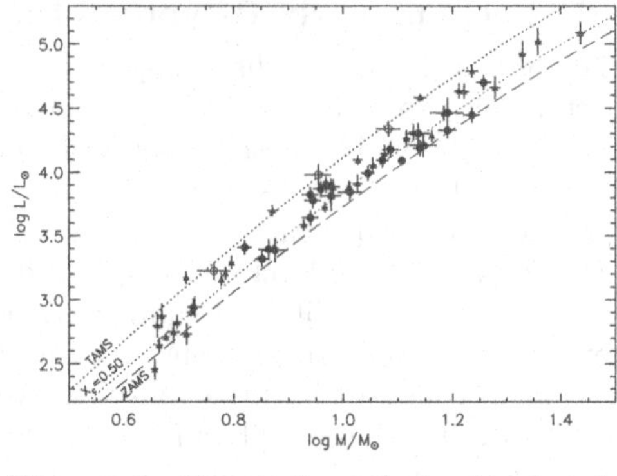

Figure 2. Mass-radius relationship for the sample stars in Fig. 1 with the same symbol encoding. Abscissa values are evolutionary masses. The ZAMS and two additional loci for 50% core-H depletion and for the TAMS are indicated by the thick/thin-dotted lines from the stellar evolution code as in Fig. 1. Error bars are shown also for the detached eclipsing binaries. See NP14 for details.

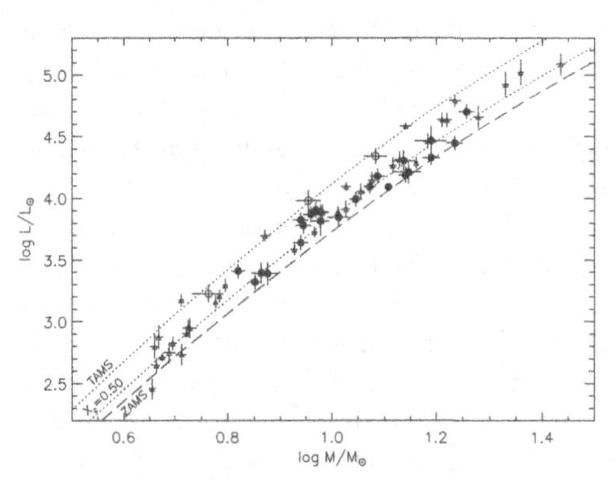

Figure 3. Mass-luminosity relationship for the sample stars in Fig. 1. Symbol and loci encoding are the same as in Figs. 1 and 2 (NP14).

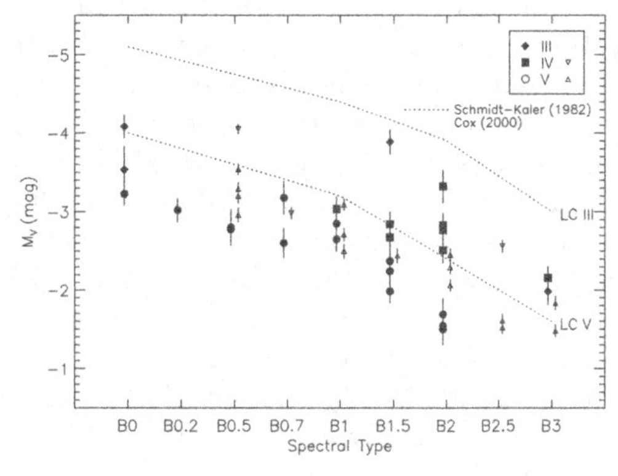

Figure 4. Absolute visual magnitudes of the sample stars in Fig. 1, encoding their luminosity class according to the legend, vs. spectral type. Older reference values are also indicated (NP14).

and the stellar wind like CMFGEN (Hillier & Miller 1998) and FASTWIND (Puls *et al.* 2005) are suited for their study.

4. Fast-rotating stars

We encounter several limitations in the analysis of intrinsically fast-rotating stars. If the star rotates so fast that its shape is affected (oblate), the assumptions of plane-parallel or even spherical symmetry are no longer valid on the global scale. Also the temperature will vary between the poles and the equator. Depending on the star's inclination, we observe spectral lines formed at different atmospheric conditions, with different local temperatures. This is clearly evident in cases of fast-rotators seen pole-on, where the lines are sharp, but nevertheless it is difficult to fit the whole spectrum with one temperature only and the assumption of establishing ionization balance for all available species simultaneously is no longer valid. If the projected rotational velocity is large, only a few spectral lines can be analyzed and many blends have to be accounted for self-consistently. If the S/N ratio is not high enough, the definition of the line continuum is one of the largest sources of systematics because the spectral lines get smeared out and appear weaker (see Korn *et al.* 2005 for a discussion). The analysis of fast-rotating stars is challenging and the limitations have to be established on a case-by-case basis. One of the most difficult tasks, when the stars have large projected rotational velocities, is identifying systematic asymmetries that can be caused by a companion in a spectroscopic binary (or multiple) system. Stellar parameters and chemical abundances derived from these kind of objects should be treated extremely carefully, because they are prone to large systematic errors. Maeder *et al.* (2014) provide in their Table 1 an example of a re-assessment of a sub-sample of objects from the FLAMES Massive Star Survey studied by Hunter *et al.* (2009), resulting in the identification of stars with previously unnoticed double or asymmetric spectral lines, which cause their natural exclusion from the interpretation of results because they were analyzed with standard techniques.

5. Be stars

Be stars are main-sequence or subgiant stars of spectral type B characterized by Balmer emission (e.g., Hα, but also other Balmer and eventually metal lines) that originates in a circumstellar disk. As most Be-stars are fast-rotators, the same limitations exposed in Sect. 4 apply to their analysis. Additionally, the clear signatures of lines formed in the disk cannot be reproduced self-consistently with any stellar atmosphere code at present. However, there are cases where the photospheric lines can be successfully reproduced, constraining reliably the stellar parameters and even metal abundances. An exploration of the extent of applicability of our spectral models and methods to such stars is ongoing.

6. Mid- and late B-type stars

Mid- and late B-type stars present fewer metal lines and less elements with different ionization stages traced in the spectra (1 or 2), in contrast to hotter stars (4 to 6). In many cases, in addition, the stars have a large projected rotational velocity, which can turn their analysis even more challenging. The identification of a companion forming a spectroscopic binary system, by resolving line-asymmetries is also a challenge for many of these stars (Zwintz *et al.*, subm.). Many objects show chemical peculiarities due to atmospheric diffusion, others present normal chemical abundances.

7. Chemically peculiar, pulsating and magnetic stars

Chemically peculiar stars are stars with distinctly unusual metal abundances in their surface layers. The stars present selective diffusion of different elements in their atmospheres causing some elements to show higher or lower abundances in their outer layers. Helium-strong stars constitute the hottest and most massive chemically peculiar stars of the upper Main Sequence. The chemical abundances and metallicities of such stars should be accounted for iteratively in every step of the analysis by computing dedicated model spectra, because their peculiar abundances are not considered in pre-computed grids of model spectra. An example of the consequence of not taking the peculiar abundances into account in the stellar parameter and chemical composition determination, in contrast to a self-consistent analysis is shown in Figs. 5 and 6, adapted from Przybilla *et al.* (2016). Analyses assuming solar helium abundances, in contrast to the enhanced

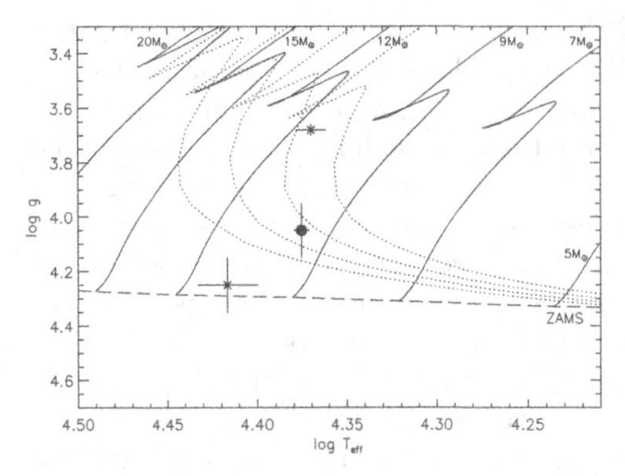

Figure 5. CPD−57 3509 in the Kiel diagram. Black dot: result from our self-consistent analysis. St. Andrew's crosses: results discussed in the literature at higher and lower lower gravity assuming solar helium abundance. Adapted from Przybilla *et al.* (2016).

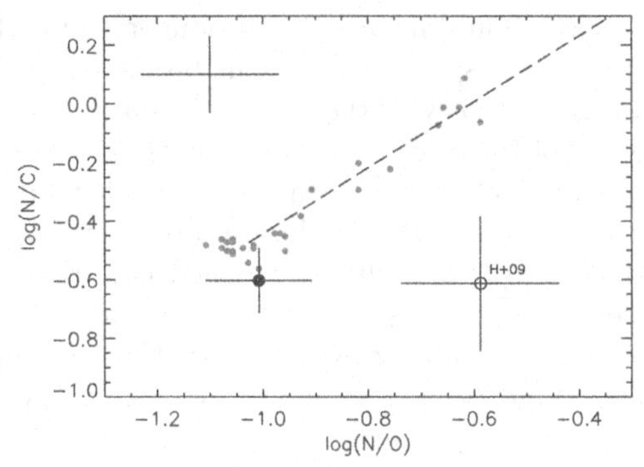

Figure 6. CPD−57 3509 in the N/O-N/C diagram (by mass). Black dot: Przybilla *et al.* (2016). Open circle: Hunter *et al.* (2009). Grey dots: early B-type stars from NP12. Dashed line: analytical approximation to the nuclear path for CN-cycled material using initial values from NP12.

helium abundance present in the star, can result in different atmospheric parameters, as seen in Fig. 5 and consequently in different chemical abundances, as shown in Fig. 6. The same methodology used by Przybilla *et al.* (2016) was consequently applied to other He-strong stars by Gonzalez *et al.* (2017), Castro *et al.* (2017), Briquet *et al.* (in prep.).

Other less-massive chemically peculiar stars (mid- to late B-type) like He-weak or HgMn stars may pose an extra challenge in their spectral analysis caused by elemental abundance stratification in their atmospheres. Clear signatures of He abundance stratification are noticed e.g. in κ Cancri, even when the modeled spectra are computed in non-LTE, whereas oxygen seems not to be stratified in non-LTE, therefore the ionization balance using lines of O I and O II formed in different depths in the atmosphere is met when considering non-LTE line formation Maza *et al.* (2014).

Strongly-pulsating stars present spectral lines with noticeable asymmetries. As model atmosphere codes usually do not include modes of stellar pulsations in their computations, such asymmetries cannot be reproduced typically, but see Irrgang *et al.* (2016). We notice, however, that some spectral lines are more symmetric because they are formed in regions where the pulsations affect the atmosphere less. In a case-by-case study, it is possible to derive atmospheric parameters through the analysis of the most symmetric lines. An example can be found in Briquet *et al.* (2011). Note, however, the difference in $\log g$ resulting from the spectroscopic and from the asteroseismic analysis in that case, which should be further investigated. We were also able to reproduce the spectra of several magnetic stars, with magnetic field strengths ranging from very weak to very strong. Some examples are discussed in Fossati *et al.* (2015), Przybilla *et al.* (2016), Gonzalez *et al.* (2017), Castro *et al.* (2017), Briquet *et al.* (in prep.), within the "B fields in OB stars" (BOB) collaboration.

8. Multiple systems

The spectral line fitting of the triple system HD 164492C in the Trifid Nebula, which is formed by a close spectroscopic binary composed by a normal early B-type star and a chemically peculiar late B-type star and a tertiary He-strong star, has been successfully realized by us in Gonzalez *et al.* (2017). The number of parameters involved in the spectral fitting is much larger than in the case of single stars. The challenge of the analysis

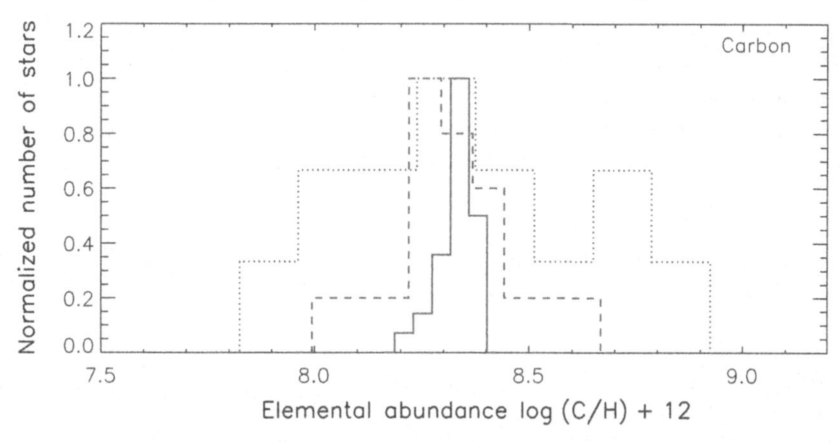

Figure 7. Present-day carbon abundances derived from early B-type stars in the Galaxy. Blue dashed line: apparently normal single stars. Blue dotted line: previously unnoticed SB2 candidates. The stars are distributed in the Galactic disc covering 4 to 15 kpc Galactocentric distance. For comparison, in red full line are carbon abundances of normal single early B-type stars in the solar neighborhood as derived by NP12.

consists in the constraint that the spectroscopic solution should provide consistent ages, spectroscopic distances, mass ratio of the primary system and flux scaling factors. And additionally, because of the chemically peculiar composition of the He-strong star, dedicated grids of model spectra have to be computed to derive parameter combinations for $T_{\rm eff}$, $\log g$, microturbulence, projected rotational velocity, macroturbulence, helium abundance and individual metal abundances per star. Within the whole procedure, several parameters are interrelated with each other, therefore the analysis is intensively iterative. In the particular case of HD 164492C, the Balmer lines do not allow to distinguish the contribution of the two brighter components, but the He I lines and the stronger Si III lines show the contribution of the faster-rotator He-strong star on the spectral line wings. The parameters of the faintest star (with a flux contribution of about 4-5%) cannot be determined from the composite spectrum. Instead, the disentangled spectrum can provide us with some constraints on stellar parameters. This kind of analysis is feasible to be done when time series spectra are available, which is time-consuming for faint objects.

9. Present-day carbon abundances in the Galactic disk as an application of accurate quantitative spectroscopy in the Galaxy

We have analyzed new high-resolution spectra taken with UVES@VLT of early B-type stars distributed along the Galactic disk covering Galactocentric distances ranging from 4 to 15 kpc. The sample was selected from the literature, and the high-quality data allowed us to identify new SB2 candidates. Preliminary carbon abundances are presented in Fig. 7. The whole sample has been analyzed assuming them as single stars, as it was done in previous work. However, if we separate stars with previously unnoticed SB2 signatures (blue dotted line, where the chosen analysis approach is inappropriate), the carbon abundance distribution is much broader than the sample of apparently single stars (blue dashed line), which still should present an intrinsic spread due to effects of the Galactic abundance gradient. Carbon abundances of apparently single early B-type stars in the solar neighborhood (red full line) as derived in NP12, are also plotted for comparison.

This is a clear case where – besides improvements in the spectral modeling and analysis technique – the data quality (spectral resolution, S/N ratio and wavelength coverage) is decisive for avoiding systematic errors. In this case, the source of systematics lies in the previously unrecognized composite nature of some spectra, which need to be excluded from the analysis based on standard assumptions.

Acknowledgements

MFN acknowledges support by the Austrian Science Fund through the Lise Meitner program N-1868-NBL.

References

Briquet, M., *et al.* 2011, *A&A*, 527, A112
Butler, K. & Giddings, J. R. 1981, *Newsletter on Analysis of Astronomical Spectra, no. 9*, University College of London
Castro, N., *et al.* 2017, *A&A*, 597, L6
Cox, A. N. 2000, *Allen's astrophysical quantities*, New York, AIP Press, Springer
Ekström, S., *et al.* 2012, *A&A*, 537, A146
Fossati, L., *et al.* 2015, *A&A*, 574, A20
Giddings, J. R. 1981, *Ph.D. thesis*, University of London
Gonzalez, J. F., *et al.* 2017, *MNRAS*, 467, 437
Hillier, D. J. & Miller, D. L. 1998, *ApJ*, 496, 407
Hubeny, I. & Lanz, T. 1995, *ApJ*, 439, 875
Hunter, I., *et al.* 2009, *A&A*, 496, 841
Irrgang, A., *et al.* 2016, *A&A*, 591, L6
Korn, A. J., Nieva, M. F., Daflon, S., & Cunha, K., 2005, *ApJ*, 633, 899
Kurucz, R. 1993, *CD-ROM No. 13*, Cambridge, Mass.: Smithsonian Astrophysical Observatory
Maeder, A., *et al.* 2014, *A&A*, 565, A39
Maza, N. L., Nieva, M.-F., & Przybilla, N. 2014, *A&A*, 572, A112
McSwain, M. V., Huang, W., & Gies, D. R. 2009, *ApJ*, 700, 1216
Nieva, M.-F. 2013, *A&A*, 550, A26
Nieva, M. F. & Przybilla, N. 2007, *A&A*, 467, 295
Nieva, M.-F. & Przybilla, N. 2012, *A&A*, 539, A143
Nieva, M.-F. & Przybilla, N. 2014, *A&A*, 566, A7
Przybilla, N., Butler, K., Becker, S. R., & Kudritzki, R. P. 2006, *A&A*, 445, 1099
Przybilla, N., *et al.* 2010, *A&A*, 517, A38
Przybilla, N., *et al.* 2016, *A&A*, 587, A7
Puls, J., *et al.* 2005, *A&A*, 435, 669

The Lives and Death-Throes of Massive Stars
Proceedings IAU Symposium No. 329, 2016
J.J. Eldridge, J.C. Bray, L.A.S. McClelland & L. Xiao, eds.

doi:10.1017/S1743921317003258

OWN Survey: a spectroscopic monitoring of Southern Galactic O and WN-type stars

Rodolfo H. Barbá[1], **Roberto Gamen**[2], **Julia I. Arias**[1] **and Nidia I. Morrell**[3]

[1]Departamento de Física y Astronomía, Universidad de La Serena, Chile
email: rbarba@userena.cl

[2]Instituto de Astrofísica de La Plata, CCT La Plata-CONICET, La Plata, Argentina

[3]Las Campanas Observatory, Observatories of the Carnegie Institution of Washington, La Serena, Chile

Abstract. We summarize the status and results of the *OWN Survey*, a high-resolution monitoring program of Southern Galactic O- and WN-type stars, after twelve years of observing campaign.

Keywords. binaries: close, binaries: spectroscopic, stars: early-type, stars: Wolf-Rayet, surveys

---

1. Introduction

The *High-resolution spectroscopic monitoring survey of Southern Galactic O- and WN-type Stars (OWN Survey)* project has its foundational ideas in the school of stellar spectroscopy that is consolidated in the Observatorio Astronómico de La Plata (Argentina). This school grew specifically in the study of early-type stellar atmospheres, massive stars and binary systems. It is worth mentioning the pioneering work of those years carried out by Dr. Jorge Sahade in the area of interacting binaries, Dr. Virpi Niemelä on massive binaries, and Dr. Hugo Levato in binary systems in open clusters, together with a strong group of students and collaborators.

It should be noted that the *OWN Survey* was born about 13 years ago, as a result of the stimulating talks with Dr. Virpi Niemela, leader of the "Massive stars Research Group" at La Plata, and with the dedicated work of Dr. Roberto Gamen, at that time, postdoc at the Astronomy Group of the Universidad de La Serena.

The idea of the spectroscopic monitoring survey arises in the lack of full knowledge of the multiplicity status of many optically bright O-type stars in the Milky Way. By the year 2005, it was known that the distribution of binary periods in massive stars differed markedly from that of periods of low- and intermediate-mass stars. Mason *et al.* (1998, 2009) show a period distribution for O-type stars which is clearly bimodal, in contrasts with the period distribution of stars of solar type stars (Duquennoy & Mayor, 1991). Since then it was suspected that this bimodality could be a consequence of a strong observational bias produced by the limitations of the techniques used in the detection of binaries in massive stars. Such bimodality posed the existence of a gap in the period distribution, expanding from periods of one month and to 20,000 years. It is illustrative, to bear in mind the ignorance about period distribution of massive binary stars has periods has an enormous impact in modeling of massive star formation and evolution.

With the aim of contributing to the knowledge of the multiplicity status of massive stars is that *OWN Survey* was born in 2005. Our approach is clearly empirical from the point of view of the determination of radial velocities (RVs) of a selected sample of massive stars during timescales of years. Parallel to the last years, considerable progress

Table 1. Instrument configurations used in *OWN Survey*

Observatory	Telescope	Spectrograph	Resolution
CASLEO	2.15 m "Jorge Sahade"	REOSC	15,000
Las Campanas	2.5 m "du Pont"	Echelle	48,000
La Silla	2.2 m	FEROS	48,000
CTIO	1.5 m	BME	50,000

has been made in the knowledge about the multiplicity status of massive stars using different observational techniques, such as the case of spectroscopy (Chini *et al.* 2012; Kobulnicky *et al.* 2012; Sana *et al.* 2012; Almeida *et al.* 2017), adaptive optics (Turner *et al.* 2008; Close *et al.* 2012), speckle interferometry (Mason *et al.* 1998, 2009, "lucky imaging" (Maíz Apellániz 2010; Barbá *et al.* 2017), optical interferometry (Sana *et al.* 2011, 2013, 2014; Sánchez-Bermúdez *et al.* 2013, 2014), Fine Guidance Sensor on board the Hubble Space Telescope (Nelan *et al.* 2014; Aldoretta *et al.* 2015), among others.

2. Goals of OWN Survey

The main goal of the *OWN Survey* is to set the multiplicity status for bright southern O- and WN-type stars through spectroscopic and radial velocity monitoring. Also, detection of spectroscopic variability is an additional outcome, for example as the case of the Oe star HD 120678, which underwent a shell-like episode in 2008 documented by the OWN survey (Gamen *et al.* 2012), the discovery of strong spectral variations in the magnetic Of?p star CPD -28 2561 (Wade *et al.* 2015), or those variations produced in interacting binaries. Another important goal is to establish a set of massive stars without close companions useful for the testing stellar models for single stars, as is the case of abundance studies in stars without interactions with close companion (e.g. Martins *et al.* 2016).

The star sample for the *OWN Survey* has been defined based on the first version of the Galactic O star Catalog" (GOSC v1.0, Maíz Apellániz 2004), and the "VIIth Catalogue of Galactic Wolf-Rayet stars" and annexes (van der Hucht 2001). That sample consists of 180 O-type stars and 58 WN-type stars without a clear evidence of binarity and stars with scarce RV information. At the present, the sample of O-stars under spectroscopic monitoring by the *OWN Survey* has been expanded to 205 stars, to include some additional targets monitored in the "Galactic O-star Spectroscopic Survey" (GOSSS, Sota *et al.* 2011, 2014; Maíz Apellániz *et al.* 2012, 2016) and the "Near-Infrared Photometric Monitoring of Galactic Star Forming Regions" (NIP, Barbá *et al.* 2011).

Observations are carried out at different facilities in the Chile and Argentina. Table 1 shows the instrumental configuration used and their typical resolutions. After twelve years of monitoring, we have obtained more than 7,000 high-resolution spectra, most of them having signal-to-noise over 200 (Fig. 1. The whole sample of 205 O-stars has been observed at least in five different epochs, having 50% of the sample 17 spectra or more. Additionally, in the framework of our project, we have observed an important number of known binaries in order to improve their orbital solutions and also to detect apsidal motions. This parameter is a very good proxy to derive absolute masses of the binary components, using stellar structure models (e.g. Ferrero *et al.* 2013; Ferrero 2016). Thus, the total number of stars monitored is almost 300.

Some additional O-type stars are observed less frequently, using using intermediate-resolution spectrographs. This also holds for the sample of 58 WN-type stars.

The *OWN Survey* is actually a project tightly coordinated with other similar survey projects, and thus sharing spectroscopic databases and results with the GOSSS (Sota

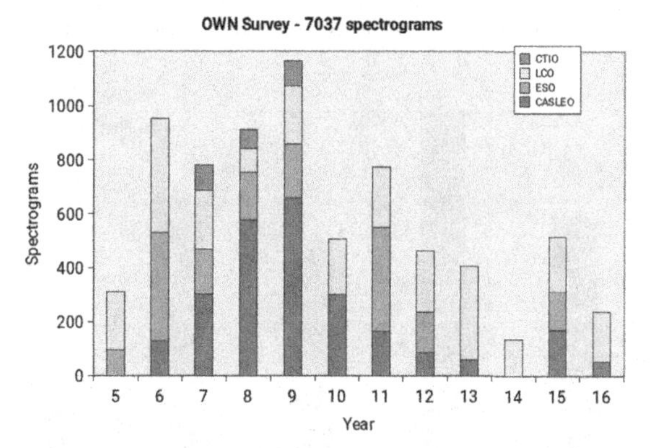

Figure 1. Number of high-resolution spectrograms obtained from different facilities as a function of time.

et al. 2011; Maíz Apellániz *et al.* 2012), IACOB (Simón-Díaz *et al.* 2011) and CAFÉ-BEANS (Negueruela *et al.* 2015).

3. Some results of the survey

From the sample of 205 southern O-stars without a clear spectroscopic signature of binarity, we have determined that 114 stars show RV variations larger than 20 km s^{-1} (typical errors about $1 - 2$ km s^{-1}), 59 stars show RV variations in the interval $10 - 20$ km s^{-1}, while 43 stars show variations below 10 km s^{-1}. It is interesting to note that 63 stars display RV variations over 50 km s^{-1}, indicative of an important number of binaries in the sample. Binaries can also contribute to the interval of $10 - 20$ km s^{-1} RV variations. Simón-Díaz (2017) has called the attention to the fact that pulsations in massive stars can potentially be confused with binary motion. Hence, low amplitude RV variations due to binarity should be defined through systematic spectroscopic monitoring, with increased cadence. This issue has a large impact in the completeness of our knowledge about systems with low-mass ratios and/or orbital inclinations.

The main result of the *OWN Survey* is the detection of 39 SB1, 47 SB2, and 16 SB3 systems, while 60 stars present RV variations larger than 10 km s^{-1}, and 43 stars can be classified as "single". We have defined as single those stars with RV variations below 10 km s^{-1}, and without a clear periodic signal. From our observations of these 102 spectroscopic binaries, we have determined 85 RV orbits, 55 of which are derived for first time. Figure 2 summarizes the main results of the *OWN Survey* compared to the RV information available for those 205 O-stars before 2005.

In the case of 58 WN-type stars comprised in the survey, most of the results have been presented in the PhD thesis of A. Collado (2014). We have determined that 22 WN-stars present RV variations over 30 km/s, while in 16 stars, variations are below that limit. We have derived the binary status for 20 systems (18 new orbits), splitted in equal amounts between SB1 and SB2. Taking into account the all known WR binaries, we have counted "only" 68 binary systems with determined periods among a total of 634 WR stars listed in the WR Catalogue compiled by P. Crowther (Rosslowe & Crowther 2015).

There are mostly two classes of WN-type stars in our sample: those very massive objects, like WR25 or WR21a, and those likely descendant from O-stars, like WR29 or WR36. In both types of WN stars, the detection of binaries is a more complicated task than in O-type stars due to the presence of strong emission lines, specially in the second group of evolved WN-type stars, where also intrinsic variations in the profiles hinder

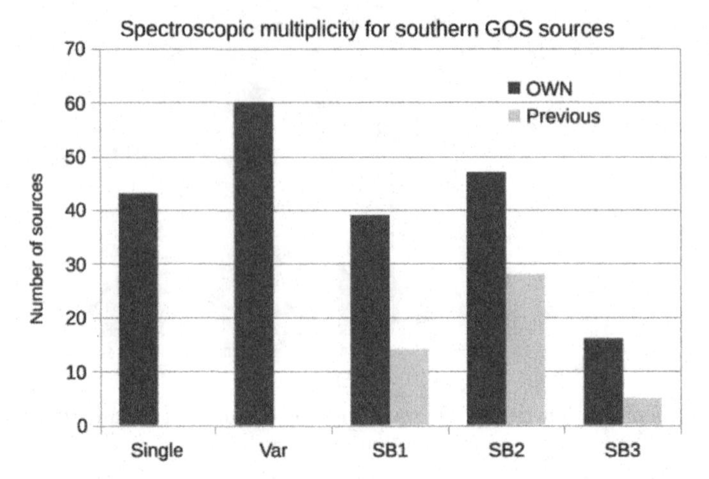

Figure 2. Spectroscopic multiplicity for southern bright GOSC sources. "Previous data" refer to systems known before 2005, starting year of OWN Survey.

the RV variations due to binary motion. Also, we must to take into account that the WR stars in our sample are in general fainter than the O-stars, therefore the spectral resolutions used in our monitoring are a factor of 10 or 20 lower than those used for the monitoring of O-type stars. This is an additional observational issue to take into account when evaluating our knowledge of the binary status of WN-tye stars. Consequently, it is early to conclude any evolutive scenario from the binary frequency in O- and WN-type stars due to the different completeness in our knowledge of both kind of objects. We shall come back to this point in the next section.

4. About binary star parameter distributions

We have collected from the literature the available information for all the known Galactic O-type binaries and combined it with our discoveries, bringing the number of known orbits to 220, a good figure for statistical studies. Figure 3 shows the period distribution for all the known Galactic O-type binaries. We have also highlighted the Southern systems and the eclipsing binary systems. The first evident conclusion is that for periods of a few days the number of Southern binaries doubles or even triplicates the number of Northern binaries. This should be compared to the 0.67 value expected for the North to South number ratio of O-type stars as derived from the GOSC (Fig. 3 right). For very long periods (thousands of days), the ratio is higher (0.75) due to observational efforts by astronomers using different techniques as enumerated in the Introduction. Therefore, we can conclude that the Southern Galactic O-type stars are more or less evenly surveyed for multiplicity in order to cover the Mason's gap in binary period distribution. Meanwhile, in the Northern hemisphere the picture is not the same, dedicated high-resolution spectroscopic monitoring surveys of massive stars are not common, and this is the niche that CAFÉ-BEANS and IACOB surveys are exploring.

The distribution of mass ratios (q) for all SB2 systems (Fig.4) in the sample considered shows a more or less even distribution between 0.3 to 1.0, except for small differences between bins, which may be explained as statistical fluctuations due to the low numbers of known SB2 systems. Comparing with the "Southern" sample, we can note that the distributions are essentially the same, with a very tiny peak for $q \sim 0.5$. For the eclipsing binaries seems to have a maximum in the bin $q = .5$, but it needs to be analyzed if in close binaries this mass-ratio is favored. Also, the sharp drop in the number of binaries

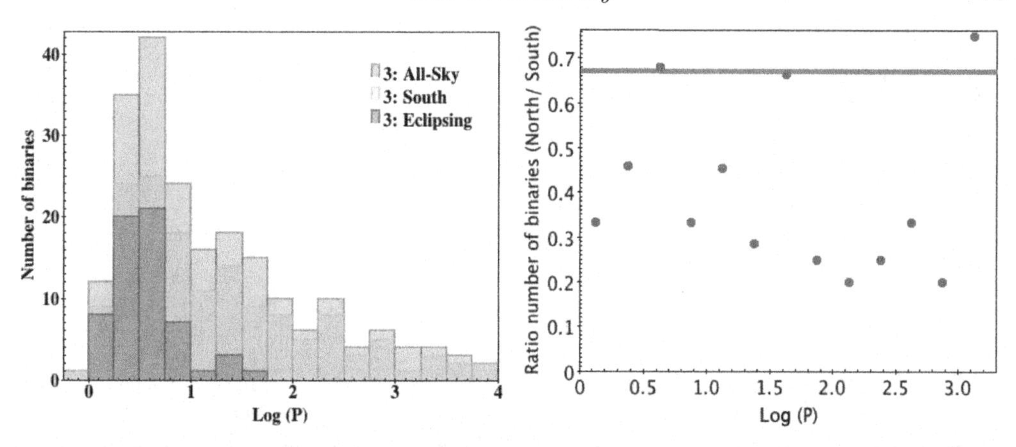

Figure 3. Left: Period distribution for known Galactic binaries. Right: North to South number ratio, the blue line represents the ratio derived for all O-stars in both hemispheres.

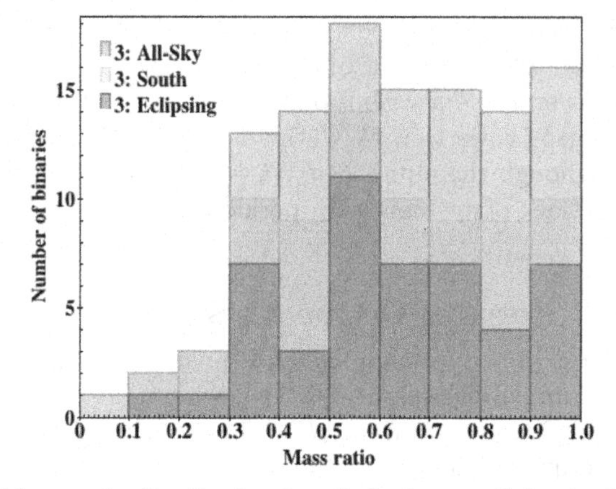

Figure 4. Mass-ratio distribution for all the known Galactic O-type stars.

with $q < 0.3$ needs to be explored. It could be related to an observational bias, because smaller mass-ratio also points to smaller luminosity-ratio, and then larger difficulties in spectroscopically detecting the secondary component. There may exist also systems with moderately long periods (weeks or months), where one of the component is a very fast rotator. In those cases, the line profiles of the fast rotator are immersed inside the continuum of the narrow-lined component, and then very hard to separate. These issue needs to be explored through the monitoring of some short period SB1 systems with highest achievable resolution and signal-to-noise.

The period-eccentricity distribution for all O-type binary systems is presented in Figure 5. The lack of systems with very high eccentricities for periods of a few days is expected because the limit imposed for the size of stars are of the same order of the orbital separations at periastron passage. Taking into account some specific spectral types, we can note that late-type supergiants show an enhancement for periods between 3 and 50 days and eccentricities larger than 0.3, compared with main-sequence binaries. This fact perhaps is indicative of evolutive signatures in binaries, for example, SB1 systems with compact companions, as in the case of HD 74194, a high-mass SB1 fast X-ray transient with a $P = 9.54$ days and $e = 0.63$ (Gamen *et al.* 2015). Moreover, it is interesting to note the relatively low number of binaries with periods longer than 100 days and eccentricities lower than 0.2. This effect in part could be explained by the difficulties in detecting

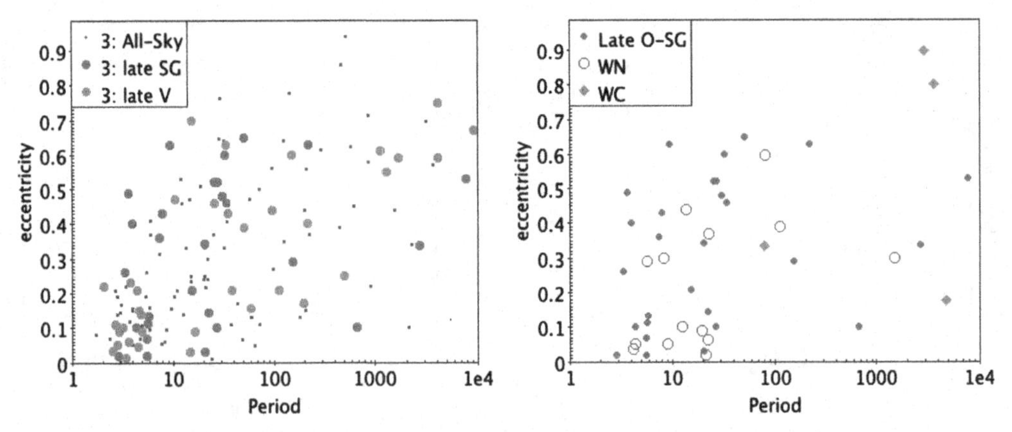

Figure 5. Eccentricity-period distribution for all Galactic O-type binaries (right), and including the WR stars (left).

low amplitude binaries with long periods, but also could be related to the mechanism involved in the formation of this kind of binaries.

The comparison with the $P-e$ distribution between late O-type supergiants and WN- and WC-type stars (Fig. 5 shows that WN stars follow a similar distribution to that of O-type supergiants. Although the number of WC-type stars with known eccentricities is low, it is noticeable the lack of systems with periods below 80 days.

5. The multiplicity status of O-type stars

The multiplicity status of O-type stars should be studied considering the differences in spectral types and luminosity classes if we wish to understand how to the systems are formed and how they evolve. Sana *et al.* (2012) concluded that the binary interaction dominates the evolution of massive stars and their calculations propose that 24% of all the systems do merge. Therefore, beyond the observational biases, we must to take into account that binary interaction could affect the true number of binaries in each spectral type and luminosity class. We have started the study of the distribution of binaries taking into account both parameters. A zero order approach for the analysis can be done counting the number of spectroscopic binaries per luminosity class. Table 2 displays the relative frequency of spectroscopic binaries for 411 stars (205 from the *OWN Survey*) for four different luminosity classes, i.e. supergiants, bright giants, giants, and main-sequence plus subgiants. Obviously, this is a coarse approach to describe the evolutionary status of these stars. For example, an O2 If* star (i.e. HD 93129 A) is more related to main-sequence stars than an O8.5 Ib-II(f)p star (i.e. HD 74194). Taking into account this caveat, we can note that the relative numbers of "single" stars show maxima for luminosity classes I, and II, reinforcing the idea that some supergiants could be merger products from the evolution of close binaries, or others became single stars after the disruption of a binary due to the evolution of the former primary and ulterior explosion as supernova, or the dynamical ejection from a massive cluster. Another interesting feature is the high number of SB2 systems for the luminosity class III. This definition of this luminosity class is based on the appearance the HeII 4686 absorption line. We can speculate that the origin of this observational feature, can be related to the presence of interacting binaries, where line profiles may be distorted. There is also the possibility that some of these "SB2 systems" are not real binaries, but fast rotators with profile variations induced by pulsations.

Table 2. Frequency of the spectroscopic multiplicity status for 411 Galactic O-type stars in function of luminosity.

Status	I/a/ab/b	II	III	IV-V-Vz
Single	0.55	0.63	0.31	0.40
SB1	0.23	0.17	0.18	0.20
SB2	0.17	0.11	0.37	0.29
SB3	0.05	0.09	0.14	0.11

Table 3. Spectroscopic multiplicity status for late O-type and early B-type supergiants

SpT	No. of Stars	No. of SB	Frequency
O7	20	11	0.57
O8	19	9	0.52
O9	63	24	0.38
B0	260	11	0.04

We can consider that the main-sequence stars in the spectral range between O5–O9.7 have masses in the range between 30–40 to 13–16 $M_\odot$ (c.f. Martins *et al.* 2005). Typical evolutionary scenarios (e.g. Brott *et al.* 2011, Ekström *et al.* 2012) for single stars in this mass range predict that the O-stars will evolve as late-O or early B-type supergiants in a timescale of few mega-years (although hypergiants may have different origin). It would thus be of great interest to compare the multiplicity status of these supergiants. Such status is relatively well known for late O-type supergiants but this is not the case for early B-type supergiants. We have compiled the information about the multiplicity status of early B-type supergiants surveying in detail the literature for more than 1,500 stars extracted from the Skiff's recompilation of spectroscopic classification of stars (Skiff 2014). The results are surprising, only 63 stars are classified as spectroscopic binaries. Table 3 shows the spectroscopic binary frequencies derived from our studies and the literature for late-O and B0 supergiants and bright giants. The sharp drop in the number of binaries between spectral types O9.7 and B0 is noticeable. It may be produced by an observational bias due to the lack of dedicated spectroscopic monitoring programs for B0 supergiants, but also we must to explore the possibility of evolutive scenarios in action like binary mergers, binary disruptions, etc.

6. Perspectives on multiplicity of massive stars

When the *OWN Survey* program started, one of our expectations was detecting systems with periods of tens of days or even longer, but one of our surprises was the discovery of bright close binaries in hierarchical triple systems as the cases of HD 92206 C (Campillay *et al.* 2007) or Herschel 36 (Arias *et al.* 2010).

Sana *et al.* (2012) have suggested that about 70% of all massive binaries interact with their companions, leading to mergers in one-third of cases. Therefore, the high number of spectroscopic hierarchical triple systems composed by a close binary and a massive companion in a wider orbit opens new evolutive scenarios, which have not been explored yet. For example, in an important number of cases we should to consider a double interacting scenario, where the close binary system evolves to a merger and then interacts with the third companion. Obviously, the different scenarios will depend on the different system parameters, such as mass-ratio of the close binary, mass of the tertiary companion, separation between different components, orbit eccentricities, spins, etc.

The *OWN Survey* has brought new interesting observational facts about the complexity of the massive binary stars.

Acknowledgments

RHB thanks support from project FONDECYT Regular No 1140076.

References

Aldoretta, E. J. *et al.* 2015, *AJ*, 149, 26.
Almeida, L. A. *et al.* 2017, *A&A*, 598, A84.
Arias, J. I. *et al.* 2010, *ApJ*, 710, L30.
Barbá, R. H. *et al.* 2011, *Bol. Asoc. Arg. Astron.*, 54, 85.
Barbá, R. H. *et al.* 2017, *in preparation.*
Brott, I. *et al.* 2011, *A&A*, 530, A115.
Campillay, A. *et al.* 2007, *VI Reunion Anual Sociedad Chilena de Astronomia (SOCHIAS), held in Valparaiso, Chile, 7-9 Nov. 2016*, 63.
Chini, R., Hoffmeister, V. H., Nasseri, A., Stahl, O., & Zinnecker, H. 2012, *MNRAS*, 424, 1925.
Close, L. M. *et al.* 2012, *ApJ*, 749, 180.
Collado, A. 2014, *PhD. Thesis (Universidad Nacional de San Juan, Argentina).*
Duquennoy, A. & Mayor, M. 1991, *A&A*, 248, 485.
Ekström, S. *et al.* 2012, *A&A*, 537, A146.
Ferrero, G., Gamen, R., Benvenuto, O., & Fernández-Lajús, E. 2013, *MNRAS*, 433, 1300.
Ferrero, G. 2016, *PhD. Thesis (Universidad Nacional de La Plata, Argentina).*
Gamen, R. *et al.* 2012, *A&A*, 546, A92.
Gamen, R. *et al.* 2015, *A&A*, 583, L4.
Kobulnicky, H. A. *et al.* 2012, *ApJ*, 756, 50.
Maíz Apellániz, J., Walborn, N. R., Galué, H. A., & Wei, L. H. 2004, *ApJS*, 151, 103.
Maíz Apellániz, J. 2010, *A&A*, 518, A1.
Maíz Apellániz, J. 2012, *ASPC*, 465, 484.
Maíz Apellániz, J. *et al.* 2016, *ApJS*, 224, 4.
Martins, F., Schaerer, D., & Hillier, D. J. 2005, *A&A*, 436, 1049.
Martins, F., Simón-Díaz, S., Barbá, R. H., Gamen, R. C., & Ekström, S. 2016, *(*A&A), in press (2016arXiv161105223M).
Mason, B. D., Gies, D. R., Hartkopf, W. I., Bagnuolo, W. G. Jr., ten Brummelaar, Th., & McAlister, H. A. 1998, *AJ*, 115, 821.
Mason, B. D., Hartkopf, W. I., Gies, D. R., Henry, T. J., & Helsel, J. W. 2009, *AJ*, 137, 3358.
Negueruela, I. *et al.* 2015, *Highlights of the Spanish Astrophysics*, 8, 524.
Nelan, E. P., Walborn, N. R., Wallace, D. J., Moffat, A. F. J., Makidon, R. B., Gies, D. R., & Panagia, N. 2014, *AJ*, 128, 323.
Rosslowe, C. K. & Crowther, P. A. 2015, *MNRAS*, 447, 2322.
Sana, H. *et al.* 2011, *ApJ*, 740, L43.
Sana, H. *et al.* 2012, *Sci*, 337, 444.
Sana, H. *et al.* 2013, *A&A*, 553, A131.
Sana, H. *et al.* 2014, *ApJS*, 215, 15.
Sánchez-Bermúdez, J., Schödel, R., Alberdi, A., Barbá, R. H., Hummel, C. A., Maíz Apellániz, J., & Pott, J.-U. 2013, *A&A*, 554, L4.
Sánchez-Bermúdez, J., Alberdi, Schödel, R., Hummel, C. A., Arias, J. I., Barbá, R. H., Maíz Apellániz, J., & Pott, J.-U. 2014, *A&A*, 567, A21.
Simón-Díaz, S., Castro, N., García, M., Herrero, A., & Markova, N. 2011, *Bull. Soc. Roy. Sci. Liege*, 80, 514.
Simón-Díaz, S. 2017, *IAU Symposium*, 329, this volume.
Skiff, B. A. 2014, *Vizier Online Data Catalogue: Catalogue of Stellar Spectral Classification.*
Sota, A. *et al.* 2011, *ApJS*, 193, 24.
Sota, A., Maíz Apellániz, J., Morrell, N. I., Barbá, R. H., Walborn, N. R., Gamen, R. C., Arias, J. I., & Alfaro, E. J. 2014, *ApJS*, 211, 10.
Turner, N. H., ten Brummelaar, Th.A., Roberts, L. C., Mason, B. D., Hartkopf, W. I., & Gies, D. R. 2008, *AJ*, 136, 554.
van der Hucht, K. A. 2001, *New Astron.*, 45, 135.
Wade, G. A. *et al.* 2015, *MNRAS*, 450, 2822.

The Lives and Death-Throes of Massive Stars
Proceedings IAU Symposium No. 329, 2016
J.J. Eldridge, J.C. Bray, L.A.S. McClelland & L. Xiao, eds.

doi:10.1017/S1743921317002976

Resolving the mass loss from red supergiants by high angular resolution

Keiichi Ohnaka

Instituto de Astronomía, Universidad Católica del Norte, Avenida Angamos 0610, Antofagasta, Chile
e-mail: `k1.ohnaka@gmail.com`

Abstract. Despite its importance on late stages of the evolution of massive stars, the mass loss from red supergiants (RSGs) is a long-standing problem. To tackle this problem, it is essential to observe the wind acceleration region close to the star with high spatial resolution. While the mass loss from RSGs is often assumed to be spherically symmetric with a monotonically accelerating wind, there is mounting observational evidence that the reality is much more complex. I review the recent progress in high spatial resolution observations of RSGs, encompassing from the circumstellar envelope on rather large spatial scales (~100 stellar radii) to milliarcsecond-resolution aperture-synthesis imaging of the surface and the atmosphere of RSGs with optical and infrared long-baseline interferometers.

Keywords. stars: mass loss, stars: imaging, stars: atmospheres, (stars:) supergiants, (stars:) circumstellar matter, infrared: stars, techniques: high angular resolution, techniques: interferometers, techniques: polarimetric

1. Introduction

The mass loss in the red supergiant (RSG) phase significantly affects the evolution of massive stars. For example, the RSG mass loss is considered to be a key to constraining the mass of the progenitors of supernovae (SNe) type IIP, which are the most common core-collapse SNe (e.g., Smartt 2015). Despite such importance, the mass loss from RSGs is one of the long-standing problems in stellar astrophysics. It is often argued that dust grains form in the atmosphere or in the circumstellar envelope of RSGs and the radiation pressure on the dust grains can drive the mass loss. However, it is not yet well understood where and how dust forms in RSGs, and it is possible that the mass loss is driven by some yet-to-be identified physical process. As Harper (2010) notes, there is no self-consistent theoretical model for the RSG mass loss at the moment.

High spatial resolution observations of RSGs provide us with valuable information to tackle the mass loss problem. However, even the closest RSG Betelgeuse has an angular diameter of only 42.5 mas (Ohnaka *et al.* 2011). This angular diameter is much smaller compared to the angular resolution of conventional ground-based imaging of ~1″, which is limited by seeing. With adaptive optics, it is possible to achieve angular resolution corresponding to the diffraction limit of the telescope, which is 17 mas and 50 mas at 0.55 μm and 2 μm with an 8 m telescope, respectively. In the mid-infrared it is possible to achieve the diffraction limit (300 mas at 10 μm) without adaptive optics if the seeing is better than ~0.6″ in the visible . This means that the structures of the circumstellar envelope of nearby RSGs on large spatial scales (~100 stellar radii) can be resolved by diffraction-limited mid-infrared imaging. The diffraction-limited adaptive optics imaging in the visible and near-infrared can resolve the structures of the circumstellar environment close to the star, within several stellar radii. However, milliarcsecond angular resolution is required to spatially resolve the detailed structures of the wind acceleration region and

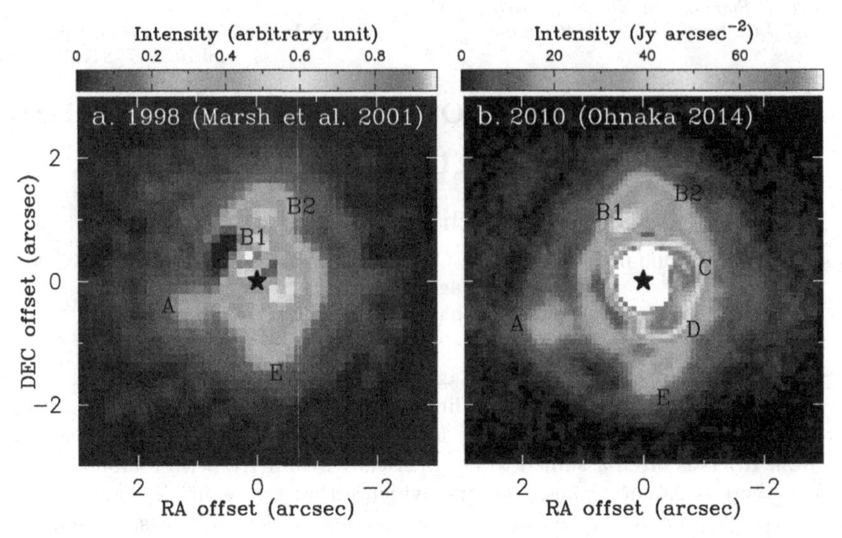

Figure 1. Clumpy dust clouds toward the RSG Antares detected in the mid-infrared. **a:** The 20.8 μm image obtained by Marsh *et al.* (2001). The central star is subtracted. **b:** The 17.7 μm image obtained by Ohnaka (2014) with the central star subtracted. The individual dust clouds are labeled. In both panels, the intensity is shown in the linear scale.

the stellar surface of RSGs, which is possible only by optical and infrared long-baseline interferometry.

2. Circumstellar envelope up to ~100 stellar radii

It is often assumed that the stellar winds from RSGs are spherically symmetric and are monotonically accelerating until they reach terminal velocities of 20–40 km s^{-1}. The spatially resolved 4.6 μm CO emission of the prototypical RSG Betelgeuse shows an overall spherical envelope with mildly clumpy structures within 3″, which corresponds to 140 stellar radii (Smith *et al.* 2009). The mid-infrared imaging of Betelgeuse by Kervella *et al.* (2011) shows the clumpy structures more clearly. The images obtained from 7.76 to 19.5 μm show that the circumstellar envelope is approximately spherical within ~1″ (~47 stellar radii) but prominent clumpy structures. The images also show emission extending up to ~2″ (~94 stellar radii) in the south and northwest.

Ohnaka (2014) detected similar clumpy structures in the circumstellar envelope of another well-studied RSG, Antares, based on the diffraction-limited 17.7 μm image. As Fig. 1 (right) shows, the image reveals six clumpy dust clumps located at 0.8–1.8″ (43–96 stellar radii) and compact emission within 0.5″ (27 stellar radii) around the star. Moreover, comparison of this image taken in 2010 and the 20.8 μm image taken in 1998 by Marsh *et al.* (2001) (Fig. 1 left) shows the outward motions of the individual dust clouds. The observed proper motions of the dust clouds (with respect to the central star) of 0.2–0.6″ in 12 years translate into the velocities of 13–40 km s^{-1} projected onto the plane of the sky. The distances and velocities of the dust clouds may not be explained by a simple monotonically accelerating wind but suggest that the individual clouds may be ejected at different velocities from the beginning, although the projection effect does not allow us to draw a definitive conclusion.

While the mass loss from Betelgeuse and Antares is roughly spherically symmetric with small-scale clumpy structures superimposed, some RSGs show deviation from spherical

Figure 2. The 9 μm image predicted by the dust torus model of the RSG WOH G64 in the Large Magellanic Cloud (Ohnaka *et al.* 2008). The dust torus is seen nearly pole-on.

symmetry on large spatial scales. For example, the RSG μ Cep shows elongation on angular scales of 0.5 to 5$''$ (de Wit *et al.* 2008; Shenoy *et al.* 2016), while the extremely dusty RSG VY CMa shows a bipolar outflow with much more complex structures overlaid (Monnier *et al.* 1999; Smith *et al.* 2001).

High angular resolution observations also allow us to probe the morphology of the circumstellar envelope of RSGs beyond the Milky Way Galaxy. The mid-infrared (8–13 μm) interferometric observations of the RSG WOH G64 in the Large Magellanic Cloud using the MIDI instrument at the Very Large Telescope Interferometer (VLTI) were the first (and are still the only) study to spatially resolve an individual star in an extragalactic system (Ohnaka *et al.* 2008). The measured angular size (15–20 mas at 8 μm and $\sim$25 mas at 13 μm) does not show noticeable position angle dependence. The 2-D radiative transfer modeling of the interferometric data and the spectral energy distribution (SED) revealed the presence of a geometrically and optically thick dust torus seen nearly pole-on with an inner boundary radius of 15 ± 5 stellar radii (see Fig. 2). While the luminosity of WOH G64 previously estimated based on the spherical models suggested an initial mass of 40 $M_\odot$, the current evolutionary theory predicts that such a high mass star does not reach the RSG phase. However, the luminosity re-estimated with the presence of the dust torus taken into account is about a half of the previously estimated value. This newly estimated luminosity suggests an initial mass of 25 $M_\odot$, instead of 40 $M_\odot$, and brings the star's position on the H-R diagram in much better agreement with the evolution theory.

3. Circumstellar environment close to the star

To understand the origin of the clumpy structures seen in the circumstellar envelope, it is necessary to study the circumstellar environment close to the star, within several stellar radii. The advent of extreme adaptive optics combined with polarimetric imaging allows us to probe the circumstellar dust environment in great detail. With the unpolarized

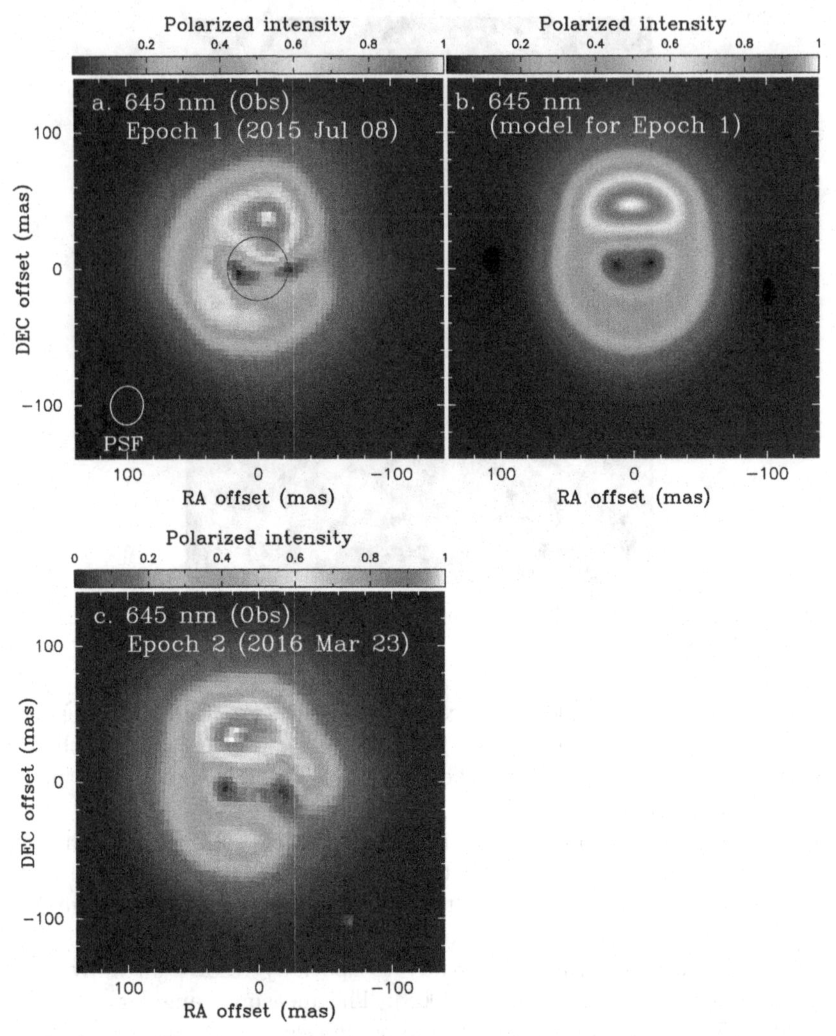

Figure 3. Clumpy dust clouds of the Mira star W Hya forming close to the photosphere detected with the VLT/SPHERE-ZIMPOL instrument (Ohnaka *et al.* 2016, 2017). **a:** Polarized intensity map obtained at 645 nm on 2015 July 8, when the star was at pre-maximum light. The black circle represents the size of the star measured with the VLTI/AMBER instrument. **b:** Model image at 645 nm for the 2015 data. **c:** Polarized intensity map obtained at 645 nm on 2016 March 23, when the star was at minimum light. In all panels, north is up, and east is to the left.

direct light from the bright central star is effectively suppressed, polarimetric imaging is very powerful for detecting faint scattered light from dust forming close to the star.

Ohnaka *et al.* (2016) observed the nearby, well-studied Mira star W Hya from 645 to 820 nm with the VLT/SPHERE-ZIMPOL instrument and detected clumpy dust clouds within ~50 mas (~2 stellar radii), revealing that dust formation takes place very close to the star (Fig. 3a). The clumpy dust formation may be induced by large convective cells as predicted by the 3-D simulations for AGB stars (Freytag & Höfner 2008). The Monte-Carlo radiative transfer modeling of the observed polarized intensity maps suggests the presence of an optically thin shell consisting of large (0.4–0.5 μm) grains of corundum (Al_2O_3), forsterite (Mg_2SiO_4), or enstatite ($MgSiO_3$), as shown in Fig. 3b. The inner boundary radius of the shell is 1.9–2.0 stellar radii, where the dust temperature reaches 1500 K.

When these observations took place, the Mira star W Hya was at pre-maximum light, just before the brightest phase. The follow-up observations of W Hya reveal significant time variations in the morphology of the dust clouds. As shown in Fig. 3c, the polarized intensity map obtained 8.5 months later at minimum light shows the formation of a new dust cloud approximately in the west and the disappearance of the dust clouds in the southwest and southeast. The degree of linear polarization measured at minimum light, 13–18%, is noticeably higher than that observed at pre-maximum light. Interestingly, the Monte Carlo radiative transfer modeling shows that the minimum-light data can be explained by a shell of 0.1 μm grains of corundum, forsterite, or enstatite, in marked contrast to the large grain sizes of 0.4–0.5 μm found at pre-maximum light. Perhaps small grains might just have started to form at minimum light in the wake of a shock induced by the large-amplitude stellar pulsation, while the pre-maximum light phase might have corresponded to the phase of efficient grain growth.

Kervella *et al.* (2016) carried out visible polarimetric imaging observations of Betelgeuse with SPHERE-ZIMPOL and detected polarimetric signal resulting from dust scattering within three stellar radii. This suggests that dust forms close to the star in RSGs as well, which may play an important role in accelerating the stellar wind.

4. Imaging the surface and atmosphere of RSGs

Since Schwarzschild (1975) predicted that the convective cells on the surface of RSGs can be as large as the radius of the star, high spatial observations have been carried out to detect such large spots. Relatively recent observations include the aperture-synthesis image of Betelgeuse obtained at 1.6 μm by Haubois *et al.* (2009). They detected two spots, which have a size of 1/4–1/2 of the radius of the star with an intensity contrast of 5–10%. Monnier *et al.* (2014) present images of VY CMa reconstructed at 1.61, 1.67, and 1.73 μm from the data taken with the VLTI/PIONIER instrument. The images reveal an extended atmosphere elongated in the E-W direction and two bright spots with a size of the stellar radius on the stellar surface. Baron *et al.* (2014) present the 1.65 μm aperture-synthesis images of two RSGs, T Per and RS Per, in the Double Cluster obtained with the Center for High-Angular Resolution Astronomy (CHARA) array. While T Per shows a bright spot with a size comparable to the stellar radius, RS Per shows a large dark region near the limb. The contrast of the surface structures is approximately 15% in both cases. These spots are considered to represent large convective cells, which were predicted by Schwarzschild (1975) and modeled by 3-D convection simulations (e.g., Chiavassa *et al.* 2010). However, Montargès *et al.* (2016) show that the interferometric data of Betelgeuse obtained at 1.6 μm suggest surface structures more inhomogeneous and with a higher contrast than predicted by the 3-D convection simulations.

5. Velocity-resolved aperture-synthesis imaging of RSGs

If we combine milliarcsecond-resolution aperture-synthesis imaging with high spectral resolution, we can have a data cube over individual molecular or atomic lines. Then it is possible to extract the spatially resolved spectrum at each spatial position from the data cube. The spatially resolved spectra of individual lines allow us to measure the gas velocity at each position over the surface of stars as well as in the atmosphere, just as routinely done in solar observations.

Ohnaka *et al.* (2009) detected the signatures of inhomogeneous gas motions in the atmosphere of Betelgeuse based on the interferometric data obtained over the 2.3 μm CO lines with the VLTI/AMBER instrument with a spectral resolution of up to 12000. Ohnaka *et al.* (2011) present 1-D aperture-synthesis images of Betelgeuse in the individual CO lines with an angular resolution of 9.8 mas. The spectral resolution was high enough to have $\sim$10 wavelength points within each CO line profile. The 1-D images reconstructed in the blue wing and the line center of the CO line profiles show an atmosphere asymmetrically extending to $\sim$1.3 stellar radii. However, the 1-D images in the red wing do not show the extended component. The observed different appearance of the star across the CO line profiles can be explained by an inhomogeneous velocity field in the extended atmosphere. For example, we assume that a gas clump in front of the star is moving toward the observer (i.e., moving upward with respect to the star), while the gas outside the clump is moving away from the observer (i.e., moving downward). The gas clump in front of the star produces blueshifted absorption, while we expect redshifted absorption from the gas outside the clump. The modeling of Ohnaka *et al.* (2009, 2011) suggests vigorous upwelling and downdrafting motions of a gas clump as large as the radius of the star at velocities of 10–30 km s^{-1}. The spatially resolved spectra over the extended atmosphere extracted from the reconstructed data cube show prominent emission, while those inside the limb of the star show absorption, which is exactly what we expect from the Kirchhoff's law (Ohnaka 2013). Ohnaka *et al.* (2013) succeeded in spatially resolving a similar, inhomogeneous velocity field in Antares based on VLTI/AMBER data. These studies demonstrate that it is now feasible to map the 2-D velocity field from the wavelength shifts of the spatially resolved spectra in a straightforward manner.

One may attribute the vigorous, inhomogeneous motions observed in Betelgeuse and Antares to large convective cells. However, Ohnaka *et al.* (2013) demonstrate that the density of the extended atmosphere at 1.3 stellar radii estimated from the observed data is $\sim 10^{-14}$ g cm^{-3}, which is higher than that predicted by the current 3D convection simulations (Chiavassa *et al.* 2010) by 6 to 11 orders of magnitude. This means that the atmosphere of Betelgeuse and Antares extending to 1.3–1.5 stellar radii cannot be explained by convection alone. Arroyo-Torres *et al.* (2015) suggest that radiation pressure on molecular lines may drive the mass loss in RSGs, although it is still necessary to construct self-consistent models and compare to the observed data.

6. Prospects

A next step is to apply the velocity-resolved imaging to different atomic and molecular lines forming at different atmospheric heights. This enables us to obtain a 3-D view of the atmospheric dynamics. The "tomographic velocity-resolved aperture-synthesis imaging" is crucial for understanding how the energy and momentum needed to accelerate the stellar winds are transferred from the deep photosphere to the outer atmosphere.

The second generation VLTI instruments GRAVITY (Eisenhauer *et al.* 2008) and MATISSE (Lopez *et al.* 2014) will be important for further probing the wind acceleration of RSGs. The GRAVITY instrument, which operates between 2 and 2.4 μm and combines four telescopes, will allow us to carry out aperture-synthesis imaging much more efficiently, although its spectral resolution of 4000 is three times lower than that of AMBER. The MATISSE instrument, which will allow us to carry out aperture-synthesis imaging in the thermal infrared (3–13 μm) for the first time, is essential for directly probing the dust formation region. The combination of these high-angular resolution observations will provide us with a comprehensive picture of the complex atmosphere and circumstellar envelope of RSGs and help us solve the long-standing problem of the mass loss.

References

Arroyo-Torres, B., Wittkowski, M., Chiavassa, A., *et al.* 2015, *A&A*, 575, A50
Baron, F., Monnier, J.D., Kiss, L.L., *et al.* 2014 *ApJ*, 785, 46
Chiavassa, A., Haubois, X., Young, J. S., *et al.* 2010, *A&A*, 515, A12
de Wit, W. J., Oudmaijer, R. D., Fujiyoshi, T., *et al.* 2008, *ApJ*, 685, L75
Eisenhauer, F., Perrin, G., Brandner, W., *et al.* 2008, *SPIE Proc.*, 7013, 70132A
Freytag, B. & Höfner, S. 2008, *A&A*, 483, 571
Harper, G. M. 2010, *ASP. Conf. Ser*, 425, 152
Haubois, X., Perrin, G., Lacour, S., *et al.* 2009, *A&A*, 508, 923
Kervella, P., Perrin, G., Chiavassa, A., *et al.* 2011, *A&A*, 531, A117
Kervella, P., Lagadec, E., Montargès, M., *et al.* 2016, *A&A*, 585, A28
Lopez, B., Lagarde, S., Jaffe, W., *et al.* 2014, *The Messenger*, 157, 5
Marsh, K. A., Bloemhof, E. E., Koerner, D. W., & Ressler, M. E. 2001, *ApJ* 548, 861
Monnier, J. D., Tuthill, P. G., Lopez, B., *et al.* 1999, *ApJ*, 512, 351
Monnier, J. D., Berger, J.-P., Le Bouquin, J.-B., *et al.* 2014, *SPIE Proc*, 9146, 91461Q
Montargès, M., Kervella, P., Perrin, G., *et al.* 2016, *A&A*, 588, A130
Ohnaka, K. 2013, *EAS Publications Series*, 60, 121
Ohnaka, K. 2014, *A&A*, 568, A17
Ohnaka, K., Driebe, T., Hofmann, K.-H., Weigelt, G., & Wittkowski, M. 2008, *A&A*, 484, 371
Ohnaka, K., Hofmann, K.-H., Benisty, M., *et al.* 2009, *A&A*, 503, 183
Ohnaka, K., Weigelt, G., Millour, F., Hofmann, K.-H., Driebe, T., Schertl, D., Chelli, A., Massi, F., Petrov, R., & Stee, Ph. 2011, *A&A*, 529, A163
Ohnaka, K., Hofmann, K.-H., Schertl, D., Weigelt, G., Baffa, C., Chelli, A., Petrov, R., & Robbe-Dubois, S. 2013, *A&A*, 555, A24
Ohnaka, K., Weigelt, G., & Hofmann, K.-H. 2016, *A&A*, 589, A91
Ohnaka, K., Weigelt, G., & Hofmann, K.-H. 2017, *A&A*, 597, A20
Schwarzschild, M. 1975, *ApJ*, 195, 137
Shenoy, D., Humphreys, R. M., Jones, T. J., *et al.* 2016, *AJ*, 151, 51
Smartt, S. J. 2015, *PASA*, 32, 16
Smith, N., Humphreys, R., Davidson, K., *et al.* 2001, *AJ*, 121, 1111
Smith, N., Hinkle, K. H., & Ryde, N. 2009, *AJ*, 137, 3558

The Lives and Death-Throes of Massive Stars
Proceedings IAU Symposium No. 329, 2016
J.J. Eldridge, J.C. Bray, L.A.S. McClelland & L. Xiao, eds.

doi:10.1017/S174392131700326X

The Young and the Massive: Stars at the upper end of the Initial Mass Function

Saida M. Caballero-Nieves[1,2] **and P. A. Crowther**[2]

[1]Department of Physics & Space Sciences, Florida Institute of Technology
150 West University Blvd, Melbourne, FL 32901, USA
email: scaballero@fit.edu

[2]Dept. of Physics & Astronomy, University of Sheffield, Hicks Building, Hounsfield RD
Sheffield S3 7RH, United Kingdom

Abstract. The upper mass limit of stars remains an open question in astrophysics. Here we discuss observations of the most massive stars (greater than 100 solar masses) in the local universe and how the observations fit in with theoretical predictions. In particular, the Large Magellanic Cloud plays host to numerous very massive stars, making it an ideal template to study the roles that environment, metallicity, and multiplicity play in the formation and evolution of the most massive stars. We will discuss the work that is instrumental in laying the groundwork for interpreting future observations by James Webb of starburst regions in the high redshift universe.

Keywords. stars: early-type, stars: mass function, stars: fundamental parameters (masses)

---

1. Introduction to Very Massive Stars

Very Massive Stars (VMS) are defined as stars born at the upper end of the Initial Mass Function (IMF). They have initial masses, M_i, greater than $100 M_{\odot}$. These stars are very rare but play an important role over a large range of scales in our Universe. Their strong stellar winds along with their subsequent supernova enrich the interstellar medium and contribute to the chemical evolution of galaxies. In the high redshift Universe, very massive stars dominate the light in unresolved clusters of starburst galaxies. Therefore, they play a crucial role in creating population synthesis models to fit the observations.

VMS provide an interesting observational challenge. The most common stellar population synthesis (Leitherer *et al.* 1999) and stellar evolutionary codes (Meynet & Maeder 2000) do not include stars with initial masses exceeding 100 or 120 $M_{\odot}$, respectively. This introduces inconsistencies in modelling starburst regions in the distant universe. In this paper, we discuss the observational evidence for stars that exceed 100 $M_{\odot}$ and make the case for the need to include such stars in future population synthesis codes.

For further reading, we recommend Martins (2015), an excellent review on the observational properties of VMS, Zinnecker & Yorke (2007), which remains a seminal review on the formation of massive stars, and references therein.

2. VMS in the Milky Way

The most obvious place to find VMS is at the core of young, dense and massive clusters. Weidner & Kroupa (2006) concluded from observations that there exists a well-defined relation between the most massive star in a cluster and the cluster mass. This means that the more massive a cluster, the more massive stars you can form. In addition, one needs to observe these clusters at early stages ($\lesssim$ 3 Myr) to successfully catch the most massive members before they evolve. In this section we provide highlights of Galactic VMS stars in clusters and in some cases, more isolated regions.

2.1. *VMS in Cluster cores*

The starburst cluster NGC 3603 is a HII region, rich in high mass stars. The very young age (1-2 Myr; Melena *et al.* 2008), based on the presence of hydrogen-rich and nitrogen-rich main sequence Wolf-Rayet stars (WNh) along with a young O star population, makes it an ideal cluster to search for VMS. It contains two VMS in the core: A1 and B. In the case of A1, it is a binary where the two components have masses of 116 $M_\odot$ + 89 $M_\odot$ (Schnurr *et al.* 2008) and are among the most massive stars dynamically weighed.

The massive open cluster, Arches Cluster, was only discovered in the mid-nineties (Cotera 1995, Figer 1995, Nagata *et al.* 1995) due to high foreground extinction towards the Galactic Center. Based on the cluster mass estimated of $\sim 4{-}6 \times 10^4 M_\odot$ estimated by Figer (2005)), the Arches cluster has a high enough mass content to form VMS. They also determined from the IMF an upper mass limit cutoff of 150 $M_\odot$. Oey & Clarke (2005) arrived to a similar conclusion using a statistical approach with a sample of clusters and OB associations. However, both of these results are limited by the extinction laws used (as is the case for the Arches Cluster) and the slope of the initial mass function (IMF) for stars greater than 10 $M_\odot$. In this high mass range, radiation pressure dominates with a shallower mass-luminosity relationship, such that for $\mathrm{L} \propto \mathrm{M}^\alpha$, α tends towards unity as M tends towards infinity. Zero age main sequence VMS approach the Eddington limit, but do not initially exceed it.

Like NGC 3603, the most massive members of the Arches cluster are found in its core. The spectroscopy study by Martins *et al.* (2008) estimated that the WNh population in the core of the Arches were around $120 M_\odot$. The revised interpretation by Crowther *et al.* 2010 using AO-based VLT photometry, found that two stars, F6 and F9, have initial masses of 185 $M_\odot$ and 165 $M_\odot$, respectively, exceeding the $150 M_\odot$ upper limit. Schneider *et al.* (2014) attribute the WNh stars to binary stellar mergers, giving an older age estimate of $\sim$ 3.5 Myr than that of NGC 3603. The complicated nature of the Arches Cluster do not make it the ideal candidate to study the stellar upper mass limit cut off.

2.2. *VMS in large star forming regions*

The prior two cases we looked at considered VMS in cluster environments. Here we note at least two cases of known VMS stars outside density cluster cores in intermediate dense regions of major massive star forming regions. This implies that you can make stars with $\mathrm{M}_{ini} > 100 M_\odot$ outside dense cluster cores.

One of the most enigmatic stars in the Milky Way is η Carinae, a Luminous Blue Variable (LBV). It has gone through two eruptive mass loss events in the last two hundred years. It is believed to have $\mathrm{M}_{current} \sim 120 M_\odot$ (Hillier *et al.* 2010), but lost $\sim 20 M_\odot$ in the last 170 years (Morris *et al.* 1999), falling into the category of VMS. It is also part of a binary system that is likely to play a role in the eruption events. η Car is usually associated with the dense star cluster Trumpler 16. However, unlike the VMS discussed previously typically found in the cluster center, it sits on the periphery of the cluster. This interesting star is discussed further in this proceedings by A. Damineli, K. Hamaguchi, M. Kiminki, T. Madura, S. Owocki and N. Smith.

In the nearby OB association, Cyg OB2, dwells a peculiar B hypergiant. Thought to be a quiescent LBV or a precursor to an LBV, Cyg OB2-12 is estimated to be a 120 $M_\odot$ star (Clark *et al.* 2012), with a high X-ray luminosity indicating that it is in a colliding wind system. The history of the association from dynamical studies (Wright *et al.* 2014) and the lack of mass segregation (Parker & Wright 2016), indicates that Cyg OB2 has

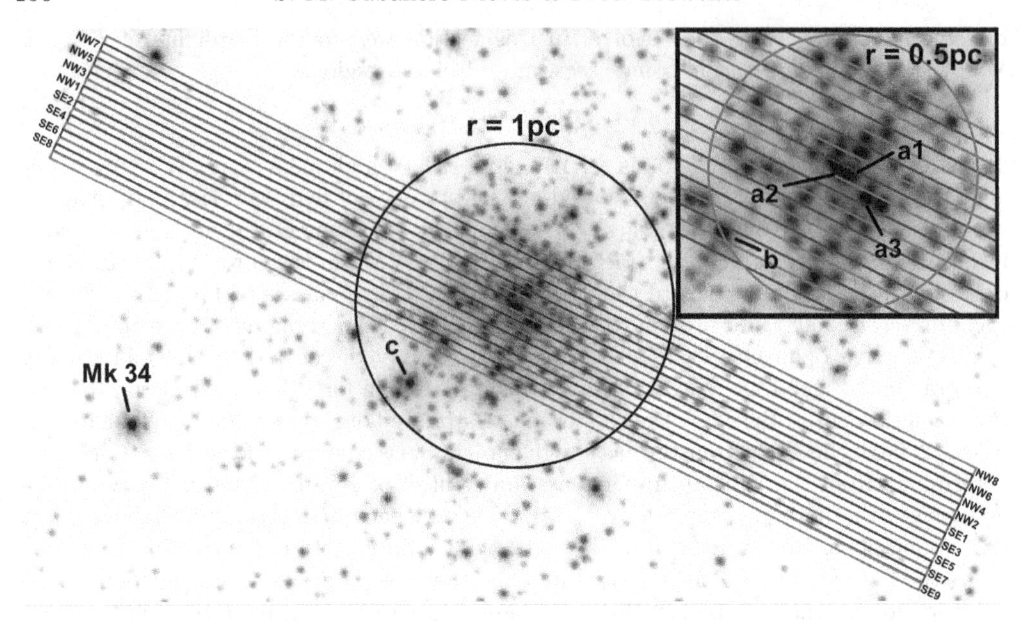

Figure 1. HST/STIS slits (52"×0".2) superimposed upon an F336W WFC3/UVIS image of R136 (de Marchi *et al.* 2011), oriented with North up and East to left, together with a circle of radius 4."1 (1 parsec) and identification of the acquisition star Melnick 34. The active slit length for MAMA observations is the central 25". The zoom highlights the central region, including identification of individual slits and the integrated R136a cluster (2."05 radius circle, centered upon R136a1 equivalent to 0.5 parsec at the distance of the LMC). Adapted from Crowther *et al.* (2016).

always been a relatively low density environment. Thus, high densities are not necessary to form VMS, but make it more likely to find VMS.

3. VMS in the LMC: R136

The star forming region 30 Doradus, or the Tarantula nebula, in the Large Magellanic Cloud (LMC) is the brightest star forming region in the Local Group. It is rich in VMS with $\sim$ 20 stars scattered throughout the region (Table 8; Crowther *et al.* 2016). The region is a conglomeration of different bursts of star formation and the stellar content cannot be treated as being coeval.

At the core of the nebula sits the starburst cluster R136, once thought to be a supermassive star (2,500 $M_{\odot}$; Cassinelli *et al.* 1981). It was finally resolved as a cluster with interferometry (Weigelt *et al.* 1985) and Hubble imaging (Hunter *et al.* 1995). R136 is ideally situated compared to Milky Way clusters, with little extinction along the line of sight and resolvable. The core is the source of feedback to the larger region through the ultraviolet (UV) radiation of its most massive members. It serves as a resolved template of starburst regions in distant galaxies.

The central parsec of R136 has a long standing history of being difficult to resolve its individual members. However it is crucial to understand both the integrated properties of the cluster itself and the individual members. R136 contains several VMS stars, among them the brightest is thought to be the most massive star known, R136a1 ($M_{ini} \sim 300 M_{\odot}$; Crowther *et al.* 2010). The mass of the R136a1 is still a subject of uncertainty (see M. M. Díez, this proceedings), and hence the stellar upper mass limit remains an open question.

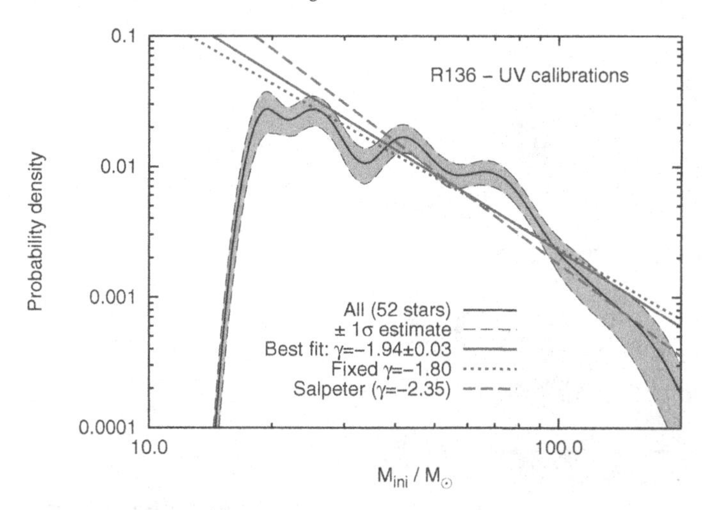

Figure 2. Initial mass function based on the UV derived parameters for the central stars in R136 using the bayesian tool BONNSAI (Schneider *et al.* 2014). The best fit IMF of R136 has a shallower slope for the most massive stars than the standard Salpeter slope. (F. Schneider, private communication).

Crowther *et al.* (2016) use Hubble STIS to create an Integral Field Spectrum (IFS) of the entire central parsec to determine both the individual properties of the stellar content along with the combined cluster properties (see Figure 1). Here we discuss the results and some of the implications from this study.

Massive stars are well known to love company, and are often found with at least one companion that will influence their evolution (Sana *et al.* 2012). In preliminary results of the multiplicity of the O-star sample in R136, we find that about 35% of them show a significant radial velocity variation that could be due to a companion. It therefore follows that these VMS could be found in binary systems too. For mass estimates, we have to ascertain that we are accounting for the correct number of stars in the luminosity we measure. We consider specifically R136a1 and the possibility that it is a binary star at different scales. To really affect the mass estimates, we are looking for two near equal mass stars, or near equal brightness, that are very close together. Schnurr *et al.* (2009) carried out a spectroscopic study and determined that in the period range they considered, R136a1 appeared single. It showed no radial velocity variations that could be attributed to a companion with $P \lesssim 44$ days. They found that R136c did seem to show radial velocity variations. This is consistent with the high X-ray luminosity observed for this star.

The X-ray properties of the central stars in R136 can also indicate binarity. High X-ray fluxes can be attributed to colliding winds, i.e. two stars with strong stellar winds causing a shock region where the winds meet, emitting X-rays. Crowther *et al.* (2010) found that the X-ray flux from the center of R136 precluded pairs of equal mass VMS in close binary orbits within a few hundred AU.

To explore the regions at wider separations where the colliding winds will not create a bright X-ray shock region or show an radial velocity variability, we need high spatial resolution observations. Hubble's interfertometer, the Fine Guidance Sensor is the best instrument to further constrain the parameter space for determining binarity of these systems. Figure 3 shows the raw interferometric signal of the region, with each 'S'

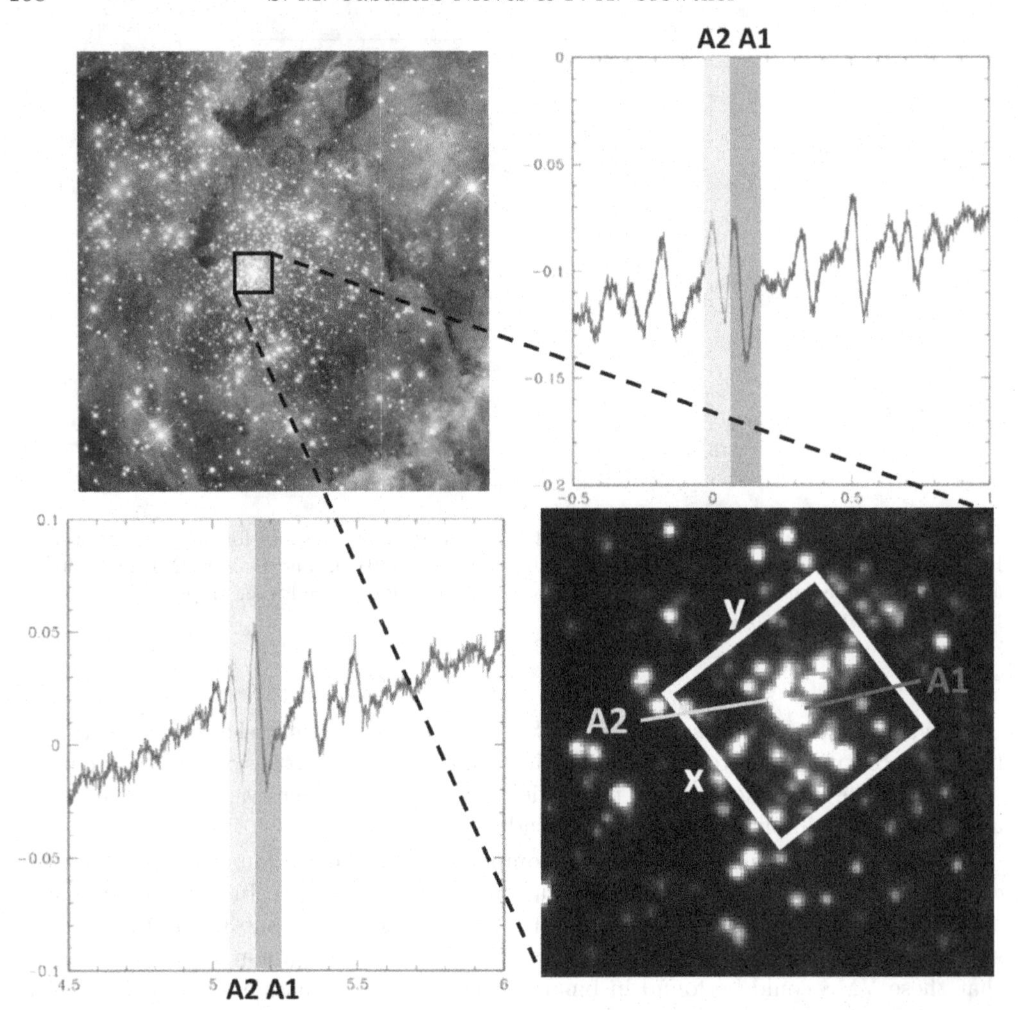

Figure 3. The super star cluster R136 at the center of 30 Doradus (top left). The cluster contains the 9 VMS, including R136a1 the most massive star known (bottom right). FGS scans in 2 perpendicular directions (top right & bottom left) allows for an unprecedented close up of this dense region.

corresponding to a star in the field. The density of the region will make disentangling the signal for stars at the center of R136 exceptionally challenging.

Crowther *et al.* (2016) provide a complete census of the most massive stars in the core of R136. Figure 2 (F. Schneider, private communication) shows the initial mass function based on the UV derived parameters for the central stars in R136 using the bayesian tool BONNSAI (Schneider *et al.* 2014). This very preliminary result hints at the possibility that the IMF of R136 has a shallower slope for the most massive stars than the standard Salpeter slope.

4. VMS in the high redshift Universe

From Crowther *et al.* (2016) we find that a handful of VMS contribute significantly to the integrated UV spectrum of the entire cluster. In 30 Doradus, $\sim$ 30 VMS contribute significantly to the total ionizing output (Doran *et al.* 2013). A third of those VMS are found in the center of R136. Of course, the rest-frame UV spectra of star forming galaxies in the early universe are redshifted to the near-infrared where JWST is exceptionally sensitive. Therefore, to interpret the rest-frame UV spectrum of distant starbursts, the presence of VMS needs to be considered (see Crowther *et al.*, these proc). Indeed, some stellar population synthesis codes (e.g. BPASS; Stanway *et al.* 2016) now account for VMS and close binary evolution.

References

Cassinelli, J. P., Mathis, J. S., & Savage, B. D. 1981, *Science*, 212, 1497
Clark, J. S., Najarro, F., Negueruela, I., *et al.* 2012, *A&Ap*, 541, A145
Cotera, A. S. 1995, Ph.D. Thesis
Crowther, P. A., Schnurr, O., Hirschi, R., *et al.* 2010, *MNRAS*, 408, 731
Crowther, P. A., Caballero-Nieves, S. M., Bostroem, K. A., *et al.* 2016, *MNRAS*, 458, 624
De Marchi, G., Paresce, F., Panagia, N., *et al.* 2011, *ApJ*, 739, 27
Doran, E. I., Crowther, P. A., de Koter, A., *et al.* 2013, *A&Ap*, 558, A134
Figer, D. F. 1995, Ph.D. Thesis
Figer, D. F. 2005, *Nature*, 434, 192
Hillier, D. J., Davidson, K., Ishibashi, K., & Gull, T. 2001, *ApJ*, 553, 837
Hunter, D. A., Shaya, E. J., Holtzman, J. A., *et al.* 1995, *ApJ*, 448, 179
Köhler, K., Langer, N., de Koter, A., *et al.* 2015, *A&Ap*, 573, A71
Leitherer, C., Schaerer, D., Goldader, J. D., *et al.* 1999, *ApJS*, 123, 3
Martins, F., Hillier, D. J., Paumard, T., *et al.* 2008, *A&Ap*, 478, 219
Martins, F. 2015, Very Massive Stars in the Local Universe, 412, 9
Melena, N. W., Massey, P., Morrell, N. I., & Zangari, A. M. 2008, *AJ*, 135, 878
Meynet, G. & Maeder, A. 2000, *A&Ap*, 361, 101
Morris, P. W., Waters, L. B. F. M., Barlow, M. J., *et al.* 1999, *Nature*, 402, 502
Nagata, T., Woodward, C. E., Shure, M., & Kobayashi, N. 1995, *AJ*, 109, 1676
Parker, R. J. & Wright, N. J. 2016, *MNRAS*, 457, 3430
Oey, M. S. & Clarke, C. J. 2005, *ApJL*, 620, L43
Sana, H., de Mink, S. E., de Koter, A., *et al.* 2012, *Science*, 337, 444
Schneider, F. R. N., Izzard, R. G., de Mink, S. E., *et al.* 2014, *ApJ*, 780, 117
Schneider, F. R. N., Langer, N., de Koter, A., *et al.* 2014, *A&Ap*, 570, A66
Schnurr, O., Casoli, J., Chené, A.-N., Moffat, A. F. J., & St-Louis, N. 2008, *MNRAS*, 389, L38
Schnurr, O., Chené, A.-N., Casoli, J., Moffat, A. F. J., & St-Louis, N. 2009, *MNRAS*, 397, 2049
Stanway, E. R., Eldridge, J. J., & Becker, G. D. 2016, *MNRAS*, 456, 485
Weigelt, G. & Baier, G. 1985, *A&Ap*, 150, L18
Weidner, C. & Kroupa, P. 2006, *MNRAS*, 365, 1333
Wright, N. J., Parker, R. J., Goodwin, S. P., & Drake, J. J. 2014, *MNRAS*, 438, 639
Yusof, N., Hirschi, R., Meynet, G., *et al.* 2013, *MNRAS*, 433, 1114
Zinnecker, H. & Yorke, H. W. 2007, *ARAA*, 45, 481

The Lives and Death-Throes of Massive Stars
Proceedings IAU Symposium No. 329, 2016
J.J. Eldridge, J.C. Bray, L.A.S. McClelland & L. Xiao, eds.

doi:10.1017/S1743921317003209

The multiplicity of massive stars: a 2016 view

Hugues Sana

Institute of Astrophysics, KU Leuven,
Celestijnlaan 200D, 3001 Leuven, Belgium
email: hugues.sana@kuleuven.be

Abstract. Massive stars like company. Here, we provide a brief overview of progresses made over the last 5 years by a number of medium and large surveys. These results provide new insights on the observed and intrinsic multiplicity properties of main sequence massive stars and on the initial conditions for their future evolution. They also bring new interesting constraints on the outcome of the massive star formation process.

Keywords. stars: early-type – binaries (including multiple): close – binaries: spectroscopic – stars: individual (VFTS352, RMC145) – binaries: general

1. Introduction

The presence of close companions introduces new physics that has the potential to affect all stages of stellar evolution, from the pre-main sequence phase, to the type of end-of-life explosions and to the properties and orbital evolution of double-compact objects that may ultimately lead to detectable gravitational wave bursts. In this short paper, we highlight major observational advances that have occurred in the last five years, with no ambition to provide a complete historical review. More information on works published prior to 2012 can be found in e.g., Mason et al. (2009), Sana & Evans (2011) and references therein. As in our previous report, we will focus on the initial multiplicity conditions, i.e. that of main-sequence massive stars.

As described by, e.g., Mason et al. (1998), Sana & Evans (2011) and Moe & Di Stefano (2016), four or five orders of magnitudes in separation need to be investigated to cover the entire parameter space relevant for the formation and evolution of massive stars (Fig. 1). This can only be achieved by a combination of instrumental techniques that all have their own regime of optimal sensitivities (masses, brightnesses, separation, flux contrasts, mass-ratios, ...). Recent advances is long-baseline interferometry – mostly the magnitude limit and efficiency of the observations (see Sect. 3) – have allowed us to bridge the spectroscopic and imaging regimes (the so-called interferometric gap, Mason et al. (1998)). However, this is currently only possible for distances of up to ~4 kpc as, farther away, most massive stars become too dim for current interferometric instrumentation.

In this short overview, we will focus on the spectroscopic (Sect.2) and interferometric (Sect. 3) regimes, with emphasis on surveys that have not been discussed elsewhere in these proceedings (see the contributions of Barbá, Simón-Díaz and Vink). We will follow the terminology adopted in Sana et al. (2014):

- **The number of observed targets** or **sample size** (N),

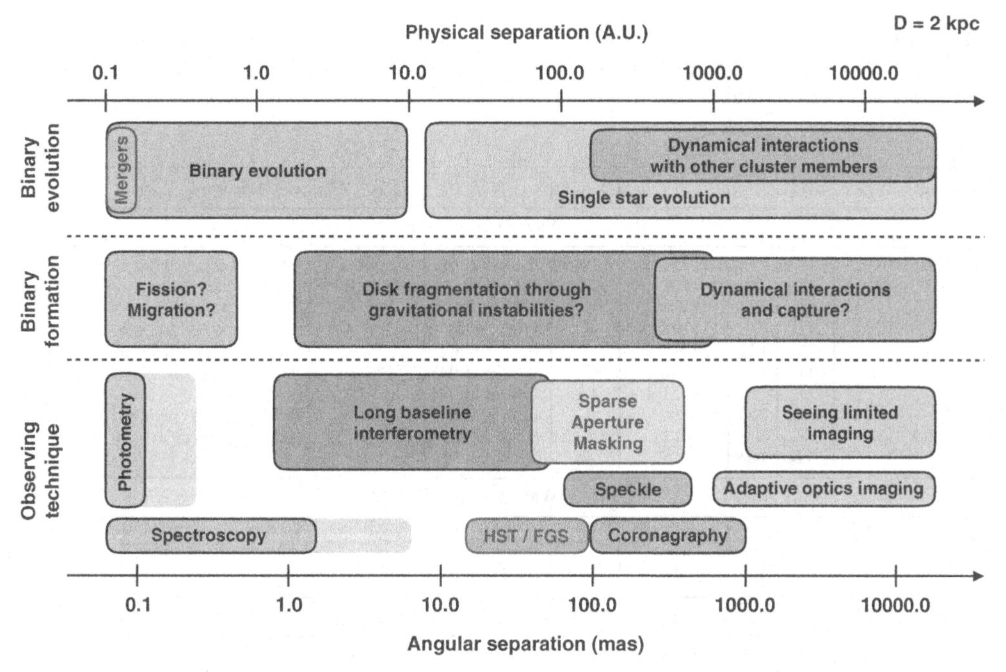

Figure 1. Sketch of the typical physical and angular separations covered by O-type binaries. A distance of 2 kpc is adopted, as representative of most galactic multiplicity surveys. Domains of sensitivity for various instrumental techniques and relevant parameter space of formation and evolution scenarios are indicated. Lightly shaded areas indicate regions where spectroscopy and photometry are in principle sensitive but where the detection likelihood drops dramatically.

- **The number of multiple systems** ($N_{\rm m}$): number of observed central objects with at least one companion,
- **The fraction of multiple systems** or **multiplicity fraction**: $f_{\rm m} = N_{\rm m}/N$,
- **The number of companions** ($N_{\rm c}$): total number of companions observed around a given sample of N central objects,
- **The fraction of companions** or **companion fraction** ($f_{\rm c} = N_{\rm c}/N$): average number of companions per central object.

Depending on the context, these quantities can be restricted to sub-categories, such as spectroscopic binaries systems (SBs) (Sect. 2) or to specific separation ranges (Sect. 3).

2. Spectroscopic surveys

2.1. *Galactic surveys*

In the last few years, two kinds of Galactic surveys have brought our understanding of massive star multiplicity to a new level:

(i) surveys focusing on specific young clusters (Sana et al. (2012)) or OB association (Cyg OB2; Kiminki & Kobulnicky (2012); Kobulnicky et al. (2014)). These surveys have acquired a large number of observational epochs per target, allowing them to obtain orbital solutions for most of the detected spectroscopic binaries in their sample. They can further be considered as volume limited surveys in their own targeted regions.

Table 1. Overview of recent spectroscopic surveys and their observed ($f_{\rm SB}^{\rm obs}$) and bias-corrected ($f_{\rm SB}^{\rm corrected}$) binary fraction. The sample size (N), number of detected binaries ($N_{\rm SB}$) and number of binaries with an orbital solution ($N_{\rm SBO}$) are also given. Error bars are computed following binomial statistics.

Region/Survey	SpT	N	$N_{\rm SB}$	$N_{\rm SBO}$	$f_{\rm SB}^{\rm obs}$	$f_{\rm SB}^{\rm corrected}$	Reference
Milky Way: O-type stars[a]							
BESO Survey	O	243	166	–	0.68 ± 0.03	–	Chini et al. (2012)
Young Clusters	O	71	40	34	0.56 ± 0.06	0.69 ± 0.09	Sana et al. (2012)
GOSSS	O	194	97 (low)	–	0.50 ± 0.03	–	Sota et al. (2014)
	O		117 (high)	–	0.60 ± 0.03	–	
IACOB[b]	O	141	66	–	0.47 ± 0.04	–	Simón-Díaz (these proceedings)
Clusters/Assoc	O	161	68 (low)	68	0.42 ± 0.04	–	Aldoretta et al. (2015)
	O		96 (high)	68	0.60 ± 0.04	–	
OWN	O	205	102	85	0.50 ± 0.03	–	Barbá (these proceedings)
Cyg OB2	O	45	23	23	0.51 ± 0.07	–[c]	Kobulnicky et al. (2014)
	OB	*128*	*48*	*48*	*0.38 ± 0.04*	*0.55*[c]	
Milky Way: B-type stars							
Cyg OB2	B0-2	83	25	25	0.30 ± 0.05	–[c]	Kobulnicky et al. (2014)
BESO Survey	B0-3	226	105	–	0.46 ± 0.03	–	Chini et al. (2012)
	B4-9	353	67	–	0.19 ± 0.02	–	
Large Magellanic Cloud							
30Dor/VFTS	O	360	124	–	0.35 ± 0.03	0.51 ± 0.03	Sana et al. (2013)
30Dor/TMBM	O			79	–	≈ 0.58	Almeida et al. (2017)
30Dor/VFTS	B0-3	408	102	–	0.25 ± 0.02	0.58 ± 0.11	Dunstall et al. (2015)

Notes: [a]The O-type samples in the Milky have strong overlap, so that the quoted measurements should not be viewed as being independent. [b]Numbers only concerns O stars with more than 3 observational epochs. The IACOB survey covers another ~ 40 Northern O stars, but with fewer epochs so far. [c]Kobulnicky et al. (2014) do not provide separate completeness correction for the O- and B-type sub-samples in Cyg OB2.

(ii) Galaxy-wide surveys with various spectral resolving power ($R = \lambda/\delta\lambda$) and number of epochs, and which are mostly magnitude limited, e.g.:

- **The Galactic O-Star Spectroscopic Survey** (GOSSS; Sota et al. (2011, 2014), Maíz Apellániz et al. (2016)): a (mostly) single-epoch survey at $R \approx 2500$, which main objective is to bring to a firm ground the spectral classification of galactic O stars down to $B = 13$;

- **The BESO Survey** of Galactic O and B stars (Chini et al. (2012)): a $R = 50\,000$ multi-epoch campaign that has allowed to search for radial-velocity (RV) variations in the observed sample;

- **The IACOB survey** (Simón-Díaz *et al.* (2015) and references therein, see also these proceedings): multi-epoch high-resolution spectroscopic survey of Northern Galactic O- and B-type stars active since 2008;

- **The OWN survey** (Barbá et al. (2014), see also Barbá in these proceedings): a high-resolution spectroscopic monitoring of Southern O and WN stars with enough multi-epoch observations to measure the orbital properties of most detected binaries.

2.2. *The Tarantula region*

Simultaneously to these groundbreaking observational campaigns in the Milky Way, the **VLT-Flames Tarantula Survey** (VFTS, Evans et al. (2011); see Vink *et al.*, these proceedings) has systematically investigated about 800 massive OB and WR stars in the 30 Dor region in the Large Magellanic Cloud. Using 6-epochs spectroscopic at moderate resolving power $R \sim 7000$, the VFTS – and its sequel program, the **Tarantula Massive Binary Monitoring** (TMBM, Almeida et al. (2017)) — has provided direct observational constraints on the massive stars multiplicity properties in a metallicity environment that is reminiscent of that of more distant galaxies at $z \sim 1 - 2$.

Recent results of the VFTS are described in Vink *et al.* (these proceedings). Here we provide more insight into the TMBM project. TMBM was designed to obtain multi-epoch spectroscopy of O-type binaries in the 30 Dor region with sufficient time sampling to measure the orbital properties of objects with orbital periods of up to about one year. TMBM targeted 60% of every O-type binary candidates detected in the VFTS (see Sana et al. (2013)) – i.e., 93 out of 152 – and obtained orbital solution for 78 of them. While specific objects are still being analysed, results so far have allowed to identify two very interesting binaries, for which (quasi)-chemically homogeneous evolution ((q)CHE) seems to be required to explain their observed properties. In both cases, (q)CHE may affect the binary evolutionary path, by preventing the merging (VFTS 352) or Roche lobe overflow (R145):

- **VFTS 352** is an overcontact 29+29 $M_{\odot}$ short period system ($P_{\rm orb} = 1.12$ d) that shows evidence of enhanced internal mixing possibly putting both components on a (q)CHE track (Almeida *et al.* (2015)). Ongoing abundance measurements may help to validate the existence of such an alternative scenario (Mandel & de Mink (2016)) to produce massive black holes with properties similar to that at the origin of the recent gravitational wave detection (Abbott et al. (2016), Mandel & de Mink (2016)).

- **RMC 145 (aka R145)** is a WN6h + O3.5If*/WN7 system with a high eccentricity ($e = 0.78$) and rather low inclination ($i = 39 \pm 6^{\rm o}$). Once suggested to contain a 300 $M_{\odot}$ star, Shenar et al. (2017) showed that the current masses were most likely of the order of 80 $M_{\odot}$. Best estimates for initial masses yield values of 105 and 90 $M_{\odot}$ for the primary and secondary component respectively. Comparison with evolution tracks and the system high eccentricity suggest (q)CHE, allowing the system to have avoided episodes of mass-transfer so far.

2.3. *The spectroscopic binary fraction*

The previously described spectroscopic surveys of O and B stars, which have samples of tens to hundreds of stars, yield detected spectroscopic binary fractions ($f_{\rm SB}^{\rm obs}$) that range from $f_{\rm SB}^{\rm obs} \sim 0.25$ for the B-type stars in the 30 Dor region to about $f_{\rm SB}^{\rm obs} \sim 0.7$ for O stars in the Milky Way (see Table 1). Much of this range can be explained by the nature and quality of the data (signal-to-noise ratio, number of epochs, ...) that directly impact the RV accuracy achieved by these surveys, hence the probability to detect significant Doppler shifts. Overall, all galactic O-star surveys yield a detected spectroscopic binary fraction $f_{\rm SB}^{\rm obs} > 0.5$, setting a firm lower limit on the intrinsic multiplicity fraction. The most optimistic estimates even reach $f_{\rm SB}^{\rm obs} \sim 0.6$ to 0.7 before any bias correction.

The overall binary detection probability of these surveys are difficult to estimate because it requires the knowledge of the parent orbital distributions of the targeted sample – which are of course ill-constrained – and a model of the sensitivity of each survey. Recent estimates yield an overall detection probability of spectroscopic binaries with $P_{\rm orb}$ up to ≈ 10 years of the order of 0.65 to 0.8 for O-type stars (Sana et al. (2012, 2013); Kobulnicky et al. (2014)). Corresponding intrinsic binary fractions are then in the range of $f_{\rm SB}^{\rm corrected} \approx 0.5$ to 0.7. It remains to be seen whether the differences observed between the Milky Way and 30 Dor surveys are indeed genuine, possibly resulting from environmental or from stellar (binary) evolution effects.

For B-type stars, $f_{\rm SB}^{\rm obs}$ rather lays in the range of 0.30 to 0.46. For these stars, the detection rate is expected to be lower owing to fainter objects and smaller Doppler

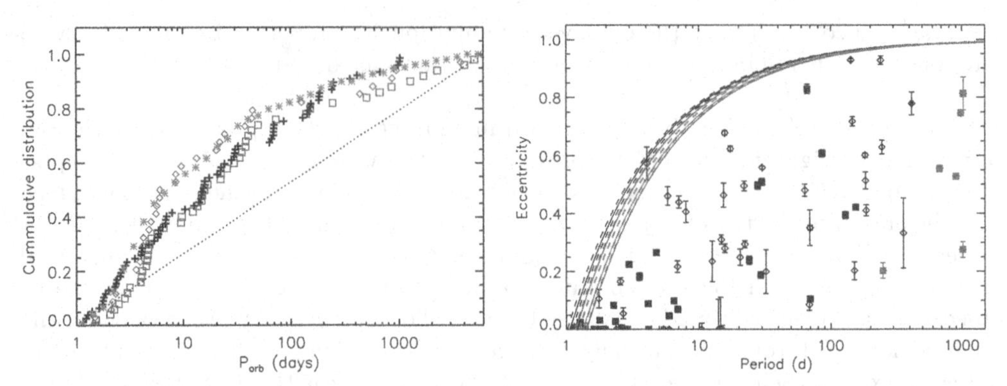

Figure 2. Left. Observed cumulative distributions of the orbital periods from different surveys: the young clusters (Sana et al. (2012), red diamonds), Cyg OB2 (Kobulnicky et al. (2014), blue squares), 30 Dor/TMBM (Almeida et al. (2017), black crosses) and OWN (Barbá, these proceedings; private communication of the period histogram). The dotted line indicates a flat distribution in $\log_{10} P_{\rm orb}$. Right. Eccentricity vs. period diagram for the TMBM O-star sample. Filled and open symbols indicates SB2 and SB1 systems, respectively. The red hexagons are for TMBM low-priority targets, i.e. stars showing ΔRV$<$ 20 km s^{-1} in VFTS. Dotted lines show the locus at which the two stars would touch at periastron for a representative range of stellar-masses and mass-ratios. Eccentricity measurement for the SB1 system at $P \sim 4$ d and $e \sim 0.6$ is biased and the data point should be neglected (see Almeida et al. (2017)).

shifts, on average, resulting from smaller masses involved. Detailed considerations of the observational biases are needed to better quantify the intrinsic B-type star multiplicity fraction. Interestingly, measurements in the 30 Dor region show that, once biases are taken into account, the O- and the B-type binary fractions are compatible within errors, with both mass ranges yielding $f_{\rm SB}^{\rm corrected} \approx 0.58$ (admittedly errors remain quite large for B stars, see Table 1).

2.4. *The orbital periods*

The initial orbital periods – and their overall distribution – are one of the most important quantities to properly predict the pre-supernova evolution of binary systems. Indeed, $P_{\rm orb}$ is the prime parameter that defines whether a system will interact through mass exchange. It sets the timing of the interaction and, to a large degree, its nature. Currently, only a few surveys have gathered enough observational epochs on a sufficiently large sample of binaries to construct observed period distributions. Constraints on the period distribution can in principle be obtained from modelling the distribution of the RV-variations (e.g., Sana et al. (2013); Dunstall et al. (2015)), but this requires assumptions on, e.g., the range of orbital periods (see discussion in Almeida et al. (2017)).

So far, three published surveys bring direct constraints on the orbital period distributions. These are displayed in Fig. 2 and show a large degree of similarities, despite the rather different environments in terms of stellar density and metallicity. Further more, preliminary results of the OWN survey (Barbá, these proceedings) confirm and improve the statistics for Galactic massive stars. Future work will allow to investigate whether the small differences seen in Fig. 2 are statistically significant or whether they result from different observational biases, sample sizes and selection effects. In the former case, these may hold clues to the formation process and early dynamical evolution of massive stars as suggested by, e.g., Kobulnicky et al. (2014).

The systematic and long-term nature of the OWN and VFTS/TMBM surveys have allowed to detect and characterise an unprecedented number of long period systems. Thanks to these efforts, the period-eccentricity diagram, once showing a dearth of

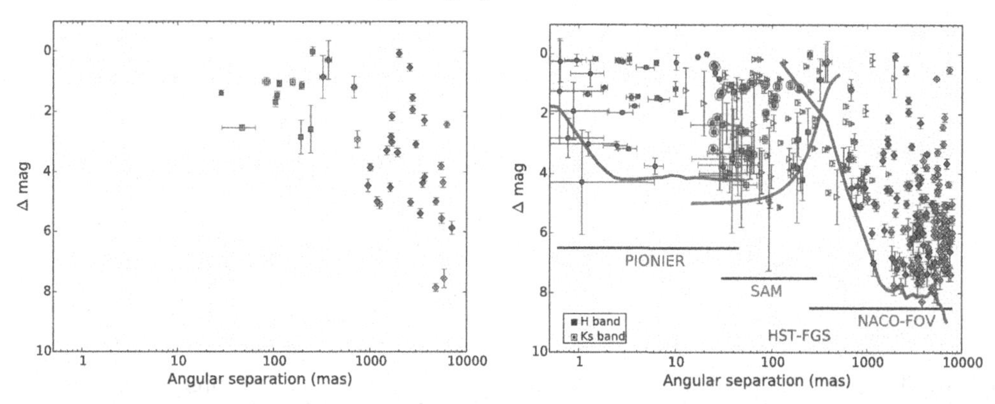

Figure 3. Left. Companions to the SMASH+ targets resolved before 2011. Right. Companions detected by the SMASH+ Survey (PIONIER: blue, NACO-SAM: green, NACO-FOV: red) in the H- (filled symbols) and Ks-bands (dotted-open symbols), and with the HST-FGS (purple) in the V-band (open symbols). Plain lines show the median sensitivity limits of the different legs of the SMASH+ survey. Companions detected both by the SMASH+ and the HST-FGS surveys appear twice in the figure. Based on data from Sana et al. (2014) and Aldoretta et al. (2015).

long-period low-eccentricity systems (see e.g., Sana et al. (2012)), is now well populated (Fig. 2 for TMBM, see also Barbá, these proceedings). The presence of these systems argues against dynamical capture and suggests that most of these spectroscopic binaries have been created during the formation process.

3. High-angular resolution surveys

Two important Galactic surveys have allowed to systematically bridge the interferometric gap that was separating the spectroscopic and imaging domains (Mason et al. (1998), see also Fig. 1). Both surveys show that the interferometric gap contains a significant number of companions and confirm previous suspicions that *the end product of massive star formation is a multiple system.*

- **The Southern MAssive Star at High angular resolution** (SMASH+, Sana et al. (2014)) is an ESO Large Program that targeted O-type stars in the Southern sky ($\delta < 0^o$). SMASH+ combined long baseline interferometry (VLTI/PIONIER; 117 objects), aperture masking (VLT/NACO-SAM; 162 objects) and adaptive optics (VLT/NACO-FOV; 162 objects) to search for companions with separations in the range of 1 to 8000 mas. 264 companions were detected, of which almost 200 were previously unresolved. After excluding runaway stars that show no resolved companions, 55% of the 96 targets observed with the full suite of instruments revealed a companion in the range of 1 to 200 mas. Combined with existing spectroscopy (see Sect. 2), the SMASH+ results yield a multiplicity fraction at angular separation $\rho < 8"$ of $f_{\rm m} = 0.90 \pm 0.03$ and an averaged number of companions of $f_{\rm c} = 2.1 \pm 0.2$. Interestingly, all dwarfs stars in the SMASH+ sample have a bound companion within ~ 100 AU.

- **The HST fine guidance sensor survey** (HST-FGS, Aldoretta et al. (2015)) observed 224 O- and B-type stars in both hemispheres. The survey allowed to resolved 58 multiple systems (incl. 43 new detections) with typical separation in the range of ≈ 20 to 1000 mas. Focusing on the 214 stars in clusters and associations, the HST-FGS resolved a companion for 31% of them. The authors performed their own literature review (see

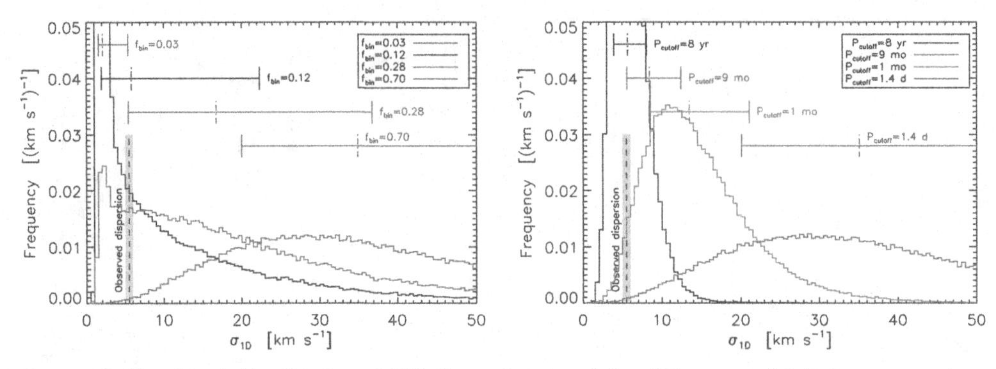

Figure 4. Simulated distribution of RV-dispersion (σ_{1D}) for different multiplicity properties of the parent populations. Left. The binary fraction is varied. Right. The minimum orbital period (P_{cutoff}) in the period distribution is varied. The shaded area (yellow) shows the RV-dispersion observed in the Ramírez-Tannus *et al.* sample. Figures adapted from Sana *et al.* (2017).

also Table 1) and obtained $0.5 < f_m < 0.7$ and $0.7 < f_c = 1.7$. The HST-FGS survey nicely covers the sensitivity gap between the NACO-SAM and FOV techniques in the SMASH+ survey and shows that companions are also found at these separations.

4. News from M17: clues to the origin of massive binaries?

Current constraints on the multiplicity of massive stars show that about half of them belong to binary systems with orbital periods of less than one month. This corresponds to separations below $\sim$ 1 AU. The existence of such tight binaries is currently not predicted by massive star formation theories. In this context, the recent Xshooter spectroscopic campaign towards a dozen young massive stars and mYSOs in the 1 Myr-old M17 region by Ramírez-Tannus *et al.* (Ramírez-Tannus *et al.* (2017), see also these proceedings) is worth mentioning. Most objects were observed only once. While no double-lined profiles were detected, the sample shows an intriguingly low RV-dispersion (σ_{1D}), of the order of 5.5 km s^{-1} (Fig. 4) only while values of 20 to 40 km s^{-1} would have been expected for the typical multiplicity properties.

Subsequent Monte-Carlo analysis confirms that such a low σ_{1D} is indeed incompatible with the multiplicity properties of well characterised massive star populations such as those presented in Table 1. Our results point either to a very low binary fraction ($f_{SB} \sim 0.12$ only) or to a lack of short period binaries ($P_{cutoff} > 9$ months). The first scenario would require to invoke that star formation has had a different outcome in M17 than in other OB-star regions in the Milky Way. The latter scenario may suggest that binaries first form at larger separations and then quickly harden to meet the observational constraints derived from slightly older populations (Table 1). Such migration scenario may be driven by interaction with other companions in a multiple systems, or with remnant of the accretion disk. The cut-off period may then be related to physical length-scales representative of the bloated pre-main-sequence stellar radii or of their accretion disks. The full discussion is to be found in Ramírez-Tannus *et al.* (2017) and Sana *et al.* (2017).

5. Final word

The results obtained in the last 5 years have definitely established the importance of massive star multiplicity. New large surveys have resulted in unprecedentedly high multiplicity and companion fraction, with values as large as $f_m > 0.9$ and $f_c > 2$ being obtained before bias-correction when considering the entire range of separations. Of particular interest is the relative similarities of the multiplicity properties measured so far for different samples, despite different environments, sample ages and even metallicities. The one exception is the results from the M17 region (see Sect. 4). More work is definitely required to further explore the possible effects of age and metallicity on the multiplicity properties. As a final note, we insist on the critical need for observers to characterise the various observational biases as carefully as possible, including selection effects and sensitivity limits. While time consuming to establish, these quantities are indeed critical to retrieve the intrinsic multiplicity properties from the observations. A posteriori attempts to combine different samples may turn more complicated without this specific information at hand (see e.g., Moe & Di Stefano (2016)).

Ackowledgments

The author is thankful to Rodolfo Barbá and Simón-Díaz for sharing OWN and IACOB results prior to publication (OWN results were used in the preliminary comparison shown in Fig. 2) and to colleagues and collaborators for many stimulating discussions.

References

Abbott, B. P., Abbott, R., Abbott, T. D., et al. 2016, Physical Review Letters, 116, 061102
Aldoretta, E. J., Caballero-Nieves, S. M., Gies, D. R., et al. 2015, AJ, 149, 26
Almeida, L. A., Sana, H., de Mink, S. E., *et al.* 2015, ApJ, 812, 102
Almeida, L. A., Sana, H., Taylor, W., et al. 2017, A&A, 598, A84
Barbá, R., Gamen, R., Arias, J. I., et al. 2014, in Revista Mexicana de Astronomia y Astrofisica Conference Series, Vol. 44, 148–148
Chini, R., Hoffmeister, V. H., Nasseri, A., Stahl, O., & Zinnecker, H. 2012, MNRAS, 424, 1925
Dunstall, P. R., Dufton, P. L., Sana, H., et al. 2015, A&A, 580, A93
Evans, C. J., Taylor, W. D., Hénault-Brunet, V., et al. 2011, A&A, 530, A108
Kiminki, D. C. & Kobulnicky, H. A. 2012, ApJ, 751, 4
Kobulnicky, H. A., Kiminki, D. C., Lundquist, M. J., et al. 2014, ApJS, 213, 34
Maíz Apellániz, J., Sota, A., Arias, J. I., et al. 2016, ApJS, 224, 4
Mandel, I. & de Mink, S. E. 2016, MNRAS, 458, 2634
Mason, B. D., Gies, D. R., Hartkopf, W. I., et al. 1998, AJ, 115, 821
Mason, B. D., Hartkopf, W. I., Gies, D. R., Henry, T. J., & Helsel, J. W. 2009, AJ, 137, 3358
Moe, M. & Di Stefano, R. 2016, arXiv: 1606.05347
Ramírez-Tannus, M. C., Kaper, L., de Koter, A., *et al.* 2017, A&A, in press (arXiv:1704.08216)
Sana, H. & Evans, C. J. 2011, in IAU Symposium, ed. C. Neiner, G. Wade, G. Meynet, & G. Peters, Vol. 272, 474
Sana, H., de Mink, S. E., de Koter, A., et al. 2012, Science, 337, 444
Sana, H., de Koter, A., de Mink, S. E., et al. 2013, A&A, 550, A107
Sana, H., Le Bouquin, J.-B., Lacour, S., et al. 2014, ApJS, 215, 15
Sana, H., Ramírez-Tannus, M. C., de Koter, A., *et al.* 2017, A&A, 599, L9
Shenar, T., Richardson, N. D., Sablowski, D. P., et al. 2017, A&A, 598, A85
Simón-Díaz, S., Negueruela, I., Maíz Apellániz, J., *et al.* 2015, Highlights of Spanish Astrophysics VIII, 576
Sota, A., Maíz Apellániz, J., Walborn, N. R., et al. 2011, ApJS, 193, 24
Sota, A., Maíz Apellániz, J., Morrell, N. I., et al. 2014, ApJS, 211, 10

The Lives and Death-Throes of Massive Stars
Proceedings IAU Symposium No. 329, 2016
J.J. Eldridge, J.C. Bray, L.A.S. McClelland & L. Xiao, eds.

doi:10.1017/S1743921317002964

Progenitors of binary black hole mergers detected by LIGO

Konstantin Postnov and Alexander Kuranov

Sternberg Astronomical Institute, Moscow M.V.Lomonosov State University
13, Universitetskij pr., 119234 Moscow, Russia
email: pk@sai.msu.ru

Abstract. Possible formation mechanisms of massive close binary black holes that can merge in the Hubble time to produce powerful gravitational wave bursts detected during advanced LIGO O1 science run are briefly discussed. The pathways include the evolution from field low-metallicity massive binaries, the dynamical formation in globular clusters and primordial black holes. Low effective black hole spins inferred for LIGO GW150914 and LTV151012 events are discussed. Population synthesis calculations of the expected spin and chirp mass distributions from the standard field massive binary formation channel are presented for different metallicities (from zero-metal Population III stars up to solar metal abundance). We conclude that that merging binary black holes can contain systems from different formation channels, discrimination between which can be made with increasing statistics of mass and spin measurements from ongoing and future gravitational wave observations.

Keywords. black hole physics, stars: evolution, (stars:) binaries (including multiple): close

---

1. Introduction and historical remarks

The epochal discovery of the first gravitational wave source GW150914 from coalescing binary black hole (BH) system (Abbott *et al.* 2016d) not only heralded the beginning of gravitational wave astronomy era, but also stimulated a wealth of works on fundamental physical and astrophysical aspects of the formation and evolution of binary BHs. The LIGO detection of GW150914 and of the second robust binary BH merging event GW151226 (Abbott *et al.* 2016c) enables BH masses and spins before the merging, the luminosity distance to the sources and the binary BH merging rate in the Universe to be estimated (Abbott *et al.* 2016b). Astrophysical implications of these measurements were discussed, e.g., in Abbott *et al.* (2016e,a).

This discovery of gravitational waves from coalescing binary BHs was long awaited. Evolution of massive binary systems was elaborated in the 1970s to explain a rich variety of newly discovered galactic X-ray binaries (van den Heuvel & Heise 1972; Tutukov *et al.* 1973). Formation of two relativistic compact remnants (neutron stars (NSs) or black holes) naturally followed from the binary evolution scenario (Tutukov & Yungelson 1973; Flannery & van den Heuvel 1975). At the dawn of the LIGO Project, Tutukov and Yungelson (Tutukov & Yungelson 1993) calculated, using the standard assumptions of massive binary evolution, the expected galactic merging rate of binary NSs and BHs. They pointed out that although the galactic merging rate of binary NSs is much larger than that of binary BHs, their detection rates by gravitational-wave interferometers can be comparable due to the strong dependence of the characteristic GW amplitude h_c on the total mass $M = M_1 + M_2$ of the coalescing binaries, $h_c \sim M^{5/2}$. A few years later, independent population synthesis calculations by the Scenario Machine code were reported in a series of papers (Lipunov *et al.* 1997b,c,a). They showed that in a wide range of possible BH formation parameters (masses,kick velocities) and under standard assumptions of the massive star evolution, the detection rate of binary BH mergings should be much higher than that of binary NSs, and the first LIGO event should most likely to be a binary BH merging. Interestingly, the mean BH masses known at that time from dynamical measurements in galactic BH X-ray binaries were about 10 $M_\odot$, which forced (cautiously) the authors of (Lipunov *et al.* 1997a) to fix the parameter $k_{BH} = M_{BH}/M_c$, where M_c is the mass of the star before the collapse, around ~ 0.3 (see Fig. 4 in that paper) in order to produce the chirp mass of coalescing binary BHs around $15M_\odot$. Taking $k_{BH} = 1$, one immediately obtains the BH masses around 30-40 $M_\odot$, which seemed outrageously high at that time.

Starting from the end of the 1990s, various groups have used different population synthesis codes to calculate the merging rates of double compact objects (see especially many papers by the Polish group based on the StarTrack code (Belczynski *et al.* 2002; Dominik *et al.* 2012)), yielding a wide range of possible BH-BH merging rates (see e.g. Table 6 in Postnov & Yungelson (2014)). Clearly, the degeneracy of binary evolution and BH formation parameters has been so high (Abadie *et al.* 2010) that only real observations could narrow the wide parameter range.

2. Standard scenario of binary BH formation

The standard scenario of double BH formation from field stars is based on well-recognized evolution of single massive stars Woosley *et al.* (2002). To produce a massive BH with $M \simeq 10M_\odot$ in the end of evolution, the progenitor star should have a large mass and low mass-loss rate. The mass-loss rate is strongly dependent on the metallicity, which plays the key role in determining the final mass of stellar remnant (see Spera *et al.* (2015) and N. Yusof's contribution in this conference). The metallicity effects were included in the population synthesis calculations (Dominik *et al.* 2013), and the most massive BHs were found to be produced by the low-metallicity progenitors. Here early metal-free Population III stars provide an extreme example, see calculations by Kinugawa *et al.* (2014); Hartwig *et al.* (2016). After the discovery of GW150914, several independent population synthesis calculations were performed to explain the observed masses of binary BH in GW150914 and the inferred binary BH merging rate $\sim 9-240$ Gpc^{-3}yr^{-1} (Abbott *et al.* 2016b) (see, among others, e.g. Belczynski *et al.* 2016; Eldridge & Stanway 2016; Lipunov *et al.* 2017).

In addition to the metallicity that affects the intrinsic evolution of the binary components, the most important uncertainty in the binary evolution is the efficiency of the common envelope (CE) stage which is required to form a compact double BH binary merging in the Hubble time. The common envelope stage remains a highly debatable issue. For example, in recent hydro simulations (Ohlmann *et al.* 2016) a low CE efficiency was found, while successful CE calculations were reported by other groups (see, e.g., N. Ivanova contribution at this conference). Another recent study (Pavlovskii *et al.* 2017) argues that it is possible to reconcile the BH formation rate through the CE channel taking into account the stability of mass transfer in massive binaries in the Hertzsprung gap stage, which drastically reduces the otherwise predicted overproduction of binary BH merging rate in some population synthesis calculations. Also, the so-called stable 'isotropic re-emission' mass transfer mode can be realized in high-mass X-ray binaries with massive BHs, thus helping to avoid the merging of the binary system components in the common envelope (van den Heuvel *et al.* 2017). This stable mass transfer mode can explain the surprising stability of kinematic characteristics observed in the galactic microquasar SS433 (Davydov *et al.* 2008).

Of course, much more empirical constraints on and hydro simulations of the common evolution formation and properties are required, but the formation channel with common envelope of binary BHs with properties similar to GW150914 remains quite plausible.

3. Other scenarios

To avoid the ill-understood common envelope stage, several alternative scenarios of binary BH formation from massive stars were proposed. For example, in short-period massive binary systems chemically homogeneous evolution due to rotational mixing can be realized. The stars remain compact until the core collapse, and close binary BH system is formed without common envelope stage (Mandel & de Mink 2016; de Mink & Mandel 2016; Marchant *et al.* 2016). In this scenario, a pair of nearly equal massive BHs can be formed with the merging rate comparable to the empirically inferred from the first LIGO observations.

Another possible way to form massive binary BH system is through dynamical interactions in a dense stellar systems (globular clusters). This scenario was earlier considered by Sigurdsson & Hernquist (1993). In the core of a dense globular clusters, stellar-mass BH form multiple systems, and BH binaries are dynamically ejected from the cluster. This mechanism was shown to be quite efficient in producing 30+30 $M_\odot$ merging binary BHs (Rodriguez *et al.* 2016b), and binary BH formed in this way can provide a substantial fraction of all binary BH mergings in the local Universe (Rodriguez *et al.* 2016a).

Finally, there can be more exotic channels of binary BH formation. For example, primordial black holes (PBHs) formed in the early Universe can form pairs which could be efficient sources of gravitational waves (Nakamura *et al.* 1997). After the discovery of GW150914, the interest to binary PBHs has renewed (Bird *et al.* 2016). Stellar-mass PBHs can form a substantial part of dark matter in the Universe (Carr *et al.* 2016). The PBHs formed at the radiation-dominated stage can form pairs like GW150914 with the merging rate compatible with empirical LIGO results, being only a small fraction of all dark matter (Eroshenko 2016; Sasaki *et al.* 2016). Different class of PBHs with a universal log-normal mass spectrum produced in the frame of a modified Affleck-Dine supersymmetric baryogenesis (Dolgov & Silk 1993; Dolgov *et al.* 2009) were shown to be able to match the observed properties of GW150914 Blinnikov *et al.* (2016).

4. Low spins of BH in GW150914 and LTV151012 events

In the framework of general relativity, a BH is fully characterized by its mass M and dimensionless angular momentum $a = J/M$ (in geometrical units $G = c = 1$) (the possible BH electric charge is negligible in real astrophysical conditions). The LIGO observations enable measurements of both masses of the coalescing BH components, M_1 and M_2, and the chirp mass that determines the strength of gravitational wave signal $\mathcal{M} = (M_1 M_2)^{3/5}/M^{1/5}$. From the analysis of waveforms at the inspiral stage, individual BH spins before the merging are poorly constrained, but their mass-weighted total angular momentum parallel to the orbital angular momentum, χ_{eff}, can be estimated with good accuracy (Abbott *et al.* 2016b)†. The O1 LIGO detections suggest that the most massive GW150914 and (less certain) LTV151012 have very low $\chi_{eff} \simeq 0$.

This observational fact has important evolutionary implications (see Kushnir *et al.* (2016); Hotokezaka & Piran (2017)). It suggests a very slow rotation of BH progenitors,

† The parameter $\chi_{eff} = (M_1\chi_1 + M_2\chi_2)/M$, where $\chi_i = a_i \cos\theta_i$ with θ_i being the angle between the angular momentum of the i-th BH and orbital angular momentum of the binary system.

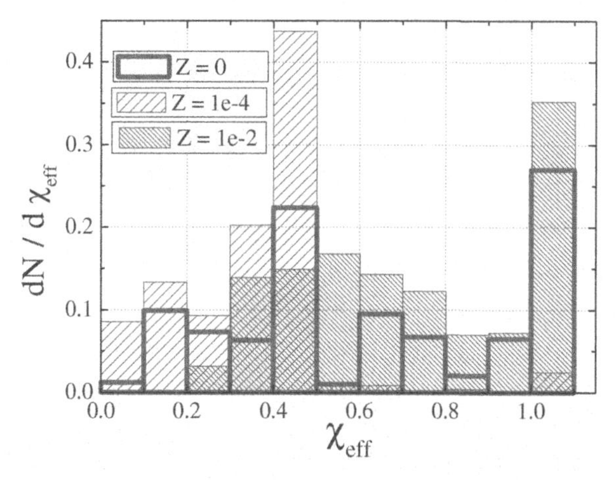

Figure 1. Distribution of the effective binary BH spin parameter χ_{eff} before the merging for different stellar metallicities.

which by itself strongly constraints, for example, chemically homogeneous pathways mentioned above in which tidally induced rotation of the close binary components plays the key role. Massive stars are observed to be rapid rotators. No significant angular momentum loss is expected during their evolution with low mass loss rate by stellar wind and at the pre-collapse stage as required to produce massive BHs (Spera *et al.* 2015). Note that low effective spin values can imply either small intrinsic BH spins $a \sim 0$, or unusual orientations of BH spins with respect to the orbital angular momentum at the inspiral stage. The last case can well be reconciled with the dynamical formation scenario (Rodriguez *et al.* 2016a), where the BH spins are not expected to be correlated with the orbital angular momentum. In the PBH scenario, BH spins must be zero as there are no vorticity in primordial cosmological perturbations.

Therefore, the mass-spin distribution of BHs can serve as a sensitive tool to discriminate between different astrophysical formation channels of coalescing massive binary BHs. To estimate the spin distribution of BH remnants in binaries, it is necessary to know how to treat the spin evolution of the stellar core, which is ill-understood and strongly model-dependent. One possible approach is to match theoretical predictions of the core rotation with observed period distribution of the young neutron stars observed as radio pulsars (Postnov *et al.* 2016). Initially, a star is assumed to rotate rigidly, but after the main sequence the star can be separated in two parts – the core and the envelope, with some effective coupling between these two parts. The coupling between the core and envelope rotation can be mediated by magnetic forces, internal gravity waves (see Fuller *et al.* (2015) and J.Fuller's talk at this conference), etc. The validity of such an approach was checked by direct MESA calculations of the rotational evolution of a 15 $M_\odot$ star (Postnov *et al.* 2016). It was found that the observed period distribution of young pulsars can be reproduced if the effective coupling time between the core and envelope is $\tau_c = 5 \times 10^5$ years (see Fig. 1 in Postnov *et al.* (2016)). Below we shall assume that this time is also applicable to the evolution of very massive stars leaving behind BH remnants.

Each angular momentum of the main-sequence components of the initial binary is assumed to be arbitrarily distributed in space, its absolute value being connected to the initial stellar mass using the empirical relation between the equatorial rotation velocity of a star with its mass $v_{rot} = 330 M_0^{3.3}/(15 + M_0^{3.45})$ km s^{-1} (here M_0 is in solar units). It was assumed that the rotation of the stellar envelope gets synchronized with the

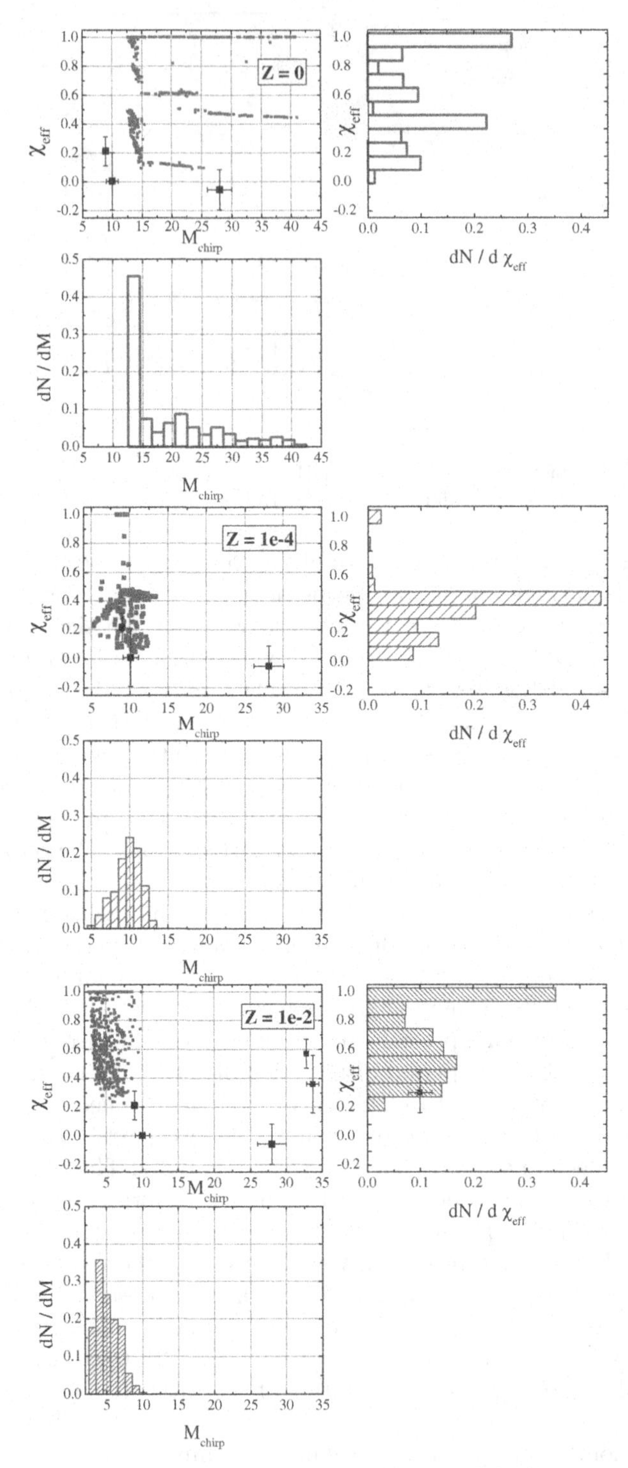

Figure 2. $\mathcal{M} - \chi_{eff}$ plane for different stellar metallicities. Filled squares show the observed BH-BH systems (Abbott *et al.* 2016b), GW150914, LTV151012 and GW151226, in order of decreasing chirp mass.

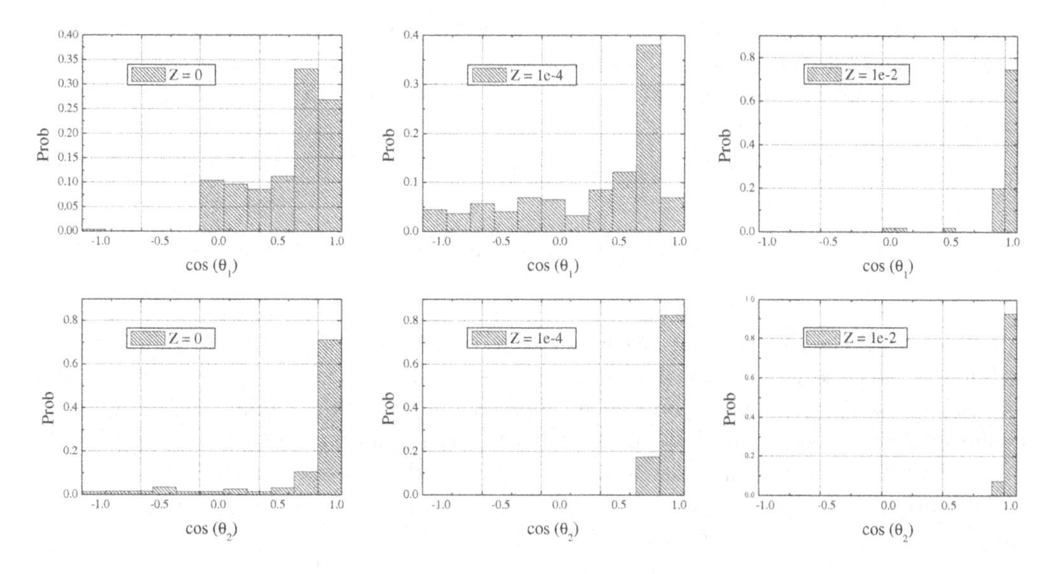

Figure 3. BH spin misalignments with orbital angular momentum (in terms of $\cos\theta$) in coalescing binary BHs for different stellar metallicities. Top and bottom row correspond to M_1 and M_2, respectively.

orbital motion with the characteristic synchronization time t_{sync}, and the process of tidal synchronization was treated as in the BSE code (Hurley *et al.* 2002). Due to intrinsic misalignment of the spin vectors of the stars with the binary orbital angular momentum $\hat{L}$, we separately treated the core-envelope coupling for the spin components parallel and perpendicular to $\hat{L}$. On evolutionary stages prior to the compact remnant formation, for each binary component we assumed that due to tidal interactions the parallel component of the stellar envelope spin $J_{||}$ gets synchronized with the orbital motion on the characteristic time t_{sync}, while the normal spin component $J_{\perp}$ of the stellar envelope decreases due to the tidal interaction in the binary system the on the same characteristic time scale, which leads to the secular evolution of the spin-orbit misalignment. The parallel component of the envelope spin also evolves due to the core-envelope interaction with the characteristic time τ_c. These processes were added to the updated BSE population synthesis code.

With these additions, the population synthesis of typically 100000 binaries per run has been carried out for different parameters of binary evolution (the common envelope stage efficiency α_{CE}, stellar metallicities etc.). No generic BH kick was assumed. The results of calculations of BH spin distributions for different stellar metallicites and for the standard CE efficiency parameter $\alpha_{CE} = 1$ are shown in Fig. 1. Evolution of zero-metallicity (primordial Population III) stars was parametrized as in Kinugawa *et al.* (2014). Fig. 2 shows the plot of coalescing binary BHs on the $\mathcal{M} - \chi_{eff}$ plane for different metallicities. BH spin misalignments with orbital angular momentum in coalescing binary BHs for different stellar metallicities are presented in Fig. 3.

A detailed analysis of these simulations will be published elsewhere (Postnov & Kuranov, in preparation), but the main conclusions can be drawn from Figs. 1-3. It is seen (expectedly) from Fig. 1 that the effective spin χ_{eff} of binary BH from field massive stars (the standard formation scenario) is distributed in a wide range, but the $\mathcal{M} - \chi_{eff}$ plot (Fig. 2) suggests that large chirp masses can hardly have $\chi_{eff} \simeq 0$. This (model) result can signal potential difficulty in explaining the most massive merging BH binaries by this formation channel only. Fig. 3 suggests that even in the absence of BH kicks, which were assumed in the present calculations, the BH spin misalignments can be quite high even for field binaries.

5. Conclusions

Presently, there are different astrophysical pathways of producing massive binary BHs that merge in the Hubble time. They can be formed from low-metallicity massive field stars, primordial Pop III remnants, can be results of dynamical evolution in dense stellar clusters or even primordial black holes. It is not excluded that all channels contribute to the observed binary BH population. For example, the discovery of very massive Schwarzschild BHs would be difficult to reconcile with the standard massive binary evolution, but can be naturally explained in the PBH scenario (Blinnikov *et al.* 2016).

As of the time of writing, another six event candidates were reported by the LIGO collaboration from the analysis of 67 days of joint operation of two LIGO interferometers during O2 run (see `http://ligo.org/news/index.php#02Apr2017update`). With the current LIGO sensitivity, the detection horizon of binary BH with masses around 30 $M_\odot$ reaches 700 Mpc. So far the statistics of binary BH merging rate as a function of BH mass as inferred from three reported LIGO O1 events is consistent with a power-law dependence, $dR/dM \sim M^{-2.5}$ (Hotokezaka & Piran 2017), which does not contradict the general power-law behavior of the stellar mass function. Clearly, more statistics of BH masses and spins inferred from binary BH mergings is required to distinguish between the possible binary BH populations which can exist in the Universe.

Acknowledgements. KP acknowledges the support from RSF grant 16-12-10519.

References

Abadie, J. *et al.* 2010, *Classical and Quantum Gravity*, 27, 173001
Abbott, B.P. *et al.* 2016a, *ApJL*, 818, L22
Abbott, B.P. *et al.* 2016b, *Physical Review X*, 6, 041015
Abbott, B.P. *et al.* 2016c, *Physical Review Letters*, 116, 241103
Abbott, B.P. *et al.* 2016d, *Physical Review Letters*, 116, 061102
Abbott, B.P. *et al.* 2016e, *ApJL*, 833, L1
Belczynski, K., Holz, D. E., Bulik, T., & O'Shaughnessy, R. 2016, *Nature* , 534, 512
Belczynski, K., Kalogera, V., & Bulik, T. 2002, *ApJ*, 572, 407
Bird, S. *et al.* 2016, *Physical Review Letters*, 116, 201301
Blinnikov, S., Dolgov, A., Porayko, N. K., & Postnov, K. 2016, *JCAP*, 11, 036
Carr, B., Kühnel, F., & Sandstad, M. 2016, *Phys. Rev. D*, 94, 083504
Davydov, V. V., Esipov, V. F., & Cherepashchuk, A. M. 2008, *Astronomy Reports*, 52, 487
de Mink, S. E. & Mandel, I. 2016, *MNRAS*, 460, 3545
Dolgov, A. & Silk, J. 1993, *Phys. Rev. D*, 47, 4244
Dolgov, A. D., Kawasaki, M., & Kevlishvili, N. 2009, *Nuclear Physics B*, 807, 229
Dominik, M. *et al.* 2012, *ApJ*, 759, 52
Dominik, M. *et al.* 2013, *ApJ*, 779, 72
Eldridge, J. J. & Stanway, E. R. 2016, *MNRAS*, 462, 3302
Eroshenko, Y. N. 2016, *ArXiv e-prints*
Flannery, B. P. & van den Heuvel, E. P. J. 1975, *A&A*, 39, 61
Fuller, J., Cantiello, M., Lecoanet, D., & Quataert, E. 2015, *ApJ*, 810, 101
Hartwig, T. *et al.* 2016, *MNRAS*, 460, L74
Hotokezaka, K. & Piran, T. 2017, *ArXiv e-prints*

Hurley, J. R., Tout, C. A., & Pols, O. R. 2002, *MNRAS*, 329, 897
Kinugawa, T., Inayoshi, K., Hotokezaka, K., Nakauchi, D., & Nakamura, T. 2014, *MNRAS*, 442, 2963
Kushnir, D., Zaldarriaga, M., Kollmeier, J. A., & Waldman, R. 2016, *MNRAS*, 462, 844
Lipunov, V.M., Postnov, K.A., & Prokhorov, M.E. 1997a, *Astronomy Letters*, 23, 492
Lipunov, V.M., Postnov, K.A., & Prokhorov, M.E. 1997b, *New Astron.*, 2, 43
Lipunov, V.M., Postnov, K.A., & Prokhorov, M.E. 1997c, *MNRAS*, 288, 245
Lipunov, V. M. *et al.* 2017, *New Astron.*, 51, 122
Mandel, I. & de Mink, S. E. 2016, *MNRAS*, 458, 2634
Marchant, P., Langer, N., Podsiadlowski, P., Tauris, T. M., & Moriya, T. J. 2016, *A&A*, 588, A50
Nakamura, T., Sasaki, M., Tanaka, T., & Thorne, K. S. 1997, *ApJL*, 487, L139
Ohlmann, S. T., Röpke, F. K., Pakmor, R., & Springel, V. 2016, *ApJL*, 816, L9
Pavlovskii, K., Ivanova, N., Belczynski, K., & Van, K. X. 2017, *MNRAS*, 465, 2092
Postnov, K. A., Kuranov, A. G., Kolesnikov, D. A., Popov, S. B., & Porayko, N. K. 2016, *MNRAS*, 463, 1642
Postnov, K. A. & Yungelson, L. R. 2014, *Living Reviews in Relativity*, 17, 3
Rodriguez, C.L., Chatterjee, S., & Rasio, F.A. 2016a, *Phys. Rev. D*, 93, 084029
Rodriguez, C.L., Haster, C.J., Chatterjee, S., Kalogera, V., & Rasio, F.A. 2016b, *ApJL*, 824, L8
Sasaki, M., Suyama, T., Tanaka, T., & Yokoyama, S. 2016, *Physical Review Letters*, 117, 061101
Sigurdsson, S. & Hernquist, L. 1993, *Nature* , 364, 423
Spera, M., Mapelli, M., & Bressan, A. 2015, *MNRAS*, 451, 4086
The LIGO Scientific Collaboration *et al.* 2016, *ArXiv e-prints*
Tutukov, A. & Yungelson, L. 1973, *Nauchnye Informatsii*, 27, 70
Tutukov, A., Yungelson, L., & Klayman, A. 1973, *Nauchnye Informatsii*, 27, 3
Tutukov, A. V. & Yungelson, L. R. 1993, *MNRAS*, 260, 675
van den Heuvel, E. P. J. & Heise, J. 1972, *Nature Physical Science*, 239, 67
van den Heuvel, E. P. J., Portegies Zwart, S. F., & de Mink, S. E. 2017, *ArXiv e-prints*
Woosley, S. E., Heger, A., & Weaver, T. A. 2002, *Reviews of Modern Physics*, 74, 1015

The Lives and Death-Throes of Massive Stars
Proceedings IAU Symposium No. 329, 2016
J.J. Eldridge, J.C. Bray, L.A.S. McClelland & L. Xiao, eds.

doi:10.1017/S1743921317002782

What can magnetic early B-type stars tell us about early B-type stars in general?

Matthew Shultz[1], Gregg Wade[2], Thomas Rivinius[3], Coralie Neiner[4], Evelyne Alecian[5,6,7], Véronique Petit[8], Jason Grunhut[9] and the MiMeS & BinaMIcS Collaborations

[1]Department of Physics and Astronomy, Uppsala University, Box 516, Uppsala 75120
email: matthew.shultz@physics.uu.se

[2]Department of Physics, Royal Military College of Canada, Kingston, Ontario K7K 7B4, Canada

[3]ESO - European Organisation for Astronomical Research in the Southern Hemisphere, Casilla 19001, Santiago 19, Chile

[4]LESIA, Observatoire de Paris, PSL Research University, CNRS, Sorbonne Universités, UPMC Univ. Paris 06, Univ. Paris Diderot, Sorbonne Paris Cité, 5 place Jules Janssen, 92195 Meudon, France

[5]Université Grenoble Alpes, IPAG, F-38000 Grenoble, France

[6]CNRS, IPAG, F-38000 Grenoble, France

[7]LESIA, Observatoire de Paris, CNRS UMR 8109, UPMC, Université Paris Diderot, 5 place Jules Janssen, 92190, Meudon, France

[8]Department of Physics and Astronomy, University of Delaware, 217 Sharp Lab, Newark, DE 19716, USA

[9]Dunlap Institute for Astronomy and Astrophysics, University of Toronto, 50 St. George Street, Toronto, ON, M5S 3H4, Canada

Abstract. Some magnetic early B-type stars display Hα emission originating in their Centrifugal Magnetospheres (CMs). To determine the rotational and magnetic properties necessary for the onset of emission, we analyzed a large spectropolarimetric dataset for a sample of 51 B5-B0 magnetic stars. New rotational periods were found for 15 stars. We determined physical parameters, dipolar magnetic field strengths, magnetospheric parameters, and magnetic braking timescales. Hα-bright stars are more rapidly rotating, more strongly magnetized, and younger than the overall population. We use the high sensitivity of magnetic braking to the mass-loss rate to test the predictions of Vink et al. (2001) and Krtička (2014) by comparing ages t to maximum spindown ages $t_{\rm S,max}$. For stars with $M_* < 10\ M_\odot$ this comparison favours the Krtička recipe. For the most massive stars, both prescriptions yield $t \ll t_{\rm S,max}$, a discrepancy which is difficult to explain via incorrect mass-loss rates alone.

Keywords. stars: magnetic fields, stars: mass loss, stars: rotation, stars: early-type

---

1. Introduction, Observations, & Methodology

Magnetic OB stars often display Hα emission originating in their corotating magnetospheres, which arise from magnetic confinement of their ionized winds (Babel & Montmerle 1997; ud-Doula & Owocki 2002). Petit et al. (2013) showed that the emission properties of magnetic early B-type stars can be explained by the presence of a *Centrifugal Magnetosphere* (CM), in which magnetically enforced corotation provides centrifugal support against gravitational infall of the magnetically confined plasma, thus enabling accumulation of plasma to optically thick densities around stars with relatively weak winds (Townsend & Owocki 2005). A CM is expected when the Alfvén radius $R_{\rm A}$, or the maximum extent of closed magnetic loops within the magnetosphere, is greater than the Kepler corotation radius $R_{\rm K}$, or the radius at which gravitational and centrifugal forces

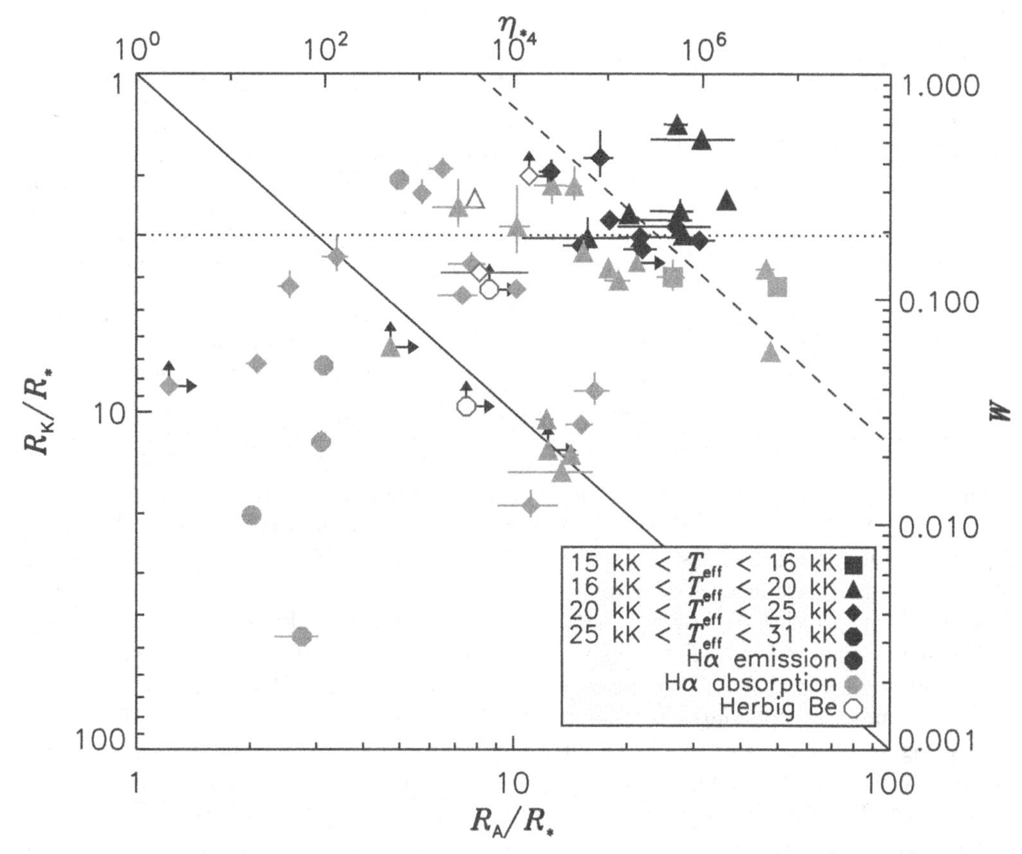

Figure 1. The rotation-magnetic confinement diagram, updated using newly determined $P_{\rm rot}$ and $B_{\rm d}$. Alternate axes give the magnetic wind confinement parameter $\eta_* = (B_{\rm d}^2 R_*^2)/(\dot{M} v_\infty)$, and the rotation parameter $W = v_{\rm rot}/v_{\rm orb}$ (where $v_{\rm orb}$ is the velocity required for a Keplerian orbit at the stellar surface, and $W = 1$ corresponds to the critical or breakup velocity). The diagnonal $R_{\rm A} = R_{\rm K}$ line divides stars with DMs (below) from those with CMs (above). The dotted line at $R_{\rm K} = 2R_*$ and the diagonal dashed line at $R_{\rm A} = 8R_{\rm K}$ divide stars with from stars without Hα emission.

are balanced. The absence of emission amongst many stars with $R_{\rm A} > R_{\rm K}$ indicates that merely possessing a CM is not a sufficient condition for the onset of Hα emission (Petit et al. 2013). However, owing to the relatively recent discovery of many of the known magnetic hot stars, their rotational periods and surface magnetic field strengths were not known, making it difficult to determine the conditions necessary for Hα emission.

To address this we assembled a sample consists of 51 main-sequence magnetic stars with spectral types between B5 and B0. The data (973 circularly polarized spectropolarimetric sequences) were obtained with ESPaDOnS at the 3.6 m Canada-France-Hawaii Telescope (CFHT), Narval at the 2 m Bernard Lyot Telescope, and HARPSpol at the European Southern Observatory (ESO) La Silla 3.6 m Telescope. The majority of the data were acquired by the Magnetism in Massive Stars (MiMeS) Large Programs (LPs). The instrumentation, data reduction, and the methodology utilized by the MiMeS LPs (in particular least-squares deconvolution of mean line profiles and measurement of the longitudinal magnetic field $\langle B_z \rangle$) are described by Wade et al. (2016). Some of the data were also acquired by the BinaMIcS LPs (Alecian et al. 2015) and the B-fields in OB stars (BOB) LP (Fossati et al. 2015), along with several CFHT and ESO PI programs.

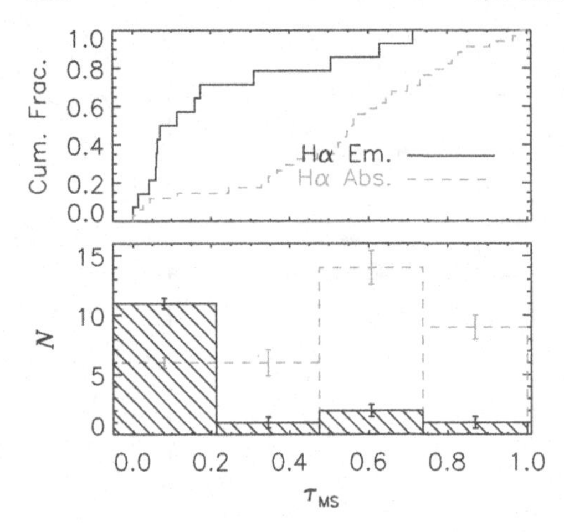

Figure 2. Cumulative distributions (top) and histograms (bottom) of $\tau_{\rm MS}$ for stars with and without Hα emission. Histogram errors were determined via Monte Carlo simulations using the formal uncertainties in $\tau_{\rm MS}$ obtained either from the HRD or cluster age MS turnoffs. Hα-bright stars are heavily concentrated amongst the youngest stars, and indeed form a majority in this age-range ($\tau_{\rm MS} < 0.25$), despite making up only 29% of the overall population. The K-S test significance for the two distributions is $\sim 10^{-3}$, indicating they are distinct distributions with high probability.

Since many of the stars possess chemical abundance spots, photometric and spectroscopic as well as magnetic variability are all rotationally modulated. Rotational periods $P_{\rm rot}$ were thus determined using Lomb-Scargle analysis of $\langle B_z \rangle$, as well as of archival photometric and/or spectroscopic time-series, when these data were available. $P_{\rm rot}$ was determined for 15 of the 18 stars for which it was previously unknown.

Stellar parameters (radii R_*, masses M_*, ages t, and fractional main sequences ages $\tau_{\rm MS}$) were determined from the Hertzsprung-Russell diagram (HRD) and the $T_{\rm eff}$-$\log g$ diagram, using the rotating evolutionary tracks calculated by Ekström et al. (2012). Cluster ages obtained from isochrone fitting to main-sequence turnoffs were used when available (15/51 stars). Stellar parameters and $P_{\rm rot}$ were used to obtain $R_{\rm K}$ (Townsend & Owocki 2005). Surface magnetic field strengths $B_{\rm d}$ were determined from $\langle B_z \rangle$ and the stellar parameters using the method developed by Preston (1967). $R_{\rm A}$ and the maximum spindown age $t_{\rm S,max}$ (ud-Doula & Owocki 2002; ud-Doula et al. 2009) were calculated using the mass-loss rates $\dot{M}$ and wind terminal velocities v_∞ obtained from the methods of both Vink et al. (2001) and Krtička (2014).

2. Results

Fig. 1 shows the rotation-magnetic confinement diagram introduced by Petit et al. (2013), updated with the newly determined magnetic and rotational parameters. Hα-bright stars are concentrated in the upper right, in the region $R_{\rm K} < 2R_*$, and $R_{\rm A} > 8R_{\rm K}$. There are no stars in this region with Hα in absorption. This demonstrates that the rotational and magnetic properties of an early B-type star are reliable predictors of whether or not it will display Hα emission.

Since rapid rotation is a necessary condition for Hα emission, but the angular momentum of strongly magnetized stars is rapidly lost due to magnetic braking (ud-Doula et al. 2009), emission should be seen primarily in young stars. Fig. 2 demonstrates that this is in fact the case. The majority of Hα-bright stars have $\tau_{\rm MS} < 0.25$. Indeed, Hα-bright stars are a majority (11/17 or 64%) of the stars in this age range, as compared to 15/51 or 29% of the overall population.

Due to magnetic braking $P_{\rm rot}$ should be longer for older stars. In Fig. 3a it can be seen that all stars with $P_{\rm rot} > 10$ d are also older than $\tau_{\rm MS} > 0.25$. Magnetic braking is highly sensitive to the mass-loss rate (ud-Doula et al. 2009), thus stars with higher masses, and stronger winds, should lose angular momentum more rapidly. As can be seen in Fig. 3a, the most slowly rotating stars in the sample are also amongst the most massive.

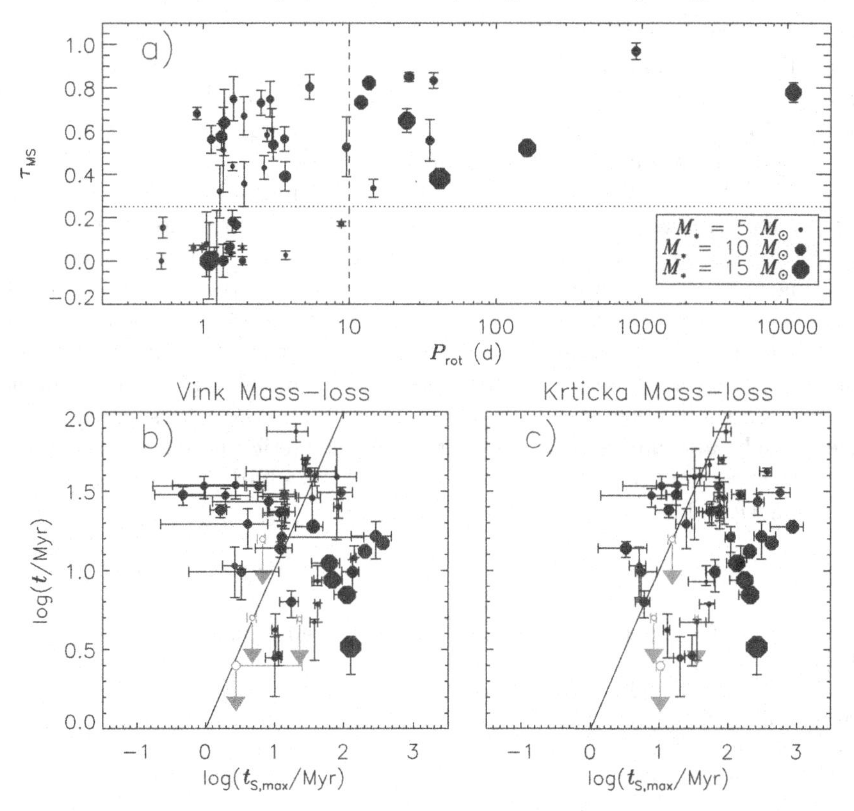

Figure 3. *a)* τ_{MS} as a function of P_{rot}. All stars with $P_{rot} > 10$ d are older than $\tau_{MS} = 0.25$. More massive stars (which have stronger winds) tend to be more slowly rotating than less massive stars of similar age. *b)* the age t inferred from the HRD, as a function of $t_{S,max}$ as computed with Vink mass-loss rates. Open symbols indicate stars with upper limits on t. The solid line shows $t = t_{S,max}$. There is poor agreement between the two timescales. *c)* as *b*, but computed with Krtička mass-loss rates. These yield a better agreement for the lower-mass stars, but cannot resolve the discrepancy for stars with $M_* \sim 15\ M_\odot$, all of which are rotating much more slowly than predicted.

The bottom panels of Fig. 3 provide a more quantitative test of magnetic braking theory via direct comparison of the stellar age t to the maximum spindown age $t_{S,max}$ (i.e. the time required for a star to spin down from critical rotation to its observed rotational velocity; ud-Doula et al. 2009). In Fig. 3b $t_{S,max}$ was calculated using Vink mass-loss, while in Fig. 3c the comparison is performed using Krtička mass-loss rates. Vink mass-loss results in $t_{S,max} \ll t$ for stars with $M_* < 10\ M_\odot$, with discrepancies up to 2 dex, i.e. the stars should be rotating much more slowly than observed. This is largely resolved by Krtička mass-loss rates. Both prescriptions yield similar solutions for higher-mass, hotter stars. For the most massive stars in the sample ($M_* \sim 15\ M_\odot$), $t_{S,max} \gg t$; resolving this, again typically around 1 dex, would require $\dot{M}$ similar to values expected for O-type stars.

3. Conclusions

Hα-bright CM-host stars are rapidly rotating, strongly magnetized, and young. Indeed, while Hα emission is fairly rare overall (occuring in about 1/4 of the population), a majority (about 2/3) of young magnetic early B-type stars possess detectable CMs (Fig. 2).

This suggests that young stellar clusters should be fertile ground for the detection of CM-host stars. The similar placement of all such stars on the rotation-magnetic confinement diagram (Fig. 1) suggests that the presence of CM emission is highly predictive of both rotational and magnetic properties. Future work will explore the sensitivity of emission strength to rotation and magnetic confinement strength.

The youth of CM-host stars, the increase in $P_{\rm rot}$ over time, and the longer $P_{\rm rot}$ typically observed for more massive stars, all serve as qualitative confirmation of rapid magnetic braking. The strong sensitivity of magnetic braking to mass-loss rates has enabled a test of competing models via comparison of t to $t_{\rm S,max}$. For stars with $M_* < 10\ M_\odot$, our results favour the mass-loss rates calculated by Krtička (2014). For more massive stars, neither Krtička nor Vink mass-loss rates are able to reconcile t with $t_{\rm S,max}$ (although Vink rates are marginally better in this mass range), and the discrepancy furthermore requires mass-loss rates similar to those of late O-type stars. As such high mass-loss rates are unlikely, alternative scenarios, e.g. rapid magnetic flux decay, strong magnetic braking on the pre-main sequence, or stellar structure modifications arising due to strong internal magnetic fields, will need to be explored.

References

Alecian E. et al., 2015, in IAU Symposium, Vol. 307, New Windows on Massive Stars, pp. 330–335
Babel J., Montmerle T., 1997, *ApJL*, 485, L29
Ekström S. et al., 2012, *A&A*, 537, A146
Fossati L. et al., 2015, *A&A*, 582, A45
Krtička J., 2014, *A&A*, 564, A70
Petit V. et al., 2013, *MNRAS*, 429, 398
Preston G. W., 1967, *ApJ*, 150, 547
Townsend R. H. D., Owocki S. P., 2005, *MNRAS*, 357, 251
ud-Doula A., Owocki S. P., 2002, ApJ, 576, 413
ud-Doula A., Owocki S. P., Townsend R. H. D., 2009, MNRAS, 392, 1022
Vink J. S., de Koter A., Lamers H. J. G. L. M., 2001, *A&A*, 369, 574
Wade G. A. et al., 2016, *MNRAS*, 456, 2

The Lives and Death-Throes of Massive Stars
Proceedings IAU Symposium No. 329, 2016
J.J. Eldridge, J.C. Bray, L.A.S. McClelland & L. Xiao, eds.

doi:10.1017/S1743921317002447

Re-examing the Upper Mass Limit of Very Massive Stars: VFTS 682, an isolated $\sim$130 $M_{\odot}$ twin of R136's WN5h core stars.

M. M. Rubio-Díez[1], F. Najarro[1], M. García[1], J. O. Sundqvist[1,2]

[1]Centro de Astrobiología, CSIC-INTA, Madrid, Spain
email: mmrd@cab.inta-csic.es

[2]Instituut voor Sterrenkunde, KU Leuven, Celestijnenlaan 200D, 3001 Leuven, Belgium

Abstract. Recent studies of WNh stars at the cores of young massive clusters have challenged the previously accepted upper stellar mass limit ($\sim$150 $M_{\odot}$), suggesting some of these objects may have initial masses as high as 300 $M_{\odot}$. We investigated the possible existence of observed stars above $\sim$150 $M_{\odot}$ by *i)* examining the nature and stellar properties of VFTS 682, a recently identified WNh5 very massive star, and *ii)* studying the uncertainties in the luminosity estimates of R136's core stars due to crowding. Our spectroscopic analysis reveals that the most massive members of R136 and VFTS 682 are very similar and our K-band photometric study of R136's core stars shows that the measurements seem to display higher uncertainties than previous studies suggested; moreover, for the most massive stars in the cluster, R136a1 and a2, we found previous magnitudes were underestimated by at least 0.4 mag. As such, luminosities and masses of these stars have to be significantly scaled down, which then also lowers the hitherto observed upper mass limit of stars.

1. Introduction

One of the key questions still to be addressed by the theory of very massive stars (VMS) formation is the existence and quantification of an upper stellar mass cutoff (Zinnecker *et al.* 2007). In 2005, Figer and Oey & Clark proposed an UPper Mass Limit (UPML) for massive stars around 150 $M_{\odot}$. However, this limit has been challenged by the massive stars located in the core of the young massive cluster R136 in the Large Magellanic Cloud. In particular, the initial masses estimated for the most deeply embedded stars in the core, a1, a2 and a3, by Crowther *et al.* (2010) (hereafter, CRW10) and Crowther *et al.* (2016) (hereafter, CRW16), range between 165-320 $M_{\odot}$.

Another question regards the nature of such massive objects, assuming they are single. Being located at the dense cores of young clusters, deep analysis of such VMS can be severely affected by crowding, which then can directly influence the photometric accuracy and, thus, the inferred luminosities and stellar masses.

Interestingly, in 2011 a new young WN5h VMS ($\sim$150 $M_{\odot}$), called VFTS 682 (#682), was discovered 30 pc from the young massive cluster, R136 (Evans *et al.* 2011, Bestenlehner *et al.* 2011; hereafter, BES11). Being in isolation, the aforementioned crowding problem disappears. Moreover, BES11 noted that this star shows a great resemblance in the optical with one of the members of R136's core stars, a3. In other words, #682 may be a key object for further examining the possible existence of VMS with initial masses above 150 $M_{\odot}$.

Thus, we obtained and analyzed VLT/XShooter full optical to NIR spectroscopic observations of #682, as well as available archival data at different wavelengths (FLAMES

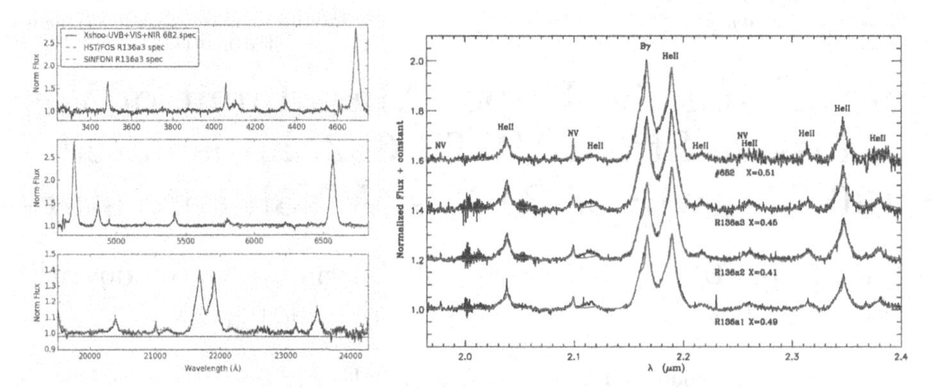

Figure 1. *Left:* a3 UV HST/FOS (blue) and VLT/SINFONI K-band (red) spectra compared to UVB and NIR XShooter #682 spectra (black). *Right:* From top to bottom, CMFGEN model fits (magenta) to XShooter and SINFONI K-band spectra of #682, a3, a2 and a1, respectively.

spectra). Simultaneously, we compared these data with available archive observations of a3 (FOS spectra; SINFONI/K-band spectral cubes).

2. Analyzing VFTS 682

Our detailed quantitative spectroscopy analysis by CMFGEN provided an excellent fit to the observed 3000 to 25000 Å of the spectrum of #682, together with an improved determination of the stellar properties and metal abundances ($T^{*}_{\tau 10}$ = 51.2 kK; log $L/L_{\odot}$=6.48; v_{∞} = 2350 Kms^{-1}; f = 0.1; log $\dot{M}$ = -4.53; X_H = 0.51) with respect to BES11. Using evolutionary tracks by Köhler *et al.* (2015), we estimated a $M_{current}$ ∼130 $M_{\odot}$ and $M_{initial}$ ∼ 145 $M_{\odot}$ for #682 (table 2).

In addition, the spectral resemblance between XShooter spectra of #682 and FOS optical and SINFONI K-band spectra of a3 is quite clear: they are almost spectroscopic twins (see fig.1-left). Furthermore, by comparing K-band spectra of #682, a1, a2 and a3†, we observed that the stars are very similar, being the hydrogen mass fraction in these four stars slightly different‡ (see fig.1-right). Moreover, according to CRW10 and CRW16, a1, a2 and a3 also have the same T_{eff} and extinction parameters. However, their luminosity estimates, and therefore their stellar masses, are quite different. Thus, the main difference between these stars comes from their estimated K magnitudes. In order to understand this apparent contradiction, we revisited the SINFONI K-band estimates of R136's core stars.

3. Revisiting SINFONI R136 data

We performed spectrophotometry of the SINFONI spectral data cubes of the most massive members of R136's core, i.e. c, b, a3, a2 and a1. We used a 2MASS filter curve and two different absolute flux calibrators, the standard star (STD) of the night and the star c (K = 11.34 ± 0.08, Campbell *et al.* 2010). We also included b and c in our study and used c as calibrator star for consistency with the procedure followed by CRW10.

The spectra of the stars were extracted for aperture radii from 1×FWHM (∼4 pixels) to 2×FWHM (∼11 pixels), except for a1 and a2. Because of their proximity, the higher the aperture radius, the higher the contamination by the closer companion. Therefore, for a1 and a2 only spectra with aperture radii as large as 7 px were extracted.

† SINFONI K-band observations cover R136's core stars b, c, a3, a2 and a1.
‡ The Br_{γ} to HeII lines ratio is very sensitive to changes in the Hydrogen mass-fraction

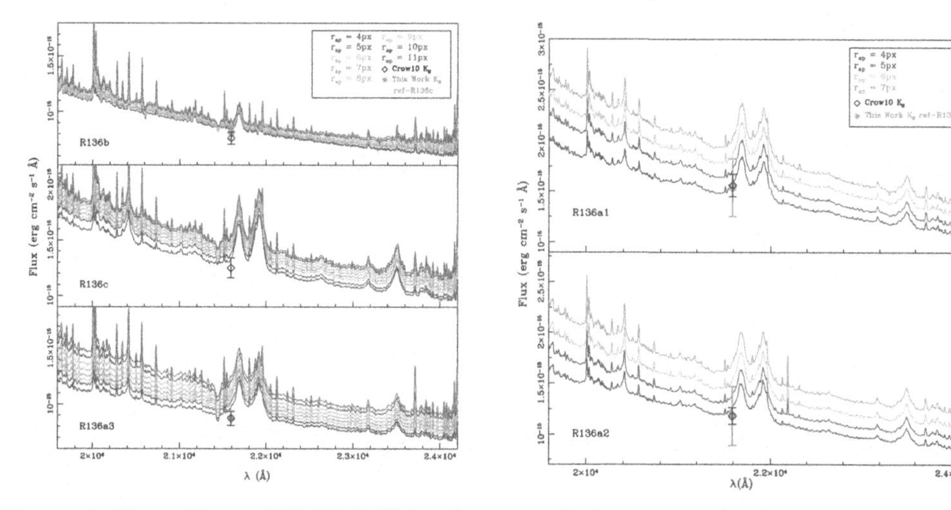

Figure 2. Flux calibrated SINFONI K-band spectra of c, b, a3, a2 and a1 for the noBS approach. Each color represents the flux-calibrated spectra extracted with different aperture radii. Black symbols represent photometric values by CRW10. Yellow symbols represent our computed K values an aperture radius of ~4 pixels and c as flux calibrator.

In order to analyze how crowding affects flux estimates, we applied two different approaches, not subtracting (noBS) and subtracting (BS) the background estimated from an annulus aperture.

Thus, in the noBS approach (fig.2) we found not only that the flux of the stars increases with aperture radius but also that this increment is larger for those stars placed in areas with higher crowding. We recovered CRW10's K-band estimates of b, a1, a2 and a3, for the smallest aperture radii and using c as flux calibrator (fig.2). On the other hand, when appliying the BS approach the flux converges with increasing aperture radius, as expected in aperture photometry. However, this is true only for the "most" isolated stars in the sample, i.e. b, c and a3 (fig.3-left). For the most embedded stars, a1 and a2, the flux not only never converges with aperture radius but also shows an irregular behaviour, which is very likely due to crowding (fig.3-right). In sight of this, it is not possible to compute reliable K-band estimates for a1 and a2, but just to provide lower limits to their K magnitudes.

We concluded that the most reliable K-band estimates are those obtained with the BS approach and using the STD star as absolute flux calibrator. We discarded c as reliable flux calibrator since it is suspected to be a binary system (Townsley *et al.* 2006, Schnurr *et al.* 2009), and also has a high uncertainty in its estimated K magnitude (Campbell *et al.* 2010).

We present the estimated K magnitudes in this work for R136's core stars compared with previous published values (CRW10) in table 1. Note that as crowding increases (from c to a2), not only the uncertainty in the measurement increases but also the difference between the K values estimated in this work and by CRW10. For a1 and a2, the K magnitudes in table 1 are actually lower limits (upper limits to their luminosities), corresponding to the computed values for an aperture radius of 4 pixels, for which the estimated flux will be less contaminated by the closer companion.

4. Discussion and Conclusions

The major findings in this work are that, in general, the uncertainty in the flux measurements heavily depends on the followed methodology and the used flux calibrator, and

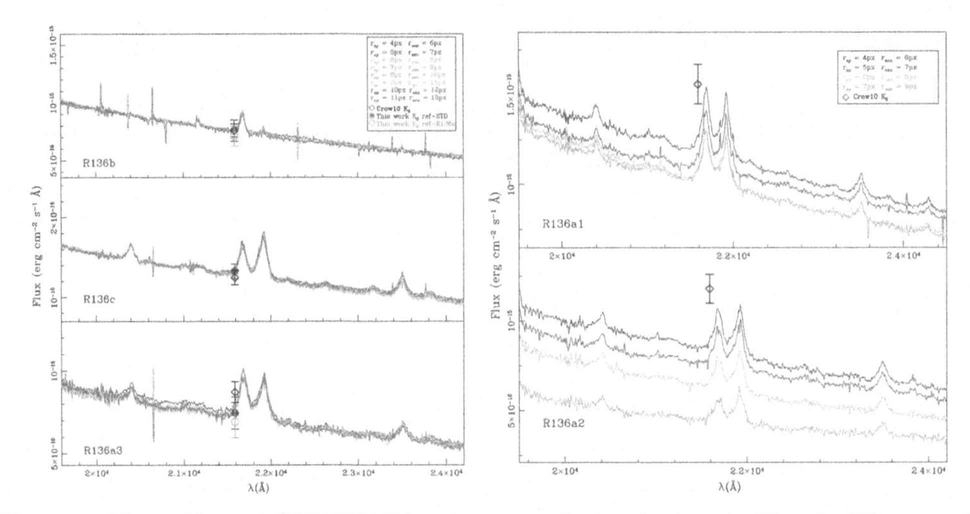

Figure 3. Flux calibrated SINFONI K-band spectra of c, b, a3, a2 and a1 for the BS approach. Each color represents the flux-calibrated spectra for different aperture and annulus radii. Black symbols represent previous estimates by CRW10. Yellow and green symbols represent our computed K values, using c and the STD star as flux calibrators, respectively.

Table 1. K_S magnitudes derived in this work for R136's core stars compared with previous values by CRW10.

Source	This work	CRW10
R136c	11.27 ± 0.09	11.34 ± 0.08
R136b	11.89 ± 0.11	11.88 ± 0.08
R136a3	11.91 ± 0.20	11.73 ± 0.08
R136a2*	$\gtrsim 11.89 \pm 0.30$	11.40 ± 0.08
R136a1*	$\gtrsim 11.51 \pm 0.30$	11.10 ± 0.08

* Corresponding to the smallest aperture radius.

that neglecting background contamination in crowded regions can lead to underestimating magnitudes. In particular, we found that the K magnitudes previously derived for the R136's core stars studied in this work were underestimated, by 0.2 magnitudes in the case of a3, and by 0.4 magnitudes, at least, for a2 and a1. To check how reliable our magnitude corrections are and if they are consistent with the measurements at other wavelengths, we compared the estimated K-band flux in this work with their UV fluxes in the literature (Hunter *et al.* 1995, Heap *et al.* 1994, de Marchi *et al.* 2011, CRW16). If a1, a2 and a3 have the same spectral type (WN5h), temperature and extinction (CRW10, CRW16), then the ratio between the fluxes of the stars in UV is the same as their flux ratio in K-band. Of the three stars, a3 is least affected by crowding† and its measured fluxes at UV and 5550Å are relatively stable ($\mathrm{F1500A}^{a3}_{Heap1994} \approx \mathrm{F1500A}^{a3}_{CRW16}$; $\mathrm{WFC2}^{a3}_{Hunter1995} \approx \mathrm{WFC3}^{a3}_{deMarchi2013}$). Therefore, we decided to use a3 as the reference star. Following this rationale, the UV a3/a1 and a3/a2 flux ratios are 0.71 and 1.24, respectively (CRW16). Thus, if $K(a3) = 11.91$, we would obtain a K-band value of 11.54 and 12.14 for a1 and a2, respectively. Note that whereas the K magnitude estimated this way for a1 is very close to our computed lower limit of 11.51, for a2 this value is even higher than our lower

† Multi-wavelength images show a3 as less contaminated by crowding than a1 and a2 (HST/WFC2-WFC3, MAD/H-K; SINFONI/K-band)

limit estimate of 11.89 (table 1). In other words, our K-band values are quite consistent, while, probably, the lower limit of the K magnitude for a2 is still underestimated.

Finally, our K-band corrections for a3, a2 and a1 lead to the determination of lower luminosities than previous estimates, by 0.1, 0.14 and 0.3 dex for a3, a2, a1, respectively. Using evolutionary tracks by Köhler *et al.* (2015), these corrections in luminosity translate into lower current masses and, therefore, into considerably lower initial masses than previous estimates for a1, a2 and a3 (table 2).

Source	log $L/L_\odot$	M_{curr} ($M_\odot$)	M_{ini} ($M_\odot$)	τ (Myr)	Ref.
#682	**6.48 ± 0.2**	**131 ± 25**	**147 ± 29**	**1.44 ± 0.25**	**This work**
R136a3	**6.48 ± 0.2**	**123 ± 24**	**141 ± 30**	**1.70 ± 0.16**	**This work**
	6.58 ± 0.09	175 ± 35	180 ± 35	1.5 ± 0.2	CRW16
R136a2	**$\lesssim$ 6.49**	**$\lesssim$ 120**	**$\lesssim$ 140**	**1.74 ± 0.55**	**This work**
	6.63 ± 0.09	190 ± 35	195 ± 35	1.6 ± 0.2	CRW16
R136a1	**$\lesssim$ 6.64**	**$\lesssim$ 171**	**$\lesssim$ 194**	**1.2 ± 0.42**	**This work**
	6.94 ± 0.09	315 ± 55	325 ± 50	0.8 ± 0.2	CRW16

Table 2. Luminosities, ages (τ) and current and initial masses estimated in this work and by CRW16.

In summary, our analyses reveal that the most massive members of R136 and #682 are very similar stars, a3 and #682 are basically twins, and confirm that crowding can severely affect flux estimates. In particular, we find that previous SINFONI K-band magnitudes of R136's core stars were underestimated. Our K magnitude corrections result in lower luminosities and therefore in lower stellar masses. As such, for #682, a3, and a2, we estimated initial masses similar or lower than 150 $M_\odot$, whereas for a1 we derived an initial mass lower than 195 $M_\odot$ (and, actually, the crowding issues discussed above suggest it is quite likely even smaller). In conclusion, the hitherto observed upper mass limit for massive stars, as large as ~325 $M_\odot$, has to be significantly scaled down, and the violation of the UPML by a1 needs more evidence to be settled.

References

Bestenlehner, J. M., Vink, J. S., Gräfener, G. *et al.* 2011, *A&A*, 530, L14
Campbell, M. A., Evans, C. J., Mackey, A. D., *et al.* 2010, *MNRAS*, 405, 421
Crowther, P. A., Schnurr, O., Hirschi, R., *et al.* 2010, *MNRAS*, 408, 731
Crowther, P. A., Caballero-Nieves, S. M., Bostroem, K. A., *et al.* 2016, *MNRAS*, 458, 624
De Marchi, G., Paresce, F., Panagia, N., *et al.* 2011, *ApJ*, 739, 27
Evans, C. J., Taylor, W. D., Hénault-Brunet, V., *et al.* 2011, *A&A*, 530, A108
Figer, D. F. 2005, *Nature*, 434, 192
Heap, S. R., Ebbets, D., Malumuth, E. M, *et al.* 1994, *ApJL*, 435, L39
Hunter, D. A., Shaya, E. J., Holtzman, J. A., *et al.* 1995, *ApJ*, 448, 179
Köhler, K., Langer, N., de Koter, A., *et al.* 2015, *A&A*, 573, A71
Oey, M. S., & Clarke, C. J.2005, *ApJL*, 620, L43
Schnurr, O., Chené, A.-N., Casoli, J., *et al.* 2009, *MNRAS*, 397, 2049
Townsley, L. K., Broos, P. S., Feigelson, E. D., *et al.* 2006 *AJ*, 131, 2164
Zinnercker, H., & Yorke, H. W.2007, *ARAA*, 45, 481

The Lives and Death-Throes of Massive Stars
Proceedings IAU Symposium No. 329, 2016
J.J. Eldridge, J.C. Bray, L.A.S. McClelland & L. Xiao, eds.

doi:10.1017/S1743921317001831

New runaway O-type stars in the first Gaia Data Release

Jesús Maíz Apellániz[1], Rodolfo H. Barbá[2], Sergio Simón-Díaz[3,4], Ignacio Negueruela[5] and Emilio Trigueros Páez[1,6]

[1]Centro de Astrobiología, CSIC-INTA, Spain
email: jmaiz@cab.inta-csic.es
[2]Universidad de La Serena, Chile
[3]Instituto de Astrofísica de Canarias, Spain
[4]Universidad de La Laguna, Spain
[5]Universidad de Alicante, Spain
[6]Universidad Complutense de Madrid, Spain

Abstract. We have detected 13 new runaway-star candidates of spectral type O combining the TGAS (Tycho-Gaia Astrometric Solution) proper motions from Gaia Data Release 1 (DR1) and the sample from GOSSS (Galactic O-Star Spectroscopic Survey). We have also combined TGAS and Hipparcos proper motions to check that our technique recovers many of the previously known O-type runaways in the sample.

Keywords. astrometry, stars:early-type, Galaxy:kinematics and dynamics

1. Introduction

On 14 September 2016 the first Gaia Data Release (DR1) was presented (Brown *et al.* 2016). Gaia DR1 includes parallaxes and proper motions from TGAS (Tycho-Gaia Astrometric Solution) for the majority (but not all) of the Tycho-2 stars. The excluded Tycho-2 stars include all of the very bright objects but also some dimmer ones. TGAS proper motions exist for a significantly larger number of stars than for Hipparcos and, for the stars in common between both catalogs, they are more precise.

The Galactic O-Star Spectroscopic Survey (GOSSS, Maíz Apellániz 2011) is obtaining $R \sim$2500, high-S/N, blue-violet spectroscopy of all optically accessible Galactic O stars. To this date, three survey papers (Sota *et al.* 2011, 2014; Maíz Apellániz *et al.* 2016) have been published with a total of 590 O stars. Several additional hundreds have already been observed and will be published in the near future.

2. Data and methods

Our initial plan with Gaia DR1 was to analyze the parallaxes in order to increase the meager number of useful trigonometric distances available for O stars (van Leeuwen 2007; Maíz Apellániz *et al.* 2008). However, the TGAS parallaxes for O stars provide little new information, as the brightest O stars are not included and only one star, AE Aur, has $\pi_o/\sigma_\pi > 6$, where π_o is the observed parallax and σ_π is the parallax uncertainty. It should be remembered that, in general, $<d> \neq 1/\pi_o$, that is, the inverse of the observed parallax is not an unbiased estimator of the trigonometric distance (Lutz & Kelker 1973; Maíz Apellániz 2001, 2005).

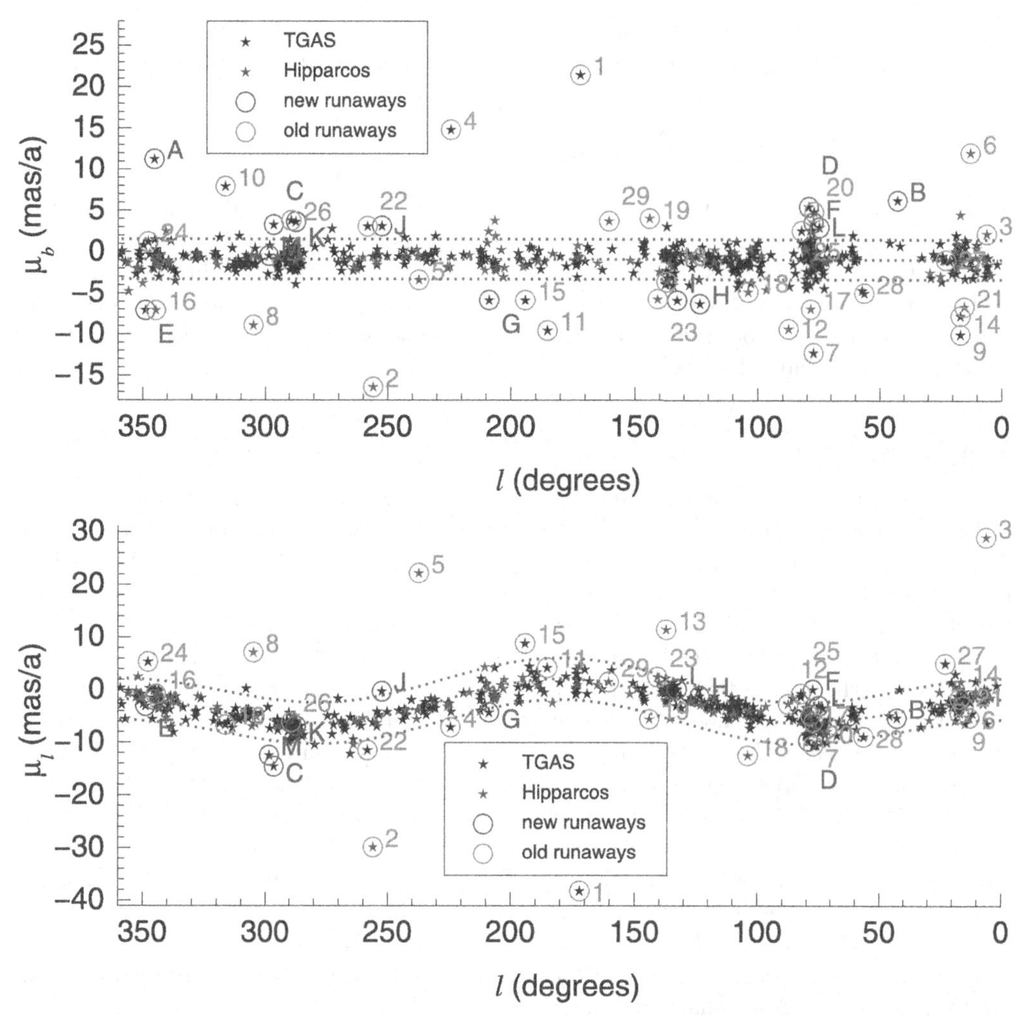

Figure 1. Galactic latitude (top) and longitude (bottom) TGAS (black) and Hipparcos (blue) proper motions for the O-star sample in the sample. Circles and IDs identify runaway candidates, both new (red, letters) and previously known (orange, numbers). See Table 1 for ID correspondences. The dotted green lines represent the functions and 2σ deviations used to detect runaway candidates.

On the other hand, the TGAS proper motions proved to have useful information. By cross-matching TGAS and GOSSS (including unpublished objects) we found 525 Galactic O stars with proper motions, of which we discarded 5 due to their large uncertainties. For the unmatched GOSSS stars we searched for Hipparcos proper motions and discovered another 96 objects, of which 7 were discarded for the same reason. That left us with a total of 520 + 89 = 609 Galactic O stars with good TGAS or Hipparcos proper motions (of those, 427 are in the published GOSSS papers).

The proper motions in RA (μ_α) and declination (μ_δ) were transformed into their equivalents in Galactic latitude (μ_b) and longitude (μ_l). A robust mean for μ_b (reflecting the solar motion in the vertical direction), $<\mu_b>$, and a robust standard deviation, σ_{μ_b}, were calculated. For μ_l we robustly fitted a functional form $f(l) = a_0 + a_1 \cos l + a_2 \cos 2l$ and we also calculated the robust standard deviation, σ_{μ_l}, from the fit. Results for stars with good TGAS or Hipparcos proper motions are shown in Fig. 1.

To detect runaway stars we computed the normalized difference (in standard deviations) of the difference between the observed proper motions and the fitted ones i.e.:

$$\Delta = \sqrt{\left(\frac{\mu_b'}{\sigma_{\mu_b}}\right)^2 + \left(\frac{\mu_l'}{\sigma_{\mu_l}}\right)^2},$$

where $\mu_b' = \mu_b - <\mu_b>$ and $\mu_l' = \mu_l - f(l)$ are the corrected proper motions, and sorted the results from largest to smallest (Table 1). This 2-D method is simpler than a full computation of the 3-D velocities (e.g. Tezlaff *et al.* 2011) and can yield false positives and negatives (see below), but it has the advantage of being self-contained and, therefore, less prone to errors introduced by the required external measurements in the 3-D method (distances and radial velocities).

3. Results

The runaway star candidates detected by our 2-D method are shown in Table 1, divided in previously known and new objects. The cut in Δ is the same in both cases and was empirically established at 3.5 (the stars at the top of the lists, AE Aur and HD 155 913, have values of 27.6 and 10.1, respectively) by comparing our results with those of Tetzlaff *et al.* (2011).

To check for false negatives in our list, we searched Tetzlaff *et al.* (2011) for runaway candidates with $P_{v_{\rm pec}} > 0.5$ missing in Table 1 but present in our sample. There are 33 objects missing but, of those, 30 were detected by Tetzlaff *et al.* (2011) based mainly on their radial velocities as they have (a) $P_{v_{\rm r,pec}} > P_{v_{\rm t,pec}}$ and (b) $P_{v_{\rm t,pec}} < 0.5$. One of the remaining three objects is HD 93 521, the highest - by far - latitude Galactic O star ($b = 52^{\rm o}$), which is difficult to detect in a 2-D method designed for objects near the Galactic Plane. The other two, HD 108 and HDE 227 018, have TGAS proper motions with significantly smaller uncertainties and closer to the mean values than the Hipparcos values, which were the ones used by Tetzlaff *et al.* (2011). Hence, a 3-D reanalysis would likely reduce their $P_{v_{\rm t,pec}}$. Therefore, we conclude that our method correctly picks up those runaway stars with large tangential velocities but, as expected, misses some which are moving mostly in a radial direction.

What about false positives? The final answer will lie, of course, in future work, but there is a good reason why the new 13 objects had not been detected before as runaways. Eight of them do not have Hipparcos proper motions and the remaining five were not included in Tetzlaff *et al.* (2011). Another indirect evidence in favor of the reality of the runaway condition for the new candidates is that in several cases it is possible to trace back the past motion of the star through its corrected proper motion to a cluster or an association as its possible origin (Table 1). Note that three of the new candidates are in the Cygnus region of the Galactic Plane (ALS 11 244, HDE 229 232 AB, and HD 192 639). For HD 46 573 we detect a bow shock in WISE images whose relative position with respect to the star is consistent with the corrected proper motion (Fig. 2).

4. Future work

Our plans for the incoming years are:

• Calculate extinction corrections (both $E(4405 - 5495)$ and R_{5495}) for all the stars in the sample with CHORIZOS (Maíz Apellániz 2004) using the Maíz Apellániz *et al.* (2014) family of extinction laws in order to obtain accurate spectroscopic parallaxes and compare them with the Gaia trigonometric parallaxes.

Table 1. Previously known O-type runaway stars (left) and new candidates (right). Each list is sorted by Δ, the normalized deviation from the mean latitude and longitude proper motions for their Galactic longitude. An ID (a number for previously known objects, a letter for new ones) is provided for each star in order to identify it in Fig. 1. The T/H flag indicates the origin of the proper motions (TGAS or Hipparcos, respectively). All new candidates have TGAS proper motions. Reference codes are listed in the bibliography. GP stands for Galactic Plane. The corrected proper motions are in mas/a.

Previously known				New candidates				
ID	Name	T/H	Ref.	ID	Name	μ'_b	μ'_l	Possible origin
1	AE Aur	T	H01	A	HD 155 913	12.14	0.17	NGC 6322
2	ζ Pup	H	H01	B	ALS 18 929	7.14	−1.15	GP, 10.6° away
3	ζ Oph	H	H01	C	HD 104 565	4.16	−8.80	GP, 4.0° away
4	HD 57 682	T	M04	D	ALS 11 244	6.38	−3.34	Cyg OB2
5	μ Col	H	H01	E	HD 155 775	−6.20	−1.10	—
6	HD 157 857	H	M04	F	HDE 229 232 AB	4.50	6.38	NGC 6913
7	Y Cyg	T	M05a	G	HD 46 573	−4.97	−4.66	GP, 2.6° away
8	HD 116 852	H	M04	H	BD +60 134	−5.38	2.04	Cas OB7
9	V479 Sct	T	R02	I	HD 12 323	−4.96	2.13	Per OB1
10	HD 124 979	T	T11	J	CPD -34 2135	4.00	4.74	—
11	HD 36 879	T	M04	K	HD 94 024	4.52	−0.64	Carina Nebula
12	68 Cyg	H	T11	L	HD 192 639	3.90	3.28	Dolidze 4
13	HD 17 520 A	H	M04	M	AB Cru	0.42	−6.83	—
14	BD -14 5040	T	G08					
15	HD 41 997	T	M04					
16	HD 152 623 AaAbB	H	M05b					
17	HD 201 345	H	T11					
18	λ Cep	H	H01					
19	α Cam	H	M05b					
20	HD 192 281	T	M04					
21	HD 175 876	H	T11					
22	HD 75 222	T	T11					
23	HD 14 633 AaAb	H	T11					
24	HD 153 919	T	M04					
25	HD 195 592	T	T11					
26	HD 96 917	T	T11					
27	BD -08 4617	T	M04					
28	9 Sge	T	M04					
29	ξ Per	H	H01					

- Extend the analysis to the rest of the OB stars. This will allow us to search for additional runaway stars and redo the study of the spatial distribution of OB stars in the solar neighborhood of Maíz Apellániz (2001) with the much better Gaia data.
- Use the multiepoch OWN (Barbá *et al.* 2010 and contribution by the same author in these proceedings), IACOB (Simón-Díaz *et al.* 2015 and contribution by the same author in these proceedings), and CAFÉ-BEANS (Negueruela *et al.* 2015) data to obtain radial velocities for OB stars corrected for binarity as a necessary step to accurately calculate their 3-D velocity.
- Expand the GOSSS sample by observing new stars.
- Incorporate the results from the new Gaia Data Releases.

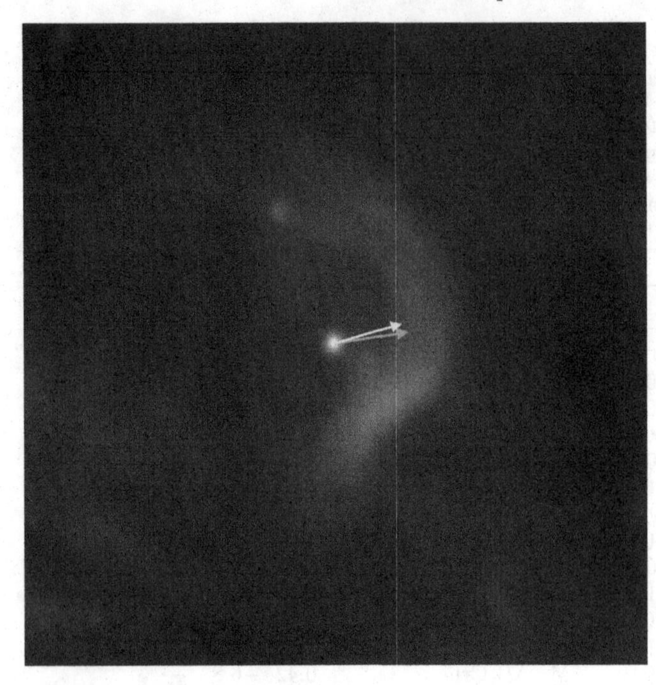

Figure 2. WISE W3+W2+W1 RGB mosaic of HD 46 573. The green arrow indicates the TGAS original proper motion and the yellow arrow the corrected proper motion (see text). Note how the bow shock correctly aligns with the proper motion. The field size is 6.′87 × 6.′87 with N to the top and E to the left.

References

Barbá, R. H. *et al.* 2010, *RvMxAA (CS)*, 38, 30
Brown, A. G. A. *et al.* 2016, *A&A*, 595, A2
Gvaramadze, V. V. & Bomans, D. J. 2008, *A&A*, 490, 1071 (G08)
Hoogerwerf, R., de Bruijne, J. H. J., & de Zeeuw, P. T. 2001, *A&A*, 365, 49 (H01)
Lutz, T. E. and Kelker, D. H. 1973, *PASP*, 85, 573
Maíz Apellániz, J. 2001, *AJ*, 121, 2737
Maíz Apellániz, J. 2004, *PASP*, 116, 859
Maíz Apellániz, J. 2005, *ESASP*, 576, 179
Maíz Apellániz, J., Alfaro, E. J., & Sota, A. 2008, *arXiv*, 0804.2553
Maíz Apellániz, J. *et al.* 2011, *Highlights of Spanish Astrophysics*, VI, 467
Maíz Apellániz, J. *et al.* 2014, *A&A*, 564, A63
Maíz Apellániz, J. *et al.* 2016, *ApJS*, 224, 4
Mdzinarishvili, T. G. 2004, *Astrophysics* 47, 155 (M04)
Mdzinarishvili, T. G. & Chargeishvili, K. B. 2005, *A&A*, 431, L1 (M05a)
Meurs, E. J. A., Fennell, G., & Norci, L. 2005, *ApJ*, 624, 307 (M05b)
Negueruela, I. *et al.* 2015, *Highlights of Spanish Astrophysics*, VIII, 524
Ribó, M. *et al.* 2002, *A&A*, 384, 954 (R02)
Simón-Díaz, S. *et al.* 2015, *Highlights of Spanish Astrophysics*, VIII, 576
Sota, A. *et al.* 2011, *ApJS*, 193, 193
Sota, A. *et al.* 2014, *ApJS*, 211, 10
Tetzlaff, N., Neuhäuser, R., & Hohle, M. M. 2011, *MNRAS*, 410, 190 (T11)
van Leeuwen, F. 2007, *A&A*, 474, 653

The Lives and Death-Throes of Massive Stars
Proceedings IAU Symposium No. 329, 2016
J.J. Eldridge, J.C. Bray, L.A.S. McClelland & L. Xiao, eds.

doi:10.1017/S1743921317003040

The evolution of magnetic fields in hot stars†

Mary E. Oksala[1,2], Coralie Neiner[2], Cyril Georgy[3], Norbert Przybilla[4], Zsolt Keszthelyi[5,6], Gregg Wade[5], Stéphane Mathis[7,2], Aurore Blazère[8,2], Bram Buysschaert[2,9]

[1]Department of Physics, California Lutheran University, 60 West Olsen Road #3700, Thousand Oaks, CA 91360, USA
email: moksala@callutheran.edu

[2] LESIA, Observatoire de Paris, PSL Research University, CNRS, Sorbonne Universités, UPMC Univ. Paris 06, Univ. Paris Diderot, Sorbonne Paris Cité, 5 place Jules Janssen, 92195 Meudon, France

[3] Geneva Observatory, University of Geneva, chemin des Maillettes 51, 1290 Sauverny, Switzerland

[4] Institut für Astro- und Teilchenphysik, Universität Innsbruck, Technikerstr. 25/8, 6020, Innsbruck, Austria

[5] Department of Physics, Royal Military College of Canada, PO Box 17000 Station Forces, Kingston, ON K7K 0C6, Canada

[6] Department of Physics, Engineering Physics and Astronomy, Queens University, 99 University Avenue, Kingston, ON K7L 3N6, Canada

[7] Laboratoire AIM Paris-Saclay, CEA/DRF - CNRS - Université Paris Diderot, IRFU/SAp Centre de Saclay, 91191 Gif-sur-Yvette, France

[8] Institut d'Astrophysique et de Géophysique, Université de Liège, Quartier Agora (B5c), Allée du 6 août 19c, 4000 Sart Tilman, Liège, Belgium

[9] Instituut voor Sterrenkunde, KU Leuven, Celestijnenlaan 200D, 3001, Leuven, Belgium

Abstract. Over the last decade, tremendous strides have been achieved in our understanding of magnetism in main sequence hot stars. In particular, the statistical occurrence of their surface magnetism has been established (~10%) and the field origin is now understood to be fossil. However, fundamental questions remain: how do these fossil fields evolve during the post-main sequence phases, and how do they influence the evolution of hot stars from the main sequence to their ultimate demise? Filling the void of known magnetic evolved hot (OBA) stars, studying the evolution of their fossil magnetic fields along stellar evolution, and understanding the impact of these fields on the angular momentum, rotation, mass loss, and evolution of the star itself, is crucial to answering these questions, with far reaching consequences, in particular for the properties of the precursors of supernovae explosions and stellar remnants. In the framework of the BRITE spectropolarimetric survey and LIFE project, we have discovered the first few magnetic hot supergiants. Their longitudinal surface magnetic field is very weak but their configuration resembles those of main sequence hot stars. We present these first observational results and propose to interpret them at first order in the context of magnetic flux conservation as the radius of the star expands with evolution. We then also consider the possible impact of stellar structure changes along evolution.

Keywords. techniques: polarimetric, stars: magnetic fields, stars: early-type, supergiants, stars: evolution.

---

† Based on observations obtained at the Canada-France-Hawaii Telescope (CFHT) operated by the National Research Council of Canada, the Institut National des Sciences de l'Univers of the CNRS of France, and the University of Hawaii, and at the European Southern Observatory (ESO), Chile (program ID 094.D-0274A, 094.D-0274B, 095.D-0155A, 096.D-0072A, and 097.D-0156A).

1. Introduction

Large-scale surveys of hot, massive stars, such as MiMeS (Wade *et al.* 2016) and BOB (Morel *et al.* 2015), have recently revealed that ~10% of hot stars host magnetic fields on the main sequence (Grunhut & Neiner, 2015; Grunhut *et al.* 2017, see also Wade *et al.*, these proceedings). Typically, the structure and strength of these fields are quite homogeneous, with dipole-dominated structure and strength ranging from ~0.3 - 20 kG. Similar magnetic field structure and strength are seen in massive pre-main sequence (PMS) stars, primarly Herbig Ae/Be stars, indicating that magnetic PMS massive stars are the progenitors of their main sequence (MS) counterparts (Alecian *et al.* 2013). Presently lacking, however, is information about the presence of such fields in the post-main sequence stellar evolution phases.

A realistic picture of what happens to these magnetic fields as the star begins its evolution off the MS and toward the red side of the HR diagram is important to understand not only the impact of the field on the star's evolution, but also to determine how the star's evolution can impact the magnetic field strength and structure. The role of magnetism at evolved phases is essential, as it impacts mass loss, stellar angular momentum redistribution and loss, internal mixing of nucleosynthetic products, differential rotation, and convection (e.g., Maeder *et al.* 2014). Ultimately, the most massive stars will end their lives as supernovae and understanding the path from the MS to this end point will give insight into the different circumstances that will affect the final state, including the field's influence on the energy budget of the explosion and the final rotation and convection properties.

Despite the wealth of spectropolarimetric data obtained in the last decade, until recently the only two known cases of evolved hot stars with detected magnetic fields were the O9.5I star ζ Ori Aa, a barely-evolved O-type star located in a binary system with a polar field strength of ~140 G (Blazère *et al.* 2015), and the B1.5II star ϵ CMa, evolved just to the end of the MS and hosting a weak field of a few tens of gauss (Fossati *et al.* 2015). No strongly evolved magnetic hot stars was known.

The observed magnetic fields of evolved hot stars are expected to weaken due to magnetic flux conservation, in which an enlarged stellar radius causes a decrease in the surface magnetic field. The surface magnetic field observed should change with time (t) according to the relation: $B(t) = \frac{B_{MS} R_{MS}^2}{R(t)^2}$, where the subscript MS indicates the radius (R) and polar field strength (B) at a particular point on the main sequence. If we know the distribution of magnetic field strengths expected for various spectral types on the MS (Shultz *et al.*, these proceedings) and combine this with radius predictions from stellar evolution models, we can roughly predict the secular variation of the measurable magnetic field throughout the star's evolution, if we assume that magnetic flux conservation is the only appreciable effect (Keszthelyi *et al.*, these proceedings). However, Fossati *et al.* (2016) indicate that other processes may act to enhance the decline in observed magnetic field strength, acting as a magnetic decay. Moreover, these evolved hot stars begin to develop subsurface convective regions and dynamo action, that could interact with the fossil magnetic field, causing further changes.

2. BRITEpol Survey

The BRITE Constellation of nano-satellites (Weiss *et al.* 2014) was launched with the goal of time-series photometric observations of every star with magnitude V $\leqslant$ 4. In response to this substantial endeavor, a coinciding magnitude-limited survey was planned

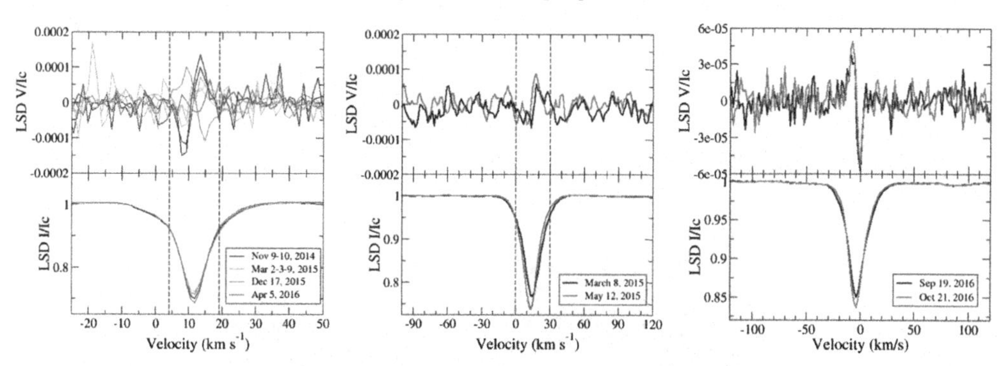

Figure 1. LSD Stokes I (bottom) and V (top) profiles of the magnetic giant and supergiant stars ι Car (A7Ib; left), HR 3890 (A8Ib; middle), and 19 Aur (A5II; right).

to observe each BRITE target using the technique of spectropolarimetry, to deduce the incidence of magnetism across all spectral types.

The BRITE spectropolarimetric (BRITEpol) survey (Neiner *et al.* 2016) aims to observe ~600 stars with 3 high-resolution spectropolarimetric instruments: HARPSpol (R~100,000) at the 3.6-m ESO telescope in Chile, Narval (R~68,000) at Téléscope Bernard Lyot (TBL) in France, and ESPaDOnS (R~68,000) at the Canada-France-Hawaii Telescope (CFHT) in Hawaii. By observing a single circular polarization spectrum or Stokes V spectrum, the presence of a magnetic field can be detected, followed up with a second observation for confirmation. The survey detected 52 magnetic stars of a wide variety of spectral types ranging from M-type stars to O-type stars, from PMS stars to evolved stars. In particular, two magnetic evolved hot stars were discovered: ι Car and HR 3890. Currently, follow up monitoring of the most interesting magnetic stars, including all hot stars, is in progress.

2.1. ι *Car*

ι Car (HR 3699) is an A7Ib supergiant, for which we determined $T_{\rm eff} = 7500 \pm 150$ K and $\log(g) = 1.85 \pm 0.1$. Stellar evolution models (Georgy *et al.* 2013) indicate the star has evolved off the MS, and is currently either in its first crossing of the HR diagram or on the blue loop, but in either case clearly evolved. The star is found to have an age of 19-56 Myr, and an enlarged radius of 50-70 $R_\odot$.

The observed Stokes V spectrum of ι Car, when analyzed with the multi-line technique Least-Squares Deconvolution (LSD; Donati *et al.* 1997), revealed an obvious magnetic signature (Fig. 1, left panel). This initial analysis allows us to derive a current dipolar field strength of $B_{\rm pol} \geqslant 3$ G. Based on the star's current and past radius estimates and the equation in Sect. 1, the magnetic dipolar field strength that would have been measurable when the star was still on the MS is $B_{\rm MS} \sim 350 - 800$ G, if we consider that the surface field strength is changed purely by magnetic flux conservation. This range is consistent with the typical magnetic field strengths currently observed in massive stars on the MS. Repeated observations of the star over a period of ~ 1.5 years indicate a rotation period of ~ 2 years based on the modulation of the magnetic field signature (see Fig. 1).

2.2. *HR 3890*

HR 3890 is an A8Ib supergiant with preliminary parameters determined by this study as $T_{\rm eff} = 7500 \pm 150$ K and $\log(g) = 1.00 \pm 0.1$. Stellar evolution models paired with these parameters indicate a post-MS evolutionary state, on its initial crossing of the HR diagram. The stellar age is estimated to be between 6-12 Myr, and the star has a radius

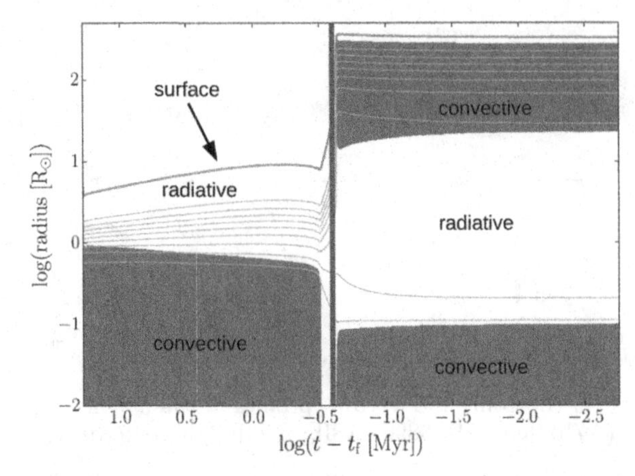

Figure 2. Kippenhahn diagram showing the evolution of the structure of the magnetic supergiant star ι Car. Convective zones are indicated in purple, while radiative zones are in white. The red line indicates the surface of the star and the thin grey lines show the mass distribution. The vertical blue line shows the current position of the star in this diagram. Figure adapted from Neiner *et al.* (in preparation).

between 190-300 $R_\odot$. The LSD average Stokes V profile shown in Fig. 1 (middle panel) reveals a magnetic signature that corresponds to a dipolar magnetic field strength of $B_{\rm pol} \geqslant 1$ G. If we consider again the relationship for the conservation of magnetic flux, and the estimated MS and current radii of HR 3890, the MS dipolar field strength would be in the range $B_{\rm MS} \sim 1450-2150$ G, again consistent with our current known properties of magnetism of MS massive stars. Additional observations are necessary to identify any variation in the Stokes V signature.

3. LIFE Survey

To fill the void of known magnetic evolved hot stars, study the evolution of their fossil magnetic fields along the stellar evolution sequence, and understand the impact of these magnetic fields on the angular momentum, rotation, mass loss, and evolution of the star itself, we have started a project called LIFE (the Large Impact of magnetic Fields on the Evolution of hot stars) to observe a statistical sample of bright (V<8), evolved (luminosity classes I, II, and III), OBA stars with ESPaDOnS at CFHT. We aim to reach very high sensitivity (below 1 G for the longitudinal field measurements) to detect weak fossil fields and possible dynamo fields. The goal is to detect more magnetic evolved hot stars and characterize the properties of the magnetic population, to test the magnetic flux conservation scenario, search for signatures of dynamos, field decay or enhancement, etc. The observations of 11 LIFE targets so far led to the detection of a first magnetic star: 19 Aur.

3.1. *19 Aur*

19 Aur (HR 1740) is an A5II giant, with a reported $T_{\rm eff}$ between 6300 and 8600 K and $\log(g)$ between 1.7 and 2.2 (e.g. Lyubimkov *et al.* 2010, Soubiran *et al.* 2016). These stellar parameters indicate an estimated $M \sim 7$ $M_\odot$ and $R \sim 35$ $R_\odot$. The LSD average Stokes V profile (Fig 1, right panel) reveals a clear magnetic signature corresponding to a dipolar magnetic field strength of $B_{\rm pol} \geqslant 6$ G. Magnetic flux conservation predicts that the MS field strength of 19 Aur would have been $B_{\rm MS} \sim 500$ G, typical of MS stars.

4. Impact of internal structure evolution

As massive stars evolve, convective regions appear at the top of the radiative zone below the stellar surface (see Fig. 2). In these regions, dynamo fields could develop (see Augustson *et al.*, these proceedings) and interact with the initial, already present, fossil magnetic field. Using MHD simulations, Featherstone *et al.* (2009) showed that the interactions between a dynamo and fossil field inside a star may increase the strength of the dynamo and modify the obliquity of the fossil field. Whether such effects indeed occur will have to be tested once the LIFE project has provided a sufficient number of magnetic evolved hot stars.

5. Conclusions

At present there are only a handful of known magnetic evolved hot stars. Their surface magnetic fields are very weak, often of only a few gauss. At first order, these observed field strengths agree with the basic assumption of magnetic flux conservation during stellar evolution. The LIFE project aims to expand our knowledge of magnetism in evolved hot stars by obtaining high quality observations of a larger number of giants and supergiants to obtain statistical information, test the flux conservation scenario, search for signatures of complex dynamo fields, and identify other potential field decay and enhancement effects.

References

Alecian, E., Wade, G. A., Catala, C., Grunhut, J. H., Landstreet, J. D., Bagnulo, S., Böhm, T., Folsom, C. P., *et al.*, 2013, MNRAS, 429, 1001

Blazère, A., Neiner, C., Tkachenko, A., Bouret, J.-C., & Rivinius, T., 2015, A&A, 582, A110

Donati, J.-F., Semel, M., Carter, B. D., Rees, D. E., & Collier Cameron, A., 1997, MNRAS, 291, 658

Featherstone, N. A., Browning, M. K., Brun, A. S., & Toomre, J., 2009, ApJ, 705, 1000

Fossati, L., Castro, N., Morel, T., Langer, N., Briquet, M., Carroll, T. A., Hubrig, S., Nieva, M. F., *et al.*, 2015, A&A, 574, A20

Fossati, L., Schneider, F. R. N., Castro, N., Langer, N., Simón-Díaz, S., Müller, A., de Koter, A., Morel, T., *et al.*, 2016, A&A, 592, A84

Georgy, C., Ekström, S., Granada, A., Meynet, G., Mowlavi, N., Eggenberger, P., & Maeder, A., 2013, A&A, 553, A24

Grunhut, J. H. & Neiner, C., 2015, in K. N. Nagendra, S. Bagnulo, R. Centeno, & M. Jesús Martínez González (eds.), IAU Symposium, Vol. 305, pp 53-60

Grunhut, J. H., Wade, G. A., Neiner, C., Oksala, M. E., Petit, V., Alecian, E., Bohlender, D. A., Bouret, J.-C., *et al.*, 2017, MNRAS, 465, 2432

Lyubimkov, L. S., Lambert, D. L., Rostopchin, S. I., Rachkovskaya, T. M., & Poklad, D. B., 2010, MNRAS, 402, 1369

Maeder, A., Przybilla, N., Nieva, M.-F., Georgy, C., Meynet, G., Ekström, S., & Eggenberger, P., 2014, A&A, 565, A39

Morel, T., Castro, N., Fossati, L., Hubrig, S., Langer, N., Przybilla, N., Schöller, M., Carroll, T., *et al.*, 2015, in G. Meynet, C. Georgy, J. Groh, & P. Stee (eds.), IAU Symposium, Vol. 307, pp 342-347

Neiner, C., Wade, G., Marsden, S., & Blazère, A., 2016, arXiv:1611.03285

Soubiran, C., Le Campion, J.-F., Brouillet, N., & Chemin, L., 2016, A&A, 591, A118

Wade, G. A. and Neiner, C. and Alecian, E. and Grunhut, J. H. and Petit, V. and Batz, B. d. and Bohlender, D. A. and Cohen, D. H., *et al.*, 2016, MNRAS, 456, 2

Weiss, W. W., Moffat, A. F. J., Schwarzenberg-Czerny, A., Koudelka, O. F., Grant, C. C., Zee, R. E., Kuschnig, R., Mochnacki, S. , *et al.*, 2014, in Guzik, W. J. Chaplin, G. Handler, & A. Pigulski (eds.), IAU Symposium, Vol. 301, pp 67-68

The Lives and Death-Throes of Massive Stars
Proceedings IAU Symposium No. 329, 2016
J.J. Eldridge, J.C. Bray, L.A.S. McClelland & L. Xiao, eds.

doi:10.1017/S1743921317002459

Magneto-asteroseismology of massive magnetic pulsators

B. Buysschaert[1,2], C. Neiner[1] and C. Aerts[2,3]

[1]LESIA, Observatoire de Paris, PSL Research University, CNRS,
Sorbonne Universités, UPMC Univ. Paris 06, Univ. Paris Diderot, Sorbonne Paris Cité,
5 place Jules Janssen, F-92195 Meudon, France
email: bram.buysschaert@obspm.fr

[2]Instituut voor Sterrenkunde, KU Leuven,
Celestijnenlaan 200D, 3001 Leuven, Belgium

[3]Dept. of Astrophysics, IMAPP, Radboud University Nijmegen,
6500 GL Nijmegen, the Netherlands

Abstract. Simultaneously and coherently studying the large-scale magnetic field and the stellar pulsations of a massive star provides strong complementary diagnostics suitable for detailed stellar modelling. This hybrid method is called magneto-asteroseismology and permits the determination of the internal structure and conditions within magnetic massive pulsators, for example the effect of magnetism on non-standard mixing processes. Here, we overview this technique, its requirements, and list the currently known suitable stars to apply the method.

Keywords. stars: magnetic fields, stars: oscillations (including pulsations), stars: interiors, stars: rotation

1. Introduction

Thanks to the efforts of large surveys to detect magnetic fields in stars, such as MiMeS (Wade *et al.*, 2016), BinaMIcS (Alecian *et al.*, 2015), the BOB campaign (Morel *et al.*, 2015), and the BRITE spectropolarimetric survey (Neiner *et al.*, 2016), the total number of known magnetic massive stars continues to increase. More than 55 massive stars are now known to show a clear magnetic signature in their Stokes V measurements. However, the full extent of the effects of such a large-scale magnetic field being present at the stellar surface remains unexplored. This magnetic field can, for example, lead to abundance inhomogeneities at the stellar surface or confine circumstellar material in a magnetosphere. Additionally, it is expected that the stellar structure, and thus also the stellar evolution, will be influenced by this surface magnetic field.

Fortunately, some of these magnetic massive stars also host stellar pulsations. These permit us to probe the internal conditions at multiple layers, enabling a direct comparison with stellar structure and evolution models. This provides a unique way of studying the impact of magnetism on, for example, the physics of non-standard mixing processes. Furthermore, magneto-asteroseismology provides strong observational constraints, since information about the stellar surface and environment is retrieved from the magnetic field, whereas information about the interior comes from pulsations.

2. Magnetic massive stars

Magnetic fields are only detected for seven to ten percent of all studied massive hot (spectral type OB) stars, and the field occurrence does not depend on the spectral type (Grunhut & Neiner 2015). Because these magnetic fields seem to be stable over long timescales and their strength does not seem to correlate with known stellar properties, it is assumed that they are of fossil origin and are frozen into the radiative envelope of the stars. The fields are those of the birth molecular cloud, partly trapped inside the pre-main sequence star during the cloud collapse phase, possibly further enhanced by a dynamo in the early fully convective stellar phase (Neiner *et al.*, 2015a). Typically, the polar field strength ranges from about a hundred Gauss up to several kiloGauss. However, some weaker fields, below 100 G, have recently been detected (e.g. Fossati *et al.*, 2015).

The stellar magnetic field influences many different regions of the star with various effects. In the deep interior of the star, the field influences the internal mixing of the star and this affects the size of the convective core overshooting region, changing the lifetime of the star by decreasing the amount of fuel for nuclear burning. Magnetic stars can also confine their stellar winds, due to their strong magnetic fields, into a magnetosphere, which slows down the rotational velocity of the star. This process is called magnetic braking and its effect is observed by the low (projected) rotational velocities for strongly magnetic stars. This magnetic braking is essentially an efficient transport of angular momentum (Mathis & Zahn 2005; Maeder & Meynet 2014). At the stellar surface, the magnetic fields can create and sustain areas of chemical over- or under-abundances and/or large temperature differences, which are called spots.

3. Pulsating massive stars

Depending on the type of pulsations and accounting for differences in global stellar properties, massive pulsators are classified as different types (see Aerts *et al.*, 2010 for a monograph on asteroseismology). The most massive main-sequence stars (having spectral type O9 – B2) are labeled β Cep pulsators. They mainly oscillate with low-order pressure modes, having periods of several hours. Less massive pulsating stars (spectral type B2 – B9) are called Slowly Pulsating B-type (SPB) stars, which pulsate with high-order gravity modes with a period of the order half to a few days. Both type of pulsations are driven by the heat mechanism, related to the iron opacity bump (e.g. Dziembowski *et al.*, 1993). Since their theoretical instability regions overlap, hybrid β Cep/SPB pulsators are expected around spectral type B2 and have been observed. Moreover, the occurrence of stellar pulsations seems to be uncorrelated to the presence of large-scale magnetic fields. Thanks to the advent of dedicated space-missions, such as MOST (Walker *et al.*, 2003), CoRoT (Baglin *et al.*, 2006), Kepler/K2 (Borucki *et al.*, 2010; Howell *et al.*, 2014), and the BRITE constellation (Weiss *et al.*, 2014), the number of known pulsating massive stars has drastically increased over the last decade.

To be able to successfully relate the stellar pulsations to the internal properties, by means of comparing them to detailed stellar and seismic models, the geometry of the pulsation mode needs to be known. This is nearly impossible from white-light photometry, except when rotational splitting is detected and / or for SPB pulsators where regular period patterns permit a direct approach to perform the mode identification. Therefore, one generally studies the line profile variations (LPVs) seen in metallic absorption lines or the amplitude ratios from multi-color photometry to unravel the mode geometry (e.g. De Cat *et al.*, 2005).

Table 1. Known magnetic massive pulsators, having N detected pulsation modes.

	Star	N	Type	$P_{\rm rot}$ [d]	$B_{\rm pol}$ [G]	Magnetic characterization	Binary?	SpT	References
1	HD 43317	> 100	both	0.90	~ 900	dip.; $i \in [20, 50]°$; $\beta \in [70, 86]°$		B3IV	(1), (2)
2	β Cen Ab	< 17	β Cep		~ 250		Y	+ B1III	(3), (4)
3	β Cep	5	β Cep	12.0	~ 300	dip.; $i \sim 60°$; $\beta \sim 95°$	Y	B0III +	(5), (6)
4	V2052 Oph	3	β Cep	3.64	~ 400	dip.; $i \sim 70°$; $\beta \sim 35°$		B2IV/V	(7), (8), (9)
5	β CMa	3	β Cep		< 30			B1II/III	(10), (11), (12)
6	16 Peg	3	β Cep	1.44	~ 500	dip.; $i \sim 70°$; $\beta \sim 70°$		B3V	(13), (14), (15)
7	ϵ Lup A	'LPV bump'	β Cep		~ 600		Y	B2V +	(16), (17)
	ϵ Lup B	> 2	β Cep		~ 300		Y	+ B3V	(16), (18)
8	ξ^1 CMa	1	β Cep	2.18	~ 600			B1V	(19), (20)
9	HD 96446	1	β Cep	23.4	~ 7500			B2IIIp	(21), (22), (23)
10	ζ Cas	1?	SPB	5.37	~ 150	dip.; $i \sim 30°$; $\beta \sim 105°$		B2IV/V	(24), (25)
11	σ Lup	1?	SPB	3.09	~ 300	dip.; $i \sim 60°$; $\beta \sim 90°$		B2III	(26), (14)
12	ϕ Cen	'LPV bump'	β Cep	1.14	~ 900			B2IV	(27), (28)

Notes:
(1): Briquet *et al.*, 2013; (2) Pápics *et al.*, 2012; (3): Alecian *et al.*, 2011; (4): Pigulski *et al.*, 2016; (5): Henrichs *et al.*, 2013; (6): Telting *et al.*. 1997; (7): Neiner *et al.*, 2012a; (8): Handler *et al.*, 2012; (9): Briquet *et al.*, 2012; (10): Fossati *et al.*, 2015; (11): Mazumdar *et al.*, 2006; (12): Shobbrook *et al.*, 2006; (13): Henrichs *et al.*, 2009; (14) Koen & Eyer 2002; (15): De Cat *et al.*, 2007; (16): Shultz *et al.*, 2015; (17): Schrijvers *et al.*, 2002; (18): Uytterhoeven *et al.*, 2005; (19): Hubrig *et al.*, 2006; (20): Williams 1954; (21): Borra & Landstreet 1979; (22): Järvinen *et al.*, 2017; (24): Neiner *et al.*, 2012b; (24): Briquet *et al.*, 2016; (25): Neiner *et al.*, 2003; (26): Henrichs *et al.*, 2012; (27): Alecian *et al.*, 2014; (28): Telting *et al.*, 2006.

4. Magnetic massive pulsators

4.1. *Magneto-asteroseismology*

We speak of magneto-asteroseismology when we study magnetic pulsators in a coherent manner, combining both the magnetometric and seismic studies. Each independent study contains information or constrains observables from a specific layer of the star. From asteroseismology, we gain information on the density, composition, and chemical mixing in multiple internal layers (depending on the number of studied frequencies). Additionally, when rotationally split pulsation modes are observed, the internal rotation profile can be retrieved (e.g. Triana *et al.*, 2015). From magnetometry, surface properties are determined, related to the chemical composition, including spots, and the magnetic field, such as its geometry, obliquity, and strength. Magnetic studies also provide constraints about the wind geometry and the circumstellar environment. Moreover, the stellar rotation period and the inclination angle towards the observer are also retrieved.

However, the seismic information is only available when two (or more) stellar pulsations have been observed and their mode order and degree have been determined, since one performs a differential study between each probed internal layer. In addition, the complete magnetic characterisation can only be performed when the rotation period can be determined from the study of the field over the complete rotation period. At present, only 12 magnetic hot pulsators are known, of which only a few have a fully characterized magnetic field and well studied stellar pulsations (see Table 1). Furthermore, magneto-asteroseismic studies have only been successfully performed for β Cep (Shibahashi & Aerts 2000) and V2052 Oph (Briquet *et al.*, 2012) so far, in part because such a study is observationally demanding.

4.2. *V2052 Oph*

V2052 Oph is a magnetic β Cep pulsator, for which 3 stellar pulsation modes have been detected and studied from ground-based spectroscopy and multi-color photometry (Briquet *et al.*, 2012, Handler *et al.*, 2012). Its magnetic field was deduced to be dipolar with a strength of $\sim$ 400 G, inclined 35° to the rotation axis (Neiner *et al.*, 2012a). From detailed seismic modelling, Briquet *et al.*, (2012) showed that the convective core overshooting region is small, with no extra mixing in spite of the relatively large rotational velocity. This clearly illustrated the effect of the magnetic field: the inhibition of chemical mixing processes by the magnetic field. This result is in agreement with theoretical criteria (e.g. Spruit 1999; Mathis & Zahn 2005), which predict that a surface magnetic field of ~ 70 G is sufficient to limit the overshooting in this star. V2052 Oph is now considered as a prime example of what magneto-asteroseismology can achieve.

4.3. *Magnetic splitting*

Similar to rotation, the (internal) magnetic field can modify the stellar pulsations by lifting some of its degeneracy (e.g. Hasan *et al.*, 2005, and references therein). Instead of just one pulsation frequency, a multiplet of frequencies is then observed, where the number of frequency peaks is governed by the mode geometry, field geometry and rotation velocity, the size of the constant frequency splitting by the strength of the magnetic field and the rotation velocity, and the amplitude of the individual peaks by the geometry of the magnetic field. This effect is known as magnetic splitting, and it was proposed as a possible explanation for the observed frequency pattern of β Cep (Shibahashi & Aerts 2000). Note, however, that these authors assumed a 6 d rotation period, instead of the now known 12 d. In practice, the magnetic splitting is difficult to observe, because of the very small expected frequency difference between the peaks. However, it may lead to a wrong mode identification when unaccounted for. The current best candidate to detect magnetic splitting is HD 43317, since this star displays two close frequency patterns (Pápics *et al.*, 2012).

4.4. *Distorted Stokes spectra*

Stellar pulsations can distort spectral absorption lines. The same effect is also observed in line-averaged spectra, such as those generated by the Least-Squares Deconvolution method using in spectropolarimetric studies; both for Stokes I (intensity) and Stokes V (polarization) profiles. Luckily, this effect seems to be minimal for most spectropolarimetric observations of pulsating magnetic stars, as the exposure times are carefully tailored to remain well below the dominant or shortest pulsation period. However, in some cases, the pulsation geometry or the overall effect of the LPVs can be so large that LPVs have to be accounted for when determining the longitudinal magnetic field measurements. Currently, this is done by constructing and fitting surface averaged synthetic spectra to the observations, for which the surface has been distorted by both the stellar pulsations and the magnetic field. This concept was successfully demonstrated for β Cep, employing the PHOEBE 2.0 pre-alpha code (Neiner *et al.*, 2015b).

5. Conclusions

To study magnetic massive pulsators accurately, one needs to account for stellar pulsations when performing the magnetic observations and analysis. Conversely, the magnetic field might complicate the mode identification of the stellar pulsations. However, carefully combining both studies, strong complementary observational constraints, e.g. the amount of chemical mixing, the rotation period or the inclination angle, lead to a tightly confined seismic and stellar model. This permits to study the internal properties of such stars and the effect a large-scale magnetic field has on them. It is only thanks to recent surveys and dedicated space missions that the sample of suitable targets has become sufficient to start to perform magneto-asteroseismology on a more regular basis.

References

Aerts, C., Christensen-Dalsgaard, J., & Kurtz, D. W. 2010, *Asteroseismology, Astronomy and Astrophysics Library. ISBN 978-1-4020-5178-4. Springer Science+Business Media B.V.*
Alecian, E., Kochukhov, O., Neiner, C., *et al.* 2011, *A&A*, 536, L6
Alecian, E., Kochukhov, O., Petit, P., *et al.* 2014, *A&A*, 567, A28
Alecian, E., Neiner, C., Wade, G. A., *et al.* 2015, *IAUS*, 307, 330
Baglin, A., Auvergne, M., Boisnard, L., *et al.* 2006, *COSP*, 36, 3749
Borra, E. F. & Landstreet, J. D. 1979, *ApJ*, 228, 809
Borucki, W. J., Koch, D., Basri, G., *et al.* 2010, *Science*, 327, 977
Briquet, M., Neiner, C., Aerts, C., *et al.* 2012, *MNRAS*, 427, 483
Briquet, M., Neiner, C., Leroy, B., *et al.* 2013, *A&A*, 557, L16
Briquet, M., Neiner, C., Petit, P., *et al.* 2016, *A&A*, 587, A126
De Cat, P., Briquet, M., Daszyńska-Daszkiewicz, J., *et al.* 2005, *A&A*, 4, 1013
De Cat, P., Briquet, M., Aerts, C., *et al.* 2007, *A&A*, 463, 243
Dziembowski, W. A., Moskalik, P., & Pamyatnykh, A. A. 1993, *MNRAS*, 265, 588
Fossati, L., Castro, N., Morel, T., *et al.* 2015, *A&A*, 574, A20
Grunhut, J. H. & Neiner, C. 2015, *IAUS*, 305, 53
Handler, G., Shobbrook, R. R., Uytterhoeven, K., *et al.* 2012, *MNRAS*, 424, 2380
Henrichs, H. F., Neiner, C., Schnerr, R. S., *et al.* 2009, *IAUS*, 259, 393
Henrichs, H. F., Kolenberg, K., Plaggenborg, B., *et al.* 2012, *A&A*, 545, A119
Henrichs, H. F., de Jong, J. A., Verdugo, E., *et al.* 2013, *A&A*, 555, A46
Hasan, S. S., Zahn, J. P., & Christensen-Dalsgaard, J. 2005, *A&A*, 444, L29
Howell, S. B., Sobeck, C., Haas, M., *et al.* 2014, *PASP*, 126, 398
Hubrig, S., Briquet, M., Schöller, M., *et al.* 2006, *MNRAS*, 369, L61
Järvinen, S. P., Hubrig, S., & Ilyin, I. 2017, *MNRAS*, 464, L85
Koen, C. & Eyer, L. 2002, *MNRAS*, 331, 45
Maeder, A. & Meynet, G. 2014, *ApJ*, 793, 123
Mathis, S. & Zahn, J. P. 2005, *A&A*, 440, 653
Mazumdar, A., Briquet, M., & Desmet, M., Aerts. C. 2006, *A&A*, 459, 589
Morel, T., Castro, N., Fossati, L., *et al.* 2015, *IAUS*, 307, 342
Neiner, C., Geers, V. C., Henrichs, H. F., *et al.* 2003, *A&A*, 406, 1019
Neiner, C., Alecian, M., Briquet, M., *et al.* 2012a, *A&A*, 537, A148
Neiner, C., Landstreet, J. D., Alecian, E., *et al.* 2012b, *A&A*, 546, 44
Neiner, C., Mathis, S., Alecian, E., *et al.* 2015a, *IAUS*, 305, 61
Neiner, C., Briquet, M., Mathis, S., & Degroote, P. 2015b, *IAUS*, 307, 443
Neiner, C., Wade, G., Marsden, S., & Blazère, A. 2016, *ArXiv e-prints*, 1611.03285
Pápics, P. I., Briquet, M., Baglin, A., *et al.* 2012, *A&A*, 542, A55
Pigulski, A., Cugier, H., Popowicz, A., *et al.* 2016, *A&A*, 588, 55
Schrijvers, C., Telting, J. H., & De Ridder, J. 2002, *ASPC*, 259, 204
Shibahashi, H. & Aerts, C. 2000, *ApJ*, 531, L143
Shobbrook, R. R., Handler, G., & Lorenz, D., Mogorosi. D 2006, *MNRAS*, 369, 171
Shultz, M., Wade, G. A., & Alecian, E., BinaMIcS collaboration 2015, *MNRAS*, 454, L1
Spruit, H. C. 1999, *A&A*, 349, 189
Telting, J. H., Aerts, C., & Mathias, P. 1997, *A&A*, 322, 493
Telting, J. H., Schrijvers, C., Ilyin, I. V., *et al.* 2006, *A&A*, 452, 945
Triana, S. A., Moravveji, E., Pápics, P. I., *et al.* 2015, *ApJ*, 810, 16
Uytterhoeven, K., Harmanec, P., Telting, J. H., & Aerts, C. 2005, *A&A*, 440, 249
Wade, G. A., Neiner, C., Alecian, E., *et al.* 2016, *MNRAS*, 456, 2
Walker, G., Matthews, J., Kuschnig, R., *et al.* 2003, *PASP*, 115, 1023
Weiss, W. W., Rucinski, S. M., Moffat, A. F. J., *et al.* 2014, *PASP*, 126, 573
Williams, A. D. 1954, *PASP*, 66, 200

The Lives and Death-Throes of Massive Stars
Proceedings IAU Symposium No. 329, 2016
J.J. Eldridge, J.C. Bray, L.A.S. McClelland & L. Xiao, eds.

doi:10.1017/S1743921317002952

X-ray diagnostics of massive star winds

L. M. Oskinova[1], R. Ignace[2], D. P. Huenemoerder[3]

[1]Institut für Physik und Astronomie, Universität Potsdam, Germany
[2]Department of Physics and Astronomy, East Tennessee State University, TN 37663, USA
[3]Massachusetts Institute of Technology, Kavli Institute for Astrophysics and Space Research, 70 Vassar St., Cambridge, MA 02139, USA

Abstract. Observations with powerful X-ray telescopes, such as XMM-Newton and Chandra, significantly advance our understanding of massive stars. Nearly all early-type stars are X-ray sources. Studies of their X-ray emission provide important diagnostics of stellar winds. High-resolution X-ray spectra of O-type stars are well explained when stellar wind clumping is taking into account, providing further support to a modern picture of stellar winds as non-stationary, inhomogeneous outflows. X-ray variability is detected from such winds, on time scales likely associated with stellar rotation. High-resolution X-ray spectroscopy indicates that the winds of late O-type stars are predominantly in a hot phase. Consequently, X-rays provide the best observational window to study these winds. X-ray spectroscopy of evolved, Wolf-Rayet type, stars allows to probe their powerful metal enhanced winds, while the mechanisms responsible for the X-ray emission of these stars are not yet understood.

Keywords. stars: atmospheres, stars: early-type, stars: mass loss, stars: winds, outflows

---

1. Introduction

Stars across the Hertzsprung-Russell diagram (HRD) emit X-rays. The quiescent X-ray luminosity of coronal type stars, like our own Sun, is less than one per cent of their bolometric luminosity, $L_{\rm x}/L_{\rm bol}\lesssim 10^{-2\ldots-3}$. The X-ray luminosity of massive stars ($M\gtrsim 9\,M_\odot$) constitutes even smaller fraction of their bolometric luminosity with only $L_{\rm x}/L_{\rm bol}\approx 10^{-7}$ (Pallavicini *et al.* 1981). Yet, despite this small output, X-rays provide invaluable diagnostics of massive star winds.

While it is not yet firmly established which physical processes lead to the generation of hot plasma emitting X-rays, the observational properties of X-ray emission from massive stars are well known (Waldron & Cassinelli 2007). The X-ray spectra are dominated by strong emission lines, and are consistent with being produced by fast expanding hot optically thin plasma. Spectral diagnostics indicate that the hot plasma observed in the X-ray band is permeated with the cool stellar wind best observed at UV wavelengths.

The classical diagnostics of OB star winds is provided by the modeling of their UV spectra. In stars with strong winds, the resonance lines of metal ions usually exhibit P Cygni type profiles. Fitting these lines with a model, e.g. based on the Sobolev approximation (Hamann 1981), provides information on the product of mass-loss rate, $\dot{M}$, and the ionization fraction. Only when the ionization fraction is well constrained, one can reliably estimate mass-loss rates. It was, however, noticed early that X-ray photons in stellar wind serve as additional source of ionization of metal ions, chiefly via Auger process (Cassinelli & Olson 1979). This effect is often referred to as "superionization". Therefore, to measure mass-loss rates, especially from stars with weaker winds, the calculations of ionization balance shall account for the influence of X-rays. Oskinova *et al.* (2011) studied the UV spectra of a sample of B-type stars. The observed spectra were modeled with the non-LTE stellar atmosphere model PoWR that accounts for the

X-ray radiation. From comparison between models and observations, it was found that the winds of the stars in our sample are quite weak. The wind velocities do not exceed the escape velocity. The X-rays strongly affect the ionization structure of these winds. But this effect does not significantly reduce the total radiative acceleration. Even when X-rays are accounted for, there is still sufficient radiative acceleration to drive a stronger mass-loss than empirically inferred from fitting the UV resonance lines. These findings are in line with conclusions reached by Prinja (1989).

2. The "weak wind problem"

Weak winds are also encountered in late O-type dwarfs (Marcolino *et al.* (2009). Lucy (2012) suggested a model explaining the low mass-loss rates empirically derived for OB dwarfs. He proposed that shock-heating increases the temperature of the gas, leading to a temperatures of a few MK. The single component flow coasts to high velocities as a pure coronal wind. The model predicts that the bulk of stellar wind is hot. Only some small fraction of the wind remains cold and is embedded in the hot wind.

To study the wind of a typical O9V star, we analyzed high-resolution X-ray spectra of μ Col (Huenemoerder *et al.* 2012). The analysis of the spectra did not reveal any significant traces of absorption of X-rays by the cool wind component, in agreement with Lucy's prediction. The analyses of line ratios of He-like ions revealed that the hot matter is present already in the inner wind. On the other hand, the X-ray emission lines are broadened up to terminal wind velocity, v_∞. The shape of the X-ray emission line profiles can be well described as originating from a hot plasma expanding with a usual β-velocity law. Moreover, the emission measure of the hot wind component is quite large, exceeding that of the cool wind. Considered together, these findings point out that the winds of O-dwarfs likely are predominantly in the hot phase, while the cool gas seen in the optical and UV constitutes only a minor wind fraction. Hence, the best observational window for studies of OB dwarfs is provided by X-rays.

3. X-ray pulsations in massive stars

Monitoring X-ray observations allow to study wind variability. Observations of pulsating B-type stars provide excellent means to investigate the links between stellar wind and stellar interior. Young B0-B2 type stars that are still burning hydrogen in their cores oscillate with periods of a few hours and are known as β Cephei-type variables (Dziembowski & Pamiatnykh 1993).

Recently, it has been shown that the X-ray emission from the magnetic star ξ^1 CMa pulsates in phase with the optical (Oskinova *et al.* 2014). Strong phase dependent variability was also detected in the high-resolution X-ray spectrum. The variability in N VII λ24.8 Å line was attributed to changes in wind ionization structure with stellar pulsational phase. Spectral diagnostics revealed that the hot X-ray emitting plasma is located very close to the stellar photosphere. The physical mechanism causing the X-ray pulsations is not yet known, but one may speculate that surface magnetic fields may be involved in coupling the wind to subphotospheric layers. Coming X-ray and UV observations of a representative sample of β Cephei variables will shed more light on this question.

4. X-ray variability of OB supergiants

Coherent and periodic variability is commonly observed in the UV lines of OB supergiants (Kaper *et al.* 1999). This variability is likely explained by the existence of corotating interaction regions (CIRs) in stellar winds (Mullan 1984, Hamann *et al.* 2001). Cranmer & Owocki (1996) showed that CIRs could result from bright stellar spots. Ramiaramanantsoa *et al.* (2014) detected corotating bright spots on ξ Per and suggested that they are generated via a breakout of a global magnetic field generated by subsurface convection. The CIRs may also be triggered by the (non)radial pulsations of the stellar surface (Lobel & Blomme 2008).

Oskinova *et al.* (2001) found the X-ray light-curve of the O-type dwarf ζ Oph being modulated on the rotation time scale, with a period similar to the one observed in UV lines. Similar conclusions on X-ray variability in O stars were reached by Massa *et al.* (2014) who analyzed X-ray observations of ξ Per (O7.5III) obtained with the *Chandra* X-ray telescope and contemporaneous Hα observations. The X-ray flux was found to vary by $\sim 15\%$, but not in phase with the Hα variability. The observations were not long enough to establish periodicity.

Among the O-stars best monitored in X-rays is the O4 supergiant ζ Pup. The *XMM-Newton* X-ray telescope observed it from time to time during a decade. Analysis of the obtained light curves revealed variations with an amplitude of $\sim 15\%$ on a time scale longer than 1 d, while no coherent periodicity was detected (Nazé *et al.* 2013). Interestingly, Howarth & Stevens (2014) reported a period of $P = 1.78$ d in optical photometry of this star, which they attributed to stellar pulsations.

In another O-type supergiant, λ Cep (O6.5I(n)), the X-ray flux also varies by $\sim 10\,\%$ on timescales of days, possibly modulated with the same period as the Hα emission (Rauw *et al.* 2015). The analysis of archival *XMM-Newton* observations of ζ Ori (O7I) shows that the X-ray variability of this star has similar properties to that of λ Cep and ζ Pup. X-ray variability with analogous character was found in δ Ori (O9.5II+B1V) from *Chandra* observations, where periodic fluctuations (not associated with its binary period) were identified (Nichols *et al.* 2015). Even Wolf-Rayet (WR) stars with very strong winds show modulations in their X-ray emission on the rotation timescale (Ignace *et al.* 2013). Summarizing, the evidence for X-ray variability on rotation time-scale is accumulating. This points to an association between X-ray emission, large scale structures in the stellar wind, and stellar spots.

5. X-ray emission lines in spectra of OB supergiants

Over the last decade, the analysis of emission lines observed in X-ray spectra has become a common tool for stellar wind studies. The profile of an emission line originating from an optically thin expanding shell was derived by Macfarlane *et al.* (1991). It was shown that continuum absorption in the stellar wind leads to characteristically blue-shifted and skewed line profiles, and suggested that the "skewness" of a line could be used to estimate the wind column density. This model was further extended and applied to X-ray emission line profiles (Waldron & Cassinelli 2001, Owocki & Cohen 2001, Ignace 2001). Stellar wind clumping was included in the modeling by Feldmeier *et al.* (2003). They showed shown that clumping reduces the effective wind opacity and affects the shape of the emission line profiles. In general, lines emerging from clumped winds are less skewed than those produced in homogeneous winds even when the mass-loss rates are the same (Oskinova *et al.* 2006).

Besides continuum absorption and wind clumping, the shape of an X-ray emission line also depends on the assumed velocity field (as illustrated in Figure 1), wind geometry, abundances, ionization balance, hot plasma distribution, onset of X-ray emission and resonance scattering in the hot plasma. This incomplete list shows that, contrary to

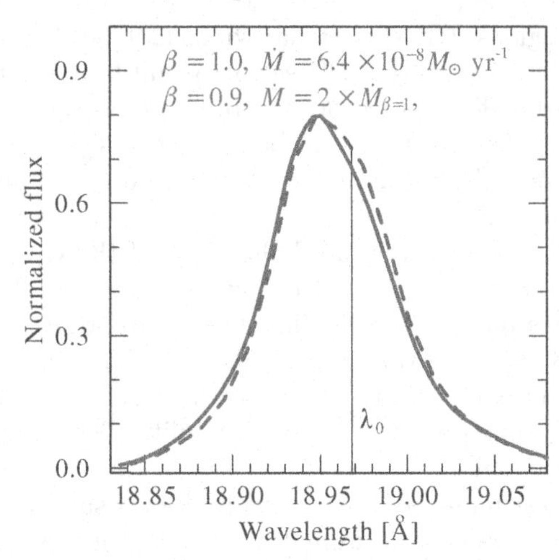

Figure 1. Illustration of the degeneracy of the X-ray emission line profiles with respect to the velocity field. The vertical line indicates the central wavelength of the O VIII Lα line. Two model lines are shown, each of them computed using the same "universal" mass-absorption coefficient κ and stellar parameters as suggested by Cohen *et al.* (2014). The blue-dotted profile is based on a smooth wind model with $\beta = 1$, onset of the X-ray emission at $R_0 = 1.33\,R_*$, and $\dot{M} = 6.3 \times 10^{-8} M_\odot\,\text{yr}^{-1}$. The red-solid line is a smooth wind model, but now for $\beta = 0.9$, $R_0 = 1.26\,R_*$, and $\dot{M} = 1.2 \times 10^{-7} M_\odot\,\text{yr}^{-1}$, i.e. twice as high.

claims in the literature, it is hardly possible to derive just one parameter, such as mass-loss rate, from fitting the shapes of X-ray emission lines by simplistic models.

On the other hand, X-ray spectroscopy complemented by the analysis of UV and optical spectra using non-LTE stellar atmosphere models is a valuable tool to constrain stellar wind properties (Hervé *et al.* 2013, Puebla *et al.* 2016). E.g., Shenar *et al.* (2015) applied PoWR non-LTE stellar atmospheres for the analysis of X-ray, optical, and UV spectra of the O star δ Ori. It was possible to reproduce the UV and optical spectra as well as the X-ray emission lines consistently, and thus derive realistic wind parameters.

6. X-ray emission from blue hypergiants and Wolf-Rayet stars

Among those stars with the most powerful stellar winds are the blue hypergiants and the WR stars. X-ray spectroscopy provides interesting insights to their stellar winds.

Recently, we analyzed the X-ray spectrum of one of the most massive and luminous stars in the Milky Way, Cyg OB2-12 ((B3Ia$^+$), obtained with the *Chandra* X-ray telescope. The analysis was complemented by modeling, using the PoWR atmosphere model (Oskinova *et al.*, in prep.).

It was shown that the X-ray spectrum of Cyg OB2-12 is produced in a hot plasma with temperatures in excess of 15 MK. Given that the stellar wind of this star is rather slow, $v_\infty \approx 400\,\text{km}\,\text{s}^{-1}$, it is difficult to explain high temperature by intrinsic wind shocks. From the ratio of fluxes in forbidden and intercombination lines in He-like ions followed that that the X-ray emitting plasma has quite a high density. Furthermore, the broadening of X-ray emission lines exceeds the one due to the stellar wind velocity. Taken together, these facts are best explained by a colliding winds scenario. Further support to the binary hypothesis is provided by the recent identification of a companion in the Cyg OB2-12 system (Maryeva *et al.* 2016, and ref. therein).

The low-resolution X-ray spectra of WR stars are well described by thermal plasmas with temperatures between 1 MK up to 50 MK (Ignace *et al.* 2003, and ref. therein).

High-resolution X-ray spectra are available so far only for one single WR star - EZ CMa (WR 6) (Oskinova *et al.* 2012). Their analysis shows that hot plasma exists far out in the wind. The X-ray emission lines are broad and strongly skewed (just as predicted by the Macfarlane *et al.* (1991) and Ignace (2001) models) being consistent with plasma expanding with constant velocity. The abundances derived from the X-ray spectra are strongly non-solar, in accordance with the advanced evolutionary state of WR stars (Huenemoerder *et al.* 2015). How X-rays are generated in winds of single WR stars is not yet understood. It is possible that, like in O type stars, X-ray generation is an intrinsic ingredient of the stellar wind driving (Gayley 2016).

To conclude, X-ray observations provide important diagnostics of massive stars and their winds at nearly all evolutionary stages.

References

Cassinelli, J. P. & Olson, G. L., 1979, *ApJ*, 229, 304
Cohen, D. H., Wollman, E. E., Leutenegger, M., *et al.*, 2014, *MNRAS*, 439, 908
Cranmer, S. R. & Owocki, S. P., 1996, *ApJ*, 462, 469
Dziembowski, W. A. & Pamiatnykh, A. A., 1993, *MNRAS*, 262, 204
&Feldmeier, A.; Oskinova, L.; Hamann, W.-R., 2003, A&A, 403, 217
Gayley, K. G., 2016, *AdSpR*, 58, 719
Hamann, W.-R., 1981, *A&A*, 93, 353
Hamann, W.-R., Brown, J. C., Feldmeier, A., *et al.*, 2001, *A&A*, 378, 946
&Hervé, A.; Rauw, G.; Nazé, Y., 2013, *A&A*, 551, 83
Howarth, I. D. & Stevens, Ian R., 2013, *MNRAS*, 445, 287
Huenemoerder, D. P., Oskinova, L. M., Ignace, R., *et al.*, 2012, *ApJ*, 756, 34
Huenemoerder, D. P., Gayley, K. G., Hamann, W.-R., *et al.*, 2015, *ApJ*, 815, 29
Ignace, R., 2001, *ApJ*, 549, 119
&Ignace, R.; Oskinova, L. M.; Brown, J. C., 2003, *A&A*, 408, 353
Ignace, R., Gayley, K. G., Hamann, W.-R., *et al.* 2013, *ApJ*, 775, 29
Kaper, L., Henrichs, H. F., Nichols, J. S., *et al.*, 1999, *A&A*, 344, 312
Lobel, A. & Blomme, R., 2008, *ApJ*, 678, 408
Lucy, L. B., 2012, *A&A*, 544, 120
Macfarlane, J. J.; Cassinelli, J. P.; Welsh, B. Y., *et al.*, 1991, *ApJ*, 380, 564
Marcolino, W. L. F., Bouret, J.-C., Martins, F., *et al.*, 2009, *A&A*, 498, 837
Maryeva, O. V., Chentsov, E. L., Goranskij, V. P., *et al.*, 2016, *MNRAS*, 458, 491
Massa, D., Oskinova, L., Fullerton, A. W., *et al.*, 2014, *MNRAS*, 441, 2173
Mullan, D. J., 1984, *ApJ*, 283, 303
Nazé, Y., Oskinova, L. M., & Gosset, E., 2013, *ApJ*, 763, 143
Nichols, J., Huenemoerder, D. P., Corcoran, M. F., *et al.*, 2015, *ApJ*, 809, 133
Oskinova, L. M., Clarke, D., & Pollock, A. M. T., 2001, *A&A*, 378, 21
Oskinova, L. M., Feldmeier, A., & Hamann, W.-R., 2006, *A&A*, 372, 313
Oskinova, L. M., Todt, H., Ignace, R., *et al.*, 2011, *MNRAS*, 416, 1456
Oskinova, L. M., Gayley, K. G., Hamann, W.-R., *et al.*, 2012, *ApJ*, 747, 25
Oskinova, L. M., Nazé, Y., Todt, H., *et al.*, 2014, *Nature Communications*, 5, 4024
Owocki, S. P. & Cohen, D. H., 2001, *ApJ*, 559, 1108
Puebla, R. E., Hillier, D. J., Zsargó, J., *et al.*, 2015, *MNRAS*, 456, 2907
Ramiaramanantsoa, T., Moffat, A. F. J., Chené, A.-N., *et al.*, 2014, *MNRAS*, 441, 910
Rauw, G., Hervé, A., Nazé, Y., *et al.*, 2015, *A&A*, 580, 59
Shenar, T., Oskinova, L., Hamann, W.-R., *et al.*, 2015, *ApJ*, 809, 135
Pallavicini, R., Golub, L., Rosner, R., *et al.*, 1981, *ApJ*, 248, 279
Prinja, R. K., 1989, *MNRAS*, 241, 721
Waldron, W. L. & Cassinelli, J. P., 2001, *ApJ*, 548, 45
Waldron, W. L. & Cassinelli, J. P., 2007, *ApJ*, 668, 456

The Lives and Death-Throes of Massive Stars
Proceedings IAU Symposium No. 329, 2016
J.J. Eldridge, J.C. Bray, L.A.S. McClelland & L. Xiao, eds.

doi:10.1017/S1743921317002393

Taking the Measure of Massive Stars and their Environments with the CHARA Array Long-baseline Interferometer

Douglas R. Gies

Center for High Angular Resolution Astronomy, Georgia State University, P.O. Box 5060, Atlanta, Georgia 30302-5060, U.S.A.
email: gies@chara.gsu.edu

Abstract. Most massive stars are so distant that their angular diameters are too small for direct resolution. However, the observational situation is now much more favorable, thanks to new opportunities available with optical/IR long-baseline interferometry. The Georgia State University Center for High Angular Resolution Astronomy Array at Mount Wilson Observatory is a six-telescope instrument with a maximum baseline of 330 meters, which is capable of resolving stellar disks with diameters as small as 0.2 milliarcsec. The distant stars are no longer out of range, and many kinds of investigations are possible. Here we summarize a number of studies involving angular diameter measurements and effective temperature estimates for OB stars, binary and multiple stars (including the σ Orionis system), and outflows in Luminous Blue Variables. An enlarged visitors program will begin in 2017 that will open many opportunities for new programs in high angular resolution astronomy.

1. The CHARA Array

Georgia State University operates the Center for High Angular Resolution Astronomy (CHARA) Array at Mount Wilson Observatory, a six-telescope interferometer designed for studies of stars and their environments at milliarcsec (mas) angular scales. The CHARA Array's six 1-m aperture telescopes are arranged in a Y-shaped configuration yielding 15 interferometric baselines from 33 to 331 meters as well as 10 independent closure phases. These include the longest OIR baselines yet implemented anywhere in the world and permit resolutions at the sub-mas level. The facility's primary components and sub-systems are described fully by ten Brummelaar *et al.* (2005), and they include telescopes, evacuated light pipes, optical path length compensation, and beam management. The final step of beam combination is made with a set of different combiners that were built primarily by members of the "CHARA Collaboration" that includes groups (and their PIs) from l'Observatoire de Paris (Vincent Coudé du Foresto), University of Michigan (John Monnier), University of Sydney (Peter Tuthill), the Australian National University (Michael Ireland), l'Observatoire de la Côte d'Azur (Denis Mourard), and the University of Exeter (Stefan Kraus). These combiners are optimized for different wavelength bands and numbers of telescopes (Table 1).

The Array is primarily used to measure interferometric visibility, a quantity that is directly related to the Fourier transform of the spatial intensity of an object in the sky. Measurements at lower spatial frequencies are related to the angular diameter of a star (and its shape if measurements are available with multiple telescope pairs), while higher spatial frequencies provide important diagnostics of the distribution of spatial intensity (for use in measuring limb darkening and spot distributions, for example). In addition,

Table 1. CHARA Array Beam Combiners

Combiner	# Tel.	Band	Typ. Limit. Mag.	Best Perf. Mag.	Spectral Resol. Power	Reference
Currently Operating						
CLASSIC	2	H or K band	7.0	8.5	Broad band	ten Brummelaar *et al.* (2005)
CLIMB	3	H or K band	6.0	7.0	Broad band	ten Brummelaar *et al.* (2013)
JouFLUOR	2	K	3.5	5.0	Broad band	Scott *et al.* (2013)
VEGA (hi-res)	2 or 3	2×7 nm, in 480-850 nm	4.0	5.0	30000 30000	Mourard *et al.* (2011)
VEGA (med-res)	2 or 3	2×35 nm, in 480-850 nm	6.5	7.5	6000 6000	Mourard *et al.* (2011)
MIRC	6	H	4.0	6.0	42	Monnier *et al.* (2006)
PAVO	2	630-900 nm	7.0	8.0	30	Ireland *et al.* (2008)
Planned						
MIRCX	6	$J + H$	7.0	9.0	42	Kraus *et al.* (2016)
MYSTIC	6	K	7.5	8.5	42	Monnier *et al.* (2016)
SPICA	6	550-850 nm	8.9	11	100	Mourard (2016)

closure phase measurements from three or more telescopes are related to asymmetries in the spatial distribution and are critical in image reconstruction from interferometry. I show in Figure 1 the range in possible measurements for the Classic two-telescope beam combiner. The x-axis shows the angular diameter of an object in the sky (measured in milli-arcseconds) and the y-axis shows the magnitude of the object in the near-infrared H-band (in the absence of interstellar extinction). The dark, solid lines show how the angular diameters vary with magnitude for stellar effective temperatures of $T_{\rm eff} = 3500$, 7000, and 25000 K. Hot objects have much larger surface intensities than cooler ones, so that the hotter O- and B-type stars appear smaller in the sky than cooler stars do at a given apparent magnitude. The gray diagonal lines show the apparent semimajor axis of binary stars of total mass 0.9, 2.3, and $34 M_\odot$ for M0 V, G0 V, and B0 V pairs, respectively, over a range in distance. Interferometers like the CHARA Array are able to resolve and map astrometric orbits for all kinds of close, spectroscopic binaries. The representative sizes of several kinds of objects that host gaseous disks are also shown in Figure 1. I discuss some of the remarkable observational results on massive stars from the CHARA Array in the next two sections.

2. Resolving massive star photospheres and outflows

The CHARA Array provides access to stars all along the main sequence. A measurement of the angular diameter together with a parallax yields the physical radius, and the effective temperature is derived from the angular diameter and spectral energy distribution. Thus, CHARA measurements lead directly to an observational Hertzsprung-Russell diagram that provides a test for theoretical models. Work has steadily progressed from cooler stars (Boyajian *et al.* 2015), through the A-type stars (Jones *et al.* 2015), and into the B-star domain (Maestro *et al.* 2013). Kathryn Gordon (GSU) is currently investigating the angular diameters of more massive stars in a sample that includes six O-type stars. Many of the early-type stars are rapid rotators, and the first science paper from CHARA highlighted the rotationally distorted shape of the B-star Regulus (McAlister *et al.* 2005). The CHARA/MIRC combiner has been used in multi-telescope

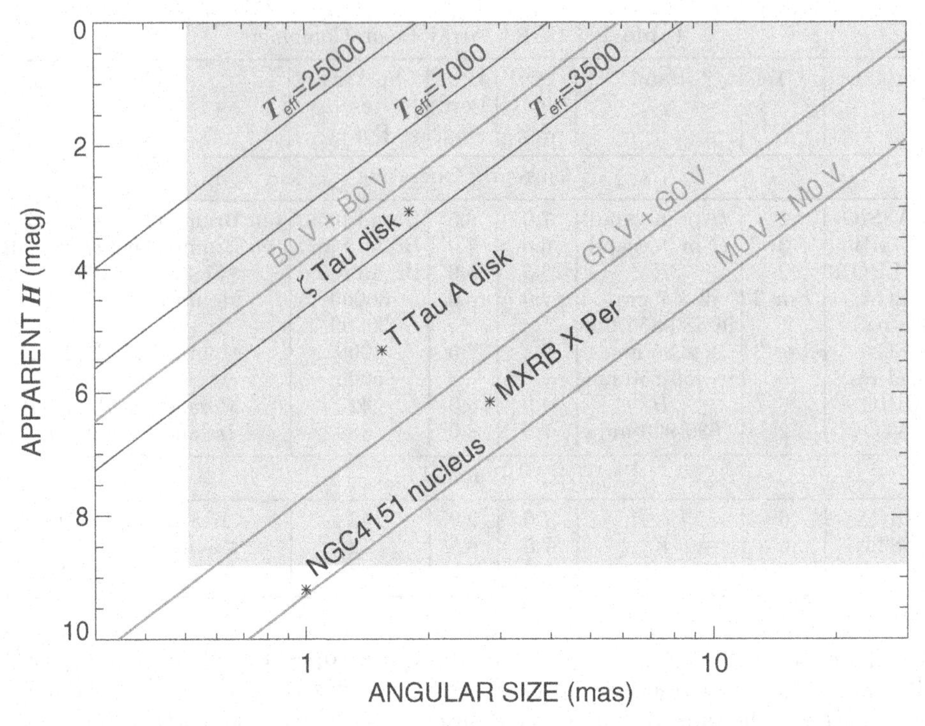

Figure 1. The approximate range in size and magnitude of targets accessible to CHARA Classic in H-band. Dark lines show diameters of stars of different temperature and gray lines show the semi-major axis of a 10 day binary over a range in distance, and these may be translated horizontally according to $(P/10)^{2/3}$. Note that the magnitude limit given here is conservative, and it is possible to go fainter in excellent seeing conditions.

configurations to obtain the first images of rapid rotators that clearly show the the hotter poles and cooler equators that result from gravity darkening (Monnier *et al.* 2007; Zhao *et al.* 2009; Che *et al.* 2011). Jones *et al.* (2016) modeled Array observations of the low inclination, rapid rotator, κ And that is the A0 V, host star of a directly imaged, low-mass planetary companion.

Massive O- and B-type stars evolve to become Cepheids, red supergiants, and luminous blue variable (LBV) stars, and investigations are underway of the properties of these older objects. Mérand *et al.* (2006) discovered extended circumstellar envelopes surrounding the Cepheids Polaris and δ Cep using very accurate visibility measurements from CHARA/FLUOR. These halos have radii several times larger than the star and contribute a few percent of the K-band flux. Additional observations by Mérand *et al.* (2007) indicate that circumstellar envelopes are often found around pulsating Cepheids, but are absent among similar non-pulsating supergiants. This indicates that the halos may be the result of pulsation-driven mass loss. Gallenne *et al.* (2012) measured diameters, distances, and pulsation modes in the classical Cepheids FF Aql and T Vul. Gallenne *et al.* (2013, 2015) detected the faint orbiting companions of a number of Cepheids enabling the determination of their relative brightness, orbital elements, precise and independent distance estimates, and masses. Combined with GAIA distances, these measurements are required to reduce eventually the systematic error of the Cepheid-based extragalactic distance scale.

The CHARA Array is ideal for studying convective structures in red supergiants, and Baron *et al.* (2014) presented some of the first results from CHARA/MIRC in the

H-band. The reconstructed images of two red supergiants show some evidence for the predicted large scale cells, and Ryan Norris (GSU) is now studying an enlarged sample. Richardson *et al.* (2013) used CHARA/MIRC to observe the supergiant P Cygni, one of the prototypes of the class of LBVs. Their reconstructed image of P Cygni clearly shows the spherical halo of flux created by the star's dense stellar wind. Richardson *et al.* (2016) found that the candidate LBV star MWC 314 is an interacting binary undergoing massive Roche lobe overflow that completely envelopes the mass gainer, perhaps what β Lyr experienced at an earlier epoch.

3. Massive binary stars

Angular orbits from CHARA Array data provide masses and distances for close and wide binaries with orbital periods from days to centuries. For example, Schaefer *et al.* (2016) determined an astrometric orbit for the σ Ori Aa,Ab binary that they combined with the spectroscopic orbit to yield very accurate masses and distance to this Ori OB1 association member (with an uncertainty of only 0.3%). Zhao *et al.* (2008) imaged the famous interacting binary β Lyr at four epochs covering approximately half the 12.9-day orbit. Those images clearly reveal the bright and elongated mass donor and thick disk surrounding the mass gainer comprising β Lyr. This represents the first direct detection of the flux of the gas torus obscuring the mass gainer star, after a century of attempts to detect the gainer by spectroscopy. The extreme mass transfer rates that occur near minimum separation in an interacting binary may lead to the formation of a circumbinary disk that is fed through leaky gas outflows through the outer axial Lagrangian points. Such mass loss processes may produce a dense and opaque circumbinary disk such as that observed in the ϵ Aur system. This enigmatic 27.1-yr eclipsing system was the subject of an observing campaign at CHARA that spanned the recent eclipse. Kloppenborg *et al.* (2010, 2015) presented images of the eclipse event made from CHARA/MIRC showing the silhouette of the optically-thick disk passing in front of the F-type supergiant primary.

The mass gainer in an interacting binary may be spun up to near the critical rate, and some of the rapidly rotating Be stars are probably the result of binary mass transfer. The former mass donor star may be stripped of its envelope, and its core will appear as a faint and hot subdwarf star (Peters *et al.* 2016). The companion of the Be star ϕ Per is one such case. Mourard *et al.* (2015) detected and mapped the orbit of the faint companion, and their CHARA/VEGA observations of the Be disk show that the gas moves in the same sense as the orbit of the companion. This finding supports the hypothesis that the Be star was spun up by mass and angular momentum transfer, so that the angular momentum being lost into the Be disk reflects the original orbital angular momentum of the mass donor. These remarkable results speak eloquently of the power interferometric imaging possesses in studying such classical systems. Binary stars continue to be a touchstone for astrophysics, renewed with each major advance in measurement technology.

4. Open access to CHARA

CHARA recently received funding from the U.S. National Science Foundation Mid-Scale Innovations Program to (1) provide open access to the community for 50 to 75 nights per year, and (2) provide an on-line data pipeline and archive of CHARA observations. Time will be allocated through the U.S. National Optical Astronomy Observatory Time Allocation Committee review beginning in 2017B (August to Dec 2017). Plans are underway for community workshops to help guest observers develop new programs and to acquire the knowledge needed to obtain observations and reduce and analyze the data. We anticipate that the new open access program will bring exciting new ideas for observational work and will help make optical long baseline interferometry part of the basic toolbox for investigations in stellar astrophysics.

Acknowledgements

This material is based upon work supported by the National Science Foundation under Grants No. AST-1211929 (McAlister), AST-1411654 (Gies), and AST-1636624 (ten Brummelaar).

References

Baron, F., Monnier, J. D., Kiss, L. L., *et al.* 2014, *ApJ*, 785, 46
Boyajian, T., von Braun, K., Feiden, G. A., *et al.* 2015, *MNRAS*, 447, 846
Che, X., Monnier, J. D., Zhao, M., *et al.* 2011, *ApJ*, 732, 68
Gallenne, A., Mérand, A., Kervella, P., *et al.* 2015, *A&A*, 579, A68
Gallenne, A., Mérand, A., Kervella, P., *et al.* 2016, *MNRAS*, 461, 1451
Gallenne, A., Monnier, J. D., Mérand, A., *et al.* 2013, *A&A*, 552, A21
Ireland, M. J., Mérand, A., ten Brummelaar, T. A., *et al.* 2008, *Proc. SPIE*, 7013, 701324
Jones, J., White, R. J., Boyajian, T., *et al.* 2015, *ApJ*, 813, 58
Jones, J., White, R. J., Quinn, S., *et al.* 2016, *ApJ* (Letters), 822, L3
Kloppenborg, B., Stencel, R., Monnier, J. D., *et al.* 2010, *Nature*, 464, 870
Kloppenborg, B., Stencel, R., Monnier, J. D., *et al.* 2015, *ApJS*, 220, 14
Kraus, S., Anugu, N., Davies, C., & Monnier, J., 2016, *CHARA Meeting*, http://www.chara.gsu.edu/files/2016Meeting/kraus.MIRCx.pdf
Maestro, V., Che, X., Huber, D., *et al.* 2013, *MNRAS*, 434, 1321
McAlister, H. A., ten Brummelaar, T. A., Gies, D. R., *et al.* 2005, *ApJ*, 628, 439
Mérand, A., Aufdenberg, J. P., Kervella, P., *et al.* 2007, *ApJ*, 664, 1093
Mérand, A., Kervella, P., Coudé du Foresto, V., *et al.* 2006, *A&A*, 453, 155
Monnier, J., Le Bouquin, J.-B., Jocou, L., *et al.* 2016, *CHARA Meeting*, http://www.chara.gsu.edu/files/2016Meeting/Monnier_MYSTIC_public_1.pdf
Monnier, J. D., Pedretti, E., Thureau, N., *et al.* 2006, *Proc. SPIE*, 6268, 62681P
Monnier, J. D., Zhao, M., Pedretti, E., *et al.* 2007, *Science*, 317, 342
Mourard, D. 2016, *CHARA Meeting*, http://www.chara.gsu.edu/files/2016Meeting/Mourard.pdf
Mourard, D., Bério, P., Perraut, K., *et al.* 2011, *A&A*, 531, A110
Mourard, D., Monnier, J. D., Meilland, A., *et al.* 2015, *A&A*, 577, A51
Peters, G. J., Wang, L., Gies, D. R., & Grundstrom, E. D. 2016, *ApJ*, 828, 47
Richardson, N. D., Moffat, A. F. J., Maltais-Tariant, R., *et al.* 2016, *MNRAS*, 455, 244
Richardson, N. D., Schaefer, G. H., Gies, D. R., *et al.* 2013, *ApJ*, 769, 118
Schaefer, G. H., Hummel, C. A., Gies, D. R., *et al.* 2016, *AJ*, 152, 213
Scott, N. J., Millan-Gabet, R., Lhomé, E., *et al.* 2013, *Journal of Astronomical Instrumentation*, 2, 40005
ten Brummelaar, T. A., McAlister, H. A., Ridgway, S. T., *et al.* 2005, *ApJ*, 628, 453
ten Brummelaar, T. A., Sturmann, J., Ridgway, S. T., *et al.* 2013, *Journal of Astronomical Instrumentation*, 2, 40004
Zhao, M., Gies, D., Monnier, J. D., *et al.* 2008, *ApJ* (Letters), 684, L95
Zhao, M., Monnier, J. D., Pedretti, E., *et al.* 2009, *ApJ*, 701, 209

The Lives and Death-Throes of Massive Stars
Proceedings IAU Symposium No. 329, 2016
J.J. Eldridge, J.C. Bray, L.A.S. McClelland & L. Xiao, eds.

doi:10.1017/S1743921317002642

The Red Supergiant Content of the Local Group

Philip Massey[1], Emily Levesque[2], Kathryn Neugent[1], Kate Evans[1,3], Maria Drout[4] and Madeleine Beck[5]

[1]Lowell Observatory, 1400 W Mars Hill Road, Flagstaff, AZ 86001 and
Dept. Physics & Astronomy, Northern Arizona University, Flagstaff, AZ 86011-6010
email: phil.massey@lowell.edu, kneugent@lowell.edu

[2]Dept. Astronomy, University of Washington, Box 351580, Seattle, WA 98195 USA
email: emsque@uw.edu

[3]REU participant, 2015; California Institute of Technology, 1200 E, California Blvd, Pasadena, CA 91125 USA
email: kevans@caltech.edu

[4]Observatories of the Carnegie Institution for Science, 813 Santa Barbara St., Pasadena, CA 91101, USA
email: mdrout@carnegiescience.edu

[5]REU participant, 2016; Wellesley College, 106 Central Street, Wellesley, MA 02481 USA
email: mbeck4@wellesley.edu

Abstract. We summarize here recent work in identifying and characterizing red supergiants (RSGs) in the galaxies of the Local Group.

Keywords. stars: early type, supergiants, stars: Wolf-Rayet

---

1. Introduction

In the Olden Days, the term "massive star" was synonymous with "hot [massive] star," and topics at these conferences were restricted to O-type stars and Wolf-Rayet (WR) stars. Gradually massive star research has expanded to include the red supergiants (RSGs). Stars with initial masses $8 - 30M_\odot$ will evolve into RSGs, and these are the progenitors of many core-collapse SNe. The amount of mass loss during the RSG phase will affect the subsequent evolution (if any) of these stars, with some evolving back to the blue side of the H-R diagram (HRD), and possibly even becoming WRs (see, e.g., Meynet *et al.* 2015).

Studying RSGs in relation to other evolved massive stars provides an exacting test of our understanding of massive star evolution, as these advanced stages act as a "sort of magnifying glass, revealing relentlessly the faults of calculations of earlier phases" (Kippenhahn & Weigert 1990). For instance, the relative number of WRs and RSGs should be a very sensitive function of the initial metallicity of the host galaxy, as first noted by Maeder *et al.* (1980). In addition, the physical properties of RSGs (effective temperatures, luminosities) provide an important check on the models.

Knowledge of the upper mass (luminosity) limit to becoming a RSG is also necessary to understand the relevance of long-period binaries on the evolution of massive stars, and in particular, the formation of WRs. Garmany *et al.* (1980) found that the close binary frequency of O-type stars is 35-40%, a number that has been confirmed by several recent studies (e.g., Sana *et al.* 2012, 2013). High angular resolution surveys and long-term radial velocity studies have shown that if long period systems (as great as

10+ years) are included, the binary frequency is significantly higher, possibly 60% or more (Caballero-Nieves *et al.* 2014, Aldoretta *et al.* 2015, Sana *et al.* 2012). But, how much do these long period systems actually matter—do the components ever interact, or do they evolve as single stars within close sight of each other? The answer to that depends upon how large the stars become. Let's take Betegeuse as a "typical" RSG; it has a radius (very roughly) of 5 AU. Imagine it had an OB star companion (it doesn't): the stars would be in contact if the period were 2 years (assuming circular orbits) from Kepler's 3rd law, assuming masses of $15M_{\odot}$+$15M_{\odot}$. But, a 5 or 10 year period would be totally uninteresting in terms of mass loss or mass transfer, unless the orbit were highly eccentric. And, above some mass ($20M_{\odot}$? $30M_{\odot}$?) we don't expect stars to become RSGs—they will instead maintain relatively small radii as they evolve to the WR phase. We can then see that it's important to measure what this limit is, so we can better understand how significant the binary channel actually is in the formation of WRs.

In this short contribution, we'll explain how we find RSGs in nearby galaxies, how we characterize their properties, and what we plan to do next. Studying these stars in nearby galaxies allows us to explore stellar evolution as a function of metallicity. It's been recognized for 40 years that mass-loss plays a significant role in the evolution of massive stars. As a result, we expect there to be galaxy-to-galaxy differences in the relative numbers of WN- and WC-type WRs, and the relative numbers of WRs and RSGs. This is because these stars will be born with different metallicities, and stellar wind mass-loss is driven through radiation pressure in highly ionized metal lines. Nearby star-forming galaxies can thus serve as our laboratories for carrying out such comparisons, as their metallicities (z) range over a factor of 30×, from the low-metallicity Sextans A and B galaxies ($z = 0.06\times$ solar), to the SMC and LMC ($z = 0.27\times$ and $0.47\times$ solar, respectively), to M33 (with a metallicity gradient going from $\sim$ solar in the center to SMC-like in the outer regions, to M31 (1.6× solar).

2. Finding RSGs

When we look at a nearby galaxy, we are looking through a sea of foreground stars in our own Milky Way. The extent of this problem was first described by Massey (1998): years earlier, Humphreys & Sandage (1980) had compared the distributions of blue and red stars in M33. It was a bit of a mystery why the former was clumpy, while the latter was smooth. Massey (1998) showed that there was significant contamination of the red star sample by foreground red dwarfs in our own Milky Way.

In the Magellanic Clouds the contamination of potential RSGs by foreground dwarfs can be nearly eliminated by the use of proper motions. In their study of LMC yellow and red supergiants, Neugent *et al.* (2012) showed that once they eliminated stars with proper motions greater than 15 mas yr^{-1} (about the limit of believability of cataloged proper motions pre-GAIA), the vast remainder of the red stars have radial velocities consistent with LMC membership. RSGs in the LMC will have V magnitudes of 9.5 to 12.5; foreground M0 dwarfs (say) in this magnitude range will be only 25-100 pc distant.

For the more distant members of the Local Group, such as M31 and M33, proper motions are not useful: at V=17-20 (the magnitude range of their RSGs), foreground red dwarfs will be at larger distances (400-1500 pc). Instead, Massey (1998) showed that one could use B-V, V-R two-color diagrams to separate supergiants from foreground stars. At these colors, model atmospheres suggested that B-V would become primarily a surface gravity indicator due to line blanketing in the blue, while V-R remains a temperature indicator. Radial velocities can be used to demonstrate the success of this, as shown in Fig. 1.

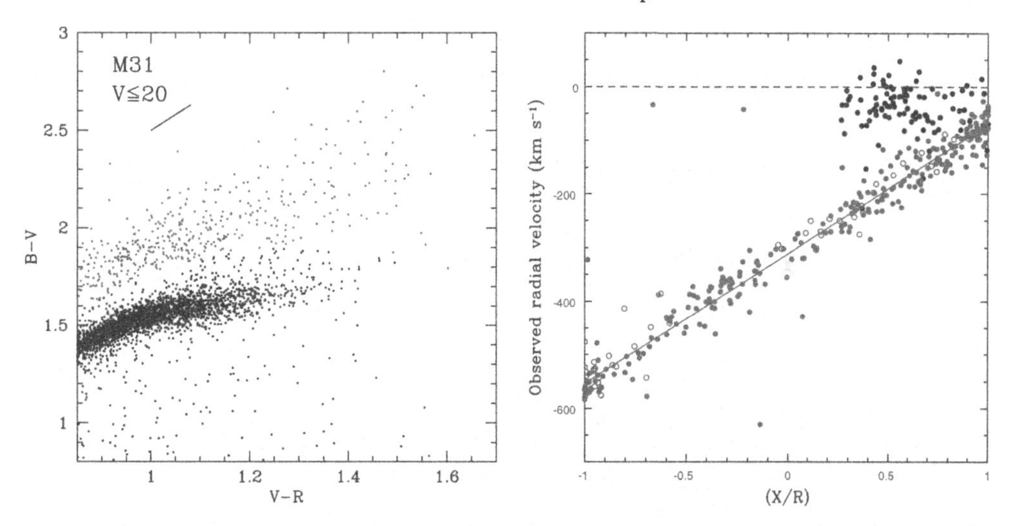

Figure 1. Left: A two color diagram of red stars seen towards M31. Stars expected to be RSGs on the basis of their B-V colors are colored red, and foreground stars, black. Right: The radial velocity of some of the sample stars. For the most part, stars we expect to be RSGs (red) follow the rotation curve of M31 (shown as a red line), while expected foreground stars (black) cluster around zero velocity. X/R denotes the position within M31. From Massey & Evans (2016).

3. Physical Properties

When our group began studying RSGs, we found that there was a bit of a problem: RSGs were considerably cooler and more luminous than the evolutionary tracks could produce (see, e.g., Massey & Olsen 2003). In fact, the locations of these stars was in the Hayashi forbidden zone (Hayashi & Hoshi 1961), where stars were no longer in hydrostatic equilibrium. Usually in such cases the problem lies with the theory, and indeed Maeder & Meynet (1987) had demonstrated that how far over the tracks extended to cooler temperatures depended upon how convection and mixing were treated. Still, what if the "observations" were wrong? After all, we don't actually observe effective temperatures and bolometric luminosities; instead, we obtain photometry and spectral types and convert these to physical properties.

Levesque *et al.* (2005) revisited this issue using a new generation of model atmospheres, and found (to their relief) that the new effective temperature scale was significantly warmer. Indeed, when they plotted the location of their stars in the HRD, there was near-perfect agreement with the evolutionary models, both in terms of the effective temperatures and upper luminosities. Furthermore, extension of this work to lower and higher metallicity environments (Levesque *et al.* 2006, Levesque & Massey 2012, Massey *et al.* 2009, Massey & Evans 2016, Drout *et al.*, 2012) have consistently shown that the change in the effective temperature scale tracks the expected metallicity-dependent change in the Hayashi limit. This was dramatically shown by Drout *et al.* (2012): as we move outwards in M33 to lower metallicities, the positions of RSGs shift to warmer temperatures.

When we are done with such studies, we can then compare the resulting HRD to that expected on the basis of the evolutionary tracks. An example is shown in Fig. 2 (right) for RSGs in M31. In this case the tracks plotted are for solar metallicity (from Ekström *et al.* 2012), which is a bit low for the metallicity of M31, but are the highest metallicity modern tracks that are available at present. We see very good agreement for the most part. The two highest luminosity stars in the figure may or may not be members; see discussion in Massey & Evans (2016).

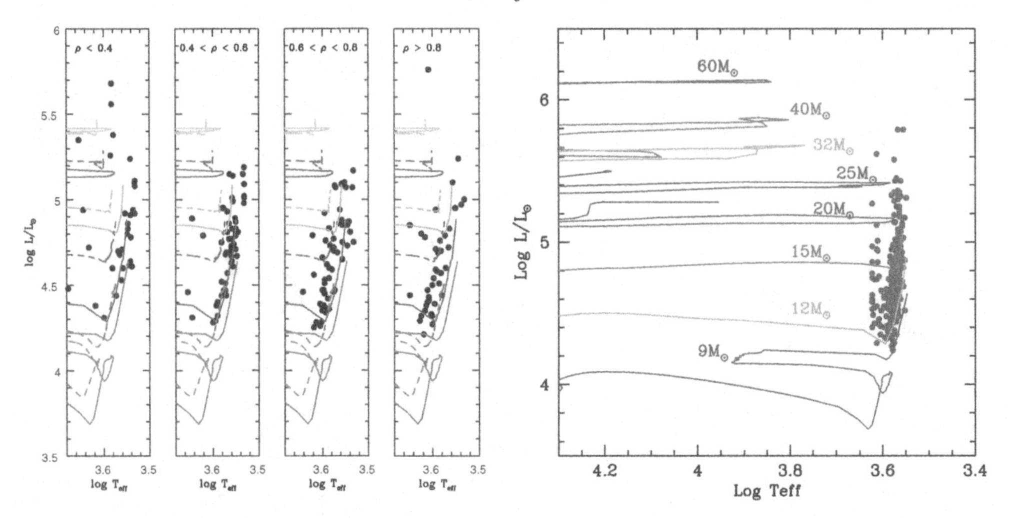

Figure 2. Left: The change in effective temperatures of RSGs in M33 with galactocentric distance ρ. At lower metallicities (greater ρ) RSGs are warmer. This is in accord with the change of the Hayashi limit with metallicity. From Drout *et al.* (2012). Right: The HRD of RSGs in M31. From Massey & Evans (2016).

4. Other Cool Things That We've Found

Along the way, we have made some other interesting discoveries. An inspection of Fig. 1 (right) reveals one star ($X/R \sim -0.15$) whose radial velocity is 300 km s^{-1} discrepant with that expected from M31's rotation curve. Was it just bad data? We had four nice MMT spectra of the star, all yielding the same result. Furthermore, the star is located about 4.5 kpc from M31's disk, about the distance a star could travel at 300 km s^{-1} (assuming a transverse peculiar velocity similar to its excess radial velocity) in 10 Myr, about the lifetime of a massive star in the RSG stage. Evans & Massey (2015) identify this star as the first runaway RSG, and the first massive extragalactic runaway star.

One of the consequences of the shifting of the Hayashi limit to warmer temperatures at lower metallicity is that the spectral types of RSGs also shift: in the Milky Way, the average RSG is about an M2 I, in the LMC it's about a M1 I, and in the SMC most RSGs are K5-7 I (Levesque *et al.* 2007). Massey *et al.* (2007) were therefore surprised to find an M4 I star in the SMC, a star which was also one of the brightest RSGs, HV 11423. The next year they reobserved the star, only to find it was then a K0-1 I. Did they just mess up? Inspection of archival spectral showed that it been caught in an M4.5-5 state on an ESO VLT spectrum taken years earlier, although no one had apparently found this remarkable at the time. Levesque *et al.* (2007) subsequently identified seven LMC and four additional SMC RSGs that were also later than expected for spectral types in their host galaxies; all of these (including HV 11423) show large V-band variability, and change their effective temperatures by 3-4% on the timescales of months. At their coolest, these stars are well into the Hayashi forbidden zone.

We have no understanding of what these stars are from an evolutionary point of view, the finding was intriguing enough for Anna Żytkow to contact us and suggest that perhaps these stars might be the long-sought after Thorne-Żytkow Objects (TŻOs). TŻOs were first proposed Thorne & Żytkow (1975, 1977); these are RSGs that have neutron star cores and hence a variety of very interesting nuclear reactions. Such objects might be the primary source of the proton-rich nuclei in the Universe, solving a long-standing mystery (Cannon 1993). The only problem was that no viable TŻO candidate had ever been identified. None of the Levesque-Massey variables prove to show the weird elemental enrichment expected of TŻO, but another RSG, selected for being too cool, did, HV 2112 (Levesque *et al.* 2014)†.

5. Where Do We Go From Here?

We have provided a quick snap-shot here of where we stand in the identification of RSGs in the Local Group, and what we have learned from these stars. We are working on characterizing complete samples of RSGs in the Magellanic Clouds, Sextans A and B, and other Local Group galaxies. Hopefully by the time of the next massive stars conference we will be able to report on clean relative numbers of RSGs and WRs throughout the galaxies of the Local Group, and compare these to both the Geneva single-star and BPASS (Eldridge & Stanway 2016) binary evolutionary models.

Acknowledgement

Support through the National Science Foundation (AST-1612874) is gratefully acknowledged.

References

Aldoretta, E. J., Caballero-Nieves, S.. M., Gies, D. R. *et al.* 2015, *AJ*, 149, 26
Caballero-Nieves, S. M., Nelan, E. P., Gies, D. R. *et al.* 2014, *AJ*, 147, 40
Cannon, R. C. 1993, *MNRAS*, 263, 817
Drout, M. R., Massey, P., & Meynet, G. 2012, *ApJ*, 750, 97
Ekström, S., Georgy, C., Eggenberger, P. *et al.* 2012, *A&A*, 537, A146
Eldridge, J. J. & Stanway, E. R. 2016, *MNRAS*, 462, 3302
Evans, K. A. & Massey, P. 2015, *AJ*, 150, 149
Garmany, C. D., Conti, P. S., & Massey, P. 1980, *ApJ*, 242, 1063
Hayashi, C. & Hoshi, R. 1961, *PASJ*, 13, 442
Humphreys, R. M., & Sandage, A.1980, *ApJS*, 44, 319
Kippenhahn, R. & Weigert, A. 1990, *Stellar Structure and Evolution* (Berlin: Springer-Verlag), 192
Levesque, E. M. & Massey, P. 2012, *AJ*, 144, 2
Levesque, E. M., Massey, P., Olsen, K. A. G., & Plez, B. 2007, *ApJ* 667, 202
Levesque, E. M., Massey, P., Olsen, K. A. G., *et al.* 2005, *ApJ*, 628, 973
Levesque, E. M., Massey, P., Olsen, K. A. G., *et al.* A. 2006, *ApJ*, 645, 1102
Levesque, E. M., Massey, P., Żytkow, A. N., & Morrell, N. 2014, *MNRAS*, 443, L94
Maccarone, T. J. & de Mink, S. E. 2016, *MNRAS*, 458, 1
Maeder, A., Lequeux, J., & Azzopardi, M. 1980, *A&A*, 90, L17
Maeder, A. & Meynet, G. 1987, *A&A*, 182, 243
Massey, P. 1998, *ApJ*, 501,153
Massey, P. & Evans, K. A. 2016, *ApJ*, 826, 224
Massey, P., Levesque, E. M., Olsen, K. A. G., Plez, B., & Skiff, B. A. 2007, *ApJ*, 660, 301
Massey, P. & Olsen, K. A. G. 2003, *AJ*, 126, 2867
Massey, P., Silva, D. R., & Levesque, E. M. 2009, *ApJ*, 703, 420
Meynet, G., Chomienne, V., Ekström, S. *et al.* 2015, *A&A*, 575, A60
Neugent, K. F., Masey, P., Skiff, B., & Meynet, G. 2012, *ApJ*, 749, 177
Sana, H., de Mink, S. E., de Koter, A. *et al.* 2012, *Sci*, 337, 444
Sana, H, de Koter A. de Mink, S. E. *et al.* 2013, *A&A*, 550, A107
horne, K. S. & Żytkow, A. N. 1975, *ApJ*, 199, L19
horne, K. S. & Żytkow, A. N. 1977, *ApJ*, 212, 832
Worley, C. C., Irwin, M. J., Tout, C. A. *et al.* 2016, *MNRAS*, 459, L31.

† Maccarone & de Mink (2016) suggest that HV 2112 is not a member of the SMC, but this was based on an unfortunate reliance on the SPM4 proper motion catalog, which is known to contain significant errors; other, more reliable catalogs show null proper motions, and the claim was quickly refuted by Worley *et al.* (2016). GAIA will of course answer this definitively.

The Lives and Death-Throes of Massive Stars
Proceedings IAU Symposium No. 329, 2016
J.J. Eldridge, J.C. Bray, L.A.S. McClelland
& L. Xiao, eds.

doi:10.1017/S1743921317003350

A high-contrast imaging survey of nearby red supergiants

Peter Scicluna[1], Ralf Siebenmorgen[2], Joris Blommaert[3], Francisca Kemper[1], Roger Wesson[4] and Sebastian Wolf[5]

[1]Academia Sinica, Institute of Astronomy and Astrophysics, 11F Astronomy-Mathematics Building, NTU/AS campus, No. 1, Section 4, Roosevelt Rd., Taipei 10617, Taiwan, Republic of China
email: `peterscicluna@asiaa.sinica.edu.tw`

[2]European Southern Observatory, Karl-Schwarzschild-Str. 2, 85748, Garching b. München, Germany

[3]Astronomy and Astrophysics Research Group, Dep. of Physics and Astrophysics, V.U. Brussel, Pleinlaan 2, 1050, Brussels, Belgium

[4]Dept. of Physics & Astronomy, University College London, Gower Street, London WC1E 6BT, UK

[5]Institute of Theoretical Physics and Astrophysics, Universität zu Kiel, Leibnizstr. 15, 24118, Kiel, Germany

Abstract. Mass-loss in cool supergiants remains poorly understood, but is one of the key elements in their evolution towards exploding as supernovae. Some show evidence of asymmetric mass loss, discrete mass-ejections and outbursts, with seemingly little to distinguish them from more quiescent cases. To explore the prevalence of discrete ejections and companions we have conducted a high-constrast survey using near-infrared imaging and optical polarimetric imaging of nearby southern and equatorial red supergiants, using the extreme adaptive optics instrument SPHERE, which was designed to image planets around nearby stars. We present the initial results of this survey, including the detection of large (500 nm) dust grains in the ejecta of VY CMa and a candidate dusty torus aligned with the maser ring of VX Sgr. We briefly speculate on the consequences for our understanding of mass loss in these extreme stars.

Keywords. stars: individual (VY CMa, VX Sgr), stars: mass loss, supergiants

---

1. Introduction

After massive stars (M $>$ 8 $M_\odot$) leave the main sequence, they undergo periods of enhanced mass loss before exploding as supernovae. This mass loss is a key factor in determining the evolution of the stars through phases including Red Supergiant (RSG), Yellow Hypergiant (YHG), Luminous Blue Variable (LBV), and Wolf-Rayet (WR) (Georgy *et al.* 2013; Smith 2014; Meynet *et al.* 2015; Georgy & Ekström 2015). A proper understanding of mass loss is therefore crucial for determining the post-main sequence evolution of massive stars and for linking classes of supernova progenitors to classes of supernovae, which remains as yet poorly understood (Georgy 2012; Groh *et al.* 2013; Ekström *et al.* 2013; Georgy *et al.* 2013; van Loon 2013).

For stars with initial masses $\lesssim 30\,M_\odot$, a large fraction of the mass loss occurs during the RSG phase. The mechanisms driving mass-loss in this phase remain a matter of debate (van Loon *et al.* 2005; Harper*et al.* 2009), in particular the origins of mass-loss asymmetries, variability and eruptions. Various mechanisms, which may not necessarily be independent, have been invoked, including non-radial pulsations, convection, binarity, and magnetic activity (e.g. Smith *et al.* 2001; Humphreys *et al.* 2007).

However, RSGs are, by definition, bright targets, and answering these questions requires high dynamic range and high angular resolution, while avoiding saturation, if the ejecta is to be separated from the central source. Numerous advances in high-contrast imaging have been driven by the need to image and characterise extrasolar planets, culminating in the development of "extreme adapative optics" (XAO) instruments, such as SPHERE (Beuzit *et al.* 2008), GPI (Macintosh *et al.* 2006) and SCExAO (Martinache & Guyon 2009), which combine coronography with a variety of differential-imaging techniques to achieve contrasts as high as 10^{-7} within $0\farcs5$.

Answering outstanding questions regarding mass-loss from evolved massive stars requires a systematic approach, in which a large sample of supernova progenitors are homogeneously analysed. Hence, we are conducting a high contrast imaging and polarimetry survey of nearby southern and equatorial evolved massive stars, primarily red supergiants, exploiting the capabilities of the new high-contrast, planet-hunting instrument SPHERE, some initial findings from which are included here.

2. Target selection

Our target selection is limited by two criteria. First, the extreme adaptive optics (XAO) exploited by modern high-contrast imagers require very bright central point sources to use as natural guide stars (NGS) for optimal AO correction. In the case of SPHERE, the AO correction begins to degrade for NGS with $m_{\rm R} > 9$. The instrument is only available in Service Mode for NGS with $m_{\rm R} \lesssim 11$ which provides high Strehl under a wide range of conditions, although it has been demonstrated that the system operates successfully with an improvement in Strehl for stars as faint as $m_{\rm R} \approx 14$ (Xu *et al.* 2015). Secondly, we require targets sufficiently close to be able to resolve structures in the circumstellar envelope with SPHERE. Given the luminosities of red and yellow supergiants, these criteria restrict us to Galactic sources on low-extinction lines-of-sight, typically within a few kpc or in nearby OB associations.

We therefore selected a sample of 18 bright galactic RSGs to observe with SPHERE, from the spectroscopically confirmed samples of Levesque *et al.* (2005) and Verhoelst *et al.* (2009), focusing on those whose 2MASS and IRAS photometry indicate excess emission, while four additional YHGs were selected based on literature searches. These targets cover a range of luminosities and mass-loss rates to ensure that we sample a useful amount of parameter space.

3. Optical Imaging Polarimetry of VY CMa and VX Sgr

3.1. *Grain sizes in VY CMa*

In Scicluna *et al.* (2015) we demonstrated some techniques for the survey on the archetypal extreme RSG VY CMa, and we summarise the results here. VY CMa was observed in December 2014 using SPHERE/ZIMPOL in active polarisation compensation mode with fast polarisation modulation. We used V- and I-band filters with a classical Lyot coronagraph to suppress the core of the PSF. The data were reduced using the pipeline to yield reduced images for the Stokes' I, Q and U components for each detector in each filter. These data were combined using python to produce total intensity, linearly-polarised intensity and polarisation angle for each filter.

Figure 1 shows the images obtained in both polarised and unpolarised light. The polarisation images show a clear centro-symmetric pattern, indicating that the polarisation is caused by the scattering of light by dust grains. As a result, we can use the fractional

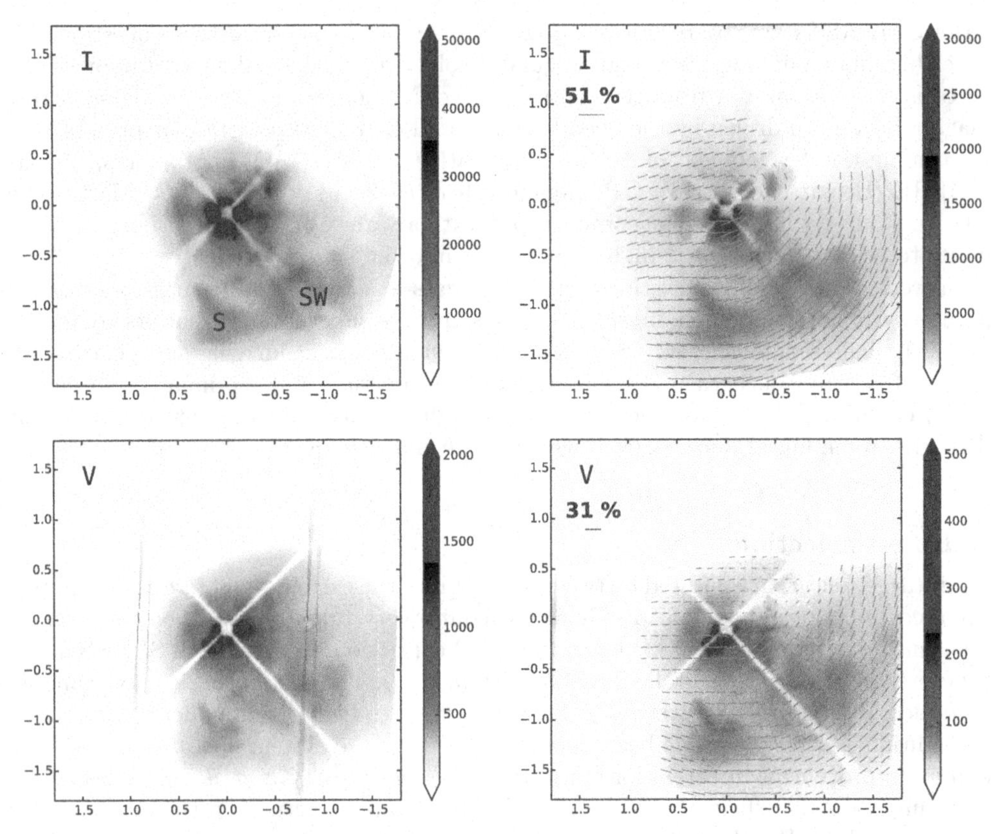

Figure 1. SPHERE observations of VY CMa based on Scicluna *et al.* (2015), with offsets in arc seconds and intensity scale in arbitrary units. *Left*: Total intensity. The locations of the South Knot and Southwest Clump are marked with 'S' and 'SW' respectively. The vertical stripes visible in the V-band data are a detector artifact resulting from the readout mode that cancels out in the polarisation data. *Right*: Polarised intensity. The overlaid vectors show the polarisation fraction and direction.

polarisation and the intensity ratio to constrain the properties of the scattering dust, in particular the size distribution of the grains. The very high maximum polarisation degree ($\sim 50\%$) indicates the presence of grains with radii similar to the wavelength of the observations ($\geqslant$100 nm), so we calculate the maximum likelihood values of the minimum ($a_{\rm min}$) and maximum ($a_{\rm max}$) radii for an MRN-like grain size distribution (Mathis *et al.* 1977), assuming oxygen-deficient silicates (Ossenkopf *et al.* 1992). To improve the constraints, we also incorporate the published Hα–polarisation measurements of Jones *et al.* (2007).

We performed detailed fits to two regions of the ejecta, the South Knot and the Southwest Clump, whose 3D positions and motions are known (Humphreys *et al.* 2007). We calculated a grid of grain-size distributions to find maximum likelihood values of $a_{\rm min}$ and $a_{\rm max}$ over a relevant parameter space. For the South Knot, a number of good solutions exist for $a_{\rm min} \geqslant 0.25\,\mu$m with $a_{\rm max}$ only slightly larger than $a_{\rm min}$, i.e. approximately monodisperse. The best solution lies at $a_{\rm min} = 0.55\,\mu$m and $a_{\rm max} = 0.58\,\mu$m, with an average size $\int an(a)\mathrm{d}a / \int a\mathrm{d}a = \langle a \rangle = 0.56\,\mu$m. The Southwest Clump shows similar behaviour. However with the maximum likelihood model ($a_{\rm min} = 0.38\,\mu$m, $a_{\rm max} = 0.48\,\mu$m, average $0.42\,\mu$m) requires a broader range of grain sizes.

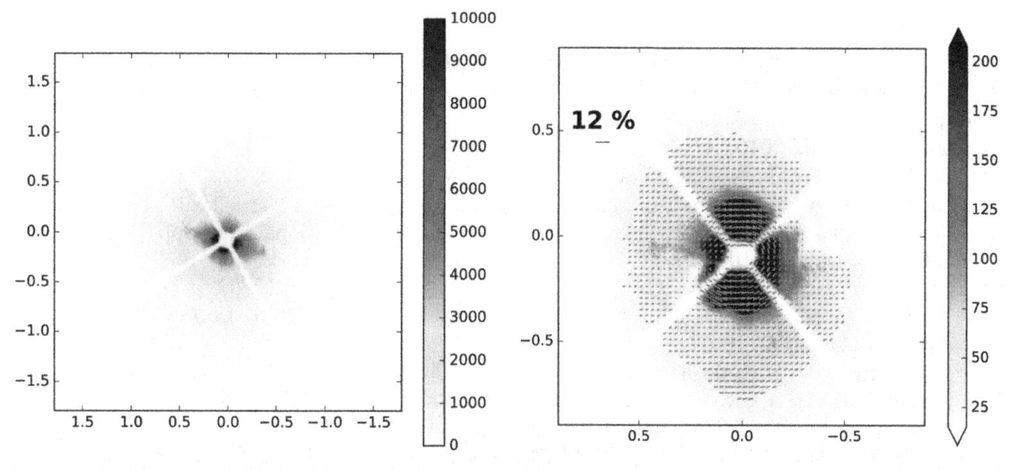

Figure 2. SPHERE observations of VX Sgr at 820 nm, with offsets in arcseconds. *Left*: Total intensity, which is likely dominated by the halo resulting from the imperfect AO correction. *Right*: Polarised intensity, zoomed to show only the central arc second.

Although grains larger than 0.1 μm have been suggested to explain a number of observations of RSGs and of VY CMa in particular, these observations directly confirm grains in this size range, approximately 50 times larger than the average size of interstellar medium (ISM) dust (Mathis *et al.* 1977). As RSGs emit the bulk of their radiation at wavelengths of a few micron, sub-micron dust grains can receive a significant amount of radiation pressure by scattering, rather than absorbing, the stellar emission (Höfner 2008; Bladh & Höfner 2012). This has been found to be an effective mechanism for driving mass loss in oxygen-rich AGB stars (Norris *et al.* 2012), and further work exploring the role of these grains in RSGs will be valuable (e.g. Haubois *et al.*, this volume).

3.2. *Morphology of VX Sgr*

VX Sgr was observed on 2015-09-11 with ZIMPOL, using V-band and 820 nm narrow-band filters. These data were reduced using the same procedure as the data for VY CMa, and the data for the 820 nm filter are shown in Fig. 2.

The total intensity shows little structure, with most of the emission probably arising from the stellar PSF and imperfect AO correction. However, the polarisation data clearly detects extended structure in the circumstellar shell for the first time. The polarisation vectors produce a clear centro-symmetric pattern with polarisation fraction of several per cent, which, although not conclusive alone, is characteristic of scattering from a dusty circumstellar disc or torus seen nearly face on.

SiO maser emission at 7 mm clearly traces a ring on 1-10 mas scales (Su *et al.* 2012); the coherent velocity structure of this ring is suggestive of an equatorial outflow within 3 $R_\star$ seen pole-on. This aligns with the expected location of the inner radius of the candidate disc. Such a torus, however, is difficult to explain without invoking either rapid rotation or the presence of a companion. No point-like sources are visible in our images and the interferometric imaging from Chiavassa *et al.* (2010) did not reveal close-in companions, while hot companions would be obvious from their photospheric spectral signatures. On the other hand, if a companion had previously been accreted it might have spun the star up, providing the necessary rotational energy to eject or form a torus (e.g. Collins *et al.* 1999).

References

Beuzit, J.-L., *et al.*, 2008, in Ground-based and Airborne Instrumentation for Astronomy II. p. 701418, doi:10.1117/12.790120

Bladh, S., Höfner S., 2012, *A&A*, 546, A76

Chiavassa, A., *et al.*, 2010, *A&A*, 511, A51

Collins, T. J. B., Frank, A., Bjorkman, J. E., Livio, M., 1999, *ApJ*, 512, 322

Ekström S., Georgy, C., Meynet, G., Groh, J., Granada, A., 2013, in Kervella, P., Le Bertre T., Perrin, G., eds, EAS Publications Series Vol. 60, EAS Publications Series. pp 31–41 (`arXiv 1303.1629`), doi:10.1051/eas/1360003

Georgy C., 2012, *A&A*, 538, L8

Georgy, C., Ekström S., 2015, in EAS Publications Series. pp 41–46 (`arXiv 1508.04656`), doi:10.1051/eas/1571007

Georgy, C., Ekström S., Saio, H., Meynet, G., Groh, J., Granada, A., 2013, in Kervella, P., Le Bertre T., Perrin, G., eds, EAS Publications Series Vol. 60, EAS Publications Series. pp 43–50 (`arXiv 1301.2978`), doi:10.1051/eas/1360004

Groh, J. H., Meynet, G., Ekström S., 2013, *A&A*, 550, L7

Harper, G. M., Richter, M. J., Ryde, N., Brown, A., Brown, J., Greathouse, T. K., Strong, S., 2009, *ApJ*, 701, 1464

Höfner S., 2008, *A&A*, 491, L1

Humphreys, R. M., Helton, L. A., Jones, T. J., 2007, *AJ*, 133, 2716

Jones, T. J., Humphreys, R. M., Helton, L. A., Gui, C., Huang, X., 2007, *AJ*, 133, 2730

Levesque, E. M., Massey, P., Olsen, K. A. G., Plez, B., Josselin, E., Maeder, A., Meynet, G., 2005, *ApJ*, 628, 973

Macintosh, B., *et al.*, 2006, in Society of Photo-Optical Instrumentation Engineers (SPIE) Conference Series. p. 62720L, doi:10.1117/12.672430

Martinache, F., Guyon, O., 2009, in Techniques and Instrumentation for Detection of Exoplanets IV. p. 74400O, doi:10.1117/12.826365

Mathis, J. S., Rumpl, W., Nordsieck, K. H., 1977, *ApJ*, 217, 425

Meynet, G., *et al.*, 2015, *A&A*, 575, A60

Norris, B. R. M., *et al.*, 2012, *Nature*, 484, 220

Ossenkopf, V., Henning, T., Mathis, J. S., 1992, *A&A*, 261, 567

Scicluna, P., Siebenmorgen, R., Wesson, R., Blommaert, J. A. D. L., Kasper, M., Voshchinnikov, N. V., Wolf, S., 2015, *A&A*, 584, L10

Smith, N., 2014, *ARA&A* 52, 487

Smith, N., Humphreys, R. M., Davidson, K., Gehrz, R. D., Schuster, M. T., Krautter, J., 2001, *AJ*, 121, 1111

Su, J. B., Shen, Z.-Q., Chen, X., Yi, J., Jiang, D. R., Yun, Y. J., 2012, *ApJ*, 754, 47

Verhoelst, T., van der Zypen N., Hony, S., Decin, L., Cami, J., Eriksson, K., 2009, *A&A*, 498, 127

Xu, S., Ertel, S., Wahhaj, Z., Milli, J., Scicluna, P., Bertrang, G. H.-M., 2015, *A&A*, 579, L8

van Loon J. T., 2013, in Kervella P., Le Bertre T., Perrin, G., eds, EAS Publications Series Vol. 60, EAS Publications Series. pp 307–316 (`arXiv 1303.0321`), doi:10.1051/eas/1360036

van Loon J. T., Cioni, M.-R. L., Zijlstra, A. A., Loup, C., 2005, *A&A*, 438, 273

The Lives and Death-Throes of Massive Stars
Proceedings IAU Symposium No. 329, 2016
J.J. Eldridge, J.C. Bray, L.A.S. McClelland & L. Xiao, eds.

doi:10.1017/S1743921317002794

The metallicity dependence of WR winds

R. Hainich, T. Shenar, A. Sander, W.-R. Hamann[1] and H. Todt

Institut für Physik und Astronomie, Universität Potsdam,
Karl-Liebknecht-Str. 24/25, D-14476 Potsdam, Germany
email: rhainich@astro.physik.uni-potsdam.de

Abstract. Wolf-Rayet (WR) stars are the most advanced stage in the evolution of the most massive stars. The strong feedback provided by these objects and their subsequent supernova (SN) explosions are decisive for a variety of astrophysical topics such as the cosmic matter cycle. Consequently, understanding the properties of WR stars and their evolution is indispensable. A crucial but still not well known quantity determining the evolution of WR stars is their mass-loss rate. Since the mass loss is predicted to increase with metallicity, the feedback provided by these objects and their spectral appearance are expected to be a function of the metal content of their host galaxy. This has severe implications for the role of massive stars in general and the exploration of low metallicity environments in particular. Hitherto, the metallicity dependence of WR star winds was not well studied. In this contribution, we review the results from our comprehensive spectral analyses of WR stars in environments of different metallicities, ranging from slightly super-solar to SMC-like metallicities. Based on these studies, we derived empirical relations for the dependence of the WN mass-loss rates on the metallicity and iron abundance, respectively.

Keywords. Wolf-Rayet stars, Magellanic Clouds, stellar atmospheres, stellar winds, mass-loss, metallicity

---

1. Introduction

The line-driven winds of hot stars are expected to be metallicity (Z) dependent, since they are predominately driven by the interaction of photons with millions of iron lines in the extreme UV. While this mechanism has been studied both empirically and theoretically for OB-type stars, a thorough investigation for Wolf-Rayet (WR) stars was pending for a long time. In the WR phase, strong and powerful stellar winds are capable of shedding a significant amount of the stellar mass, determining the conditions under which these objects might or might not explode as supernovae (SNe, see e.g. Heger *et al.* 2003, Dessart *et al.* 2011, Groh *et al.* 2013). Therefore, it is urgent to understand how the properties of WR winds change as a function of Z. This is all the more important in the light of the recent gravitational wave discoveries (Abbot *et al.* 2016a, b). First empirical studies of the metallicity dependence of WR stars were performed by Crowther (2006) and Nugis *et al.* (2007), while theoretical approaches were presented by Gräfener & Hamann (2008) and Vink & de Koter (2005).

In a series of paper, we presented analyses of an unprecedented number of WR stars of the nitrogen sequence (WN stars), covering metallicities from solar down to the Small Magellanic Cloud (SMC). In the SMC, we investigated the complete population, consisting of seven potentially single stars (Hainich *et al.* 2015) and five binary systems (Shenar *et al.* 2016). The Large Magellanic Cloud (LMC) sample encompasses almost all putative single stars (Hainich *et al.* 2014), while the analysis of the binary systems is currently underway. Solar metallicity is covered by the analyses of Galactic WN stars (Hamann *et al.* 2006, Martins *et al.* 2008, Liermann *et al.* 2010, Oskinova *et al.* 2013).

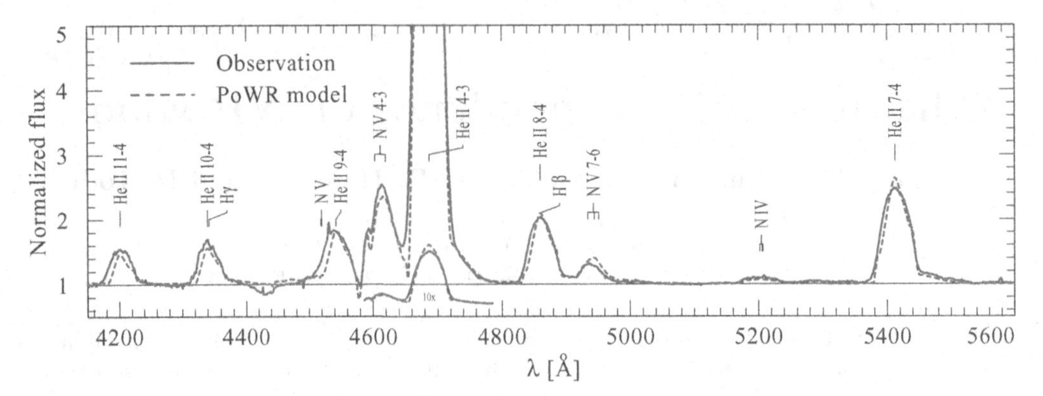

Figure 1. Spectral fit for WR 18, an early-type Galactic WN star. The observation is depicted as a solid blue line, while the PoWR model is shown as a red dashed line.

These comprehensive data sets allow us to investigate the relation between the mass-loss rate and the metallicity for WN stars, as presented in this contribution.

2. Stellar atmosphere models

The spectral analyses employed in this work were mostly performed with the Potsdam Wolf-Rayet (PoWR) model atmospheres. The PoWR code is a state-of-the-art stellar atmosphere code (see e.g. Sander 2017, these proceedings) that, among other things, allow for deviations from LTE, complex model atoms, iron line-blanketing, wind inhomogeneities, and a consistent treatment of the quasi-hydrostatic domain. The statistical equations are solved simultaneously with the radiative transfer that is calculated in the co-moving frame, while energy conservation is ensured.

The analysis of complete WN populations is only feasible by means of large model grids that we have computed for a variety of metallicities. These model grids, as well as additional grids for WC and OB-type stars, are publicly available via our PoWR website† (Todt *et al.* 2015). For a detailed description of grid based analyses, we refer to Sander *et al.* (2012) and Hainich *et al.* (2014). We note that additional models with adjusted abundances and terminal velocities were calculated for a subset of the stars discussed in this work. A typical fit of the normalized optical line spectrum is presented in Fig. 1.

3. The mass-loss rate as a function of metallicity

A comparison between the spectra of typical WN stars from different metallcity regimes clearly reveals that the emission line strengths are declining with decreasing metallicity, indicating a profound dependency of the wind strength on Z. To derive an empirical relation between the mass-loss rate and the metallicity, we utilized the "modified wind momentum" (Kudritzki *et al.* 1995, Puls *et al.* 1996, Kudritzki *et al.* 1999), which is defined as $D_{\rm mom} = \dot{M} v_\infty R_*^{1/2}$. Here, we neglect a potential Z dependency of the terminal wind velocity v_∞, since this dependence is weak, if present at all (Niedzielski *et al.* 2004). Consequently, $D_{\rm mom}$ is expected to show the same metallicity dependence as the mass-loss rate.

A tight relation is predicted between $D_{\rm mom}$ and the luminosity L, the so-called wind-momentum luminosity relation (WLR, Kudritzki *et al.* 1999). This allows us to implicitly

† `www.astro.physik.uni-potsdam.de/PoWR`

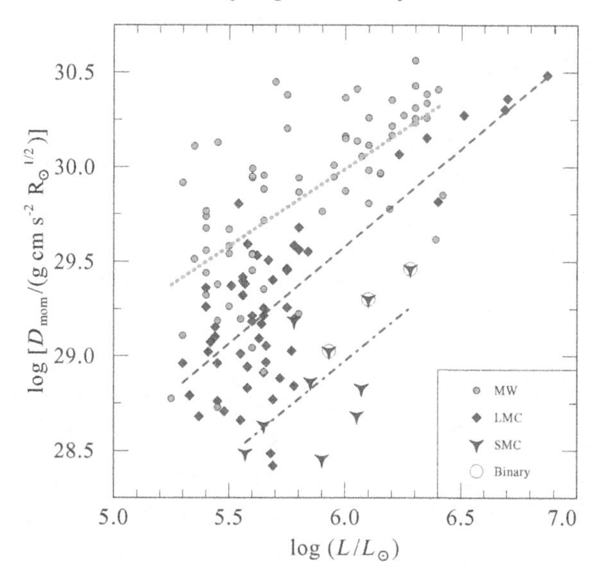

Figure 2. Modified wind momentum as a function of the luminosity for the galactic WN stars (green symbols), LMC WN stars (red symbols), and SMC WN stars (blue symbols). The lines are the WLRs fitted to the three populations.

Table 1. Coefficients of the wind-momentum luminosity relations $D_{mom} = D_0 \cdot L^\alpha$ fitted to the WN populations in the MW, LMC, and SMC, respectively.

	$\log D_0$	α
MW	25.1 ± 0.7	0.8 ± 0.1
LMC	23.4 ± 0.6	1.0 ± 0.1
SMC	22.9 ± 2.6	1.0 ± 0.4

account for the dependence of the wind strength on the luminosity, while utilizing the WLRs of the distinct WN population to determine the $\dot{M}$-Z-relation. The modified wind momentum of the WN stars in the MW, LMC, and SMC are plotted in Fig. 2 as a function of the luminosity. This figure also depicts the three WLRs fitted to the different WN populations. The coefficients of the respective relations are given in Table 1.

Since the slope of the WLR fitted to the Galactic population is slightly different compared to the other two WLRs, we have to either evaluate these equations at a specific luminosity (e.g. $\log L/L_\odot = 5.9$), or assume the same slope for all three WLRs. In fact, both methods give the nearly identical results. For Fig. 3, we have evaluated the WLRs (shown in Fig. 2) at a luminosity of $\log(L/L_\odot) = 5.9$. The differences between the WLRs on the D_{mom}-scale are plotted over the difference in the metallicities of the corresponding host galaxies. A linear fit to these data points reveals the metallicity dependence of D_{mom} and, consequently, of the mass-loss rate:

$$\dot{M} \propto Z^{1.2\pm0.1} \,. \tag{3.1}$$

We note that the quoted uncertainty only represents the statistical error of the linear regression.

Since the main contribution to the radiative acceleration of the line driven winds of massive stars is provided by interaction of the stellar radiation field with the multitude of iron lines in the extreme UV, we alternatively investigated the dependence of the wind momentum specially on the iron abundance X_{Fe}. Due to the lack of high resolution

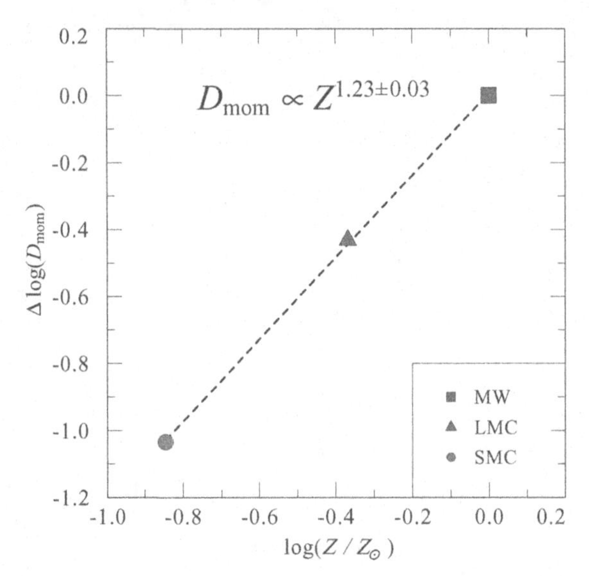

Figure 3. Offsets between the wind-momentum luminosity relations plotted over the difference between the metallicities of the corresponding galaxies.

and high signal-to-noise UV data for most of the MW WN stars, individual iron abundances are not available for most of these objects. Therefore, we used the iron abundance measured for the Galactic stellar population as a proxy for the iron abundance of the individual WN stars. In the outer part of the our Galaxy, the iron abundance shows a well established gradient (Hayden *et al.* 2015), while it seems to be rather constant and slightly supersolar in the inner part (Cunha *et al.* 2007, Najarro *et al.* 2009, Ryde & Schultheis 2015). Consequently for stars with a galactocentric distance of less than 6 kpc, we assume an iron abundance of $X_{\rm Fe} = 0.0014$ (mass fraction), while for stars outwards of this radius, it is reduced according to the gradient in the Galactic iron abundance.

Based on these assignments, the Galactic WN population can be separated in two groups, one with a high and one with a relatively low iron abundance. As for the metallicity, the LMC, SMC, and the two Galactic samples can be evaluated by means of their WLRs. In Fig. 4, the differences between the WLRs on the $D_{\rm mom}$-scale are plotted over the iron abundances of the respective populations. Fitting these data points with a linear regression provides the following relation between the mass-loss rate of WN stars and their iron abundance:

$$\dot{M} \propto X_{\rm Fe}^{1.5\pm0.1} \; . \tag{3.2}$$

4. Implications

Based on comprehensive spectral analyses of the WN stars in the MW, LMC, and SMC, we determine the dependence of the mass-loss rate on the environments' total metallicity and, alternatively, specifically the iron abundance. The latter proves to be steeper than the former, pointing to the importance of iron for the line driving. However, the uncertainty of this relation is large.

In comparison to what was found for OB-type stars (see e.g. Vink *et al.* 2007, Mokiem *et al.* 2007), the metallicity dependence of the WN stars is found to be significantly steeper: $\dot{M}_{\rm OB} \propto Z^{0.8}$ vs. $\dot{M}_{\rm WN} \propto Z^{1.2}$. This probably reflects the growing importance of multiple scattering for the wind driving with higher metallicity.

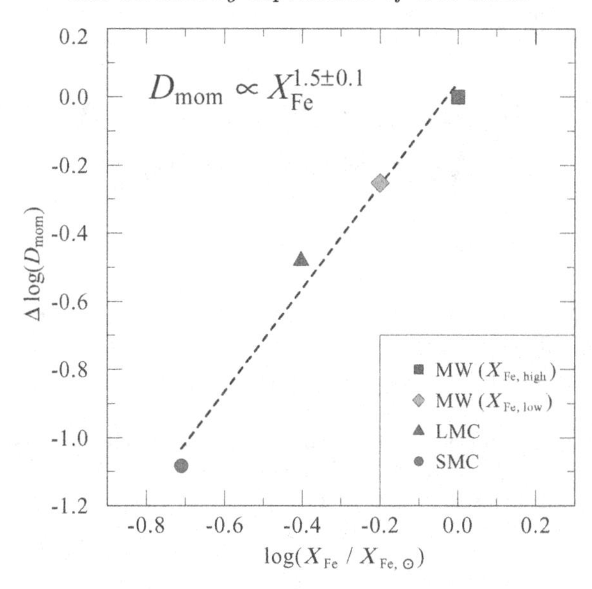

Figure 4. Differences between the wind-momentum luminosity relations plotted over the differences between the iron abundance of the massive stars in the corresponding galaxies.

References

Abbott, B. P., Abbott, R., Abbott, T. D., *et al.* 2016, *Phys. Rev. Lett.*, 116, 061102

Abbott, B. P., Abbott, R., Abbott, T. D., *et al.* 2016, *Phys. Rev. Lett.*, 116, 241103

Crowther, P. A. 2006, in *ASP-CS*, Vol. 353, *Stellar Evolution at Low Metallicity: Mass Loss, Explosions, Cosmology*, ed. H. J. G. L. M. Lamers, N. Langer, T. Nugis, & K. Annuk, 157

Cunha, K., Sellgren, K., Smith, V. V., *et al.* 2007, *ApJ*, 669, 1011

Dessart, L., Hillier, D. J., Livne, E., *et al.* 2011, *MNRAS*, 414, 2985

Gräfener, G. & Hamann, W.-R. 2008, *A&A*, 482, 945

Groh, J. H., Georgy, C., & Ekström, S. 2013, *A&A*, 558, L1

Hainich, R., Pasemann, D., Todt, H., *et al.* 2015, *A&A*, 581, A21

Hainich, R., Rühling, U., Todt, H., *et al.* 2014, *A&A*, 565, A27

Hamann, W.-R., Gräfener, G., & Liermann, A. 2006, *A&A*, 457, 1015

Hayden, M. R., Bovy, J., Holtzman, J. A., *et al.* 2015, *ApJ*, 808, 132

Heger, A., Fryer, C. L., Woosley, S. E., Langer, N., & Hartmann, D. H. 2003, *ApJ*, 591, 288

Kudritzki, R.-P., Lennon, D. J., & Puls, J. 1995, in *Science with the VLT*, ed. J. R. Walsh & I. J. Danziger, 246

Kudritzki, R. P., Puls, J., Lennon, D. J., *et al.* 1999, *A&A*, 350, 970

Liermann, A., Hamann, W.-R., Oskinova, L. M., Todt, H., & Butler, K. 2010, *A&A*, 524, A82

Martins, F., Hillier, D. J., Paumard, T., *et al.* 2008, *A&A*, 478, 219

Mokiem, M. R., de Koter, A., Vink, J. S., *et al.* 2007, *A&A*, 473, 603

Najarro, F., Figer, D. F., Hillier, D. J., Geballe, T. R., & Kudritzki, R. P. 2009, *ApJ*, 691, 1816

Niedzielski, A., Nugis, T., & Skorzynski, W. 2004, *AcA*, 54, 405

Nugis, T., Annuk, K., & Hirv, A. 2007, *Baltic Astronomy*, 16, 227

Oskinova, L. M., Steinke, M., Hamann, W.-R., *et al.* 2013, *MNRAS*, 436, 3357

Puls, J., Kudritzki, R.-P., Herrero, A., *et al.* 1996, *A&A*, 305, 171

Ryde, N. & Schultheis, M. 2015, *A&A*, 573, A14

Sander, A., Hamann, W.-R., & Todt, H. 2012, *A&A*, 540, A144

Sander, A., Todt, H., Hainich, R., & Hamann, W.-R. 2014, *A&A*, 563, A89

Shenar, T., Hainich, R., Todt, H., *et al.* 2016, *A&A*, 591, A22

Todt, H., Sander, A., Hainich, R., *et al.* 2015, *A&A*, 579, A75

Vink, J. S., de Koter, A., & Lamers, H. J. G. L. M. 2001, *A&A*, 369, 574

Vink, J. S. & de Koter, A. 2005, *A&A*, 442, 587

The Lives and Death-Throes of Massive Stars
Proceedings IAU Symposium No. 329, 2016
J.J. Eldridge, J.C. Bray, L.A.S. McClelland & L. Xiao, eds.

doi:10.1017/S1743921317002629

The Evolutionary Status of WN3/O3 Wolf-Rayet Stars

Kathryn F. Neugent[1,2], Phil Massey[1,2], D. John Hillier[3] and Nidia I. Morrell[4]

[1]Lowell Observatory, 1400 W Mars Hill Road, Flagstaff, AZ 86001
email: `kneugent@lowell.edu`, `phil.massey@lowell.edu`

[2]Department of Physics and Astronomy, Northern Arizona University, Flagstaff, AZ, 86011-6010

[3]Department of Physics and Astronomy, University of Pittsburgh, Pittsburgh, PA 15260
email: `hillier@pitt.edu`

[4]Las Campanas Observatory, Carnegie Observatories, Casilla 601, La Serena, Chile
email: `nmorrell@lco.cl`

Abstract. As part of a multi-year survey for Wolf-Rayet stars in the Magellanic Clouds, we have discovered a new type of Wolf-Rayet star with both strong emission and absorption. While one might initially classify these stars as WN3+O3V binaries based on their spectra, such a pairing is unlikely given their faint visual magnitudes. Spectral modeling suggests effective temperatures and bolometric luminosities similar to those of other early-type LMC WNs but with mass-loss rates that are three to five times lower than expected. They additionally retain a significant amount of hydrogen, with nitrogen at its CNO-equilibrium value (10× enhanced). Their evolutionary status remains an open question. Here we discuss why these stars did not evolve through quasi-homogeneous evolution. Instead we suggest that based on a link with long-duration gamma ray bursts, they may form in lower metallicity environments. A new survey in M33, which has a large metallicity gradient, is underway.

Keywords. stars: Wolf-Rayet, Magellanic Clouds

---

1. Introduction

A few years ago we began a search for Wolf-Rayet (WR) stars in the Magellanic Clouds (Massey *et al.* 2014, 2015, 2016). As I write this proceeding, we have just finished our fourth year of observations and have imaged the entire optical disks of both the Large Magellanic Cloud (LMC) and the Small Magellanic Cloud (SMC). We will likely have a few candidate WRs left to spectroscopically confirm after this season of imaging, but so far our efforts have been quite successful. We have discovered four WN-type stars, one WO-type star, eleven Of-type stars, and ten stars that appear to belong to an entirely new class of WR star.

The ten stars, as shown in Figure 1 (left), exhibit strong WN3-like emission line features as well as O3V star absorption lines. While one might instinctively think these are WN3 stars with O3V binary companions, this cannot be the case. They are faint with $M_V \sim -2.5$. An O3V by itself has an absolute magnitude of $M_V \sim -5.5$ (Conti 1988) so these stars cannot contain an O-type star companion. We call these stars WN3/O3s where the "slash" represents their composite spectra. As mentioned previously, we have found ten of these stars, or 6.5% of the LMC WR population. As Figure 1 (right) shows, all of their spectra are nearly identical.

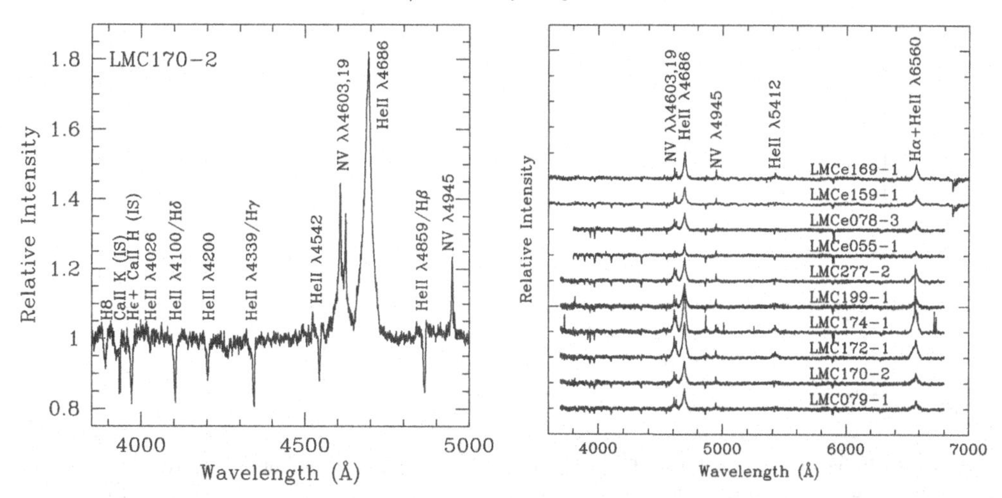

Figure 1. Left: The spectrum of LMC170-2, a WN3/O3 Wolf-Rayet star. Notice both the strong emission and absorption lines. Right: The spectra of all ten WN3/O3s.

2. Modeling Efforts and Physical Parameters

Given that these stars are not WN3 stars with O3V companions, we were interested to see if we could model their spectra as single stars. We obtained medium dispersion optical spectra for all ten WN3/O3s in addition to high-dispersion optical spectra for one WN3/O3, UV spectra for three WN3/O3s, and NIR spectra for one WN3/O3. For LMC170-2 we obtained spectra from all four sources giving us coverage from 1000-25000Å. To model these stars we turned to CMFGEN, a spectral modeling code that contains all of the complexities needed to model hot stars near their Eddington limits (Hillier & Miller 1998). Using CMFGEN, we began by modeling the spectrum of LMC170-2.

Using the parameters given in Table 1, we obtained a good fit to both the emission and the absorption lines for LMC170-2. Hainich *et al.* (2014) recently modeled most of the previously known WN-type WRs in the LMC using PoWR and thus we can compare the star's physical parameters with those of more "normal" WN-type LMC WRs. We can additionally compare the parameters with those of LMC O3Vs. Table 1 shows that the abundances and He/H ratio are comparable with LMC WN3s. Figure 2 (left) shows that while the temperature is a bit on the high side of what one would expect for a LMC WN, it is still within the expected temperature range. However, the biggest surprise is the mass-loss rate. As is shown in Table 1, the mass loss rate of the WN3/O3s is more similar to that of an O3V than of a normal LMC WN. This is shown visually in Figure 2 (right).

Table 1. Physical parameters of WN3/O3s, WNs, and O3Vs in the LMC[a]

	WN3/O3	WN3	O3V
$T_{\rm eff}\,(K)$	100,000	80,000	48,000
$\log \frac{L}{L_\odot}$	5.6	5.7	5.6
$\log \dot{M}^b$	-5.9	-4.5	-5.9
He/H (by #)	1.0	1.0-1.4	0.1
N (by mass)	10× solar	5-10× solar	0.5× solar
M_V	-2.5	-4.5	-5.5

a: Hainich *et al.* (2014); Massey *et al.* (2013)
b: $\log \dot{M}$ assumes a clumping filling factor of 0.1

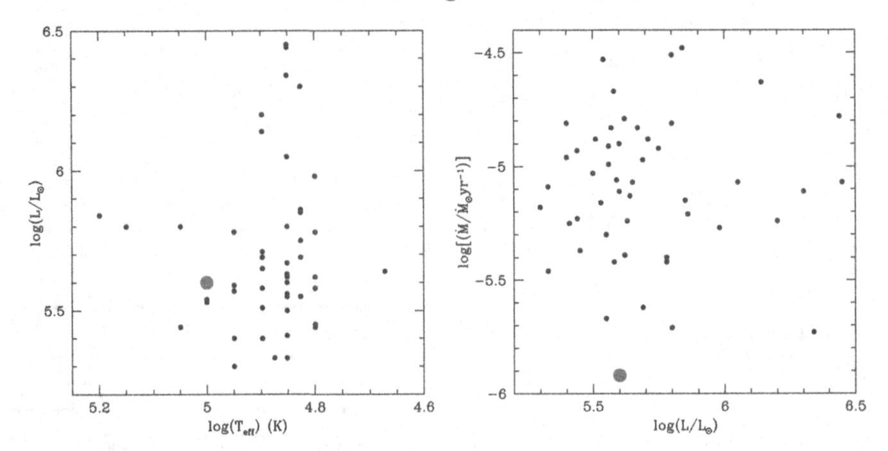

Figure 2. Left: A comparison of the bolometric luminosity vs. effective temperature for LMC170-2. Right: A comparison of the mass-loss rate vs. bolometric luminosity for LMC170-2. The small black circles represent the early-type LMC WNs (WN3s and WN4s) analyzed by Hainich *et al.* (2014). LMC170-2 is represented as a large red circle.

While all the spectra look visually similar, we still wanted to determine a range of physical parameters for these WN3/O3s. Thus, we modeled all 10 of them using CMFGEN. As discussed above, we obtained UV data for three of our stars. Based on these data, we determined that C IV $\lambda1550$ was not present in the spectra of our stars. This again confirms that there is not an O3V star within the system. Additionally, it points to a high temperature as this line disappears at an effective temperature at 80,000 K. As an upper limit, as the effective temperature increases above 110,000 K, both He II $\lambda4200$ and O VI $\lambda1038$ become too weak. Thus, the overall temperature regime is between 80,000 - 110,000 K. However, in practice, our models only varied between 100,000 - 105,000 K. Most of the other physical parameters all stayed relatively consistent and are thus well constrained as is shown in Table 2. However, the exception is again the mass-loss rate. Figure 3 shows the range in the mass-loss rates for our WN3/O3s. While most of them are quite low (like those of O3Vs), a few of them are on the low end of some other early-type LMC WNs. In particular, there are three LMC WNs with low mass-loss rates and similar luminosities to the WN3/O3s. However, these three stars are visually much brighter than the WN3/O3s and thus we do not believe that they are the same type of star. We are still planning on studying them further.

Table 2. Physical parameter range of WN3/O3s

$T_{\rm eff}\,(K)$	100,000 - 105,000
$\log \frac{L}{L_\odot}$	5.6
$\log \dot{M}$*	-6.1 - -5.7
He/H (by #)	0.8 - 1.5
N (by mass)	5-10× solar

*stellar wind parameters: clumping filling factor of 0.1, terminal velocity of 2,600 km s^{-1}, and $\beta = 1$.

3. Evolutionary Status

Now that we have a good handle on their physical parameters, we are currently investigating possible WN3/O3 progenitors as well as their later stages of evolution. Figure 4 shows where the WN3/O3s are located within the disk of the LMC. If they were all

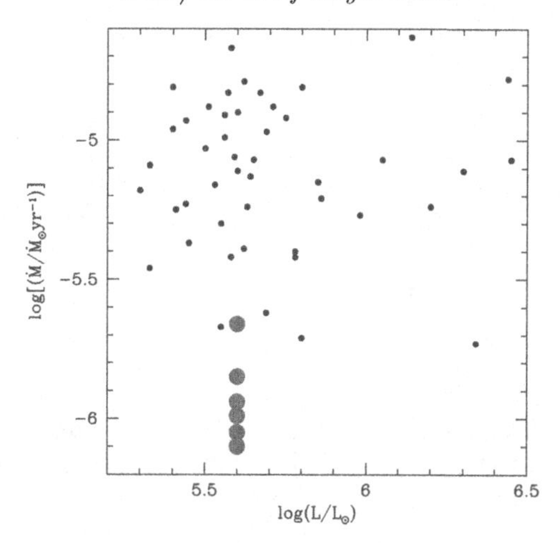

Figure 3. A comparison of the mass-loss rate vs. luminosity for all ten WN3/O3s. The small black circles represent the early-type LMC WNs (WN3s and WN4s) analyzed by Hainich *et al.* (2014). The WN3/O3s are represented as large red circles.

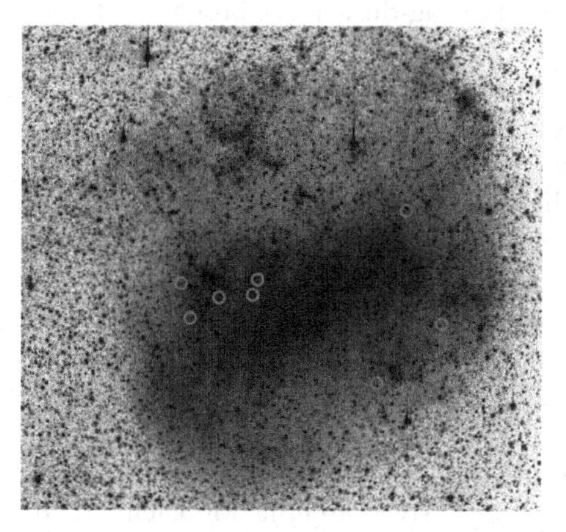

Figure 4. The locations of the WN3/O3s within the LMC.

grouped together, we might assume that they formed out of the same stellar nursery. Instead, they are pretty evenly spaced across the LMC.

There are other non-binary WRs with hydrogen absorption lines denoted as WNha stars. Martins *et al.* (2013) has argued that the majority of these stars evolved through quasi-homogeneous evolution. In this case, the stars have high enough rotational velocities that the material in the core mixes with the material in the outer layers. This creates CNO abundances near equilibrium, much as we find in the WN3/O3s. However, the WN3/O3s have relatively low rotational velocities ($V_{\rm rot} \sim 150$ km s^{-1}) and extremely low mass-loss rates. It is difficult (if not impossible) to imagine an scenario where a star could begin its life with a large enough rotational velocity (typically ~ 250 km s^{-1}) to produce homogeneous evolution and later slow down to 150 km s^{-1} given the small mass-loss rates (Song *et al.* 2016). Instead, a homogeneous star must have either a larger rotational velocity or a larger mass-loss rate. So, at this point we can rule out quasi-homogeneous evolution.

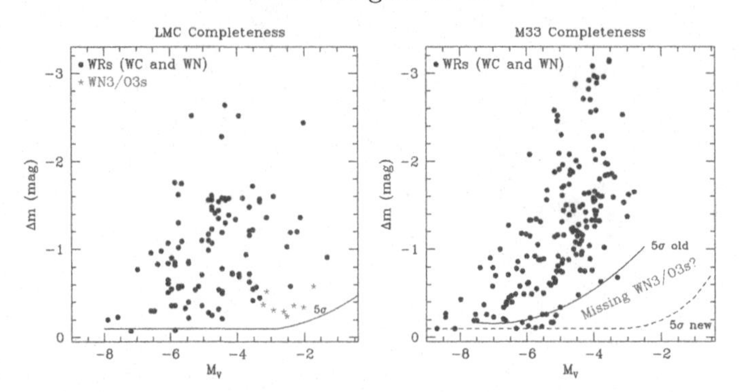

Figure 5. Magnitude difference (emission minus continuum filter) vs. absolute magnitude showing that our previous M33 surveys did not go deep enough to detect WN3/O3s.

4. Next Steps

Drout *et al.* (2016) looked at various types of supernovae progenitors and found that WN3/O3s, with their low mass-loss rates and high wind velocities, might be the previously-unidentified progenitors of Type Ic-BL supernovae. A subset of these Type Ic-BL supernovae then turn into long-duration gamma ray bursts which are preferentially found in low metallicity environments (Vink *et al.* 2011). Thus, we are currently in the process of investigating any metallicity dependence for the formation of WN3/O3s.

So far we have only found them in the low metallicity LMC. Given the high number of WRs currently known in the Milky Way, we would expect to have found at least a few of them. However, these dim WRs with strong absorption lines have not been found elsewhere. To investigate any metallicity dependence, we have decided to begin another search for WRs in M33 which has a strong metallicity gradient. We previously conducted a search for WRs in M33 but our previous survey simply did not go deep enough, as Figure 5 shows. So, there may be an entire yet undiscovered population of WN3/O3s within M33.

We continue to discuss the evolutionary status of these WN3/O3s with our theoretician colleagues while also observationally attempting to determine where these stars come from and what they turn into. By searching for them in M33 we should be able to constrain their metallicity dependence and gain further insight into their evolution.

Support for this work from the National Science Foundation (AST-1612874) and the Space Telescope Science Institute (GO-13780) is gratefully acknowledged.

References

Conti, P. S. 1988, NASA Special Publication, 497, 119
Drout, M. R., Milisavljevic, D., Parrent, J., Margutti, R., *et al.* 2016, ApJ, 821, 57
Hainich, R., Rühling, U., Todt, H., *et al.* 2014, A&A, 565, A27
Hillier, D. J. & Miller, D. L. 1998, ApJ, 496, 407
Martins, F., Depagne, E., Russeil, D., & Mahy, L. 2013, A&A, 554, 23
Massey,, P., Neugent, K. F., Hillier, D. J., & Puls, J. 2013, ApJ, 768, 6
Massey, P., Neugent, K. F., & Morrell, N. 2015, ApJ, 807, 81
Massey, P., Neugent, K. F., & Morrell, N. 2016, in prep.
Massey, P., Neugent, K. F., Morrell, N., & Hillier, D. J. 2014, ApJ, 788, 83
Song, H. F., Meynet, G., Maeder, A., & Ekström, Eggenberger, P. 2016, A&A, 585, 120
Vink, J. S., Gräener, G., & Harries, T. J. 2011, A&A, 536, L10

The Lives and Death-Throes of Massive Stars
Proceedings IAU Symposium No. 329, 2016
J.J. Eldridge, J.C. Bray, L.A.S. McClelland & L. Xiao, eds.

doi:10.1017/S1743921317003210

The Most Massive Heartbeat: Finding the Pulse of ι Orionis

Herbert Pablo[1], Noel Richardson[2], Jim Fuller[3,4], Anthony F. J. Moffat[1], BEST and Ritter Observing team

[1]Département de physique and Centre de Recherche en Astrophysique du Québec (CRAQ), Université de Montréal, C.P. 6128, Succ. Centre-Ville, Montréal, Québec, H3C 3J7, Canada
email: hpablo@astro.umontreal.ca

[2]Ritter Observatory, Department of Physics and Astronomy, The University of Toledo, Toledo, OH 43606-3390, USA

[3]TAPIR, Walter Burke Institute for Theoretical Physics, Mailcode 350-17, Caltech, Pasadena, CA 91125, USA

[4]Kavli Institute for Theoretical Physics, Kohn Hall, University of California, Santa Barbara, CA 93106, USA

Abstract. ι Orionis is a massive binary system consisting of O9III + B1 III/IV stars. Though the system has been well studied, much about its fundamental properties have been difficult to determine. In this paper we report on the discovery of the heartbeat phenomenon in ι Orionis making it the most massive heartbeat system currently known. Using this phenomenon we have found empirical values for the masses and radii of both components. Moreover, we report the detection of tidally induced oscillations in an O-type star for the first time. These discoveries open a new avenue for exploring asteroseismology in massive stars.

Keywords. (stars:) binaries (including multiple): close, stars: fundamental parameters (classification, masses, radii), stars: oscillations (including pulsations), stars: individual: ι Ori

---

1. Introduction

Understanding stellar evolution is essential as it is stars which are the building blocks on which galaxies, and by extension, the universe as a whole is built. It is even more important that we understand these processes in massive stars as these stars will explode and pollute the interstellar medium with metals. Unfortunetely, massive stars have comparatively short lifetimes compared to lower mass stars, making ensemble statistical studies extremely difficult. Another avenue of exploration is through the use of asteroseismology. By studying the oscillations produced within stars, we can interpret their interior structure, and from that their evolutionary state. However, this channel has also been largely restricted as there are only handful of O stars which are known to pulsate (see Buysschaert *et al.* 2015; Pablo *et al.* 2015, and references therein).

Further complicating matters is that almost all massive stars are born as binary stars (Sana *et al.* 2012, Sana *et al.* 2014, Aldoretta *et al.* 2015) and it has been shown that only 24 % will not interact or merge during their lives. This makes it imperative that we include binary systems when trying to understand evolution in massive stars. Studying binary systems, though, is not a simple proposition. While it is true that in such systems we are able to determine fundamental parameters, this typically requires close eclipsing binary systems. Such constraints have led to only $\approx$ 50 O star binaries for which fundamental parameters are known (Gies *et al.* 2012). These constraints are also problematic as many such binaries will have had interactions making it difficult to determine the evolutionary history of the system.

Virtually all these problems could be mitigated, however, by the discovery of a new class of binary systems. This class first discovered with the *Kepler* space telescope (Thompson *et al.* 2012), are commonly labeled heartbeat systems, as their light curves' bear qualitative similarities to the "normal sinus rhythm" signal of an electrocardiogram. This peculiar heartbeat occurs in eccentric binary systems, which show strong evidence of ellipsoidal variation only at periastron, when the stars are closest. Due to the strong dependence of this heartbeat phenomenon on inclination it is possible to use such systems to derive fundamental parameters, even in the absence of eclipses. A further consequence of the stars' close proximity at periastron, is the ability to induce tidally excited oscillations (TEOs) in their companion. This affords us the opportunity to do asteroseismology within binary systems.

The only drawback is the lack of known systems. Due to the high binary fraction of O stars and their short lifetimes (and consequently higher eccentricities), this would seem to be due largely to selection bias. Because this phenomenon is often only a few parts per thousand, the heartbeat phenomenon is very difficult to observe from the ground in addition to requiring long time-baseline photometry. The only space missions which have heretofore observed this effect, *CoRoT* and *Kepler*, have looked at a total of 6 O stars (though this will undoubtedly change with K2). However in 2013, a new network of nanosatellites, called *BRIght Target Explorer* (BRITE)-Constellation, was launched to address this sort of problem. Its mission is to look at the brightest stars in the sky with high-precision, long time-baseline observations Weiss et. al, (2014). As these stars are often the most intrinsically luminous stars it is an excellent resource for studying massive star variability, such as the heartbeat phenomenon.

In these proceedings we will give an overview of the results of Pablo *et al.* (2017, accepted) who have identified ι Ori a massive O9 III + B1 III/IV eccentric binary ($e = 0.76$) with $P_{\rm orb} = 29.13376$ d as the most massive heartbeat star ever weighed. Using light curves from two separate BRITE observing runs, in addition to radial velocities from Marchenko *et al.* (2000) as well as new observations from the 1 m Ritter observatory telescope (University of Toledo) we were able to obtain a full binary solution. In addition, we were able to identify 5 separate TEOs, the first ever observed in an O star. Finally we will discuss the implications of this work as it relates massive star evolution.

2. Overview

Binary Simulation and fitting.

The prevalence of TEOs in heartbeat systems is extremely useful for asteroseismic analysis, but has a single important drawback when attempting to find a binary solution: the oscillations are not canceled out when the light curve is phased on the orbital period as they occur at multiples of the orbital frequency. This is only an issue at periastron, but it is a significant one as the frequencies (given in detail in Pablo *et al.* (2017)) are as large as 10 % of the heartbeat effect. Therefore, the TEOs which had the largest influence on the phased light curve were removed using a least squares fit on the phased light curve, where the heartbeat itself is masked. This is explained in greater detail in Pablo *et al.* (2017). Once the oscillations were removed the system was run through the binary simulation code PHOEBE (Prsa & Zwitter 2005). First, the system was fit using a Levenburg-Marquardt algorithm to determine a set of realistic parameters which resulted in an adequate fit to both the light and radial velocity curves. These values were used to initialize our Monte Carlo Markov Chains (MCMC). We then used these chains to probe

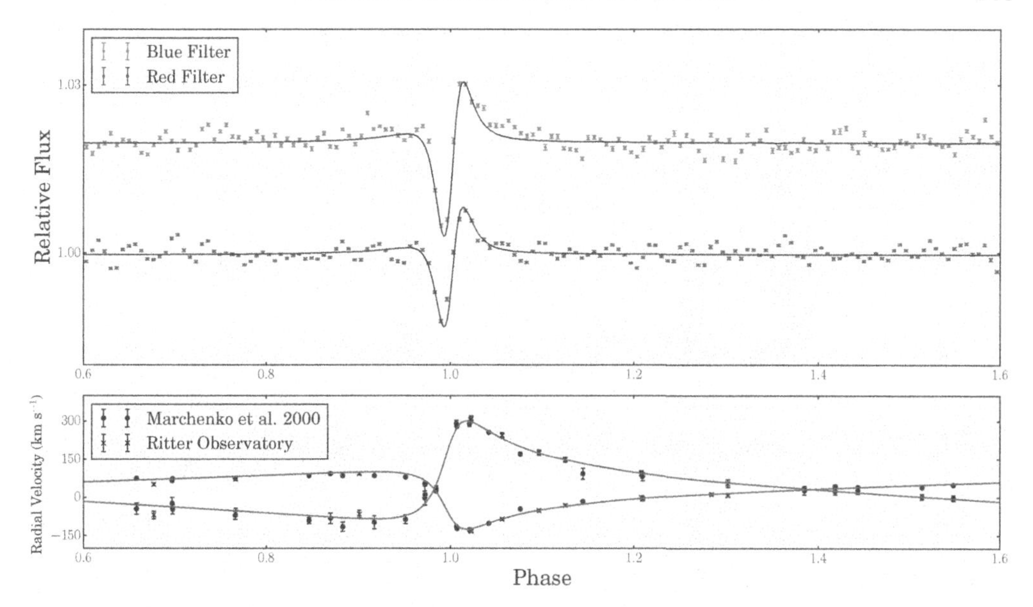

Figure 1. Simulation of 50 percentile values (black) derived from the simultaneous MCMC fit of phase-binned BRITE light curves (top) and radial velocities (bottom). The frequencies have not been removed, to demonstrate the presence of TEOs.

the parameter space, identifying degeneracies as well as determining the extent of the global minimum for error calculation. Our MCMC implementation is achieved through the Python package EMCEE (Foreman-Mackey *et al.* 2013). Because this procedure is computationally expensive several parameters were not fit, namely period and time of periastron as no noticeable change could be seen in the new data from the values given in Marchenko *et al.* (2000). The fit is given in Fig. 1.

This equates to a primary star with parameters $M_\mathrm{p} = 23.18^{+0.57}_{-0.53} M_\odot$, $R_\mathrm{p} = 9.10^{+0.12}_{-0.10} R_\odot$ and a secondary with parameters $M_\mathrm{s} = 13.44^{+0.30}_{-0.30} M_\odot$, $R_\mathrm{s} = 4.94^{+0.16}_{-0.23} R_\odot$. When compared with the expected values given for O stars (Martins 2005) and B stars (Nieva & Pryzbilla 2014) the only noticeable difference is in the luminosity class. The primary values put it likely as a luminosity class IV as its parameter values fall between that of class III and V. For the secondary, while it is possible that it has luminosity class of IV, there is no discernible difference in its parameters from that of a class V star of the same spectral type. The only other unusual quantity is the temperature of the secondary, $T_\mathrm{eff} = 18319^{+531}_{-758}$ K. This value is almost 10000 K lower than would be expected from a B1 star. While it is clear from the MCMC calculations that this value is more favorable and is independent of range or initial value, it is also clear that this temperature is highly unlikely. The most likely explanation, therefore is that the two BRITE filters are not enough to get an accurate value of the temperature in the absence of eclipses. Instead, there is likely a degeneracy with another parameter which was not fit. See Table 2 in Pablo *et al.* (2017) for a full list of binary parameters.

Fourier Analysis & Tidally Excited Oscillations.

Due to a 6 month gap between two separate campaigns on the Orion field, the frequency analysis was done on each campaign separately as well as on data from each filter. This resulted in 4 separate datasets from which frequencies were determined. In total 5 frequencies were found to be multiples of the orbital frequency $f_\mathrm{orb} = 0.0343244(2)$ d^{-1}. The frequencies, shown in Fig. 2, are at 6, 23, 25, 27, 33 f_orb. As can be seen these frequencies appear in more than one color and campaign. The complete list of frequencies

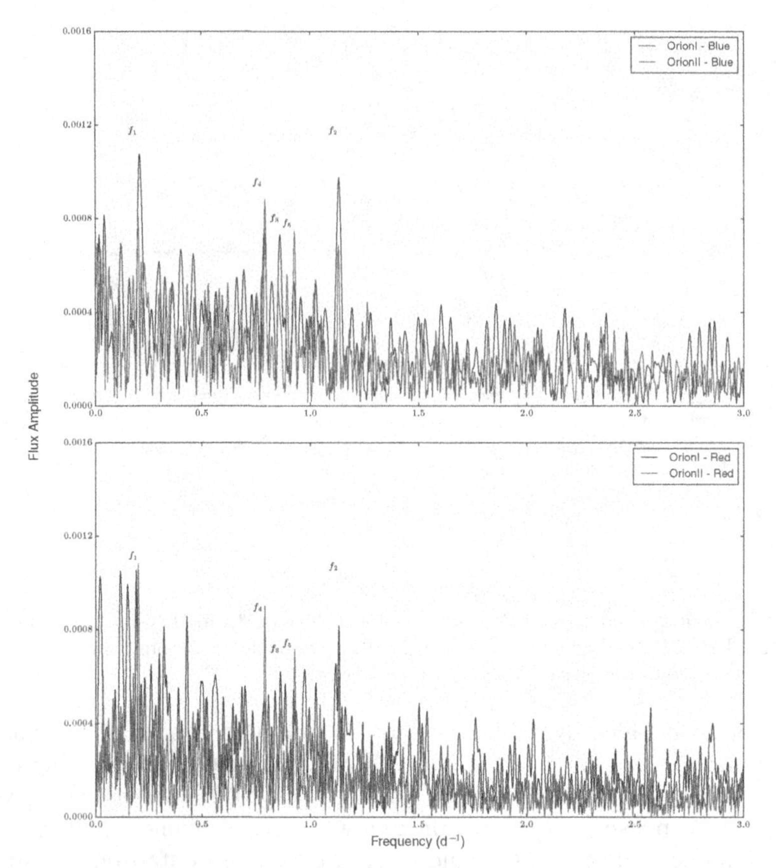

Figure 2. Discrete Fourier spectrum, from the blue (top) and red (bottom) filter of the OrionI and OrionII BRITE observing campaigns. The tidally excited oscillations are marked.

and their parameters (see Pablo *et al.* (2017), Tab. 3) also suggest that the amplitudes and phases are largely stable between observation epochs.

From the binary parameters, stellar models were calculated through the us of the MESA stellar evolution code (Paxton *et al.*, 2011, Paxton *et al.*, 2013, Paxton *et al.*, 2015). Based on these models oscillation modes were found using the non-adiabatic version of the GYRE pulsation code (Townsend & Teitler 2013). The oscillations produced were able to adequately reproduce 4 of the 5 frequencies as tidally induced $l = 2$ modes. The one frequency which does not fit the models is the 6th harmonic, which is at a significantly lower frequency than all others. It is possible that this frequency just happens to be close to an orbital harmonic but is not tidally induced. However, its persistence across datasets makes this idea unlikely. A possible explanation is that the pulsation arises from non-linear interactions between TEOs, a common occurrence in heartbeat stars (Fuller *et al.*, 2012, Hambleton *et al.*, 2013, O'Leary *et al.*, 2014, Guo *et al.* 2016). These models have also led to the estimation of a tidal dissipation rate, $Q_{\rm tide} = 4 \times 10^4$, for the primary star. While this result is model dependent, it represents the first empirical estimate of the tidal dissipation rate in an O star.

3. Implications

The determination of fundamental parameters for ι Ori is significant, if only because there are very few O type systems for which these parameters are known. What makes the system even more interesting, though, is that these parameters are known in a system with clear pulsations. This gives us the ability to constrain and test asteroseismic properties, which has been underutilized in massive stars simply due to the dearth of systems. While this analysis is still in its infancy ι Ori has already allowed for unprecedented asteroseismic calculations in massive stars, and these results will only improve as more data are obtained.

Beyond the results of ι Ori are the avenues this opens up for the study of massive star evolution. Heartbeat stars provide us the opportunity to study the interior structure of massive stars in a manner which, up to now, has never been done. What's more these systems are not rare. While ι Ori is the only system which has currently been published several more are known in the BRITE sample including ϵ Lupi, the only massive binary where both components are magnetic (Pablo *et al.* in preparation). Moreover, it is likely that several systems exist in the literature which have been overlooked as heartbeat systems were only discovered in 2011. Over the coming years analysis of these heartbeat systems should redefine our knowledge of how massive stars evolve, as they not only give us access to asteroseismology, but provide fundamental parameters for non-interacting massive stars from which to constrain many other properties such as mass loss rates and magnetism.

References

Aldoretta, E. J., Caballero-Nieves, S. M., Gies, D. R., Nelan, E. P., & Wallace, D. J., 2015, *AJ*, 149, 26

Buysschaert, B., Aerts, C., Bloemen, S., Debosscher, J., & Neiner, C., 2015, *MNRAS*, 453, 89

Foreman-Mackey, D., Hogg, D. W., Lang, D., & Goodman, J., 2013, *PASP*, 125, 306

Fuller, J., & Lai, D., 2012, *MNRAS*, 420, 3126

1995, *Meteoritics*, 30, 490

Guo, Z., Gies, D. R., & Fuller, J. 2016, *ApJ*, 834, 59

Hambleton, K. M., Kurtz, D. W., Prša, A., Guzik, J. A., Pavlovski, K., *et al.*, 2013, *MNRAS*, 434, 925

Marchenko, S. V., Rauw, G., Antokhina, E. A., Antokhin, I. I., Ballereau, D. *et al.* 2000, *MNRAS*, 317, 333

O'Leary, R. M. & Burkart, J. 2014, *MNRAS*, 440, 3036

Pablo, H., Richardson, N. D., Moffat, A. F. J., Corcoran, M., Shenar, T. *et al.* 2015, *ApJ*, 809, 134

Pablo, H., Richardson, N. D., Fuller J., Moffat, A. F. J., Rowe, J. *et al.* 2017, MNRAS, accepted

Paxton, B., Bildsten, L., Dotter, A., Herwig, F., Lesaffre, P. *et al.* 2011, *ApJs*, 192, 3

Paxton, B., Cantiello, M., Arras, P., Bildsten, L., Brown, E. F. *et al.* 2013, *ApJs*, 208, 4

Paxton, B., Marchant, P., Schwab, J., Bauer, E. B., Bildsten, L. *et al.* 2015, *ApJs*, 220, 15

Sana, H., de Mink, S. E., de Koter, A., Langer, N. Evans, C. J. *et al. Science*, 337, 444

Sana, H., Le Bouquin, J.-B., Lacour, S., Berger, J.-P., Duvert, G. *et al.* 2014, *ApJs*, 215, 15

Thompson, S. E., Everett, M., Mullally, F., Barclay, T., Howell, S. B. *et al.* 2012, *ApJ*, 2012, 753

Townsend, R. H. D. & Teitler, S. A. 2013, *MNRAS*, 435, 3406

Weiss, W. W., Rucinski, S. M., Moffat, A. F. J., Schwarzenberg-Czerny, A., Koudelka, O. F. *et al.* 2014, *PASP*, 126, 573

The Lives and Death-Throes of Massive Stars
Proceedings IAU Symposium No. 329, 2016
J.J. Eldridge, J.C. Bray, L.A.S. McClelland & L. Xiao, eds.

doi:10.1017/S1743921317002630

The wind-wind collision hole in eta Car

A. Damineli[1], M. Teodoro[2,3], N. D. Richardson[9], T. R. Gull[2], M. F. Corcoran[2,3], K. Hamaguchi[2,3], J. H. Groh[5], G. Weigelt[6], D. J. Hillier[7], C. Russell[2], A. Moffat[4], K. R. Pollard[8] and T. I. Madura[10]

[1]Inst. de Astron., Geofísica e Ciências Atmosféricas, Univ. de São Paulo, R. do Matão 1226, São Paulo 05508-900, Brazil
email: augusto.damineli@iag.usp.br

[2]Astroph. Sci. Division, Code 660, NASA Goddard Space Flight Center, Greenbelt, MD 20771, USA

[3]USRA, 7178 Columbia, MD 20146, USA

[4]Départ. de Physique, Univ. de Montréal, CP 6128, Succursale: Centre-Ville, Montréal, QC, H3C 3J7, Canada

[5]School of Physiscs, Trinity College Dublin, The Un. of Dublin, Dublin 2, Ireland

[6]Max-Planck-Institut for Radioastronomy, Auf dem Hügel 69, D-53121 Bonn, Germany

[7]Depart. of Phys. and Astr., Univ. of Pittsburgh, 3941 O'Hara Street, Pittsburgh, PA 15260, USA

[8]Department of Physics and Astronomy, University of Canterbury, Chirstchurch, New Zealand

[9]Ritter Observ., Depart. of Phys. and Astr., The University of Toledo, Toledo, OH 43606-3390, USA

[10]San Jose State University, Depart. of Physics and Astronomy, San Jose, CA, USA

Abstract. Eta Carinae is one of the most massive observable binaries. Yet determination of its orbital and physical parameters is hampered by obscuring winds. However the effects of the strong, colliding winds changes with phase due to the high orbital eccentricity. We wanted to improve measures of the orbital parameters and to determine the mechanisms that produce the relatively brief, phase-locked minimum as detected throughout the electromagnetic spectrum. We conducted intense monitoring of the He II λ4686 line in η Carinae for 10 months in the year 2014, gathering $\sim$300 high S/N spectra with ground- and space-based telescopes. We also used published spectra at the FOS4 SE polar region of the Homunculus, which views the minimum from a different direction. We used a model in which the He II λ4686 emission is produced by two mechanisms: a) one linked to the intensity of the wind-wind collision which occurs along the whole orbit and is proportional to the inverse square of the separation between the companion stars; and b) the other produced by the 'bore hole' effect which occurs at phases across the periastron passage. The opacity (computed from 3D SPH simulations) as convolved with the emission reproduces the behavior of equivalent widths both for direct and reflected light. Our main results are: a) a demonstration that the He II λ4686 light curve is exquisitely repeatable from cycle to cycle, contrary to previous claims for large changes; b) an accurate determination of the longitude of periastron, indicating that the secondary star is 'behind' the primary at periastron, a dispute extended over the past decade; c) a determination of the time of periastron passage, at $\sim$4 days after the onset of the deep light curve minimum; and d) show that the minimum is simultaneous for observers at different lines of sight, indicating that it is not caused by an eclipse of the secondary star, but rather by the immersion of the wind-wind collision interior to the inner wind of the primary.

Keywords. stars: individual (η Carinae) — stars: massive — binaries: general — stars: circumstellar matter — atomic data — atomic processes — radiation mechanisms: general — plasmas

1. Introduction

Eta Carinae is one of the most luminous evolved stars in the local Universe (Davidson & Humphreys 1997). It became famous after the great eruption in the 1840's, which created the beautiful bipolar Homunculus flow and there were additional smaller mass ejections. The discovery that it is long period eccentric binary (Damineli 1996, Damineli, Conti & Lopes 1997) provided the potential to determine the mass of the companions and to explain a number of complex, variable features. One very interesting is the connection between the sharp photometric peaks during the great eruption and timing of periastron passages, indicating that at periastron, there is strong interaction between the companion stars (Damineli 1996, Prieto *et al.* 2014). Moreover, the shape of the homunculus displays features formed when the ejecta was a thousand times smaller than at the time of the eruption (Steffen *et al.* 2014).

In practice, variations of the system do not easily lead to the physical parameters: both companions have strong winds, with especially the primary's known to be strongly variable, leading to strong and variable wind-wind collision (WWC), which give rise to many complex and variable phenomena like free-free and free-bound emission, radio emission, high and low energy emission lines. A recent discovery (Steiner & Damineli 2004) of a "burst" in the He II λ4686 line, because of its sharp and phase-locked behaviour (Teodoro *et al.* 2012), turned out to be a key phenomenon to study the binary system. As shown by Madura & Owocki (2010) and Madura *et al.* (2013), the 'bore hole' effect can account for the escaping of UV radiation from the inner interacting binary to the external regions, through the cavity (filled by the sparser secondary wind) produced by the WWC. 3D SPH simulations are necessary to calculate the opacity to each direction. Such a tool is critical to explore the nature of the 'low excitation events'. As shown by Ishibashi *et al.* (1999), the minimum in the X-ray light curve is too long to be explained by an eclipse, even when adding a disk around the primary star. The radio light-curve also shows that the minimum is not restricted to our direction (Duncan & White 2003). The radio nebula grows and shrinks continuously, reaching a minimum simultaneously with all other features. It could be that an eclipse is possible involved in the minimum, but it is not the main or the sole mechanism.

Results in this paper are described in details by Teodoro *et al.* (2016).

2. The 2014.6 monitoring campaign of the He II λ4686 line in η Carinae

Observations of the He II 4686Å emission were done with HST/STIS beginning after the periastron event in 2009 through 2015 for a total of twenty one intervals (Gull *et al.* 2016, Mehner *et al.* 2015). Even more observations with ground-based telescopes were carried out during 2014, with 114 spectra from CTIO/Chiron, 90 spectra from OPD/LNA, 37 from SOAR/Goodman, 26 from MJUO/Hercules and 19 from CASLEO/REOSC. The group of amateurs SASER (Southern Astro Spectroscopy Email Ring) contributed 46 spectra. We also used 9 spectra from the Hexapod Telescope/BESO collected in critical times after the 2009.1 periastron. The resolution of the spectra varied among the facilities employed, but S/N was $>$ 200 per resolving element for the vast majority of the data. The typical uncertainty of the EW (equivalent width) measurements was $\sim$2 $\AA$ when inter–comparing contemporary data from different observers. For such a faint line, in a rich emission line spectrum, the placement of the continuum is challenging. Although it is impossible to obtain an unbiased measurement of the line intensity for an isolated spectrum, the variability can be derived in a robust way by a careful rectification of the stellar continuum (using a flux standard star) and using always the same wavelength windows (blue: 4600 ± 5 Å; red: 4742 ± 1 Å) and line integration (4675–4694 Å). Results are insensitive to instrumental setup for ground-based telescopes, for which the slit samples a region of 1–3 *Å* projected on the sky. The HST/STIS spectra, with spatial sampling was $\sim$ 0.1 arcsec, excludes considerable extended wind and nebular structures, which contribute considerable continuum, thus decreasing the equivalent width of the 4686Å feature. Fortunately thirteen of the HST/STIS observations were mapped across the 2x2 arcsec^2 region centered on eta Car. Comparison of the integrated mapped spectra were found to be compatible with measures of the ground-based spectra Teodoro *et al.* 2016).

We used several different procedures to derive a period that better fits the light-curve using the He II λ4686 line and also other features measured by different authors, like broad-band photometry, X-rays and radio flux, obtaining P= 2022.7 ± 0.3 d Damineli *et al.* (2008). It is important to notice that such a high accuracy (1 part in 10 000) opens the possibility to detect changes in the period due to the mass loss in future periastron passages.

The EW He II λ4686 light-curve resembles that of X-rays. It stays at a fairly constant level ($\sim$ 50 *mÅ*) for about 4.5 years up to phase=0.9. The He II EW slowly rises for $\sim$4 months, then rapidly grows for 2 months, reaching a peak of $\sim$ 3.5 *Å*. The fading phase reaches zero intensity in a timescale as short as 2 weeks. Different from X-rays, it stays at minimum for just $\sim$1 week and recovers with phase-locked behaviour, to reach another local peak (P3) $\sim$30 days after the minimum. The light curve then returns to the low level of $\sim$50 m*Å* at $\sim$2 months after the minimum. The similarity between the X-ray and He II λ4686 light curves when approaching the minimum and their large differences after that indicate that different mechanisms are controlling these two light curves in the recovery phase.

3. The light curve minimum seen from two directions

Comparison of He II λ4686 as seen from different view points provides information on the periastron and symmetry, which is tilted out of the sky plane at 45 deg latitude (Gull *et al.*, 2009, Madura *et al.* 2013). Our vantage point views the orbital axis at 45 deg latitude. The FOS4 position reflects light from the orbital polar region - 90 deg latitude - (Mehner *et al.* 2015). With a time-delay of 17 days, derived from the Homunculus shape and expansion rate (Smith 2006, both light curves show coincident temporal events. Once this correction is done, the time of the minimum of both light curves are in close coincidence in time. The immediate conclusion is that the 'event' cannot be an eclipse, as otherwise already hinted by the radio light curve. The simple interpretation is that the 'event' is produced by the blanketing of the UV ionizing radiation when the secondary star penetrates the inner dense wind of the primary. The cavity inside the WWC shock cone is wrapped by the orbital motion to the point that it no longer enables the radiation emitted at its apex to escape in any direction. This happens when the secondary star reaches periastron so that this critical point can be determined from the combination between the direct and reflected light curves.

The orbital orientation (specially the longitude of periastron) can be derived from modelling the He II λ4686 light curve. We used a model in which the intensity of emission goes as $1/D^2$ (D is the distance between the companion stars), following Fahed *et al.* (2011). Since the light curve grows/fades at a steeper pace around periastron, there must be an additional source which is present just when the apex of the WWC is at $\sim$1 month from periastron. We assume that such emission comes from recombination of He^{2+} ionized from He^{+} by the UV and soft X-rays from the WWC apex. This demands

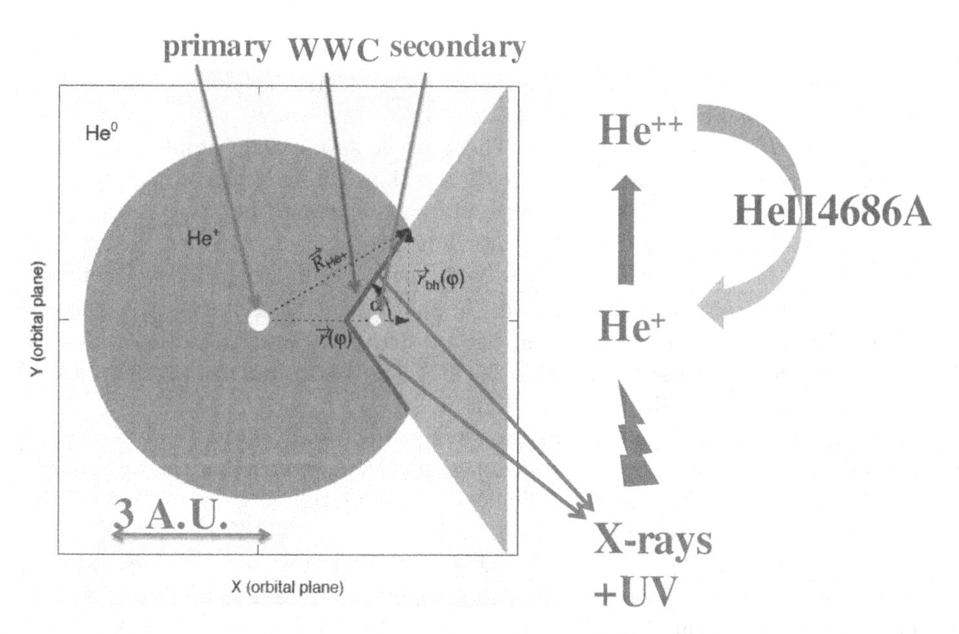

Figure 1. The bore hole effect, following Madura & Owocki (2010). When the WWC apex gets close and enters inside the He^+ sphere, it is illuminated by a powerful extreme-UV radiation source and is ionized to He^{2+}. Recombination to He^+ produces He II λ4686 photons. This accounts for the steep increase of the He II λ4686 light-curve for about ±1 month on either side of periastron. This emission adds up to the $1/D^2$ effect that modulates the He II λ4686 emission along the entire orbit as the distance (D) between the companion stars changes.

a huge reservoir of He^+, as is predicted by Hillier *et al.* (2011) model for the atmosphere of the primary star. When the WWC apex penetrates that sphere (R=3 AU) a luminous source of He II λ4686 is ignited and grows vey fast as the star approaches periastron. Both sources of emission add up and peak symmetrically around periastron. Electron scattering in the primary's wind produces opacity and the visibility of the He II λ4686 source gets modulated according to the orbital orientation relative to the observer. In addition to the primary star wind, the cavity inside the WWC shock cone also impacts the opacity. This is the 'bore hole' mechanism (Madura and Owocki 2010). We used 3D SPH simulations to model the He II II λ4686 light curve from the direct and reflected views. A very narrow range of periastron longitudes ($\omega = 234\,\text{deg}$–$252\,\text{deg}$) were allowed and the orbital inclination also could be constrained. It is interesting to see how good the fit of both light curves was. Even minor details, like the smaller amplitude of the light curve in reflected light, was well reproduced.

It is also noticeable that the time of periastron passage derived from this model is in close agreement with the time of minimum for hard X-rays. The fact that the recovery from the minimum in X-rays does not repeat from cycle to cycle, indicates that there additional mechanisms at work (Corcoran *et al.* 2010). For example, some X-ray emission is produced far from the WWC apex or it may be subject to density fluctuations due to clumping in the primary's wind.

4. Conclusions

Periodicity. The period length defined by the He II λ4686 line during the last 3 cycles is 2022.9 ± 1.6 d. When combined with other features across the electromagnetic spectrum (Damineli *et al.* 2008) it is better constrained to P= 2022.7 ± 0.3 d.

Time of periastron passage. The periastron passage occurs ~4 days after the He II λ4686 line reaches its minimum (EqW~0) at JD=2,456,874.1±2.3 days.

Stability of the He II λ4686 light-curve. There are no statistically significant changes in the line intensities when comparing the 2009.1 and the 2014.6 events contrary to previous claims (Mehner *et al.* 2015). The P2 peak could have increased by up to ~30 percent in the 2014.6 periastron as compared to the previous one.

Ingredients to reproduce the He II λ4686 light curve. The light curve is well reproduced by adopting for the emission a $1/D^2$ law like in Fahed *et al.* (2011) along the whole orbit of WR140, plus a "bore hole" component like in Madura & Owocki (2010) which works when the WWC is very close to or inside the He^+ sphere around the primary star. The absorption is calculated by 3D SPH simulations.

The orbital parameters. The simultaneous fit to direct and reflected light curves gives the longitude of periastron as $\omega = 234 - 252$ deg, in such a manner that the secondary star is "behind" the primary at periastron. The orbital inclination angle is $i = 135 - 153$ deg. The orbital ecccentricity was adopted as $e = 0.9$.

Verification from high spatial resolution. These orbital parameters are in excellent agreement with high spatial resolution observations reported by Gull *et al.* (2016) with HST (50 *mas* resolution) showing the high excitation lines at the opening of the WWC cavity. Weigelt *et al.* (2016) used the AMBER camera at the ESO/VLTI (6 *mas* resolution) to show that the WWC cavity is opened in our direction along most of the orbit, also in agreement with our orbital orientation.

References

Corcoran M.F., Hamaguchi, K., Pittard, J.M. *et al.* 2010 *ApJ*, 725, 1528
Damineli A., 1996, *ApJ* (Letters), 460, L49
Damineli A., Conti P. S., Lopes D. F., 1997, *New Astron.*, 2, 107
Damineli A., *et al.* 2008, *MNRAS*, 384, 1649
Davidson K., Humphreys R. M., 1997, *ARAA*, 35, 1
Duncan R.A & Whithe, S.M. 2003 2003, *MNRAS*, 338, 425
Fahed R., *et al.*, 2011, *MNRAS*, 418, 2
Gull T., *et al.*, 2009, *MNRAS*, 396,1308
Gull T., *et al.*, 2016, *MNRAS*, 462, 3196
Hillier D. J., Davidson K., Ishibashi K., Gull T., 2001, *ApJ*, 553, 837
Ishibashi, K., Corcoran, M. F., *et al.* 1999,*ApJ*, 524, 983
Madura T. I., Owocki S. P. 2010,*RMAA*, 38, 52
Madura T. I., *et al.*, 2013, *MNRAS*, 436, 3820
Mehner A., *et al.*, 2015, *A&A*, 578, 122
Prieto *et al.*, 2014 *ApJ*, 787, L8
Smith N., 2006, *AJ*, 644, 1151
Steffen W., *et al.*, 2014, *MNRAS*, 442, 3316
Steiner J. E., Damineli A., 2004, *ApJ* (Letters), 612, L133
Teodoro M., *et al.*, 2012, *ApJ*, 746, 73
Teodoro M., Damineli, A., *et al.* 2016, *ApJ*, 819, 131
Weigelt G., *et al.*, 2016, *A&A*, 594, 106

Session 4: Theory of stellar evolution & atmospheres: beyond standard physics, rotation, duplicity, mass loss and magnetic fields and instabilities

The Lives and Death-Throes of Massive Stars
Proceedings IAU Symposium No. 329, 2016
J.J. Eldridge, J.C. Bray, L.A.S. McClelland & L. Xiao, eds.

doi:10.1017/S1743921317003179

Evolution models of red supergiants

Cyril Georgy

Department of Astronomy, University of Geneva,
Chemin des Maillettes 51, 1290 Versoix, Switzerland
email: `cyril.georgy@unige.ch`

Abstract. The red supergiant (RSG) phase is a key stage for the evolution of massive stars. The current uncertainties about the mass-loss rates of these objects make their evolution far to be fully understood. In this paper, we discuss some of the physical processes that determine the duration of the RSG phase. We also show how the mass loss affect their evolution, and can allow for some RSGs to evolve towards the blue side of the Hertzsprung-Russell diagram. We also propose observational tests that can help in better understanding the evolution of these stars.

Keywords. stars: evolution, stars: interiors, stars: luminosity function, stars: mass loss, supergiants, convection

1. Introduction

The red supergiant (RSG) phase is an important phase of the life of massive stars. Indeed, a significant fraction of all massive stars will spend some time in this stage. Moreover, some of them will finish their life as a RSG, and explode as a type II-P supernova (SN), which represents about half of the total number of core-collapse SNe (Smith *et al.* 2011). The RSG phase occurs because at the end of the main sequence (MS), the core contraction will lead to a huge extension of the envelope (the so-called "mirror effect", see e.g. Kippenhahn & Weigert 1990). This makes the star evolve redwards in the Hertzsprung-Russell diagram (HRD), up to the point where the effective temperature of the star is $\log(T_{\rm eff}) \sim 3.5 - 3.6$. However, the modelling of the crossing of the HRD is strongly dependent on the way several physical processes are implemented into stellar evolution codes (e.g. convection, or rotation-induced instabilities).

During the RSG phase, the major uncertainty from the point of view of stellar evolution modelling is the mass loss. Mass loss from RSGs is still not understood on a theoretical point of view, so there is so far no equivalent to the radiative wind theory (e.g. Castor *et al.* 1975), established for hot stars, for this cool phase. The mass-loss rates of RSGs rely thus on observational determinations, which can vary over orders of magnitude at a given luminosity or effective temperature (see below).

The choice of weak or strong mass-loss rates during the RSG phase affects dramatically the modelling of the advanced stages of a massive star. In some cases, it is possible to make the star leaves the RSG branch and starts a new crossing of the HRD, becoming again a blue supergiant (BSG). In this paper, we illustrate some of these difficulties in modelling RSG, and how it is affected by the treatment of convection and mass-loss in stellar evolution codes. In section 2, we discuss briefly the first crossing of the HRD at the end of the MS. In section 3, we show the impact of the choice of the mass-loss rates during the RSG phase on the subsequent evolution. Finally, we discuss a few observational tests that could be used to test stellar evolution codes in section 4.

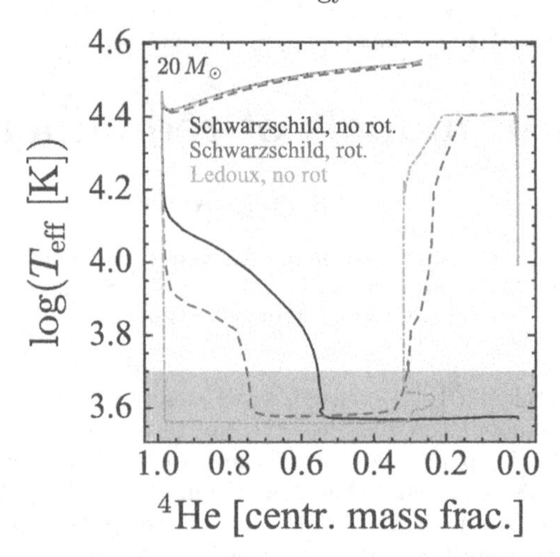

Figure 1. Tracks of $20\,M_\odot$ stars at solar metallicity in the $\log(T_{\rm eff})$ vs. central helium mass fraction plane, with different inputs physics. The solid line is a non-rotating model computed with the Schwarzschild criterion for convection. The long-dashed one includes the effect of rotation (initial velocity corresponding to 0.4 the critical one). The dot-dashed line is a non-rotating model computed with the Ledoux criterion for convection. The ZAMS is located on the top. The models then proceed along the MS towards the left, up to the point where the central helium mass fraction reaches almost 1. The tracks then cross the HRD, entering the red zone of the plot, illustrating the RSG region. At the end of their life, the rotating and Ledoux model cross again the HRD, reaching higher effective temperature, while the non-rotating Schwarzschild model remains in the RSG region.

2. First crossing of the HRD

Figure 1 illustrates the sensitivity of the first crossing of the HRD on the implemented physics inside a stellar evolution code in the $\log(T_{\rm eff})$ vs. central helium mass fraction plane. The figure shows the tracks of three different models of solar metallicity $20\,M_\odot$ stars: a non-rotating model computed with the Schwarzschild criterion for convection (solid line), a rotating model with the same criterion for convection (long-dashed line), and a non-rotating model computed with the Ledoux criterion for convection (dot-dashed line). For all three models, the ZAMS is located on the top, where all three curves are merged. The evolution proceeds towards the left, the central He mass fraction increasing to 1 during the MS. Then the tracks cross the HRD with different behaviour, reaching the region highlighted in red, corresponding to the effective temperatures of RSGs. The crossing of the HRD occurs very differently according to the physical assumptions that are made: the non-rotating Schwarzschild model crosses slowly, becoming a RSG only when it has burned about 45% of its central helium content. On the other hand, the model computed with the Ledoux criterion crosses directly the HRD before starting burning helium as a RSG.

These different behaviours are linked to the position and activity of an intermediate convective zone appearing on top of the former hydrogen-burning core at the end of the MS. The way both rotation and convection are implemented influences this intermediate convective shell. When it is active enough, it can sustain the luminosity of the star for a while, making the crossing occurs slowly. When it is less active, the star contracts more quickly, producing a fast crossing of the HRD.

These behaviours have important consequences on various aspects of the post-main sequence evolution of the star. It changes the time spent in the blue or red side of the HRD. It will thus affect the blue- to red-supergiant ratio in stellar population, ratio

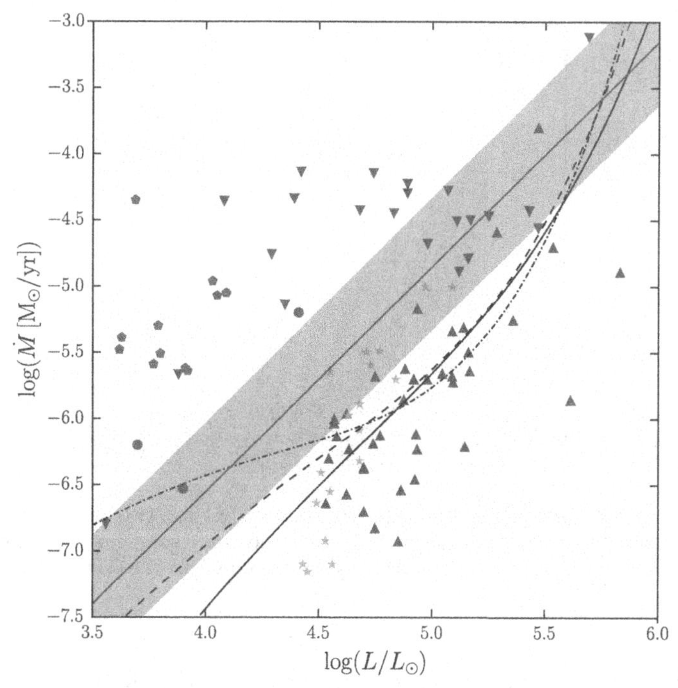

Figure 2. Mass-loss rates of RSGs deduced by different authors: Mauron & Josselin (2011, dark blue up-triangles), M-type (red down-triangles), MS-type (green circles), and carbon RSGs (purple pentagons) from van Loon *et al.* (2005) and RSGs from Beasor & Davies (2016, cyan stars). The three curves are mass-loss rates computed with the de Jager *et al.* (1988) recipe, for effective temperature of $\log(T_{\rm eff}) = 3.5$ (solid line), 3.6 (dashed line) and 3.7 (dot-dashed line). The shaded area shows the mass-loss rate used in the Geneva stellar evolution code (from observations from Crowther 2001): standard rate follows the bottom boundary of the shaded area, the middle line represents a mass-loss rate multiplied by a factor of 3, and the top boundary by a factor of 10.

which is still poorly reproduced by stellar models as of today (Langer & Maeder 1995; Eggenberger *et al.* 2002, see also Eldridge *et al.* 2008). By changing the duration of the RSG phase, it modifies the total amount of mass lost during this stage. Also, the way the HRD is crossed affects the time at which a possible Roche-lobe overflow can occur in case of a binary system.

3. Impact of the mass-loss rates on the advanced stages of massive star life

Figure 2 shows the mass-loss rates deduced observationally from different samples of RSG (Mauron & Josselin 2011; van Loon *et al.* 2005; Beasor & Davies 2016). From this plot, we see that the classical mass-loss recipes used in stellar evolution codes (often from de Jager *et al.* 1988) fit well the bulk of RSGs with the lowest mass-loss rates. However, some stars exhibit a much higher rate (more than 1 order of magnitude). It is not clear yet why so high mass-loss rates are present: is it linked to a peculiar evolutionary phase of the star? Is the duration of this high mass-loss rate phase long enough to have an impact on the total mass lost during the RSG phase? How to include it in stellar evolution codes?

This has motivated various studies, showing how an increased mass-loss rate during the RSG phase affects the subsequent evolution of the star. First of all, an increased mass-loss rate allows for some massive stars (typically with an initial mass

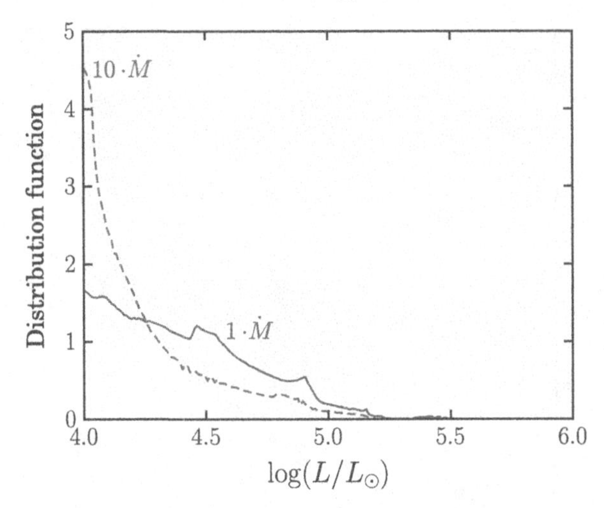

Figure 3. Luminosity distribution function for single RSG for two different mass-loss rates: computed with models using the standard mass-loss rate (blue solid line) and using a mass-loss rate increased by a factor of 10 during the RSG phase (red dashed line). From Georgy & Ekström (2016).

higher than about $20\,M_\odot$) to cross the HRD for a second time after the RSG phase (Vanbeveren *et al.* 1998a,b). This is a possible solution for the so-called "red supergiant problem" (Smartt *et al.* 2009, see also Georgy *et al.* 2012; Walmswell & Eldridge 2012): no type II-P supernova (SN) progenitor is found with an initial mass higher than about $17\,M_\odot$, while higher mass RSGs are known, which should explode as a type II-P SN. A solution to this problem is thus that higher initial mass RSGs evolve towards the blue side of the HRD before exploding.

In Georgy (2012), we have discussed the impact of an increased mass-loss rate during the RSG phase for stars with an initial mass of about $12 - 15\,M_\odot$, and shown that their end-point in the HRD is located inside the yellow supergiant region. Their very small hydrogen-rich envelope would make it appears as a type IIb or II-L SN at the time of the explosion. In case high $\dot{M}$ are possible for RSGs in this mass range, this could be an alternative scenario to the classical binary channel scenario (e.g. Podsiadlowski *et al.* 1993; Claeys *et al.* 2011; Benvenuto *et al.* 2013) for producing yellow supergiant SN progenitors.

Meynet *et al.* (2015a) have studied in more details the impact a higher $\dot{M}$. They have shown that using a large mass-loss rate indeed decreases the time spent as a RSG. Interestingly, however, it appears that the total amount of mass lost during the RSG phase is almost not affected by the choice of the mass-loss rate for RSG stars evolving back to blueside of the HRD.

The change in the duration of the RSG phase at different mass for different mass-loss rates has, as a direct consequence, a modification of the luminosity distribution function of such stars. This is illustrated in Fig. 3. Indeed, as discussed in Meynet *et al.* (2015a), a high mass-loss rate tends to remove the Cepheid loops in the intermediate mass star regime, making them spending more time in the RSG phase. On the other hand, a higher mass-loss rate will decrease the duration of the RSG phase for stars with a mass of about $20 - 25\,M_\odot$, by making them evolving back to the blue side of the HRD. This makes the luminosity distribution more peaked at lower luminosity for higher mass-loss rates during the RSG phase.

4. A second population of BSGs?

As mentioned in the previous section, a high mass-loss rate during the RSG phase favours a blueward evolution. This occurs when the mass of the core becomes an important fraction of the total mass of the star (typically more than about 60%, see Giannone 1967). In case such kind of evolutionary path would exist, we can wonder whether there is a way to distinguish those BSGs coming back from the red side of the HRD after the RSG phase (hereafter group 2 BSGs) and the classical BSGs that are stars on their first crossing of the HRD (hereafter group 1 BSGs).

In Saio *et al.* (2013), we have shown that group 1 and group 2 BSGs have very different pulsational properties. Indeed, due to the strong mass loss during the RSG phase, group 2 stars have a much higher luminosity to mass ratio, favouring excitation of pulsations inside these stars. The period of radial pulsations in our models of group 2 BSGs are qualitatively in good agreement with the observed period of variability of α Cygni variables, indicating that this class of variable stars could be post-RSG stars. However, these models, computed with the Schwarzschild criterion for convection, have a surface chemical composition strongly processed by CNO cycle, with for example N/C ratio up to 60. This is not compatible with the observed value of about 2-3 for the α Cygni variable stars Rigel and Deneb (Przybilla *et al.* 2010).

In a following study (Georgy *et al.* 2014), we have shown that using the Ledoux criterion instead of the Schwarzschild one improves a lot the situation. Indeed, the position of the intermediate hydrogen-burning convective zone is shifted at a higher position inside the star with the Ledoux criterion. This prevents the transport strongly CNO-processed material close enough to the surface. However, we still have some difficulties in reproducing the observed surface gravity for these stars (see also Saio *et al.* 2015).

Another interesting property of BSGs is that they lie in a tight linear relation in the bolometric magnitude vs $\log(g/T_{\rm eff}^4)$ plane (called the "flux-weighted gravity – luminosity relation", FWLR, see Kudritzki *et al.* 2003). In Meynet *et al.* (2015b), we have studied how population of stars containing both group 1 and group 2 BSGs fits this relation. It appeared that the FWLR is not compatible with a lot of objects of the group 2. However, the general agreement between our models and the observed FWLR is quite good. The next step in this direction will be to include the effects of binary evolution in this diagram.

5. Conclusions

In this short paper, we have discussed the way a massive star crosses the HRD at the end of the MS to become a RSG, and show the sensitivity of this crossing on the way convection or rotation is implemented inside stellar evolution code. Once the star is a RSG, its mass-loss rate is a key ingredient in determining what will be its future evolution. In the most extreme cases, the star loses enough mass to cross the HRD a second time, becoming again a BSG. We propose different ways to observationally constrain the physics (particularly, the physics of convection and the average mass-loss rate during the RSG phase) in the models: the luminosity distribution of RSG, the pulsational properties of BSGs and the FWLR can provide useful informations about the RSG stage.

References

Beasor, E. R. & Davies, B. 2016, *MNRAS* 463, 1269
Benvenuto, O. G., Bersten, M. C., & Nomoto, K. 2013, *ApJ* 762, 74
Castor, J. I., Abbott, D. C., & Klein, R. I. 1975, *ApJ* 195, 157

Claeys, J. S. W., de Mink, S. E., Pols, O. R., Eldridge, J. J., & Baes, M. 2011, *A&A* 528, A131

Crowther, P. A. 2001, in D. Vanbeveren (ed.), *The Influence of Binaries on Stellar Population Studies*, Vol. 264 of *Astrophysics and Space Science Library*, p. 215, Kluwer Academic Publishers, Dordrecht

de Jager, C., Nieuwenhuijzen, H., & van der Hucht, K. A. 1988, *A&AS* 72, 259

Eggenberger, P., Meynet, G., & Maeder, A. 2002, *A&A* 386, 576

Eldridge, J. J., Izzard, R. G., & Tout, C. A. 2008, *MNRAS* 384, 1109

Georgy, C. 2012, *A&A* 538, L8

Georgy, C. & Ekström, S. 2016, *IAU Focus Meeting* 29, 454

Georgy, C., Ekström, S., Meynet, G., *et al.* 2012, *A&A* 542, A29

Georgy, C., Saio, H., & Meynet, G. 2014, *MNRAS* 439, L6

Giannone, P. 1967, *ZAp* 65, 226

Kippenhahn, R. & Weigert, A. 1990, *Stellar Structure and Evolution*, Stellar Structure and Evolution, XVI, 468 pp. 192 figs. Springer-Verlag Berlin Heidelberg New York. Also Astronomy and Astrophysics Library

Kudritzki, R. P., Bresolin, F., & Przybilla, N. 2003, *ApJ* 582, L83

Langer, N. & Maeder, A. 1995, *A&A* 295, 685

Mauron, N. & Josselin, E. 2011, *A&A* 526, A156

Meynet, G., Chomienne, V., Ekström, S., *et al.* 2015a, *A&A* 575, A60

Meynet, G., Kudritzki, R.-P., & Georgy, C. 2015b, *A&A* 581, A36

Podsiadlowski, P., Hsu, J. J. L., Joss, P. C., & Ross, R. R. 1993, *Nature* 364, 509

Przybilla, N., Firnstein, M., Nieva, M. F., Meynet, G., & Maeder, A. 2010, *A&A* 517, A38+

Saio, H., Georgy, C., & Meynet, G. 2013, *MNRAS* 433, 1246

Saio, H., Georgy, C., & Meynet, G. 2015, in G. Meynet, C. Georgy, J. H. Groh, & Ph. Stee (ed.), *IAU Symposium*, Vol. 307 of *IAU Symposium*, pp 230–231

Smartt, S. J., Eldridge, J. J., Crockett, R. M., & Maund, J. R. 2009, *MNRAS* 395, 1409

Smith, N., Li, W., Filippenko, A. V., & Chornock, R. 2011, *MNRAS* 412, 1522

van Loon, J. T., Cioni, M.-R. L., Zijlstra, A. A., & Loup, C. 2005, *A&A* 438, 273

Vanbeveren, D., De Donder, E., van Bever, J., van Rensbergen, W., & De Loore, C. 1998a, *New A* 3, 443

Vanbeveren, D., De Loore, C., & Van Rensbergen, W. 1998b, *A&A Rev.* 9, 63

Walmswell, J. J. & Eldridge, J. J. 2012, *MNRAS* 419, 2054

The Lives and Death-Throes of Massive Stars
Proceedings IAU Symposium No. 329, 2016
J.J. Eldridge, J.C. Bray, L.A.S. McClelland & L. Xiao, eds.

doi:10.1017/S1743921317003398

Common envelope: progress and transients

Natalia Ivanova

Department of Physics, University of Alberta, Edmonton AB T6G 2E1 Canada
email: nata.ivanova@ualberta.ca

Abstract. We review the fundamentals and the recent developments in understanding of common envelope physics. We report specifically on the progress that was made by the consideration of the recombination energy. This energy is found to be responsible for the complete envelope ejection in the case of a prompt binary formation, for the delayed dynamical ejections in the case of a self-regulated spiral-in, and for the steady recombination outflows during the transition between the plunge-in and the self-regulated spiral-in. Due to different ways how the recombination affects the common envelope during fast and slow spiral-ins, the apparent efficiency of the orbital energy use can be different between the two types of spiral-ins by a factor of ten. We also discuss the observational signatures of the common envelope events, their link a new class of astronomical transients, Luminous Red Novae, and to a plausible class of very luminous irregular variables.

Keywords. binaries: close, hydrodynamics, stars: outflows

---

1. Introduction: the energy sources and the energy sinks

Common-envelope events (CEEs) are fate-defining episodes in the lives of close binary systems. During a common envelope phase, the outer layers of one of the stars expand to engulf the companion, and two stars start to temporarily orbit within their shared envelope. This pivotal binary changeover ends with a luminosity outburst, leaving behind either a significantly shrunk binary, or a single merged star. These episodes are believed to be vital for the formation of a wide range of extremely important astrophysical objects, including X-ray binaries, close double-neutron stars, the potential progenitors of Type Ia supernovae and gamma-ray bursts, and double black holes that could produce gravitational waves (for more details on overall importance of the CEEs, as well as on many aspects of the involved physics, see the review in Ivanova *et al.* 2013).

The outcomes of the CEEs are believed to fall into two main divergent categories – either a close binary formation, or a merger of the two stars into a single star. The boundary between the outcomes is usually found by comparing the available energy source (the energy difference between the orbital energies before and after the CEE, $\Delta E_{\rm orb}$), and the required energy expense (the energy required to displace the envelope to infinity, $E_{\rm bind}$). This is known as the energy formalism (Webbink 1984; Livio & Soker 1988):

$$\alpha \Delta E_{\rm orb} = E_{\rm bind} = \frac{G m m_{\rm env}}{\lambda R} \tag{1.1}$$

Here, α is the efficiency of the use of the orbital energy, and it can be only less than one. m is the mass of the donor star – the star whose expanded envelope has formed the CE, $m_{\rm env}$ is the mass of that envelope, R is the radius of the donor. The parameter λ relates the envelope's binding energy $E_{\rm bind}$ as integrated from the stellar structure with its parameterized form.

This famous equation, while seems to be transparent and straightforward, buries a lot of not yet fully understood physics. For example, there are plenty of uncertainties in how to determine $E_{\rm bind}$. That includes such questions as what is the boundary between the core and the ejected envelope, whether the thermal energy can be converted effectively in the mechanical energy of the envelope, and whether the out-flowing envelope should be evaluated using Bernoulli integral which is inclusive of the P/ρ term in addition to thermal energy (Dewi & Tauris 2000; Deloye & Taam 2010; Ivanova & Chaichenets 2011; Ivanova 2011a). For instance, the uncertainty in the boundary may lead to an order of magnitude uncertainty in the value of $E_{\rm bind}$, and, consequently, same uncertainty in the orbital separations of a post-CEE binary (Ivanova 2011a).

Another deficiency of the classic energy formalism is that, by design, the Equation (1.1) implies that the kinetic energy of the ejected envelope at infinity is zero (or, in other words, is substantially smaller than the two considered energies). However, as has been shown recently for the case of low-mass giants, if the entire envelope has been successfully ejected, that envelope can carry away between 20% and 55% of the released orbital energy, mainly in the form of the kinetic energy, and, to a lesser degree, in the form of the thermal energy (Nandez & Ivanova 2016).

Considering three fundamental energies – gravitational potential energy, thermal energy of the envelope and kinetic energy – CEEs were studied using different three-dimensional (3D) hydrodynamic codes. Universal evolution of a CEE in 3D simulations is to start a plunge-in of the companion, during which the binary orbit shrinks strongly on the timescale comparable to the initial binary orbital period. By the end of the plunge-in, the strength of all frictional interactions between the shrunken binary and the inflated envelope is strongly reduced. The binary settles into a slow spiral-in with a minuscule orbital dissipation rate (Ricker & Taam 2008; De Marco *et al.* 2011; Ricker & Taam 2012; Passy *et al.* 2012; Nandez *et al.* 2015; Ohlmann *et al.* 2016a; Staff *et al.* 2016). Independently the type of employed code, only partial envelope ejections had been obtained. It showed clearly that something essential is missing, and the missing piece is neither the type of the code, nor the resolution, but should be related to physics that has not been yet taken into account.

Indeed, there are other, "non-fundamental", ways in which the energy can be generated or lost during a CEE.

One of the sources of energy is due to accretion on a companion while it swirls inside the common envelope. Energy comes from the release of the potential energy of the accreted material while it reaches the surface of the companion, in the form of heat and radiation. If the companion does not accept all the accreted material, some energy may be released back via jets. Jets inject the kinetic energy back to the common envelope, inflating "bubbles" and helping to remove the common envelope this way (Akashi & Soker 2016; Shiber *et al.* 2017). The total input from this energy source depends on the mass of the companion, on the mass accretion rate, and the time during which the accretion takes place. To find the accretion rate, a common way in the past was to use Bondi-Hoyle-Lyttleton prescription (Hoyle & Lyttleton 1939; Bondi & Hoyle 1944; Bondi 1952). It has been found however that in 3D simulations the accretion rate onto the companion is significantly smaller than the Bondi-Hoyle-Lyttleton prescription would provide (Ricker & Taam 2012). On the other hand, more recent, albeit simplified, studies of the accretion during a CEE, have found accretion rates that approach the Bondi-Hoyle-Lyttleton prescription (MacLeod & Ramirez-Ruiz 2015a,b). It is not clear if one of the accepted simplifications, or the differences in the considered stellar models in 3D studies and in simplified studies, have led to the striking difference in the accretion rates. The time on which this energy can be generated efficiently can be as small as the initial orbital period, as this dictates the timescale of the plunge-in. After the plunge, a shrunken binary clears out its neighborhood and may avoid continued accretion. Whatever accretion rate would be eventually found to be correct, the case when the accretion source of energy can become comparable to the binding energy of the envelope is likely limited to the case when a companion, while accreting at its Eddington rate, is spiraling-in to a very large donor, $\sim 1000 R_{\odot}$.

The role of the magnetic field has also been contemplated. Magnetic fields were found to strongly shape the outflows from the common envelopes (Nordhaus & Blackman 2006; Nordhaus *et al.* 2007). For little-bound envelopes of AGB stars, these magnetic outflows has been argued to help to unbind the entire envelope, although the complete ejection was not directly obtained in simulations. For low-mass red giants, the presence of the magnetic field was determined to be dynamically irrelevant for a common envelope ejection, despite strong amplification of the magnetic fields (Ohlmann *et al.* 2016b).

In some CEEs, if a non-degenerate companion has initially failed to eject the common envelope, and has to merge with the donor's core, the companion's material can trigger an explosive nucleosynthesis on the outer parts of the core of the evolved donor. This can lead to the explosive ejection of not just the envelope, but also of both the hydrogen and the helium layers (Ivanova 2002; Podsiadlowski *et al.* 2010).

The energies listed above are not guaranteed to be present in all CEEs. However, there is one source of energy which is naturally present in all the cases – the recombination energy. It is important that four phases of a CEE, qualitatively different in the involved dominant physical processes and the timescales, are currently distinguished: (a) loss of corotation, (b) plunge-in (this is the stage which is often mistaken for a CEE as a whole), (c) self-regulating spiral-in (this stage only takes place if the plunge-in did not lead to complete envelope ejection); (d) termination of the self-regulating phase, with either a delayed dynamical ejection, or a nuclear ejection, or with a merger (for qualitative definitions of the phases, see Ivanova 2011b; Ivanova *et al.* 2013, and quantitative definitions can be found in Ivanova & Nandez 2016). Recently, it was shown that during the loss of corotation, a substantial fraction of the initial envelope mass can be lost before the CEE enters the dynamical phase during which the energy formalism is applicable. The mass is lost while the the donor overfills its Roche lobe, but the expanded envelope does not yet go beyond L_2/L_3 points, and the phase can last for thousand of years (Pavlovskii & Ivanova 2015; Pavlovskii *et al.* 2017). The self-regulating phase also could last thousand of years (Meyer & Meyer-Hofmeister 1979; Podsiadlowski 2001). At this timescale the radiative energy loss from the common envelope surface is becoming large enough to affect the overall energy budget (Ivanova 2002). As it has appeared, the recombination energy plays an important, while varying, role during most of stages of a CEE.

2. The role of the recombination energy

As the common envelope expands and its material cools down, the ionized plasma can recombine, releasing binding energy which is usually referred to as recombination energy, $\Delta E_{\rm rec}$. We note that as cooling continues, formation of molecules can take place, also releasing energy, but here we will not consider the energy related to molecule formation. Recombination energy was suggested to be helpful for ejecting outer stellar layers even before the concept of a CEE, to say nothing of the energy formalism (e.g., Lucy 1967; Roxburgh 1967). Binary population synthesis studies have shown that the inclusion of the recombination energy in the energy formalism, as a part of the envelope's internal energy, provides the best fits to the observations of subdwarf B stars (Han *et al.* 1994, 2002). On the other hand, it has been argued that the recombination energy cannot help ejecting the CE, as the most of the recombination energy would leave the envelope immediately in a form of radiation, as opacity in the envelope might be too low to effectively reprocess energy released in photons of specific wavelengths (Soker & Harpaz 2003). We note that this restriction is indeed valid if the optical depth of the layer where the recombination takes place is small (it is close to the photosphere), and the layer itself is very thin, so the released photons can not be used to heat the envelope material.

The amount of the recombination energy that is stored in an envelope of the mass $m_{\rm env}$ prior the start of a CEE, neglecting the ionization of the elements others than hydrogen and helium, can be evaluated as follows:

$$\Delta E_{\rm rec} \approx 2.6 \times 10^{46} \times \frac{m_{\rm env}}{M_\odot} {\rm ergs} \times (X f_{HI} + Y f_{HeII} + 1.46 Y f_{HeI}) \tag{2.1}$$

Here X is the hydrogen mass fraction, Y is the helium mass fraction, f_{HI} is the fraction of hydrogen that becomes neutral, f_{HeI} is the fraction of helium that becomes neutral, and f_{HeII} is the fraction of helium that becomes only singly ionized. With a typical value for helium content and assuming complete recombination from initially completely ionized material, the released energy can be as high as $\Delta E_{\rm rec} \approx 3 \times 10^{46} \times m_{\rm env}/M_\odot$ erg. Comparing this energy with the binding energy as in the Equation (1.1), one can see that once the radius of the star exceeds $R \gtrsim 127 R_\odot/\lambda$, a star can be said to have positive total energy even before the start of the CEE, i.e. it is unbound. However, first of all, the release of this energy has to be triggered. Second, this energy should not escape in a form of radiation, but be reprocessed by the envelope itself.

2.1. *Recombination during a plunge-in phase*

As was mentioned above, an unavoidable outcome of 3D simulations of a CEE with a hydrodynamic code that did not include the recombination energy in the adopted equation of state is to obtain a plunge-in phase, to eject a part of the envelope, to inflate the remaining bound envelope well above the binary orbit, and to start a slow spiral-in, during which the depletion of the binary orbit is becoming too small to be further treated by a hydrodynamic code (Ricker & Taam 2008; De Marco *et al.* 2011; Ricker & Taam 2012; Passy *et al.* 2012; Nandez *et al.* 2015; Ohlmann *et al.* 2016a; Staff *et al.* 2016). The primary reason for this outcome is the decoupling of the shrunken binary orbit from the remaining inflated envelope, as both gravitational or viscous drags are becoming too small (Ivanova & Nandez 2016). On the other hand, the very first attempt to include the recombination energy in the equation of state have resulted in the complete common envelope ejection (Nandez *et al.* 2015).

This very first study, where the common envelope was completely ejected, have considered the formation of the specific double-white dwarf (DWD) binary WD 1101+364, a well-measured binary system that has $P_{\rm orb} = 0.145$ d, and a mass ratio of $q = M_1/M_2 = 0.87 \pm 0.03$, where $M_1 \simeq 0.31 M_\odot$ and $M_2 \simeq 0.36 M_\odot$ are the masses of the younger and older WDs, respectively (Marsh *et al.* 1995). DWD binaries are the best test-site for CEE as their younger white dwarfs must have been formed during a CEE, and their pre-CEE binary separations are strongly restricted by the well known core-radius relation of low-mass giants, albeit there is a fairly small dependence on the total giant mass (van der Sluys *et al.* 2006). Several simulations performed to form WD 1101+364 using the allowed range of the initial binaries and using the equation of state that did not include the recombination energy, also did not unbind the envelope (Nandez *et al.* 2015). The analysis has shown that the binding energy of the remaining bound envelope could be easily overcome by the release of the recombination energy, if the recombination energy release will be triggered at the right time. This is exactly what the simulations with the recombination energy taken into account have shown (Nandez *et al.* 2015).

The physics of the complete envelope ejection can be understood via introduction of the *recombination radius* – the radius at which the released specific recombination energy is larger than the local specific potential energy (for more detail, see Ivanova & Nandez 2016). Usually hydrogen starts its recombination when all helium is already recombined; in this case this radius is $r_{\rm rec,H} \approx 105 R_\odot \times m_{\rm grav}/M_\odot$. Here $m_{\rm grav}$ is the mass within the recombination radius – this mass includes the companion, the core of the donor, and the mass envelope within $r_{\rm rec,H}$.

During a CEE, at first, the frictional forces dissipate energy from the binary orbit and dump the same energy into the common envelope. This leads to the first dynamical ejection of a fraction of the envelope, and it is the ejection that is present in all the types of 3D simulations, independent of the equation of state or the adopted method.

If a still bound envelope has been dynamically expanded beyond of the envelope's recombination radius, its material is doomed to be ejected to infinity via the recombination outflows on a dynamical timescale, leading to a prompt binary formation (Ivanova & Nandez 2016). If the envelope expansion beyond the recombination radius is slow (only a small fraction of the envelope has been expanded beyond the recombination radius on a dynamical timescale), a transition to a slow spiral-in takes place. In this case, recombination leads to *steady* recombination-powered outflows, the mass lass through these outflows can be slowly accelerating, as $m_{\rm grav}$ decreases during the continuing mass loss (Ivanova & Nandez 2016). We note that it has been proposed, but not yet verified against the 3D outcomes, that in the case when *steady* outflows are established, the envelope's enthalpy rather than the envelope's thermal energy determines the outcome (Ivanova & Chaichenets 2011). During the transition to a slow-spiral-in, the remaining bound envelope can also "fall" back on its parabolic trajectory. Such a fallback triggers another partial envelope ejection that acts on a dynamical time and is presumably powered by the compression ionisation and then recombination of the helium layer (Ivanova & Nandez 2016).

Let us now consider the efficiency of the use of the recombination energy. It has been found that the structure of ionisation zones in an expanded common envelope is drastically different from the same in unperturbed stars. The zones of partial ionisation of helium and hydrogen, i.e. where f_{HI}, f_{HeI} and f_{HeII} are changing from 0 to 1, are very thick in mass each – e.g., they can reach $\sim 0.5 M_\odot$ in a low-mass giant. Hydrogen is still 1% ionized at an optical depth of 100 or more (Ivanova *et al.* 2015; Ivanova & Nandez 2016), although a smaller degree of ionisation can remain in some cases closer to the photosphere. The recombination energy therefore can be well reprocessed. Notably, the recombination energy of helium has absolutely no chance for escape in a form of radiation and all can be used for the envelope expansion (Ivanova *et al.* 2015).

2.2. *Recombination during a self-regulated spiral-in*

During a self-regulated spiral-in, the energy transfer throughout the common envelope, the nuclear energy generation, and the energy losses from envelopes surface are becoming important both for the energy budget and for the thermal structure of the shared envelope. At the same time, the orbital period of the shrunken binary is becoming substantially smaller than the dynamical timescale of the inflated envelope, mandating a 3D hydrodynamic code to switch to a timestep which is extremely small if compared to the timescale on which the envelope evolves. As a result of these complications, no existing 3D hydrodynamic code is capable of following the self-regulated spiral-in (we note that the first step towards treating the convection properly has been made recently by

Ohlmann *et al.* 2017). There is a 3D study that specifically investigated how the plunge-in transits, via a slow spiral-in, into the self-regulated spiral-in (Ivanova & Nandez 2016). However, the simulations had to end by the time when the thermal timescale processes could become important. Instead of 3D, a common approach for studying a self-regulated spiral-in is to use an one-dimensional (1D) stellar code, modifying it to *mimic* CEE conditions, with a number of simplifications which could be different from study to study (the pioneering studies are Taam *et al.* 1978; Meyer & Meyer-Hofmeister 1979, and many thereafter).

In one of the 1D studies, it has been found that during the self-regulated spiral-in, after the envelope has been inflated, a delayed dynamical instability initiating pulsations of growing amplitude takes place (Ivanova 2002; Han *et al.* 2002). These growing pulsations might lead to a delayed dynamical ejection of the envelope, although the ejection itself was not obtained.

Ivanova *et al.* (2015) have explored 1D CEE evolution for a low-mass giant in a systematic way, by introducing a constant "heating" source of the two types – uniform heating throughout the envelope, and a shell-type at the base of the envelope. As a reaction to the artificial "heating", the envelope readjusts by expanding to its new "equilibrium" radius – the radius at which the inflated star radiates away the amount of energy that it receives from both the artificial heating and the shell nuclear burning – and is cooling down. Double ionized helium starts its recombination. This recombination is becoming energetically important and can produce an even higher rate of the energy input than the artificial heating. The recombination zones of once ionized helium and hydrogen propagate inwards in mass. With high heating rates and quick initial envelope expansion, outer layers start moving faster than their local escape velocity. For moderate heating rates, the envelope expands to its "equilibrium" radius but is becoming unstable. Ivanova *et al.* (2015) determined that, due to the expansion of the zones of partial ionisation of hydrogen and helium in mass, the envelope's pressure-weighted Γ_1 becomes less than 4/3, and almost the entire envelope becomes dynamically unstable.

In further studies, using a similar approach for an artificial heating source while using a 1D stellar code that includes hydrodynamic terms, Clayton *et al.* (2017) found that the heated envelopes, if not dynamically ejected at high heating rates, also become unstable and start to experience non-regular pulsations, with the periods between 3 and 20 years. Some pulsations lead to the ejection of a fraction of the envelope, with up to 10% of the envelope mass escaping per ejection episode. These ejections have a nature similar to the shell-triggered ejections found earlier in 3D studies of slow spiral-ins (Ivanova & Nandez 2016).

2.3. *Recombination and the outcomes of CEEs*

Two families of the outcomes are expected.

If a CEE has resulted in a prompt binary formation, the "classical" α that relates the initial donor and the final orbit, as in the Equation (1.1), can be as large as one or even a bit more that one. The revised energy formalism that taken into account the energy that the ejected material carried away and the recombination energy can be found in Nandez & Ivanova (2016). This revision of the energy formalism is based on the fits of 3D simulations of CEEs for the grid of initial binaries with low-mass giant donors and low-mass white dwarfs.

If a CEE has resulted in a self-regulated spiral-in, the "classical" α is only about 0.05-0.25, and the envelope's material is lost in semi-regular recombination-triggered pulsations with an interval between the ejections of 3-20 years (Clayton *et al.* 2017). We note however that this value of α does not yet take into account that some material has been ejected "dynamically" before the self-regulated spiral-in has started, and more studies are needed.

2.4. *Appearance of the CEEs*

All CEEs, including those that end up as mergers, are accompanied by a dynamical ejection of at least some envelope material. As plasma expands, it cools down and starts recombination. Before the recombination starts, gas expansion is adiabatic. As most of envelope material has initially about the same entropy, the location at which the recombination starts is also similar for the ejected material. Opacities below the place where gas recombines are high, while above they are low, at least until the cooled gas can form dust. This recombination front appears as a "photosphere" that hides beneath it the common envelope, for as long as there is plasma to be recombined. Once all material have recombined, it reveals the common envelope. This model of Wavefront of Cooling and Recombination (WCR) has been proposed by Ivanova *et al.* (2013). It utilizes an analytical model of Popov (1993), proposed for hydrogen envelope cooling in Type II supernovae during the plateau phase.

This WCR model explains naturally curious observational features of the new class of transients, Luminous Red Novae:

- Large "apparent" size and luminosities, plateau phase for the light-curve.
- "Red" color (temperature of the object is about 5000K).
- Fast decline of luminosities (timescale of the decline is a fraction of the plateau time, and it is much smaller than the inferred dynamical timescale of the object)
- Spectroscopic velocities, which are few hundreds of km/s, are larger than the expansion rate of the "effective" radius, which are less than a hundred of km/s

Ivanova *et al.* (2013) have shown that the range of the expected plateau time and luminosities for stellar mergers is consistent with the observed ranges for LRNe, and that the rate at which LRNe are observed can also be provided by the stellar mergers. Some attempts are made to fit the observed light-curves of LRNe. To fit V1309 Sco outburst (Tylenda *et al.* 2011), Ivanova *et al.* (2013) have used Popov's analytical model, for which velocities and the mass of the ejecta were provided by 3D simulations (detail of 3D simulations are in Nandez *et al.* 2014). The light-curve of M31 2015 LRN was fitted with the merger of a binary system in which the primary star is a $3 - 5.5 M_\odot$ sub-giant branch star with radius of $30 - 40 R_\odot$ (MacLeod *et al.* 2017).

If a CEE has entered into self-regulated spiral-in, the common envelope object appears as a luminous pulsation variable (note that an LRN-type outburst is expected to precede this). On the Hertzsprung-Russell diagram, the pulsations swirl around the equilibrium point, the position of which is dictated by the heating rate. Depending on the heating rate, that point can be located at $\log_{10} T_{\rm eff} \approx 3.4 - 3.5$ (while $\log_{10} T_{\rm eff}$ during the pulsation can be changing between 3.2 and 3.7) and at $\log_{10}(L/L_\odot) \approx 4.0 - 4.4$ (while $\log_{10}(L/L_\odot)$ can vary by up to 500 times between the minimum luminosity during the pulsation, and the maximum luminosity). The pulsations are not symmetric with time, and the time that a heated envelope spends at higher than equilibrium luminosity is much smaller than the time it takes for the star to be "re-heated" back to its equilibrium value (for examples of light-curves, see Clayton *et al.* 2017).

However, if a CEE had neither resulted in a clean merger, nor had entered in a self-regulated spiral-in, the observational signatures are less understood. While the first dynamical ejection can provide an LRN-type outburst, further outflows take places when some initially available recombination energy has been processed to unbind the envelope. This may change the observed luminosities, presence of the plateau, and the timescale of the outbursts. No self-consistent 3D modeling of a CEE leading to a binary formation inclusive of radiative energy loss have been done yet, and is the important subject of future studies.

References

Akashi, M. & Soker, N. 2016, *MNRAS*, 462, 206

Bondi, H. & Hoyle, F. 1944, *MNRAS*, 104, 273

Bondi, H. 1952, *MNRAS*, 112, 195

Clayton, M., Podsiadlowski, P., Ivanova, N., & Justham, S. 2017, *MNRAS*, accepte, arXiv:1705.08457

Deloye, C. J. & Taam, R. E. 2010, *ApJ* (Letters), 719, L28

De Marco, O., Passy, J.-C., Moe, M., *et al.* 2011, *MNRAS*, 411, 2277

Dewi, J. D. M. & Tauris, T. M. 2000, *A&A*,360, 1043

Han, Z., Podsiadlowski, P., & Eggleton, P. P. 1994, *MNRAS*, 270, 121

Han, Z., Podsiadlowski, P., Maxted, P. F. L., Marsh, T. R., & Ivanova, N. 2002, *MNRAS*, 336, 449

Hoyle, F. & Lyttleton, R. A. 1939, Proceedings of the Cambridge Philosophical Society, 34, 405

Ivanova, N. 2002, Ph.D. Thesis,

Ivanova, N. 2011a, *ApJ*, 730, 76

Ivanova, N. 2011b, Evolution of Compact Binaries, 447, 91

Ivanova, N. & Chaichenets, S. 2011, *ApJ* (Letters), 731, L36

Ivanova, N., Justham, S., Avendano Nandez, J. L., & Lombardi, J. C. 2013, *Science*, 339, 433

Ivanova, N., Justham, S., Chen, X., *et al.* 2013, *A&ARv*, 21, 59

Ivanova, N., Justham, S., & Podsiadlowski, P. 2015, *MNRAS*, 447, 2181

Ivanova, N. & Nandez, J. L. A. 2016, *MNRAS*, 462, 362

Livio, M. & Soker, N. 1988, *ApJ*, 329, 764

Lucy, L. B. 1967, *AJ*, 72, 813

MacLeod, M. & Ramirez-Ruiz, E. 2015a, *ApJ* (Letters), 798, L19

MacLeod, M. & Ramirez-Ruiz, E. 2015b, *ApJ*, 803, 41

MacLeod, M., Macias, P., Ramirez-Ruiz, E., *et al.* 2017, *ApJ*, 835, 282

Marsh, T. R., Dhillon, V. S., & Duck, S. R. 1995, *MNRAS*, 275, 828

Meyer, F. & Meyer-Hofmeister, E. 1979, *A&A*,78, 167

Nandez, J. L. A., Ivanova, N., & Lombardi, J. C., Jr. 2014, *ApJ*, 786, 39

Nandez, J. L. A., Ivanova, N., & Lombardi, J. C. 2015, *MNRAS*, 450, L39

Nandez, J. L. A. & Ivanova, N. 2016, *MNRAS*, 460, 3992

Nordhaus, J. & Blackman, E. G. 2006, *MNRAS*, 370, 2004

Nordhaus, J., Blackman, E. G., & Frank, A. 2007, *MNRAS*, 376, 599

Ohlmann, S. T., Röpke, F. K., Pakmor, R., & Springel, V. 2016a, *ApJ* (Letters), 816, L9

Ohlmann, S. T., Röpke, F. K., Pakmor, R., Springel, V., & Müller, E. 2016b, *MNRAS*, 462, L121

Ohlmann, S. T., Roepke, F. K., Pakmor, R., & Springel, V. 2017, *A&A*

Passy, J.-C., De Marco, O., Fryer, C. L., *et al.* 2012, *ApJ*, 744, 52

Pavlovskii, K. & Ivanova, N. 2015, *MNRAS*, 449, 4415

Pavlovskii, K., Ivanova, N., Belczynski, K., & Van, K. X. 2017, *MNRAS*, 465, 2092

Podsiadlowski, P. 2001, Evolution of Binary and Multiple Star Systems, 229, 239

Podsiadlowski, P., Ivanova, N., Justham, S., & Rappaport, S. 2010, *MNRAS*, 406, 840

Popov, D. V. 1993, *ApJ*, 414, 712

Ricker, P. M. & Taam, R. E. 2008, *ApJ* (Letters), 672, L41

Ricker, P. M. & Taam, R. E. 2012, *ApJ*, 746, 74

Roxburgh, I. W. 1967, *Nature*, 215, 838

Shiber, S., Kashi, A., & Soker, N. 2017, *MNRAS*, 465, L54

Soker, N. & Harpaz, A. 2003, *MNRAS*, 343, 456

Staff, J. E., De Marco, O., Wood, P., Galaviz, P., & Passy, J.-C. 2016, *MNRAS*, 458, 832

Taam, R. E., Bodenheimer, P., & Ostriker, J. P. 1978, *ApJ*, 222, 269

Tylenda, R., Hajduk, M., Kamiński, T., *et al.* 2011, *A&A*,528, A114

van der Sluys, M. V., Verbunt, F., & Pols, O. R. 2006, *A&A*,460, 209

Webbink, R. F. 1984, *ApJ*, 277, 355

The Lives and Death-Throes of Massive Stars
Proceedings IAU Symposium No. 329, 2016
J.J. Eldridge, J.C. Bray, L.A.S. McClelland & L. Xiao, eds.

doi:10.1017/S1743921317003246

Clumping in stellar winds and interiors

Götz Gräfener

Argelander Institut für Astronomie, Universität Bonn,
Auf dem Hügel 71, 53121 Bonn, Germany
email: goetz@astro.uni-bonn.de

Abstract. The uncertain clumping properties pose a major problem for the quantitative analysis and the modelling of hot star winds. New results suggest that also the outer envelopes of massive stars may be affected by clumping, with important consequences for their observable radii and ionising properties. In this talk I will discuss how clumping is incorporated in stellar interior and wind/atmosphere models, how current theoretical results compare with observations, and what we can learn from a combination of stellar interior models and winds.

---

1. Introduction

Over the last twenty years it became clear that the radiatively-driven winds of hot massive stars are structured and inhomogeneous. This is in stark contrast to stellar wind and atmosphere models, where these outflows are often treated as homogeneous. The presence of inhomogeneities has a strong influence on the opacities and the radiative transfer in stellar winds. It thus affects the modelling and diagnostics of stellar winds, and introduces large uncertainties in theoretical predictions and empirical determinations of the mass-loss rates of massive stars.

More recent works started to investigate the influence of inhomogeneities on the inflated envelopes of massive stars, as they occur when very massive stars (VMS) on the main-sequence, or evolved stars in the Wolf-Rayet (WR) phase, are approaching the Eddington limit.

In the following the current status of research on the inhomogeneities in stellar winds and envelopes is reviewed, and it is discussed how a connection of wind and envelope models may help to obtain more information on their uncertain clumping properties.

2. The origin of clumping

The presence of inhomogeneities has been predicted in a variety of theoretical studies of stellar winds and envelopes. In stellar winds, Doppler shifts introduced by velocity gradients lead to the so-called line-driven instability (Owocki *et al.* 1988), resulting in wind shocks. The shock-heated material can reach temperatures of several MK and is believed to be the origin of the observed X-Ray emission of single OB stars (Feldmeier *et al.* 1997). Further details on X-Ray wind diagnostics are discussed, e.g., by Oskinova (this volume).

Instabilities in stellar envelopes have been predicted by Glatzel (2008, 2009), for the non-linear regime of strange-mode pulsations (Saio *et al.* 1998; Glatzel & Kaltschmidt 2002). The latter occur in cavities, which are typically formed in inflated stellar envelopes (cf. Sect. 4.1). 3D models by Jiang *et al.* (2015) show that these systems can form a pronounced clumpy and porous structure (cf. also Jiang this volume).

Clumping and porosity have a strong effect on the mean opacity in astrophysical plasmas. There is a general trend that the increased densities within clumps, compared to a homogeneous medium, lead to an increase of the opacity (given in cm^2/g). Porosity, on the other hand, can lower the effective mean opacity, as, depending on the detailed clump geometry, velocity, and optical depth, radiation may leak through gaps between clumps. In the following we will discuss the effect of clumping and porosity on the diagnostics and the modelling of stellar winds and envelopes.

3. Clumping in stellar winds

3.1. *Wind diagnostics*

The empirical determination of stellar wind densities mainly relies on the analysis of optical recombination lines, such as Hα. Because recombination is a two-particle process the emissivity of these lines scales with ρ^2, where ρ denotes the enhanced clump density. The total line strength scales with the spatial mean $< \rho^2 >$. If we introduce a clumping factor such that the clump density $\rho = D \times < \rho >$ and the inter-clump medium is void, the derived empirical mass-loss rates scale with $\dot{M} \propto 1/\sqrt{D}$, where D is usually ill-constrained. This introduces large uncertainties in empirically derived mass-loss rates. The scaling with $\dot{M} \propto 1/\sqrt{D}$ is valid as long as individual clumps are optically thin in the diagnostic lines.

For an empirical determination of clumping factors it is necessary to compare empirical mass-loss rates based on $< \rho^2 >$-diagnostics with values derived by different means. This was first done for WR stars, whose strong emission lines show electron-scattering wings whose strength scales linearly with $< \rho >$ (Hillier 1984). Because of the presence of the electron-scattering wings, the clumping factors in WR winds are at least roughly known. The relative weakness of the electron-scattering wings typically implies clumping factors of the order of $D = 10$ (cf. also Hamann & Koesterke 1998), corresponding to mass-loss reductions of the order of 3.

Fullerton *et al.* (2006) found a much higher mass-loss discrepancy of a factor 10 (corresponding to clumping factors of the order of $D = 100$) for a large sample of Galactic O stars, based on the strength of P V P-Cygni absorption profiles obtained with the Far Ultraviolet Spectroscopic Explorer (FUSE). This extreme result implied that the wind mass-loss rates of O stars are much weaker than thought before, and started a broad discussion of the relevance of stellar-wind mass loss for massive stars (e.g. Smith & Owocki 2006; Smith 2014). In the meantime it has become clear that the P V absorption profiles are also affected by porosity. This occurs when individual clumps become optically thick in the diagnostic lines, and radiation leaking through gaps between clumps affects the line profiles. Oskinova *et al.* (2007) pointed out that this results in much more moderate mass-loss reductions for O stars. The effects of porosity are further enhanced by discontinuities in the velocity fields of stellar winds (velocity-porosity; Owocki 2008).

While the inclusion of optically-thin clumping is now a standard in spectroscopic analyses, the effects of porosity have only been studied by combining standard non-LTE model atmospheres with 3D Monte-Carlo models (Sundqvist *et al.* 2010, 2011; Šurlan *et al.* 2012, 2013). These studies all suggest moderate mass-loss reductions. Sundqvist *et al.* (2014) pointed out that doublet ratios, i.e., the relative strength of different doublet components in UV resonance lines, can serve as a useful diagnostic means for clumping and porosity. In a homogeneous medium the doublet ratios should simply reflect the relative line strength of the components, as it follows from atomic physics. In a clumped and porous medium, also the clump geometry affects the line formation. For the most extreme case that individual clumps are optically thick and both doublet components

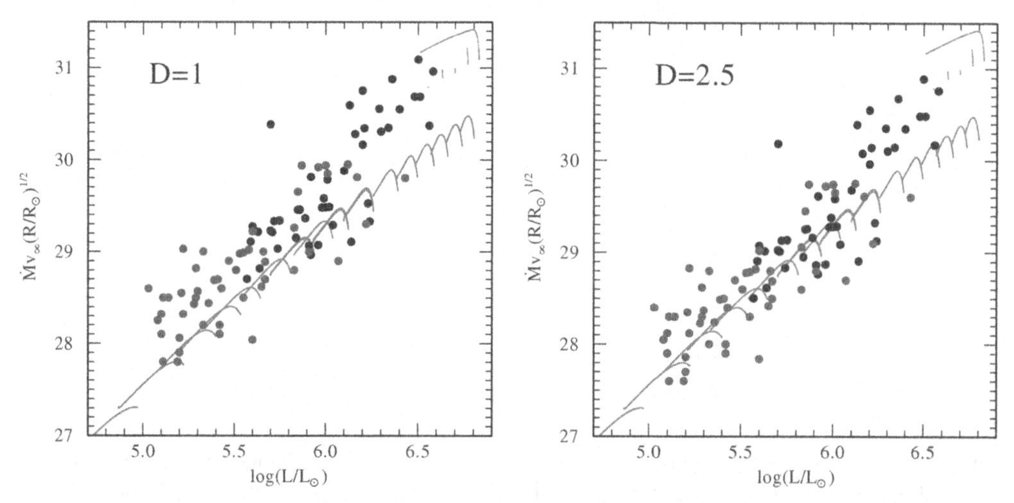

Figure 1. Wind momentum – luminosity relation for O-Of-WNh stars from the VLT-FLAMES Tarantula Survey. Filled circles indicate empirical results for O dwarfs (red), O giants and supergiants (blue), and the Of – WNh transition (black). The plot combines the results of three studies (see text). Shown are all results, regardless of sample overlaps. Upper limits, as they have been derived for some O-type stars with weak winds, have been skipped. The left panel shows the empirical results without clumping ($D = 1$). In the right panel the empirical results have been shifted according to an adopted clumping factor $D = 2.5$. Grey lines indicate wind momenta for O-type stars that have been extracted from evolutionary tracks without rotation from Brott *et al.* (2011); Köhler *et al.* (2015).

are equally saturated, the line profiles are *only* affected by the clump geometry, and both doublet components appear equally strong.

The fact that the observed P V doublet ratios, in many cases, do not follow atomic physics is a direct sign of the existence of porosity, implying that the mass-loss reductions claimed by Fullerton *et al.* (2006) are too high. Sundqvist *et al.* (2014) further developed an effective opacity formalism for porous two-component media, and demonstrated that there are two solutions for the mass-loss rate, depending on whether an observed line profile is dominated by the clump, or inter-clump medium. This result likely resolves previous controversies about porous O star mass-loss rates from Monte-Carlo modeling.

The uncertainty in the clumping factors of O stars is still a major problem for the determination of empirical mass-loss rates. For this reason the clumping factor D is usually treated as an open parameter. The best way forward in the understanding of clumping and porosity in stellar winds will be to include an appropriate effective opacity algorithm in non-LTE atmosphere models, and to perform detailed UV + optical studies of large stellar samples.

3.2. *Empirical mass-loss rates*

The most recent empirical results from optical analyses within the VLT-FLAMES Tarantula survey (Evans *et al.* 2011) are presented in Fig. 1. The plot shows results from three different studies of O dwarfs (Sabín-Sanjulián *et al.* 2017) in red, O giants and supergiants (Ramírez-Agudelo *et al.* 2017) in blue, and the Of – WNh transition regime of the most massive stars (Bestenlehner *et al.* 2014) in black. We compare the results with theoretical mass-loss rates that are extracted from stellar evolutionary models of Brott *et al.* (2011); Köhler *et al.* (2015), and which are based on the mass-loss estimates of Vink *et al.* (2000, 2001). First of all, ignoring the effects of clumping (left panel in Fig. 1), the

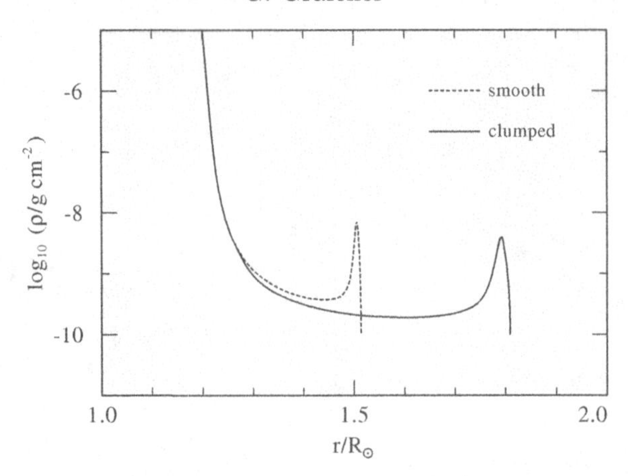

Figure 2. Density structure of inflated stellar envelopes with and without clumping from Gräfener *et al.* (2012). The clumped model is computed with a clumping factor $D = 2$.

theoretical mass-loss rates match the empirical ones for O dwarfs, but are systematically lower than the ones for O giants/supergiants.

In the right panel of Fig. 1 a clumping factor of $D = 2.5$ is adopted, introducing a shift of the empirical values by $\sqrt{D}$ so that the bulk of the O star data matches the theoretical mass-loss rates. This means that, if the real clumping factors are higher than 2.5, the mass-loss rates in stellar evolution models need to be corrected downwards by a factor $\sqrt{D/2.5}$.

A second aspect in Fig. 1 are the enhanced mass-loss rates of VMS in the WNh phase. Wind models predict a strong dependence on the Eddington factor Γ_e in this regime Gräfener & Hamann (2008), with a kink between optically thin O star winds and optically thick WNh stars (Vink *et al.* 2011). The strong dependence on Γ_e is supported by empirical studies of VMS in the Galactic center and the LMC (Gräfener *et al.* 2011; Bestenlehner *et al.* 2014). The WR mass-loss relations in evolutionary models (visible in the upper right corners in the diagrams in Fig. 1), which are based on the classical surface-chemistry criteria, fail to reproduce the observations. The extreme importance of VMS $\gtrsim 100\,M_\odot$ for the ionising and mechanical feedback in young star-forming regions has been pointed out by Doran *et al.* (2013). A correct inclusion of these objects in evolutionary models and population synthesis computations thus seems to be crucial (cf. also Wofford *et al.* 2014; Gräfener & Vink 2015; Crowther *et al.* 2016).

4. Clumping in stellar envelopes

Clumping in stellar envelopes is a relatively new topic that has been brought forward by Gräfener *et al.* (2012). Clumping turns out to be of major importance for inflated stellar envelopes because their structure is dominated by the topology of the Rosseland-mean opacity in the stellar interior (given as a function of density ρ and temperature T, or alternatively, gas pressure $P_{\rm gas}$ and radiation pressure $P_{\rm rad}$). In the following we will discuss how clumping and porosity can affect the envelope structure, and how stellar winds may connect to such envelopes.

4.1. *Stellar envelope inflation*

The envelope inflation effect is a radial inflation of the outer stellar envelope of stars near the Eddington limit (with classical Eddington factors of the order of

$\Gamma_e = \chi_e L/(4\pi cGM) \approx 0.5$, where χ_e denotes the free-electron opacity). The effect has been studied with chemically homogeneous models for WR stars and massive stars on the ZAMS (Ishii *et al.* 1999; Petrovic *et al.* 2006; Gräfener *et al.* 2012; Ro & Matzner 2016), and its relevance during the main-sequence evolution of massive stars at different metallicities has been discussed by Sanyal *et al.* (2015, 2017).

The underlying reason for the envelope inflation effect lies in the topology of the Fe opacity peak near $\sim$160 kK. As the envelope solution has to cross the region of the Fe opacity peak at some point, and the Fe opacity can become extremely high, the star struggles to avoid a super-Eddington situation in this region. In case that convection is efficient enough, this can be done by lowering the radiative flux in the region of high opacity. If convection is inefficient, the opacity $\chi(\rho, T)$ near the Fe-peak needs to be lowered. The only way to do this for given T, is to go to low densities ρ. Close to the Eddington limit, the topology of the opacity peak thus leads to the formation of a low-density region within the stellar envelope near 150 kK. Due to the relation between temperature and optical depth (approximately with $T^4 \propto \tau$), and $\tau \propto \rho\,\Delta R$, a low density automatically implies a large radial extension ΔR. Above the low-density zone the density increases again, so that a cavity is formed (cf. Fig. 2). Gräfener *et al.* (2012) described this effect in an analytical approach which revealed the existence of a stability limit, reminiscent of the S-Dor instability strip in the upper HRD. This makes the inflation effect one of the best candidates to explain the radius variations LBVs (cf. also Fig. 3).

Because of the importance of the density at the Fe opacity peak, the enhanced density in clumps has a major effect on the radii of stars in this regime. As long as individual clumps stay optically thin, clumping can be implemented in stellar models analogous to stellar atmosphere models, by simply computing the opacities $\chi(\rho, T)$ for a higher clump density $D \times \rho$. This approach has been used by Gräfener *et al.* (2012) to reproduce the observed radii of H-free WR stars. The adopted clumping factors were of the order of $D = 10$, and thus comparable to those found in WR-type stellar winds. This way it was possible to resolve the long-standing "WR radius problem" that the observed radii of WR stars are much larger (by up to a factor 10) than the ones predicted by classical stellar structure models (cf. Fig. 3).

As the inflation effect is expected to be highly dependent on metallicity (Ishii *et al.* 1999; Petrovic *et al.* 2006), it can also account for the fact that late WR subtypes are tendentially found in high-metallicity environments such as the Galactic center, while early subtypes dominate in low-metallicity environments such as the LMC. This is also supported by the study of McClelland & Eldridge (2016) who used the same clumping approach to explain the properties of the WR populations in the Galaxy and LMC (see also McClelland this volume).

It is important to note that the research on the inflation effect is currently still in its infancy. The recent 3D simulations by Jiang *et al.* (2015) have shown that inflated envelopes may not only be clumped, but also porous. If clumps become optically thick in the continuum, the radiative flux tends to avoid high-density material, leading to an effective reduction of the Rosseland-mean opacity (cf. Shaviv 1998, 2001; Owocki *et al.* 2004). Furthermore, studies by Petrovic *et al.* (2006); Ro & Matzner (2016) have shown that the dynamics of the envelope material in presence of a stellar wind can inhibit the inflation effect. To obtain a more realistic view on the inflation effect it will thus be necessary to include clumping and porosity in dynamic models stellar structure models.

4.2. *Connection with stellar winds*

Apart from purely dynamical effects, optically-thick stellar winds also impose boundary conditions on stellar envelopes because of their back warming. Gräfener & Vink (2013)

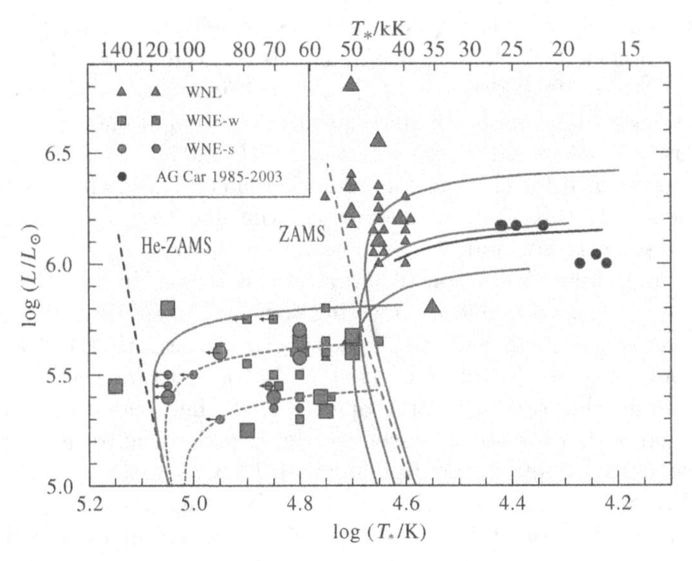

Figure 3. Hertzsprung-Russel diagram of the Galactic WN stars, and the LBV AG Car. Red/blue symbols indicate observed HRD positions of WN stars from Hamann *et al.* (2006) (blue: with hydrogen ($X > 0.05$); red: hydrogen-free ($X < 0.05$)). Black symbols indicate the HRD positions of AG Car throughout its S Dor Cycle from 1985–2003, according to Groh *et al.* (2009). Large symbols refer to stars with known distances from cluster/association membership. The symbol shapes indicate the spectral subtype (see inlet). Arrows indicate lower limits of $T_\star$ for stars with strong mass loss. The observations are compared to stellar structure models from Gräfener *et al.* (2012) (blue with hydrogen, red hydrogen-free), and for AG Car ($X = 0.36$, black line). The dashed red lines indicate models for which clumping factors of 4 and 16 have been adopted to match the observed radii of H-free WR stars.

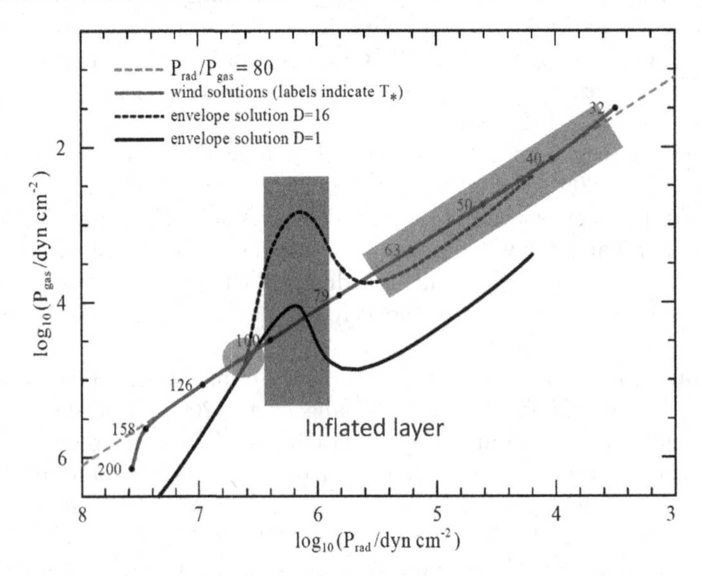

Figure 4. Conditions at the sonic point for a 14 $M_\odot$ helium star with Galactic metallicity. The solid and dashed black lines indicate stellar envelope solutions with different clumping factors $D = 1$ and 16 in the sub-surface layers. The conditions imposed by the optically thick stellar wind are indicated in blue. Labels indicate the stellar temperature $T_\star$ of the wind solutions. The red shaded areas indicate possible wind-envelope connections below (red circle) and above (red bar) the inflated envelope layers (grey).

estimated the densities and temperatures at the sonic point of optically-thick radiatively-driven winds, and found that these correspond to very high, and almost constant ratios of $P_{\rm rad}/P_{\rm gas} \approx 80$. These high ratios reflect the ratio of the terminal wind velocity over the sonic speed v_∞/a, and are *independent* of wind clumping. They are most likely the reason why optically-thick winds tend to occur only near the Eddington limit.

Fig. 4 demonstrates, for the example of a He star with $14\,M_\odot$ and solar metallicity, that clumping can help to achieve a match between envelope and wind solutions at the sonic point. The reason is that clumping shifts the envelope solution to lower densities, and thus helps to achieve higher ratios of $P_{\rm rad}/P_{\rm gas}$. For the un-clumped envelope solution (black solid curve in Fig. 4) a match between envelope and wind can only be achieved below the inflated layer (indicated as grey shaded area). For the clumped solution (dashed curve) such a match is also possible above the inflated zone. While these considerations are currently only of qualitative value, they show that the radii and mass-loss rates of stars with optically-thick winds may be the result of a very complex interplay between envelope, wind, and the clumping properties of the material in the envelope/wind transition region.

5. Conclusions

Clumping and porosity have strong influence on the effective opacities in stellar winds and envelopes. In particular, they affect the empirical determination of stellar wind mass-loss rates, and the radial extension of inflated stellar envelopes. A good knowledge of clumping and porosity is thus crucial for the understanding of the mass loss, and the radii of massive stars.

The current lack of accurate knowledge of clumping factors and porosity demand for the inclusion of new approaches for the description of clumpy and porous media in stellar atmosphere and stellar structure models, and their exposition to empirical tests.

References

Bestenlehner, J. M., Gräfener, G., Vink, J. S., *et al.* 2014, *A&A* 570, A38

Brott, I., de Mink, S. E., Cantiello, M., *et al.* 2011, *A&A* 530, A115

Crowther, P. A., Caballero-Nieves, S. M., Bostroem, K. A., *et al.* 2016, *MNRAS* 458, 624

Doran, E. I., Crowther, P. A., de Koter, A., *et al.* 2013, *A&A* 558, A134

Evans, C. J., Taylor, W. D., Hénault-Brunet, V., *et al.* 2011, *A&A* 530, A108

Feldmeier, A., Kudritzki, R.-P., Palsa, R., Pauldrach, A. W. A., & Puls, J. 1997, *A&A* 320, 899

Fullerton, A. W., Massa, D. L., & Prinja, R. K. 2006, *ApJ* 637, 1025

Glatzel, W. 2008, in A. Werner & T. Rauch (ed.), *Hydrogen-deficient stars*, Vol. 391 of *Astronomical Society of the Pacific Conference Series*, p. 307, San Francisco: Astronomical Society of the Pacific

Glatzel, W. 2009, *Communications in Asteroseismology* 158, 252

Glatzel, W. & Kaltschmidt, H. O. 2002, *MNRAS* 337, 743

Gräfener, G. & Hamann, W.-R. 2008, *A&A* 482, 945

Gräfener, G., Owocki, S. P., & Vink, J. S. 2012, *A&A* 538, A40

Gräfener, G. & Vink, J. S. 2013, *A&A* 560, A6

Gräfener, G. & Vink, J. S. 2015, *A&A* 578, L2

Gräfener, G., Vink, J. S., de Koter, A., & Langer, N. 2011, *A&A* 535, A56

Groh, J. H., Hillier, D. J., Damineli, A., *et al.* 2009, *ApJ* 698, 1698

Hamann, W.-R., Gräfener, G., & Liermann, A. 2006, *A&A* 457, 1015

Hamann, W.-R. & Koesterke, L. 1998, *A&A* 335, 1003

Hillier, D. J. 1984, *ApJ* 280, 744

Ishii, M., Ueno, M., & Kato, M. 1999, *PASJ* 51, 417

Jiang, Y.-F., Cantiello, M., Bildsten, L., Quataert, E., & Blaes, O. 2015, *ApJ* 813, 74
Köhler, K., Langer, N., de Koter, A., *et al.* 2015, *A&A* 573, A71
McClelland, L. A. S. & Eldridge, J. J. 2016, *MNRAS* 459, 1505
Oskinova, L. M., Hamann, W.-R., & Feldmeier, A. 2007, *A&A* 476, 1331
Owocki, S. P. 2008, in W.-R. Hamann, A. Feldmeier, & L. M. Oskinova (eds.), *Clumping in Hot-Star Winds*, p. 121
Owocki, S. P., Castor, J. I., & Rybicki, G. B. 1988, *ApJ* 335, 914
Owocki, S. P., Gayley, K. G., & Shaviv, N. J. 2004, *ApJ* 616, 525
Petrovic, J., Pols, O., & Langer, N. 2006, *A&A* 450, 219
Ramírez-Agudelo, O. H., Sana, H., de Koter, A., *et al.* 2017, *ArXiv e-prints* 1701.04758
Ro, S. & Matzner, C. D. 2016, *ApJ* 821, 109
Sabín-Sanjulián, C., Simón-Díaz, S., Herrero, A., *et al.* 2017, *ArXiv e-prints* 1702.04773
Saio, H., Baker, N. H., & Gautschy, A. 1998, *MNRAS* 294, 622
Sanyal, D., Grassitelli, L., Langer, N., & Bestenlehner, J. M. 2015, *A&A* 580, A20
Sanyal, D., Langer, N., Szécsi, D., -C Yoon, S., & Grassitelli, L. 2017, *A&A* 597, A71
Shaviv, N. J. 1998, *ApJ* 494, L193
Shaviv, N. J. 2001, *ApJ* 549, 1093
Smith, N. 2014, *ARAA* 52, 487
Smith, N. & Owocki, S. P. 2006, *ApJ* 645, L45
Sundqvist, J. O., Puls, J., & Feldmeier, A. 2010, *A&A* 510, A11
Sundqvist, J. O., Puls, J., Feldmeier, A., & Owocki, S. P. 2011, *A&A* 528, A64
Sundqvist, J. O., Puls, J., & Owocki, S. P. 2014, *A&A* 568, A59
Šurlan, B., Hamann, W.-R., Aret, A., *et al.* 2013, *A&A* 559, A130
Šurlan, B., Hamann, W.-R., Kubát, J., Oskinova, L. M., & Feldmeier, A. 2012, *A&A* 541, A37
Vink, J. S., de Koter, A., & Lamers, H. J. G. L. M. 2000, *A&A* 362, 295
Vink, J. S., de Koter, A., & Lamers, H. J. G. L. M. 2001, *A&A* 369, 574
Vink, J. S., Muijres, L. E., Anthonisse, B., *et al.* 2011, *A&A* 531, A132
Wofford, A., Leitherer, C., Chandar, R., & Bouret, J.-C. 2014, *ApJ* 781, 122

The Lives and Death-Throes of Massive Stars
Proceedings IAU Symposium No. 329, 2016
J.J. Eldridge, J.C. Bray, L.A.S. McClelland
& L. Xiao, eds.

doi:10.1017/S1743921317003362

Recent advances in non-LTE stellar atmosphere models

Andreas A. C. Sander

Institut für Physik & Astronomie, Universität Potsdam,
Karl-Liebknecht-Str. 24-25, 14476 Potsdam, Germany
email: ansander@astro.physik.uni-potsdam.de

Abstract. In the last decades, stellar atmosphere models have become a key tool in understanding massive stars. Applied for spectroscopic analysis, these models provide quantitative information on stellar wind properties as well as fundamental stellar parameters. The intricate non-LTE conditions in stellar winds dictate the development of adequate sophisticated model atmosphere codes. The increase in both, the computational power and our understanding of physical processes in stellar atmospheres, led to an increasing complexity in the models. As a result, codes emerged that can tackle a wide range of stellar and wind parameters.

After a brief address of the fundamentals of stellar atmosphere modeling, the current stage of clumped and line-blanketed model atmospheres will be discussed. Finally, the path for the next generation of stellar atmosphere models will be outlined. Apart from discussing multi-dimensional approaches, I will emphasize on the coupling of hydrodynamics with a sophisticated treatment of the radiative transfer. This next generation of models will be able to predict wind parameters from first principles, which could open new doors for our understanding of the various facets of massive star physics, evolution, and death.

Keywords. hydrodynamics, methods: data analysis, methods: numerical, radiative transfer, stars: atmospheres, stars: early-type, stars: fundamental parameters, stars: mass loss, stars: winds, outflows, stars: Wolf-Rayet

---

1. Introduction

A few decades ago, it was Edwin Salpeter who asked Dimitri Mihalas a question that some might consider to be offensive, namely "Why in world would anyone want to study stellar atmospheres?". A proper answer to this question is nothing less than the justification of an important keystone in modern astrophysics, and thus D. Mihalas spent several pages on it (Hubeny & Mihalas 2014). Repeating all the arguments from this book would already exceed the page limit of these proceedings, but essentially the bottom line is that the stellar atmosphere is all we actually see from a star, and its spectrum is usually the only information we get.

With the recent advances in asteroseismology and the advent of gravitational wave astronomy this will change for some types of objects, but so far clearly not for the majority of stars. Thus, understanding the spectrum is the only way to obtain information about them. However, in order to reproduce the spectrum, a proper modeling of the stellar atmosphere is necessary. Given sufficient observations, stellar atmospheres can unveil the stellar and wind parameters of a star, such as $T_{\rm eff}$, $\log g$, L, v_{∞}, and $\dot{M}$, provide its chemical composition, and give insights into its interaction with the environment. Studying stellar atmospheres is therefore prerequisite for a plethora of applications and analyses.

2. Modeling stellar atmospheres

The atmospheres of hot and massive stars are especially challenging in terms of modeling requirements. Their extreme non-LTE situation has to be taken into account and their population numbers can only be determined by solving the high-dimensional system of equations describing statistical equilibrium. Their winds require the calculation of the radiative transfer in an expanding atmosphere, either by using the so-called "Sobolev approximation" (Sobolev 1960) or by solving the radiative transfer in the comoving frame (CMF, Mihalas *et al.* 1975).

For proper modeling, all significant elements in the stellar atmosphere have to be described by detailed model atoms. This is especially challenging for the elements of the iron group which have thousands of levels and millions of line transitions, leading to the so-called "blanketing" effect. Since an explicit treatment is impossible, basically all modern atmosphere codes use a superlevel concept, going back to Anderson (1989) and Dreizler & Werner (1993). This is often combined either with some kind of opacity distribution function (ODF), where the detailed cross sections are not conserved, but resorted to give the correct total superlevel cross-section, or an opacity sampling technique where the complex detailed cross sections are sampled on the frequency grid.

In an expanding, non-LTE environment, also the determination of the electron temperature stratification is not trivial. Going back to the ideas of Unsöld (1951, 1955) and Lucy (1964), temperature corrections can be obtained from the equation of radiative equilibrium and it's integral describing the conservation of the total flux. Alternatively, one can also obtain the corrections from calculating the electron thermal balance going back to Hummer & Seaton (1963). Since each of these methods have their strengths and weaknesses, stellar atmosphere codes usually use a combination.

Eventually, the construction of a proper stellar atmosphere model is not just the sum of the tasks mentioned in this section. In fact, the tasks are highly coupled, outlining the complexity of the problem. In CMF radiative transfer, there is an essential coupling in space, while there is an intrinsic coupling in frequency in the rate equations and temperature corrections. Due to the huge dimensionality and the non-linear character of the problem, the only way is therefore an iterative algorithm in order to establish the consistent solution for all quantities (e.g. radiation field, population numbers, temperature stratification).

Unless the stellar wind is very dense, the vast majority of photons originate in the quasi-hydrostatic layers below the supersonic wind. For an easier description, some early models have treated these two regimes separately, thereby allowing for simplifying approximations in each of the domains. However, this "core-halo" approach has limits and introduces an artificial boundary. Since the beginning of the 1990s, the concept of "unified model atmospheres" became more common, where the quasi-hydrostatic and wind regime are described within the same model atmosphere.

3. State of the art CMF atmospheres

The basic scheme for current, state of the art model atmospheres describing stars with stationary winds can be summarized as follows: the stellar and wind parameters of a star with an expanding atmosphere are given as input parameters for the calculation of an atmosphere model. After assuming a certain starting approach, the equations of statistical equilibrium and radiative transfer are iteratively solved where such iteration must be "accelerated" by a suitable algorithm (e.g. Hamann 1987). In addition, the temperature stratification has to be updated in parallel. When the total changes in this iterative scheme drop below some prespecified level, the atmosphere model is considered to be converged. In a last step, the emergent spectrum (in the observer's frame) is calculated based on the converged model atmosphere.

Regarding the techniques, current, state of the art model atmospheres share the following properties: the models can account for many elements, including the iron group. There is no artificial boundary between the subsonic (quasi-hydrostatic) and supersonic (wind) regime. For the latter, a predescribed velocity law $v(r)$ is adopted while a more detailed treatment is applied to the quasi-hydrostatic part. Density inhomogeneities (aka "clumping") can be tackled in an approximate way. For more details see Gräfener (2017, these proceedings). For the detailed radiative transfer, typically either a Monte Carlo (MC) or a comoving frame (CMF) approach is implemented.

In the CMF approach, the radiative acceleration a_{rad} is obtained by an evaluation of the integral

$$a_{\mathrm{rad}}(r) = \frac{4\pi}{c}\frac{1}{\rho(r)}\int\limits_{0}^{\infty} \kappa_{\nu}(r)H_{\nu}(r)\mathrm{d}\nu \tag{3.1}$$

in the comoving frame. While approximate treatments following CAK (named after Castor, Abbott, & Klein 1975) or it's later extensions have their strength in reducing the description of a a_{rad} into a (semi-)analytical form, allowing a fast calculation, they neglect effects like multiple scattering and essentially break down for dense winds. In contrast, the CMF calculation implicitly includes various effects and thus works for all line-driven winds, ranging from classical OB stars to Luminous Blue Variables (LBVs) and Wolf-Rayet (WR) stars. Even low-mass stars with line-driven winds, such as O subdwarfs or WR-type central stars of planetary nebulae, can be calculated in this way.

4. Recent advances in state of the art atmospheres

The current generation of stellar atmosphere models has reached high complexity. Updating or extending the model codes is therefore often a non-trivial task. Apart from the new features eventually visible to the general user, a lot of technical aspects have to be taken into account and the developers have to check whether the original premises made for their code still hold in all of the current (and intended) applications. This includes technical and physical aspects, such as start approximations, boundary conditions, blanketing treatment, the description of $v(r)$ in the subsonic domain, microturbulence, and clumping, the calculation of the emergent spectrum, or the accuracy and completeness of the atomic data. Some of these tasks are implemented at a fundamental level of the code and adjusting them can therefore result in a significant amount of work, often not or just partly visible to the general audience. In order to shed a bit more light on the great amount of work that has been performed in this field on all scales during approximately the last decade, the following list of advances does not only cover such updates which are immediately interesting for the user, but also some more technical aspects as far as they were documented. The codes are listed in alphabetical order:

CMFGEN (Hillier 1987, 1990, 1991; Hillier & Miller 1998; diagnostic: UV, optical, IR) uses a detailed CMF radiative transfer, fully accounting for line blanketing. It received a major extension in the last decade by the inclusion of the time-dependent terms in the rate equations and the addition of time-dependent radiative transfer modules in order to allow for the calculation of supernovae spectra (Dessart & Hillier 2010). Regarding stars, the concept of "hydrostatic iterations" was introduced (Martins & Hillier 2012) where the velocity description is updated a few times in order to better match with the hydrostatic equation in the subsonic part. For the emergent spectrum, rotational broadening can now be considered (Hillier *et al.* 2012) and the code can handle depth-dependent Doppler and Stark broadening. In 2016, the option to handle H^- opacity has been added. To speed up the calculations of a single model, CMFGEN can also make use of more than one core, using partial code parallelization.

FASTWIND (Santolaya-Rey *et al.* 1997; Puls *et al.* 2005; diagnostic: optical, IR) focusses on calculation speed and therefore performs a split into so-called "explicit" and "background" elements. Originally, only the explicit elements are treated with a CMF approach, while background elements are tackled with Sobolev. The line blanketing is approximated, leading to an additional performance gain. The more recent developments of FASTWIND are described in Rivero González *et al.* (2011, 2012). The photospheric line acceleration is now properly treated and, in contrast to earlier versions, important background lines are now also treated in the CMF. A new version that is able to treat all lines in the CMF and thus also extends the diagnostic range to the UV is in development and described by Puls (2017, these proceedings). Furthermore, FASTWIND is now also able to consider X-ray emission from wind-embedded shocks (Carneiro *et al.* 2016).

Krtička & Kubat (Krtička & Kubat 2004, 2009; Krtička 2006; no emergent spectra) have a code that is focussed on predicting mass-loss rates by solving the hydrodynamical equation together with the rest of the atmosphere iteration. While their code originally relied on the Sobolev approximation, the line force is now also calculated in the CMF (Krtička & Kubat 2010). Their calculations can furthermore account for turbulent broadening of the line profiles.

PHOENIX (Hauschildt 1992; Hauschildt & Baron 1999, 2004; diagnostic: UV, optical, IR) is a CMF-based code, commonly applied in the modeling of cool stars and supernovae, but in principle also suitable for hot stars (see, e.g. Hauschildt 1992). So far unmet by other codes, the developers of PHOENIX have started with a 3D branch of their code in addition to their standard 1D branch (see Sect. 5.1). Some 3D test results also help to improve their 1D branch, e.g. in the case of obtaining limb darkening coefficients.

PoWR (Hamann 1985, 1986; Hamann & Schmutz 1987; Koesterke *et al.* 2002; Gräfener *et al.* 2002; diagnostic: UV, optical, IR) also uses the CMF radiative transfer and fully accounts for line blanketing. Originally developed for Wolf-Rayet stars, it has since been significantly extended to be nowadays applicable to any hot star. For an accurate photospheric density stratification, the quasi-hydrostatic domain is now treated self-consistently, i.e. the velocity and density stratification are constantly updated in the course of the iteration (Sander *et al.* 2015). For a proper calculation of the emergent spectrum, the formal integral accounts for line broadening, including the linear and quadratic Stark effect, rotation (Shenar *et al.* 2014), and mircoturbulent broadening. On the more technical side, the blanketing treatment has been updated and the superlevels now strictly separate the different parities. Furthermore, the temperature correction can now alternatively be obtained via thermal balance instead of radiative equilibrium.

In order to predict mass-loss rates, the PoWR code has also a recently added branch to calculate hydrodynamically consistent models (see also Sect. 5.2.) Although based on the ideas of the first efforts from Gräfener & Hamann (2005, 2008) for Wolf-Rayet models, the technical details of the new method differ significantly in detail.

WM-basic (Pauldrach 1987, Pauldrach *et al.* 1994, 2001; diagnostic: UV, optical) applies the Sobolev approximation for the line transfer. It can account for an EUV and X-ray shock source function and is able to calculate hydrodynamically consistent models. In the past decade, WM-basic has also been extended to calculates supernova "snapshot" spectra (e.g. Pauldrach *et al.* 2012). Furthermore, with the inclusion of stark broadening

(Kaschinski *et al.* 2012, Pauldrach *et al.* 2014), the diagnostic range was extended to the optical regime.

An alternative to CMF-based codes are *Monte Carlo codes*, where the radiative acceleration is obtained by following energy or photon packages throughout the atmosphere. Basically all current approaches have built up on the concepts of Friend & Abbott (1986). Using so-called "moving reversing layers", mass-fluxes are obtained as eigenvalues by Lucy (2007). While already outlined in Lucy & Solomon (1970), it requires the modern generation of computers to calculate a larger range of models and study details like the influence of the microturbulence, metallicity or the potential roots of the so-called "weak-wind problem". Consequently, this approach was not just improved in technical details, such as improving the lower boundary condition or updating the line list, but foremost applied to various parameter ranges (Lucy 2010a, 2010b, 2012). A different approach using MC techniques was taken by Müller & Vink (2008), who found solutions for the velocity field with the help of the Lambert W-function by adopting a semi-analytic description for the radiative acceleration in the (isothermal) wind. This allows to achieve local hydrodynamical consistency in their models, in contrast to the earlier mass-loss predictions from Vink *et al.* (1999, 2000, 2001) where only a global consistency was guaranteed. The new approach was extended to 2D and applied to rotating massive O stars in Müller & Vink (2014), allowing to study effects like oblateness and gravity darkening. A new MC-based code is also currently in development by Noebauer & Sim (2015).

5. What's next?

When discussing the potential next steps for non-LTE stellar atmosphere models, things such as 2D/3D calculations, hydrodynamical consistency, consistent X-ray treatment, multi-component winds or non-monotonic velocity fields are coming to mind. For the sake of time (or page) limitation, only the first two will be discussed below. But apart from this "wishlist" from the user's point of view one has also to be aware of the more imperceptive challenges in the development process which only become visible if we take the developer's point of view. To identify and tackle "problematic" parameter regimes in the codes is already a task in itself. Finding "good" compromises between accuracy, numerics, and computational performance is also a constant topic for code developers as we see more and more applications coming up where whole grids of models are required and manual checks for each model become practically impossible. On top of these technical aspects, there are scientific challenges like a better description of turbulence and clumping. And of course there is one of the biggest underlying challenges introducing unknown uncertainties basically in any model, namely the atomic data. This is not just a question of completeness – also more unapparent aspects like the details of superlevel approximations or the handling of ionization cross-sections can have an effect on the results. While finding good constraints on these is one important future task, it might be even more important, especially for the user, to simply keep in mind that there are systematic errors and simplifying assumptions and thus one should not overestimate the precision of the derived spectra.

5.1. *2D and 3D approaches*

Given the significant computational effort when using a CMF radiative transfer, much more work in the field of multi-D approaches, which are necessary for studying non-spherical stars or structures, has been done with Monte Carlo models. As mentioned in the previous section, Müller & Vink (2014) modeled an axially symmetric rotating star and found an equatorial decrease in the mass-loss, implying that the total mass-loss

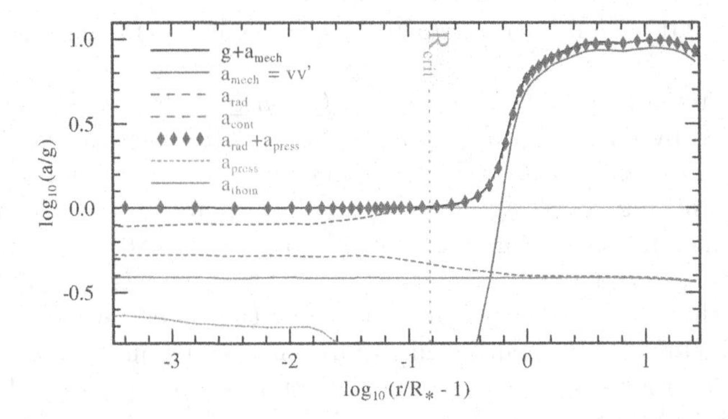

Figure 1. Acceleration stratification for an O supergiant model: The red diamonds ($a_{\rm rad} + a_{\rm press}$) match the black curve ($g + a_{\rm mech}$), illustrating that the HD Eq. (5.1) is fulfilled.

rate would be lower than in the 1D-spherical case, thereby contradicting the predictions from Maeder & Meynet (2000). On the other hand, there is the very sophisticated and ambitious idea of a full 3D CMF radiative transfer. This has been started by Hauschildt & Baron (2006) using their PHOENIX code, now termed PHOENIX/3D for this branch. In a series of currently 11 papers, they present the ideas and methods of their 3D branch and show several test calculations where they compare 3D results to their 1D counterparts. While their efforts and progress is impressive, it also shows that detailed 3D non-LTE atmospheres will not become a standard tool in the near future, since the current calculations require large supercomputers as soon as complex ions are used. In Hauschildt & Baron (2014), they presented a small test case using only 62 non-LTE levels and a very large case using 4686 non-LTE levels. For a single iteration, i.e. one the solution of the 3D radiative transfer plus one solution of the statistical equations etc., already the small case requires about half a year on a single core using an Intel Xeon E5420 CPU with 2.50 GHz clock-speed. For the large, but much more realistic case this value rises to 4300 years for a single iteration, thus leading to the effect that Hauschildt & Baron (2014) decide to give the total linear calculation time, i.e. CPU time until model convergence, as 15 μHubble or roughly 215 000 years. While the corresponding wall clock times could be lowered significantly by parallelization, the number of cores required to bring this down to the order of days would be enormous.

5.2. *Hydrodynamically self-consistent atmosphere models*

While nowadays stellar atmosphere models are mostly used to measure the wind parameters ($v(r), \dot{M}$) of a star, a new generation of models is designed to predict these. The key ingredient for this task is the inclusion of the hydrodynamic (HD) equation

$$v\left(1 - \frac{a_{\rm s}^2}{v^2}\right)\frac{{\rm d}v}{{\rm d}r} = a_{\rm rad}(r) - g(r) + 2\frac{a_{\rm s}^2}{r} - \frac{{\rm d}a_{\rm s}^2}{{\rm d}r}, \tag{5.1}$$

which has to be fulfilled at all depths in a self-consistent atmosphere model. The predictablity of the wind parameters is achieved due to the additional constraint introduced by the hydrodynamic Eq. (5.1) and its critical point. The requirement to have a smooth transition of $v(r)$ through the critical point can be translated into a condition for the mass-loss rate $\dot{M}$. Thereby the model predicts this fundamental wind parameter purely from the given stellar parameters.

While the idea is relatively simple and goes back to Lucy & Solomon (1970), its actual implementation is not. Early efforts using a pure CMF line force implementation were made by Pauldrach et al. (1986) and a Sobolev-based implementation became part of WM-basic with Pauldrach et al. (2001). The concept is also applied in the more theoretical works of Krtička & Kubát (2004), first via Sobolev and later in Krtička & Kubát (2010) with a CMF approach. The first complete implementation into a CMF-based analysis code was done by Gräfener & Hamann (2005) using the PoWR code. Their implementation technique, based on a generalized force multiplier concept, was successfully applied to a WC and later also to a grid of WN models (Gräfener & Hamann 2008). A new implementation utilizing a different technique was recently added to the PoWR code (Sander *et al.* 2015b, 2017), finally allowing to also calculate hydrodynamically consistent models for OB stars. An example for an O supergiant model based on ζ Pup can be seen in Fig. 1. Recently, J. Puls and J. Sundqvist have also started to implement consistent hydrodynamics into their new version of FASTWIND.

6. Summary

Compared to earlier decades, the current generation of non-LTE expanding stellar atmosphere models has a significantly increased applicability, ranging from OB stars via transition stages like LBVs or Of/WN stars up to the classical Wolf-Rayets. Moreover, they are applicable for hot low-mass stars, such as central stars of planetary nebulae. With more codes allowing for a such a broad range of applications, more benchmarking will be possible, allowing to cross-check results between different codes and methods and identify problems. PoWR, CMFGEN, and FASTWIND can also treat X-rays from wind-embedded shocks, thereby extending the diagnostic range into this wavelength regime.

A completely different branch has been opened by CMFGEN, PHOENIX, and WM-basic with their option to calculate so-called supernova "snapshot" spectra. By adding the time-dependent terms in the rate equations and the radiative transfer, atmosphere codes can be used to analyze supernova spectra and study their evolution.

In the years to come, hydrodynamically consistent models will provide a new generation of model atmospheres which can predict wind parameters from a given set of stellar parameters, thereby opening up a third branch next to MC and CAK-like techniques. Due to the local consistence of the models and the possibility to calculate emergent spectra, the results can immediately be cross-checked with observations. Apart from a Sobolev-based implementation in WM-basic at the beginning of the century, HD-consistent models can now also be calculated with PoWR and potentially with FASTWIND.

Although mainly used for cool stars, significant efforts to obtain a 3D CMF radiative transfer have been made with PHOENIX/3D. Unfortunately, massive parallelization is required to reach manageable wall clock times for 3D models, and thus 1D model atmospheres will remain the standard tool for massive stars in the near future.

References

Anderson, L. S. 1989, *ApJ*, 339, 558
Carneiro, L. P., Puls, J., Sundqvist, J. O., & Hoffmann, T. L. 2016, *A&A*, 590, A88
Castor, J. I., Abbott, D. C., & Klein, R. I. 1975, *ApJ*, 195, 157
Dessart, L. & Hillier, D. J. 2010, *MNRAS*, 405, 2141
Dreizler, S. & Werner, K. 1993, *A&A*, 278, 199
Friend, D. B. & Abbott, D. C. 1986, *ApJ*, 311, 701
Gräfener, G. & Hamann, W.-R. 2005, *A&A*, 432, 633
Gräfener, G. & Hamann, W.-R. 2008, *A&A*, 482, 945
Gräfener, G., Koesterke, L., & Hamann, W.-R. 2002, *A&A*, 387, 244
Hamann, W.-R. 1985, *A&A*, 148, 364

Hamann, W.-R. 1986, *A&A*, 160, 347
Hamann, W.-R. 1987, *Num. rad. transfer*, 35 (ed. W. Kalkofen, Cambridge: Univ. Press, 1987)
Hamann, W.-R. & Schmutz, W. 1987, *A&A*, 174, 173
Hauschildt, P. H. 1992, *J. Quant. Spec. Radiat. Transf.*, 47, 433
Hauschildt, P. H., Allard, F., & Baron, E. 1999, *ApJ*, 512, 377
Hauschildt, P. H. & Baron, E. 2004, *A&A*, 417, 317
Hauschildt, P. H. & Baron, E. 2006, *A&A*, 451, 273
Hauschildt, P. H. & Baron, E. 2014, *A&A*, 566, A89
Hillier, D. J. 1990, *A&A*, 231, 116
Hillier, D. J. 1991, *A&A*, 247, 455
Hillier, D. J., Bouret, J.-C., Lanz, T., & Busche, J. R. 2012, *MNRAS*, 426, 1043
Hillier, D. J. & Miller, D. L. 1998, *ApJ*, 496, 407
Hubeny, I. & Mihalas, D. 2014, *Theory of Stellar Atmospheres* (Princeton: Univ. Press, 2014)
Hummer, D. G. & Seaton, M. J. 1963, *MNRAS*, 125, 437
Kaschinski, C. B., Pauldrach, A. W. A., & Hoffmann, T. L. 2012, *A&A*, 542, A45
Koesterke, L., Hamann, W.-R., & Gräfener, G. 2002, *A&A*, 384, 562
Krtička, J. 2006, *MNRAS*, 367, 1282
Krtička, J. & Kubát, J. 2004, *A&A*, 417, 1003
Krtička, J. & Kubát, J. 2009, *MNRAS*, 394, 2065
Krtička, J. & Kubát, J. 2010, *A&A*, 519, A50
Lucy, L. B. 1964, *SAO Special Report*, 167, 93
Lucy, L. B. 2007, *A&A*, 468, 649
Lucy, L. B. 2010a, *A&A*, 512, A33
Lucy, L. B. 2010b, *A&A*, 524, A41
Lucy, L. B. 2012, *A&A*, 543, A18
Lucy, L. B. & Solomon, P. M. 1970, *ApJ*, 159, 879
Maeder, A. & Meynet, G. 2000, *A&A*, 361, 159
Martins, F. & Hillier, D. J. 2012, *A&A*, 545, A95
Mihalas, D., Kunasz, P. B., & Hummer, D. G. 1975, *ApJ*, 202, 465
Müller, P. E. & Vink, J. S. 2008, *A&A*, 492, 493
Müller, P. E. & Vink, J. S. 2014, *A&A*, 564, A57
Noebauer, U. M. & Sim, S. A. 2015, *MNRAS*, 453, 3120
Pauldrach, A., Puls, J., & Kudritzki, R. P. 1986, *A&A*, 164, 86
Pauldrach, A. W. A., Hoffmann, T. L., & Hultzsch, P. J. N. 2014, *A&A*, 569, A61
Pauldrach, A. W. A., Hoffmann, T. L., & Lennon, M. 2001, *A&A*, 375, 161
Pauldrach, A. W. A., Kudritzki, R. P., Puls, et al. 1994, *A&A*, 283, 525
Pauldrach, A. W. A., Vanbeveren, D., & Hoffmann, T. L. 2012, *A&A*, 538, A75
Puls, J. 1987, *A&A*, 184, 227
Puls, J., Urbaneja, M. A., Venero, R., et al. 2005, *A&A*, 435, 669
Rivero González, J. G., Puls, J., & Najarro, F. 2011, *A&A*, 536, A58
Rivero González, J. G., Puls, J., Najarro, F., & Brott, I. 2012, *A&A*, 537, A79
Sander, A., Shenar, T., Hainich, R., et al. 2015, *A&A*, 577, A13
Sander, A., Hamann, W.-R., Hainich, R., et al. 2015, *Wolf-Rayet Stars: Proceedings*, p.139-142 (ed. W.-R. Hamann, A. Sander, & H. Todt, Universitätsverlag Potsdam, 2015)
Sander, et al. 2017, http://adsabs.harvard.edu/abs/2017arXiv170408698S
Santolaya-Rey, A. E., Puls, J., & Herrero, A. 1997, *A&A*, 323, 488
Shenar, T., Hamann, W.-R., & Todt, H. 2014, *A&A*, 562, A118
Sobolev, V. V. 1960, *Moving envelopes of stars* (Cambridge, Harvard University Press, 1960)
Unsöld, A. 1951, *Naturwissenschaften*, 38, 525
Unsöld, A. 1955, *Physik der Sternatmosphären* (Berlin, Springer, 1955. 2. Aufl.)
Vink, J. S., de Koter, A., & Lamers, H. J. G. L. M. 1999, *A&A*, 350, 181
Vink, J. S., de Koter, A., & Lamers, H. J. G. L. M. 2000, *A&A*, 362, 295
Vink, J. S., de Koter, A., & Lamers, H. J. G. L. M. 2001, *A&A*, 369, 574

The Lives and Death-Throes of Massive Stars
Proceedings IAU Symposium No. 329, 2016
J.J. Eldridge, J.C. Bray, L.A.S. McClelland & L. Xiao, eds.

doi:10.1017/S1743921317002563

Massive stars in advanced evolutionary stages, and the progenitor of GW150914

Wolf-Rainer Hamann, Lidia Oskinova, Helge Todt, Andreas Sander, Rainer Hainich, Tomer Shenar and Varsha Ramachandran

Institut für Physik und Astronomie, Universität Potsdam, Germany

Abstract. The recent discovery of a gravitational wave from the merging of two black holes of about 30 solar masses each challenges our incomplete understanding of massive stars and their evolution. Critical ingredients comprise mass-loss, rotation, magnetic fields, internal mixing, and mass transfer in close binary systems. The imperfect knowledge of these factors implies large uncertainties for models of stellar populations and their feedback. In this contribution we summarize our empirical studies of Wolf-Rayet populations at different metallicities by means of modern non-LTE stellar atmosphere models, and confront these results with the predictions of stellar evolution models. At the metallicity of our Galaxy, stellar winds are probably too strong to leave remnant masses as high as $\sim$30 $M_\odot$, but given the still poor agreement between evolutionary tracks and observation even this conclusion is debatable. At the low metallicity of the Small Magellanic Cloud, all WN stars which are (at least now) single are consistent with evolving quasi-homogeneously. O and B-type stars, in contrast, seem to comply with standard evolutionary models without strong internal mixing. Close binaries which avoided early merging could evolve quasi-homogeneously and lead to close compact remnants of relatively high masses that merge within a Hubble time.

Keywords. stars: atmospheres, stars: early-type, stars: evolution, stars: fundamental parameters, Hertzsprung-Russell diagram, stars: mass loss, stars: winds, outflows, stars: Wolf-Rayet

---

1. Introduction

At the 14th of September 2015, the advanced LIGO detectors registered for the first time a gravitational wave (Abbott *et al.* 2016). According to the analysis of the waveform, this wave testified the event of two merging black holes (BHs) of $36^{+5}_{-4}\,M_\odot$ and $29^{+4}_{-4}\,M_\odot$ at a distance of about 400 Mpc. The immediate conclusion, and even the prediction prior to the measurement, was that such heavy BHs can only form by stellar evolution at low metallicity, where the mass-loss due to stellar winds is low and hence the stellar remnants can be more massive (Belczynski *et al.* 2016). Still heavily debated is whether such BHs form separately in dense clusters and then combine into a close pair by dynamical interactions, or whether they evolve as close binaries all the time. Since both scenarios have their problems, a primordial origin has also been suggested (see Postnov, these proceedings).

2. Galactic Wolf-Rayet stars

Massive stars may end their life in a gravitational collapse while being in the red-supergiant (RGS) phase or as Wolf-Rayet (WR) stars. The sample of putatively single and optically un-obscured Galactic WR stars has been comprehensively analyzed with increasing sophistication (cf. Fig. 1). It became clear that the WN stars (i.e. the WR stars of the nitrogen sequence) actually form two distinct groups. The very luminous WNs with $\log L/L_\odot > 6$ are slightly cooler than the zero-age main sequence and typically still

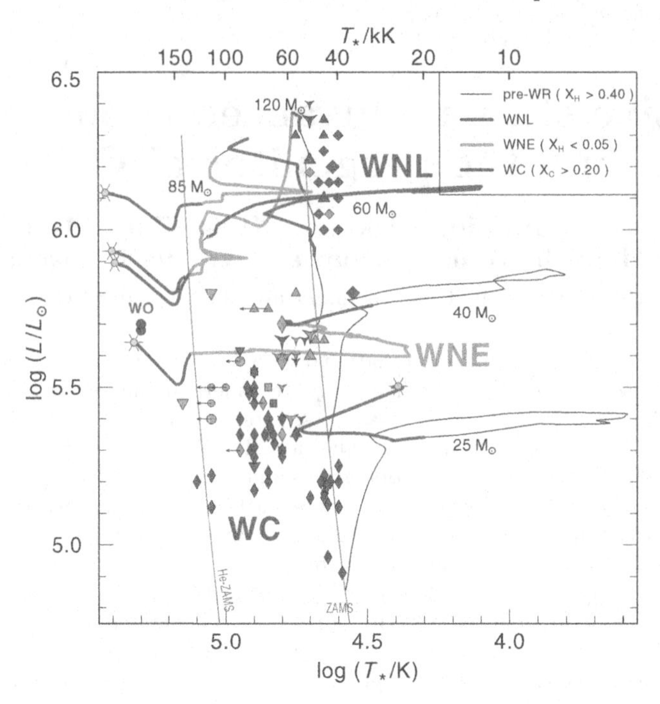

Figure 1. HRD of single WR stars in the Galaxy. The discrete symbols represent analyzed stars (Hamann *et al.* 2006, Sander *et al.* 2012), while the tracks are from the Geneva group accounting for rotationally induced mixing (Georgy *et al.* 2012).

contain hydrogen in their atmosphere (often termed WNL for "late"). In contrast, the less luminous WNE stars are hotter ("early" WN subtypes) and typically hydrogen free. The WR stars of the carbon sequence (WC) are composed of helium-burning products and share their location in the Hertzsprung-Russell diagram (HRD) with the WNE stars.

From this empirical HRD one can deduce the evolutionary scenario (Sander *et al.* 2012). The WNL stars evolve directly from O stars of very high initial mass ($> 40\,M_\odot$). In the mass range $20 - 40\,M_\odot$ the O stars first become RSGs and then WNE stars and finally WCs. Stars with initially less then $\approx 20\,M_\odot$ become RSGs and explode there as type II supernova before having lost their hydrogen envelope.

Evolutionary tracks still partly fail to reproduce this empirical HRD quantitatively, despite of all efforts, e.g. with including rotationally induced mixing. The WNE and WC stars are observed to be much cooler than predicted; this is probably due to the effect of "envelope inflation" (Gräfener *et al.* 2012). Moreover, the mass (and luminosity) range of WNE and WC stars is not covered by the post-RSG tracks. Evolutionary calculations depend sensitively on the mass-loss rates $\dot{M}$ that are adopted as input parameters. Empirical $\dot{M}$ suffer from uncertainties caused by wind inhomogeneities: when clumping on small scales ("microclumping") is taken into account, lower values for $\dot{M}$ are derived from observed emission-line spectra. Large-scale inhomogeneities ("macroclumping"), on the other hand, can lead to underestimating mass-loss rates (Oskinova *et al.* 2007).

Due to the open questions of mixing and the true $\dot{M}$, it is still uncertain which is the highest BH mass that can be produced from single-star evolution at Galactic metallicity. For instance, the luminosities of the two WO stars included in Fig. 1 would correspond to masses as high as $\approx 20\,M_\odot$ if they were chemically homogeneous, while the displayed evolutionary track for initially $40\,M_\odot$ ends with only $12\,M_\odot$ at core collapse.

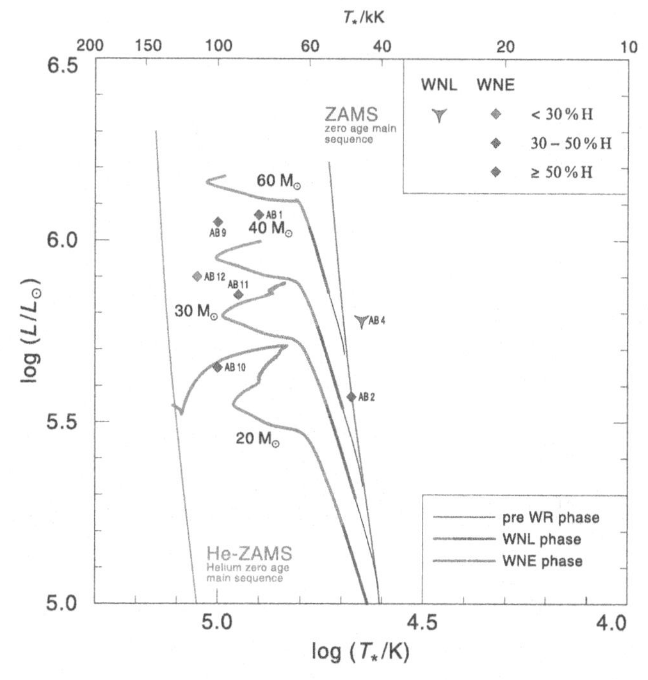

Figure 2. WN stars in the SMC; the discrete symbols are from the analyses of the stars as labeled, while the tracks from Brott *et al.* (2011) adopt very strong rotational mixing (initial $v_{\rm rot} > 500\,{\rm km/s}$). Colors code in both cases for the hydrogen mass fraction (see inlet). From Hainich *et al.* (2015)

3. Massive stars at low metallicity

The population of massive stars depends critically on their metallicity Z. This becomes obvious, e.g., from the WR stars in the Small Magellanic Cloud (SMC) where Z is only about 1/7 of the solar value. In contrast to the Galaxy, *all* putatively single WN stars in the SMC show a significant fraction of hydrogen in their atmosphere and wind, like the Galactic WNL stars. However, the WN stars in the SMC are all hot and compact, located in the HRD (Fig. 2) between the zero age main sequence for helium stars (He-ZAMS) and the H-ZAMS (or at least, in two cases, close the latter). Such parameters cannot be explained with standard evolutionary tracks, unless very strong internal mixing is assumed which makes the stars nearly chemically homogeneous. Corresponding tracks are included in Fig. 2. Quantitatively, they still do not reproduce the observed hydrogen mass fractions.

Stellar winds from hot massive stars are driven by radiation pressure intercepted by spectral lines. The literally millions of lines from iron and iron-group elements, located in the extreme UV where the stellar flux is highest, play a dominant role. Hence a metallicity dependence is theoretically expected. For O stars, such Z dependence is empirically established (e.g. Mokiem *et al.* 2007). For WN stars, Hainich *et al.* (2015) found a surprisingly steep dependence, probably due to the multiple-scattering effect (see also Hainich, these proceedings).

Hence, the lower mass-loss for massive O stars in the SMC, compared to the Galaxy, might reduce the angular-momentum loss and thus maintain the rapid rotation which causes the mixing and quasi-homogeneous evolution to the WR regime. Alternatively, one might speculate that the low $\dot{M}$ in the SMC is insufficient to remove the hydrogen envelope, and thus prevents the formation of single WR stars. This would imply that

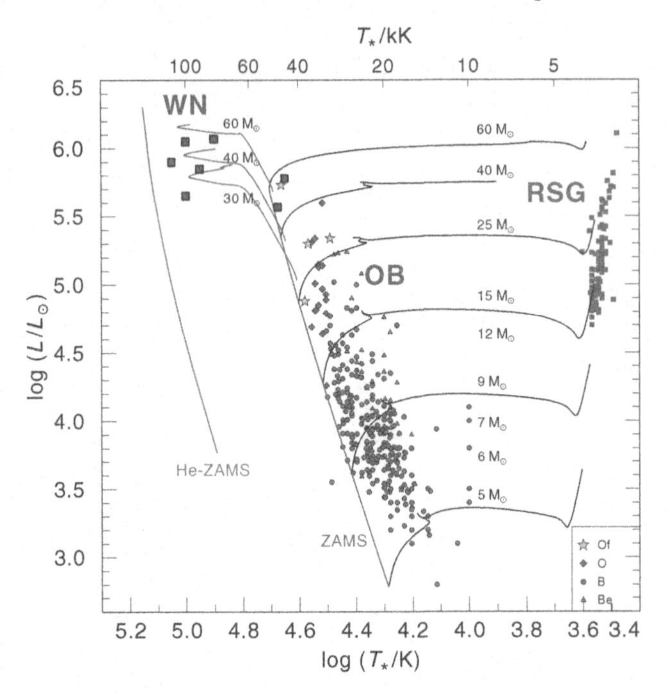

Figure 3. Massive stars in the SMC. The discrete symbols are from our analyses of about 300 O and B type stars (Ramachandran *et al.* in prep.), and the red supergiants are from Massey *et al.* (2003). The WN stars are the same as in Fig. 2, as are the quasi-homogeneous evolutionary tracks. The "normal" tracks are also from Brott *et al.* (2011) but with slower initial rotation ($v_{\rm rot} = 300\,{\rm km/s}$).

the observed single WNs have all formed through the binary channel, possibly as merger products.

But what happens with SMC stars of slightly lower mass? We have analyzed about 300 OB stars in the region of the supergiant shell SGS 1 (Ramachandran *et al.* in prep.). Their HRD positions are included in Fig. 3, together with "normal" tracks with less rotational mixing. As the comparison shows, the O and B-type stars with initial masses below $30\,M_\odot$ are consistent with "normal" evolution to the RSG stage. Only the more massive Of stars might also be consistent with quasi-homogenous evolution, as are the WN stars discussed above.

4. Massive binaries

In their majority, massive stars are born in binary systems. Marchant *et al.* (2016) suggested a scenario of "massive overcontact binary (MOB) evolution" that could lead to a tight pair of massive black holes as observed in the GW events. Two massive stars which are born as tight binary would evolve fully mixed due to their tidally induced fast spin and interaction. They would swap mass several times, making their masses about equal, but under lucky circumstances they might avoid early merging.

Figure 4 shows two such evolutionary tracks from Marchant (priv. comm.). In both examples, the tracks end at core collapse with a pair of $34+34\,M_\odot$ objects. We have calculated synthetic spectra for representative points along the evolutionary tracks (marked by asterisks in Fig. 4) and found that such spectra would look unspectacular if observed; in the advanced stages, the stars would appear as WN-type (with hydrogen, like those in

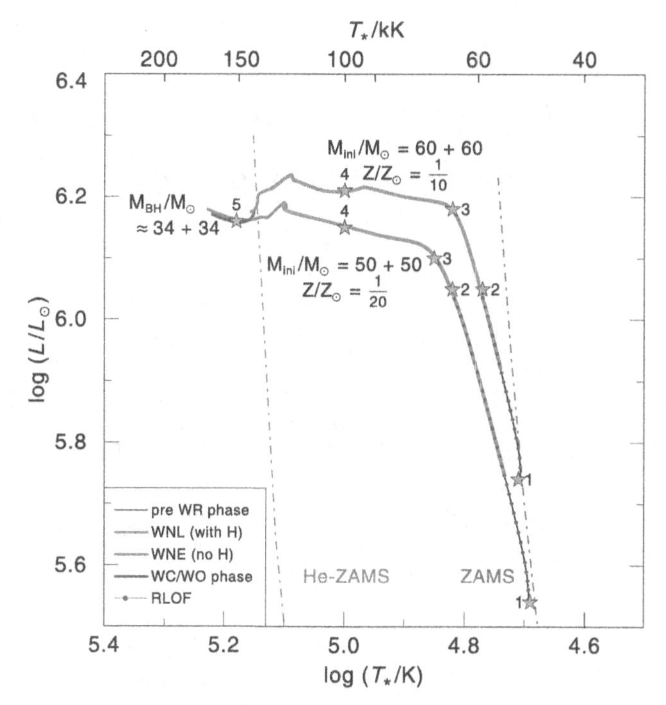

Figure 4. Tracks for "massive overcontact binary evolution (MOB)" (from Marchant, priv. comm.). Due the mass exchange, both binary components have equal masses and therefore evolve identically. One of the two tracks shown is for a binary of initially $50 + 50\,M_\odot$ and metallicity of 1/20 solar, and the another one for $60 + 60\,M_\odot$ and metallicity of 1/10 solar. The asterisks mark positions for which we calculated representative synthetic spectra (Hainich *et al.* in prep.).

the SMC discussed above) or, towards the end of the track for $60\,M_\odot$, as a hot WC type, always with otherwise weak metal lines due to the low abundances. The only characteristic differences compared to single stars would be the doubled luminosity and, if the orbital inclination is favorite, the radial-velocity variation in the double-lined spectrum with short period.

Acknowledgements

We thank Pablo Marchant for providing the MOB evolutionary tracks.

References

Abbott, B. P., Abbott, R., Abbott, T. D., *et al.*, 2016a, *PhRvL*, 116, 061102
Belczynski, K., Holz, D. E., Bulik, T., & O'Shaughnessy, R., *Nature*, 534, 512
Brott, I., de Mink, S. E., Cantiello, M., *et al.*, 2011, *A&A*, 530, A115
Georgy, C., Ekström, S., Meynet, G., *et al.*, 2012, *A&A*, 542, A29
Gräfener, G., Owocki, S. P., & Vink, J. S., 2012, *A&A*, 238, A40
Hainich, R., Pasemann, D., Todt, H., *et al.*, 2015, *A&A*, 581, A21
Hamann, W.-R., Gräfener, G., & Liermann, A., 2006, *A&A*, 457, 1015
Marchant, P., Langer, N., Podsiadlowski, P., Tauris, T. M., & Moriya, T. J., 2016, *A&A*, 588, A50
Massey, P. & Olsen, K. A. G., 2003, *AJ*, 126, 2867
Mokiem, M. R., de Koter, A., Vink, J. S., *et al.*, 2007, *A&A*, 473, 603
Oskinova, L. M., Hamann, W.-R., & Feldmeier, A., 2007, *A&A*, 476, 1331
Sander, A., Hamann, W.-R., & Todt, H., 2012, *A&A*, 540, A144

The Lives and Death-Throes of Massive Stars
Proceedings IAU Symposium No. 329, 2016
J.J. Eldridge, J.C. Bray, L.A.S. McClelland & L. Xiao, eds.

doi:10.1017/S1743921317002800

Properties of O dwarf stars in 30 Doradus

Carolina Sabín-Sanjulián[1,2] and the VFTS collaboration

[1]Departamento de Física y Astronomía, Universidad de La Serena, Av. Cisternas 1200 Norte, La Serena, Chile

[2] Instituto de Investigación Multidisciplinar en Ciencia y Tecnología, Universidad de La Serena, Raúl Bitrán 1305, La Serena, Chile
email: cssj@dfuls.cl

Abstract. We perform a quantitative spectroscopic analysis of 105 presumably single O dwarf stars in 30 Doradus, located within the Large Magellanic Cloud. We use mid-to-high resolution multi-epoch optical spectroscopic data obtained within the VLT-FLAMES Tarantula Survey. Stellar and wind parameters are derived by means of the automatic tool IACOB-GBAT, which is based on a large grid of FASTWIND models. We also benefit from the Bayesian tool BONNSAI to estimate evolutionary masses. We provide a spectral calibration for the effective temperature of O dwarf stars in the LMC, deal with the mass discrepancy problem and investigate the wind properties of the sample.

Keywords. Galaxies: Magellanic Clouds Stars: atmospheres Stars: early-type Stars: fundamental parameters

---

1. Introduction

The VLT-FLAMES Tarantula Survey (VFTS, Evans *et al.* 2011) is an ESO large programme that has obtained mid-to-high resolution multi-epoch optical spectra of hundreds of O-type stars in 30 Doradus, the largest H II region in the Local Group, located in the Large Magellanic Cloud. VFTS was developed aiming at the study of rotation, binarity and wind properties of an unprecedented number of massive stars within the same star-forming region. Within this project, a huge effort has been made in the study of O-type stars, including the analysis of multiplicity (Sana *et al.* 2013), spectral classifications (Walborn *et al.* 2014) and the distribution of rotational velocities (Ramírez-Agudelo *et al.* 2013, 2015). These studies are essential for the estimation of stellar parameters and chemical abundances of the complete O-type sample, which will allow to investigate fundamental questions in stellar and cluster evolution. To date, the quantitative study of the O-type sample is about to be complete (see Sabín-Sanjulián *et al.* 2014, 2017; Ramírez-Agudelo *et al.* 2017), as well as the determination of nitrogen abundances (Grin *et al.* 2017, Simón-Díaz *et al.* 2017.)

In this work, we study the properties of a sample of O stars close to the ZAMS, those with luminosity classes IV and V.

2. Observations and data sample

We have used spectroscopic data obtained by means of the Medusa mode of the FLAMES spectrograph at the VLT (Paranal, Chile). In addition, *I*-band images obtained with the *HST* (GO12499, P. I.: D. J. Lennon) were utilized to evaluate possible contamination in the Medusa fibres.

We have selected a sample of 105 likely single and unevolved O stars based on the multiplicity properties derived by Sana *et al.* (2013) and the spectral classifications by

Walborn *et al.* (2014). Our sample includes O stars with luminosity classes V and IV, as well as Vz, V-III (uncertain classification) and III-IV (interpolation between III and IV).

3. Quantitative spectroscopic analysis

The quantitative analysis of the spectroscopic data for our sample of O dwarfs was performed by means of the IACOB Grid-Based Automatic Tool (IACOB-GBAT, see Simón-Díaz *et al.* 2011, Sabín-Sanjulián *et al.* 2014). IACOB-GBAT is based on a grid of ~190 000 precomputed FASTWIND stellar atmosphere models (Santolaya-Rey et al. 1997, Puls *et al.* 2005) and a χ^2 algorithm that compares observed and synthetic H and He line profiles.

Absolute magnitudes in the V-band calculated by Maíz-Apellániz *et al.* (2014) and projected rotational velocities by Ramírez-Agudelo *et al.* (2013) were used.

We estimated mean values and uncertainties for effective temperature $T_{\rm eff}$, surface gravity g, helium abundance $Y(He)$, stellar radius R, luminosity L, spectroscopic mass $M_{\rm sp}$ and wind-strength Q-parameter (defined as $Q = \dot{M}(R\,v_\infty)^{-3/2}$, where $\dot{M}$ is the mass-loss rate and v_∞ the wind terminal velocity). Evolutionary masses ($M_{\rm ev}$) were obtained by means of the Bayesian tool BONNSAI (Schneider *et al.* 2014), which used evolutionary models by Brott *et al.* (2011) and Köhler *et al.* (2015).

Most of the stars in our sample showed strong nebular contamination, but only 11 critical cases had to be analyzed using nitrogen lines (HHeN). This diagnostic was also used for 9 cases with too weak or inexistent He I lines. We could only provide upper limits for $\log Q$ for about 70% of the sample due to thin winds. In addition, possible/confirmed contamination in fibre was present in several stars, which could have altered the determination of physical parameters. A few cases were found with too low helium abundances and too high gravities, a possible indication of undetected binarity.

4. Spectral calibration

Figure 1 represents our $T_{\rm eff}$ scale as a function of spectral types. We perform a linear fit to the points excluding O2 stars, since they show indications of undetected binarity and/or contamination in fibre. When comparing with the most recent and complete spectral calibration for O dwarfs in the LMC before this work (Rivero-González *et al.* 2012a,b, who used HHeN diagnostics to analyse optical spectra of 25 stars, including 16 dwarfs), we found that there is an excellent agreement between both scales for spectral types later than O4. In the earliest regime, Rivero-González *et al.* utilized a quadratic fit obtaining hotter temperatures than ours, although the estimates for our O2 stars agree with their calibration. However, we cannot reach a firm conclusion about the necessity of changing the slope in the O2-O4 regime due to the small number of observed stars and their very likely binarity/multiplicity. A larger sample of O-type stars earlier than O4, as well as information on their possible composite nature are necessary.

5. Mass discrepancy

The mass discrepancy problem consists on a lack of agreement between spectroscopic and evolutionary masses. Since its identification by Herrero *et al.* (1992), several explanations have been proposed but it remains unresolved due to the diversity of results in different environments.

We compare our derived spectroscopic and evolutionary masses in Figure 2. We find a certain trend in the distribution: for masses below 20 $M_\odot$, we can only find stars with $M_{\rm ev} > M_{\rm sp}$, i.e., a positive discrepancy, while for higher masses both positive and negative

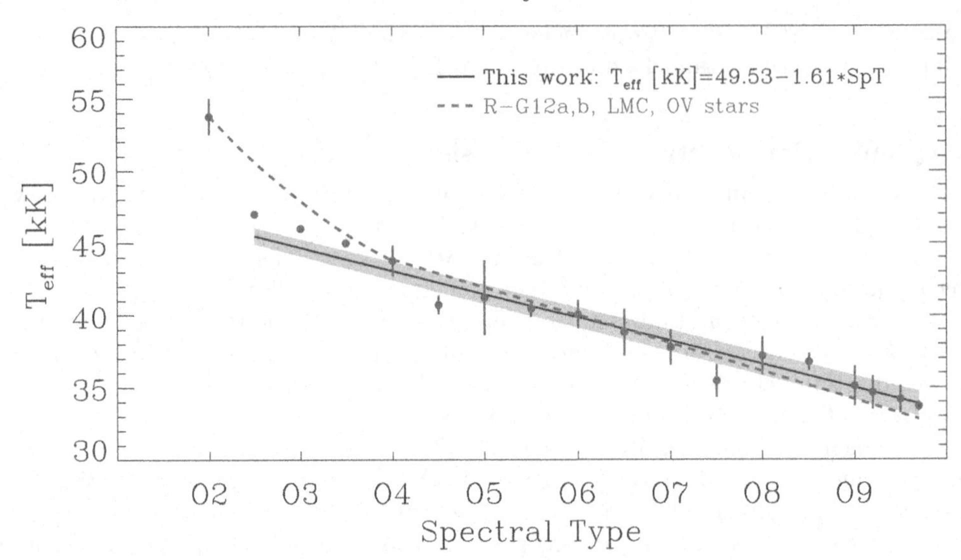

Figure 1. Effective temperature calibration compared to that derived by Rivero-González *et al.* (2012a,b). For each spectral type, the dark grey circles and bars represent the mean value and standard deviation of our T_{eff} estimates. Our linear fit is represented by the black line, and the grey zones indicate the associated uncertainties. O2 stars have been excluded from the fit due to their possible contamination in fibre and/or undetected binarity.

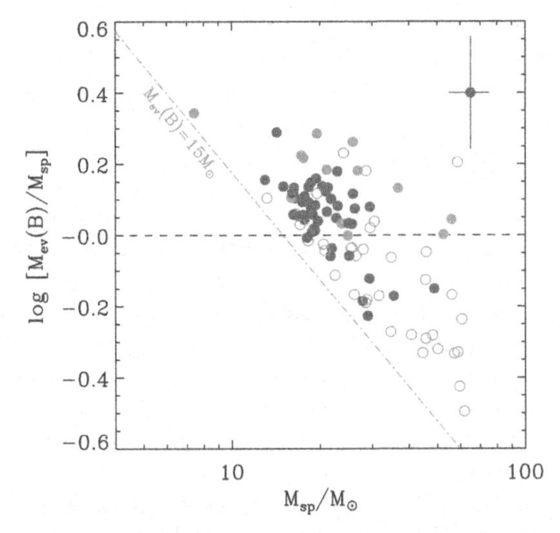

Figure 2. Logarithmic ratio of evolutionary masses from BONNSAI ($M_{ev}(B)$) to spectroscopic masses (M_{sp}). Grey filled symbols correspond to stars analyzed with nitrogen lines and no indications of binarity or contamination in fibre. Open symbols are those stars with possible undetected binarity or contamination in fibre. Typical error bars are given in the upper-right corner. The dashed line indicates points where $M_{ev}(B) = 15\,M_{\odot}$. O2 stars are excluded.

discrepancies are present. Nevertheless, this trend could be explained by a selection bias in our sample. All the O dwarfs in this work have masses systematically above 15 $M_{\odot}$ (see indication in Fig. 2). As a consequence, all stars with a negative mass discrepancy and masses below 15 $M_{\odot}$ are not present. To investigate this effect and reach a clearer conclusion about the mass discrepancy in the VFTS sample, results from the ongoing quantitative analysis of the early B-dwarfs should be included.

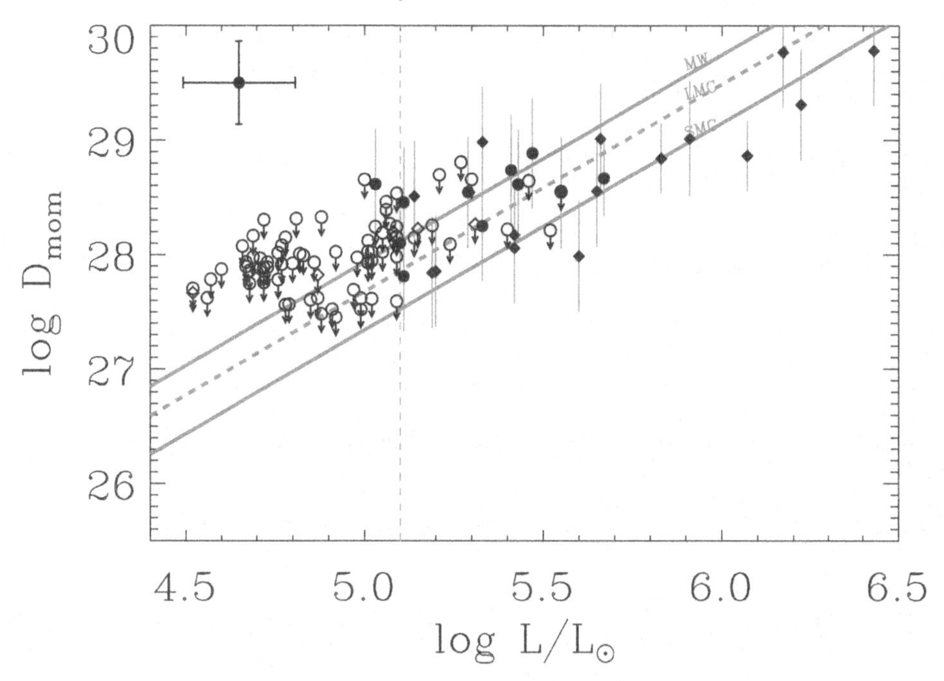

Figure 3. Wind-momentum – Luminosity Relationship (WLR) for our sample compared with theoretical predictions from Vink *et al.* (2001) for the metallicities of the Milky Way and the Magellanic Clouds. Open symbols represent stars with upper limits on $\log Q$ (hence $D_{\rm mom}$). Diamonds correspond to stars analyzed using nitrogen lines. Typical uncertainties are shown in the upper-left corner, and the vertical dashed line indicates $\log L/L_\odot = 5.1$.

6. Wind properties

In this study, we estimated the logarithm of the wind-strength Q-parameter (see Sect. 3) of the sample stars. If terminal velocities were available, $\log Q$ could be used to estimate mass-loss rates. Nevertheless, we lack UV spectroscopic data to derive directly terminal velocities. For this reason, we have followed the approach used by Mokiem *et al.* (2007a,b) to estimate the escape velocity $v_{esc} = 2gR(1-\Gamma)^{1/2}$ and the terminal velocity $v_\infty/v_{esc} = 2.65\,Z^{0.13}$, where Γ is the Eddington factor and Z the metallicity corresponding to the LMC. The wind momentum was calculated via $D_{\rm mom} = \dot{M}v_\infty R^{1/2}$, and is represented in Figure 3 as a function of stellar luminosity. Two different behaviours are present in the distribution:

• At luminosities higher than $\log L/L_\odot \sim 5.1$, winds were constrained for most of the stars. Most of them show consistency with the linear trend predicted by Vink *et al.* (2001) within the error bars. However, a large dispersion is found, showing some stars above the theoretical WLR for Galactic metallicity and some others below the prediction for the SMC. The first case could be explained by unclumped wind models and large uncertainties on the estimated terminal velocities, but no suitable explanation can be found for the second one.

• At luminosities lower than $\log L/L_\odot \sim 5.1$, only upper limits could be given for $\log Q$ and therefore $D_{\rm mom}$. To constrain such thin winds, UV spectroscopy is necessary.

References

Brott I., *et al.*, 2011, A&A, 530, A115

Evans C. J., *et al.*, 2011, A&A, 530, A108

Grin, N. J., Ramirez-Agudelo, O. H., de Koter, A., *et al.* 2017, A&A, 600, A82
Herrero A., Kudritzki R. P., Vilchez J. M., Kunze D., Butler K., Haser S., 1992, A&A, 261, 209
Köhler K., *et al.*, 2015, A&A, 573, A71
Maíz Apellániz J., *et al.*, 2014, A&A, 564, A63
Mokiem M. R., *et al.*, 2007, A&A, 465, 1003
Mokiem M. R., *et al.*, 2007, A&A, 473, 603
Puls J., Urbaneja M. A., Venero R., Repolust T., Springmann U., Jokuthy A., Mokiem M. R., 2005, A&A, 435, 669
Ramírez-Agudelo O. H., *et al.*, 2013, A&A, 560, A29
Ramírez-Agudelo O. H., *et al.*, 2015, A&A, 580, A92
Ramírez-Agudelo, O. H., Sana, H., de Koter, A., *et al.* 2017, A&A, 600, A81
Rivero González J. G., Puls J., Najarro F., Brott I., 2012, A&A, 537, A79
Rivero González J. G., Puls J., Massey P., Najarro F., 2012, A&A, 543, A95
Sabín-Sanjulián C., *et al.*, 2014, A&A, 564, A39
Sabín-Sanjulián, C., Simón-Díaz, S., Herrero, A., *et al.* 2017, A&A, 601, A79
Sana H., *et al.*, 2013, A&A, 550, A107
Santolaya-Rey A. E., Puls J., Herrero A., 1997, A&A, 323, 488
Schneider F. R. N., Langer N., de Koter A., Brott I., Izzard R. G., Lau H. H. B., 2014, A&A, 570, A66
Simón-Díaz S., Castro N., Herrero A., Puls J., Garcia M., Sabín-Sanjulián C., 2011, JPhCS, 328, 012021
Vink J. S., de Koter A., Lamers H. J. G. L. M., 2001, A&A, 369, 574
Walborn N. R., *et al.*, 2014, A&A, 564, A40

The Lives and Death-Throes of Massive Stars
Proceedings IAU Symposium No. 329, 2016
J.J. Eldridge, J.C. Bray, L.A.S. McClelland
& L. Xiao, eds.

doi:10.1017/S1743921317002769

Dynamo Scaling Relationships

Kyle Augustson[1], Stéphane Mathis[1], Sacha Brun[1] and Juri Toomre[2]

[1]Laboratoire AIM Paris-Saclay, CEA/DRF – CNRS – Université Paris Diderot, IRFU/SAp
Centre de Saclay, F-91191 Gif-sur-Yvette Cedex, France
[2]JILA, University of Colorado – Boulder, Colorado, USA 80309

Abstract. This paper provides a brief look at dynamo scaling relationships for the degree of equipartition between magnetic and kinetic energies. Two simple models are examined, where one that assumes magnetostrophy and another that includes the effects of inertia. These models are then compared to a suite of convective dynamo simulations of the convective core of a main-sequence B-type star and applied to its later evolutionary stages.

1. Introduction

The effects of astrophysical dynamos can be detected at the surface and in the environment of many magnetically-active objects, such as stars (e.g., Christensen *et al.* 2009; Donati & Landstreet 2009; Donati 2011; Brun *et al.* 2015). Yet predicting the nature of the saturated state of such turbulent convective dynamos remains quite difficult. Nevertheless, one can attempt to approximate the shifting nature of those dynamos. There may be the potential to identify a few regimes for which some global-scale aspects of stellar dynamos might be estimated with only a knowledge of the basic parameters of the system. For instance, consider how the magnetic energy of a system may change with a modified level of turbulence and also how rotation may influence it. Establishing the global-parameter scalings of convective dynamos, particularly with stellar mass and rotation rate, is useful given that they provide an order of magnitude approximation of the magnetic field strengths generated within the convection zones of stars as they evolve from the pre-main-sequence to a terminal phase. This could be especially useful in light of the recent evidence for magnetic fields within the cores of red giants, pointing to the existence of a strong core dynamo being active in a large fraction of main-sequence, intermediate-mass stars (Fuller *et al.* 2015; Cantiello *et al.* 2016; Stello *et al.* 2016). In turn, such estimates place constraints upon transport processes, such as those for angular momentum.

2. Scaling of Magnetic and Kinetic Energies

Convective flows often possess distributions of length scales and speeds that are peaked near a single characteristic value. One estimate of these quantities in stellar convection zones assumes that the energy containing flows possess a kinetic energy proportional to the stellar luminosity (L) that is approximately $\mathrm{v}_{\mathrm{rms}} \propto (2L/\rho_{\mathrm{CZ}})^{1/3}$ (Augustson *et al.* 2012) , where ρ_{CZ} is the average density in the convection zone. However, such a mixing-length velocity prescription only provides an order of magnitude estimate (e.g., Landin *et al.* 2010). Since stars are often rotating fairly rapidly, their dynamos may reach a quasi-magnetostrophic state wherein the Coriolis acceleration also plays a significant part in balancing the Lorentz force. Such a balance has been addressed and discussed at length in Christensen (2010), Brun *et al.* (2015), and Augustson *et al.* (2016).

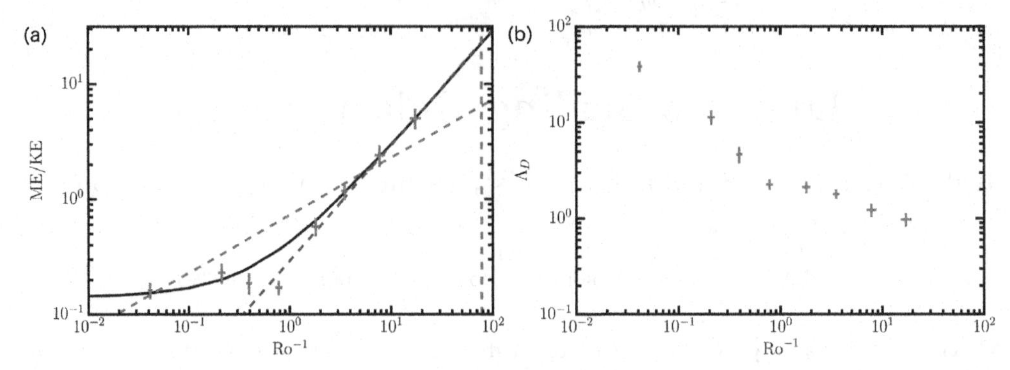

Figure 1. (a) The scaling of the ratio of magnetic to kinetic energy (ME/KE), with data from Augustson *et al.* (2016). The black curve indicates the scaling defined in Equation (2.1), with $\beta = 0.5$. The blue dashed line is for magnetostrophy ($\beta = 0$). The green dashed line represents the buoyancy-work-limited dynamo scaling, where $\mathrm{ME/KE} \propto \mathrm{Ro}^{-1/2}$. The red dashed line indicates the critical Rossby number of the star, corresponding to its rotational breakup velocity. (b) The scaling of the dynamic Elsasser number (Λ_D) with inverse Rossby number in simulations from Augustson *et al.* (2016). The uncertainty of the measured Rossby number and energy ratio or dynamic Elsasser number that arises from temporal variations are indicated by the size of the cross for each data point.

In Augustson *et al.* (2017), it is shown that one can derive a scaling relationship based upon the vorticity equation. In particular, integrating the enstrophy equation and ignoring any loss of enstrophy through the boundary requires that

$$\boldsymbol{\nabla}\times\left[\rho\mathbf{v}\times\boldsymbol{\omega} + 2\rho\mathbf{v}\times\boldsymbol{\Omega} + \frac{\mathbf{J}\times\mathbf{B}}{c} + \boldsymbol{\nabla}\cdot\sigma\right] = 0.$$

Thus, the primary balance is between inertial, Coriolis, Lorentz, and viscous forces. Scaling the derivatives as the inverse of a characteristic length scale ℓ and taking fiducial values for the other parameters in the above equation yields $\mathrm{ME/KE} \propto 1 + \mathrm{Re}^{-1} + \mathrm{Ro}^{-1}$, when divided through by $\rho \mathrm{v}_{\mathrm{rms}}^2/\ell^2$. Here the Reynolds number is taken to be $\mathrm{Re} = \mathrm{v}_{\mathrm{rms}}\ell/\nu$. However, the leading term of this scaling relationship is found to be less than unity, at least when assessed through simulations. Replacing it with a parameter to account for dynamos that are subequipartition leaves

$$\mathrm{ME/KE} \propto \beta(\mathrm{Ro}, \mathrm{Re}) + \mathrm{Ro}^{-1}. \tag{2.1}$$

Here β is unknown apriori as it depends upon the intrinsic ability of the non-rotating system to generate magnetic fields, which in turn depends upon the specific details of the system such as the boundary conditions and geometry of the convection zone.

For a subset of dynamos, like those discussed in Augustson *et al.* (2016), Equation (2.1) may hold. Such dynamos are sensitive to the degree of rotational constraint on the convection and upon the intrinsic ability of the convection to generate a sustained dynamo. The inertial term, in particular, may permit a minimum magnetic energy state to be achieved, bridging the subequipartition slowly rotating dynamos to the rapidly rotating magnetostrophic regime, where $\mathrm{ME/KE} \propto \mathrm{Ro}^{-1}$. For low Rossby numbers, or large rotation rates, it is possible that the dynamo can reach superequipartition states where $\mathrm{ME/KE} > 1$. Indeed, it may be much greater than unity, as is expected for the Earth's dynamo (e.g., Roberts & King 2013).

Consider the data for the evolution of a set of MHD simulations using the Anelastic Spherical Harmonic code presented in Augustson *et al.* (2016). These simulations attempt to capture the dynamics within the convective core of a 10 $M_\odot$ B-type star. Given the choices of rotation rates for this suite of simulations, they have nearly three decades of

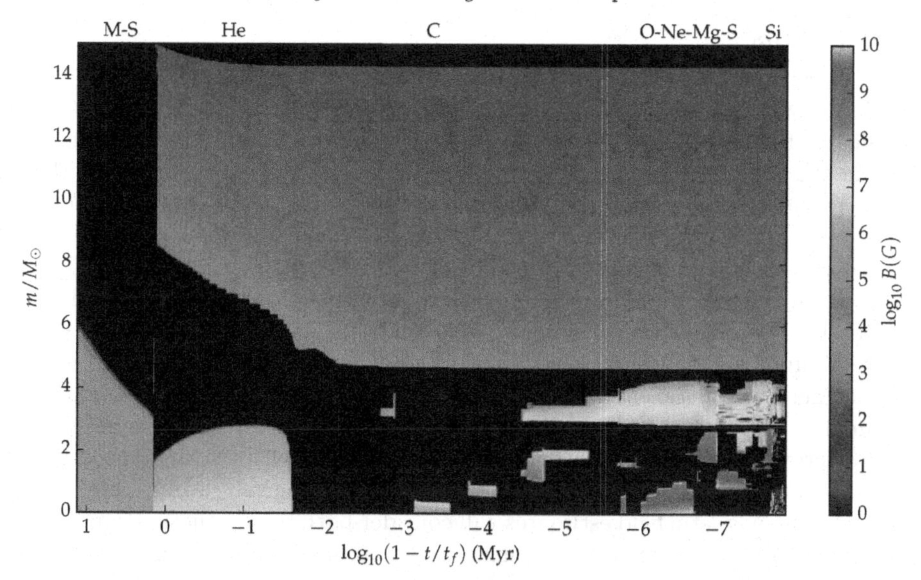

Figure 2. A magnetic Kippenhahn diagram showing the evolution of the equipartition magnetic field for a 15 $M_\odot$ star. The abscissa show the time remaining in Myr before the iron core infall that occurs at t_f. The burning phase of the core is indicated at the top of the diagram.

coverage in Rossby number, as shown in Figure 1(a). In that figure, the force-based scaling given in Equation (2.1) is depicted by the black curve (where ME/KE $\propto 0.5+\mathrm{Ro}^{-1}$). This scaling does a reasonable job of describing the nature of the superequipartition state for a given Rossby number. These simulated convective core dynamos appear to enter a regime of magnetostrophy for the four cases with the lowest average Rossby number, where the scaling for the magnetostrophic regime is denoted by the dashed blue line in Figure 1(a). This transition to the magnetostrophic regime can be better understood through Figure 1(b), which shows the dynamic Elsasser number ($\Lambda_D = {\mathrm{B_{rms}}}^2/(8\pi\rho_0\Omega_0\mathrm{v_{rms}}\ell)$, where ℓ is the typical length scale of the current density $\mathbf{J}$). So as Λ_D approaches unity, the balance between the Lorentz and the Coriolis forces also approaches unity, indicating that the dynamo is close to magnetostrophy.

The scaling relationship between the magnetic and kinetic energies of convective dynamos in turn provide an estimate of the rms magnetic field strength in terms of the local rms velocity and density at a particular depth in a convective zone. Therefore, these relationships permit the construction of magnetic Kippenhahn diagrams that show the equipartition magnetic field, which is estimated based on the mixing length velocities achieved in stellar evolution models, as shown in Figure 2 for a 15 $M_\odot$ star. During the main sequence, the magnetic field generated by the dynamo running in the convective core has an estimated rms strength of about 10^6 Gauss, which is consistent with the simulations described in Augustson *et al.* (2016). Likewise, during the helium-burning phase, the equipartition magnetic field rises to about 10^7 Gauss. During subsequent burning phases, the field amplitude continues to rise largely due to the increasing density of the convective regions where it eventually reaches 10^{10} Gauss during the oxygen-neon and silicon burning stages. The density dependence of the equipartition magnetic field can be seen more directly in the scaling $B \propto \rho_{\mathrm{CZ}}^{1/6} L^{1/3}$, which follows from the scaling of the mixing length velocity discussed above, and by noting the surface luminosity of the star does not change significantly during these late-stage burning phases.

3. Conclusions

As discussed in Augustson *et al.* (2017) and Augustson (2017), there appear to be two scaling laws for the level of equipartition of magnetic and kinetic energies that are applicable to stellar systems, one in the high magnetic Prandtl number regime and another in the low magnetic Prandtl number regime. Within the context of the large magnetic Prandtl number systems, the ratio of the magnetic to the kinetic energy of the system scales as $\mathrm{ME}/\mathrm{KE} \propto \beta + \mathrm{Ro}^{-1}$, where β depends upon the details of the non-rotating system, as mentioned above in §2 and in Augustson *et al.* (2016). For low magnetic Prandtl number and fairly rapidly rotating systems, such as the geodynamo and rapidly rotating low-mass stars, another scaling relationship may be more applicable. This scaling relies upon a balance of buoyancy work and magnetic dissipation and it yields a ratio of magnetic to kinetic energy that scales as the inverse square root of the convective Rossby number (Davidson 2013; Augustson *et al.* 2017). In either case, it is likely that the magnetic energy can grow to be near or above equipartition with the kinetic energy, which allows the estimation of the magnetic energy at various stages of evolution as shown in Figure 2. Future magnetic field estimates will consider both the magnetic Prandtl number and the star's rotational evolution, utilizing angular momentum transport techniques such as those discussed in Amard *et al.* (2016). Yet, more work is needed to establish more robust scaling relationships that cover a greater range in both magnetic Prandtl number and Rossby number. Likewise, numerical experiments should explore a larger range of Reynolds number and level of supercriticality. Indeed, as in Yadav *et al.* (2016), some authors have already attempted to examine such an increased range of parameters for the geodynamo. Nevertheless, to be more broadly applicable in stellar physics, there is a need to find scaling relationships that can bridge both the low and high magnetic Prandtl number regimes that are shown to exist within main-sequence stars. The authors are currently working toward this goal, as will be presented in an upcoming paper.

Acknowledgements

K. C. Augustson and S. Mathis acknowledge support from the ERC SPIRE 647383 grant. A. S. Brun acknowledges funding by ERC STARS2 207430 grant, INSU/PNST, CNES Solar Orbiter, PLATO and GOLF grants, and the FP7 SpaceInn 312844 grant. J. Toomre thanks the NASA TCAN grant NNX14AB56G for support.

References

Amard, L., Palacios, A., Charbonnel, C., Gallet, F., & Bouvier, J. 2016, A&A, 587, A105
Augustson, K. 2017, ArXiv e-prints
Augustson, K., Mathis, S., & Brun, A. S. 2017, ArXiv e-prints
Augustson, K. C., Brown, B. P., Brun, A. S., Miesch, M. S., & Toomre, J. 2012, ApJ, 756, 169
Augustson, K. C., Brun, A. S., & Toomre, J. 2016, ApJ, 829, 92
Brun, A. S., García, R. A., Houdek, G., Nandy, D., & Pinsonneault, M. 2015, SSR, 196, 303
Cantiello, M., Fuller, J., & Bildsten, L. 2016, ApJ, 824, 14
Christensen, U. R. 2010, SSR, 152, 565
Christensen, U. R., Holzwarth, V., & Reiners, A. 2009, Nat, 457, 167
Davidson, P. A. 2013, Geophysical Journal International, 195, 67
Donati, J.-F. 2011, in IAU Symposium, Vol. 271, , 23–31
Donati, J.-F., & Landstreet, J. D. 2009, ARAA, 47, 333
Fuller, J., Cantiello, M., Stello, D., Garcia, R. A., & Bildsten, L. 2015, Science, 350, 423
Landin, N. R., Mendes, L. T. S., & Vaz, L. P. R. 2010, A&A, 510, A46
Roberts, P. H., & King, E. M. 2013, Reports on Progress in Physics, 76, 096801
Stello, D., Cantiello, M., Fuller, J., *et al.* 2016, Nat, 529, 364
Yadav, R. K., Gastine, T., Christensen, U. R., Wolk, S. J., & Poppenhaeger, K. 2016, PNAS, 113, 12065

The Lives and Death-Throes of Massive Stars
Proceedings IAU Symposium No. 329, 2016
J.J. Eldridge, J.C. Bray, L.A.S. McClelland & L. Xiao, eds.

doi:10.1017/S1743921317003441

The First 3D Simulations of Carbon Burning in a Massive Star

A. Cristini[1] †, C. Meakin[2,3], R. Hirschi[1,4], D. Arnett[3], C. Georgy[5,1] and M. Viallet[6]

[1]Astrophysics Group, Keele University, Lennard-Jones Laboratories, Keele, ST5 5BG, UK
[2]Karagozian & Case, Inc., 700 N. Brand Blvd. Suite 700, Glendale, CA, 91203, USA
[3]Department of Astronomy, University of Arizona, Tucson, AZ 85721, USA
[4]Kavli IPMU (WPI), The University of Tokyo, Kashiwa, Chiba 277-8583, Japan
[5]Geneva Observatory, University of Geneva, Maillettes 51, 1290 Versoix, Switzerland
[6]Max-Planck-Institut für Astrophysik, Karl Schwarzschild Strasse 1, Garching, D-85741, Germany

Abstract. We present the first detailed three-dimensional hydrodynamic implicit large eddy simulations of turbulent convection for carbon burning. The simulations start with an initial radial profile mapped from a carbon burning shell within a $15\,M_\odot$ stellar evolution model. We considered 4 resolutions from 128^3 to 1024^3 zones. These simulations confirm that convective boundary mixing (CBM) occurs via turbulent entrainment as in the case of oxygen burning. The expansion of the boundary into the surrounding stable region and the entrainment rate are smaller at the bottom boundary because it is stiffer than the upper boundary. The results of this and similar studies call for improved CBM prescriptions in 1D stellar evolution models.

Keywords. Convection, hydrodynamics, turbulence, methods: numerical, stars: interiors.

One-dimensional (1D) stellar evolution codes are currently the only way to simulate the entire lifespan of a star. This comes at the cost of having to replace complex, inherently three-dimensional (3D) processes, such as convection, rotation and magnetic activity, with generally simplified mean-field models. An essential question is "how well do these 1D models represent reality?" Answers can be found both in empirical and theoretical work. On the empirical front, we can investigate full star models, by comparing them to observations of stars under a range of conditions. On the theoretical side, multi-dimensional simulations can be used to test 1D models under astrophysical conditions that are difficult to recreate in terrestrial laboratories. We present here the latter; three-dimensional hydrodynamic simulations of carbon shell burning in a $15\,M_\odot$ star.

The reasons for choosing carbon burning as opposed to other burning regions were: this phase of stellar evolution has never been studied before; cooling is dominated by neutrino losses, allowing radiative effects to be neglected (very high Péclet[1] number) which reduces the computational cost of the simulations; the initial composition and structure profiles of the shell are simpler than the more advanced stages as the composition in the region where the shell forms has been homogenised by the preceding convective helium burning core; and finally this shell plays an important role in setting the final mass of the iron core prior to the core-collapse event. Choosing a shell as opposed to a core burning region also affords the simulation of two distinct convective boundaries rather than one.

We prepared the initial conditions by calculating a $15\,M_\odot$, solar metallicity, non-rotating 1D stellar evolution model until the end of the oxygen burning phase using the

† Email: a.j.cristini@keele.ac.uk
1 The Péclet number is the ratio of heat transfer through conduction to heat transfer through advective motions.

Geneva stellar evolution code (GENEC; Eggenberger et al. 2008). The hydrodynamic simulations were calculated from the structure given by this stellar model during the growth of the carbon burning shell. These simulations were calculated using the Prometheus MPI (PROMPI; Meakin and Arnett 2007) code which solves the inviscid Euler equations using a finite-volume Eulerian solver which utilises the piecewise parabolic method of Colella and Woodward (1984). We chose to model the domain within a plane-parallel, Cartesian geometry. This 'box-in-a-star' approach allows us to maximise the effective resolution at the convective boundaries, and ease the difficult Courant time scale at the inner boundary of the grid. More details on the PROMPI code and the model set-up can be found in Cristini et al. (2016*a*).

Simulations of turbulence involve some kind of initialisation of turbulent motions, followed by a transient phase whereby the global motion settles down into a quasi-steady state of turbulence. Initial test calculations of carbon burning revealed that the time-scale for this relaxation to the quasi-steady state was long, and therefore simulations of the quasi-steady state over time-scales that are statistically significant (several convective turnovers) would not be possible given our available computational resources. We therefore decided to boost the nuclear energy generation rate by a factor of 1000 in order to match that of oxygen burning; this reduces both the relaxation time and the convective turnover time. Such an artificial boost in luminosity does not affect the structure for the following reasons: hydrostatic equilibrium is still maintained; the entropy and composition profiles in the convective region remain flat; and the structure in the stable regions away from the boundary are determined by the evolutionary history of the model and are unaffected by the turbulence.

To test the dependence of our results on numerical resolution we simulated the carbon shell at four different resolutions. These models are named according to their resolution: lrez - 128^3, mrez - 256^3, hrez - 512^3 and vhrez - 1024^3. The temporal evolution of the global (integrated over the convective zone) specific kinetic energy for all of the models is presented in the left panel of Fig. 1. The first $\sim$1000 seconds of evolution is characterised by the initial transient associated with the onset of convection. By $\sim$1250 s, all of the models settle into the quasi-steady state of turbulence, characterised by semi-regular pulses in kinetic energy occurring on a time scale of the order of the convective turnover time. These pulses are associated with the formation and eventual breakup of semi-coherent, large-scale eddies or plumes that traverse a good fraction of the convection zone before dissipating. It is a phenomena that is typical of stellar convective flow (Meakin and Arnett 2007; Arnett and Meakin 2011*a*,*b*; Viallet et al. 2013; Arnett et al. 2015). Although these simulations do not sample a large number of convective turnover times (between $\sim$2 and $\sim$6), resolution trends are still apparent.

Some aspects of our models are sensitive to the grid resolution. At the lower convective boundaries of our models a spurious spike in dissipation appears at all resolutions (see figs. 7 and 8 in Cristini et al. 2016*a*). This spike appears to be numerical and undermines the statistical analysis, although the general behaviour of the numerical dissipation is sane, and the discrepancy is localised. The spike reduces in amplitude and width with increasing resolution, suggesting convergence to a physically relevant solution. Our resolution study shows that a radial resolution of 512 zones is sufficient to resolve the upper boundary but a resolution of roughly 1500 zones is needed to fully resolve the lower boundary.

The qualitative description of convection and CBM is very different from that which describes the parameterisations that are used in stellar evolution models. The velocity magnitude, $\sqrt{v_r^2 + v_y^2}$ (where v_r and v_y are the radial and horizontal velocities, respectively) of the hrez model is shown in the right panel of Fig. 1. Entrainment events

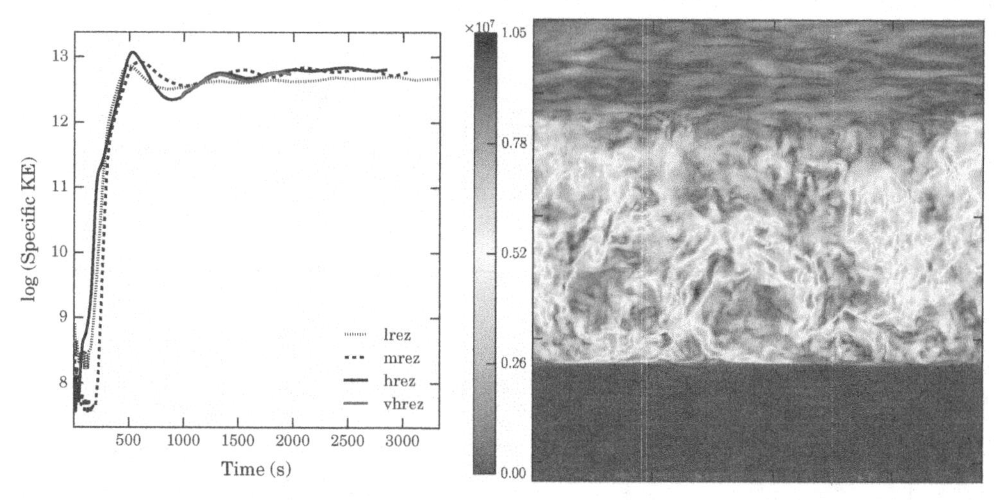

Figure 1. *Left:* Temporal evolution of the global specific kinetic energy: thin dashed - lrez; thick dashed - mrez; black solid - hrez; red solid - vhrez. The quasi-steady state begins after approximately 1,000 s. *Right:* Vertical cross-section of the radial and horizontal component of the velocity vector field, (v_r, v_y). The colour-map represents the velocity magnitude in cm s^{-1}. This snapshot was taken at 2,820 s into the hrez simulation. Figure taken from Cristini et al. (2016*a*).

(similar to those found for oxygen burning, see e.g. fig. 23 in Meakin and Arnett 2007) can be seen in the convection zone (see e.g. bottom left of convective zone where material from below the convective zone is entrained upwards or top corners of the convective zones where the material is entrained from the top stable layer). Strong flows can be seen in the centre of the convective region and shear flows can be seen over the entire convective region. These shear flows have the greatest impact at the convective boundaries, where composition and entropy are mixed between the convective and radiative regions.

Turbulent entrainment at both boundaries pushes the boundary position over time into the surrounding stable regions. In order to calculate the boundary entrainment velocities, first the convective boundary positions must be determined in the simulations. In the 3D simulations, the boundary is a two-dimensional surface and is not spherically symmetric as in 1D stellar models. Thus the convective boundary position must be estimated. In order to do this we first map out a two-dimensional horizontal boundary surface, $r_{j,k} = r(j,k)$, for $j = 1, n_y$; $k = 1, n_z$, where n_y and n_z are the number of grid points in the horizontal y and z directions. We estimate that the radial position of the boundary at each horizontal coordinate coincides with the position where the average atomic weight, $\bar{A}$, is equal to the mean value of $\bar{A}$ between the convective and corresponding radiative zones, this threshold composition value is denoted as A_{th}. The boundary position at each time-step is then approximated as the horizontal mean, $\overline{r}_{j,k}$ (henceforth denoted as $\overline{r}$), over the boundary surface. We define the error in the estimated boundary position as the standard deviation (σ) from the horizontal surface mean, $\overline{r}$.

The average atomic weight, $\bar{A}$, is used as an input variable in the Helmholtz equation of state (Timmes and Arnett 1999), and so as the models evolve we update the value of A_{th} and hence calculate new boundary surface positions, $r_{j,k}$. Our method is a valid but not unique way in which to calculate the boundary positions (e.g. Sullivan et al. 1998; Fedorovich, Conzemius and Mironov 2004; Meakin and Arnett 2007; Liu and Ecke 2011;

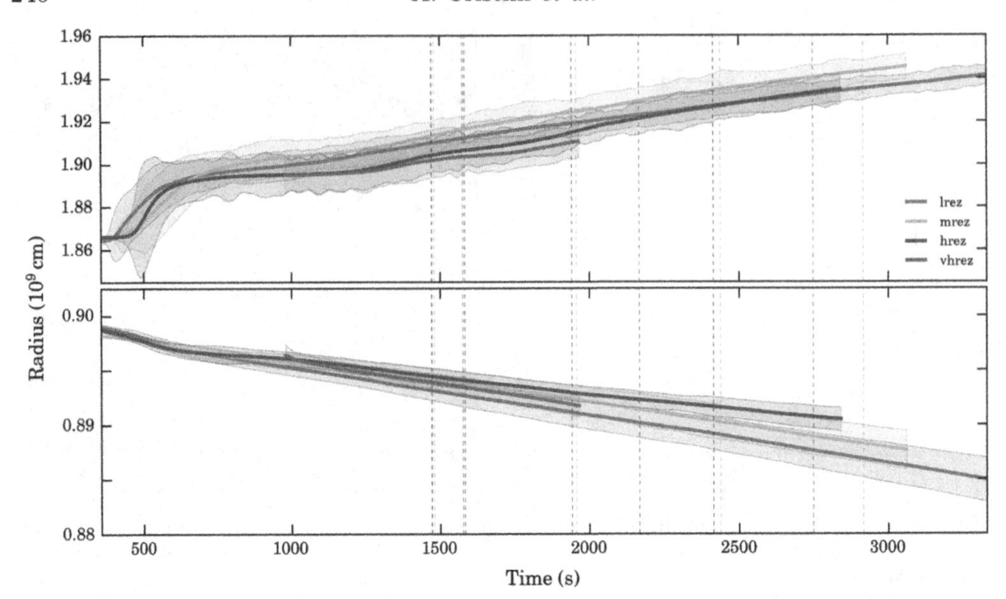

Figure 2. Time evolution of the mean radial position of the upper convective boundary (top panel) and lower convective boundary (bottom panel), averaged over the horizontal plane for all four resolutions. Shaded envelopes are twice the standard deviation from the boundary mean. Vertical dotted lines indicate convective turnover times, taken from the beginning of the quasi-steady state estimated at ~1000s. The shaded areas represent the variation in the boundary height due to the fact that the boundary is not a flat surface. This can be compared to the surface of the ocean not being flat due to the presence of waves. Figure taken from Cristini et al. (2016*a*).

Sullivan and Patton 2011; van Reeuwijk, Hunt and Jonker 2011; Garcia and Mellado 2014; Gastine, Wicht and Aurnou 2015).

The variation in time of the average surface position, $\overline{r}$, of both boundaries is shown for all models in Fig. 2. Positions are shown as solid lines and twice the standard deviation as the surrounding shaded envelopes. These small envelopes are due to the vertical extent of the boundary surface, which is not flat. Following the initial transient (> 1000 s) a quasi-steady expansion of the convective shell proceeds, convective turnovers are indicated by vertical dashed lines for each resolution in Fig. 2.

In conclusion, 1D stellar evolution models should include CBM at all convective boundaries, and turbulent entrainment should be accounted for in the advanced stages of massive star evolution. At the lower boundary of our models, which is stiffer, the entrainment is slower and the boundary width is narrower. This confirms the dependence of entrainment and mixing on the stiffness of the boundary. Since the boundary stiffness will vary both in time and with the convective boundary considered, a single constant parameter is probably not going to correctly represent the dependence of the mixing on the instantaneous convective boundary properties. We suggest the use of the bulk Richardson number (Cristini et al. 2016*b*) in new prescriptions in order to include this dependence.

The authors acknowledge support from EU-FP7-ERC-2012-St Grant 306901. RH acknowledges support from the World Premier International Research Centre Initiative (WPI Initiative), MEXT, Japan. This work used the Extreme Science and Engineering Discovery Environment (XSEDE), which is supported by National Science Foundation grant number OCI-1053575. CM and WDA acknowledge support from NSF grant 1107445 at the University of Arizona. The authors acknowledge the Texas Advanced Computing

Center (TACC) at The University of Texas at Austin (http://www.tacc.utexas.edu) for providing HPC resources that have contributed to the research results reported within this paper. MV acknowledges support from the European Research Council through grant ERC-AdG No. 341157-COCO2CASA. This work used the DiRAC Data Centric system at Durham University, operated by the Institute for Computational Cosmology on behalf of the STFC DiRAC HPC Facility (www.dirac.ac.uk). This equipment was funded by BIS National E-infrastructure capital grant ST/K00042X/1, STFC capital grants ST/H008519/1 and ST/K00087X/1, STFC DiRAC Operations grant ST/K003267/1 and Durham University. DiRAC is part of the National E-Infrastructure.

References

Arnett, W. D. and C. Meakin. 2011*a*. "Toward Realistic Progenitors of Core-collapse Supernovae." *ApJ* 733:78.

Arnett, W. D. and C. Meakin. 2011*b*. "Turbulent Cells in Stars: Fluctuations in Kinetic Energy and Luminosity." *ApJ* 741:33.

Arnett, W. D., C. Meakin, M. Viallet, S. W. Campbell, J. C. Lattanzio and M. Mocák. 2015. "Beyond Mixing-length Theory: A Step Toward 321D." *ApJ* 809:30.

Colella, P. and P. R. Woodward. 1984. "The Piecewise Parabolic Method (PPM) for Gas-Dynamical Simulations." *J. Comput. Phys.* 54:174–201.

Cristini, A., C. Meakin, R. Hirschi, D. Arnett, C. Georgy and M. Viallet. 2016*a*. "3D Hydrodynamic Simulations of Carbon Burning in Massive Stars." *ArXiv e-prints* .

Cristini, A., C. Meakin, R. Hirschi, D. Arnett, C. Georgy and M. Viallet. 2016*b*. "Linking 1D evolutionary to 3D hydrodynamical simulations of massive stars." *Phys. Scr.* 91(3):034006.

Eggenberger, P., G. Meynet, A. Maeder, R. Hirschi, C. Charbonnel, S. Talon and S. Ekström. 2008. "The Geneva stellar evolution code." *Ap&SS* 316:43–54.

Fedorovich, E., R. Conzemius and D. Mironov. 2004. "Convective Entrainment into a Shear-Free, Linearly Stratified Atmosphere: Bulk Models Reevaluated through Large Eddy Simulations." *J. Atmos. Sci.* 61:281.

Garcia, J. and J. Mellado. 2014. "The two-layer structure of the entrinament zone in the convective boundary layer." *J. Atmos. Sci.* pp. 1935–1955.

Gastine, T., J. Wicht and J. M. Aurnou. 2015. "Turbulent Rayleigh-Bénard convection in spherical shells." *J. Fluid Mech.* 778:721–764.

Liu, Y. and R. E. Ecke. 2011. "Local temperature measurements in turbulent rotating Rayleigh-Bénard convection." *Phys. Rev. E* 84(1):016311.

Meakin, C. A. and D. Arnett. 2007. "Turbulent Convection in Stellar Interiors. I. Hydrodynamic Simulation." *ApJ* 667:448–475.

Sullivan, P. P., C.-H. Moeng, B. Stevens, D. H. Lenschow and S. D. Mayor. 1998. "Structure of the Entrainment Zone Capping the Convective Atmospheric Boundary Layer." *J. Atmos. Sci.* 55:3042–3064.

Sullivan, P. P. and E. G. Patton. 2011. "The Effect of Mesh Resolution on Convective Boundary Layer Statistics and Structures Generated by Large-Eddy Simulation." *J. Atmos. Sci.* 68:2395–2415.

Timmes, F. X. and D. Arnett. 1999. "The Accuracy, Consistency, and Speed of Five Equations of State for Stellar Hydrodynamics." *ApJS* 125:277–294.

van Reeuwijk, M., G. R. Hunt and H. J. Jonker. 2011. Direct simulation of turbulent entrainment due to a plume impinging on a density interface. In *Journal of Physics Conference Series*. Vol. 318 of *Journal of Physics Conference Series* p. 042061.

Viallet, M., C. Meakin, D. Arnett and M. Mocák. 2013. "Turbulent Convection in Stellar Interiors. III. Mean-field Analysis and Stratification Effects." *ApJ* 769:1.

The Lives and Death-Throes of Massive Stars
Proceedings IAU Symposium No. 329, 2016
J.J. Eldridge, J.C. Bray, L.A.S. McClelland
& L. Xiao, eds.

doi:10.1017/S1743921317002587

Effect of a Dipole Magnetic Field on Stellar Mass-Loss

Chris Bard[1] and Richard Townsend[2]

[1]NPP Fellow, NASA Goddard Space Flight Center
8800 Greenbelt Rd., Greenbelt, MD, USA
email: christopher.bard@nasa.gov

[2]Dept. of Astronomy, University of Wisconsin-Madison,
475 N. Charter St., Madison, WI, USA
email: townsend@astro.wisc.edu

Abstract. Massive star winds greatly influence the evolution of both their host star and local environment though their mass-loss rates, but current radiative line-driven wind models do not incorporate any magnetic effects. Recent surveys of O and B stars have found that about ten percent have large-scale, organized magnetic fields. These massive-star magnetic fields, which are thousands of times stronger than the Sun's, affect the inherent properties of their own winds by changing the mass-loss rate. To quantify this, we present a simple surface mass-flux scaling over the stellar surface which can be easily integrated to get an estimate of the mass-loss rate for a magnetic massive star. The overall mass-loss rate is found to decrease by factors of 2-5 relative to the non-magnetic CAK mass-loss rate.

Keywords. stars: magnetic fields; stars: winds, outflows; stars: mass-loss; stars: early-type

1. Introduction

Recent surveys of magnetic massive stars have found that 5-10% of OB stars have detectable magnetic fields (Wade *et al.* 2014; Morel *et al.* 2015). These magnetic fields have significant effects on their host star. In order to quantify how these fields affect the stellar wind, Bard & Townsend (2016) undertook a steady-state analysis of a Castor *et al.* (1975) (hereafter CAK) line-driven wind flowing along dipole magnetic field lines for both the optically-thick formulation of the radiative acceleration (a la CAK) and the more general limit (Owocki *et al.* 1988). For this simplest case, a point source of radiation was assumed such that the line acceleration was solely radial. This assumption of "radially-streaming" photons can be relaxed by treating the star as a finite disk. The transformation between these two cases is provided by the so-called "finite-disk correction factor" (fdcf). Numerical solutions of a CAK wind modified with the fdcf (so-called "mCAK" models) (Friend & Abbott 1986; Pauldrach *et al.* 1986) found that the fdcf had a large effect on the stellar mass-loss rate, reducing it by roughly half.

In this proceeding, we will focus on the effects of a dipole field on the finite-disk corrected CAK wind and present a simple scaling relation for the surface mass-flux of a magnetic massive star. We will leave the full mathematical details of this derivation to a forthcoming paper, though they can currently be found in Bard (2016).

2. Surface Mass-Flux Scaling

We provide here a simple scaling which roughly reproduces the more detailed model of Bard (2016) to within an $\approx (25\%, 15\%, 7\%, 3\%, 0\%)$ underestimate for critical rotation

Table 1. Stellar and wind parameters used in this proceeding to represent a typical magnetic B-type star with a centrifugal magnetosphere and an O-type star with a dynamical magnetosphere. Values are identical to those used in Bard & Townsend (2016).

Type	M_*	R_p	$T_{\rm eff}$	α	$\Gamma_{\rm el}$	$\bar{Q}$	B_p	η_*
B	9.0 $M_\odot$	4.5 $R_\odot$	21000 K	0.56	9.27×10^{-3}	1025.14	11 kG	4.29×10^{5}
O	50 $M_\odot$	19 $R_\odot$	41860 K	0.6	0.5	500	3.715 kG	100

fractions $\omega = (0.65, 0.5, 0.35, 0.2, 0.0)$. This scaling assumes an aligned dipole magnetic field and a general CAK line force:

$$\dot{m}_*(\theta) \approx \mu_B^{1+1/\alpha} f^{1/\alpha} \Sigma_f \aleph^{1-1/\alpha} \dot{m}_{\rm CAK}, \tag{2.1}$$

with θ the surface co-latitude, α the CAK power-law index and f is the finite-disk correction factor evaluated at the pole ($\theta = 0$). This scaling relation does require the calculation of the polar f for each star, but $f \approx 0.6 - 0.7$ can be used in a pinch.

With R_p the polar radius of the star, the CAK surface mass-flux is

$$\dot{m}_{\rm CAK} = \frac{\dot{M}_{\rm CAK}}{4\pi R_p^2} = \frac{L_*}{4\pi R_p^2 c^2} \frac{\alpha}{1-\alpha} \left(\frac{\bar{Q}\Gamma_{\rm el}}{1-\Gamma_{\rm el}} \right)^{(1-\alpha)/\alpha}, \tag{2.2}$$

with $\bar{Q}$ the Gayley (1995) Q-parameter and $\Gamma_{\rm el} \equiv \kappa_e L_*/(4\pi c G M_*)$ is the Eddington parameter. μ_B is the surface tilt of the magnetic field with respect to the stellar surface normal,

$$\mu_B = \frac{2\cos\theta - \sin\theta (R'_*/R_*)}{\sqrt{(1+3\cos^2\theta)(1+(R'_*/R_*)^2)}}. \tag{2.3}$$

The stellar radius of a rotating star is given by

$$\frac{R_*}{R_p} = \frac{3}{\omega \sin\theta} \cos\left[\frac{\pi + \arccos(\omega \sin\theta)}{3} \right], \tag{2.4}$$

with its derivative $R'_* = dR_*/d\theta$ defined as

$$\frac{1}{R_p}\frac{dR_*}{d\theta} = \frac{\cot\theta \sin\{\frac{1}{3}\left[\pi + \arccos(\omega\sin\theta)\right]\}}{1-\omega^2\sin^2\theta} - \frac{3\cot\theta\csc\theta\cos\{\frac{1}{3}\left[\pi + \arccos(\omega\sin\theta)\right]\}}{\omega}. \tag{2.5}$$

For $\omega = 0$, $R_* = R_p$ and $R'_* = 0$. The rotation effect parameter is

$$\aleph \equiv 1 - \frac{12\cos^3\left[\frac{1}{3}(\pi + \cos^{-1}(\omega\sin\theta))\right]}{\omega\sin\theta}, \tag{2.6}$$

and the so-called "optically-thin correction" parameter is defined as

$$\Sigma_f \equiv \frac{\left| 1 - \alpha - \left[\frac{1-\left(\frac{\chi_0 \aleph}{\mu_B f}\right)^{1/\alpha - 1}}{1-\left(\frac{\chi_0 \aleph}{\mu_B f}\right)^{1/\alpha}} \right] \right|}{\alpha}. \tag{2.7}$$

Finally, we have

$$\chi_0 = (1-\Gamma_{\rm el})/(\Gamma_{\rm el}\bar{Q}). \tag{2.8}$$

For O-stars with high mass-loss rates, the "optically-thick" version of the CAK

Table 2. Mass-loss rates (in units of $10^{-9}\ M_\odot/\,\mathrm{yr}$) for our example B-type star ($\eta_* = 4.28\times10^5$). "No B" indicates a mCAK-type mass-loss rate calculated from a non-rotating radial flow with spherical divergence. The other mass-loss rates are calculated from a dipole magnetosphere with the given rotation fraction ω. "Optically-Thick" indicates the mass-loss calculated from using the optically-thick CAK radiative acceleration; the rest use the general version. "Open" is the mass-loss into open field lines ($L > R_c$). "Disk" is the mass-loss into field lines with a centrifugally supported disk ($R_K < L < R_c$). Numbers in parentheses next to a mass-loss rate represent the ratio of that particular rate to the "General" mass-loss at its rotation fraction ω.

	No B	$\omega = 0.0$	0.2	0.35	0.5	0.65	0.8
Optically-Thick	0.89	0.373	0.373	0.372	0.371	0.365	0.349
General	0.605	0.253	0.253	0.252	0.252	0.248	0.236
Open	...	0.021(0.08)	0.020(0.08)	0.020(0.08)	0.019(0.075)	0.017(0.07)	0.015(0.06)
Disk	...	...	0.090(0.36)	0.138 (0.54)	0.178(0.7)	0.208(0.84)	0.221(0.93)
Effective	...	0.021(0.08)	0.11(0.44)	0.157(0.62)	0.196(0.77)	0.225(0.91)	0.236(0.999)

line-acceleration can be used instead of the general version. In this case, Σ_f can be set to 1 and the remaining terms of the above scaling relation are unchanged.

3. Magnetic Effects on Mass-Loss Rate

The estimated stellar mass-loss rate is found by integrating the mass-flux scaling over the stellar surface:

$$\dot{M}_{\mathrm{global}} = \int \dot{m}_r \, dA = 2\pi \int R_*^2 \, \mu_B \, \dot{m}_* d\mu \ , \tag{3.1}$$

with $\mu = \cos\theta$. Following Bard & Townsend (2016), we define "open", "disk", and "effective" mass-loss rates based on the behavior of the wind after it flows away from the stellar surface. The "effective" mass-loss is simply the mass lost to wind flowing along open field lines out of the magnetosphere ("open") or into a centrifugally-supported disk ("disk"). Plasma which does not escape the magnetosphere nor settles into a disk eventually falls back to the stellar surface, so it is never "lost". We define open field lines as having dipole shell radii (L) larger than the closure radii (R_C) as derived from MHD simulations:

$$R_C \approx R_p + 0.7[R_p(0.3 + \eta_*^{1/4}) - R_p], \tag{3.2}$$

with the usual "wind magnetic confinement parameter"

$$\eta_* \equiv \frac{B_{\mathrm{eq}}^2 R_*^2}{\dot{M}_{B=0} v_{\infty,B=0}} \tag{3.3}$$

(ud-Doula & Owocki 2002). Lines with a centrifugally-supported disk are defined with $L > R_K$, where the Kepler radius

$$R_K = \frac{GM_*}{v_\phi^2} = \omega^{-2/3} R_p, \tag{3.4}$$

demarcates the beginning of the centrifugally-supported disk.

To illustrate the effect of a magnetic field on stellar mass-loss rates, we integrate the full finite-disk-corrected mass-flux model from Bard (2016) for two prototypical magnetic massive stars (Table 1). One star represents a B-star with a centrifugal magnetosphere,

Table 3. Same as Table 2, except for an O-type star with $\eta_* = 100$. All mass-loss rates are given in 10^{-6} $M_\odot$/yr. Numbers in parentheses next to a mass-loss rate represent the ratio of that particular rate to the "General" mass-loss with the same rotation fraction.

	No B	$\omega = 0.0$	0.2	0.35	0.5	0.65	0.8
Optically-Thick	3.34	1.50	1.50	1.50	1.49	1.46	1.39
General	3.26	1.46	1.46	1.46	1.45	1.42	1.35
Open	...	0.40(0.27)	0.36(0.25)	0.37(0.25)	0.34(0.24)	0.34(0.24)	0.32(0.24)
Disk	...	...	0.25(0.17)	0.51 (0.35)	0.76 (0.52)	0.94 (0.66)	1.03 (0.76)
Effective	...	0.40(0.27)	0.62(0.42)	0.88(0.60)	1.10(0.76)	1.28(0.9)	1.35 (0.999)

and the other represents an O-star with a dynamical magnetosphere. The results are presented in Tables 2 and 3. We find that the overall effect of a dipole magnetic field is to reduce the "effective" mass-loss rate by roughly a factor of 2 (at $\omega = 0.8$) to 5 (at $\omega = 0.2$) compared to the CAK-estimated mass-loss rates for non-magnetic, non-rotating stars. This result implies that magnetic massive stars will both evolve differently and have a disparate impact on their interstellar environments compared their non-magnetic counterparts. Models of magnetospheric emission will also be affected, though the results here are more applicable to centrifugal magnetospheres with stronger fields than to dynamical magnetospheres with weaker fields.

References

Bard, C. 2016, PhD. Thesis, University of Wisconsin-Madison
Bard, C. & Townsend, R. 2016, *MNRAS*, 462, 3672
&Castor, J. I. and Abbott, D. C. and Klein, R. I. 1975, *ApJ*, 195, 157
&Friend, D. B. and Abbott, D. C. 1986, *ApJ*, 311, 701
&Morel, T. and Castro, N. and Fossati, L., *et al.* 2015, *IAU Symposium*, 307, 342
&Pauldrach, A. and Puls, J. and Kudritzki, R. P. 1986, *A&A*, 164, 86
&Owocki, S. P. and Castor, J. I. and Rybicki, G. B. 1988, *ApJ*, 335, 914
&ud-Doula, A. and Owocki, S. P. 2002, *ApJ*, 576, 413
&Wade, G. A. and Grunhut, J. and Alecian, E., *et al.* 2014, *IAU Symposium*, 302, 265

The Lives and Death-Throes of Massive Stars
Proceedings IAU Symposium No. 329, 2016
J.J. Eldridge, J.C. Bray, L.A.S. McClelland & L. Xiao, eds.

doi:10.1017/S174392131700309X

New Insights into the Puzzling P-Cygni Profiles of Magnetic Massive Stars

Christiana Erba[1], Alexandre David-Uraz[1,2], Véronique Petit[1,2] and Stanley P. Owocki[1]

[1]Deptartment of Physics and Astronomy, Bartol Research Institute, University of Delaware, Newark, DE, 19716, USA

[2]Department of Physics and Space Sciences, Florida Institute of Technology, Melbourne, FL 32904, USA

Abstract. Magnetic massive stars comprise approximately 10% of the total OB star population. Modern spectropolarimetry shows these stars host strong, stable, large-scale, often nearly dipolar surface magnetic fields of 1 kG or more. These global magnetic fields trap and deflect outflowing stellar wind material, forming an anisotropic magnetosphere that can be probed with wind-sensitive UV resonance lines. Recent HST UV spectra of NGC 1624-2, the most magnetic O star observed to date, show atypically unsaturated P-Cygni profiles in the C IV resonant doublet, as well as a distinct variation with rotational phase. We examine the effect of non-radial, magnetically-channeled wind outflow on P-Cygni line formation, using a Sobolev Exact Integration (SEI) approach for direct comparison with HST UV spectra of NGC 1624-2. We demonstrate that the addition of a magnetic field desaturates the absorption trough of the P-Cygni profiles, but further efforts are needed to fully account for the observed line profile variation. Our study thus provides a first step toward a broader understanding of how strong magnetic fields affect mass loss diagnostics from UV lines.

Keywords. stars: magnetic fields, stars: mass loss, ultraviolet: stars

1. Introduction

Hot, luminous stars undergo mass loss through a steady outflow of supersonic material from the stellar surface. These radiatively driven *stellar winds* are best characterized through modeling UV resonance lines, which are sensitive to wind properties.

Recent spectropolarimetric measurements have revealed approximately 10% of massive stars host strong, stable, nearly dipolar magnetic fields with surface field strength of approximately 1kG or more (Wade *et al.* 2016, Fossati *et al.* 2015). The magnetic field channels the flow of the stellar wind along its field lines, confining the wind within closed loops and reducing the overall mass loss (see Fig. 1). This results in a *dynamic* and *structurally complex* so-called magnetosphere (ud-Doula & Owocki 2002) with observational diagnostics in the optical, UV, and X-ray regimes (Petit *et al.* 2013, David-Uraz *et al.*, these proceedings).

Fig. 2 shows HST UV spectra of NGC 1624-2, the most magnetic O star observed to date, compared with HD 93146 and HD 36861 (non-magnetic O stars of similar spectral type). In stark contrast to the strong line saturation observed in non-magnetic O stars, NGC 1624-2 and other magnetic O-type stars such as HD 57628, HD 191612, CPD -28 2561, show *atypically unsaturated* P-Cygni profiles in the C IV resonant doublet, as well as a distinct dependence on rotational phase (Grunhut *et al.* 2009, Wade *et al.* 2011a, Wade *et al.* 2012, Marcolino *et al.* 2013, Nazé *et al.* 2015, Marcolino *et al.* 2012, Petit *et al.* 2013). Therefore, the development of specialized analytical tools is necessary to interpret these profiles and derive the associated wind properties.

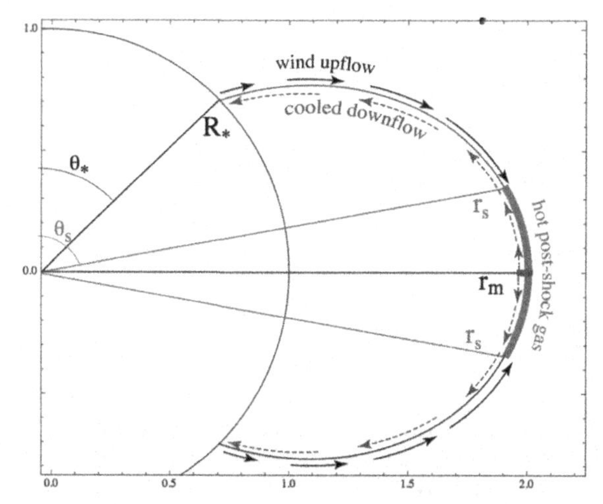

Figure 1. Cartoon depiction of the Analytic Dynamical Magnetosphere (ADM) model, with a magnetic loop intersecting the stellar surface at colatitude θ_* and reaching a maximum radius r_m at the magnetic equator. Magnetically-confined wind material from both stellar hemispheres flows up from the stellar surface (black arrows), and shocks at the magnetic equator. The shock front retreats to shock radius r_s, cooling the hot gas through the release of X-rays. The cooled post-shock material then flows back along the field lines toward the stellar surface (blue arrows). Reproduced from Owocki *et al.* (2016).

2. The Magnetically Confined Wind

In general, the dipolar axis of magnetic OB stars is not aligned with the rotational axis. Therefore, the resulting magnetospheres produce rotationally modulated variability in wind-sensitive UV line profiles and significantly reduced mass-loss rates (Wade *et al.* 2011b, Sundqvist *et al.* 2012, Petit *et al.* 2013). P-Cygni line profiles can provide a powerful tool to probe the structure of these magnetospheres. A detailed investigation of the density and velocity structure of magnetically channeled winds has been carried out using magnetohydrodynamic (MHD) simulations (ud-Doula & Owocki 2002, ud-Doula *et al.* 2008, ud-Doula *et al.* 2009). Although this work has provided unprecedented insight, these simulations remain computationally cumbersome and expensive.

The recently published Analytic Dynamical Magnetosphere (ADM) model (Owocki *et al.* 2016) provides a simplified parametric prescription corresponding to a time-averaged picture of the normally complex magnetospheric structure previously derived through full MHD simulations. As shown in Fig. 1, outflowing material leaves the stellar surface and is channeled along the field lines. At the magnetic equator, the outflowing material ("upflow") from each hemisphere collides, producing a shock. Cooling is achieved through X-ray production, allowing the wind material to flow back toward the stellar surface ("downflow"). This model shows remarkable agreement with Hα and X-ray observations. However, as yet the ADM model has not been used to explain UV line variability, particularly with respect to the observed unsaturated absorption troughs of C IV lines. This provides the unique opportunity to conduct a thorough parameter study capable of accurately reproducing observed trends without the usual computational complexities associated with multi-dimensional MHD modeling.

3. Initial Results and Future Applications

We examine the effect of non-radial, magnetically channeled wind outflows on UV line formation through the development of synthetic UV wind-line profiles which implement

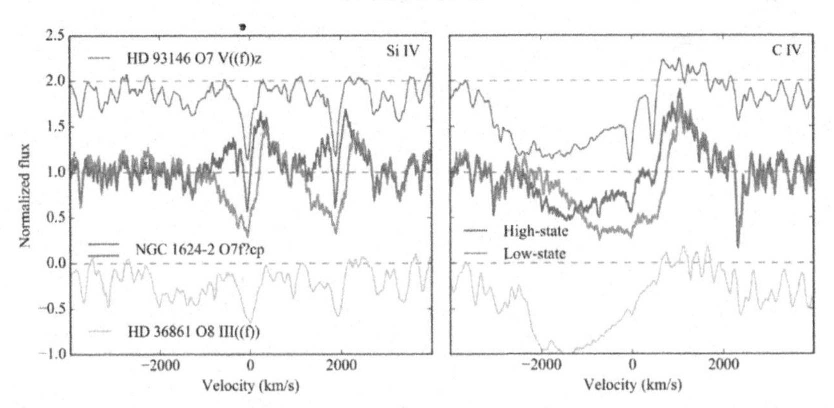

Figure 2. HST/STIS data showing the comparison of the SiIV and CIV UV resonance lines of the magnetic O-star (NGC 1624-2, blue/red, middle) with non-magnetic O-stars of similar spectral type (HD 93146, green, top, and HD 36861, yellow, bottom). The absorption troughs of the CIV lines are a particularly illustrative example of the difference between non-magnetic O-stars (where the absorption trough is fully saturated) and magnetic O-stars (where the absorption trough is clearly unsaturated). Additionally, the magnetic pole-on view (blue) and equator-on view (red) of NCG 1624-2 are shown, demonstrating the modulation present in P-Cygni profiles of the same line at different phases.

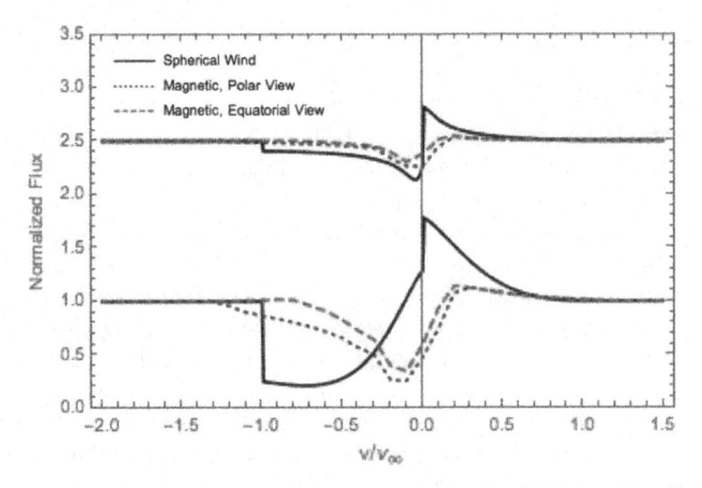

Figure 3. Three synthetic wind-line profiles at line strengths similar to the CIV and SiIV UV resonance lines. The black curve shows the P-Cygni profile of a non-magnetic stellar wind with spherical symmetry; in contrast, the blue and red curves depict the P-Cygni profiles of a magnetically-confined wind viewed respectively from the magnetic pole-on and equator-on angles. These initial results suggest the addition of a dipole field alone is enough to produce unsaturated line profiles, although it does not reproduce the rotational modulation visible in HST/STIS observations.

the ADM formalism developed in Owocki *et al.* (2016). We solve the equation of radiative transfer using a Sobolev Exact Integration (SEI) method (both in the optically thin and optically thick approximations, following Owocki & Rybicki (1985)) to produce synthetic P-Cygni profiles for comparison with observational data. For simplicity, these preliminary models only implement the wind upflow component. We also limit these initial investigations to the magnetic “pole-on” and “equator-on” viewing angles.

Initial results (see Fig. 3) show that significant desaturation of the UV wind-line absorption troughs, both in the magnetic “pole-on” and “equator-on” views, occurs naturally from the addition of a dipolar magnetic field at the stellar surface. This is in agreement with earlier studies, which used MHD simulations to synthesize UV line profiles (Marcolino *et al.* 2013) as well as observational data (see Fig. 2). This suggests that the desaturation is a direct result of the presence of a magnetic field. However, comparison with observational data also reveals a failure to reproduce the exact phase dependence of the lines, indicating the need for further investigation and additional modeling.

In comparison to other computationally expensive MHD modeling techniques, the ADM formalism allows us to provide meaningful results that can be applied across a broader spectrum of magnetic massive stars. Future models will include both the wind upflow and downflow regions, as well as the addition of a "shock retreat" boundary to account for X-ray production in the wind. Further consideration of the effects of an approximated source function (as opposed to one that is solved self-consistently) also need to be explored.

The detection and characterization of magnetic fields in massive stars currently relies heavily on spectropolarimetry, a powerful but expensive technique. Given these limitations, the future of the field of massive star magnetism might rely on indirect detection methods. Therefore, the development of robust UV diagnostics will be a critical step forward in understanding the circumstellar environment of massive stars, as well as the possible presence of an underlying field.

Bibliography

Bard, C. & Townsend, R. H. D. 2016, *MNRAS*, 462.4, 3672

Fossati, L., Castro, N., Schöller, M., Hubrig, S., Langer, N., Morel, T., Briquet, M., Herrero, A., Przybilla, N., Sana, H., *et al.* 2015, *A&A*, 582, A45

Grunhut, J. H., Wade, G. A., Marcolino, W. L. F., Petit, V., Henrichs, H. F., Cohen, D. H., Alecian, E., Bohlender, D., Bouret, J. C., Kochukhov, O., *et al.* 2009, *MNRAS*, 400.1, L94

Marcolino, W. L. F., Bouret, J. C., Walborn, N. R., Howarth, I. D., Nazé, Y., Fullerton, A. W., Wade, G. A., Hillier, D. J., & Herrero, A. 2015, *MNRAS*, 422.3, 2314

Marcolino, W. L. F., Bouret, J. C., Sundqvist, J. O., Walborn, N. R., Fullerton, A. W., Howarth, I. D., Wade, G. A., & ud-Doula, A. 2013, *MNRAS*, 431.3, 2253

Nazé, Y., Sundqvist, J. O., Fullerton, A. W., ud-Doula, A., Wade, G. A., Rauw, G., & Walborn, N. R. 2015, *MNRAS*, 452.3, 2641

Owocki, S. P., ud-Doula, A., Sundqvist, J. O., Petit, V., Cohen, D. H., & Townsend, R. H. D. 2016, *MNRAS*, 462.4, 3830

Owocki, S. P. & Rybicki, G. B. 1985, *ApJ*, 299, 265

Petit, V., Owocki, S. P., Wade, G. A., Cohen, D. H., Sundqvist, J. O., Gagné, M., Apellániz, J. M., Oksala, M. E., Bohlender, D. A., Rivinius, T., *et al.* 2013, *MNRAS*, 429.1, 398

Sundqvist, J. O., ud-Doula, A., Owocki, S. P., Townsend, R. H. D., Howarth, I. D., & Wade, G. A. 2012, *MNRAS*, 423.1, L21

Townsend, R. H. D. & Owocki, S. P. 2005, *MNRAS*, 357.1, 251

ud-Doula, A., Owocki, S. P., & Townsend, R. H. D. 2008, *MNRAS*, 385.1, 97

ud-Doula, A., Owocki, S. P., & Townsend, R. H. D. 2009, *MNRAS*, 392.3, 1022

ud-Doula, A. & Owocki, S. P. 2002, *ApJ*, 576.1, 413

Wade, G. A., Grunhut, J., Gräfener, G., Howarth, I. D., Martins, F., Petit, V., Vink, J. S., Bagnulo, S., Folsom, C. P., Nazé, Y., *et al.* 2011, *MNRAS*, 419.3, 2459

Wade, G. A., Howarth, I. D., Townsend, R. H. D., Grunhut, J. H., Shultz, M., Bouret, J. C., Fullerton, A., Marcolino, W., Martins, F., Nazé, Y., *et al.* 2011, *MNRAS*, 416.4, 3160

Wade, G. A., Apellániz, J. M., Martins, F., Petit, V., Grunhut, J., Walborn, N. R., Barbá, R. H., Gagné, M., García-Melendo, E., Jose, J., *et al.* 2012, *MNRAS*, 425.2, 1278

Wade, G. A., Neiner, C., Alecian, E., Grunhut, J. H., Petit, V., de Batz, B., Bohlender, D. A., Cohen, D. H., Henrichs, H. F., Kochukhov, O., *et al.* 2016, *MNRAS*, 456.1, 2

The Lives and Death-Throes of Massive Stars
Proceedings IAU Symposium No. 329, 2016
J.J. Eldridge, J.C. Bray, L.A.S. McClelland & L. Xiao, eds.

doi:10.1017/S1743921317002745

The evolution of magnetic hot massive stars: Implementation of the quantitative influence of surface magnetic fields in modern models of stellar evolution

Zsolt Keszthelyi[1,2], **Gregg A. Wade**[1] **and Véronique Petit**[3,4]

[1]Department of Physics, Royal Military College of Canada,
PO Box 17000 Station Forces, Kingston, ON, K7K 0C6, Canada
email: zsolt.keszthelyi@rmc.ca

[2] Department of Physics, Engineering Physics and Astronomy, Queen's University,
99 University Avenue, Kingston, ON, K7L 3N6, Canada

[3] Department of Physics and Space Sciences, Florida Institute of Technology,
150 W. University Blvd, Melbourne, FL, 32904, USA

[4] Department of Physics and Astronomy, University of Delaware,
Newark, DE, 19711, USA

Abstract. Large-scale dipolar surface magnetic fields have been detected in a fraction of OB stars, however only few stellar evolution models of massive stars have considered the impact of these fossil fields. We are performing 1D hydrodynamical model calculations taking into account evolutionary consequences of the magnetospheric-wind interactions in a simplified parametric way. Two effects are considered: i) the global mass-loss rates are reduced due to mass-loss quenching, and ii) the surface angular momentum loss is enhanced due to magnetic braking. As a result of the magnetic mass-loss quenching, the mass of magnetic massive stars remains close to their initial masses. Thus magnetic massive stars - even at Galactic metallicity - have the potential to be progenitors of 'heavy' stellar mass black holes. Similarly, at Galactic metallicity, the formation of pair instability supernovae is plausible with a magnetic progenitor.

Keywords. stars: magnetic fields, stars: mass loss, stars: evolution

1. Introduction

Surface magnetic fields are detected in about 10% of hot stars (Wade *et al.* 2016), and are understood to be of fossil origin, likely remaining from the star formation history or the pre-main sequence evolution of the star (Donati & Landstreet 2009; Braithwaite & Spruit 2015). These surface fields are known to form a magnetosphere due to the interaction with line-driven winds of hot stars (Babel & Montmerle 1997). This interaction has been extensively studied in the literature by means of magnetohydrodynamic simulations (ud-Doula & Owocki 2002; ud-Doula *et al.* 2008; ud Doula *et al.* 2009) and analytical studies (Owocki & ud-Doula 2004; Bard & Townsend 2016; Owocki *et al.* 2016).

Two major effects occur that influence both the mass loss and angular momentum loss from the star on a dynamical time scale. i) The magnetosphere channels the wind material along magnetic field lines leading to an infall of mass (ud-Doula & Owocki 2002). The confined plasma is thus deposited back to the stellar surface, and hence the effective mass-loss rates are reduced. ii) The magnetosphere reduces the surface angular momentum budget via Maxwell stresses, which leads to an efficient slow down of the surface rotation (magnetic braking). We incorporate and test these 'surface' effects caused by fossil magnetic fields in the one dimensional hydrodynamical stellar evolution code Module for Experiments in Stellar Astrophysics, MESA (Paxton *et al.* 2011, 2013, 2015).

2. Methods

2.1. *Mass-loss quenching*

When surface magnetic fields are coupled to the line-driven stellar winds of hot massive stars, then these magnetic fields are capable of channeling the mass along the field lines ('flux tubes'). To quantify this interaction, ud-Doula & Owocki (2002) introduced the magnetic confinement parameter, $\eta_\star$ (cf. their Equation 20), which describes the magnetic field energy compared to the stellar wind kinetic energy at the stellar surface,

$$\eta_\star = \frac{B_{\rm eq}^2 R_\star^2}{\dot{M}_{B=0} v_\infty}, \tag{2.1}$$

where $B_{\rm eq}$ is the surface equatorial magnetic field strength, $R_\star$ is the stellar radius, $\dot{M}_{B=0}$ is the mass-loss rate the star would have in absence of the magnetic field, and v_∞ is the wind terminal velocity. For convenience, we use the polar surface field strength $B_{\rm p}$, considering that $B_{\rm p} = 2\,B_{\rm eq}$ for a dipolar field. The mass-loss quenching effect due to surface magnetic fields occurs when the stellar wind, confined to flow along closed magnetic loops, ultimately shocks, stalls, and returns to the stellar surface. This effectively reduces the mass-loss rates. According to ud-Doula & Owocki (2002), the equatorial radius of the farthest closed magnetic loop, that is the closure radius, R_c, in a magnetized wind with a dipolar geometry at the stellar surface is of the order of the Alfvén radius R_A,

$$R_c \sim R_\star + 0.7(R_A - R_\star)\,. \tag{2.2}$$

The location of the Alfvén radius corresponds to the point in the magnetic equatorial plane where the magnetic field energy density equals the wind kinetic energy, that is $\frac{R_A}{R_\star} \approx 0.3 + (\eta_\star + 0.25)^{1/4}$. With the obtained polar field strength $B_{\rm p}(t)$, the non-magnetic mass-loss rate, terminal velocity, and the stellar radius, we calculate the Alfvén radius. It is straightforward then to obtain a parameter describing the escaping wind fraction f_B.

$$f_B = \frac{\dot{M}_{\rm eff}}{\dot{M}_{B=0}} = 1 - \sqrt{1 - \frac{R_\star}{R_c}}. \tag{2.3}$$

where $\dot{M}_{\rm eff}$ is the effective mass loss allowed by a magnetosphere. This parametric description accounts for the fraction of the stellar surface covered by open magnetic field loops. Along these loops, wind material can escape effectively, while along closed loops, material will fall back to the stellar surface.

We imposed magnetic flux conservation on the evolving models,

$$F \sim 4\pi R_\star^2 B_p = \text{constant}. \tag{2.4}$$

As a consequence, as the star evolves and the stellar radius $R_\star$ increases, the surface magnetic field strength changes according to:

$$B_p(t) = B_{p,0}\left(\frac{R_{\star,0}}{R_\star(t)}\right)^2, \tag{2.5}$$

where $B_{p,0}$ and $R_{\star,0}$ correspond to the polar field and stellar radius defined at the zero age main sequence. The final adopted mass-loss rate is obtained by scaling the current time step mass-loss rates with the escaping wind fraction f_B allowing for mass to escape only via open loops, such that:

$$\dot{M}_{\rm eff} = f_B \, \dot{M}_{B=0}, \tag{2.6}$$

where $\dot{M}_{B=0}$ refers to the non-magnetic mass-loss rates according to any applicable scheme. In our main sequence models above 12.5 kK, we adopt the Vink rates (Vink *et al.* 2000, 2001). However, uncertainties related to this scheme (or any other scheme) will have an impact on the quantitative results since the magnetic mass-loss quenching is coupled to the wind scheme.

2.2. *Magnetic braking*

We implemented magnetic braking in MESA; this will be discussed in a forthcoming publication (Keszthelyi *et al.*, in prep). We followed the prescription derived by ud Doula *et al.* (2009), and thus adopted an additional source of angular momentum loss,

$$\frac{{\rm d}J}{{\rm d}t} = \frac{2}{3} \dot{M}_{B=0} \, \Omega \, R_\star^2 \, [(\eta_\star + 0.25)^{1/4} + 0.29]^2, \tag{2.7}$$

where J is the total angular momentum, t is the time, and Ω is the surface angular velocity. It is important to note that $\dot{M}_{B=0}$ in this equation strictly means that most of the angular momentum is lost by Maxwell stresses, and not by the effective mass loss. An important issue is that the way this surface angular momentum loss propagates into the interior layers is unknown, as the effect of large-scale fossil magnetic fields on differential rotation is not well constrained. Meynet *et al.* (2011) have already implemented this prescription in the Geneva stellar evolution code, and tested two cases - solid body rotation and differential rotation. Indeed, the major problem arises from the fact that the surface angular momentum loss depends on the internal angular momentum transport mechanisms. While state-of-the-art stellar evolution modelling allows for testing both solid body and differential rotation, observational evidence on the internal rotational properties and thus on the transport mechanisms in hot stars only exists in very few cases (e.g. KIC 10526294 by Moravveji *et al.* 2015, and V2052 Oph by Briquet *et al.* 2012).

3. Results

3.1. *Models including mass-loss quenching*

We find that the evolution of the magnetic confinement parameter depends mostly on how $\dot{M}_{B=0}(T_{\rm eff})$ evolves, and in a general scenario the other parameters ($B_{\rm p}(t), R_\star, v_\infty$, respectively) compensate each other. The impact of mass-loss rates on massive star evolution, and, in particular, the dependence of $\dot{M}_{B=0}$ on $T_{\rm eff}$ has recently been discussed in a more general context by Keszthelyi *et al.* (2017). In Figure 1 the evolution of magnetic confinement parameter (solid line, left ordinate) is contrasted to the evolution of the mass-loss rate (dashed line, right ordinate). Indeed, $\eta_\star$ is most sensitive to changes in $\dot{M}_{B=0}$. In particular, the large jump in mass-loss rate at the bi-stability (shifted down to 20 kK in accordance with the suggestion from Keszthelyi *et al.* 2017 based on findings from Petrov *et al.* 2016) causes a drop in the magnetic confinement. This is powerful enough that the initially strong ($\eta_\star > 10$) and then moderate ($1 < \eta_\star < 10$) magnetic confinement rapidly becomes weak ($\eta_\star < 1$).

3.2. *The evolution of fossil magnetic fields*

A key ingredient to incorporate the effects of fossil magnetic fields in stellar evolution models is currently poorly understood: the evolution of stellar magnetic fields over

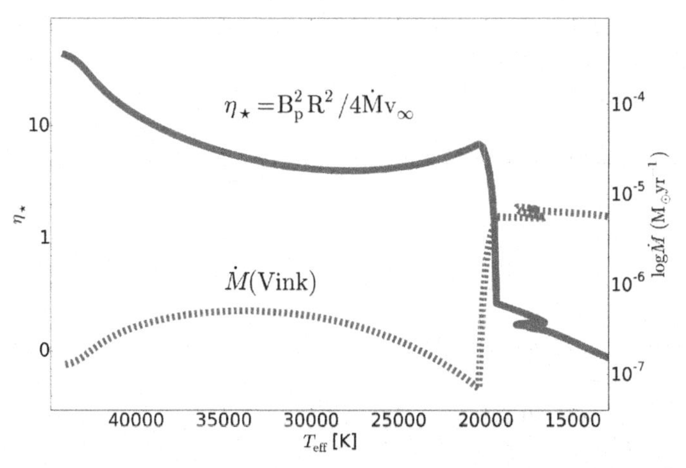

Figure 1. Shown is the evolution of the magnetic confinement parameter (solid line, left vertical axis, defined in Equation 2.1) and the effective mass-loss rates (dashed line, right vertical axis) against the effective temperature for a 40 $M_\odot$ Galactic metallicity non-rotating model.

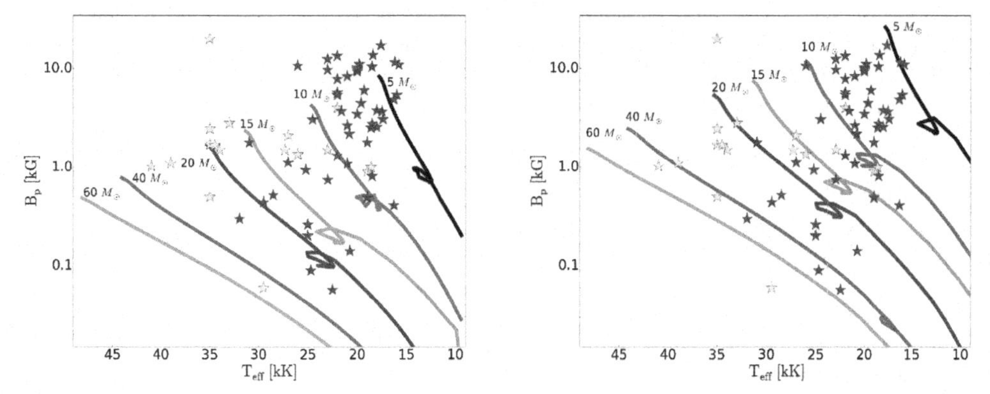

Figure 2. Plotted are the known sample of magnetic OB stars from Petit *et al.* (2013) and Shultz *et al.* (in prep). The former are denoted with yellow stars, the latter with blue stars. Evolutionary models with different initial masses are plotted. *Left*: Models adopting an initial magnetic flux of $F = 10^{27.5}$ G cm^2. This yields an initial dipole field strength in, e.g., the 40 $M_\odot$ model of $B_{\rm p} = 1$ kG. *Right*: Models adopting an initial magnetic flux of $F = 10^{28}$ G cm^2. This yields an initial dipole field strength in, e.g., the 40 $M_\odot$ model of $B_{\rm p} = 2.5$ kG.

evolutionary timescales. Adopting magnetic flux conservation (Equation 2.4) allows for calculating the first grid of models that adopt a field evolution other than a constant magnetic field strength (that is, increasing flux) as in the work by Meynet *et al.* (2011). Two major scenarios are considered. The range of observed magnetic field strengths is very large, over 3 orders of magnitude in stellar remnants (e.g. Ferrario & Wickramasinghe 2006). Interestingly, this range is consistent with the scale of field strengths in main sequence stars, and thus supports magnetic flux conservation. However, in light of theoretical considerations (e.g. Braithwaite 2008; Braithwaite & Spruit 2015) a plausible scenario of magnetic flux decay should be taken into account. Most commonly, Ohmic decay is speculated to lead to magnetic flux decay. Additionally, observations may also argue for such a trend (Landstreet *et al.* 2007; Fossati *et al.* 2016). Figure 2 shows stellar evolution models adopting magnetic flux conservation to describe the field evolution. A sample of magnetic OB stars from Petit *et al.* (2013) and Shultz *et al.* (in prep) are shown and their magnetic fluxes were calculated based on the measured magnetic field strengths and stellar radii. The MESA models are computed for two different initial flux values (left panel: $F = 10^{27.5}$ G cm^2, right panel: $F = 10^{28}$ G cm^2). Since the initial flux is the same for all models, but the stellar radius is different due to the different initial masses, the initial field strength for higher mass (larger radius) models is smaller than that of lower mass (smaller radius) models.

4. Conclusions

For the first time, we incorporated the effects of surface magnetic fields in the MESA code. These effects are based on the dynamical interactions between the magnetosphere and the stellar wind, and we conclude from our model calculations that they have a large impact on massive star evolution. Two evolutionary scenarios are discussed by Petit *et al.* (2017) and Georgy *et al.* (in press).

Acknowledgements

We acknowledge the MESA developers for making their code publicly available. GAW acknowledges Discovery Grant support from the Natural Sciences and Engineering Research Council (NSERC) of Canada. VP acknowledges support provided by the NASA through Chandra Award Number GO3-14017A issued by the Chandra X-ray Observatory Center, which is operated by the Smithsonian Astrophysical Observatory for and on behalf of the NASA under contract NAS8-03060.

References

Babel, J. & Montmerle, T. 1997, *A&A*, 323, 121
Bard, C. & Townsend, R. H. D. 2016, *MNRAS*, 462, 3672
Braithwaite, J. 2008, *MNRAS*, 386, 1947
Braithwaite, J. & Spruit, H. C. 2015, ArXiv 1510.03198
Briquet, M., Neiner, C., Aerts, C., *et al.* 2012, *MNRAS*, 427, 483
Donati, J. & Landstreet, J. D. 2009, *ARAA*, 47, 333
Ferrario, L. & Wickramasinghe, D. 2006, *MNRAS*, 367, 1323
Fossati, L., Schneider, F. R. N., Castro, N., *et al.* 2016, *A&A*, 592, A84
Keszthelyi, Z., Puls, J., & Wade, G. A. 2017, *A&A*, 598, A4
Landstreet, J. D., Bagnulo, S., Andretta, V., *et al.* 2007, *A&A*, 470, 685
Meynet, G., Eggenberger, P., & Maeder, A. 2011, *A&A*, 525, L11
Moravveji, E., Aerts, C., Pápics, P. I., Triana, S. A., & Vandoren, B. 2015, *A&A*, 580, A27
Owocki, S. P. & ud-Doula, A. 2004, *ApJ*, 600, 1004
Owocki, S. P., ud-Doula, A., Sundqvist, J. O., *et al.* 2016, *MNRAS*, 462, 3830
Paxton, B., Bildsten, L., Dotter, A., *et al.* 2011, *ApJS*, 192, 3
Paxton, B., Cantiello, M., Arras, P., *et al.* 2013, *ApJS*, 208, 4
Paxton, B., Marchant, P., Schwab, J., *et al.* 2015, *ApJS*, 220, 15
Petit, V., Keszthelyi, Z., MacInnis, R., *et al.* 2017, *MNRAS*, 466, 1052
Petit, V., Owocki, S. P., Wade, G. A., *et al.* 2013, *MNRAS*, 429, 398
Petrov, B., Vink, J., & Gräfener, G. 2016, MNRAS, 458, 1999
ud-Doula, A. & Owocki, S. P. 2002, *ApJ*, 576, 413
ud-Doula, A., Owocki, S. P., & Townsend, R. H. D. 2008, MNRAS, 385, 97
ud Doula, A., Owocki, S. P., & Townsend, R. H. D. 2009, MNRAS, 392, 1022
Vink, J. S., de Koter, A., & Lamers, H. J. G. L. M. 2000, *A&A*, 362, 295
Vink, J. S., de Koter, A., & Lamers, H. J. G. L. M. 2001, *A&A*, 369, 574
Wade, G. A., Neiner, C., Alecian, E., *et al.* 2016, *MNRAS*, 456, 2

The Lives and Death-Throes of Massive Stars
Proceedings IAU Symposium No. 329, 2016
J.J. Eldridge, J.C. Bray, L.A.S. McClelland & L. Xiao, eds.

doi:10.1017/S1743921317003404

Helium stars: Towards an understanding of Wolf–Rayet evolution

Liam A. S. McClelland and J. J. Eldridge

Department of Physics, University of Auckland
email:lmcc054@aucklanduni.ac.nz
email:j.eldridge@auckland.ac.nz

Abstract. Recent observational modelling of the atmospheres of hydrogen-free Wolf–Rayet stars have indicated that their stellar surfaces are cooler than those predicted by the latest stellar evolution models. We have created a large grid of pure helium star models to investigate the dependence of the surface temperatures on factors such as the rate of mass loss and the amount of clumping in the outer convection zone. Upon comparing our results with Galactic and LMC WR observations, we find that the outer convection zones should be clumped and that the mass-loss rates need to be slightly reduced. We discuss the implications of these findings in terms of the detectability of Type Ibc supernovae progenitors, and in terms of refining the Conti scenario.

Keywords. stars: evolution, stars: Wolf-Rayet, binaries: general

---

1. Introduction

Wolf–Rayet (WR) stars are massive helium-burning stars that, through strong mass loss, have lost all or most of their hydrogen envelopes leaving a partially or fully exposed helium core. We have generated a grid of pure helium star models at various metallicities and shall only study the evolution from onset of core-helium burning onwards.

2. Computational method

Construction of the Models. To investigate the evolution of helium stars, we have constructed a grid of hydrogen-free models. We make use of the Cambridge STARS evolutionary code. Originally developed by Eggleton (1971), it has been modified by various groups; herein, we employ the version described by Stancliffe & Eldridge (2009). We make our selection of metallicities based on the expected environments of WR stars: $Z = 0.008$ for the Large Magellanic Cloud; $Z = 0.014$ and $Z = 0.02$ being, respectively, "new" and "old" solar metallicity; and $Z = 0.04$, double "old" solar. We have utilised the solar abundance determinations from Grevesse & Sauval (1998) ("old" solar) and Asplund *et al.* (2009) ("new" solar). A comparison of evolution between models of "old" and "new" solar compositions shows very little difference, and in light of this, we prefer "old" solar abundances for use in our models.

Mass-loss Scheme. We employ a mass-loss scheme, outlined in Eldridge & Vink (2006), which is derived from Nugis & Lamers (2000):

$$\dot{M}_Z \propto \dot{M}\beta \left(\frac{Z}{Z_\odot}\right)^{\frac{1}{2}}, \qquad (2.1)$$

where $\dot{M}$ is taken from Eldridge & Vink (2006), Z is the metallicity of the model, and $Z_\odot$ is solar metallicity. To test the effect of varying the mass-loss rate on the evolution of a model, we introduce a parameter, β. We may use this parameter to estimate the evolution of the helium star model if, before the hydrogen envelope were removed, more helium burning had occurred. For example in the case of $\beta = 1$, the hydrogen envelope is removed before the beginning of helium burning, so the tracks represent the greatest possible effect of mass loss on the models. Thus, more helium mass would be lost from the WR stars. However for $\beta = 0$, the evolution towards the end of the track represents how the star would appear if the hydrogen envelope were removed near the end of helium burning.

Envelope inflation. Massive stars develop an extended envelope structure due to an increase in opacity caused by the iron-opacity peak. At stellar temperatures near the iron-opacity peak ($\log T_{\rm eff} \sim 5.2$), a small convective layer forms above a near-void region of the star. The overall effect is a radiation-driven expansion of this high opacity outer layer; a reduction of the apparent stellar temperature. The effect is more severe with increasing metallicity due to the increased abundance of iron-group elements.

We allow the material in inflated envelope to be inhomogeneous and define a clumping factor, D, as described by Gräfener, Owocki & Vink (2012). To investigate the effect of density inhomogeneities in the inflated envelopes of helium stars, we select two values for our clumping factor: 1 and 10 referring to "unclumped" and "clumped", respectively.

When referring to a particular set of models, we shall label them by the metallicity, β, and clumping factor like so: (Z, β, D). For example, (0.02, 1, 1) refers to the grid of models with $Z = 0.02$, $\beta = 1$, and a normal envelope without the clumping factor to enhance the opacity.

3. Helium giants & Wolf–Rayet stars

We may divide our helium star models into two categories: low mass ("helium-giant type"), and high mass ("Wolf–Rayet type").

Low-mass helium stars. Low-mass ($< 8\mathrm{M}_\odot$) helium star models evolve as "helium giants". A helium giant has a stellar structure that is analogous to that of a red-giant star: a dense core region with an expansive envelope (see left panel of Fig. 1).

High-mass helium stars. High-mass ($> 8\mathrm{M}_\odot$) helium star models evolve as "traditional" Wolf–Rayet stars, having characteristic high temperatures due to strong mass loss (Crowther 2007). The structure of a high-mass helium star differs from that of a low-mass helium star by the properties of its envelope: an extended region of near-constant

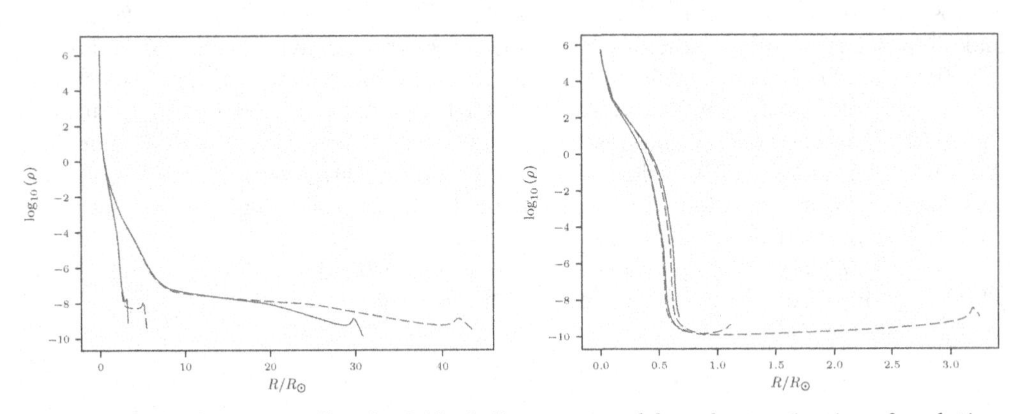

Figure 1. Left: Density profile of a 3 $\mathrm{M}_\odot$ helium star model at the termination of evolution. The blue and red lines denote $Z = 0.008$ and $Z = 0.02$, respectively. Additionally, the solid and broken lines denote a clumping factor of $D = 1$ and $D = 10$, respectively. Right: The same for a model of mass 15 $\mathrm{M}_\odot$. In all cases, $\beta = 1$.

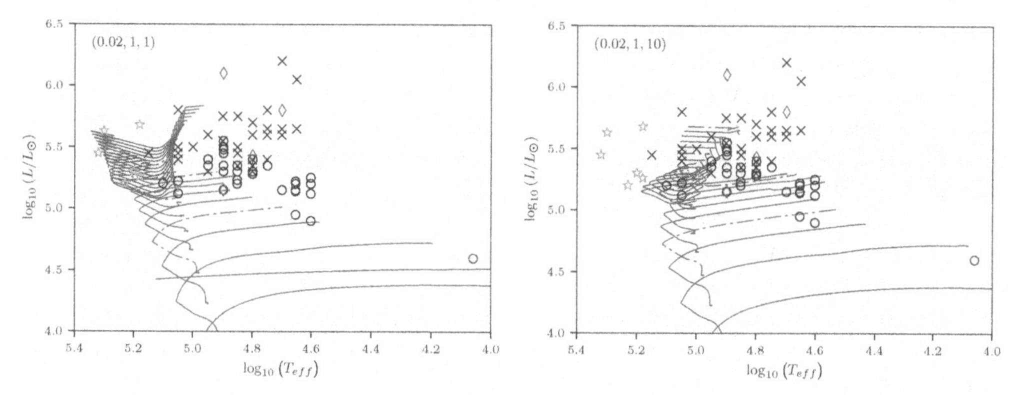

Figure 2. HR diagram of our evolved Galactic-metallicity models. For clarity, models with initial masses of 5, 10, 15 and 20 $M_\odot$ are represented by dot-dashed lines. Observed WR star locations (see text for details) are marked as follows: WN, saltires; WC, circles; WO, yellow stars; and WN/WC transition objects, blue diamonds. Progenitor iPTF13bvn is marked with a purple circle (see Eldridge & Maund 2016). All observed stars are hydrogen-free. The phase of WR mass loss is indicated, WN (solid, green line) and WC (solid, blue line); non-WR mass loss is shown with solid, black lines. Metallicity, mass-loss rate, and clumping factor are noted on the plots.

density with a large density inversion at the surface (see right panel of Fig. 1). The density inversion sits atop the extended envelope structure of the high-mass helium star models, and due to the stellar interior reaching the phase space of the iron-opacity peak Gräfener, Owocki & Vink (2012), is affected by metallicity.

4. Results

We present our results for Galactic-metallicity models in Figure 2. Observational data is taken from Hamann *et al.* (2006) and Sander *et al.* (2012), for Galactic WN and WC stars; Hainich *et al.* (2014), for Large Magellanic Cloud (LMC) WN stars; and Tramper *et al.* (2013) and Tramper *et al.* (2015), for WO stars. Also included are results from Crowther *et al.* (2002) for LMC WC stars. In this work, we attempt to reproduce the observed locations of WR stars.

Galactic WN stars. The observed early-type WN stars lie near the HeZAMS for massive helium stars and are, generally, in good agreement with helium star models of initial masses above $\approx 10\,M_\odot$. However, the agreement is not so favourable for the observed late-type WN stars. These WN stars have stellar temperatures that are far cooler than those at the HeZAMS, and their locations cannot be reproduced by using models with $\beta > 0$.

Without mass loss ($\beta = 0$), we see an interesting result: the higher mass helium star models do, indeed, cross the region of observed (hydrogen-free) late-type WN stars for solar metallicity ("old" and "new"). A small amount of mass loss will remove the outer layers of the envelope and expose the hot interior of the model; thus, without mass loss, the model swells due to inflation and the surface temperature decreases. The inclusion of envelope clumping results in cooler stellar temperatures, as expected. However, even with the inclusion of clumping, our helium star models are unable to reproduce the observed locations of the coolest Galactic WN stars if we use the standard WR mass-loss rates (i.e. $\beta = 1$).

To summarise, the large radii of the WN stars is due to either a small amount of mass loss or clumping in the envelope.

WC and WO stars. The expected locations of WC stars (marked in solid blue lines) are in very poor agreement with the positions of observed WC stars. The standard evolutionary picture of WR evolution–suggested by Sander *et al.* (2012) where the WC phase succeeds a WN phase–is clearly insufficient to explain this discrepancy. We note from Fig. 2 that low-mass helium models can reproduce the observed locations of early- and late-type WC stars.

We note the observed WO stars on the HR diagram are hotter and more luminous than the observed WC stars. A higher luminosity implies a higher stellar mass, and we indeed find the observed WO stars in a region predicted by our high-mass models. Though difficult to draw definitive conclusions due to the lack of observational evidence, we argue the standard description of WC stars used in evolutionary models is incorrect and actually applies instead to WO stars.

As can be seen in Fig. 2, the locations of the observed WC stars coincides with the locations of the low-mass helium giants. The surface composition identifies these low-mass helium giants as WN stars, not WC stars. Due to the weak mass-loss rates of the low-mass helium stars, nitrogen remains abundant on the stellar surface rather than being removed, as is the case for the higher mass stars. It is possible that the surface nitrogen may be removed by an alternative mechanism of extra mixing (Frey, Fryer & Young 2013): nitrogen is mixed into the stellar interior and removed via nuclear processing.

We note that, by including clumping in our stellar models, our models evolve closer to locations on the HR diagram where WC stars are observed. However, the coolest WC stars still arise from the same low-mass helium giants. Furthermore, WO stars–which arise from the most massive stars—have much shorter lifetimes than the stars that we suggest are WC stars. This may explain why WO stars make up only a small fraction of the WR population.

In summary, our results suggest that both WC and WO stars are represented by models with $\beta = 1$. However, they differ in terms of mass: WO stars are from the most massive of stars (an initial helium star mass of $\gtrsim 13\,\mathrm{M}_\odot$), while WC stars come from less massive stars (an initial helium star mass of $\lesssim 13\,\mathrm{M}_\odot$). Similar findings were made by Groh *et al.* (2013) and Sander *et al.* (2012). We note that adding clumping improves the agreement but we find that WO stars must be unclumped.

Binary Configurations. Radial expansion by way of a clumped envelope may affect the nature of binary interactions. The evolution of binary systems of helium stars will be analysed in a future paper. Figure 3 shows the initial conditions required for the system to experience a binary interaction—a Roche-lobe overflow. We see that, in general, helium binary systems do not interact, even with the application of enhanced inflation. For all metallicities, a WR binary system requires a period of less than a day to interact. In contrast, helium giants do interact quite readily due to their expansive envelopes.

5. Discussion

In light of our work (see McClelland & Eldridge 2016), we can draw some firm conclusions about certain aspects of WR star evolution and speculate about others. We shall now discuss our conclusive findings.

First, WO stars are what we have always considered to be WC stars in stellar models. They are the progeny of the most massive WN stars ($M_{\mathrm{He,i}} > 8\,\mathrm{M}_\odot$) that have suffered significant mass loss and are the hottest WR stars.

Second, WC stars evolve from less massive stars ($M_{\mathrm{He,i}} < 8\,\mathrm{M}_\odot$) and are unlikely to be the evolutionary end-points of the typical observed WN star. The evolution of these

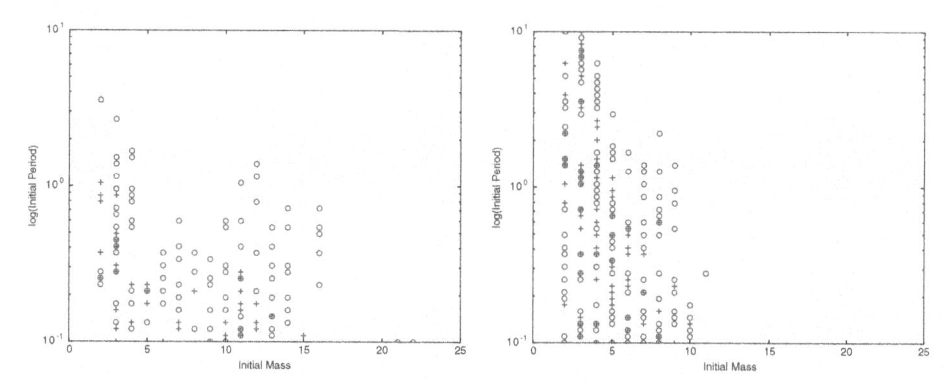

Figure 3. Initial period and initial mass of models that undergo a binary interaction. *Left to right*: $Z = 0.008$ and 0.02 respectively. Red circles indicate a clumped envelope; blue crosses indicate an unclumped envelope.

stars could be described either as an inflationary effect occurring towards the end of their lives or as them becoming helium giants.

If the stellar envelope is clumped, it can only occur for WN stars. The only way to reproduce the locations of the late-WN stars on the HR diagram with standard-evolution models is to remove all mass loss, as even slight mass loss will drive a model to higher stellar temperatures. WC and WO stars do not require enhanced inflation to reproduce their HR diagram positions. The evolutionary models predict the locations of WO stars without a problem: a hot phase consequential to high-mass WN stars. The predictions for WC stars are correctly realised for the cooler temperatures of low-mass WC stars.

References

Asplund M., Grevesse N., Sauval A. J., Scott P. 2009, *ARAA* , 47, 481
Crowther P. A., *et al.*, 2002, *A&A*, 392, 653
Crowther P. A., 2007, *ARAA*, 45, 177
Eggleton P. P. 1971, *MNRAS*, 151, 351
Eldridge J. J., Vink J. S. 2006, *A&A* , 452, 295
Eldridge, J. J. & Maund, J. R. 2016, *MNRAS*, 461, L117
Frey L. H., Fryer C. L., Young P. A., 2013, *ApJ*(Letters), 773, L7
Gräfener G., Owocki S. P., Vink J. S., 2012, *A&A*, 538, A40
Grevesse N., Sauval A. J. 1998, *Space Sci. Revs*, 85, 161
Groh J. H., Meynet G., Georgy C., Ekström S., 2013, *A&A*, 558, A131
Hainich R. *et al.*, 2014, *A&A*, 565, A27
Hamann W.-R., Gräfener G., Liermann A. 2006, *A&A*, 457, 1015
McClelland, L. A. S. & Eldridge, J. J. 2016, *MNRAS*, 459, 1505
Nugis T., & Lamers H. J. G. L. M. 2000,*A&A* , 360, 227
Sander A., Hamann W.-R., Todt H., 2012, *A&A*, 540, A144
Stancliffe R. J., Eldridge J. J. 2009, *MNRAS*, 396, 1699
Tramper F. *et al.*, 2013, *A&A*, 559, A72
Tramper F., *et al.*, 2015, in IAU Symposium, Vol. 307, IAU Symposium, pp. 144–145

Session 5: Massive stars and their supernovae as galactic building blocks and engines: Milky Way, nearby galaxies and the early Universe

The Lives and Death-Throes of Massive Stars
Proceedings IAU Symposium No. 329, 2016
J.J. Eldridge, J.C. Bray, L.A.S. McClelland
& L. Xiao, eds.

doi:10.1017/S1743921317002988

Massive infrared clusters in the Milky Way

André-Nicolas Chené[1], Sebastian Ramírez Alegría[2,3], Jordanka Borissova[3,4], Anthony Hervé[5], Fabrice Martins[6], Michael Kuhn[3,4] Dante Minniti[7,8] and the VVV Science Team

[1]Gemini Observatory, Northern Operations Center, 670 A´ohoku Place, Hilo, HI 96720, USA
email: andrenicolas.chene@gmail.com

[2]Instituto de Astronoma, Universidad Catlica del Norte, Antofagasta, Chile [3]Millennium Institute of Astrophysics, MAS, Chile

[4]Instituto de Fisica y Astronomia, Facultad de Ciencias, Universidad de Valparaiso, Av. Gran Bretana 1111, Playa Ancha, Casilla 5030, Valparaiso, Chile

[5]Astronomical Institute of the ASCR, Fričova 298, 251 65 Ondřejov, Czech Republic

[6]LUPM-UMR 5299, CNRS & Universite Montpellier, Place Eugene Bataillon, 34095 Montpellier Cedex 05, France

[7]Vatican Observatory, V00120 Vatican City State

[8]Departamento de Ciencias Fisicas, Universidad Andres Bello, Republica 220, Santiago, Chile

Abstract. Our position in the Milky Way (MW) is both a blessing and a curse. We are nearby to many star clusters, but the dust that is a product of their very existence obscures them. Also, many massive young clusters are expected to be located near, or across the Galactic Center, where the dust extinction is extreme ($A_V > 15$ mag) and can be better penetrated by infrared photons. This paper reviews the discoveries and the study of new MW massive stars and massive clusters made possible by near infrared observations that are part of the VISTA Variables in the Vía Láctea (VVV) survey. It discusses what the studies of their fundamental parameters have taught us.

Keywords. surveys, stars: distances, stars: evolution, stars: fundamental parameters, stars: Wolf-Rayet, (Galaxy:) open clusters and associations: general

---

1. Introduction

Our location within our own Galaxy gives us a unique perspective from which we can study star clusters in great detail. It is a great opportunity for making significant progress in our knowledge of stellar formation and stellar evolution. Indeed, it is commonly accepted that the majority of stars with masses $\geqslant 0.50\ M_\odot$ form in clustered environments (e.g. Lada & Lada 2003 ; de Wit *et al.* 2005), rather than individually. Consequently, the study of clusters with massive stars is the study of the birth environment of these stars. Also, young clusters can be seen as stellar evolution snap-shots (e.g. Martins *et al.* 2008, Crowther *et al.* 2010, Davies *et al.* 2012, and Liermann *et al.* 2012). When observed in their first few million years of age, the clusters' stellar population can be used to trace the evolution scenario of the most massive stars. The main challenge for reaching a significant statistical sample of clusters is the extinction caused by dust that can be found along the line-of-sight, and/or near the clusters that produced it. This is why new near infrared (NIR) images provided by the VISTA Variable in Vía Láctea (VVV) survey are key for discovering and studying new clusters that are used to reveal the mechanisms involved in the birth and the faith of massive stars.

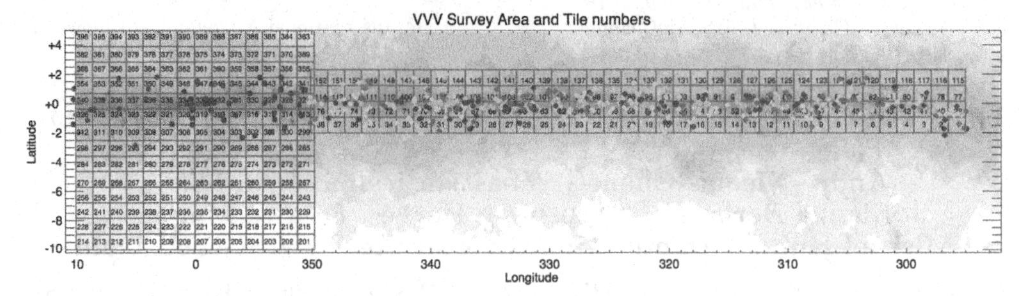

Figure 1. Area covered by the VVV survey. The VISTA tile are 1.5 deg^2 wide and each have a unique ID number. The dots mark the cluster candidates and clusters discovered by (blue) Borissova *et al.* (2011), (2014a), (red) Solin *et al.* (2014) and (yellow) Barbá *et al.* (2015).

2. VISTA Variables in Vía Láctea (VVV) survey

VVV was a key ESO survey that targeted the enigmatic central region of the Galaxy (Minniti *et al.* 2010, Saito *et al.* 2012, Hempel *et al.* 2014), and was carried out from Paranal using the 4-m Visible and Infrared Survey Telescope for Astronomy (VISTA). This successful survey recently completed after having undertaken $\sim$ 2000 hours of exposure through $ZYJHK_S$ filters. Those observations revealed an unprecedented $\sim 10^9$ point sources across 562 deg^2 of sky that spanned the Galaxy's bulge and an adjacent region of the Galactic disk. Our team has constructed color-magnitude diagrams (CMDs) of nearly 1500 bona fide and candidate clusters located in the VVV area, including an impressive new 735 candidates uncovered in the survey that are not included in the Morales *et al.* (2013) catalogue (Borissova *et al.* 2011, 2014a, Solin *et al.* 2014 and Barbá *et al.* 2015, see Fig. 1).

For more information about the survey, please visit `https://vvvsurvey.org/`. For information on how to get your own set of public data, click the link "Data Releases" at the top menu. For J, H, and K_S combined photometric catalogs, one can also search the *Nuevo Observatorio Virtual Argentino* (NOVA) at `http://nova.conicet.gov.ar`.

3. Census of the clusters with massive stars in the MW

Fig. 2 shows the distribution of young (< 10 Myrs) clusters and cluster candidates in the MW. The red and blue points are from the Karchenko *et al.* (2013) catalog, which derived fundamental parameters solely based on photometric analysis. Therefore, the uncertainty on these points is unknown and highly variable from point to point. Nevertheless, when used as a bulk, it gives a fair representation of the distribution of the young clusters that are known to date. The red points mark the clusters that were discovered using infrared light. One can appreciate that infrared allows to reach further than the optical light (the blue points). The green points mark the position of the young and massive clusters as derived from the Table 4 of Davies *et al.* (2012). The names of some of the most famous ones are written on the figure. The yellow points mark the clusters with massive stars discovered in the VVV survey. The names of the clusters that are described in more details later in these proceedings are written.

The following is a brief description of some of the most interesting recent findings.

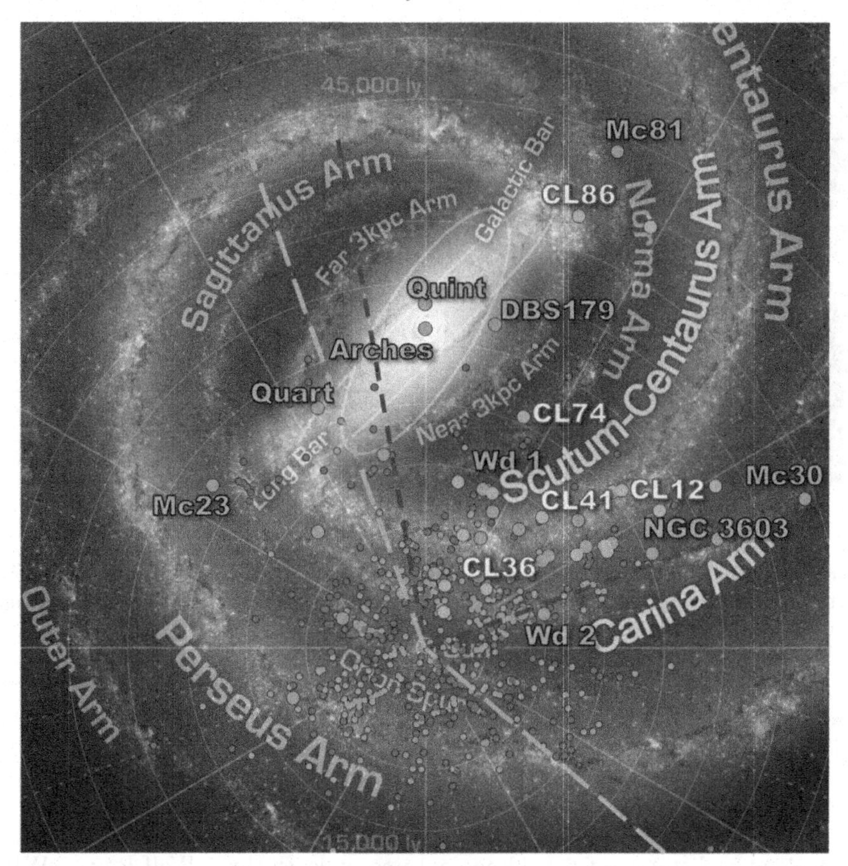

Figure 2. Artist illustration produced by NASA/JPL-CALTECH of the MW seen face-on. The Galactic Bar and different arms are identified. The small blue and red points are young (age < 10 Myrs) clusters and clusters candidates that were discovered in the optical and in the NIR light, respectively. There distances were determined by Karchenko *et al.* (2013). The big green points mark the census of known massive clusters. The big yellow points mark the clusters with massive stars that were discovered using the VVV survey images. The red dashed line indicates the area covered by the VVV survey. The green dashed lines show the wider area that will be covered by the newly allocated survey VVVX (`https://vvvsurvey.org/`).

4. The VVV clusters

4.1. *VVV CL086, an OB star cluster at the fat end of the Galactic Bar*

The position of VVV CL086 places this clusters containing numerous OB stars at the far end of the Galactic Bar (Ramírez Alegría *et al.* 2014). This cluster is the second one found in that region, after Mercer 81 (Davies *et al.* 2012).

4.2. *VVV CL012, a cluster with massive young stellar objects*

Using the host clusters of massive young stellar objects (YSO), we can determine their age, distance and reddening. Also, since VVV provides us with multi-epoch of K_S observations of the whole field over many years, we can extract the light curve of the massive YSOs (see Fig. 3). As demonstrated by Caratti o Garatti *et al.* (2016; and references therein), massive YSO show the same type of outburst as the one seen in YSO of other masses. The monitoring of the changes in brightness of YSO helps defining what is the priviledged mechanism for massive stars formation. Fig. 3 shows an example from

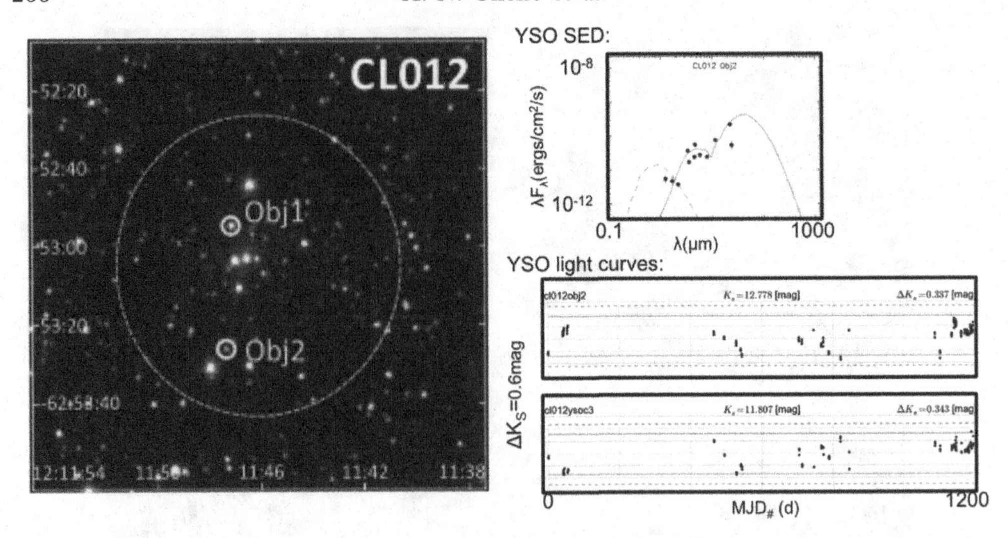

Figure 3. *Left*: False color JHK_S image from VVV. The stars marked in yellow were observed in spectroscopy. *Right*: The top shows the SED obtained from VVV, Spitzer and Apex observations. The bottom panels show the K_S light curves extracted from VVV for the two stars marked in yellow in the image at the left.

Borissova *et al.* (2016) with VVV CL012, where two massive YSOs are monitored. Also, an SED fit is made possible by combining VVV, Spitzer and APEX data.

4.3. *VVV CL074, having access to the whole massive star population*

VVV CL074, at a distance of 6 ± 1 kpc contains at least three Wolf-Rayet (WR) stars and a handful of OB stars (Chené *et al.* 2013). In addition to the VVV photometry, we collected the spectra of 11 stars. All the observed O stars are supergiants. We used the radiative transfer code CMFGEN (Hillier & Miller 1998) to determine their stellar and wind parameters. It allowed us to place the best observed stars into the HR diagram shown in Fig. 4. This first result gives an approximate age of 7 Myrs and could indicate that $\sim$40 $M_\odot$ O supergiant stars are linked to the evolved WN8 stars (Hervé *et al.* 2015). This result requires the spectra of more stars for final confirmation. Very interestingly, the O dwarfs are about 2 magnitudes fainter, which means that in a reasonable amount of time on an 8m telescope, we can characterize by using NIR spectroscopy the complete massive stellar population of the cluster. We currently have secured observing time on the VLT to achieve this in the 17A semester.

4.4. *VVV CL036, highlighting binary evolution*

VVV CL036 has a fairly peculiar massive star population (see Fig. 5). In the same cluster, we find a WN, a blue supergiant (BSG) and a red supergiant (RSG) star (Chené *et al.* 2013). These three stars cannot be coeval, if one considers only single star evolution. The only way to explain such a population is by involving a multiple system evolution for, at least, the WN star that was rejuvenated through matter exchange during binary interaction. The study of more stars in the cluster would pin down a more accurate age, defining a clearer scenario of the cluster's history. Note that the WN stars shows a WR nebula (Borissova *et al.* 2014b), which is not detectable in the optical due to the extreme reddening ($A_V \sim 24$ mag).

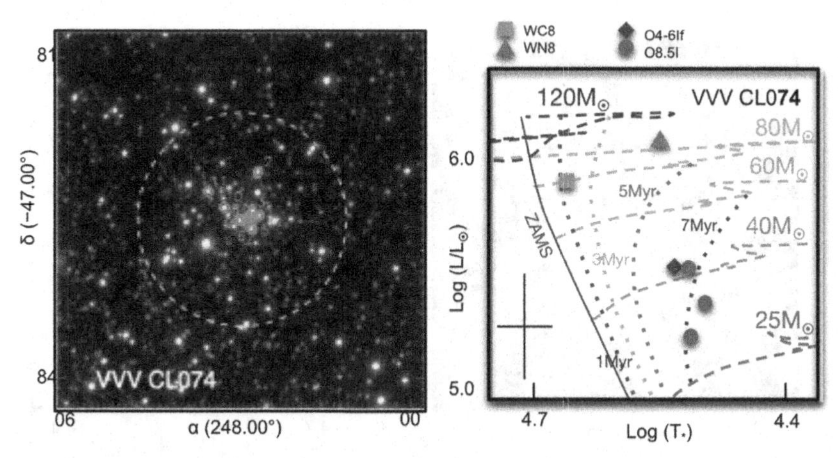

Figure 4. *Left*: False color JHK_S image from VVV. The stars marked in blue were observed in spectroscopy. *Right*: HR diagram with the brightest observed stars. The evolution tracks are from Ekström *et al.* (2012). The error bars are plotted on the bottom-left corner.

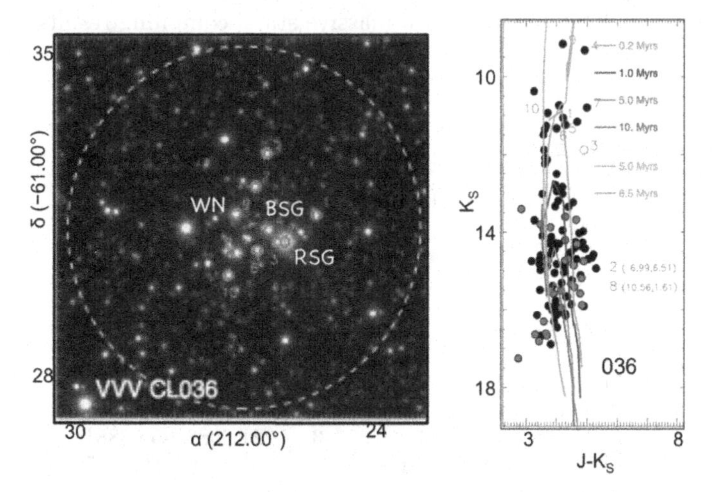

Figure 5. *Left*: False color JHK_S image from VVV. The stars marked in blue were observed in spectroscopy. *Right*: K_S vs $J - K_S$ diagram. The isochrones are from Ekström *et al.* (2012). The values for the RSG star (star#2) are off boundaries, and are written on the diagram. The WN and the BSG stars are stars #9 and #4, respectively.

4.5. *VVV CL041, a new very massive star candidate*

Very massive stars (VMS) are defined as stars with masses higher than 100 $M_\odot$. VVV CL041 contains a newly discovered VMS candidate, WR 62-2 (Chené *et al.* 2015). That star, a WN8-9h type star equivalent to what is observed in the Arches cluster (Martins *et al.* 2008), is one of the most luminous stars known to date. Applying the same spectral analysis as described for VVV CL074, we obtained a HR diagram for the 5 brightest stellar members of the cluster. The cluster's age is around 2 Myrs, and WR 62-2 might have a mass higher then 120 $M_\odot$. Of course, if it is in a binary system, then the total mass would be shared by the two components.

4.6. *Summary of the best studied VVV clusters with massive stars*

One can refer to Sebastian Ramírez Alegría's paper in these proceedings about the massive clusters in VVV. Here, we present a summary of the fundamental parameters for

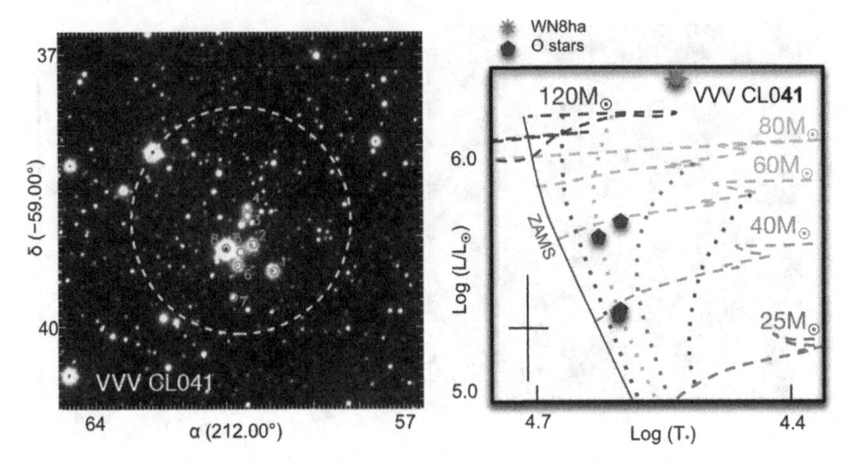

Figure 6. Same as Fig. 4, for VVV CL041.

the current census of VVV clusters with massive stars, compiling results from Borissova *et al.* (2016), Chené *et al.* (2013), (2015), Ramírez Alegría *et al.* (2014), (2016).

Table 1. Parameters of VVV clusters with massive stars

VVV	A_V (mag)	d (kpc)	Age (Myrs)	RA	DEC
CL009	5.4 ± 0.1	5 ± 1	4-6	179.0125	-63.3117
CL010	12.6 ± 0.3	8 ± 2	2-5	182.9458	-61.7733
CL011	9.4 ± 0.1	5 ± 2	3-7	183.1708	-62.7094
CL012	11.8 ± 0.3	7 ± 1	8-12	185.0583	-62.8850
CL013	11.4 ± 0.3	4 ± 2	1-5	187.1542	-62.9733
CL027	5.0 ± 0.2	6 ± 2	6-8	203.1000	-62.7275
CL028	3.7 ± 0.3	6 ± 1	20	205.0958	-61.7333
CL036	24 ± 3	2 ± 1	5-7	212.2666	-61.2661
CL041	8.0 ± 0.2	4 ± 1	1-3	221.6083	-59.3881
CL059	17.7 ± 0.3	12 ± 2	<20	241.4666	-50.7967
CL062	6.3 ± 0.3	1 ± 2	10	243.0333	-51.9689
CL073	19.3 ± 0.3	4 ± 3	<7	247.6000	-48.2167
CL074	23.4 ± 0.1	6 ± 1	4-6	248.0250	-47.8253
CL086	12.8 ± 0.1	11 ± 5	1-5	252.0625	-45.4350
CL088	15.7 ± 0.3	3 ± 2	6-8	253.1416	-44.6019
CL089	15.0 ± 0.3	2 ± 2	8-10	253.4458	-43.2675
CL099	15.2 ± 0.2	4 ± 1	4-6	258.6083	-38.1661

5. Mention of other groups work based on VVV data

There are currently more than 30 publications written based on VVV data by authors that are not member of the VVV science team. This includes massive stars related studies such as the X-ray observations of massive stars in VVV CL077 (Bodaghee *et al.* 2015), the monitoring Cepheid stars in open clusters (Chen *et al.* 2015), and the thorough study of the Dragonfish nebula de la Fuente *et al.* (2016). The later uses HST for the inner, crowded areas, and VVV for the surrounding clusters.

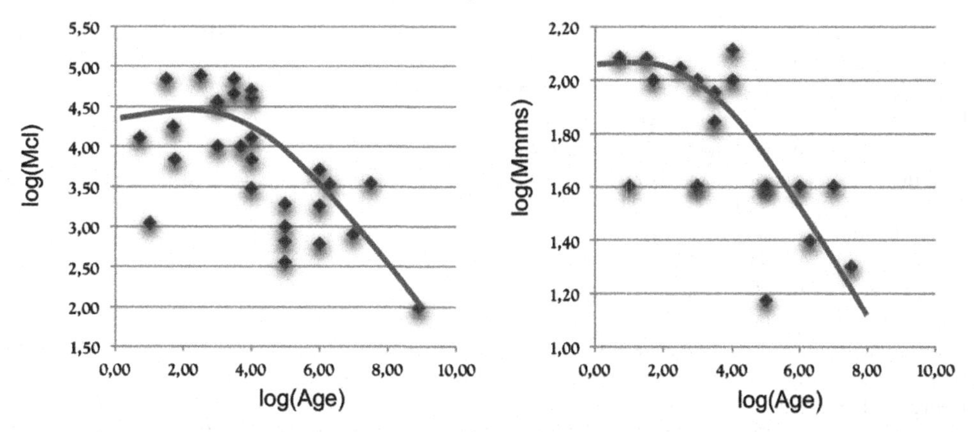

Figure 7. Mass of the cluster (Mcl) and mass of the most massive star cluster member (Mmms) as a function of age (in logarithmic scale). The blue lines show the general trends in the two cases.

6. Playing with the clusters fundamental parameters

Having access to previous studies, together with fresh results on newly discovered clusters gives a preliminary list of more than 40 clusters with massive stars. We can play the game of comparing some fundamental parameters together. We searched for correlations between the physical radius, the age, the total mass of the cluster, the mass of the most massive star in the cluster, the number of WR stars and the extinction using a Kendall's τ rank analysis. We could find only 3 real significant (anti-)correlations:

- the total mass of the cluster vs. the mass of the most massive star,
- the age vs. the mass of the most massive star,
- the age vs. the total mass of the cluster.

The first one echoes the result from Weidner *et al.* (2010), but with a much poorer statistic. The two last ones are plotted in Fig. 7. The second one may show an effect of stellar evolution, since passed an age of 4 Myrs, the mass of the most massive star decreases. It is indeed expected that the most massive members of the clusters end their life after ony a few Myrs. The last one could be showing the effect of cluster evolution, since after 3 Myrs, the total mass of the cluster seems to decrease. It is indeed suggested that clusters loose mass at early times, due to gravitational interaction or massive mass loss in the form of gas expelled during the supernova explosion of the most massive stars, or both (Portegies Zwart *et al.* 2010).

Of course, the statistics of these results are quite poor. Yet, there could be an indication that adding more clusters would clarify some phase of stellar and cluster evolution. And potentially, outliers would provide us with opportunities to unveil other star formation channels or yet unexpected cluster evolution destiny.

References

Barbá, R. H., Roman-Lopes, A., Nilo Castellón *et al.* 2015, *A&A*, 581A, 120

Bodaghee, A., Tomsick, J. A., Fornasini, F., Rahoui, F., & Bauer, F. E. 2015, *ApJ*, 801, 49

Borissova, J., Bonatto, C., Kurtev, R. *et al.* 2011, *A&A*, 532, A131

Borissova, J., Chené, A.-N., Ramírez Alegría, S. *et al.* 2014a, *A&A*, 569, A24

Borissova, J., Kumar, M. S. N., Amigo, P., Chené, A.-N., Kurtev, R., & Minniti, D. 2014b, *AJ*, 147, 18

Borissova, J., Ramrez Alegra, S., Alonso, J. *et al.* 2016,*AJ*, 152, 74

Caratti o Garatti,A., Stecklum, B., Garcia Lopez, R. *et al.* 2016, *Nature Physics*, doi:10.1038/nphys3942
Chen, X., de Grijs, R., & Deng, L. 2015, *MNRAS*, 446, 1268
Chené, A.-N., Borissova, J., Bonatto, C., *et al.* 2013, *A&A*, 549, A98
Chené, A.-N., Ramírez Alegría, S., Borissova, J., *et al.* 2015, *A&A*, 584, A31
Crowther, P. A., Schnurr, O., Hirschi, R., Yusof, N., Parker, R. J., Goodwin, S. P., & Kassim, H. A. 2010, *MNRAS*, 408, 731
Davies, B., de la Fuente, D., Najarro, F., Hinton, J. A., Trombley, C., Figer, D. F., & Puga, E. 2012, *MNRAS*, 419, 1860
Ekström, S., Georgy, C., Eggenberger, P., *et al.* 2012, *A&A*, 537, A146
de la Fuente, D., Najarro, F., Borissova, J., *et al.* 2016,*A&A*, 589A, 69
Hempel, M., *et al.* 2014, *The Messenger*, 155, 29
Hervé, A., Martins, F., Chené, A.-N., Bouret, J.-C., & Borissova, J. 2016, *NewA*, 45, 84
Hillier, D. J. & Miller, D. L. 1998, *ApJ*, 496, 407
Kharchenko, N. V., Piskunov, A. E., Schilbach, E., Röser, S., & Scholz, R.-D. 2013, *A&A*, 558A, 53
Lada & Lada 2003, *ARA&A*, 41, 57
Liermann, A., Hamann, W.-R., & Oskinova, L. M. 2012, *A&A*, 540A, 14
Morales, E. F. E., Wyrowski, F., Schuller, F., & Menten, K. M. 2013, *A&A*, 560A, 76
Martins, F., Hillier, D. J., Paumard, T., Eisenhauer, F., Ott, T., & Genzel, R. 2008, *A&A*, 478,219
Minniti, D., Lucas, P., Emerson, J., *et al.* 2010, *New A*, 15, 433
Portegies Zwart S. F., McMillan, S. L. W., & Gieles, M. 2010, *ARA&A*, 48, 431
Ramírez Alegría, S., Borissova, J., Chené, A.-N., *et al.* 2014, *A&A*, 564, L9
Ramírez Alegría, S., Borissova, J., Chené, A.-N., *et al.* 2016, *A&A*, 588A, 40
Saito, R. K., Hempel, M., Minniti, D. *et al.* 2012, *A&A*, 537A, 107
Solin, O., Haikala, L., & Ukkonen, E. 2014, *A&A*, 562, A115
Weidner, C., Kroupa, P., & Bonnell, I. A. D. 2010, *MNRAS*, 401, 275
de Wit, W. J., Testi, L., Palla, F., & Zinnecker, H. 2005, *A&A*, 437, 247

The Lives and Death-Throes of Massive Stars
Proceedings IAU Symposium No. 329, 2016
J.J. Eldridge, J.C. Bray, L.A.S. McClelland & L. Xiao, eds.

doi:10.1017/S174392131700299X

High-mass stars in Milky Way clusters

Ignacio Negueruela

Departamento de Física, Ingeniería de Sistemas y Teoría de la Señal
Escuela Politécnica Superior, Universidad de Alicante
Carretera de San Vicente del Raspeig s/n, E03690, San Vicente del Raspeig, Alicante, Spain
email: ignacio.negueruela@ua.es

Abstract. Young open clusters are our laboratories for studying high-mass star formation and evolution. Unfortunately, the information that they provide is difficult to interpret, and sometimes contradictory. In this contribution, I present a few examples of the uncertainties that we face when confronting observations with theoretical models and our own assumptions.

Keywords. open cluster and associations: general, stars: massive, formation, evolution

1. Introduction

Young open clusters are the natural laboratories where we can study high-mass star formation and evolution. This is because (most) high-mass stars are born in young open clusters, which provide us with astrophysical context for these objects. Until now, membership in an open cluster has been the only reliable way to estimate the distance to (almost all) high-mass stars. This situation is going to change dramatically in a few months with the release of accurate parallaxes by *Gaia*, but even then open clusters will give us other important parameters, such as age and interstellar extinction, that cannot be easily derived from an isolated star. These proceedings contain many examples of how young open clusters can be used to learn important facts about high-mass star formation and evolution. In this contribution, I will provide a few more examples, with special emphasis on the problems that we face to make effective use of these natural laboratories.

2. Constraints on high-mass star formation

By now, there is a general consensus on the idea that the vast majority of high-mass stars are formed in open clusters (Zinnecker & Yorke 2007 for a review). There is still, however, a heated debate about an apparently very similar proposition: *all high-mass stars form in clusters.* Going from "*almost all*" to "*all*" is not just a subtlety. Simply because they are very rare, high-mass stars should preferentially be found in the most massive clusters (Elmegreen 1983). However, if the observed initial mass function (IMF) is due to purely stochastical sampling of an underlying distribution, we should ocassionally see a high-mass star form in a low-mass environment (Maschberger & Clarke 2008). Contrarily, if the processes that form stars have a way of knowing about their environment, we would expect to see tight correlations between the mass of a cloud and the mass of the stars it forms (e.g. Weidner *et al.* 2013). Observations of clusters show that, in general, the mass of the most massive cluster member is a function of the cluster mass (Larson 1982, Weidner *et al.* 2010), but the interpretation of observations is compounded by many complicating issues, such as the effects of binarity (Eldridge 2012) or even the very definition of open cluster and IMF (Cerviño *et al.* 2013). So far, the strongest observational constraint in favour of purely statistical sampling is the detection by Oey *et al.* (2013) of a sample of 14 OB stars in the Small Magellanic Cloud (SMC) that meet strong criteria for having formed under extremely sparse star-forming conditions.

This issue is intimately linked to the ongoing discussion about the mechanisms that lead to the formation of very massive stars (very massive star is used here in the general sense of a star significantly more massive than the most numerous high-mass stars, with $M_* \sim 10-15\ M_\odot$). At present, there are two main competing theories about how such stars may form: (a) monolithic core accretion, in most respects a scaled-up version of classical low-mass formation theories, where very high opacities allow infalling material to overcome radiation pressure (e.g. Yorke & Sonnhalter 2002, Krumholz *et al.* 2009) and (b) competitive accretion, where massive stars are formed in cluster cores, benefiting from the gravitational potential of the whole cluster to accrete more material (e.g. Bonnell & Bate 2006, Smith *et al.* 2009). Observations do not provide a definite answer. On one side, there is (quite) direct evidence of disks around some moderately massive stars (up to $M_* \sim 25-30\ M_\odot$; see Beltrán & de Wit 2016 for a review), supporting the idea of monolithic collapse. On the other hand, massive stars in young clusters seem to lie in the regions of highest stellar density (Rivilla *et al.* 2014), a key prediction of the competitive accretion scenario that seems at odds with monolithic collapse.

Looking for a different approach, we can explore the environments where we find stars with the earliest spectral types (O2 – O3) in the Milky Way. These objects are not the most massive stars, which generally present WNh spectral types (e.g. Schnurr *et al.* 2008), but can safely be assumed to have masses $\gtrsim 40\ M_\odot$ (Massey *et al.* 2005). If we take all the stars with these spectral types in the GOSC v3 (Maíz Apellániz *et al.* 2013), we find 20 objects. Of these, 14 lie inside or close to some of the most massive clusters in the Milky Way, but four are in smaller clusters ($M_{\rm cl} \sim 10^3\ M_\odot$) and two appear in relative isolation within active star formation regions (Marco & Negueruela 2017). Taken at face value, this distribution seems to support the stochastic scenario, but we must exercise caution when interpreting such data.

For example, if we take a closer look to one of the two isolated objects, the O3 V((f*))z star HD 64 568, in the star-forming complex NGC 2467, we can see that there are almost no nearby OB stars (Fig. 1). However, it lies about 15′ away from the open cluster Haffner 19, which contains at least one mid-O star. There is some indirect evidence suggesting that it may be a runaway coming from the vicinity of this cluster. Interestingly, to the other side of Haffner 19, the apparently isolated mid-O star HD 64 315 illuminates the bright Hα region Sh2-311. This star was recently classified as O5.5 Vz+O7 V (Sota *et al.* 2014), but interferometric observations show that it can be resolved into at least two visual components (Tokovinin *et al.* 2010, Aldoretta *et al.* 2015). Analysis of high-resolution spectra shows that the "star" includes at least two SB2 systems with a minimum of four O-type stars (Lorenzo *et al.*, submitted). Such a complex system cannot have been ejected from a nearby cluster, and must therefore have formed in relative isolation. On the other hand, what was originally thought to be a single star appears now as a stellar aggregate that may well hide a population of lower-mass stars. This hidden multiplicity beautifully illustrates the difficulty in defining isolation or measuring the mass of a small cluster.

3. Evolution after the main sequence

Young open clusters provide most of the information that we have about high-mass star evolution. Unfortunately, this information is difficult to interpret, because most young open clusters have quite small populations of (evolved) high-mass stars. As an example, take NGC 457, a typical cluster in the Perseus Arm, with a distance estimate of 2.5 kpc

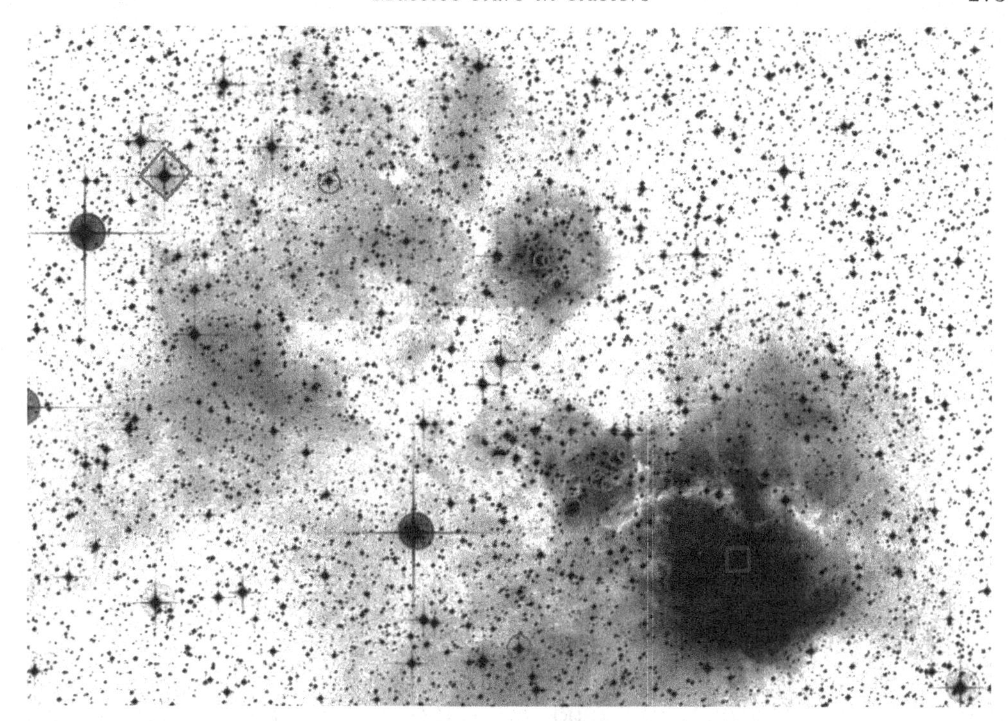

Figure 1. The position of HD 64 568 (marked by the purple diamond) relative to the star-forming complex NGC 2467 is shown on this DSS2 blue plate. The position of HD 64 315 is shown by a blue square. Other OB stars in the area are shown as red circles. The small group to the North is Haffner 18, while the larger concentration towards the centre is Haffner 19.

and more than sixty B-type members identified from photometry (Fitzsimmons 1993, and references therein). This cluster contains three supergiants: the B5 Ib star HD 7 902, the F0 Ia MK standard ϕ Cas, and the M2 Ib red supergiant (RSG) HD 236 687. With such a complement of evolved stars, it should be an ideal laboratory to constrain the evolution of moderately massive stars. More than 4° South of the Galactic Plane, and separated from all nearby OB associations, its membership should be essentially free of confusion problems. Abundance determinations for some of its members imply solar-like values for α elements (Dufton *et al.* 1994), allowing direct comparison to theoretical tracks.

But gleaning information is not always straightforward. Figure 2 shows the classification spectra of six of the brightest non-supergiant blue stars in the area of NGC 457. Of them, S153 is an extreme Be star and thus difficult to classify or analyse. Star 19, formally B1.5 III, looks more luminous than the others, but this is mostly a visual effect due to its very low projected rotational velocity (Huang & Gies 2006, who also find that S19 and S37 are likely binaries). Star 128 is about as luminous, though a fast rotator. Star 14 is slightly less luminous, and a Be star, while S37 is somewhat later, around B2 III (but this is almost certainly an SB2 binary). The spectral types of these four latter stars are consistent, and their presence at the top of the sequence suggests an age perhaps slightly younger than the ~21 Ma proposed by Huang *et al.* (2006). In a Geneva isochrone for $\sim$ 20 Ma with moderate initial $v_{\rm rot}$ at solar metallicity (Georgy *et al.* 2013), the objects at the top of the sequence have $M_{*} \sim 11\ {\rm M}_{\odot}$. For an age of 16 Ma, we find masses $\sim 13\ {\rm M}_{\odot}$, bracketting the expectations for the spectral type. But S120 is clearly earlier than any other star in the cluster. At O9.7 V, it should be a blue straggler (as Huang & Gies 2006 measure a radial velocity similar to those of other

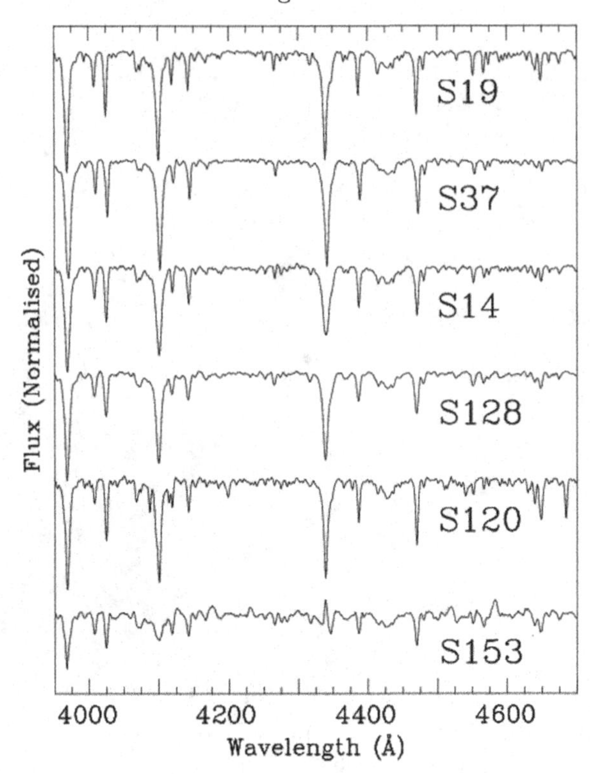

Figure 2. Classification spectra of six bright blue stars in the open cluster NGC 457. S19, S128 and S37 (and also likely the extreme Be star S153) are all giants between B1.5 and B2. S14 is a very slightly less luminous Be star, while S120 (at O9.7 V) is a blue straggler.

cluster members). Moving to the supergiants, Levesque *et al.* (2005) derive parameters for HD 236 687 from a global model fit to the spectrum, finding that it is compatible with a $\sim 12\,\mathrm{M}_\odot$ track. For HD 7 902, McErlean *et al.* (1999), using tailored model fits, derive parameters compatible with a $\sim 18\,\mathrm{M}_\odot$ star. Finally, the parameters of ϕ Cas are much more uncertain. Combining model fits with observational calibrations, Kovtyukh *et al.* (2012) derive parameters that would put it well above the $20\,\mathrm{M}_\odot$ track, but Luck (2014) finds very different values of log g (and, thence, luminosity) when using different atmosphere models.

Can we use all this apparently conflicting information to constrain stellar evolution? Is there any useful way to mix the results of different groups? All three supergiants have been analysed using different techniques and sets of models. Are the results on the same theoretical scale? Do all groups use the same assumptions? McErlean *et al.* (1999) use a distance modulus to NGC 457 0.2 mag longer than used by Levesque *et al.* 2005. This may not seem like much, and is certainly within the uncertainty of most distance determination methods, but it represents an increase of 10% in the distance. Although it cannot explain the higher mass implied for HD 7 902, it brings it closer to the value derived from the cluster giants. The position of HD 7 902 could also perhaps be explained by a very high initial rotational velocity, as stars that rotate very fast become brighter and evolve more slowly (Georgy *et al.* 2013). Another possibility (that could also explain S120) is that it gained mass in a binary while it was still on the main sequence and then followed a standard evolution (Langer 2012). The properties of ϕ Cas could require a more complex explanation. If its mass is $> 13\,\mathrm{M}_\odot$, it should not be in a Cepheid loop.

Perhaps it is a post-RSG object. There are many possible post-MS tracks, and we only have three stars clearly detached from the main sequence that may have followed very different paths. However, not all hope is lost. There is still a lot of reliable information in the cluster. We observed 15 stars close to the top of the main sequence in NGC 475 (including the six in Fig. 2) for the programme described in Marco *et al.* (2007). Of these, 12 have spectral types between B1.5 and B2.5, with luminosity classes between V and III, well matched to their relative brightness. The brightest main-sequence stars have spectral type B2 V providing a good definition of the turn-off, which, through the use of calibrations based on eclipsing binaries, we can place around 9 $M_\odot$.

We can easily check this conclusion. In Fig. 3, we plot the existing photoelectric photometry for stars in the field of NGC 457, from Pesch (1959) and Hoag *et al.* (1961). We can verify the validity of the spectral types, as the photometric turn-off coincides with the transition between class V and higher luminosities. We can then plot the corresponding isochrones from Georgy *et al.* (2013), reddened with the values from Fitzsimmons (1993). We use the 16 Ma isochrone without rotation and the 20 Ma one with $\Omega/\Omega_{\rm crit} = 0.3$ (they are almost identical). The fit is not perfect, but it is surprisingly good (we just guessed the age based on the spectral types of the brightest stars). A result is much stronger when it is based on many stars hinting at approximately the same conclusion. We can see how the blue straggler lies on the continuation of the main sequence, well to the left of the turn-off. We can also see that the analysis by McErlean *et al.* (1999) is not in error. The B5 Ib supergiant is much more luminous than the corresponding isochrone.

To fight low-number statistics, we can resort to utilising information from many different clusters, but then the difficulties increase. The metallicity of young Galactic clusters cannot simply be assumed to be solar. There is a gradient with galactocentric distance and strong local variability (see Negueruela *et al.* 2015 and references therein). Moreover, results obtained by different groups can have large systematic effects, affecting many parameters. Membership criteria may vary. Age scales change as new stellar models are developed and successive generations of isochrones are released. For these reasons, the more massive a cluster is (and thus the more high-mass stars it contains) the more useful it is as a laboratory. To continue with our example, the Perseus arm contains a few young open clusters more massive (and hence containing more evolved stars) than NGC 457, namely, NGC 7419, NGC 663, NGC 869, and NGC 884. Their combined analysis shows that blue stragglers happen frequently, but are not very numerous. It also shows that HD 7 902 is not an exception; blue supergiants (BSGs) almost always fall on tracks rather more massive than the main sequence turnoff (see also Marco *et al.* 2007). These clusters, however, offer very little information about what should be a very strong constraint on evolutionary timescales, the blue to red supergiant ratio (Eggenberger *et al.* 2002), as it varies wildly among them.

4. Massive clusters

Clusters with large populations of high-mass stars become much more reliable laboratories. The most extreme example is the starburst cluster Westerlund 1. It contains many dozen supergiants, covering all spectral types at very high luminosities (Clark *et al.* 2005). The sequence of BSGs is well populated from O9, with examples of almost all the B subtypes (Negueruela *et al.* 2010). Under these conditions, the much higher number of O9 – B2 supergiants can be taken as a very strong suggestion of a much faster evolution after $T_{\rm eff} \approx 18\,000$ K is reached. Because the population is so large, we find a high number of spectroscopic binaries, including a few eclipsing binaries, which allow a definite measurement of the masses of these stars, which lie in the $\sim 35 - 40\ M_\odot$ range

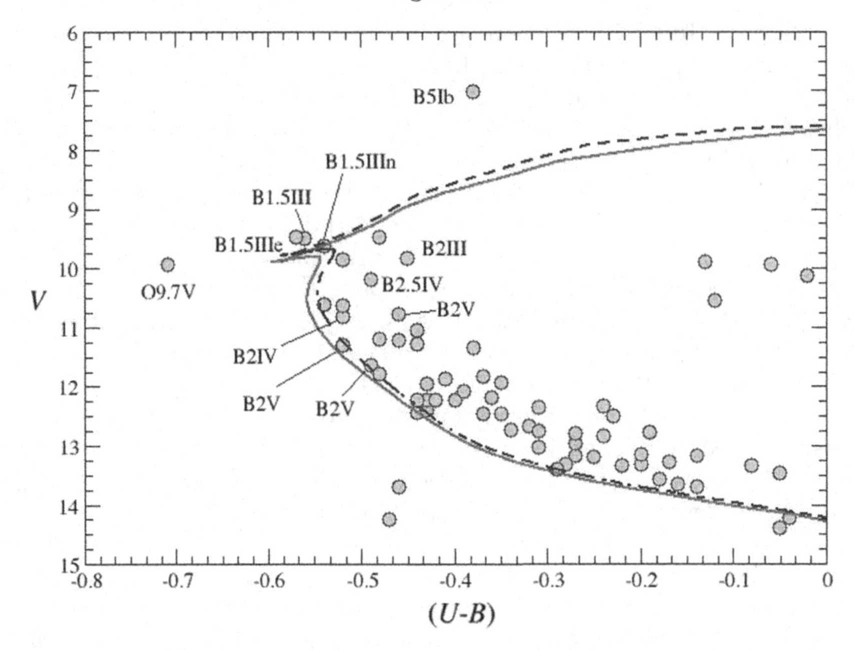

Figure 3. Photoelectric photometry of blue stars in NGC 457, with our spectral types superimposed. The continuous line is the 16 Ma isochrone for no initial rotation, while the dashed line is the 20 Ma isochrone for $\Omega/\Omega_{\rm crit} = 0.3$ from Georgy *et al.* (2013).

(Koumpia & Bonanos 2012). Moreover, again because of high statistics, we may find binary systems with similar masses, but different orbital parameters. Comparison of their properties leads strength to the concept of diverse evolutionary paths for stars in close binaries (Sana *et al.* 2012). Indeed, in Westerlund 1, we can see examples of many binary systems that represent different stages along these paths (Clark *et al.* 2014b)

Unfortunately, young massive clusters (YMCs) are rare, and this means that we generally find them at large distances and behind heavy extinction. If we follow Hanson *et al.* (2015) in defining a YMC as a cluster with $M_{\rm cl} \gtrsim 10^4$ M$_\odot$, we know less than twenty YMCs in the Milky Way. Moreover, some of the most massive clusters in the Galaxy, such as NGC 3063 or the Arches cluster, are too young to display any effects of stellar evolution. Given their scarcity, obtaining parameters for YMCs with high accuracy seems paramount, but it is not an easy task. For example, Westerlund 1 is behind more than 10 mag of extinction, and in spite of its astrophysical interest and the effort dedicated over the past ten years, its distance is still uncertain by almost a factor of two.

As an illustration of this difficulty, let us take VdBH 222, one of the less reddened YMCs with $E(B-V) \approx 2.5$, despite its location towards the Galactic Centre ($\ell = 349°$). We observed a large population of red and yellow supergiants, from whose spectra it was easy to measure the radial velocity of the cluster. With a very distinctive value of $v_{\rm LSR} = -99 \pm 4$ km s^{-1}, the Galactic rotation curve allows two possible distances for VdBH 222, one at $d \sim 6$ kpc, and the other at $d \sim 11$ kpc (Marco *et al.* 2014). Though such distance difference appears really huge, it only amounts to 1.3 mag in distance modulus. With the high value of $E(B-V)$, a small difference in the value of R_V can account for most of this gap. In addition, the shape of isochrones does not change significantly over the age range between ~ 10 and ~ 20 Ma. Only the gap in brightness between the red and blue supergiants can help determining the cluster age, but this also depends on the extinction law. We found that it was possible to fit all photometric data with a 12 Ma cluster at $\sim$ 10.5 kpc or a 18 Ma cluster at $\sim$ 6 kpc (in fact, the photometry allowed many other possibilities between these two, but the radial velocity offers further constraints). The data do not provide any direct way to choose between the two options. However, given our knowledge of Galactic structure, the second option looked rather unlikely, because it places the cluster at less than 3 kpc from the Galactic Centre, in a region supposed to be free of star formation (Marco *et al.* 2014).

However, additional data showed this to be the correct solution. Spectral monitoring of the cool supergiants revealed one of them to be a massive Cepheid. The determination of a 24.3 d period indicated that this is a $\sim 10\,M_{\odot}$ star (again, the chances of finding such an unusual star increase with cluster population), and so the cluster is definitely on the near side of the Galactic Centre (Clark *et al.* 2015). This has been confirmed by classification spectra of the brightest blue members, which shows all of them to be early-B (bright) giants, and not supergiants as would be required by a higher distance (and hence lower age). Interestingly, this means that this cluster rich in RSGs hosts no BSGs, in analogy to the Perseus arm cluster NGC 7419 (Marco & Negueruela 2013). Based on data that we are currently analysing on two other clusters rich in RSGs, this scarcity or even absence of BSGs seems to be a common theme.

Regrettably, obtaining a good characterisation of other clusters rich in RSGs is even more difficult than finding a good solution for VdBH 222, because they are hidden by higher extinction (see a list of these clusters with their parameters in Negueruela 2016). In the extreme case of RSGC1, with $A_K \approx 2.5$ mag (Davies *et al.* 2008), we are limited to work in the infrared – even J-band spectroscopy is hard to get! But even for the much less obscured Stephenson 2 (Davies *et al.* 2007), we find that spectroscopy in the optical is not feasible, while photometry in the U and B bands is beyond reach. Under this conditions, the reddening law cannot be determined with accuracy, resulting in large uncertainties in the distance to the cluster (and hence all other parameters). Fortunately, there are other tools, and we are likely to see in the near future accurate distances to both Westerlund 1 and Stephenson 2 (and perhaps other clusters) based on trigonometric parallaxes of the SiO masers detected in some of their RSGs (Verheyen *et al.* 2012). This will allow full characterisation of their properties and a much better use as laboratories of stellar evolution (see Negueruela 2016 for a discussion of these potentialities).

A great example of this kind of work can be found in the contribution by Beasor & Davies to these proceedings, where they use the large collection of RSGs in the YMC NGC 2100 (in the LMC) to present strong evidence for an evolutionary sequence among the RSGs: mass-loss rates and luminosity show a clear correlation, as has also been found in an analysis of the LMC field population by Dorda *et al.* (2016). A similar behaviour is readily seen among the red RSG population of Stepheson 2 (Clark *et al.* 2014a). Interestingly, in both clusters the range in brightness of the supergiant population (above 2 mag) is much wider than allowed by any single theoretical isochrone. Why is this? Why do RSGs and BSGs tend to avoid each other in YMCs? Why are BSGs almost always more luminous than the isochrone of their parent cluster? All these questions are presented by the simple observation of young open clusters. Their continued study and accurate characterisation in the *Gaia* era will certainly provide new constraints and new questions.

References

Aldoretta, E. J., Caballero-Nieves, S. M., Gies, D. R., *et al.* 2015, *AJ*, 149, 26
Beltrán, M. T. & de Wit, W. J. 2016, *A&AR*, 24, 6
Bonnell, I. A. & Bate, M. R. 2006, *MNRAS*, 370, 488

Cerviño, M., Román-Zúñiga, C., Luridiana, V., *et al.* 2013, *A&A*, 553, A31
Clark, J. S., Negueruela, I., Crowther, P. A., & Goodwin, S. P. 2005, *A&A*, 434, 949
Clark, J. S., Negueruela, I., & González-Fernández, C. 2014, *A&A*, 561, A15
Clark, J. S., Ritchie, B. W., Najarro, F., *et al.* 2014b, *A&A*, 565, A90
Clark, J. S., Negueruela, I., Lohr, M. E., *et al.* 2015, *A&A*, 584, L12
Davies, B., Figer, D. F., Kudritki, R.-P., *et al.* 2007, *ApJ*, 671, 781
Davies, B., Figer, D. F., Law, C. J., *et al.* 2008, *ApJ*, 676, 1016
Dorda, R., Negueruela, I., González-Fernández, C., & Tabernero, H. M. 2016, *A&A*, 592, A16
Dufton, P. L., Fitzsimmons, A., & Rolleston, W. R. J. 1994, *A&A*, 286, 449
Eggenberger, P., Meynet, G., & Maeder, A. 2002, *A&A*, 386, 576
Eldridge, J. J. 2012, *MNRAS*, 422, 794
Elmegreen, B. G. 1983, *MNRAS*, 203, 1011
Fitzsimmons, A. 1993, *A&AS*, 99, 15
Georgy, C., Ekström, S., Granada, A., *et al.* 2013, *A&A*, 553, A24
Hanson, M. M., Froebrich, D., Martins, F., *et al.* 2015, Highlights of Astronomy, 16, 410
Hoag, A. A., Johnson, H. L., Iriarte, B., *et al.* 1961, *Publications of the U.S. Naval Observatory Second Series*, 17, 344
Huang, W. & Gies, D. R. 2006, *ApJ*, 648, 580
Krumholz, M. R., Klein, R. I., McKee, C. F., *et al.* 2009, *Science*, 323, 754
Kovtyukh, V. V., Gorlova, N. I., & Belik, S. I. 2012, *MNRAS*, 423, 3268
Koumpia, E. & Bonanos, A. Z. 2012, *A&A*, 547, A30
Lamb, J. B., Oey, M. S., Werk, J. K., & Ingleby, L. D. 2010, *ApJ*, 725, 1886
Langer, N. 2012, *ARA&A*, 50, 107
Larson, R. B. 1982, *MNRAS*, 200, 159
Luck, R. E. 2014, *AJ*, 147, 137
McErlean, N. D., Lennon, D. J., & Dufton, P. L. 1999, *A&A*, 349, 553
Maíz Apellániz, J., Sota, A., Morrell, N. I., *et al.* 2013, msao.confE.198M
Marco, A. & Negueruela, I. 2013, *A&A*, 552, A92
Marco, A. & Negueruela, I. 2017, *MNRAS*, 465, 784
Marco, A., Negueruela, I., & Motch, C. 2007, *Active OB-Stars: Laboratories for Stellar and Circumstellar Physics*, 361, 388
Marco, A., Negueruela, I., González-Fernández, C., *et al.* 2014, *A&A*, 567, A73
Maschberger, Th. & Clarke, C. J. 2008, *MNRAS*, 391, 711
Massey, P., Puls, J., Pauldrach, A. W. A., *et al.* 2005, *ApJ*, 627, 477
Negueruela, I. 2016, *IAU Focus Meeting*, 29, 461
Negueruela, I., Clark, J. S., & Ritchie, B. W. 2010, *A&A*, 516, A78
Negueruela, I., Simón-Díaz, S., Lorenzo, J., Castro, N., & Herrero, A. 2015, *A&A*, 584, A77
Oey, M. S., Lamb, J. B., Kushner, C. T., Pellegrini, E. W., & Graus, A. S. 2013, *ApJ*, 768, 66
Pesch, P. 1959, *ApJ*, 130, 764
Rivilla, V. M., Jiménez-Serra, I., Martín-Pintado, J., & Sanz-Forcada, J. 2014, *MNRAS*, 437, 156
Sana, H., de Mink, S. E., de Koter, A., *et al.* 2012, *Science*, 337, 444
Sota, A., Maíz Apellániz, J., Morrell, N. I., *et al.* 2014, *ApJS*, 211, 10
Smith, R. J., Longmore, S., & Bonnell, I. 2009, *MNRAS*, 400, 1775
Schnurr, O., Casoli, J., Chené, A.-N., *et al.* 2008, *MNRAS*, 389, L38
Tokovinin, A., Mason, B. D., & Hartkopf, W. I. 2010, *AJ*, 139, 743
Verheyen, L., Messineo, M., & Menten, K. M. 2012, *A&A*, 541, A36
Weidner, C., Kroupa, P., & Bonnell, I. A. D. 2010, *MNRAS*, 401, 275
Weidner, C., Kroupa, P., & Pflamm-Altenburg, J. 2013, *MNRAS*, 434, 84
Yorke, H. W. & Sonnhalter, C. 2002, *ApJ*, 569, 846
Zinnecker, H. & Yorke, H. W. 2007, *ARA&A*, 45, 481

The Lives and Death-Throes of Massive Stars
Proceedings IAU Symposium No. 329, 2016
J.J. Eldridge, J.C. Bray, L.A.S. McClelland & L. Xiao, eds.

doi:10.1017/S1743921317002496

The VLT-FLAMES Tarantula Survey

Jorick S. Vink[1], C.J. Evans[2], J. Bestenlehner[1,3], C. McEvoy[4], O. Ramírez-Agudelo[2], H. Sana[5], F. Schneider[6] and VFTS

[1]Armagh Observatory, College Hill, BT61 9DG, Armagh, Northern Ireland
email: `jsv@arm.ac.uk` [2]ATC, Royal Observatory Edinburgh, Blackford Hill, Edinburgh, EH9 3HJ, UK [3]Departament of Physic and Astronomy University of Sheffield, Sheffield, S3 7RH, UK [4]ARC, School of Mathematics and Physics, QUB, Belfast BT7 1NN, UK [5]Institute of Astrophysics, KU Leuven, Celestijnenlaan 200D, 3001, Leuven, Belgium [6]Department of Physics, University of Oxford, Keble Road, Oxford OX1 3RH, UK

Abstract. We present a number of notable results from the VLT-FLAMES Tarantula Survey (VFTS), an ESO Large Program during which we obtained multi-epoch medium-resolution optical spectroscopy of a very large sample of over 800 massive stars in the 30 Doradus region of the Large Magellanic Cloud (LMC). This unprecedented data-set has enabled us to address some key questions regarding atmospheres and winds, as well as the evolution of (very) massive stars. Here we focus on O-type runaways, the width of the main sequence, and the mass-loss rates for (very) massive stars. We also provide indications for the presence of a *top-heavy* initial mass function (IMF) in 30 Dor.

Keywords. stars: early-type, stars: massive, stars: evolution, stars: luminosity function, mass function, stars: mass loss, stars: fundamental parameters

---

1. Introduction

Massive star evolution is important for many fields of Astrophysics including supernovae (SNe; Levesque, these proceedings). Yet, it remains largely unconstrained (Langer 2012; Meynet these proceedings). Progress can be made using high-quality observations from nearby sources, as well as from large data-sets such as VFTS (Evans *et al.* 2011) discussed here. In parallel, VFTS data are analysed using state-of-the-art model atmospheres such as CMFGEN (Hiller & Miller 1998) and FASTWIND (Puls *et al.* 2005), as well as automatic fitting tools (Sabín-Sanjulián *et al.* 2014; Bestenlehner *et al.* 2014; Ramírez-Agudelo *et al.* 2017).

In addition to this observational progress, our VFTS collaboration strives to make theoretical progress on stellar winds and evolution, and we are in the unique position to confront our new models against VFTS data. In the following, we highlight a number of recent results that we argue make a real difference to our knowledge of massive stars.

1.1. *Motivation for the Tarantula region*

The Tarantula region (30 Doradus) is the largest active star-forming region in our Local Universe for which individual spectra of the massive-star population can be obtained. Because it is the largest region, it provides a unique opportunity to study the most massive stars, including very massive stars (VMS) with masses up to 200-300 $M_{\odot}$ (Crowther *et al.* 2010; Bestenlehner *et al.* 2014; Martins 2015; Vink *et al.* 2015). This allows us to properly investigate whether the upper-IMF may be top-heavy (Schneider *et al.* 2017). Answering this question is important as these VMS that are thought to dominate the ionizing radiation and wind feedback from massive stars (Doran *et al.* 2013).

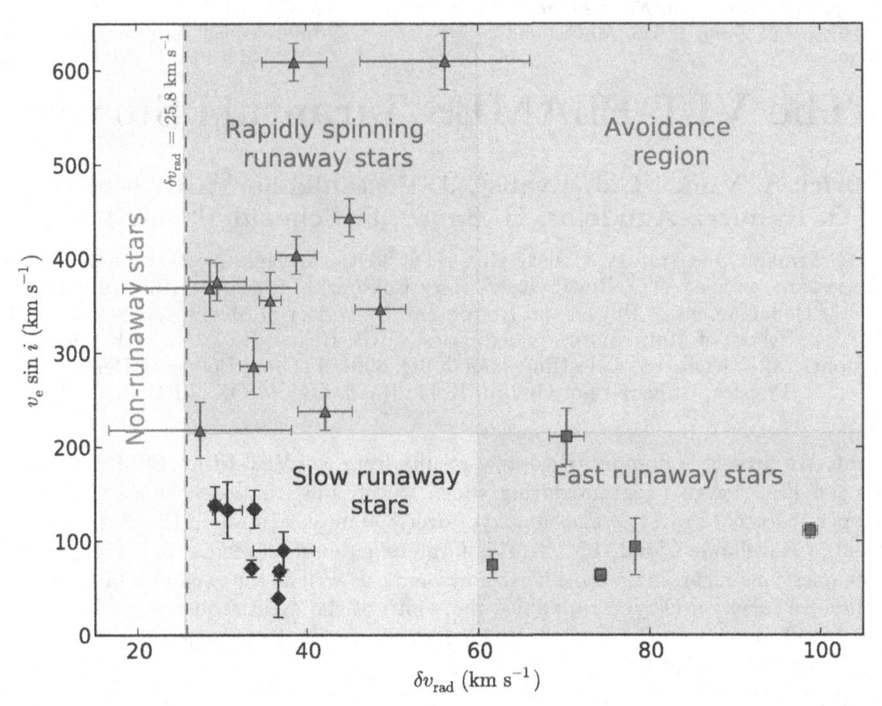

Figure 1. Rotational velocities of both slow & fast runaways (Sana *et al.* in prep.). Whilst there are slow runaways at high v sin i, and fast runaways at relatively low v sin i, there appears to be a Region-of-Avoidance for fast runaways with large v sin i. These diagnostics might enable us to disentangle the different proposed origins for runaways, as discussed in the text.

Another reason to study 30 Doradus is that testing massive star evolution *requires* large data-sets. For instance, the issue of the location of the terminal-age main sequence (TAMS) can only be addressed when the sample-size is sufficiently large to populate both the main-sequence with O-type stars (Sabín-Sanjulián *et al.* 2017; Ramírez-Agudelo *et al.* 2017) and B supergiants (McEvoy *et al.* 2015).

2. Results on binarity, rotation rates, and runaways

The aims of VFTS were to determine the stellar parameters, such as $T_{\rm eff}$, $\log g$ & $\log L$ to place our objects on the HR-diagram; the mass and $\dot{M}$ to determine the evolution & fate of massive stars; and the helium (He) and nitrogen (N) abundances to test (rotational) mixing processes (Grin *et al.* 2016; Rivero-Gonzalez *et al.* 2012). All these parameters require sophisticated atmosphere modeling, but VFTS also offered some model-independent parameters including the rotational velocities v sin i and radial velocities (RVs) thanks to the multi-epoch nature of the survey. The latter allowed us to obtain information on the $\sim$ 50% frequency in 30 Dor (Sana *et al.* 2013) and the opportunity to study the dominant mechanism for runaways (Fig. 1; Sana *et al.* in prep.).

Figure 1 might allow us to disentangle the dynamical runaway scenario (Gies & Bolton 1986) from the binary-SN kick scenario (Stone 1991), as the first scenario might produce relatively fast runaways, whilst one would expect the binary SN kick scenario to produce rapid rotators. Obviously, definitive conclusions can only be obtained when more sophisticated models become available.

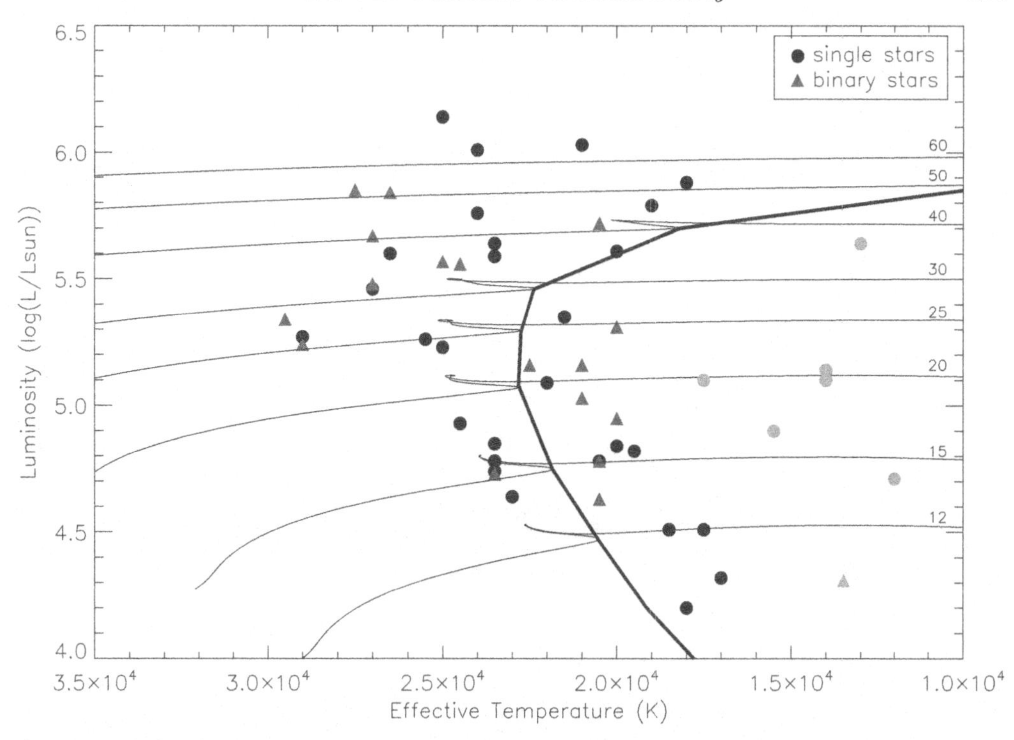

Figure 2. A zoomed-in version of the Hertzsprung-Russell diagram for single (circle) and binary (triangle) B supergiants from McEvoy *et al.* (2015). Also shown are the LMC evolutionary tracks of Brott *et al.* (2011) for $v_{\rm rot} \simeq 225$ km/s, with the initial mass (in units of the solar mass) given on the right hand side. The dark line represents the TAMS (terminal age main sequence). The stars highlighted in green are far enough from the TAMS line that they may be interpreted as core He-burning objects.

3. The width of the main-sequence and constraints on core overshooting

Figure 2 shows a zoomed-in version of the Hertzsprung-Russell diagram for both single and binary B supergiants from McEvoy *et al.* (2015). The position of the dark line indicates the position of the TAMS, with its location is determined by the value of the core overshooting parameter ($\alpha_{\rm ov}$) which is basically a "free" parameter (e.g. Vink *et al.* 2010; Brott *et al.* 2011) until astro-seismology on a large number of OB supergiants becomes available. The Brott *et al.* models employ a value of $\alpha_{\rm ov} = 0.335$, whilst the Geneva models (Georgy these proceedings) employ a smaller value. The VFTS results shown in Fig. 2 appear to suggest a *larger* value of $\alpha_{\rm ov}$ than 0.335.

Larger $\alpha_{\rm ov}$ makes bi-stability braking (BSB; Vink *et al.* 2010; Keszthelyi *et al.* 2017) feasible, which we test by showing $v \sin i$ of both VFTS and previous FLAMES-I results (Hunter *et al.* 2008) versus $T_{\rm eff}$ in Fig. 3. Note the presence of another "Region-of-Avoidance"†, where rapidly-rotating "cool" (cooler than the bi-stability location of 20 000 K; Petrov *et al.* 2016) B supergiants are simply not observed. The reason for this avoidance below 20 000 K could either involve BSB, or it might be that the slowly rotating cool B supergiants are He-burning objects (due to post red-supergiant evolution or binarity).

† The perceived lack of rapid rotators on the hot side of the diagram is not real, there are many rapidly rotating O-type stars. These O-stars are just not included here.

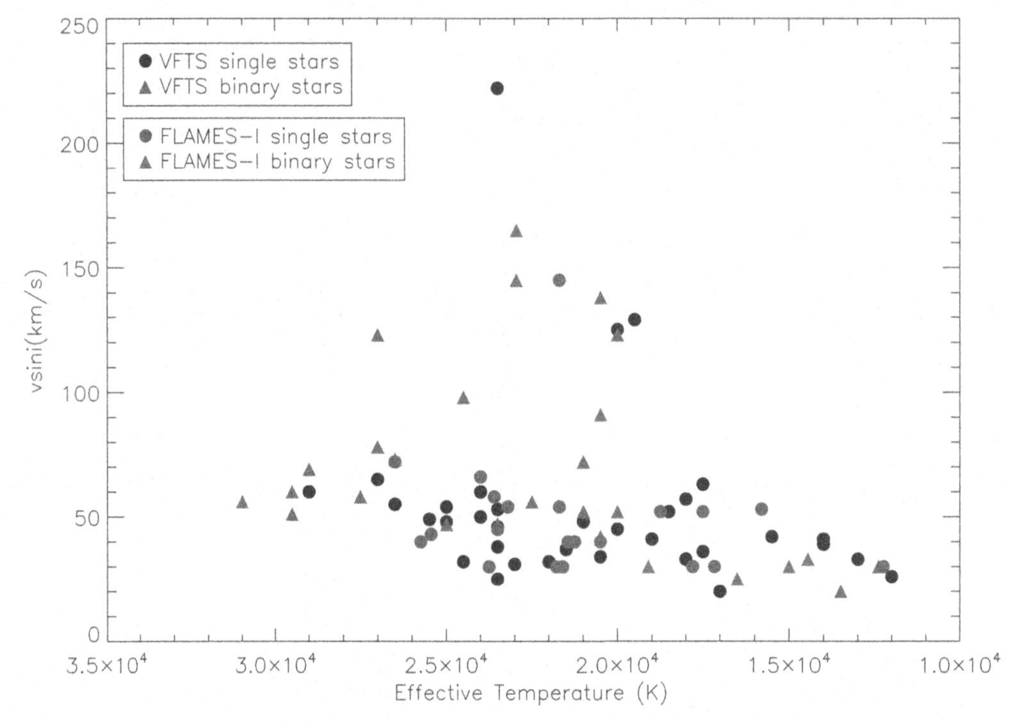

Figure 3. Effective temperatures plotted against v sin i for FLAMES–I and VFTS B-type supergiants in the LMC (see Vink *et al.* 2010; McEvoy *et al.* 2015). The perceived lack of rapid rotators on the hot side of the diagram is not real, there are many rapidly rotating O-type stars; they are just not included here.

4. The mass-loss rates

The mass-loss rates for O-type dwarfs were discussed by Sabín-Sanjulián *et al.* (2014; 2017), whilst those for the O giants and supergiants are plotted in the form of the wind-momentum-luminosity relation (WLR; Kudritzki & Puls 2000; Puls *et al.* 2008) in Fig. 4. Interestingly, the empirical WLR lies above the theoretical WLR (of Vink *et al.* 2001). Usually a discrepancy between theoretical and empirical values would be interpreted such that the theoretical rates would be too low, but here it is different, as it is widely accepted that empirical modeling is more dependent on wind clumping and porosity than theory (see Muijres *et al.* 2011 for theoretical expectations).

Indeed, it is more likely that the empirical WLR is too high, as a result of wind clumping, which has not been included in the analysis. This would imply that the empirical WLR would need to be lowered by a factor $\sqrt{D}$, where D is the clumping factor, which is as yet uncertain. However, given the model-independent (from clumping & porosity) transition mass-loss rate (Vink & Gräfener 2012; next Sect.) a value of $D \simeq 10$ (with a mass-loss rate and WLR reduction of $\sim$3) would bring the empirical WLR and theory in reasonable agreement. None of this means that the theoretical rates for lower mass-and-luminosity O stars need necessarily to be correct. Therefore, spectral analysis of large data-sets of O-stars including clumping & porosity (Surlan *et al.* 2013; Sundqvist *et al.* 2014) are needed to provide definitive answers.

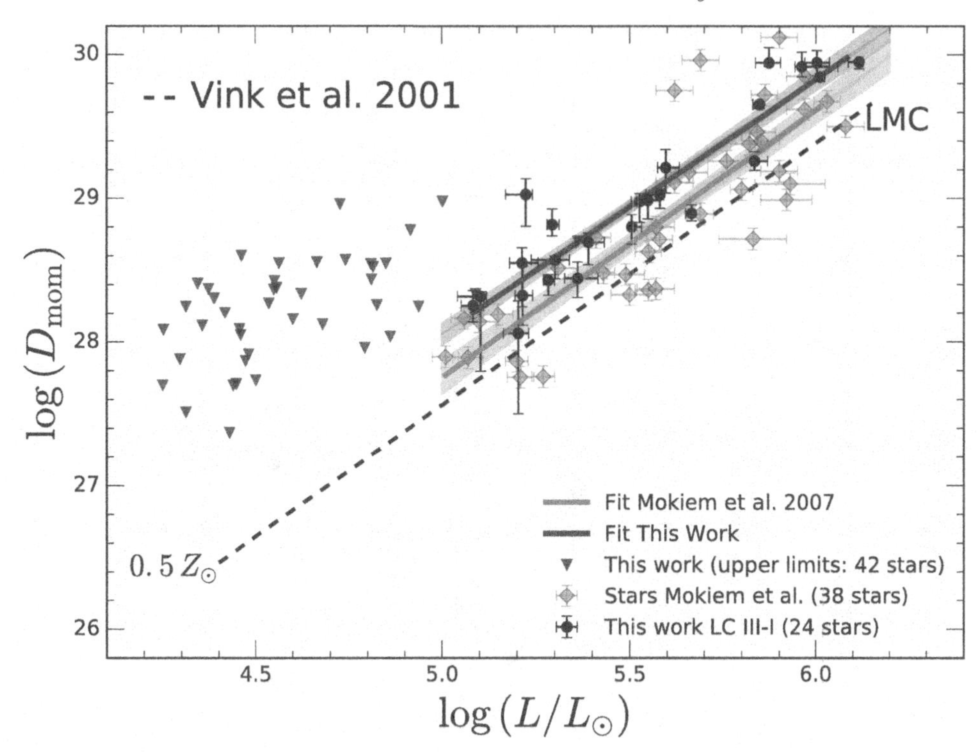

Figure 4. Modified wind momentum ($D_{\rm mom}$) versus luminosity diagram from Ramírez-Agudelo *et al.* (2017). The dashed lines indicate the theoretical predictions of Vink *et al.* (2001) for homogeneous winds. The empirical fit and Mokiem *et al.* (2007) (both for $L/L_\odot > 5.0$) are shown in shaded blue and gray bars, respectively. For stars with $L/L_\odot \leqslant 5.0$, only upper limits could be constrained.

5. Very Massive Stars

The most massive stars in VFTS were analysed by Bestenlehner *et al.* (2014), plotted in the HRD of Fig. 5. Over-plotted are VMS evolutionary tracks and the location of the ZAMS. The HRD shows the presence of 12 VMS (with $M > 100\ M_\odot$; Vink *et al.* 2015), which enables us to derive the upper-IMF of 30 Dor for the first time. Figure 6 compares the preferred value for the mass function to that of Salpeter. It is found that the slope is different to that of Salpeter (at ~85% confidence), and also that a Salpeter IMF cannot reproduce the larger number of massive stars above $30\,M_\odot$ at >99% confidence (Schneider *et al.* 2017). As this result is obtained using the largest spectroscopic data-set ever obtained, and analysed with the most sophisticated analysis tools, we consider this the most robust test to date. A top-heavy IMF would have major implications for the interpretation of spectral modelling of high-redshift galaxies, as well as the ionizing radiation and kinetic wind energy input into galaxies. Answers will strongly depend on the mass-loss rates of these VMS, as discussed next.

Figure 7 shows VFTS results of the mass-loss rates of the most massive stars in 30 Dor (Bestenlehner *et al.* 2014). Whilst at relatively low values of the Eddington value Γ, the slope of the empirical data is consistent with that for O stars, those above the crossover point are not. Here the mass-loss rate kinks upwards, with a steeper slope. The winds have become optically thick, and show WR-like spectra. Also, above this critical Γ point, the wind efficiency crosses unity, enabling a calibration of the absolute mass-loss rates for the first time (Vink & Gräfener 2012). Moreover, Bestenlehner *et al.* (2014) found

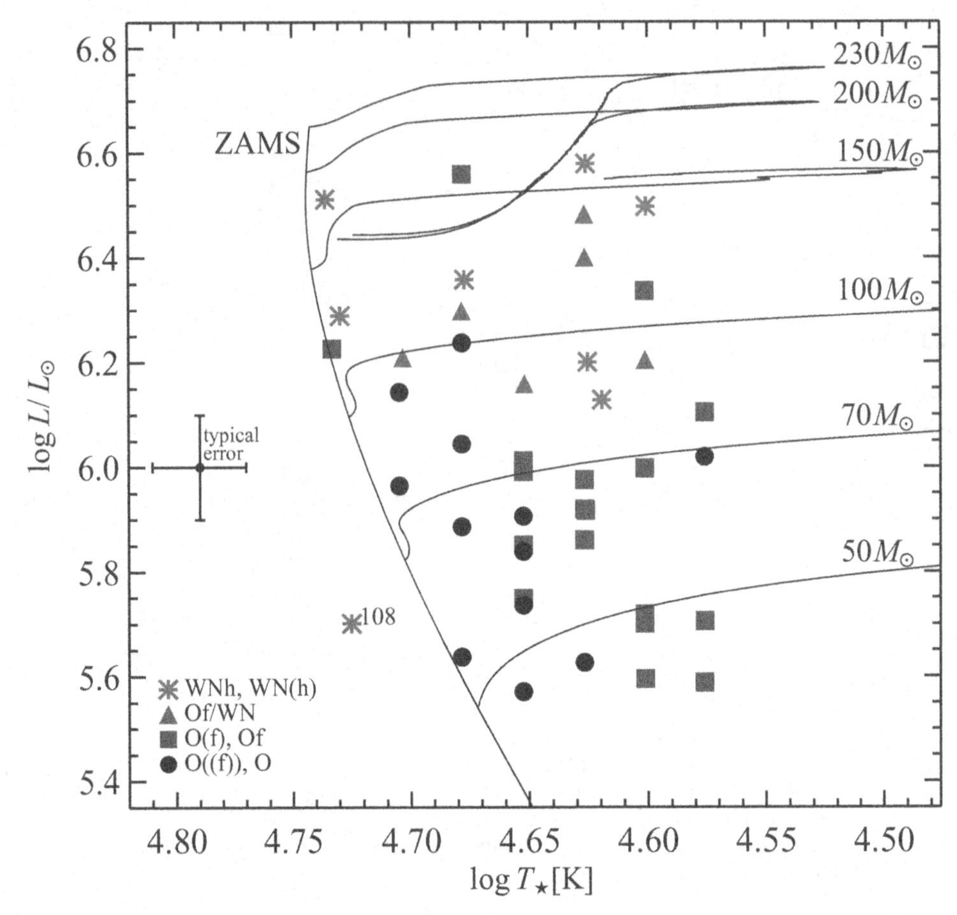

Figure 5. Distribution of spectral types of the sample of Bestenlehner *et al.* (2014) in the HR-diagram. The different symbols indicate different stellar sub-classes. Black lines indicate evolutionary tracks from Köhler *et al.* (2015) for an initial rotation rate of 300 km s^{-1} and the location of the Zero-Age Main Sequence (ZAMS).

profound changes in the surface He abundances exactly coinciding with the luminosity threshold where mass loss is enhanced. This suggests that Of/WN and WNh stars are objects whose H-rich layers have been stripped by enhanced mass-loss during their main-sequence life. Note that this mass-loss enhancement for VMS has not been included in most stellar evolution calculations, and this implies there will be many exciting surprises for extra-galactic applications of massive stars in the near future!

6. Final Words

The VFTS has conclusively shown that binaries are common in 30 Dor. With a corrected close-binary fraction of ~50% (Sana *et al.* 2013), we do not yet know whether this hints at a lower binary frequency at low metallicity, or it it is still consistent with the larger Galactic frequency of ~70% when evolutionary considerations are taken into account. Either way, we now know we require both single & binary evolutionary models to make progress. Another interesting finding is that there is a high-velocity tail present in single O-type supergiants (Ramírez-Agudelo *et al.* 2013), which is not present in the

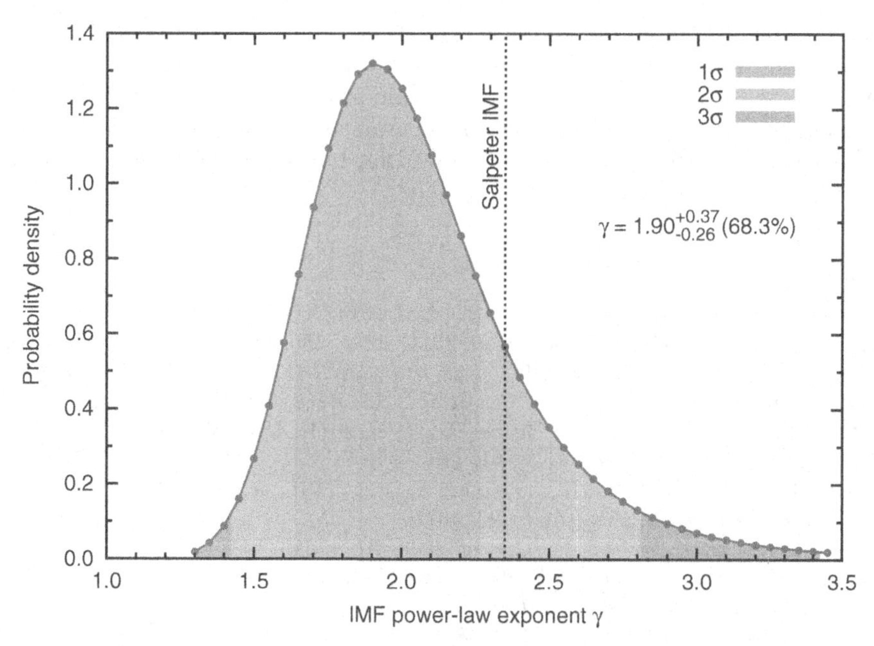

Figure 6. Probability density distribution of the slope of the IMF (Schneider *et al.* 2017).

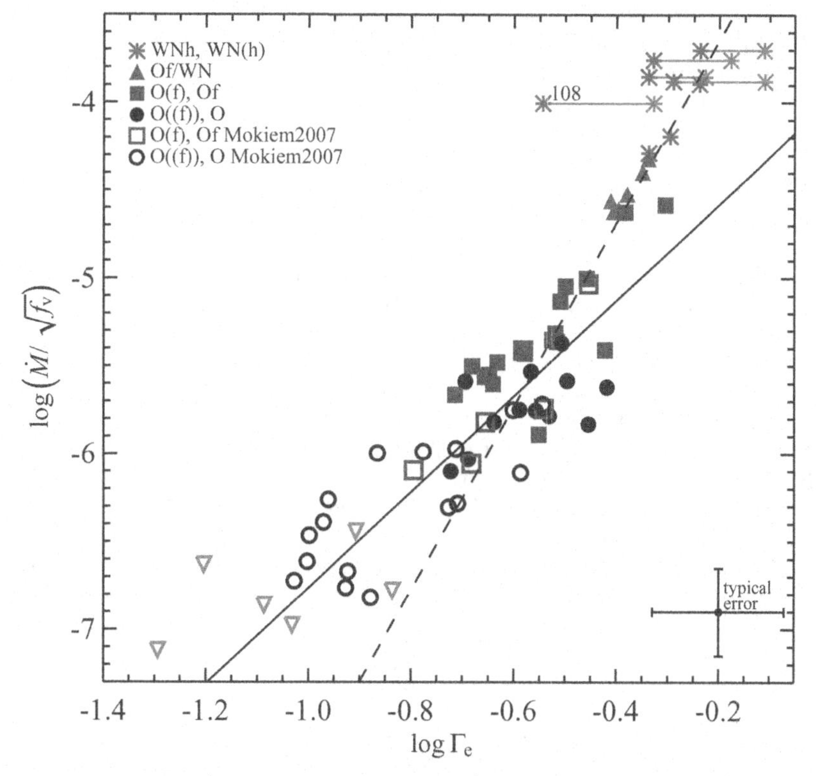

Figure 7. Unclumped $\log \dot{M}$ vs. $\log \Gamma_e$. Solid line: $\dot{M} - \Gamma$ relation for O stars. The different symbols indicate stellar sub-classes from Bestenlehner *et al.* (2014). Dashed line: the steeper slope of the Of/WN and WNh stars. The *kink* occurs at $\log \Gamma_e = -0.58$. The grey asterisks indicate the position of the stars with $Y > 0.75$ under the assumption of core He-burning. The grey upside down triangles are stars from Mokiem *et al.* (2007) which only have upper limits and are excluded from the fit. Note the presence of a kink, as predicted by Vink *et al.* (2011).

spectroscopic binaries (Ramírez-Agudelo *et al.* 2015). This suggests that binary interactions need to be accounted for to understand the underlying rotational distribution.

The VFTS results also indicate that the main-sequence needs widening. This hints at a larger value for the core overshooting parameter than usually adopted. Finally, VMS up to at least $200M_\odot$ are common in 30 Dor, but VMS mass-loss rates have been *under*estimated.

References

Bestenlehner, J. M., Gräfener, G., Vink, J. S., *et al.* 2014, *A&A*, 570, A38
Brott, I., de Mink, S. E., Cantiello, M., *et al.* 2011, *A&A*, 530, A115
Crowther, P. A., Schnurr, O., Hirschi, R., *et al.* 2010, *MNRAS*, 408, 731
Doran, E. I., Crowther, P. A., de Koter, A., *et al.* 2013, *A&A*, 558, A134
Evans, C. J., Taylor, W. D., Hénault-Brunet, V., *et al.* 2011, *A&A*, 530, A108
Gies, D. R. & Bolton, C. T. 1986, *ApJS*, 61, 419
Grin, N. J., Ramírez-Agudelo, O. H., de Koter, A., *et al.* 2016, arXiv:1609.00197
Hillier, D. J. & Miller, D. L. 1998, *ApJ*, 496, 407
Hunter, I., Lennon, D. J., Dufton, P. L., *et al.* 2008, *A&A*, 479, 541
Keszthelyi, Z., Puls, J., & Wade, G. A. 2017, *A&A*, 598, A4
Köhler, K., Langer, N., de Koter, A., *et al.* 2015, *A&A*, 573, A71
Kudritzki, R.-P. & Puls, J. 2000, *ARA&A*, 38, 613
Langer, N. 2012, *ARA&A*, 50, 107
Martins, F. 2015, *Very Massive Stars in the Local Universe*, 412, 9
McEvoy, C. M., Dufton, P. L., Evans, C. J., *et al.* 2015, *A&A*, 575, A70
Mokiem, M. R., de Koter, A., Evans, C. J., *et al.* 2007, *A&A*, 465, 1003
Muijres, L. E., de Koter, A., Vink, J. S., *et al.* 2011, *A&A*, 526, A32
Petrov, B., Vink, J. S., & Gräfener, G. 2016, *MNRAS*, 458, 1999
Puls, J., Urbaneja, M. A., Venero, R., *et al.* 2005, *A&A*, 435, 669
Puls, J., Vink, J. S., & Najarro, F. 2008, *A&Ar*, 16, 209
Ramírez-Agudelo, O. H., Simón-Díaz, S., Sana, H., *et al.* 2013, *A&A*, 560, A29
Ramírez-Agudelo, O. H., Sana, H., de Mink, S. E., *et al.* 2015, *A&A*, 580, A92
Ramírez-Agudelo, O. H., Sana, H., de Koter, A., *et al.* 2017, arXiv:1701.04758
Rivero González, J. G., Puls, J., Najarro, F., & Brott, I. 2012, *A&A*, 537, A79
Sabín-Sanjulián, C., Simón-Díaz, S., Herrero, A., *et al.* 2014, *A&A*, 564, A39
Sana, H., de Koter, A., de Mink, S. E., *et al.* 2013, *A&A*, 550, A107
Stone, R. C. 1991, *AJ*, 102, 333
Sundqvist, J. O., Puls, J., & Owocki, S. P. 2014, *A&A*, 568, A59
Šurlan, B., Hamann, W.-R., Aret, A., *et al.* 2013, *A&A*, 559, A130
Vink, J. S. & Gräfener, G. 2012, *ApJL*, 751, L34
Vink, J. S., de Koter, A., & Lamers, H. J. G. L. M. 2001, *A&A*, 369, 574
Vink, J. S., Brott, I., Gräfener, G., *et al.* 2010, *A&A*, 512, L7
Vink, J. S., Heger, A., Krumholz, M. R., *et al.* 2015, *Highlights of Astronomy*, 16, 51

The Lives and Death-Throes of Massive Stars
Proceedings IAU Symposium No. 329, 2016
J.J. Eldridge, J.C. Bray, L.A.S. McClelland & L. Xiao, eds.

doi:10.1017/S1743921317003076

The Massive stellar Population at the Galactic Center

Francisco Najarro[1], Diego de la Fuente[1], Tom R. Geballe[2], Don F. Figer[3] and D. John Hillier[4]

[1]Centro de Astrobiología (CSIC/INTA), ctra. de Ajalvir km. 4, 28850 Torrejón de Ardoz, Madrid, Spain [2]Gemini Observatory, 670 N. A'ohoku Place, Hilo, HI 96720, USA
[3]Center for Detectors, Rochester Institute of Technology, 54 Lomb Memorial Drive, Rochester, NY 14623, USA
[4]Department of Physics and Astronomy, University of Pittsburgh, 3941 O'Hara Street, Pittsburgh, PA 15260

Abstract. We present results from our ongoing infrared spectroscopic studies of the massive stellar content at the Center of the Milky Way. This region hosts a large number of apparently isolated massive stars as well as three of the most massive resolved young clusters in the Local Group. Our survey seeks to infer the presence of a possible top-heavy recent star formation history and to test massive star formation channels : clusters vs isolation.

Keywords. stars: early-type –stars: mass loss – stars: winds – stars: abundances – Galaxy: center – infrared: stars

---

1. Massive stars and massive star formation in the GC.

Hosting three of the most massive young clusters in the Local Group (Central, Arches and Quintuplet), the Galactic Center (GC) constitutes an ideal test-bed to investigate massive stars and massive star formation. Whether the latter occurs in a similar way as in the giant molecular clouds elsewhere in the Galaxy, despite the harsh environment pervaded with intense radiation fields and high densities, has been a matter of debate over the last two decades. Several studies have aimed at establishing whether the Initial Mass Functions (IMFs) and Present Day Mass functions (PDMFs) of these clusters are top-Heavy and therefore characterized by a shallower slope caused by an excess of massive stars when compared to a Salpeter distribution (Figer *et al.* 1999; Stolte *et al.* 2002; Kim *et al.* 2006; Espinoza *et al.* 2009; Bartko *et al.* 2010; Hu*ß*mann *et al.* 2012). Being the youngest of the three GC clusters, the Arches constitutes an unique laboratory to investigate the shape of the IMF and PDMF. However, the results obtained range from quite shallow slopes (Figer *et al.* 1999; Stolte *et al.* 2002; Kim *et al.* 2006) to Salpeter-like (Habibi *et al.* 2013) or half-way (Espinoza *et al.* 2009) and are strongly dependent on the extinction and transformations of the photometric systems.

Over the last decade, new ingredients have been added to the overall picture, as several studies have revealed the presence of a large number of isolated massive stars (Mauerhan *et al.* 2010a,b) at the GC which is comparable to the massive star population of each of the clusters (Figer *et al.* 1999, 2002). Such detection of apparently isolated massive stars in this region has raised a further fundamental issue - whether these "massive field stars" are results of tidal interactions among clusters, are escapees from a disrupted cluster, or represent a new mode of massive star formation in isolation (Dong *et al.* 2015). Following the numerical dynamical simulations from Harfst *et al.* (2010), and including the effects of stellar evolution and the orbit of the Arches cluster in the Galactic Center potential, Habibi *et al.* (2014) investigated the first option and found that models were able to

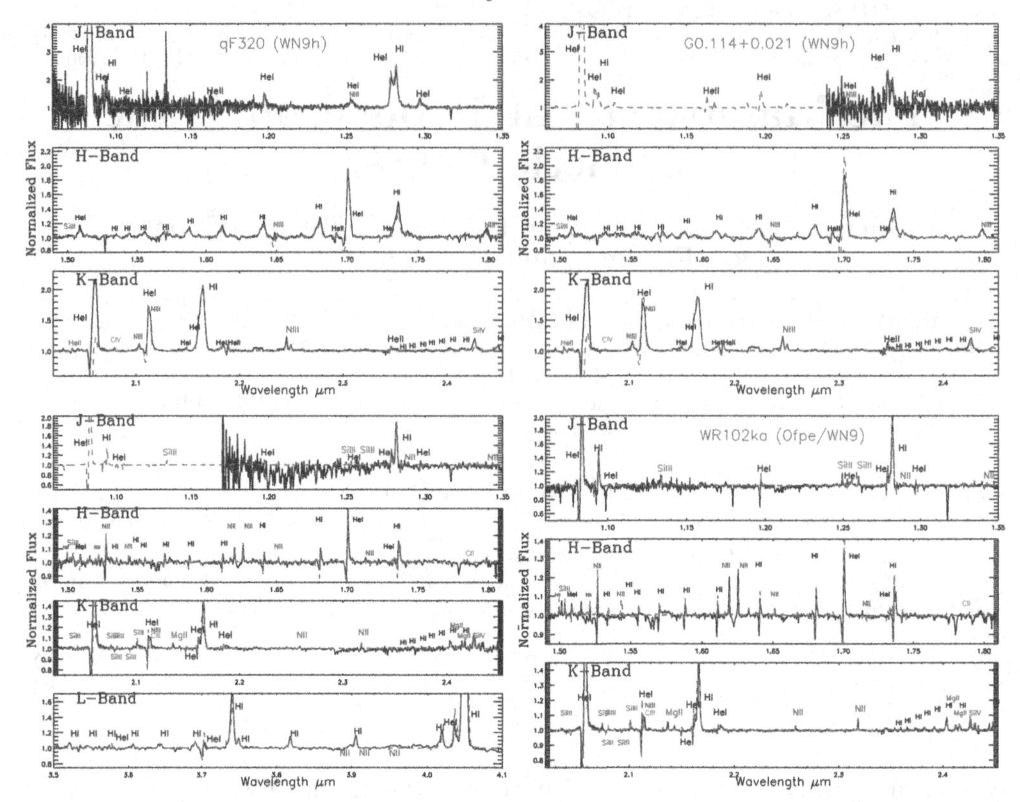

Figure 1. Top. A consistent α-element enhancement is found for the Quintuplet sources **(left)** and isolated massive stars **(right)**. **Bottom** Breaking the $T_{\rm eff}$ degeneracy in OIfpe/WN9 stars. **Bottom-feft)** Q8 in the Quintuplet Cluster. **Bottom-right)** The isolated WR102ka star (see text).

account for ~60% of the isolated sources within the central 100pc as sources drifted away from the center of the clusters. On the other hand, radial velocity measurements of a sample of eight objects in the vicinity of the Arches cluster (Dong *et al.* 2015) suggest that two of them could have been associated with the cluster while other two likely formed in isolation. The latter option was also inferred for the very luminous WR102ka star (Oskinova *et al.* 2013) from radial velocity studies and a deep integral-field K-band spectroscopic survey of its surroundings. We note, however, that radial velocity estimates of these objects (mainly OIf+ and WNh) may be subject to high uncertainties, as the spectral lines utilized in these studies are severely contaminated by the stellar winds (see below). Thus, further detailed evidence for or against these scenarios is still lacking and awaits precise proper motion measurements (currently underway) providing 3-D velocities of the sources relative to the clusters.

Confronting the stellar properties of the isolated sources, including ages and abundances with those of the putative feeding clusters (Quintuplet and Arches) through detailed spectroscopic studies, may constitute a major step to differentiate among the above scenarios. Comparison of the results of the quantitative model-atmosphere analysis to theoretical isochrones will allow us to determine if these stars were born in single co-eval cold molecular cloud event or formed over an extended (eg, 1-10Myr) period. Obtaining metal abundances from these "field" objects is crucial to understand the metal

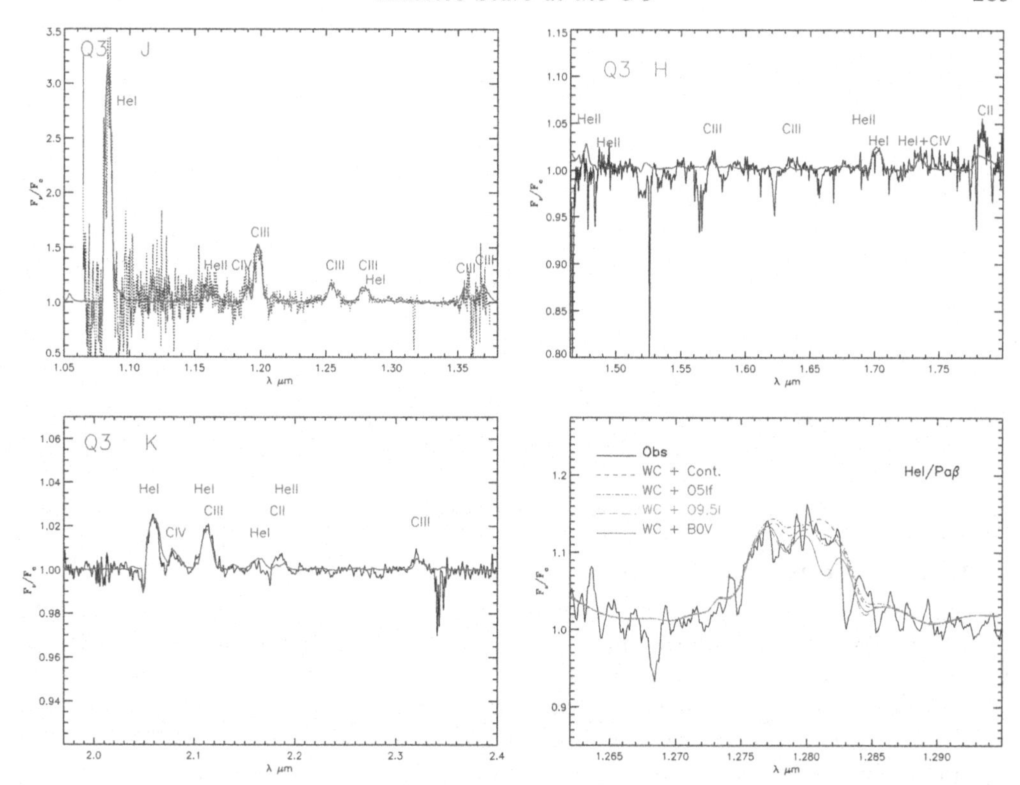

Figure 2. First detection of stellar lines in the NIR spectra of the dusty WCL proper members of the Quintuplet Cluster. J,H and K band observations and model fits for GCS4 (Q3). The bottom-right plot displays the potential of Paschenβ to trace the presence of the OB companion.

enrichment history of the GC and to test whether these isolated stars have followed a metal-enrichment scenario different than those in the GC clusters.

The Quintuplet and Arches clusters provide the stellar reference sources to perform such studies. At its present evolutionary phase (age ~4 Myr, Figer *et al.* 1999) the massive members of the Quintuplet Cluster are currently WN9-10h (plus a WN6), weak lined WC9, OIf+ stars and LBVs (Figer *et al.* 1999; Liermann *et al.* 2009), while the massive population of the younger Arches cluster is dominated by WN8-9h and OIf+ stars (Figer *et al.* 2002; Najarro *et al.* 2004; Martins *et al.* 2008). These spectral types basically encompass all the the isolated evolved massive stars identified from recent follow-up spectroscopic observations of Paα emission objects in the GC (Mauerhan *et al.* 2010a,b) which show quite similar spectral morphology to those present in the Quintuplet and some of the brightest Arches members.

Abundance analyses may as well help to distinguish between top-heavy and standard star formation in the region. In a top-heavy IMF scenario, the occurrence of a larger number of type II supernovae produce enhanced yields of α-elements, resulting in an increase of α-element vs Fe (the main suppliers of iron are type Ia SNe). Najarro *et al.* (2004, 2009) have shown that quantitative NIR spectroscopy of high-mass stars may provide estimates of both absolute abundances and abundance ratios, telling us about the global integrated enrichment history up to the present, placing constraints on models of galactic chemical evolution, and acting as clocks by which chemical evolution can be measured.

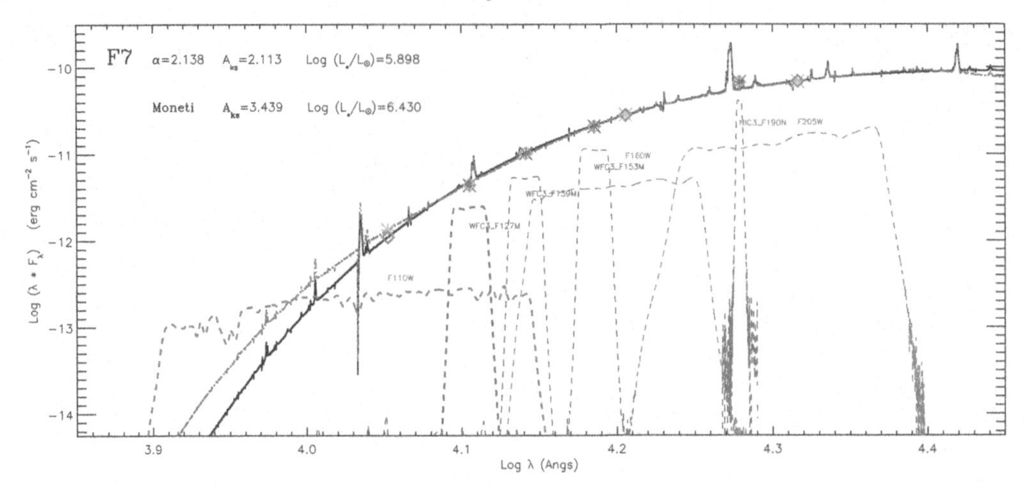

Figure 3. The importance of the extinction law to infer Luminosities and masses (see text).

2. Observations and results

Over the last five years we have been carrying an observing program at GEMINI North to obtain high S/N, medium-resolution spectra of the most massive stars in the Quintuplet (including the dusty WCL proper members) as well as isolated massive stars in the inner GC. So far, around 20 massive stars have been observed. We are currently modeling the early-type spectra with the CMFGEN code (Hillier & Miller 1998) to obtain physical and chemical properties and present below some results from our ongoing analysis:

We obtain a clear α-element enrichment from the analysis of the Quintuplet WNh stars, consistent with the results derived for the LBVs (Najarro *et al.* 2009) which denoted a clearly enhanced α/Fe=2 ratio with respect to solar (Fig.1-right).

Stellar abundances of the isolated objects (Fig.1-left) seem to show a similar trend on average, with the presence of even higher α-element enrichment in some cases.

Our new high S/H spectroscopic J, H and K data provide important diagnostic lines (N II-III, SiII-IV, CII-IV, etc), which are crucial not only for abundance determinations but also to constrain stellar properties. As an example, previously found uncertainties in $T_{\rm eff}$ ($\pm$6000 K) for the Ofpe/WN9 objects at the GC (Najarro *et al.* 1997; Martins *et al.* 2007) due to the lack of He II lines in the spectra, are drastically reduced ($\pm$1000 K)by making use of the N II/N III and Si III/Si IV ionization equilibria. This is shown in Fig. 1-bottom, where model fits are displayed for two OIfpe/WN9 stars in our sample (Q8 in the Quintuplet cluster and WR102ka, an object relatively far from the three clusters).

When available the HI and He lines at Pβ, provide an excellent He/H ratio diagnostic, allowing much more accurate He abundance determinations than those performed by means of K-Band spectra. Simple blue (He I) to red (H I) peak ratios may be used (see Pβ complex all four J-Band spectra in Fig. 1)

Radial velocity estimates, if obtained for OIf+ and WNh stars, require detailed modeling of the observed spectra (Figer *et al.* 2004). For the OIf+ stars, the He II absorption lines, which are decent diagnostics for OV and for some OI stars with weak-to-moderate stellar winds start to be filled by the stellar wind, producing an effective blue-shift as high as 80-90km s^{-1}. This may have important consequences when associating the radial velocities of these objects with the nearby gas and clusters. Further, even quantitative modeling may suffer from high uncertainties. Our fits to WR102ka making use of the full J, H and K band spectra (Fig. 1-bottom-right) reveal a radial velocity of $\sim$100km s^{-1}.

This value differs significantly from the 60km s^{-1} obtained by Oskinova *et al.* (2013) by means of only K-band spectra and with a slightly (%25) lower spectral resolution.

We detect, for the first time, the WR stellar lines in the NIR SEDs of the dusty WCL proper members in the Quintuplet Cluster (see Fig.2). The deep J band spectra, where the dust contribution weakens clearly display the presence of the He I-II and C III-IV lines of the WC9 component. Further, the huge S/N achieved in the H and K band allows to clearly identify and model the stellar lines which are severely diluted by the dust continuum. Thus, we confirm spectroscopically the WR+OB binary nature of these systems and, fitting spectrophotometric data, derive the individual contributions of the dust, WR and OB components to the total SED of the system. Figure 2 displays as well the potential of the observed Pβ line profile to infer the nature of the OB-component.

Finally, it is worth mentioning the crucial role of extinction, a skeleton on everyone's closet, and its implications on the inferred luminosities and masses of the massive stars and therefore on the IMFs (PDMFs) of the clusters. Figure 3 displays two different reddening approaches to the SED of the stellar model which best reproduces the K-band spectrum of the WNh star F7 in the Arches cluster. Both extinction laws match satisfactorily the observed HST photometry from J to K band but yield Luminosities with differ by more than 0.5dex. A much lower luminosity and stellar mass (70 vs 130$M_{\odot}$) is obtained if a simple $A_{\lambda} \propto (\lambda/\lambda_K)^{\alpha}$ relation is used for the whole NIR compared to previously utilized (Figer *et al.* 2002; Najarro *et al.* 2004) laws. By securing spectra of the O dwarfs in the Arches, for which the scatter in intrinsic K-Band luminosity is much lower than for the WNh stars, one could definitely discern between both extinction approaches.

F.N. and D.dF. acknowledge grants FIS2012-39162-C06-01, ESP2013-47809-C3-1-R and ESP2015-65597-C4-1-R

References

Bartko, H., Martins, F., Trippe, S., *et al.* 2010, *ApJ*, 708, 834
Dong, H., Mauerhan, J., Morris, M. R., Wang, Q. D., & Cotera, A. 2015, *MNRAS*, 446, 842
Espinoza, P., Selman, F. J., & Melnick, J. 2009, *A&A*, 501, 563
Figer, D. F., McLean, I. S., & Morris, M. 1999, *ApJ*, 514, 202
Figer, D. F., Najarro, F., Gilmore, D., *et al.* 2002, *ApJ*, 581, 258
Figer, D. F., Najarro, F., & Kudritzki, R. P. 2004, *ApJ* (Letters), 610, L109
Habibi, M., Stolte, A., Brandner, W., Hußmann, B., & Motohara, K. 2013, *A&A*, 556, A26
Habibi, M., Stolte, A., & Harfst, S. 2014, *A&A*, 566, A6
Harfst, S., Portegies Zwart, S., & Stolte, A. 2010, *MNRAS*, 409, 628
Hillier, D. J. & Miller, D. L. 1998, *ApJ*, 496, 407
Huβmann, B., Stolte, A., Brandner, W., Gennaro, M., & Liermann, A. 2012, *A&A*, 540, A57
Kim, S. S., Figer, D. F., Kudritzki, R. P., & Najarro, F. 2006, *ApJ* (Letters), 653, L113
Liermann, A., Hamann, W.-R., & Oskinova, L. M. 2009, *A&A*, 494, 1137
Martins, F., Genzel, R., Hillier, D. J., *et al.* 2007, *A&A*, 468, 233
Martins, F., Hillier, D. J., Paumard, T., *et al.* 2008, *A&A*, 478, 219
Mauerhan, J. C., Cotera, A., Dong, H., *et al.* 2010a, *ApJ*, 725, 188-199
Mauerhan, J. C., Muno, M. P., Morris, M. R., Stolovy, S. R., & Cotera, A. 2010b, *ApJ*, 710, 706
Najarro, F., Krabbe, A., Genzel, R., *et al.* 1997, *A&A*, 325, 700
Najarro, F., Figer, D. F., Hillier, D. J., & Kudritzki, R. P. 2004, *ApJ* (Letters), 611, L105
Najarro, F., Figer, D. F., Hillier, D. J., Geballe, T. R., & Kudritzki, R. P. 2009, *ApJ*, 691, 1816
Oskinova, L. M., Steinke, M., Hamann, W.-R., *et al.* 2013, *MNRAS*, 436, 3357
Stolte, A., Grebel, E. K., Brandner, W., & Figer, D. F. 2002, *A&A*, 394, 459

The Lives and Death-Throes of Massive Stars
Proceedings IAU Symposium No. 329, 2016
J.J. Eldridge, J.C. Bray, L.A.S. McClelland
& L. Xiao, eds.

doi:10.1017/S1743921317002484

The Tarantula Nebula as a template for extragalactic star forming regions from VLT/MUSE and HST/STIS

Paul A. Crowther[1], Saida M. Caballero-Nieves[1,2], Norberto Castro[3] and Christopher J. Evans[4]

[1]Department of Physics & Astronomy, University of Sheffield, Hounsfield Road, Sheffield, S3 7RH, UK
email: Paul.Crowther@sheffield.ac.uk

[2]Physics & Space Sciences, Florida Institute of Technology, 150 W. University Blvd, Melbourne, FL 32901, USA

[3]Department of Astronomy, University of Michigan, 1805 S.University, Ann Arbor, MI 48109, USA

[4]UK Astronomy Technology Centre, Royal Observatory Edinburgh, Blackford Hill, Edinburgh, EH9 3HJ, UK

Abstract. We present VLT/MUSE observations of NGC 2070, the dominant ionizing nebula of 30 Doradus in the LMC, plus HST/STIS spectroscopy of its central star cluster R136. Integral Field Spectroscopy (MUSE) and pseudo IFS (STIS) together provides a complete census of all massive stars within the central 30×30 parsec^2 of the Tarantula. We discuss the integrated far-UV spectrum of R136, of particular interest for UV studies of young extragalactic star clusters. Strong He II λ1640 emission at very early ages (1–2 Myr) from very massive stars cannot be reproduced by current population synthesis models, even those incorporating binary evolution and very massive stars. A nebular analysis of the integrated MUSE dataset implies an age of ~4.5 Myr for NGC 2070. Wolf-Rayet features provide alternative age diagnostics, with the primary contribution to the integrated Wolf-Rayet bumps arising from R140 rather than the more numerous H-rich WN stars in R136. Caution should be used when interpreting spatially extended observations of extragalactic star-forming regions.

Keywords. stars: early-type – stars: Wolf-Rayet – open clusters and associations: individual: NGC 2070 – galaxies: star clusters: individual: R136 – ISM: HII regions

---

1. Introduction to the Tarantula Nebula

Our interpretation of distant, unresolved star-forming regions relies heavily upon population synthesis models, including Starburst99 (Leitherer *et al.* 2014) and BPASS (Eldridge & Stanway 2012). However, such models rely on a number of key assumptions, involving the initial mass function (IMF), rotation, star formation history. In particular, the post-main sequence evolution of massive stars depends sensitively on internal mixing processes and adopted mass-loss prescriptions. Locally, close binary evolution looks to play a major role (Sana *et al.* 2012) and there is some evidence that the upper IMF extends well beyond the usual $M_{\rm max} \sim 100 M_{\odot}$ (Crowther *et al.* 2010).

Therefore, it is important to benchmark population synthesis models against empirical results from nearby, resolved star-forming regions. Within the Local Group of galaxies, the LMC's Tarantula Nebula is the closest analogue to the intensive star-forming clumps of high-redshift galaxies (Jones *et al.* 2010). Kennicutt (1984) compares the nebular properties of nearby extragalactic HII regions, including the Tarantula (alias 30 Doradus). Although the Tarantula Nebula extends over several hundred parsecs, the central ionizing

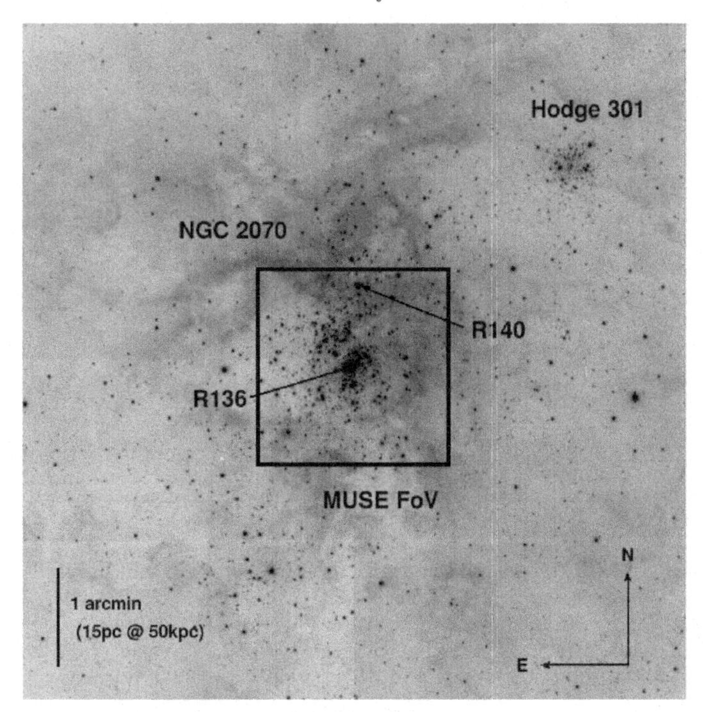

Figure 1. VLT/FORS2 R-band image of the central 6×6 arcmin^2 of the Tarantula Nebula, corresponding to 100×100 parsec at the 50kpc LMC distance, indicating NGC 2070, R136, R140, Hodge 301 and the VLT/MUSE field of view.

region, NGC 2070, spans 40 pc, with the massive, dense cluster R136† at its core (see Table 1 of Walborn 1991).

In this work, we exploit the high spatial resolution of HST to dissect the central 4×4 arcsec^2 of the R136 star cluster (Crowther *et al.* 2016) plus the new VLT large field-of-view integral field spectrograph MUSE which has observed the central 2×2 arcmin^2 region of NGC 2070 (N. Castro *et al.* in prep). Figure 1 shows a VLT/FORS2 R-band image of the central region of the Tarantula, including NGC 2070 and an older star cluster (Hodge 301) 3′ to the NW. Comparisons with population synthesis predictions are made to assess their validity for young starburst regions.

2. HST/STIS ultraviolet spectroscopy of R136

R136 is the young massive star cluster at the core of the Tarantula Nebula. Establishing the stellar content of R136 has generally required high spatial spectroscopy from HST at UV and optical wavelengths or a ground-based 8m telescope with Adaptive Optics. Massey & Hunter (1998) have used HST/FOS to observe stars in the R136 region (few parsec), while Crowther *et al.* (2016) used HST/STIS to map the core of R136, obtaining complete UV/optical spectroscopy of the central parsec. Spectroscopic analysis of the brightest 52 O stars reveals a cluster age of $1.5^{+0.3}_{-0.7}$ Myr, confirming this as one of the youngest massive clusters in the Local Group.

The integrated UV spectrum of R136 exhibits strong stellar C IV $\lambda\lambda$1548-51, N V $\lambda\lambda$1238-42 P Cygni profiles plus prominent He II λ1640 emission. Within unresolved high

† Strictly, the star cluster is R136a, with R136b and R136c representing individual very massive stars, but R136 is generally used to describe the cluster

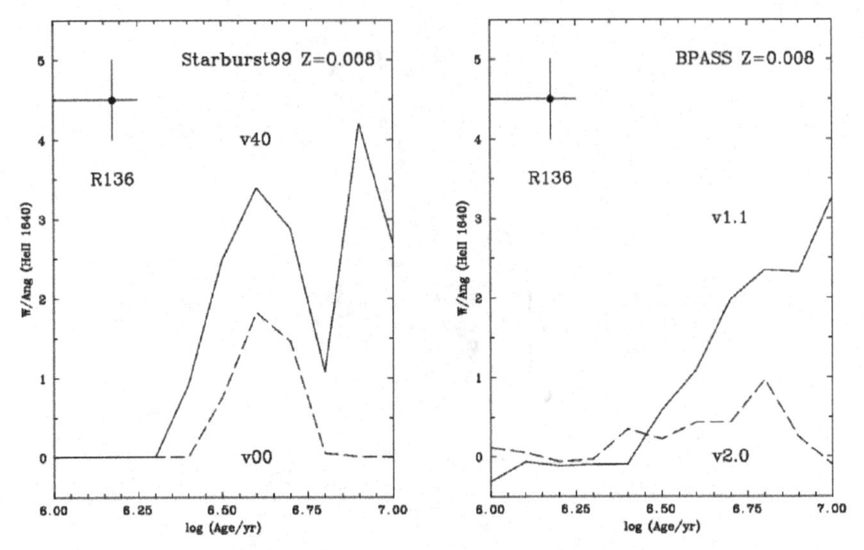

Figure 2. Comparison between observed He II λ1640 emission in R136 (Crowther *et al.* 2016) and Z = 0.008 instantaneous burst models from (left): Starburst99 (Leitherer *et al.* 2014) for non-rotating (v00, dashed) and rotating (v40, solid) Geneva single star models with $M_{\rm max} = 120 M_{\odot}$, (right): BPASS (Eldridge & Stanway 2012, Stanway *et al.* 2016) close binary models for $M_{\rm max} = 100 M_{\odot}$ (v1.1, solid) and $M_{\rm max} = 300 M_{\odot}$ (v2.0, dashed).

mass star clusters, He II λ1640 emission is usually believed to originate from classical He-burning Wolf-Rayet (WR) stars which become prominent after several Myr. The presence of strong He II emission in the young R136 indicates a different origin, namely the presence of very massive stars (VMS, $M_{\rm init} \geqslant 100 M_{\odot}$). Indeed, ∼10 VMS within the central parsec contribute one third of the integrated far-UV continuum, and dominate the observed He II emission.

Standard population synthesis models fix the upper mass limit at $M_{\rm max} \sim 100 M_{\odot}$, so fail to produce significant He II λ1640 emission until ∼2–3 Myr after the initial burst, as illustrated in the left panel of Fig 2, which compares predictions from Starburst99 based on metal-poor (Z = 0.008) Geneva models (Leitherer *et al.* 2014) with the observed strong, broad He II emission in R136 (Crowther *et al.* 2016). These predictions are based on single, non-rotating (v00) or rotating (v40) models, so exclude the contribution of close binary evolution and VMS. Close binary evolution has been included into metal-poor Z = 0.008 BPASS burst models (v1.1, Eldridge & Stanway 2012), which is presented on the right panel of Fig. 2 together with recent updates (v2.0, Stanway et al. 2016) in which very massive stars are incorporated. Weaker emission in the more recent models for ages of ⩾3 Myr results from the use of different spectral libraries.

It is apparent that both single and binary population synthesis models currently fail to reproduce the observed strong He II λ1640 in R136, indicating an incomplete treatment of rotation, binarity and mass-loss for the most massive stars in star clusters. In the nearby universe, a number of young massive star clusters also exhibit strong He II λ1640 emission (Wofford *et al.* 2014, Smith *et al.* 2016) while it is a prominent (stellar and nebular) spectral feature in the integrated rest-frame UV light of high redshift Lyman Break galaxies (Shapley *et al.* 2003). VMS and classical WR stars will initially play a major role in stellar He II λ1640 emission, with close binary evolution subsequently dominant.

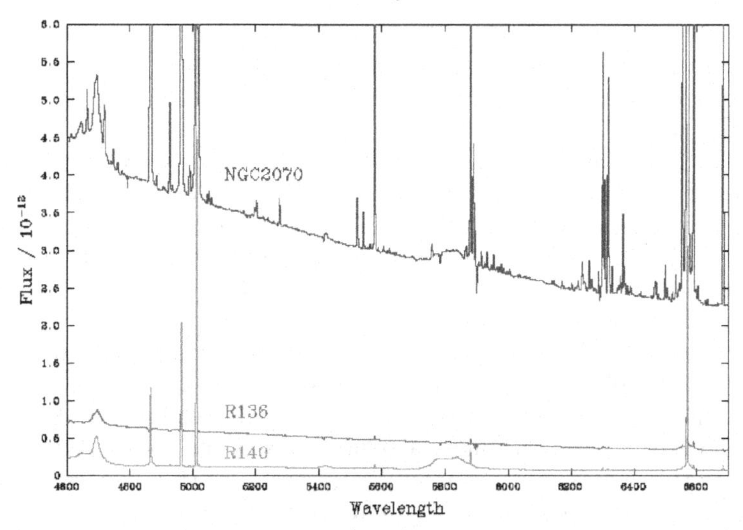

Figure 3. Integrated VLT/MUSE spectrum of NGC 2070 from N. Castro *et al.* (in prep), highlighting strong blue (He II λ4686, C III $\lambda\lambda$4647-51) and yellow (C IV $\lambda\lambda$5801-12) Wolf-Rayet features, including the individual contributions from R136 and R140.

3. VLT/MUSE integral field spectroscopy of NGC 2070

Individual massive stars within NGC 2070 – the dominant HII region within the Tarantula Nebula – have been extensively investigated from the ground (Melnick 1985, Walborn & Blades 1997, Bosch *et al.* 1999, Evans *et al.* 2011), but the advent of large field-of-view integral field spectrographs such as MUSE permit the integrated properties of nearby giant HII regions to be investigated. We have obtained four pointings of VLT/MUSE which cover the central region of NGC 2070 (Fig. 1), and provide medium resolution optical spectroscopic observations (R = 3,000, λ= 4600–9360Å) of the 2×2 arcmin^2 region, sampled at 0.3 arcsec spaxel^{-1}.

Nebular and stellar kinematics of NGC 2070 are discussed in N. Castro *et al.* (in prep), while the individual stellar content will be discussed elsewhere. Here we focus upon the integrated stellar and nebular properties of the VLT/MUSE dataset, which samples a 30×30 parsec^2 region of NGC 2070, including R136. At a distance of 10 Mpc, this region would subtend only 0.6 arcsec, typical of long-slit spectroscopy of extragalactic HII regions from large ground-based telescopes. The integrated VLT/MUSE spectrum of NGC 2070 is presented in Fig. 3, revealing prominent nebular emission features. An interstellar extinction of E(B-V) = 0.38 follows from the observed Hα/Hβ flux ratio, in good agreement with Pellegrini *et al.* (2010), while the measured Hα flux is 60% of that measured by Kennicutt *et al.* (1995) for a 3′ radius centred upon R136, such that the inferred Hα luminosity, $\log L(\mathrm{H}\alpha)$ = 39.1 erg s^{-1}, corresponds to 10% of the total emission from the Tarantula Nebula (Kennicutt 1984).

Commonly used diagnostics to estimate the age of young extragalactic star forming regions are nebular Balmer emission and the presence of WR features in the integrated light. We infer an age of ~4.5 Myr from $\log W_\lambda$ (Hα)/Å = 2.85 for NGC 2070, even though it is clear that the individual ages of massive stars within NGC 2070 span a broad range (e.g. Selman *et al.* 1999, F. Schneider *et al.* in prep) with massive star formation still ongoing in NGC 2070 (Walborn *et al.* 2013) plus Hodge 301 to the NW of NGC 2070 indicating a burst of star formation 15–25 Myr ago. Recalling Fig. 3, both the blue (He II λ4686, C III $\lambda\lambda$4647-51) and yellow (C IV $\lambda\lambda$5801-12) WR features are prominent in the integrated VLT/MUSE dataset, with W_λ(blue) ~12Å, and W_λ(yellow) ~9Å. As with

He II λ1640 in ultraviolet spectroscopy, it is usually assumed that WR bumps arise from classical He-burning stars. Indeed, for a burst age of 4–5 Myr, Starburst99 models at Z = 0.008 predict W_λ(blue)∼15Å (5Å) and W_λ(yellow)∼6Å (3Å) from v40 rotating (v00 non-rotating) models.

Therefore, at face value it appears that a single 4–5 Myr burst is reasonably consistent with the observed nebular and stellar line features in NGC 2070. However, integral field spectroscopy also permits us to consider the origin of these WR features within NGC 2070. Fig. 3 also displays the integrated spectrum of R136, which exhibits weak He II λ4686 emission from the VMS in this young ∼1.5 Myr cluster, contributing only 15% of the blue NGC 2070 WR feature, and crucially negligible emission at C IV λλ5801–12 since it is too young to host any WC stars. Indeed, the dominant source of the yellow WR feature in NGC 2070 is R140, a relatively modest group of stars including two classical WN stars and a WC star (indicated in Fig. 1, while R140 also contributes 25% of the integrated blue WR feature. The higher line luminosities of classical WR stars (in R140) with respect to H-rich WN stars (in R136) compensate for their reduced population.

Therefore caution should be used when interpreting spatially extended regions in extragalactic star forming regions. In the absence of spatially resolved spectroscopy, one would anticipate that the integrated stellar and nebular properties of NGC 2070 are dominated by the R136 region, whereas the nebular-derived age represents a composite of the young R136 cluster and older OB stars, while the stellar-derived age is biased towards WR stars possessing the highest line luminosities (R140), rather than the those within the dominant ionizing cluster (R136). By way of example, the dominant cluster in NGC 3125-A has an age of 1–2 Myr (Wofford *et al.* 2014) while a burst age of 4 Myr is inferred from its associated star-forming region from nebular and WR diagnostics (Hadfield & Crowther 2006).

References

Bosch, G., Terlevich, R., Melnick, J., & Selman, F. 1999, *A&AS* 137, 21
Crowther, P. A., Schnurr, O., Hirschi, R. *et al.* 2010, *MNRAS* 408, 731
Crowther, P. A., Caballero-Nieves, S., Boestrom, K. A. *et al.* 2016, *MNRAS* 458, 624
Eldridge, J. J. & Stanway, E. R. 2012, *MNRAS* 419, 479
Evans, C. J., Taylor, W. D., Henault-Brunet, V. *et al.* 2011, *A&A* 530, A108
Hadfield, L. J. & Crowther, P. A. 2006, *MNRAS* 368, 1822
Kennicutt, R. C. J.r 1984, *ApJ* 287, 116
Kennicutt, R. C. J.r, Bresolin, F., Bomans, D. J., Bothun, G. D., & Thompson, I. B. 1995, *AJ* 109, 594
Jones, T. A., Swinbank, A. M., Ellis, R. S., & Stark, D. P. 2010, *MNRAS* 404, 1247
Massey, P. & Hunter, D. A. 1998, *ApJ* 493, 180
Melnick, J. 1985, *A&A* 153, 235
Pellegrini, E. W., Baldwin, J. A., & Ferland, G. J. 2010, *ApJS* 191, 160
Sana, H., de Mink, S. E., de Koter, A. *et al.* 2012, *Science* 337, 444
Selman, F., Melnick, F., Bosch, G., & Terlevich, R. 1999, *A&A* 347, 532
Shapley, A. E., Steidel, C. C., Pettini, M., & Adelberger, K. L. 2003, *ApJ* 588, 65
Smith, L. J., Crowther, P. A., Calzetti, D., & Sidoli, F. 2016, *ApJ* 823, 38
Stanway, E. R., Eldridge, J. J. & Becker G. D. 2016, *MNRAS* 456, 485
Walborn, N. R. 1991, in: Haynes, R. & Milne, D. (eds.), *The Magellanic Clouds*, Proceedings of IAU Sympisium No. 148, (Dordrecht: Kluwer), p. 145
Walborn, N. R. & Blades, C. J. 1997, *ApJS* 112, 457
Walborn, N. R., Barba, R. H., & Sewilo, M. M. 2013, *AJ* 145, 98
Wofford, A., Leitherer, C., Chandar, R., & Bouret, J.-C. 2014, *ApJ* 781, 122

The Lives and Death-Throes of Massive Stars
Proceedings IAU Symposium No. 329, 2016
J.J. Eldridge, J.C. Bray, L.A.S. McClelland
& L. Xiao, eds.

doi:10.1017/S1743921317003118

Extragalactic Supergiants

Miguel A. Urbaneja[1] and Rolf P. Kudritzki[2]

[1]Institut für Astro- und Teilchen-Physik, Universität Innsbruck,
Technikerstr. 25/8, A-6020, Innsbruck, Austria
email: `Miguel.Urbaneja-Perez@uibk.ac.at`

[2]Institute for Astronomy, University of Hawaii,
2680 Woodlawn Drive, Honolulu HI 96822, USA
email: `kud@ifa.hawaii.edu`

Abstract. Blue supergiant stars of B and A spectral types are amongst the visually brightest non-transient astronomical objects. Their intrinsic brightness makes it possible to obtain high quality optical spectra of these objects in distant galaxies, enabling the study not only of these stars in different environments, but also to use them as tools to probe their host galaxies. Quantitative analysis of their optical spectra provide tight constraints on their evolution in a wide range of metallicities, as well as on the present-day chemical composition, extinction laws and distances to their host galaxies. We review in this contribution recent results in this field.

Keywords. stars: abundances, early-type, supergiants; galaxies: abundances, distances and redshifts; cosmology: distance scale.

---

1. Introduction

Blue supergiant stars with B and A spectral types are amongst the visually brightest non-transient astronomical objects. They are the descendants of stars born in the mass range between 15 to 40 $M_{\odot}$, in a short-lived evolutionary phase, after the end of the H-core burning phase. Their intrinsic brightness makes it possible to obtain high quality optical spectra of these objects in distant galaxies, enabling the study not only of these stars in different environments, but also to use them as tools to probe their host galaxies. Their optical spectra are rich in metal absorption lines from many different elements (among others, C, N, O, Mg, Si, S, Ti, Fe). As young objects, with ages of a few Myr, they provide key probes of the current chemical composition of the interstellar medium. Quantitative interpretation of these optical spectra by means of sophisticated model atmosphere/line formation codes, coupled with efficient analysis methods, allowed in recent years the study of individual supergiants in galaxies well beyond the realm of the Local Group, providing information not only on the characteristics of the stars, but also opening a window to investigate properties of their host galaxies, such as reddening, extinction laws, chemical composition and distances.

Whilst studies of these objects within the Local Group have been conducted for a while (see Kudritzki *et al.* 2008a and references therein), the jump to more distant galaxies is relatively recent, the main reasons being the availability of very efficient multi-object spectrographs attached to the generation of 10m class telescopes, in combination with advanced radiative transfer models that properly treat the physics describing the atmospheres of these objects (intense radiation fields propagating in low density environments) and the development of novel analysis techniques. This last is a crucial point, since spectral resolution has to be sacrificed in order to achieve very good signal-to-noise ratios for faint objects. Following on the first ideas presented in Kudritzki *et al.* (1995), the seminal papers by Kudritzki *et al.* (2008b) for the case of mid B to A-type supergiants, and by Urbaneja *et al.* (2005) for the case of early B-type supergiants, represent the cornerstones for the quantitative analysis of low resolution ($\lambda/\Delta\lambda \sim 1000$) optical spectra of these objects. Besides obtaining information on the stellar parameters and chemical composition of the stars in NGC 300, a mid size disk galaxy located at 1.9 Mpc, these works allowed the first detailed study of the stellar metallicity gradient in an external galaxy. More importantly, they confirmed the existence of the so-called "Flux-weighted Gravity–Luminosity Relationship" initially introduced by Kudritzki *et al.* (2003).

Blue supergiants have long been recognised as potentially distance indicators in external galaxies (see for example Hubble 1936, Tully & Wolff 1984). However, the harsh conditions in their atmospheres required a significant amount of development in the non-LTE model atmosphere techniques for them to be used reliably (Przybilla *et al.* 2006). The work by Kudritzki *et al.* (2008b) based on quantitatively analyzing high signal-to-noise optical spectra to determine stellar parameters such as effective temperature T_{eff}, gravity $\log g$ and metallicity with unprecedented accuracy and reliability, led to the detection of a tight relationship between stellar absolute bolometric magnitude $\mathrm{M_{bol}}$ and the flux-weighted gravity g/T_{eff}^4 ($\equiv g_f$), the "Flux-weighted Gravity - Luminosity Relationship (FGLR)", of the form

$$M_{\rm bol} = a(\log g_f - 1.5) + b, \tag{1.1}$$

A very basic back-of-the-envelope calculation confirms that this simple form of the FGLR can be understood as the result of (single) massive star evolution beyond the main sequence, in the crossing to the red supergiant phase at constant luminosity and constant mass. Under these simplifying assumptions, the quantity g/T_{eff}^4 remains constant during the evolution accross the Hertzsprung-Russell Diagram and the FGLR is then the result of a simple relationship between stellar luminosity L and stellar mass M, $L \sim M^x$ (see Kudritzki *et al.* 2008b, for further details). Even though the luminosity is not exactly constant during the evolution (also the exponent x decreases with increasing stellar mass), the detailed study by Meynet *et al.* (2015) combining state-of-the-art stellar evolution calculations (for single stars) with population synthesis has recently shown that the FGLR still holds.

2. The case for alternative metallicity indicators

The spatial distribution of chemical elements in galaxies represents a key tool to understand how the chemical evolution proceeds in the universe. In the case of star-forming galaxies, the classical procedure has been to use oxygen abundances derived from emission lines present in the spectra of H II regions for studies of abundance gradients in individual galaxies, and more recently the existence of a relationship between the mass of the galaxy and their *metal* content (Tremonti *et al.* 2004). Gas phase oxygen abundances have been also widely used to interpret results obtained from studies of populations of massive stars, such as the dependence of Wolf-Rayet types or the dependence of the number of Blue-to-Red supergiants with the *metallicity* of the environment (see Massey 2003 and references therein), as well as for studies of the *metallicity* dependence of the Period–Luminosity relationship of Cepheid stars, with far reaching implications for the distance scale of the universe (Freedman & Madore 2010, Bono *et al.* 2010).

All these studies relied on the analysis of the emission spectrum of H II regions because it is very easy to collect high quality spectra for these objects even in very distant galaxies, and under the key assumption that their physics was quite simple. In recent years however, it has become very clear that these objects are not as simple as originally thought, and that their physics has yet to be properly understood (Stasińska 2008).

Among others, the dependence of the derived oxygen abundances upon the choice of calibration used to interpret the ratio of some of the most prominent forbidden lines (the so-called statistical or strong-line methods, see for example Bresolin *et al.* 2009), the large discrepancies between abundances based on the derivation of the electron temperature of the gas and the ones obtained from statistical methods (both applied to the same set of spectra) or the systematic difference between abundances based on collisionally excited lines, CELs, and on recombination lines, RLs, when observed in the same object (termed in the literature as the "Abundance Discrepancy Factor", ADF, see for example Toribio San Cipriano *et al.* 2016), are sources of concern when interpreting any observed property in terms of the *metal* content of the host galaxy. Furthermore, the evolution of massive stars is at the very least the result of the interplay between rotation and mass-loss (Meynet & Maeder 2005). Whilst light elements, such as oxygen, will play a role in the acceleration of the stellar wind in the outer layers of the atmosphere, hence having a significant impact on the wind terminal velocity, the mass-loss rate is set at the base of the wind, where the radiative acceleration is provided by a myriad of iron lines (Vink *et al.* 2000, Puls *et al.* 2000). Thus, any proper interpretation of the wind properties of massive stars, such as the Wind Momentum–Luminosity Relationship (WLR, Kudritzki & Puls 2000) and its dependence with *metallicity* requires the knowledge of the iron content. Since the α-to-iron ratio (oxygen being an α-element) in any stellar system will depend not only on the original composition, but also on how the system has evolved in time (oxygen in mainly produced when massive stars die, whilst iron is mainly produced by less massive stars), there is a priori no reason for the ratio observed in the solar vicinity to apply universally (as sometimes is assumed).

The case of the *too strong* winds in the putative low *metallicity* Local Group galaxy IC 1613 illustrates this point. Recent studies of massive stars in this galaxy (among others Tramper *et al.* 2011, Herrero *et al.* 2012; see also the contribution by Miriam Garcia in these proceedings) observed an apparent over strength of the stellar winds when comparing the WLR of stars in IC1613 with Galactic and LMC and SMC counterparts, with respect to predictions provided from the theory of Radiatively Driven Winds (Kudritzki & Puls 2000). However, the interpretation of these results were based on the oxygen content of this galaxy. Here we point out that previous studies of blue type supergiants in this galaxy (Bresolin *et al.* 2007) confirmed the low oxygen abundances derived from H II regions, about 1/10th of the solar neighbourhood value. However, upon a closer inspection, it turns out that the O/Fe ratio in this galaxy is sub-solar, with the current Fe content of IC 1613 being somewhere closer to the SMC value (Tautvaišienė *et al.* 2007). A recent high resolution study of A-type supergiants in this galaxy (Urbaneja *et al.* in prep., see Fig. 1) confirms the oxygen abundances from H II regions and B-type supergiants as well as sub-solar O/Fe ratio. When the WLR results are interpreted in terms of the iron content of the different galaxies, the results fully support the theoretical predictions.

As indicated above, oxygen abundances derived from emission line spectra of ionised gas are affected by a number of systematic problems, not yet understood. One way to tackle these issues, at least in part, is through the comparison of several different abundance indicators. Whilst it is true that spectroscopy of individual stars can not reach the far distances covered with nebular studies, it is possible to study nearby systems, in order to investigate which, if any, of the nebular diagnostics is more reliable, and hence should be preferred over the others. A number of nearby galaxies have been studied in recent years with this purpose in mind. Comparative studies of H II regions and blue supergiants seem to be yielding some fruitful results. Very recently, Bresolin *et al.* (2016) presented a comparison of abundances in different galaxies (both in terms of the oxygen content and the mass of the galaxies). While still not definitive, it seems that CEL-based

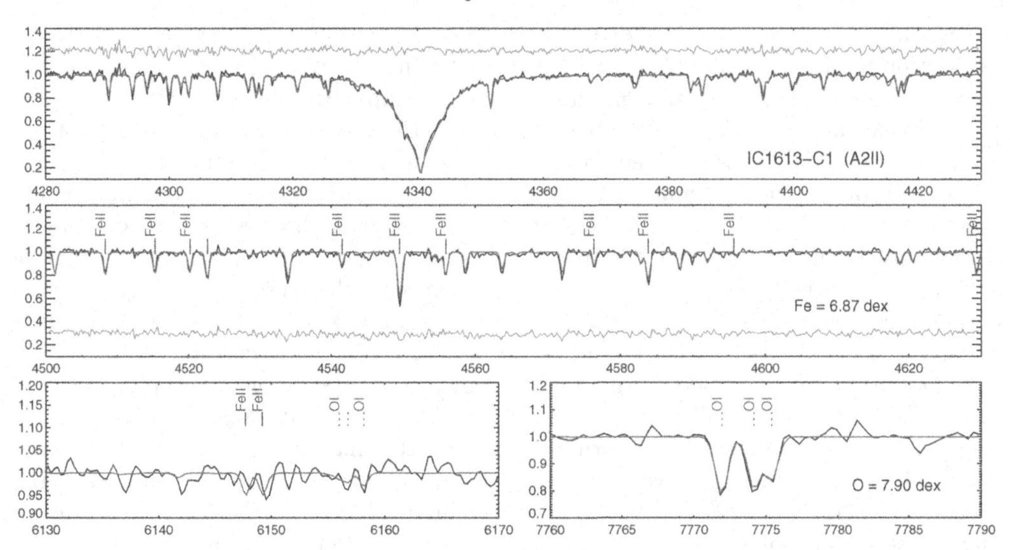

Figure 1. Observed X-Shooter (R∼9900) spectrum of an A-type supergiant in IC 1613 (black) with the final tailored model (red). These analyses confirm the low oxygen content as well as the sub-solar O/Fe ratio (Urbaneja *et al.*, in prep).

abundances (when the electron temperature of the gas is known) are fully consistent with stellar abundances for oxygen abundances below the solar neighbourhood content, whilst RL-based abundances seem to be systematically higher. On the other hand, the situation reverts for oxygen abundances above the solar neighbourhood value. This result needs to be confirmed with a larger sample of objects in this second regime. However, upon confirmation, this would help to constrain the physical origin of the ADF problem.

3. A stellar based Mass–Metallicity relationship for nearby galaxies

Based on the analysis of more than 40000 star forming galaxies, Tremonti *et al.* (2004) showed that there seems to be a relationship between the amount of oxygen in a galaxy and its mass. This "Mass–Metallicity" relationship of star-forming galaxies represents a very powerful tool to investigate the chemical evolution of the universe, by for example comparing the relationship observed at different redshifts (Zahid *et al.* 2014). As is customary, oxygen abundances derived by means of a specific strong line method were used in this work as proxy for metallicities. However, the placement of our nearest companions (the Magellanic Clouds) and the Milky Way in this relationship, when using what is known about the oxygen content of their stars, results in the oddity that these three galaxies are 3σ outliers in this relationship. This is at least in part related to the choice of the strong line indicator used to derive gas phase oxygen abundances. In a later work, Kewley & Ellison (2008) illustrated how not only the absolute abundance scale but also the shape of the relationship would depend upon the choice of the calibration. To overcome the potential problems that could appear when comparing results obtained by different authors using different strong line diagnostics, these authors provided relative calibrations that allow transformation between the different diagnostics. Thus, everything would potentially be fine when working in relative terms. However, there are applications for which an *absolute* scale is required, like for example to investigate the metallicity dependence of the PL relationship of Cepheid stars.

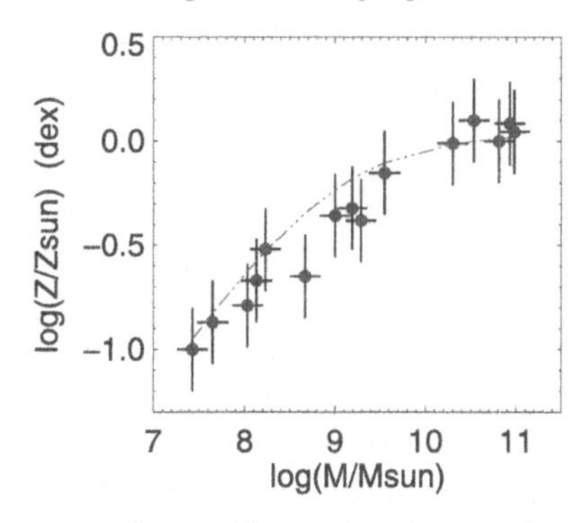

Figure 2. The "stellar" Mass–Metallicity relationship for nearby star-forming galaxies. Metallicities are obtained from the quantitative spectroscopy of individual blue supergiant stars in these galaxies.

It is possible to create a similar mass–metallicity relationship for nearby galaxies based only on stellar abundances derived from individual blue supergiant stars, albeit with a much smaller number of galaxies. This is shown in Fig. 2, where abundances derived from individual blue supergiant stars and stellar masses for the galaxies have been collected from the literature. The red line shown in the figure is not a fit to the data, but just a simple zero point adjustment of the Mass–Metallicity relationship obtained by Andrews & Martini (2013), based on staked spectra (using mass bins) of the same galaxy sample studied by Tremonti *et al.* (2004) and Kewley & Ellison (2008).

4. A Spectroscopic distance indicator: the FGLR

Along with metallicities, knowledge of distances to galaxies represent a key ingredient in our understanding of the universe. The primary stellar distance indicator, the Period–Luminosity relationship of Cepheid stars, suffers from two major problems: extinction and metallicity dependence, both of which are difficult to estimate individually with the stringent precision required to reduce the uncertainties in measuring the Hubble Constant H_0 (with this being ultimately related to constraining the equation of state of Dark Energy, Riess *et al.* 2016). Quantitative spectroscopy of Cepheids beyond the Magellanic Clouds is currently not feasible, because of the intrinsic brightness of these objects (about 3 orders of magnitude fainter than blue supergiant stars). Hence, in order to gauge the effect that the metallicity may or may not have in the PL relationship, alternative metallicity indicators are used as proxies. Mainly, gas phase oxygen abundances derived from H II regions. Therefore any interpretation in terms of *metallicity* is severely hampered by the unknown systematics discussed above. The second source of concern is related to internal extinction in the host galaxies. Whilst it is true that there are ways to try to circumvent this problems (by using "reddening-free" magnitudes, or moving to the IR domain), there is still no clear consensus on the accuracy of these methods. Moreover, the lack of information concerning the real form of the extinction law is always present (see Kudritzki & Urbaneja 2012a for a detailed discussion).

Blue supergiant stars can also contribute in this field. As young objects, they are closely related to Cepheids, and hence expected to present similar extinction values (note that we are referring here exclusively to B- and A-type supergiant stars). First, for each individual blue supergiant, information regarding its metal content is directly obtained from the analysis of the optical spectrum. At the same time, reddening as well as information about the form of the extinction curve can be obtained individually for each star, by combining

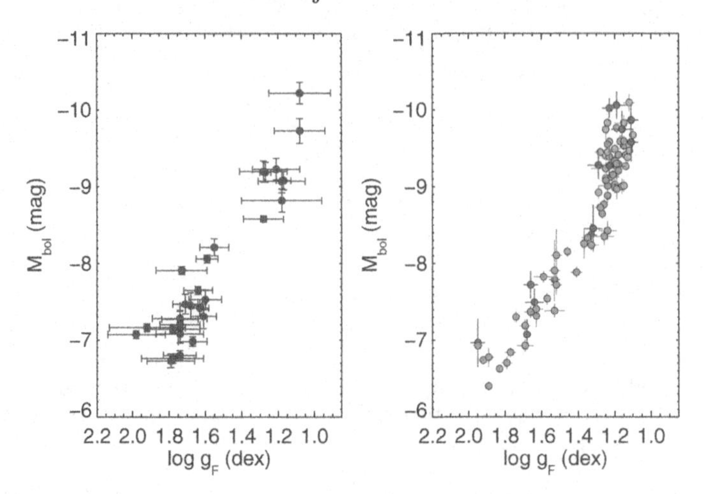

Figure 3. The FGLR of blue supergiant stars. Left–current FGLR calibrator: stars in NGC 300 (adapted from Kudritzki *et al.* 2008). Right–stars in the LMC (Urbaneja *et al.*, in prep).

the intrinsic colours predicted by the model atmosphere corresponding to the physical parameters describing the observed spectrum with precise multi-band photometric observations. This method has been for example used in several galaxies (NGC 300–Kudritzki *et al.* 2008, M33–U *et al.* 2009, M81–Kudritzki *et al.* 2012b) to investigate the differences between the observed E(B-V) values derived from blue supergiants with the foreground values customarily applied in the Cepheid's literature. The basic conclusion from these comparisons is that the concept of applying a constant foreground reddening correction for Cepheid stars is a dangerous notion (Kudritzki & Urbaneja 2012a).

Besides these kind of comparisons, blue supergiants can be used to derive distances to nearby galaxies by means of the FGLR. As discussed in the introduction, there is a tight observed correlation between the absolute bolometric magnitude and the flux-weighted gravity of blue supergiant stars (see eq. 1.1). Once this relationship is calibrated in absolute terms with stars in a galaxy with a well know distance, the "apparent" FGLR (based on apparent bolometric magnitudes) of stars in any other galaxy can be compared with the FGLR calibrator, and the difference in the intersects will directly provide the relative distance between both galaxies. FGLR based distances have been derived for a number of nearby galaxies, using NGC 300 as calibrator: WLM–Urbaneja *et al.* 2008, M33–U *et al.* 2009, M81–Kudritzki *et al.* 2012b, NGC 3109–Hosek *et al.* 2014, NGC 3621–Kudritzki *et al.* 2014, NGC 55 –Kudritzki *et al.* 2016 and M83–Bresolin *et al.* 2016. Excluding the yet to be understood large discrepancy for the case of M33 (U *et al.* 2009, see also Bonanos *et al.* 2006), the FGLR derived distances are in good agreement with distances derived by using multi-band photometry (covering the optical regime as well as the J and K bands) of Cepheid stars.

5. Blue supergiants in era of the ELTs.

The future introduction of the Extremely Large Telescopes (E-ELT, TMT, GMT) will significantly enlarge the volume of the universe in which quantitative spectroscopy of individual stars would be feasible. Whilst these new towering giants are being designed primary to work in the IR domain, some capabilities in the optical will still be present, enabling the use of individual blue supergiant stars for studies of their host galaxies well

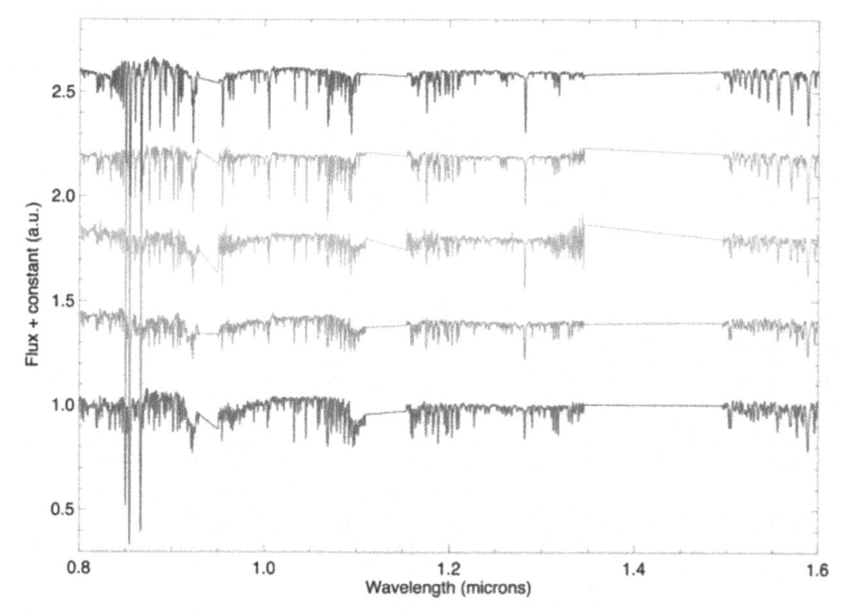

Figure 4. IR spectra of Galactic F-type supergiants from the IRTF Spectral library (Rayner *et al.* 2009, R∼2000) in the 0.8–1.6 μm spectral window. These illustrate the potential of yellow supergiants as metallicity indicators in external galaxies.

beyond the limits of current facilities. But it would be in the IR domain, thanks to the extreme AO capabilities, where these telescopes will provide the maximum efficiency. Most likely, B- to early A-type supergiants will have a diminishing impact in the extragalactic field. However, these new facilities will strengthen the relevance of red supergiants (see Ben Davies' contribution in these proceedings) and some newcomers to the extragalactic playground: yellow supergiants, with spectral types from late A to early G.

As part of the first-light instrument suite for ESO's E-ELT, the Multi-Adaptive Optics Imaging Camera for Deep Observations (MICADO, Davies *et al.* 2016) is being design to provide spectroscopic capabilities with very high spatial resolution in the 0.8–2.4 μm wavelength range, split into two settings, zYJ and HK, at a spectral resolution of R∼8000. Fig. 4 displays a sequence of real IR spectra of F-type supergiant stars from the IRTF Spectral Library (Rayner *et al.* 2009), in the 0.8–1.6 μm at a lower spectral resolution R∼2000. Many individually resolved atomic lines from several species are contained in this region. Thus, in principle, the same techniques that have been successfully employed in the optical for B- and A-type supergiants could potentially be applied for the quantitative analysis of the IR spectrum of these objects, provided that the models atmospheres (and corresponding model atoms) are fully developed for their application in this scarcely explored regime.

Acknowledgment: part of the work presented in this paper was funded by the Hochschulraumstrukturmittel (HRSM) provided by the Austrian Ministry for Research (bmwfw).

References

Andrews, B. H. & Martini, P. 2013, *ApJ*, 765, 140
Bonanos, A. Z., Stanek, K. Z., Kudritzki, R. P., *et al.* 2006, *ApJ*, 652, 313
Bono, G., Caputo, F., Marconi, M., & Musella, I. 2010, *ApJ*, 715, 277

Bresolin, F., Urbaneja, M. A., Gieren, W., Pietrzyński, G., & Kudritzki, R.-P. 2007, *ApJ*, 671, 2028
Bresolin, F., Gieren, W., Kudritzki, R.-P., *et al.* 2009, *ApJ*, 700, 309
Bresolin, F., Kudritzki, R.-P., Urbaneja, M. A., *et al.* 2016, *ApJ*, 830, 64
Davies, R., Schubert, J., Hartl, M., *et al.* 2016, *Proc SPIE*, 9908, 99081Z
Freedman, W. L. & Madore, B. F. 2010, *ARAA*, 48, 673
Herrero, A., Garcia, M., Puls, J., *et al.* 2012, *A&A*, 543, A85
Hosek, M. W., Jr., Kudritzki, R.-P., Bresolin, F., *et al.* 2014, *ApJ*, 785, 151
Hubble, E. 1936, *ApJ*, 84, 158
Kewley, L. J. & Ellison, S. L. 2008, *ApJ*, 681, 1183-1204
Kudritzki, R.-P., Lennon, D. J., & Puls, J. 1995, *Science with the VLT*, 246
Kudritzki, R.-P. & Puls, J. 2000, *ARAA*, 38, 613
Kudritzki, R. P., Bresolin, F., & Przybilla, N. 2003, *ApJL*, 582, L83
Kudritzki, R., Urbaneja, M. A., Bresolin, F., & Przybilla, N. 2008 (a), *Massive Stars as Cosmic Engines*, 250, 313
Kudritzki, R.-P., Urbaneja, M. A., Bresolin, F., *et al.* 2008 (b) *ApJ*, 681, 269-289
Kudritzki, R.-P. & Urbaneja, M. A. 2012, *APSS*, 341, 131 (a)
Kudritzki, R.-P., Urbaneja, M. A., Gazak, Z., *et al.* 2012, *ApJ*, 747, 15 (b)
Kudritzki, R.-P., Urbaneja, M. A., Bresolin, F., Hosek, M. W., Jr., & Przybilla, N. 2014, *ApJ*, 788, 56
Kudritzki, R. P., Castro, N., Urbaneja, M. A., *et al.* 2016, *ApJ*, 829, 70
Massey, P. 2003, *ARAA*, 41, 15
Meynet, G. & Maeder, A. 2005, *A&A*, 429, 581
Meynet, G., Kudritzki, R.-P., & Georgy, C. 2015, *A&A*, 581, A36
Przybilla, N., Butler, K., Becker, S. R., & Kudritzki, R. P. 2006, *A&A*, 445, 1099
Puls, J., Springmann, U., & Lennon, M. 2000, *A&APS*, 141, 23
Riess, A. G., Macri, L. M., Hoffmann, S. L., *et al.* 2016, *ApJ*, 826, 56
Stasińska, G. 2008, *Low-Metallicity Star Formation: From the First Stars to Dwarf Galaxies*, 255, 375
Tautvaišienė, G., Geisler, D., Wallerstein, G., *et al.* 2007, *AJ*, 134, 2318
Toribio San Cipriano, L., García-Rojas, J., Esteban, C., Bresolin, F., & Peimbert, M. 2016, *MNRAS*, 458, 1866
Tramper, F., Sana, H., de Koter, A., & Kaper, L. 2011, *ApJL*, 741, L8
Tremonti, C. A., Heckman, T. M., Kauffmann, G., *et al.* 2004, *ApJ*, 613, 898
Tully, R. B. & Wolff, S. C 1984, *ApJ*, 281, 67
Rayner, J. T., Cushing, M. C., & Vacca, W. D. 2009, *ApJS*, 185, 289
U, V., Urbaneja, M. A., Kudritzki, R.-P., *et al.* 2009, *ApJ*, 704, 1120
Urbaneja, M. A., Herrero, A., Bresolin, F., *et al.* 2005, *ApJ*, 622, 862
Urbaneja, M. A., Kudritzki, R.-P., Bresolin, F., *et al.* 2008, *ApJ*, 684, 118-135
Vink, J. S., de Koter, A., & Lamers, H. J. G. L. M. 2000, *A&A*, 362, 295
Zahid, H. J., Dima, G. I., Kudritzki, R.-P., *et al.* 2014, *ApJ*, 791, 130

Discussion

The Lives and Death-Throes of Massive Stars
Proceedings IAU Symposium No. 329, 2016
J.J. Eldridge, J.C. Bray, L.A.S. McClelland & L. Xiao, eds.

doi:10.1017/S1743921317002927

What can distant galaxies teach us about massive stars?

Elizabeth R. Stanway

Physics Department, University of Warwick, Gibbet Hill Road, Coventry, CV4 7AL, UK
email: e.r.stanway@warwick.ac.uk

Abstract. Observations of star-forming galaxies in the distant Universe ($z > 2$) are starting to confirm the importance of massive stars in shaping galaxy emission and evolution. Inevitably, these distant stellar populations are unresolved, and the limited data available must be interpreted in the context of stellar population synthesis models. With the imminent launch of JWST and the prospect of spectral observations of galaxies within a gigayear of the Big Bang, the uncertainties in modelling of massive stars are becoming increasingly important to our interpretation of the high redshift Universe. In turn, these observations of distant stellar populations will provide ever stronger tests against which to gauge the success of, and flaws in, current massive star models.

Keywords. galaxies: evolution, galaxies: high-redshift, stars: luminosity function, mass function.

1. Introduction

As studies elsewhere in these proceedings have shown, as many as ~70% of massive stars are expected to interact with a binary companion during their lifetimes. These stars dominate the integrated light of young stellar populations (< 100 Myr) and the effects of binary interactions are typically more pronounced at significantly sub-Solar metallicities. This presents an interesting opportunity: the low metallicity, highly star-forming galaxies we observe at high redshifts both require an understanding of massive stellar evolution for their interpretation, and represent a laboratory in which to test that understanding. This synergy has recently been recognised by a growing subset of both the extragalactic and massive stars communities, and will only become stronger as the advent of highly-multiplexed near-infrared spectroscopy from the ground is complemented by the eagerly-anticipated *James Webb Space Telescope* (JWST).

Here I review recent observational indications of the presence and influence of massive stars in the distant Universe, as well as discussing the the role of stellar modelling in their interpretation and the possible insights this synergy enables.

2. Observing the Distant Universe

Over the last twenty years, the number of spectroscopically-confirmed galaxies in the distant Universe (in this context, $z > 2$) has grown exponentially, from a mere handful to tens of thousands of sources. The primary driver of this process is the 'Lyman break technique' first applied on a large scale by Steidel *et al.* (1996), and later extended to higher redshifts (e.g. Stanway *et al.* 2003; Bouwens *et al.* 2011). This method allows the selection of high redshift galaxy candidates by their distinctive photometric colours. These arise from a strong discontinuity imposed on their rest-frame ultraviolet spectrum due to absorption by neutral hydrogen in the intergalactic medium (IGM). It preferentially

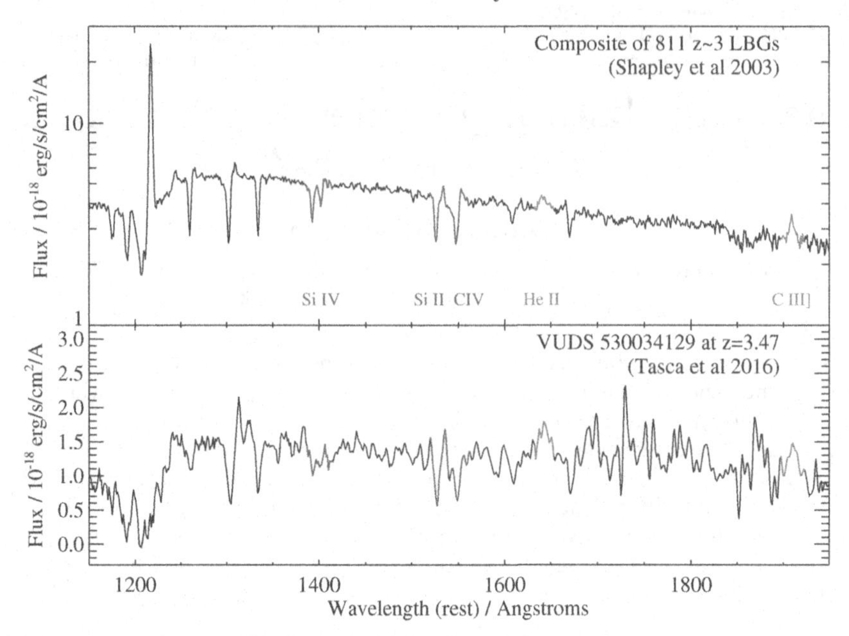

Figure 1. Example spectra of high redshift sources, highlighting the features which confirm the presence of massive stars. Indicated lines are Si IV, Si II, C IV, He II and C III] - all lines which either include a direct component from massive stars or which are powered by the ionizing radiation of massive stellar populations. Early composites of Lyman break galaxies at $z \sim 3$ (Shapley *et al.* 2003) have now been complemented by deep spectra of individual targets, the example given here being a bright $z = 3.47$ source from the VUDS survey (Tasca *et al.* 2016).

selects galaxies with high ultraviolet luminosity, and thus those with some component of on-going star formation. The redshift of these can be confirmed (for many cases) by follow-up spectroscopy of the rest-frame ultraviolet, redshifted into the observed optical. At very high redshift ($z \sim 5 - 7$) spectroscopic characterisation is often restricted to detection of an isolated Lyman-α emission line ($\lambda_{\rm rest} = 1216$Å), with perhaps a second strong UV feature (e.g. He II or CIII]) to provide confirmation in rare cases (e.g. Stark *et al.* 2015). Beyond $z \sim 7$, the rising neutral hydrogen fraction in the IGM, combined with the shift of the Lyman break into the near-infrared, means that spectroscopic confirmation is very seldom possible, but the properties of galaxies may still be inferred from their photometry (e.g. Caruana *et al.* 2014; Smit *et al.* 2014).

2.1. *Rest-frame Ultraviolet Spectroscopy*

Rest frame ultraviolet spectroscopy can be interpreted through the construction of composites which yield the 'typical' properties of galaxies in a population or subset thereof (e.g. Shapley *et al.* 2003). These have been complemented by observations of individual sources, either particularly bright, or lensed targets, or simply using extremely deep spectroscopy (e.g. Tasca *et al.* 2016). While individual sources show variation, figure 1 illustrates certain features that are common in the high redshift galaxy population. Either directly (through features arising from the stellar spectra) or indirectly (through emission from the nebular gas of H II regions), these are diagnostic of the massive stellar population in these galaxies.

Clearly the strongest and most obvious of these features is the Lyman-α emission line. A full analysis of the emission and radiative transfer of this resonantly-scattered line is beyond the scope of this review. Nonetheless, both the equivalent width distribution of this line (e.g. Malhotra & Rhoads 2002; Dijkstra & Wyithe 2007; Hashimoto *et al.* 2017), and its typical offset in velocity from interstellar absorption features in the same source (e.g. Adelberger *et al.* 2003; Shapley *et al.* 2003), are indicative of a hard radiative field, powering galaxy-scale outflows. Given the young typical ages of galaxies (a few hundred Myrs), and the lack of evidence for significant AGN activity, these observations both imply the presence of sufficient massive stars to cause powerful radiative feedback and significantly affect the evolution of their circumgalactic environments.

For populations in which we can move beyond Lyman-α (mostly at $z \lesssim 5$, although with some exceptions), the rest-UV contains other diagnostics of massive stellar populations. As figure 1 shows, the prominant absorption features of C IV 1548,1550Å, Si IV 1393,1402Å and Si II 1526Å are all seen in both composite and individual spectra. Similarly the He II 1640Å and CIII] 1907,1909Å emission features appear to be far more common in the distant galaxy population than in local star forming system. Each of these may have broad (stellar wind-driven) and narrow (nebular) components, but show strengths that are difficult to reproduce with conventional stellar populations (Shapley *et al.* 2003). The emission lines in particular, are also indicative of a far harder ionizing spectrum than that seen in local sources. A few rare He II emitting star forming regions in the local Universe are interpreted as hosting massive, often Wolf-Rayet, stars (Kehrig *et al.* 2015). Models incorporating more detailed analysis of massive stellar populations, either in terms of rotation or binary interaction, and exploring these effects at sub-Solar metallicities are proving both necessary for and successful in simultaneous fitting of these line strengths in the high redshift population (e.g. Eldridge & Stanway 2012; Steidel *et al.* 2016).

2.2. *Rest-Frame Optical Spectroscopy*

The recent advent of multi-object near-infrared spectrographs on 8-10m class telescopes has had a strong impact in this field. The rest-frame optical spectra of $z \sim 2-3$ Lyman break galaxies are now accessible in a reasonable integration time, allowing direct comparison with galaxy populations observed in the local Universe. In particular two programmes using the Keck/MOSFIRE instrument, KBSS (Steidel *et al.* 2014) and MOSDEF (Shapley *et al.* 2015), have built large samples of several hundred nebular emission line spectra in this redshift range. As figure 2 (left) illustrates, one of the key findings of these programmes has been an offset in line ratios which probe the ionization conditions of nebular gas. The Baldwin, Phillips, & Terlevich (BPT, 1981) diagram is often used to distinguish between irradiation by star formation and that arising from active galactic nuclei. This diagram makes use of two pairs of lines, each close in wavelength to minimise the effects of dust and continuum subtraction uncertainties. The [N II]/Hα ratio is primarily sensitive to the shape of the ionizing spectrum just above 1 Rydberg, while the [O III]/Hβ ratio probes the 1-3 Rydberg range. The offset in the population medians for $z \sim 2$ galaxies relative to the local $z < 0.2$ Sloan Digital Sky Survey (SDSS) star forming galaxy population has been interpreted as evidence for a harder ionizing field than that of the near-Solar metallicity local population – supporting a similar interpretation of the He II and C III] emission in the rest-frame ultraviolet (Masters *et al.* 2014; Stanway *et al.* 2014; Sanders *et al.* 2016; Steidel *et al.* 2016; Strom *et al.* 2016).

These high ionization environments are not unique to the highest redshifts. A number of $z \sim 0-0.3$ galaxy populations have been identified as analoguous to those at high redshift, in terms of their star formation and emission properties (see Greis *et al.* 2016). As figure 2 demonstrates, these also appear distinct from the bulk of the low redshift galaxy locus in the BPT diagram, and so may provide more local (and hence accessible) environments in which to test our understanding of these stellar populations.

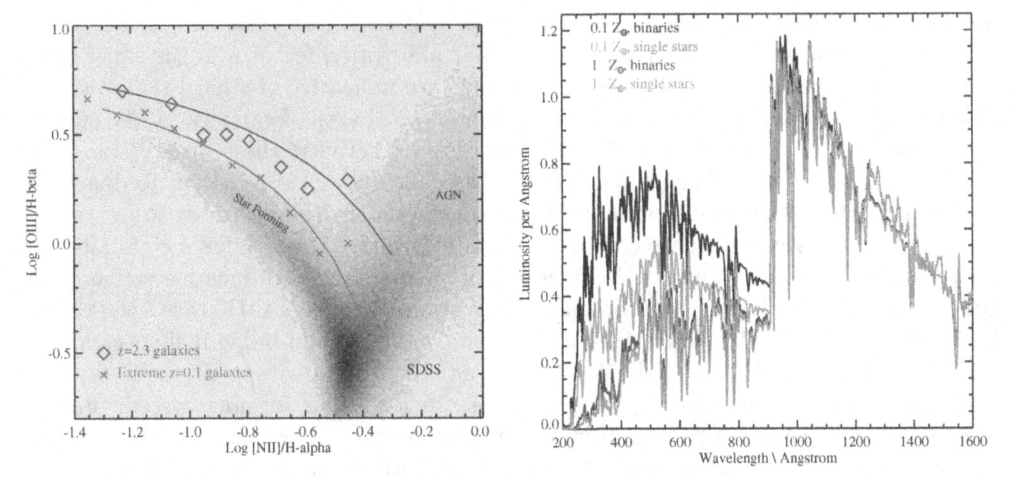

Figure 2. The importance of the hardness of the ionizing spectrum. (Left) The often-used BPT diagram which measures the hardness of the ionizing spectrum both close to 1 Rydberg (the [NII]/Hα ratio) and at 1-3 Ryd ([OIII]/Hβ). Greyscale indicates the distribution of local star forming galaxies at near Solar metallicity from the SDSS. Points and lines indicate the properties of low metallicity, intensely star forming galaxies in the local Universe (blue crosses, Greis *et al.* 2016) and at $z \sim 2.3$ (red diamonds, Strom *et al.* 2016). (Right) The effect of metallicity and binary evolution effects on the hardness of the far-ultraviolet stellar continuum. Stellar population models have formed stars at a constant rate for 30 Myr, and are taken from BPASS v2.0 (Stanway *et al.* 2016).

2.3. *Indirect Constraints*

At the highest redshifts, detailed spectroscopy is seldom possible. Nevertheless, there are strong indications that the dominant stellar population differs from that seen either at $z \sim 2-4$ or in the local Universe. The rest-frame ultraviolet photometric colours of galaxies at $z > 5$ indicate extremely blue spectral slopes; again, these are difficult to fit with most stellar population synthesis models unless a very young, very low metallicity population is invoked (e.g. Stanway, McMahon, & Bunker 2005; Finkelstein *et al.* 2012; Wilkins *et al.* 2016a). As photometry extends to the rest-frame optical (shifted to the thermal infrared, and necessarily observed from space), spectral energy distribution modelling has suggested that many distant galaxies require extremely powerful nebular emission lines - consistent with the hard ultraviolet slope (e.g. Smit *et al.* 2014).

A second indirect constraint is derived from the so-called Epoch of Reionization, a period at $z > 7$ over which the IGM was ionized by ultraviolet photons escaping from star forming galaxies. Current constraints on the evolution of the ionized hydrogen fraction require either that the luminosity function of galaxies extends to very low stellar masses, that the escape fraction of photons is unreasonably high, or that the ionizing photon output of galaxies exceeds that currently inferred from their 1500Å (rest) luminosities. All three factors are probably significant, but it appears likely that a hard ionizing spectrum, arising from evolution in the massive stellar population of typical galaxies, is a key requirement (Topping & Shull 2015; Stanway *et al.* 2016; Wilkins *et al.* 2016b; Ma *et al.* 2016).

3. The Key Role of Population Synthesis Models

When observing distant galaxies, it is very rare that individual stars, or even individual star forming regions can be resolved. At $z = 1$, the angular scale is $\sim$8 kpc per arcsec and sub-kiloparsec resolution is usually only possible from space, and for cases distorted by strong gravitational lensing. As a result, the emission that we observe is the integrated light of large stellar populations, potentially of varying age and metallicity. As the previous section demonstrates, interpreting this necessarily requires comparison with stellar population synthesis (SPS) models, which construct the expected integrated spectrum from a stellar population with known parameters (star formation history, initial mass function (IMF), metallicity etc.) for comparison with the data. A number of SPS codes are in common usage. These range from those based on purely empirical observations of nearby stellar populations to others based on theoretical stellar evolution and atmosphere models. Each contains different formalisms for evolutionary stages and the handling of complex populations, and as a result, each has different strengths and weaknesses. Recent reviews of a number of SPS codes include those by Conroy (2013) and Wofford *et al.* (2016).

Here I will focus on results from a particular SPS model set - the Binary Population and Spectral Synthesis code (BPASS, Eldridge & Stanway 2009, 2012; Stanway *et al.* 2016, Eldridge *et al.* in prep)†. This uses a library of theoretical stellar evolution models, and combines these with synthetic stellar atmospheres. Importantly, it also includes prescriptions for binary system evolutionary effects omitted in many other SPS codes, including mass loss and (a simple prescription for) rotation. At Solar metallicity, the output model spectra for continuously star forming populations are comparable to those of other codes for a Salpeter-like initial mass function, while at early ages, different IMFs and low metallicities the different model sets diverge. The effects of binary evolution tend to prolong the epoch over which very blue stars dominate the spectrum. The results presented here are from the v2.0 BPASS data release, supplemented by lower metallicity models which will be made available in BPASS v2.1, and for an IMF with a broken power law slope, with indices -1.35 at $M< 0.5\,M_\odot$ and -2.35 for $0.5\,M_\odot < M < M_{max}$.

Such models can confront the very high ionizing fluxes inferred for galaxies in the distant Universe. In figure 3 we illustrate the dramatic difference in ionizing flux output between different stellar populations and compare these to observational constraints derived from distant sources. These are usually presented using the parameter $\xi_{\rm ion} = \dot{N}_{\rm ion}/L_{\rm UV}$, i.e. the ratio between the ionizing photon production rate and the rest-frame ultraviolet continuum luminosity. While the latter can be directly observed in the distant Universe, the former must be inferred from indirect observations - primarily of the strong nebular emission lines generated from the galaxy in question, assuming that a fraction $(1 - f_{esc})$ of the ionizing continuum is absorbed by the interstellar medium (see Stark *et al.* 2015; Bouwens *et al.* 2016, for discussion). While the metallicity of stellar populations in the distant Universe are still poorly constrained, the inferred ionizing fluxes are significantly higher than those predicted for a canonical, continuously forming stellar population at near-Solar metallicity (the canonical value of Kennicutt 1998, is indicated on the figure). Instead, populations with a higher upper mass cut-off ($300\,M_\odot$ rather than the usually-assumed $100\,M_\odot$) and those which account for binary effects (e.g. through evolutionary effects or rotation) are favoured. The steady increase in $\xi_{\rm ion}$ measurements with redshift is more rapid than that expected from simple cosmic metallicity evolution for single star models, and may indicate that binary effects at low metallicity are becoming significant.

Not only the ionizing photon flux but also the hardness of its spectrum can provide insight into the massive stellar populations in distant galaxies. In figure 2 (right) we illustrate the shape as well as the strength of the Lyman continuum emission region for stellar models with the same star formation history (constant star formation over a 30 Myr

† see bpass.auckland.ac.nz

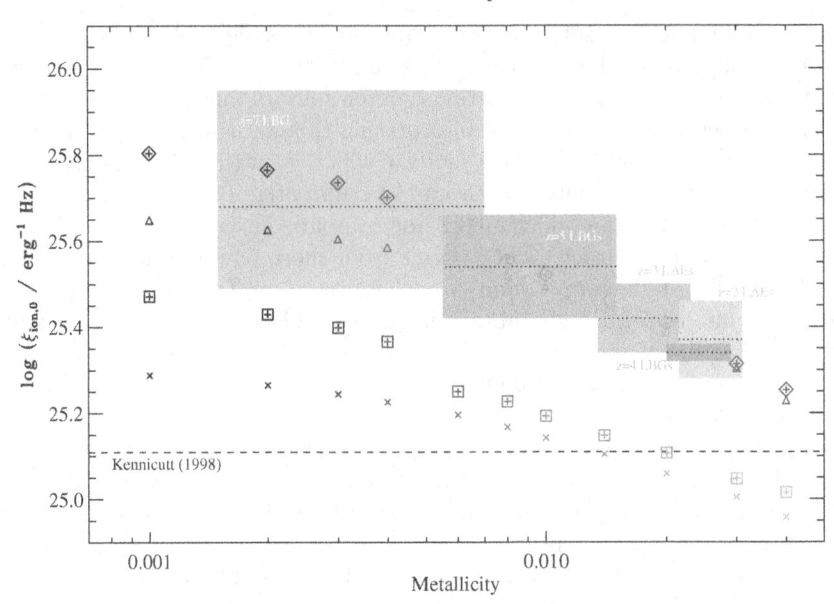

Figure 3. The dependence of ionizing photon efficiency, $\xi_{\rm ion,0}$, on metallicity, initial mass function and single vs binary star evolution, assuming $f_{esc} = 0$. At each metallicity we show the model values for binary (large symbols with '+') and single star (smaller symbols) stellar populations, and for two IMFs ($M_{\rm max}$=300 $M_\odot$, upper pair, and $M_{\rm max}$=100 $M_\odot$, lower pair). Shaded regions indicate observational constraints ($\pm 1\sigma$) on $\xi_{\rm ion,0}$ from high redshift galaxy populations from Bouwens *et al.* (2016, $z \sim 4-5$ LBGs), Matthee *et al.* (2017, $z = 2$ LAEs, using the β dust correction for consistency with Bouwens *et al.*, Nakajima *et al.* (2016, $z = 3$ LAEs) and Stark *et al.* (2015, $z \sim 7$ LBG). Metallicity is poorly constrained in high redshift observations and the observational constraints are shown with horizontal offsets for clarity, illustrating a plausible range of metallicities rather than measurements. Model values are for a 100 Myr old, continuously star forming population using BPASS v2.0 (Stanway *et al.* 2016).

interval) but with different metallicities and including or omitting the effect of stellar binaries. The shape of the ionizing spectrum is quite different, and will result in shifts in the ratios of line emission, since these probe different energy ranges. In particular, the presence of a hard (blue) ionizing spectrum in the 1-3 Rydberg range will shift galaxies vertically in the ionization-sensitive BPT diagram (see also Xiao *et al.* 2017, in these proceedings). If the observed shift in the populations of both distant galaxies and their local analogues are interpreted as a stellar population effect then models suggest that low metallicity models which incorporate binary evolution effects are required to reproduce them (although note that models with enhanced N/H ratios may also be appropriate, Shapley *et al.* 2015).

4. Implications

There is clear evidence that the rest-frame ultraviolet spectra of galaxies, and the nebular emission powered by reprocessing of the same photons, have evolved over cosmic time. This is, of course, to be expected - the volume-averaged mean metallicity of the IGM and the mean stellar population age both drop towards higher redshifts simply because the Universe itself is younger. The observed metallicity of gas clouds in the IGM evolves slowly, with the matter density in C IV absorbers dropping by a dex between $z \sim 2$ and $z \sim 6$ (D'Odorico *et al.* 2013). This is sufficient that we would expect to see significant changes in the character and influence of the massive star population over the same epoch. At the relatively low metallicities prevalent in the distant Universe, and the still lower ones inferred for the first few generations of star formation (i.e. during the Epoch of Reionization), we would expect the stars to be more massive, and to retain more of their mass due to the weakening of stellar winds. At the same time the influence of binary interactions will be important for these swollen massive stars. They are likely to exchange mass and angular momentum with binary companions, and so to rejuvenate ageing secondary stars. They may also spin up to velocities where rotational mixing influences the stellar evolution pathway. These changes are reflected in the far higher ionizing photon output seen in figure 3 at low metallicities, particularly for populations with a slightly higher limiting stellar mass than previously assumed.

Figure 3 also demonstrates why observers of distant galaxy populations are increasingly turning towards synthesis codes incorporating detailed stellar models to interpret their data. Constraints on the ionizing photon production in distant galaxies are suggesting that the effects of both IMF and multiplicity are becoming impossible to ignore in these sources. A corollary of this is that such galaxies provide a laboratory in which to test our models for massive stellar populations - they place the effects of these massive stars into a wider context. If a set of stellar population synthesis models cannot reproduce the observed properties of a distant source, we are left with three alternatives: either the observations are incorrect, or the wrong type of object is being modelled, or the models need to be reviewed and improved. In the local Universe it has rarely been necessary to answer such questions: the stellar population can frequently be resolved, or observed in sufficient detail to remove any ambiguity, while the models of near-Solar stellar populations are relatively mature and well constrained. The distant Universe, on the other hand, is now throwing up cases where searching questions can be asked of those modelling stellar populations, primary amongst them 'do we understand how massive stars form, interact and evolve at low metallicity?'.

A case in point is that of the $z = 6.6$ Lyman break galaxy CR7. This source was selected for its extremely strong Lyman-α line emission, but also shows strong He II 1640Å emission, with constraints on the non-detection of O III 1665Å and C III] 1909Å lines (Sobral *et al.* 2015). Together with its photometric colours, these characteristics have been proposed as indicative of a metal-free (Population III) stellar population, while alternative interpretations include a primordial Direct Collapse Black Hole (Pallottini *et al.* 2015). Either would be somewhat surprising, given the source's luminosity and redshift. As Bowler *et al.* (2016) discuss, stellar evolution models can go some way towards addressing this question, but are unable to resolve the issue completely and the interpretation of CR7 will remain ambiguous until further observations and improved modelling of all possible scenarios are undertaken.

While this is an isolated case, and the constraints placed on massive stellar models by galaxies (rather than the reverse) are still weak, JWST will revolutionise this field. The NIRSPEC instrument will enable deep spectroscopy of hundreds (if not thousands) of star-forming galaxies at $3 < z < 8$, from the rest-frame ultraviolet through to the rest-optical (see e.g. Giardino *et al.* 2016). As a result, it is likely to produce both more anomalous examples and much tighter constraints on the metal abundance and interstellar medium properties in galaxies in the distant Universe. Direct measurements of strong line ratios and probes of the stellar and interstellar absorption lines (as opposed to just strong line emission) should be possible on both individual and stacked sources. These will allow direct comparison with stellar population synthesis models, and test whether these recover the observed parameter spaces. Whether our understanding of the massive star population and its behaviour is sufficient to confront the wealth of observational data expected in the next few years remains to be seen. It will certainly be tested. Evidence to date is that both fields can learn from and be enriched by this synergy and it is my hope that they will continue to do so.

References

Adelberger, K. L., Steidel, C. C., Shapley, A. E., & Pettini, M., 2003, *ApJ*, 584, 45
Baldwin, J. A., Phillips, M. M., & Terlevich, R., 1981, *PASP*, 93, 5
Bouwens, R. J., *et al.*, 2011, *Nature*, 469, 504
Bouwens, R. J., *et al.*, 2016, *ApJ*, 831, 176
Bowler, R. A. A., *et al.*, 2016, arXiv, arXiv:1609.00727
Caruana, J., *et al.*, 2014, *MNRAS*, 443, 2831
Conroy, C., 2013, *ARA&A*, 51, 393
Dijkstra, M. & Wyithe, J. S. B., 2007, *MNRAS*, 379, 1589
D'Odorico, V., *et al.*, 2013, *MNRAS*, 435, 1198
Eldridge, J. J. & Stanway, E. R., 2012, *MNRAS*, 419, 479
Eldridge, J. J. & Stanway, E. R., 2009, *MNRAS*, 400, 1019
Finkelstein, S. L., *et al.*, 2012, *ApJ*, 756, 164
Giardino, G., *et al.*, 2016, *ASPC*, 507, 305
Greis, S. M. L., Stanway, E. R., Davies, L. J. M., & Levan, A. J., 2016, *MNRAS*, 459, 2591
Hashimoto, T., *et al.*, 2017, *MNRAS*, 465, 1543
Kehrig, C., *et al.*, 2015, *ApJ*, 801, L28
Kennicutt, R. C., Jr., 1998, *ARA&A*, 36, 189
Ma, X., *et al.*, 2016, *MNRAS*, 459, 3614
Malhotra, S. & Rhoads, J. E., 2002, *ApJ*, 565, L71
Masters, D., *et al.*, 2014, *ApJ*, 785, 153
Matthee, J., *et al.*, 2017, *MNRAS*, 465, 3637
Nakajima, K., *et al.*, 2016, *ApJ*, 831, L9
Pallottini, A., *et al.*, 2015, *MNRAS*, 453, 2465
Sanders, R. L., *et al.*, 2016, *ApJ*, 816, 23
Shapley, A. E., *et al.*, 2015, *ApJ*, 801, 88
Shapley, A. E., Steidel, C. C., Pettini, M., & Adelberger, K. L., 2003, *ApJ*, 588, 65
Smit, R., *et al.*, 2014, *ApJ*, 784, 58
Sobral, D., *et al.*, 2015, *ApJ*, 808, 139
Stanway, E. R., Bunker, A. J., & McMahon, R. G. 2003, *MNRAS*, 342, 439
Stanway, E. R., McMahon, R. G., & Bunker, A. J., 2005, *MNRAS*, 359, 1184
Stanway, E. R., Eldridge, J. J., & Becker, G. D. 2016, *MNRAS*, 456, 485
Stanway, E. R., Eldridge, J. J., Greis, S. M. L., *et al.* 2014, *MNRAS*, 444, 3466
Stark, D. P., *et al.*, 2015, *MNRAS*, 454, 1393
Steidel, C. C., Giavalisco, M., Pettini, M., Dickinson, M., & Adelberger, K. L., 1996, *ApJ*, 462, L17
Steidel, C. C., *et al.*, 2014, *ApJ*, 795, 165
Steidel, C. C., *et al.*, 2016, *ApJ*, 826, 159
Strom, A. L., *et al.*, 2016, arXiv, arXiv:1608.02587
Tasca, L. A. M., *et al.*, 2016, arXiv, arXiv:1602.01842
Topping, M. W. & Shull, J. M., 2015, *ApJ*, 800, 97
Wilkins, S. M., *et al.*, 2016a, *MNRAS*, 455, 659
Wilkins, S. M., *et al.*, 2016b, *MNRAS*, 458, L6
Wofford, A., *et al.*, 2016, *MNRAS*, 457, 4296

The Lives and Death-Throes of Massive Stars
Proceedings IAU Symposium No. 329, 2016
J.J. Eldridge, J.C. Bray, L.A.S. McClelland & L. Xiao, eds.

doi:10.1017/S1743921317003088

Low-metallicity (sub-SMC) massive stars

Miriam Garcia[1], Artemio Herrero[2,3], Francisco Najarro[1], Inés Camacho[2,3], Daniel J. Lennon[4], Miguel A. Urbaneja[5] and Norberto Castro[6]

[1]Centro de Astrobiología (INTA-CSIC), Departamento de Astrofísica.
Ctra. Torrejón a Ajalvir km.4, E-28850 Torrejón de Ardoz (Madrid), Spain
email: mgg@cab.inta-csic.es

[2]Instituto de Astrofísica de Canarias, 38205 La Laguna (Tenerife), Spain
[3]Departamento de Astrofísica, Universidad de La Laguna, 38206 La Laguna (Tenerife), Spain
[4]European Space Astronomy Centre (ESA/ESAC), Villanueva de la Cañada (Madrid), Spain
[5]Institut fuer Astro- und Teilchenphysik, Universitaet Innsbruck, Innsbruck, Austria
[6]Astronomy Department, University of Michigan, Ann Arbor, MI 48109, USA

Abstract. The double distance and metallicity frontier marked by the SMC has been finally broken with the aid of powerful multi-object spectrographs installed at 8-10m class telescopes. VLT, GTC and Keck have enabled studies of massive stars in dwarf irregular galaxies of the Local Group with poorer metal-content than the SMC. The community is working to test the predictions of evolutionary models in the low-metallicity regime, set the new standard for the metal-poor high-redshift Universe, and test the extrapolation of the physics of massive stars to environments of decreasing metallicity. In this paper, we review current knowledge on this topic.

Keywords. Galaxies: individual: IC 1613, NGC 3109, WLM, Sextans A – Stars: early-type – Stars: Population III – Stars: winds, outflows – Ultraviolet: stars

---

1. Introduction

Massive stars leave their imprint through the ages of the Universe and, as such, hold the key to interpret a plethora of astrophysical phenomena. The copious amount of ionizing and mechanical feedback from a population of massive stars drives locally the gas dynamics of their natal cloud, and can impact at a global level the evolution of host galaxies. Either as individuals or in binary systems, massive stars are the progenitors of the most energetic events in the Universe that can be used to probe high-redshifts: type Ibc,II supernovae (SNe) and arguably pair-instability supernovae, super-luminous supernovae and long γ-ray bursts (GRBs). The aftermath products, neutron stars and black holes, are sites of extreme physics.

In order to understand and quantify the role of massive stars in an evolving Universe, and to eventually use SNe/GRBs as lighthouses, it is necessary to describe the variation of their physical properties as a function of metallicity. The metal-poor regime is subject to a particularly growing interest in order to understand the conditions of earlier cosmic epochs, and to ultimately extrapolate the prescriptions for physical properties to the roughly metal-free Universe at the re-ionization epoch.

We do expect significant differences between metal poor ($Z \leqslant 1/10 Z_{\odot}$) massive stars and those in the Milky Way (MW) in terms that will translate into significant impact on their feedback. Radiation-driven winds (RDW) are ubiquitous in the hot stages of massive star evolution. They not only drive mechanical feedback but also regulate the evolution of the star (e.g. Meynet *et al.* 1994), and therefore the overall ionizing radiation and the final properties of the pre-SN core. Because the driving mechanism involves absorption

Table 1. Metal-poor early-type massive stars: headcount

	IC 1613	NGC 3109	WLM	Sextans A
Metallicity	1/7 $O_\odot$	1/7 $O_\odot$	1/7 $O_\odot$	1/10 $Z_\odot$
# O-B2 stars	56	44	8	12
# O-stars	26	12	2	5
# Earlier than O7	4	0	0	0

References: Azzopardi *et al.* (1988); Lozinskaya *et al.* (2002); Bresolin *et al.* (2006, 2007); Evans *et al.* (2007); Garcia & Herrero (2013); Camacho *et al.* (2016), Herrero, Garcia *et al.* in prep.

and re-emission of photons by metallic transitions RDWs are expected to decrease with decreasing metallicity, and be almost negligible at Z$\sim$1/1000$Z_\odot$ except for very bright objects (Kudritzki 2002). However, the metallicity dependence of RDWs has only been confirmed observationally down to the Small Magellanic Cloud (SMC) (Mokiem *et al.* 2007).

The evolutionary pathways may also change drastically. Metal-poor massive stars rotating sufficiently fast may bring He produced at the stellar core to the surface, mix it in fast time-scales, and experience chemical homogeneous evolution (CHE). The star will never leave the high temperature regime of the HR-diagram (HRD); the overall ionizing energy emitted through its evolution will be enhanced with respect to analog stars that, either because of a slower initial v_{rot}, or a higher metallicity powered wind, do reach the red supergiant stage (RSG). For instance, a 150$M_\odot$ undergoing CHE will double the amount of HI-ionizing photons and quadruple the HeII-ionizing photons with respect to the redwards evolving analog according to Szécsi *et al.* (2015)'s 1/50$Z_\odot$ tracks. CHE of binary massive stars within the Roche lobe is one proposed scenario to generate the double $\sim$30$M_\odot$ black hole system that produced the first detection of gravitational waves when it merged (Mandel & de Mink 2016). Yet, while some evidence on CHE exists (Martins *et al.* 2013) a sufficiently large number of stars that establishes this evolutionary pathway has not been detected.

Today, the Small Magellanic Cloud stands the reference for the metal-poor Universe. Spectral libraries of SMC stars feed population synthesis models (e.g. Leitherer *et al.* 2001). However, the SMC's 1/5$Z_\odot$ falls short of the $\sim$1/30$Z_\odot$ metallicity at the peak of star formation (Madau & Dickinson 2014) and is not valid an approximation for the roughly metal-free early Universe. In this paper we will review recent efforts to surpass the SMC frontier, and what we have learnt about sub-SMC metallicity massive stars.

2. The 1/7 $O_\odot$ galaxies IC 1613, NGC 3109 and WLM

The full development of the 8-10m telescopes and their multi-object spectrographs (MOS) in the mid-2000's enabled the discovery and study of massive stars beyond the Magellanic Clouds with good quality data. Three galaxies were identified within reach, with promising 1/7$O_\odot$† nebular abundances: IC 1613, NGC 3109 and WLM (Fig. 1).

Spectroscopic studies with the previous generation of 4m telescopes had been restricted to the brightest supergiants, LBV and WR stars, because of faint magnitude limits and crowding issues. Nowadays the 10.4m Gran Telescopio Canarias (GTC) enables R$\sim$800

† At this point, we would like to note that metallicity ($Z/Z_\odot$) is often calculated from the ratio of oxygen abundances against the Solar value only, with the implied hypothesis that the chemical mixture follows the Solar pattern. However, the relative ratios of elements (e.g. α/Fe) do vary with the chemical evolution of galaxies (as we discuss again in Sect. 2.4). For this reason, we use the notation of $O/O_\odot$ and $Fe/Fe_\odot$ to distinguish the metallicity indicator, and only use $Z/Z_\odot$ when information from at least both Fe and O is available.

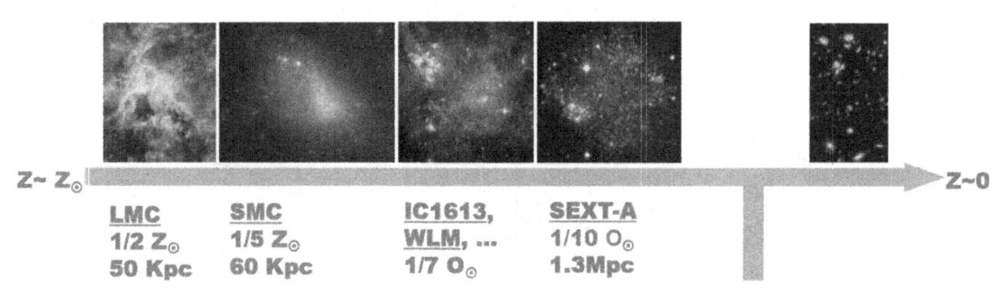

Figure 1. Road-map to the early-Universe: sequence of resolved galaxies of decreasing metal content. Characterizing massive stars in these galaxies will open the metal-poor regime. The ultimate goal is to produce prescriptions valid for the early-Universe.

identification spectroscopy of an IC 1613 O-type star at 750Kpc (V=19.6) in 1 hour. The actual exposure times per target are being decreased with detectors of ever increasing sensitivity, and divided by the multiplexing capabilities of the MOSs.

The task first required a strong exploratory effort to confirm the nature of massive star candidates with spectroscopy. The candidates could not be directly targeted with multi-slit or multi-fiber spectrographs for deep high resolution data because blue massive stars cannot be solely identified from optical colors (though some optimized methods for the blue types exist, e.g. Garcia & Herrero 2013). The high multi-plexing of VLT-VIMOS and FORS was of great help, although numbers are still scarce. Table 1 summarizes the number of sub-SMC metallicity early-type massive stars known to date.

Although great progress has been made, our studies are still limited in sample size, resolution and epochs. At the moment, we cannot tackle very interesting issues like rotational velocities, binary frequencies, or the possibility of a top-heavy initial mass function in metal-poor environments (in connection to the extremely massive, metal-free First Stars). However, the observations do allow for quantitative analyses to obtain stellar parameters, constrain the winds and enable first contrast against evolutionary models.

2.1. $T_{\rm eff}$ *vs Spectral Type calibration*

The effective temperatures ($T_{\rm eff}$) of all OBA stars studied in IC 1613, NGC 3109 and WLM are compiled in Fig. 2-left, and compared against calibrations with spectral type for various metallicities. These calibrations are useful to assign effective temperatures in sight of spectral type to first approximation, and to quantify ionizing photons emitted by unresolved populations.

The temperatures of O-stars show a large dispersion in Fig. 2-left. The scatter is of the order of the expected differences between different luminosity classes, although it likely reflects the heterogeneity of the analyzed data that either have low resolution, low signal-to-noise ratio (SNR) or both. When only giant and supergiant stars are considered, the $1/7\mathrm{O}_{\odot}$ effective temperature scale is roughly $\sim$1000 K hotter than the SMC's (Garcia & Herrero 2013).

B-supergiants show a much smaller dispersion, and roughly follow Trundle *et al.* (2007)'s calibration for the SMC. The early-A supergiants can be found at the extrapolation of this calibration. One early-B supergiant stands out with high effective temperature. This is likely a miss-classification issue, since low resolution or SNR may hinder the detection of the HeII lines, which will be weak at the latest O-types.

2.2. *The HR-diagram*

The 1/7 $\mathrm{O}_{\odot}$ massive stars are found at their expected loci in the HR-diagram (Fig. 2-right). The types that have so far been confirmed by spectroscopy include O-stars and

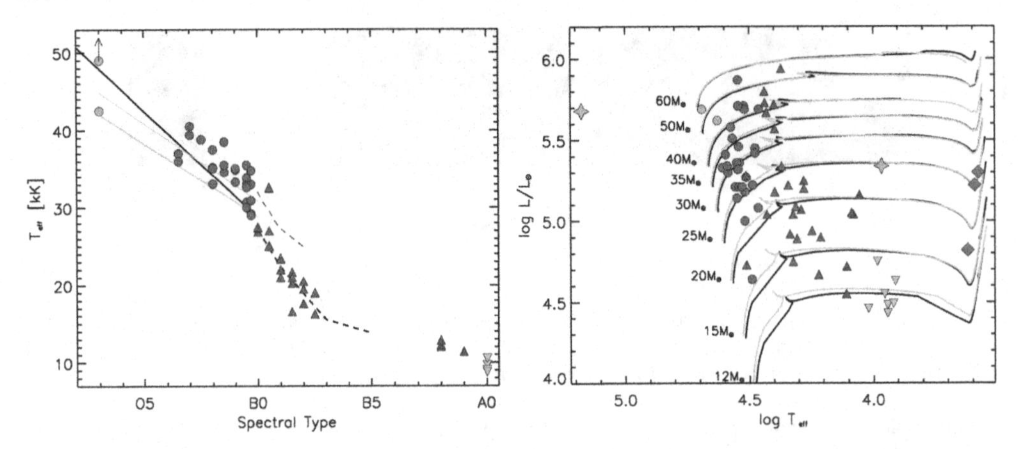

Figure 2. Properties of known stars in 1/7O$_\odot$ galaxies. **Left:** $T_{\rm eff}$ *vs* spectral type for early–type (O-A0) massive stars, compared with SMC results: Trundle *et al.* (2007)'s calibration for SMC B-stars (black dashed lines), and a linear fit to Mokiem *et al.* (2007) and Massey *et al.* (2009) parameters for O stars (black solid line). For comparison, the MW $T_{\rm eff}$ *vs* spectral type calibration by Martins *et al.* (2005) is included in gray. For all calibrations: thick lines represent supergiants and thin lines dwarfs. **Right:** stars in the HR-diagram compared with Brott *et al.* (2011a)'s SMC tracks with v_{ini}=300 km s^{-1}(black) and 0 km s^{-1}(gray). **References:** Stellar samples collected from Bresolin *et al.* (2007); Evans *et al.* (2007); Tautvaišienė *et al.* (2007); Garcia & Herrero (2013); Tramper *et al.* (2013, 2014); Hosek *et al.* (2014); Bouret *et al.* (2015); Herrero *et al.* (2010, 2012, in prep.), Camacho *et al.* in prep. **Symbols:** Circles: O-stars, early-types in cyan; Up-triangles: B-supergiants, late-types in dark-green; Down-triangles: A-supergiants; Red-rhomboids: RSGs; Orange-star: the LBVc V39; Cyan-star: the WO DR1.

BA supergiants, plus some advanced stages in IC 1613: RSGs, the oxygen Wolf-Rayet (WO) star DR1, and the LBV-candidate V39. O-type stars draw the main sequence, with only one early-O star close to the zero-age main sequence (ZAMS). B-supergiants overlap with the H-burning region of SMC tracks, which could be partly explained by the shift of the evolutionary paths to higher $T_{\rm eff}$ as metallicity decreases.

The total sample is still small but at present there is no evidence of a sequence of stars undergoing CHE, which would be located at higher temperatures than the ZAMS. Likewise, no super-luminous RSG has been detected that would support the inflation and redwards evolution scenario of slowly-rotating very metal-poor stars (Szécsi *et al.* 2015). At the moment we cannot contrast this information with rotational velocities because of the low-resolution used for the majority of the observing runs, but no high-speed rotator has been detected.

Finally, it is noteworthy that the most massive stars detected, including the WO star in IC 1613 (Tramper *et al.* 2015), have M_{ini}~60M$_\odot$.

2.3. *The Wind Momentum - Luminosity Relation (WLR)*

Mass loss is one crucial ingredient to compute the evolution and fate of massive stars. Quantifying true mass loss rates, providing prescriptions that hold in varying metallicity environments, and testing such prescriptions in metal-poor environments, is a must to be able to simulate massive stars at high-redshifts, their death and feedback.

The WLR is currently the main diagnostic tool, and compares observational measurements of the mechanical wind-momentum ($\propto \dot{M} \cdot v_\infty$) with theory. Based on the nature of RDWs, the WLR tests simultaneously the expected positive correlation of wind-momentum with stellar luminosity and its anti-correlation with metallicity.

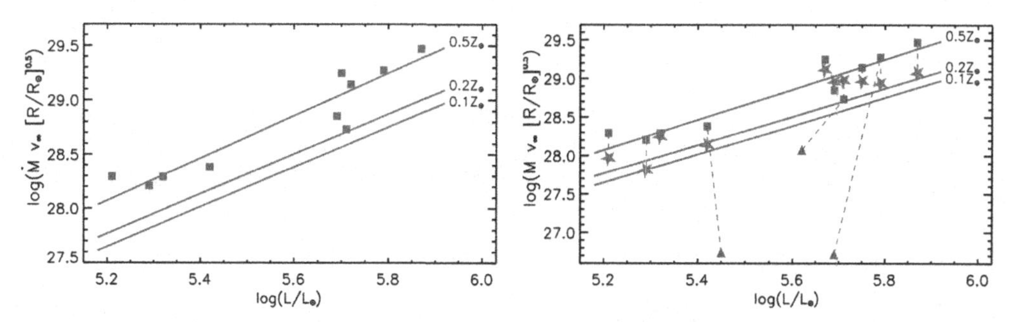

Figure 3. The 1/7O$_\odot$ Wind Momentum - Luminosity Relation. **Left:** Wind-momentum of 1/7O$_\odot$ OB-stars from optical data (Herrero *et al.* 2012; Tramper *et al.* 2011, 2014): $\dot{M}$'s were derived from H$_\alpha$, and v_∞=2.65v_{esc} (gray squares). The unclumpled observational relations for the MW, LMC and SMC (Mokiem *et al.* 2007) are shown for comparison. **Right:** The WLR recomputed with UV terminal velocities (red stars), either direct measurements of the star or a star with similar spectral type (from Garcia *et al.* 2014; Bouret *et al.* 2015). The results from a full UV+optical analysis of 3 stars by Bouret *et al.* (2015) are also included (purple triangles).

When the WLR for 1/7O$_\odot$ stars was first plotted, the community was alarmed: most of the stars were found at the loci of more metal-rich 0.5Z$_\odot$ LMC stars (Fig. 3-left). Together with the previous detection of strong P Cygni profiles in IC 1613's V39 (Herrero *et al.* 2010), and the existence of a WO in the galaxy (e.g. Tramper *et al.* 2015), these results were suggestive of stronger winds than expected in the studied galaxies. If these results were confirmed, we would need updated wind recipes, evolutionary models, feedback calculations and SN and GRB rates for very low-metallicity massive stars.

However, the *strong wind problem* can be largely explained by the lack of suitable diagnostics for both the wind and metallicity (see Sect. 2.4). The optical-only studies that reported the *strong winds* are based on H$_\alpha$ profile fitting. They are not sensitive to small mass loss rates and are degenerated to the parameter Q=$\dot{M}/(v_\infty \cdot R_\star)^{1.5}$ (Kudritzki & Puls 2000). In order to derive $\dot{M}$, and lacking a direct ultraviolet (UV) measurement, v_∞ is calculated from the escape velocity (v_∞/v_{esc}=2.65) and scaled with metallicity ($v_\infty \propto Z^{0.13}$ Leitherer *et al.* 1992). The v_∞/v_{esc}=2.65 calibration is long known to suffer a large scatter (Kudritzki & Puls 2000), but recent results also argue against a straightforward $v_\infty \propto$ Z relation (Garcia *et al.* 2014). Thus, in absence of UV data, v_∞ uncertainties add up and propagate to $\dot{M}$ and the WLR.

The installation of the Cosmic Origins Spectrograph (COS) on the HST enabled UV spectroscopy (~1150-1800Å) of OB stars out to ~1Mpc (see Sect. 3), providing crucial information on their RDWs. The wind terminal velocity can be measured from the resonance lines of NV λ1238.8,1242.8, SiIV λ1393.8,1402.8 and CIV λ1548.2,1550.8. Together with CIII λ1176 and NIV λ1718.0, these lines constrain mass loss rates even for the low $\dot{M} < 10^{-7}$M$_\odot$ yr^{-1} values expected for metal-poor stars that render H$_\alpha$ insensitive.

Fig. 3-right shows the 1/7O$_\odot$ WLR recalculated with UV derived terminal velocities by Garcia *et al.* (2014) and Bouret *et al.* (2015). The wind-momentum of most of the stars was revised downwards, and the overall trend is now closer to Mokiem *et al.* (2007)'s regression describing the SMC (0.2Z$_\odot$). Moreover, the joint UV+optical analysis of 3 stars by Bouret *et al.* (2015) yielded wind-momenta much below the 1/7Z$_\odot$ line. While the difference is partly due to the consideration of clumping in their study, the authors also propose the existence of a hot gas component in the wind that do not impact P Cygni profiles and therefore cannot be diagnosed from the UV. In any case, the WLR

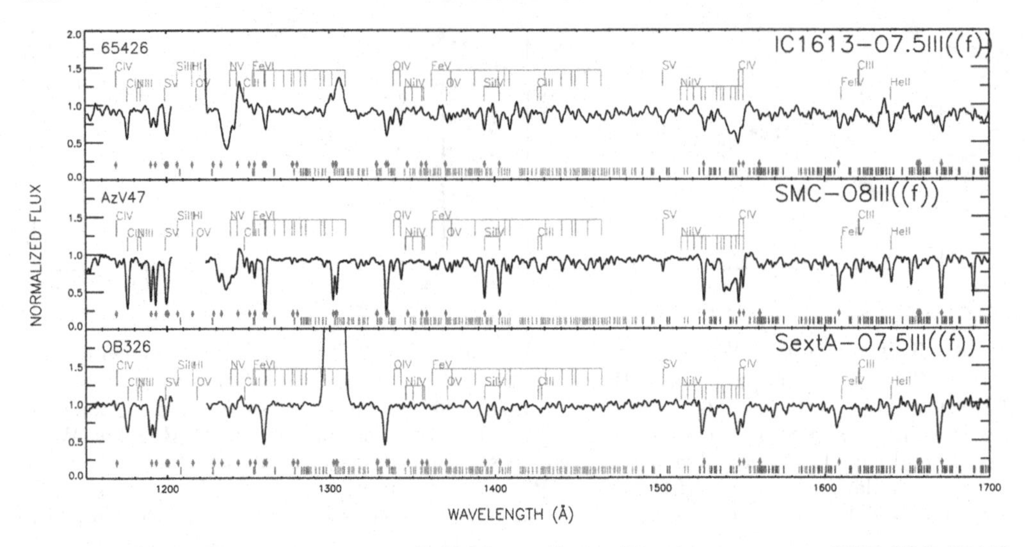

Figure 4. First UV spectroscopy at sub-SMC metallicity. The chart compares HST-COS-G140L observations of stars in IC 1613 and Sextans A (PI M. Garcia) with the STIS spectrum of a star with similar spectral type. #65426 in IC 1613 shows the strongest P Cygni profiles of CIV and NV, stronger than the SMC star. The wind profiles are much weaker in the Sextans A star, as expected with $1/10Z_\odot$ metallicity, although the terminal velocity can still be measured from CIV. The plot also illustrates striking differences in the Fe-content; the lines in the FeV forest (green ticks) are strongest in the IC 1613 star, and decrease as we move downwards in the plot, suggesting a succession of decreasing metallicity.

of the $1/7O_\odot$ galaxies is still not settled and clearly requires a larger sample of high quality spectra.

2.4. *The actual metallicity of IC 1613, NGC 3109 and WLM*

The HST-COS observations delivered an interesting by-product: the iron content of IC 1613 and WLM is similar or even larger than the $\sim 1/5Fe_\odot$ content of the SMC, superseding the $1/7Z_\odot$ value scaled from oxygen (Garcia *et al.* 2014; Bouret *et al.* 2015). Tautvaišienė *et al.* (2007) obtained a similar value from the analysis of three RSGs in IC 1613. Meanwhile Hosek *et al.* (2014) reported an $\sim 0.21Fe_\odot$ iron abundance for the massive stars of NGC 3109. The updated $\gtrsim 0.2Fe_\odot$ content of IC 1613, NGC 3109 and WLM also helps to alleviate the problem of *strong winds*, as iron is the main driver of mass loss (e.g. Vink *et al.* 2001) and a larger wind-momentum is expected.

These results reflect the known fact that the ratio of α-elements to Fe depends on the chemical evolution of galaxies, and that oxygen is not always a good proxy for metallicity. Unfortunately for the subject of this work, it also implies that *the regime of massive stars with poorer metal-content than the SMC ($\lesssim 0.1Fe_\odot$) remains unexplored.*

3. A promising $1/10Z_\odot$ galaxy: Sextans A

Located 1.3Mpc away, Sextans A is potentially the most iron-poor galaxy of the Local Group. Hosek *et al.* (2014) compiled iron and α-element abundances of nearby irregular galaxies determined from blue supergiants, thus probing the present-day metallicity while avoiding the calibration issues of nebular studies: Sextans A stands out as the most iron-poor with $[Fe/H] \lesssim -1.0$ and $[\alpha/Fe] \sim 0$.

Camacho *et al.* (2016) recently confirmed the first OB-stars in Sextans A (see Table 1) with low-resolution long-slit spectroscopy performed at the GTC. The data enabled the first determination of the stellar parameters but yielded no information on the wind or abundances (as expected). The identified stars are not very massive, M_{ini}~20-40M$_\odot$, and 4-6Myr old. The youngest ones concentrate on the main galactic-scale over-densities of neutral hydrogen, similarly to IC 1613 (Garcia *et al.* 2010), although no very-early spectral types were detected. However, the most massive stars of Sextans A may be gas- and dust-enshrouded within the HII shells, and out of reach to the depth of the observing program.

The UV spectra of some Sextans A OB-stars were obtained by pushing HST-COS to its limits (Fig. 4). The stars display wind profiles that are weak, but strong enough to set constraints on the wind properties (Garcia *et al.* in prep). The continuum depletion caused by Fe and Ni transitions is almost negligible, specially when comparing with IC 1613 or SMC stars (see Fig. 4), therefore confirming the poor Fe content of the stars.

Sextans A may become the next sub-SMC metallicity standard and important observational effort is being devised to mine this extremely interesting galaxy. Stay tuned for GTC-OSIRIS multi-object spectroscopic observations (Camacho *et al.* in prep).

4. Summary and outlook: prospects for the E-ELT and LUVOIR

The analysis of stars in the 1/7O$_\odot$ galaxies IC 1613, NGC 3109 and WLM has not revealed significant differences with SMC stars to date. This could be partly due to the very similar iron content of the four galaxies. The IC 1613, NGC 3109 and WLM stars open the window to study the effects of non-solar abundance patterns in the evolution and winds of massive stars, although such task would require much larger samples of stars.

The number of confirmed OBA stars in the true sub-SMC metallicity Sextans A is still scarce. The large distance to the galaxy, 1.3Mpc, makes observations challenging. In addition, internal extinction at the expected sites for OB stars may be non-negligible since the youngest populations often co-exist with ionized and neutral gas, and dust. Deeper observations are needed.

In fact, progress in the characterization of sub-SMC metallicity massive stars requires efforts in 2 steps. Firstly deep, low-resolution (R~ 700-2000), wide-field spectroscopic observations are key to pierce through gas and dust, reach sufficiently weak stars, and enlarge the samples. In this respect, the high multiplexing capabilities of the second-generation spectrographs on 8-10m telescopes will be crucial. We can expect significant contributions with MEGARA at the GTC (Gil de Paz *et al.* 2016) and MUSE at the VLT (Henault *et al.* 2003) in the near future.

Secondly stellar properties, wind parameters and abundances will be derived from quantitative spectroscopic analysis. The required ultraviolet observations and medium resolution (R~5000) optical spectroscopy again meet the problem of reaching stars as faint as V~20-21 and extinction, critical to the UV. It is important to stress that UV observations (optimally in the 950-1800Å range) are a necessary companion to optical spectroscopy. This range hosts both the diagnostics for the terminal velocity of OB stars and key lines to constrain shocks in the wind (Garcia & Bianchi 2004) and micro-/macro-clumping (Sundqvist *et al.* 2011).

We are pushing current observing facilities to the limit. Not only larger collecting surfaces are needed, but also source confusion cripples studies of individual objects in interesting sites farther than Sextans A.

Beyond the Local Group, I Zw18 (18.2Mpc Aloisi *et al.* 2007) is the most metal poor galaxy known to sustain significant star formation. With 1/32Z$_\odot$ (Vílchez & Iglesias-Páramo 1998), it may become the reference to study star formation in the first galaxies of the Universe and it is the ideal site to test $1/50Z_{\odot}$ stellar tracks. In fact, I Zw18 exhibits intense HeII4686 emission (Kehrig *et al.* 2015) that could be caused by Szécsi *et al.* (2015)'s TWUINs, very hot and massive stars experiencing CHE. However, I Zw18 is about 10 times farther than Sextans A and resolving its stellar population escapes current facilities.

The field will experience a tremendous boost with the arrival of two great astronomical facilities in the near/mid-term future. Multi-object spectrographs at the E-ELT (e.g. MOSAIC) will be able to reach main-sequence O-stars almost as far as 4Mpc, and resolve intricate environments like 30 Dor out to 1.5Mpc. The second key facility is NASA's likely next flagship mission LUVOIR, a 12m space telescope with UV, optical and IR capabilities. The observatory would be sensitive to unprecedentedly faint UV sources *per se*, but it will also host a multi-object spectrograph for UV observations (LUMOS) that will enable to observe entire extragalactic OB-associations in one shot. Both large facilities will be able to target individual, luminous stars in I Zw18. The synergies between E-ELT and LUVOIR will surely revolutionize the studies of truly metal-poor massive stars.

Acknowledgements

M. Garcia acknowledges funding by the Spanish MINECO *via* grants FIS2012-39162-C06-01, ESP2013-47809-C3-1-R and ESP2015-65597-C4-1-R.

References

Aloisi, A., Clementini, G., Tosi, M., *et al.* 2007, *ApJL*, 667, L151
Azzopardi, M., Lequeux, J., & Maeder, A. 1988, *A&A*, 189, 34
Bouret, J.-C., Lanz, T., Hillier, D. J., *et al.* 2015, *MNRAS*, 449, 1545
Bresolin, F., Pietrzyński, G., Urbaneja, M. A., *et al.* 2006, *ApJ*, 648, 1007
Bresolin, F., Urbaneja, M. A., Gieren, W., Pietrzyński, G., & Kudritzki, R.-P. 2007, *ApJ*, 671, 2028
Brott, I., de Mink, S. E., Cantiello, M., *et al.* 2011a, *A&A*, 530, A115
Camacho, I., Garcia, M., Herrero, A., & Simón-Díaz, S. 2016, *A&A*, 585, A82
Evans, C. J., Bresolin, F., Urbaneja, M. A., *et al.* 2007, *ApJ*, 659, 1198
Fullerton, A. W., Massa, D. L., & Prinja, R. K. 2006, *ApJ*, 637, 1025
Garcia, M. & Bianchi, L. 2004, *ApJ*, 606, 497
Garcia, M., Herrero, A., Castro, N., Corral, L., & Rosenberg, A. 2010, *A&A*, 523, A23
Garcia, M. & Herrero, A. 2013, *A&A*, 551, A74
Garcia, M., Herrero, A., Najarro, F., Lennon, D. J., & Alejandro Urbaneja, M. 2014, *ApJ*, 788, 64
Gil de Paz, A., Gallego, J., Carrasco, E., *et al.* 2016, Multi-Object Spectroscopy in the Next Decade: Big Questions, Large Surveys, and Wide Fields, 507, 103
Henault, F., Bacon, R., Bonneville, C., *et al.* 2003, *SPIE*, 4841, 1096
Herrero, A., Garcia, M., Uytterhoeven, K., *et al.* 2010, *A&A*, 513, A70
Herrero, A., Garcia, M., Puls, J., *et al.* 2012, *A&A*, 543, A85
Hosek, M. W., Jr., Kudritzki, R.-P., Bresolin, F., *et al.* 2014, *ApJ*, 785, 151
Kehrig, C., Vílchez, J. M., Pérez-Montero, E., *et al.* 2015, *ApJL*, 801, L28
Kudritzki, R.-P. & Puls, J. 2000, *ARA&A*, 38, 613
Kudritzki, R. P. 2002, *ApJ*, 577, 389
Leitherer, C., *et al.* 1992, *ApJ*, 401, 596
Leitherer, C., Leão, J. R. S., Heckman, T. M., *et al.* 2001, *ApJ*, 550, 724
Lozinskaya, T. A., Arkhipova, V. P., Moiseev, A. V., & Afanas'Ev, V. L. 2002, Astronomy Reports, 46, 16
Madau, P. & Dickinson, M. 2014, *ARA&A*, 52, 415

Mandel, I. & de Mink, S. E. 2016, *MNRAS*, 458, 2634
Martins, F., Schaerer, D., & Hillier, D. J. 2005, *A&A*, 436, 1049
Martins, F., *et al.* 2013, *A&A*, 554, A23
Massey, P., Zangari, A. M., Morrell, N. I., *et al.* 2009, *ApJ*, 692, 618
Meynet, G., Maeder, A., Schaller, G., *et al.* 1994,*A&AS*, 103, 97
Mokiem, M. R., de Koter, A., Vink, J. S., *et al.* 2007, *A&A*, 473, 603
Sundqvist, J. O., Puls, J., Feldmeier, A., & Owocki, S. P. 2011, *A&A*, 528, A64
Szécsi, D., Langer, N., Yoon, S.-C., *et al.* 2015, *A&A*, 581, A15
Tautvaišienė, G., Geisler, D., Wallerstein, G., *et al.* 2007, *AJ*, 134, 2318
Tramper, F., Sana, H., de Koter, A., & Kaper, L. 2011, *ApJL*, 741, L8
Tramper, F., Gräfener, G., Hartoog, O. E., *et al.* 2013, *A&A*, 559, A72
Tramper, F., Sana, H., de Koter, A., Kaper, L., & Ramírez-Agudelo, O. H. 2014, *A&A*, 572, AA36
Tramper, F., Straal, S. M., Sanyal, D., *et al.* 2015, *A&A*, 581, A110
Trundle, C., Dufton, P. L., Hunter, I., *et al.* 2007, *A&A*, 471, 625
Vílchez, J. M. & Iglesias-Páramo, J. 1998, *ApJ*, 508, 248
Vink, J. S., *et al.* 2001, *A&A*, 369, 574

Discussion

The Lives and Death-Throes of Massive Stars
Proceedings IAU Symposium No. 329, 2016
J.J. Eldridge, J.C. Bray, L.A.S. McClelland & L. Xiao, eds.

doi:10.1017/S1743921317003489

II Zw 40 – 30 Doradus on Steroids

Claus Leitherer[1], Janice C. Lee[1] and Emily M. Levesque[2]

[1]Space Telescope Science Institute,
3700 San Martin Drive, Baltimore, MD 21218, USA
email: leitherer@stsci.edu, jlee@stsci.edu

[2]Dept. of Astronomy, University of Washington
C327, Seattle, WA 98195, USA
email: emsque@uw.edu

Abstract. We obtained HST COS G140L spectra of the enigmatic nearby blue compact dwarf galaxy II Zw 40. The galaxy hosts a nuclear super star cluster embedded in a radio-bright nebula, similar to those observed in the related blue compact dwarfs NGC 5253 and Henize 2-10. The ultraviolet spectrum of II Zw 40 is exceptional in terms of the strength of He II 1640, O III] 1666 and C III] 1909. We determined reddening, age, and the cluster mass from the ultraviolet data. The super nebula and the ionizing cluster exceed the ionizing luminosity and stellar mass of the local benchmark 30 Doradus by an order of magnitude. Comparison with stellar evolution models accounting for rotation reveals serious short-comings: these models do not account for the presence of Wolf-Rayet-like stars at young ages observed in II Zw 40. Photoionization modeling is used to probe the origin of the nebular lines and determine gas phase abundances. C/O is solar, in agreement with the result of the stellar-wind modeling.

Keywords. galaxies: dwarf, galaxies: starburst, galaxies: stellar content

---

1. Introduction

II Zw 40 (= UGCA 116), together with I Zw 18, is the founding member of the class of blue compact dwarf (BCD) galaxies. This class was originally defined by Sargent (1970) and Sargent & Searle (1970) who demonstrated that the optical spectra of these galaxies are indistinguishable from those of extragalactic H II regions. BCDs and related dwarf starbursts have since been recognized as objects that are characterized by their blue optical colors, small sizes (< 1 kpc), and low luminosities of $M_{\rm B} > -18$ (Gil de Paz & Madore 2005). These galaxies do not typically show signs of AGN activity but have strongly enhanced star formation in a relatively pristine chemical environment. Owing to these properties, BCDs have been proposed as nearby analogs of star formation in young galaxies at high redshift (Thuan 2008). Among BCDs are some of the most metal-poor star-forming galaxies detected in the universe (Kunth & Östlin 2000). These properties make BCDs preferred targets for observations in the ultraviolet (UV) where they are generally bright due to their blue colors. II Zw 40 is one of the few exceptions, which lacked UV spectra despite its remarkable properties due to its high Galactic foreground reddening (see below).

Super star clusters are an important mode of star formation in starbursts (Meurer *et al.* 1995). II Zw 40's morphology is dominated by a central super star cluster, which is the powering source of a radio-infrared super nebula (Beck *et al.* 2013, Kepley *et al.* 2014). The radio super nebula requires a powering ionizing photon output of $\sim 10^{52.7}$ s^{-1}, which makes this region 10 times as luminous as 30 Doradus, the most luminous giant H II region in the Local Group. Radio super nebulae have also been detected in the BCDs

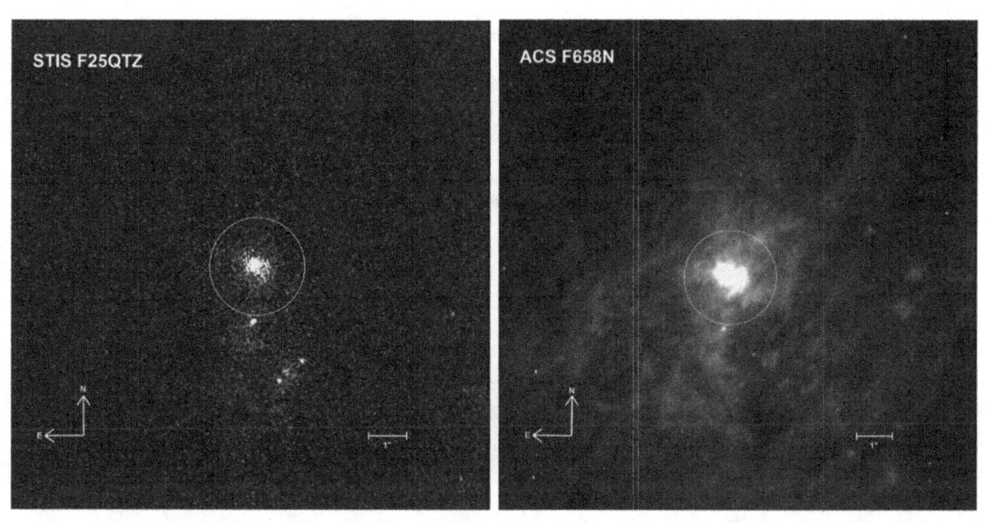

Figure 1. STIS F25QTZ (left) and ACS F658N (right) images of the super nebula. F25QTZ is centered on $\lambda \approx 1600$ Å, and F658N encompasses Hα. The circle indicates the location of the COS aperture, which covers a physical area of 135 pc diameter.

NGC 5253 and Henize 2-10 (Beck 2015). They are thought to be the earliest phases in the evolution of super star clusters, suggesting a very young age for the cluster.

2. New Data

We obtained UV spectra of the super nebula and its ionizing cluster with HST's COS using the G140L grating. The spectra cover the wavelength range 1150 - 2000 Å and have a spectral resolution of about 0.5 Å. The ionizing stars are concentrated in one super star cluster, which is essentially a single point source in the UV (see Fig. 1, left). In contrast, the optical Hα image (Fig. 1, right) shows nebular emission which is more extended inside the COS aperture, as well as diffuse emission over hundreds of parsecs. The super nebula is at the core of the galaxy whose extended (several kpc) tidal tails indicate previous interaction and merging of two dwarf galaxies (Kepley *et al.* 2014).

The processed spectrum is reproduced in Fig. 2. No reddening or redshift correction has been applied. Owing to its location close to the Galactic plane ($b = 10.8°$), II Zw 40 has a high foreground reddening of $E(B-V) = 0.73$. Despite the high dust attenuation the COS spectrum permits useful constraints on the ionizing cluster and the super nebula. Apart from the geocoronal Lyman-α 1216 and O I 1300 emission lines, the spectrum shows, among others, stellar N V 1240, C IV 1550, He II 1640 as well as nebular O III] 1666 and C III] 1909.

3. Results

We compared the observed UV spectrum of the super nebula and its ionizing cluster to synthetic spectra generated with the population synthesis code Starburst99 (Leitherer *et al.* 2014). We adopted metal-poor ($Z = 1/7^{th}$ $Z_{\odot}$) stellar evolution models with rotation (Ekström *et al.* 2012, Georgy *et al.* 2013), a Kroupa (Kroupa 2008) initial mass function (IMF), a single stellar cluster, and a distance of 11.1 Mpc (Marlowe *et al.* 1997). Comparison of the spectral slope, luminosity and stellar spectral lines provides the intrinsic reddening, cluster mass, and age, respectively. The best-fit model (see Fig. 3)

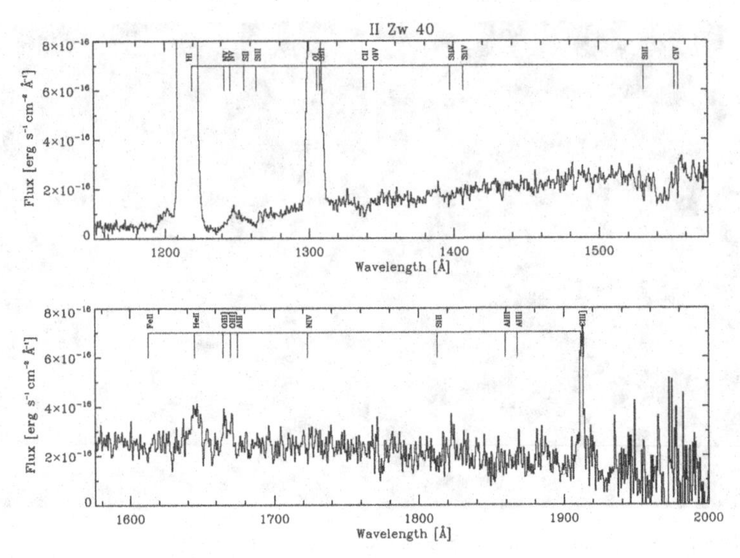

Figure 2. COS G140L spectrum of the super nebula and its ionizing star cluster with line identifications given at the top.

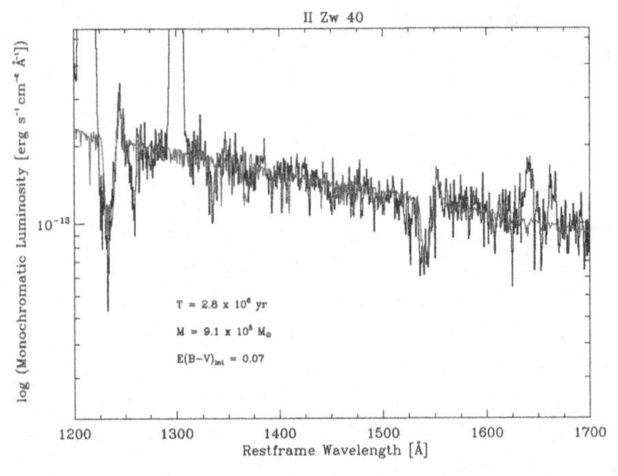

Figure 3. Restframe UV spectrum of the super nebula corrected for foreground and intrinsic reddening (black) compared to a model spectrum of a star cluster with parameters listed (blue).

corresponds to a stellar cluster with intrinsic reddening $E(B-V) = 0.07$, a mass of 9.1×10^5 $M_\odot$, and an age of 2.8×10^6 yr. The young age is constrained mainly by the strength of N V 1240 and the absence of Si IV 1400. This age is consistent with the upper limit derived in the radio from the strength of the free-free and the weakness of the synchrotron emission, suggesting few core-collapse supernovae (Kepley *et al.* 2014).

The Starburst99 models can be used to predict the ionizing luminosity from an extrapolation of the best-fit UV spectrum to below 912 Å. This leads to $\log N_{\rm L} = 52.8$ s^{-1}, in excellent agreement with the reddening-independent value derived from the radio free-free luminosity by Kepley *et al.* (2014). The ionizing luminosity and the stellar mass of the super star cluster exceed the values determined for the proto-typical local giant H II region 30 Doradus and its ionizing cluster NGC 2070 by an order of magnitude. The mass is similar to the masses of the most massive Galactic globular clusters (Gnedin & Ostriker 1997) and the super star clusters in the Antennae galaxies (Whitmore *et al.* 2010).

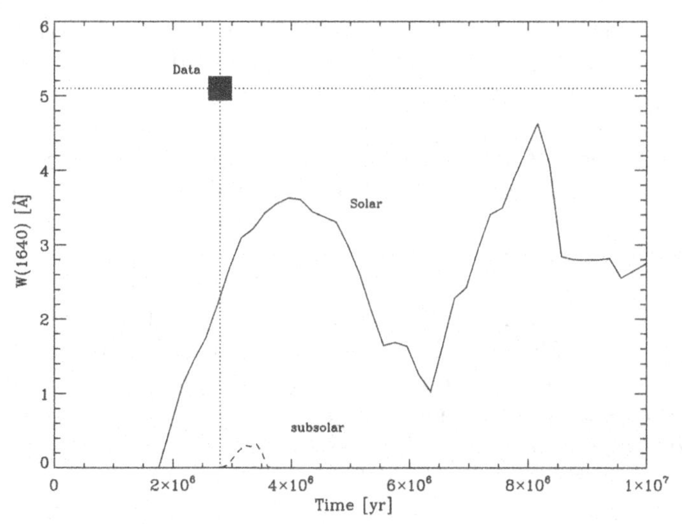

Figure 4. Stellar He II 1640 equivalent width for solar (solid) and $1/7^{th}$ solar (dashed) chemical composition predicted by evolution models with rotation. The filled square is the measured value.

It also approaches the masses of the most massive molecular clouds in the prototypical starburst galaxy M82 (Keto *et al.* 2005).

The synthetic and the observed stellar lines agree rather well in Fig. 3, with the notable exception of the broad He II 1640. This line has traditionally been associated with Wolf-Rayet (W-R) stars (Conti 1991), the late stages of massive-star evolution. Our models do not predict significant numbers of W-R stars due to the young age of the cluster so that the line is absent in the models. This is further illustrated in Fig. 4 where we compare the measured equivalent width to model predictions of rotating evolution models at solar and sub-solar chemical composition. Metal-rich populations barely reach the observed value, but only at older ages. Metal-poor populations do not produce significant He II 1640 at any age. The stars responsible for the observed He II 1640 are not accounted for in these models.

The strongest nebular line in the UV spectrum of II Zw 40 is C III] 1909 with an equivalent width of 9.5 Å. This is among the highest values observed in local star-forming galaxies and comparable to observations at high redshift (Rigby *et al.* 2015). The O III] 1666 has an equivalent width of about 3 Å, corrected for the underlying interstellar Al II 1670 absorption. Lyman-α is not observable in II Zw 40 due to its low velocity of 789 km s^{-1} and the corresponding geocoronal and Galactic contamination. Comparison of the observed C III] 1909 and O III] 1666 line strengths with photo-ionization models of Gutkin *et al.* (2016) suggests a very high ionization parameter and a near-solar C/O ratio of 0.4. This is consistent with the result of the stellar-wind modeling, where we obtained an excellent fit to C IV 1550 with solar C/O. II Zw 40's location in the C/O versus O/H diagram (Esteban *et al.* 2014) agrees with that of other dwarf galaxies, in which carbon is a primary element.

4. Implications

Our UV data nicely complement existing studies in the optical, infrared, and radio. The derived mass of the stellar cluster is of order 10^6 $M_\odot$, which approaches the maximum masses observed in young and old stellar clusters. It is also comparable to the mass of the most massive individual molecular clouds in starburst galaxies like M82, where dynamical shear is believed to set the upper mass limit. The spectrum of the super star cluster in II Zw 40 can be reproduced by a standard Kroupa-type IMF, which is essentially a Salpeter IMF in the mass range sampled. Our models include stellar masses up to 120 $M_\odot$. From analyses of individual stars, Crowther *et al.* 2016 find evidence of the initial mass function in NGC 2070 extending up to 300 $M_\odot$. Similarly, Smith *et al.* (2016) suggest the presence of massive stars with masses well above 100 $M_\odot$ in the nearby dwarf galaxy NGC 5253. In the absence of tracks for extremely massive stars in the evolution models available to us, we cannot test this suggestion in II Zw 40. However, we do find shortcomings in the existing evolution models. The UV spectrum of the super star cluster in II Zw 40 shows strong broad He II 1640 in emission. This line is most likely produced by hot, He-rich stars with strong mass loss. Such stars are commonly interpreted as "classical" W-R stars, yet they are not expected in large numbers at the young age of the cluster. Most likely, the emission is generated in the winds of very massive, mass losing core-hydrogen burning stars close to the main sequence. This stellar phase is not (yet) accounted for in the evolution models.

References

Beck, S. 2015, *International Journal of Modern Physics D*, 24, 1530002

Beck, S., Turner, J., Lacy, J., Greathouse, T., & Lahad, O. 2013, *ApJ*, 767, 53

Conti, P. S. 1991, *ApJ*, 377, 115

Crowther, P. A., Caballero-Nieves, S. M., Bostroem, K. A., *et al.* 2016, *MNRAS*, 458, 624

Ekström, S., Georgy, C., Eggenberger, P., *et al.* 2012, *A&A*, 537, A146

Esteban, C., García-Rojas, J., Carigi, L., *et al.* 2014, *MNRAS*, 443, 624

Georgy, C., Ekström, S., Eggenberger, P., *et al.* 2013, *A&A*, 558, A103

Gil de Paz, A. & Madore, B. F. 2005, *ApJS*, 156, 345

Gnedin, O. Y. & Ostriker, J. P. 1997, *ApJ*, 474, 223

Gutkin, J., Charlot, S., & Bruzual, G. 2016, *MNRAS*, 462, 1757

Kepley, A. A., Reines, A. E., Johnson, K. E., & Walker, L. M. 2014, *AJ*, 147, 43

Keto, E., Ho, L. C., & Lo, K.-Y. 2005, *ApJ*, 635, 1062

Kroupa, P. 2008, in *Pathways Through an Eclectic Universe*, ed. J. H. Knapen, T. J. Mahoney, & A. Vazdekis (San Francisco, CA: ASP), 390, 3

Kunth, D. & Östlin, G. 2000, *A&ApR*, 10, 1

Leitherer, C., Ekström, S., Meynet, G., *et al.* 2014, *ApJS*, 212, 14

Marlowe, A. T., Meurer, G. R., Heckman, T. M., & Schommer, R. 1997, *ApJS*, 112, 285

Meurer, G. R., Heckman, T. M., Leitherer, C., *et al.* 1995, *AJ*, 110, 2665

Rigby, J. R., Bayliss, M. B., Gladders, M. D., *et al.* 2015, *ApJ*, 814, L6

Sargent, W. L. W. 1970, *ApJ*, 160, 405

Sargent, W. L. W. & Searle, L. 1970, *ApJ*, 162, L155

Smith, L. J., Crowther, P. A., Calzetti, D., & Sidoli, F. 2016, *ApJ*, 823, 38

Thuan, T. X. 2008, in IAU Symp. 255, Low-metallicity Star Formation: From the First Stars to Dwarf Galaxies, ed. L. K. Hunt, S. C. Madden, & R. Schneider (Cambridge: Cambridge Univ. Press), 348

Whitmore, B. C., Chandar, R., Schweizer, F., *et al.* 2010, *AJ*, 140, 75

The Lives and Death-Throes of Massive Stars
Proceedings IAU Symposium No. 329, 2016
J.J. Eldridge, J.C. Bray, L.A.S. McClelland & L. Xiao, eds.

doi:10.1017/S1743921317003349

The very massive star content of the nuclear star clusters in NGC 5253

Linda J. Smith[1], Paul A. Crowther[2] and Daniela Calzetti[3]

[1]European Space Agency and Space Telescope Science Institute,
3700 San Martin Drive, Baltimore, MD 21218, USA
email: lsmith@stsci.edu

[2]Dept. of Physics and Astronomy, University of Sheffield, Sheffield S3 7RH, UK
email: paul.crowther@sheffield.ac.uk

[3]Dept. of Astronomy, University of Massachusetts – Amherst, Amherst, MA 01003, USA
email: calzetti@astro.umass.edu

Abstract. The blue compact dwarf galaxy NGC 5253 hosts a very young starburst containing twin nuclear star clusters. Calzetti *et al.* (2015) find that the two clusters have an age of 1 Myr, in contradiction to the age of 3–5 Myr inferred from the presence of Wolf-Rayet (W-R) spectral features. We use *Hubble Space Telescope (HST)* far-ultraviolet (FUV) and ground-based optical spectra to show that the cluster stellar features arise from very massive stars (VMS), with masses greater than 100 $M_{\odot}$, at an age of 1–2 Myr. We discuss the implications of this and show that the very high ionizing flux can only be explained by VMS. We further discuss our findings in the context of VMS contributing to He II λ1640 emission in high redshift galaxies, and emphasize that population synthesis models with upper mass cut-offs greater than 100 $M_{\odot}$ are crucial for future studies of young massive clusters.

Keywords. galaxies: dwarf – galaxies: individual (NGC 5253) – galaxies: starburst – galaxies: star clusters: general – stars: massive – stars: Wolf-Rayet

1. Introduction

NGC 5253 is a blue compact dwarf galaxy at a distance of 3.15 Mpc (Freedman *et al.* 2001, Davidge 2007). It hosts a very young central starburst; the near-absence of non-thermal radio emission suggests that very few supernovae have exploded so far (Beck *et al.* 1996). The galaxy contains many young massive clusters, which display W-R emission features, suggesting that the burst is 3–5 Myr old (Campbell *et al.* 1986, Schaerer *et al.* 1997). The metallicity of NGC 5253 is 35% solar (Monreal-Ibero *et al.* 2012). Observations at radio wavelengths show that the central region is dominated by a massive ultracompact H II region (Turner *et al.* 2000). Alonso-Herrero *et al.* (2004) used the Near Infrared Camera and Multi Object Spectrometer (NICMOS) on HST to discover a double nuclear star cluster with the two components separated by $0''.3$–$0.''4$ (≈ 5 pc). The eastern cluster coincides with the peak of the Hα emission in NGC 5253 and is the youngest optical star cluster identified by Calzetti *et al.* (1997). The western cluster is very reddened and is coincident with the ultracompact H II region (Turner *et al.* 2000).

Recently, Calzetti *et al.* (2015) presented a detailed analysis of the two nuclear clusters (#5 and #11 in their terminology) using HST imaging from the FUV to the IR. By fitting their spectral energy distributions (SED), they find that the two clusters are extremely young and massive with ages of only 1±1 Myr and masses of 7.5×10^4 and 2.5×10^5 $M_{\odot}$ for #5 and #11 respectively. In Fig. 1, we show the central region of NGC 5253 containing the clusters #5 and #11, taken from Calzetti *et al.* (2015). The young ages contradict the

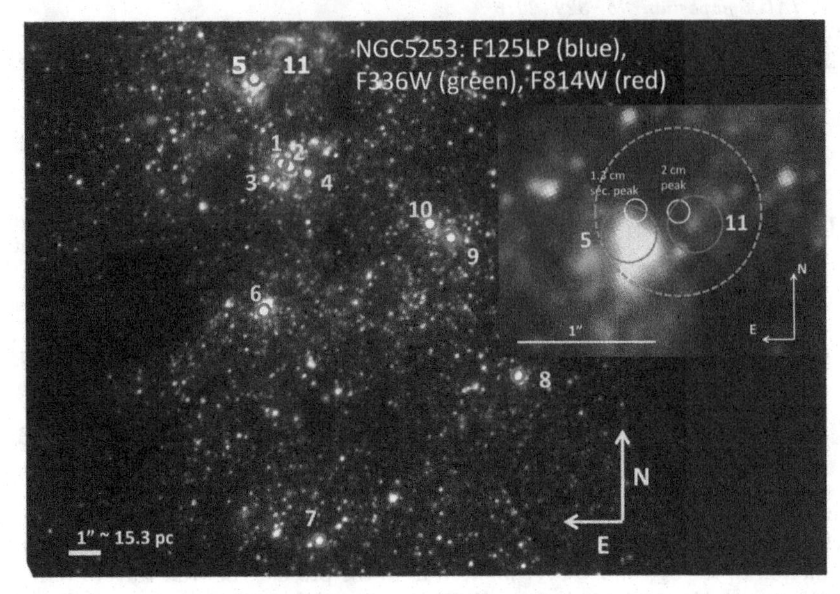

Figure 1. Three-color composite of the central 20″ × 16″ (300 × 250 pc) of NGC 5253 from Calzetti *et al.* (2015). The 11 brightest clusters from this study are identified with circles and numbered. The inset shows a detailed view of the two nuclear clusters #5 and #11, which are separated by a projected distance of 5 pc. The small circles correspond to the peaks of the radio emission (Turner *et al.* 2000). The absolute astrometric uncertainty of the HST images is comparable to the radius of the larger circles (0″.2).

values of 3–5 Myr inferred from the presence of W-R stars in cluster #5 (e.g. Monreal-Ibero *et al.* 2010). We consider here whether the W-R features could instead arise from hydrogen-rich very massive stars (VMS; masses > 100 $M_{\odot}$). The full study is described in Smith *et al.* (2016).

In a series of papers studying the core of the cluster R136 in the 30 Doradus region of the Large Magellanic Cloud, Crowther *et al.* (2010, 2016) used FUV spectra obtained with spectrographs on HST to determine the masses of the most massive stars using modelling techniques. They found that the R136 cluster is only 1.5 ± 0.5 Myr old and contains eight stars more massive than 100 $M_{\odot}$, with the most massive star (called R136a1) having a current mass of 315 ± 50 $M_{\odot}$. The four most massive stars account for one-quarter of the total ionizing flux from the star cluster. These VMS have very dense, optically thick winds and their emission-line spectra resemble W-R stars, but they are hydrogen-rich and on the main sequence.

2. Observations and Results

We obtained FUV Space Telescope Imaging Spectrograph (STIS) and Faint Object Spectrograph (FOS) spectra of cluster #5 from the Mikulski Archive for Space Telescopes (MAST). The STIS G140L spectrum covers 1150–1730 Å with a spectral resolution of 1.8 Å. The FOS G190H spectrum has a resolution of 3.8 Å and covers 1590–2310 Å. The two spectra were merged in the overlap region.

In Fig. 2, we compare the HST/STIS+FOS spectrum of cluster #5 with the integrated HST/STIS spectrum of R136a from Crowther *et al.* (2016). The similarity between the two spectra is striking. The strengths, widths and velocities of the N V, O V and C IV lines are in excellent agreement between the two spectra. The crucial VMS spectral features are the presence of blue-shifted O V λ1371 wind absorption, broad He II λ1640 emission, and

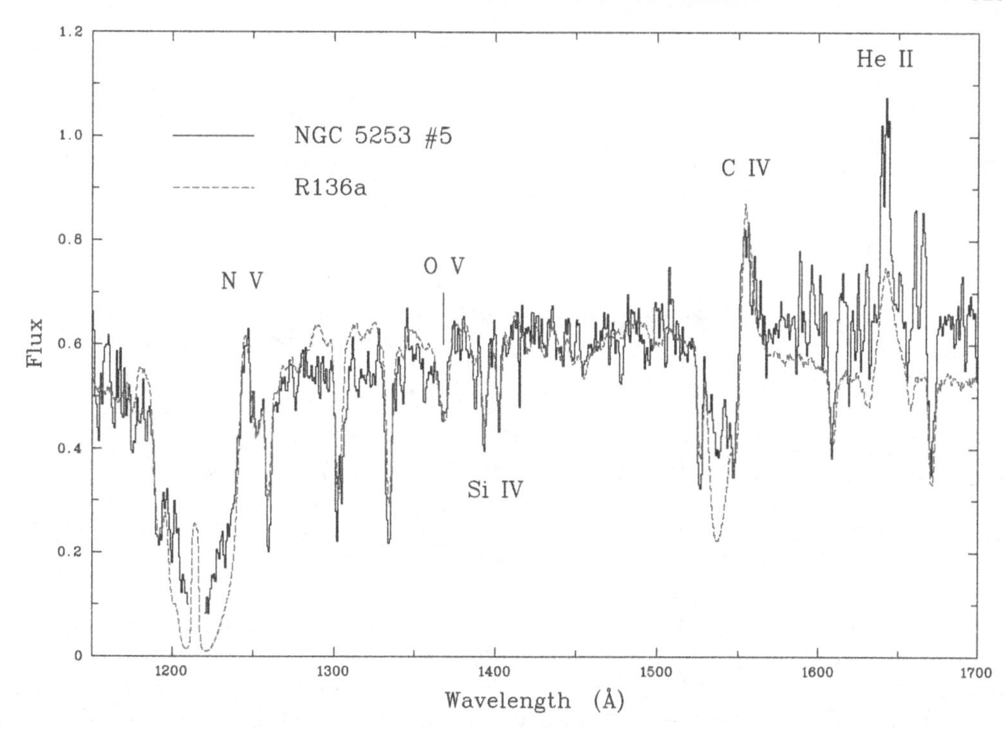

Figure 2. Integrated HST/STIS spectrum of R136a from Crowther *et al.* (2016) compared to the HST/STIS+FOS spectrum of cluster #5. The R136a spectrum has been scaled to the distance of NGC 5253. The flux is in units of 10^{-15} ergs cm^{-2} s^{-1} Å^{-1}.

the absence of a Si IV λ1400 P Cygni profile (expected in classical W-R stars). Crowther *et al.* (2016) find that 95% of the broad He II emission shown in the R136a spectrum in Fig. 2 originates solely from VMS. Thus, the presence of this feature in emission together with the O V wind absorption indicates a very young age (< 2 Myr) for cluster #5 and a mass function that extends well beyond 100 $M_{\odot}$.

We have also obtained high spectral resolution spectroscopy across the starburst core of NGC 5253 with the UV-Visual Echelle Spectrograph (UVES) at the Very Large Telescope (VLT). This spectrum covers 3100–10360 Å and thus both the blue and red W-R bumps. The red bump due to WC C IV λλ5801, 5812 emission is absent. The blue bump corresponds to the W-R emission features of N III λ3634–4641, C III λ4647–4651 and He II λ4686. We show this region in Fig. 3. Broad He II emission (1450 km s^{-1}) is the only W-R feature present. N III is present but narrow and the individual line components are resolved. They are likely to be nebular in origin, although detecting these transitions is unusual. We also detect nebular He II λ4686. The blue bump region of the VMS in R136 is dominated by strong broad He II λ4686 with N III absent (Caballero-Nieves *et al.* in prep.). From this, the optical signature of VMS is strong and broad He II λ4686 emission and N III λ3634–4641 absent.

3. Discussion and Conclusions

We have examined UV and optical spectroscopy of cluster #5 in the nucleus of NGC 5253 with the aim of reconciling the extremely young age of 1 ± 1 Myr found by Calzetti *et al.* (2015) with the presence of W-R features in the cluster spectrum. Specifically, we have investigated whether the W-R features arise from hydrogen rich, very massive stars

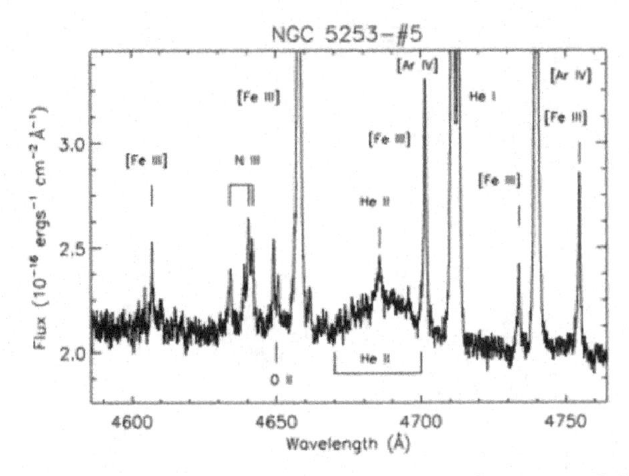

Figure 3. VLT/UVES spectrum of NGC 5253-cluster #5 in the region of the W-R blue bump.

with masses greater than 100 $M_{\odot}$. We conclude from the UV and optical comparisons with R136 that cluster #5 is very young with an age of less than 2 Myr and that the broad He II emission is produced by VMS that have dense, optically thick winds and show W-R-like emission features while they are on the main sequence.

The presence of VMS in the nuclear region of NGC 5253 can explain the very high ionizing flux of $Q(\mathrm{H\,I}) = 2.2 \times 10^{52}$ s^{-1} for the central 5 pc region (Turner & Beck 2004). Previous studies (e.g. Turner *et al.* 2015) have assumed an age of 3–5 Myr and found that the predicted ionizing fluxes from *Starburst99* (Leitherer *et al.* 1999) significantly under-predict the ionizing flux, unless a top heavy initial mass function is assumed. Calzetti *et al.* (2015) likewise find that 50% of the ionizing flux is unaccounted for using a lower age of 1 Myr and SED modelling with an upper mass cut-off of 100 $M_{\odot}$. For R136, Doran *et al.* (2013) find that $Q(\mathrm{H\,I}) = 7.5 \times 10^{51}$ s^{-1} and that the four most massive VMS produce 25% of this ionizing flux. We find that the extra ionizing flux needed for the central 5 pc of NGC 5253 at an age of 2 Myr over *Starburst99* (with an upper mass cutoff of 100 $M_{\odot}$) is the equivalent of 12 VMS. This number seems reasonable given the six times greater mass for clusters #5 and #11 compared to R136.

The FUV spectrum of cluster #5 is representative of a very young, nearby, low metallicity, nuclear starburst. It bears a striking resemblance to the UV rest-frame spectrum of Q2343-BX418 (Erb *et al.* 2010). These authors suggest this $z = 2.3$ young, low mass, low metallicity galaxy is a plausible analogue to the young, low metallicity star-forming galaxies at $z > 5$. Q2343-BX418 has broad He II λ1640 emission, and nebular emission from O III] $\lambda\lambda$1661, 1666 and C III] $\lambda\lambda$1907, 1909. The C III] equivalent width of 7.1 Å is very similar to the value of 7.7 Å that we measure for the FOS spectrum near cluster #5.

The James Webb Space Telescope (JWST) will obtain rest frame UV spectra of high redshift star-forming galaxies. These spectra may reveal the presence of VMS through the presence of broad He II λ1640 emission and O V λ1371 wind absorption, and the absence of Si IV λ1400 P Cygni emission. Population synthesis models typically have upper mass cutoffs of 100 $M_{\odot}$. For all studies near and far, it is crucial to extend these into the VMS regime to correctly account for the radiative, mechanical and chemical feedback, which will be dominated by VMS for the first 1–3 Myr in star-forming regions.

References

Alonso-Herrero, A., Takagi, T., Baker, A. J., *et al.* 2004, *ApJ*, 612, 222
Beck, S. C., Turner, J. L., Ho, P. T. P., Lacy, J. H., & Kelly, D. M. 1996, *ApJ*, 457, 610
Calzetti, D., Meurer, G. R., Bohlin, R. C., *et al.* 1997, *AJ*, 114, 1834
Calzetti, D., Johnson, K. E., Adamo, A., *et al.* 2015, *ApJ*, 811, 75
Campbell, A., Terlevich, R., & Melnick, J. 1986, *MNRAS*, 223, 811
Crowther, P. A., Schnurr, O., Hirschi, R., *et al.* 2010, *MNRAS*, 408, 731
Crowther, P. A., Caballero-Nieves, S. M., Bostroem, K. A., *et al.* 2016, *MNRAS*, 458, 624
Davidge, T. J. 2007, *AJ*, 134, 1799
Doran, E. I., Crowther, P. A., de Koter, A., *et al.*2013, *A&A*, 558, A134
Erb, D. K., Pettini, M., Shapley, A. E., *et al.* 2010, *ApJ*, 719, 1168
Freedman, W. L., Madore, B. F., Gibson, B. K., *et al.* 2001, *ApJ*, 553, 47
Leitherer, C., Schaerer, D., Goldader, J. D., *et al.* 1999, *ApJS*, 123, 3
Monreal-Ibero, A., Vílchez, J. M., Walsh, J. R., & Muñoz-Tuñón, C. 2010, *A&A*, 517, A27
Monreal-Ibero, A., Walsh, J. R., & Vílchez, J. M. 2012, *A&A*, 544, A60
Schaerer, D., Contini, T., Kunth, D., & Meynet, G. 1997, *ApJ*, 481, L75
Smith, L. J., Crowther, P. A., Calzetti, D., & Sidoli, F. 2016, *ApJ*, 823:38
Turner, J. L., Beck, S. C., Benford, D. J., *et al.* 2015, *Nature*, 519, 331
Turner, J. L. & Beck, S. C. 2004, *ApJ*, 602, L85
Turner, J. L., Beck, S. C., & Ho, P. T. P. 2000, *ApJ*, 532, L109

The Lives and Death-Throes of Massive Stars
Proceedings IAU Symposium No. 329, 2016
J.J. Eldridge, J.C. Bray, L.A.S. McClelland & L. Xiao, eds.

doi:10.1017/S1743921317003374

High angular resolution radio and infrared view of optically dark supernovae in luminous infrared galaxies

Seppo Mattila[1], Erkki Kankare[2], Erik Kool[3,5], Cristina Romero-Cañizales[4], Stuart Ryder[5] and Miguel Perez-Torres[6]

[1]Tuorla observatory, University of Turku, 21500 Kaarina, Finland,
email: sepmat@utu.fi

[2]Astrophysics Research Centre, Queen's University Belfast, Belfast, UK

[3]Department of Physics & Astronomy, Macquarie University, Sydney, Australia

[4]Millennium Institute of Astrophysics & Universidad Diego Portales, Chile

[5]Australian Astronomical Observatory, Sydney, Australia

[6]Instituto de Astrofisica de Andalucia, Granada, Spain

Abstract. In luminous and ultraluminous infrared galaxies (U/LIRGs), the infall of gas into the central regions strongly enhances the star formation rate (SFR), especially within the nuclear regions which have also large amounts of interstellar dust. Within these regions SFRs of several tens to hundreds of solar masses per year ought to give rise to core-collapse supernova (SN) rates up to 1-2 SNe every year per galaxy. However, the current SN surveys, almost exclusively being ground-based seeing-limited and working at optical wavelengths, have been blinded by the interstellar dust and contrast issues therein. Thus the properties and rates of SNe in the nuclear environments of the most prolific SN factories in the Universe have remained largely unexplored. Here, we present results from high angular resolution observations of nearby LIRGs at infrared and radio wavelengths much less affected by the effects of extinction and lack of resolution hampering the optical searches.

Keywords. supernovae: general, dust extinction, galaxies: nuclei, galaxies: starburst, infrared: galaxies, instrumentation: adaptive optics, instrumentation: high angular resolution

---

1. Introduction

The current very wide-field supernova (SN) searches are increasing substantially the SN statistics locally covering a large fraction of the whole sky every few nights and in a less biassed manner than was possible before. However, most of the local searches are working at optical wavelengths and are limited by the ground-based seeing making the detection of SNe challenging especially within the dusty and often bright and complex nuclear (<500 pc) regions of the most strongly star-forming galaxies. Therefore, the rates and properties of the SNe occurring within the nuclear regions of galaxies have remained largely unexplored.

Deep infrared (IR) surveys will enable studies of the evolution of SN rates extending beyond the peak of the cosmic star formation history (SFH). As core-collapse SNe come from massive short-lived stars, their explosion rate directly reflects the on-going massive star formation rate (SFR) at the explosion sites, and thus CCSNe provide an independent way to probe the SFRs in their host galaxies and also the cosmic SFH (e.g., Dahlén *et al.* 1999, Madau & Dickinson 2014). Core-collapse SNe can thus provide a very useful consistency check on the cosmic SFH, independent from many assumptions and biases with the more conventional methods that are based on galaxy luminosities. Measuring the evolution of the core-collapse SN rate with redshift is of major importance for determining the universal history of metal production and thus for all galaxy formation and evolution models. Furthermore, the SN behaviour in such regions might be extreme as the interaction of the SN ejecta with the dense nuclear environment may convert a substantial fraction of the SN kinetic energy into radiation producing more luminous and slowly declining events.

2. Supernovae in luminous infrared galaxies

Despite the impressive statistics, the current SN rates may still suffer from significant systematic omissions. This was illustrated by Horiuchi *et al.* (2011) claiming a significant deficit of core-collapse SNe detected as a function of redshift compared to the expectations from the cosmic SFR. In fact, a large fraction of the massive star formation especially at intermediate and high redshifts (Magnelli *et al.* 2011) took place in luminous and ultraluminous infrared galaxies (LIRGs and ULIRGs, respectively). The starburst-dominated LIRGs form several tens to hundreds of solar masses of stars per year. This corresponds to core-collapse SN rates up to several SNe yr^{-1} per galaxy, some hundred times higher than in "ordinary" spiral galaxies like the Milky Way. However, even in the local Universe this entire SN population has been almost completely neglected by the previous SN searches. A cross-correlation between the Asiago Supernova Catalog and the IRAS Revised Bright Galaxy Sample reveals about 45 confirmed core-collapse SNe discovered in these systems since year 2000 which is just the tip of the iceberg, the total intrinsic number being over 4000 SNe over this period (Kool *et al.* 2017, submitted). Furthermore, only a small number of such SNe have been followed-up and studied in detail (e.g., Kankare *et al.* 2014; Romero-Cañizales *et al.* 2014; Kangas *et al.* 2016).

The properties and the rates of SNe in the nuclear environments of the most prolific SN factories in the Universe have thus remained largely unexplored. A very promising approach to detect and study SNe within the dust obscured nuclear regions of starburst galaxies, LIRGs and ULIRGs is working at IR or radio wavelengths where the dust extinction is strongly reduced (e.g., Mattila *et al.* 2013). However, it has become very clear that high spatial resolution is also critically important for a successful detection and study of SNe within the bright and complex nuclear regions of U/LIRGs (e.g. Mattila *et al.* 2007, Perez-Torres *et al.* 2007). At IR wavelengths this can be achieved either by space telescopes or by ground-based Adaptive Optics (AO) imaging used to compensate for the blurring by a turbulent atmosphere. The combination of near-IR and AO instrumentation provides a spatial resolution ($\sim$0.1 arcsec) which is 10$\times$ better than typically available under ground-based natural seeing conditions and is comparable to that attainable with space telescopes. At radio wavelengths even higher angular resolution ($<$10 milliarcsec) is available thanks to the Very Long Baseline Interferometry (VLBI) imaging techniques. Radio observations are completely free from the effects of dust allowing studies of the SN population within the innermost nuclear regions of nearby U/LIRGs (e.g. Herrero-Illana *et al.* 2012).

Figure 1 shows examples of our high angular resolution near-IR and radio observations of the nearby LIRG Arp 299 (Kankare *et al.* 2014; Perez-Torres *et al.* 2009). The near-IR image comes from our AO observations and shows the entire merging system of galaxies with the locations of a number of circumnuclear SNe indicated. The radio image comes from VLBI observations and shows the innermost 150 pc regions of the eastern component of Arp 299 with a number of resolved compact radio sources, consistent with radio SNe and SN remnants. To have hope of detecting these events one clearly needs a high spatial

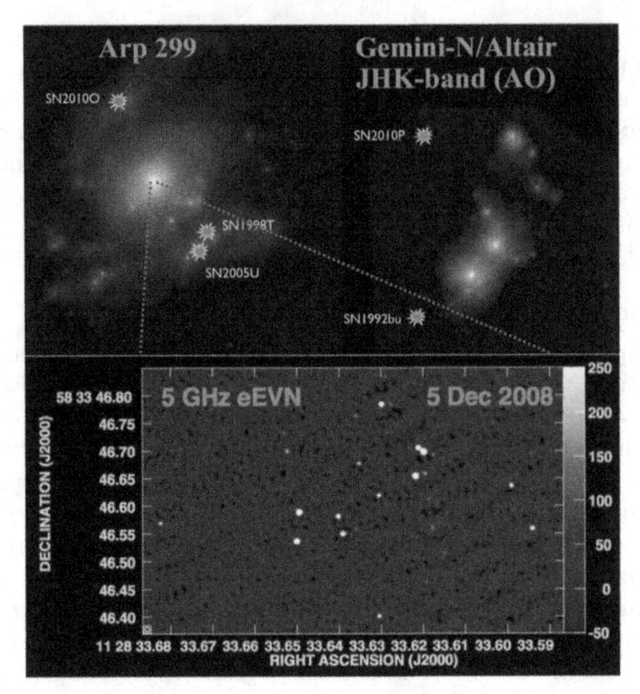

Figure 1. Top: Colour combined near-IR JHK-band image (FWHM = 0.1 arcsec) of the luminous infrared galaxy Arp 299 obtained with the Gemini-North telescope with the Altair/NIRI AO system with laser guide star (Kankare *et al.* 2014). The locations of a number of circumnuclear SNe detected at optical or near-IR wavelengths are indicated (for details see Mattila *et al.* 2012). Bottom: Interferometric radio image (FWHM = 8.5 mas) at 5 GHz of the central 150 pc nuclear regions of the eastern component of Arp 299 obtained with the electronic European VLBI Network (eEVN). The radio image shows a number of compact sources most of which are young radio SNe and SN remnants consistent with the high SFR within the nuclear regions of Arp 299 (Perez-Torres *et al.* 2009; Romero-Cañizales *et al.* 2011).

resolution and the use of either IR or radio observations to tackle the dust extinction. However, at larger distances even a 10 milliarsec angular resolution is not sufficient to resolve individual components that might be blended, and/or be embedded in diffuse emission (e.g., Romero-Cañizales *et al.* 2012).

3. AO assisted near-IR SN searches in luminous infrared galaxies

Our near-IR AO assisted programmes have already yielded the best available SN search dataset of nearby LIRGs. In our Gemini-North programme we used the laser guide star AO for repeatedly imaging a sample of 8 nearby LIRGs in near-IR K-band over a 2.5 year period. Follow-up observations in JHK bands allowed constraining the SN types and line-of-sight extinction. During the programme we discovered/confirmed a total of 6 nuclear SNe (Ryder *et al.* 2014) in good agreement with the expectations. For example, SN 2008cs detected in the circumnuclear regions of IRAS 17138-1017 was observed to suffer from a very high host galaxy extinction of $A_V = 16$ therefore making it only detectable at IR and radio wavelengths (Kankare *et al.* 2008). Furthermore, SNe 2010cu and SN 2011hi were discovered at only ~180pc and ~380pc (0.4 and 0.8 arcsec) galactocentric distances in the LIRG IC 883 and our follow-up observations allowed estimating their extinction of $A_V = 0$ and 5-7, respectively (Kankare *et al.* 2012). High resolution radio follow-up observations provided upper limits for the radio luminosities in agreement with their

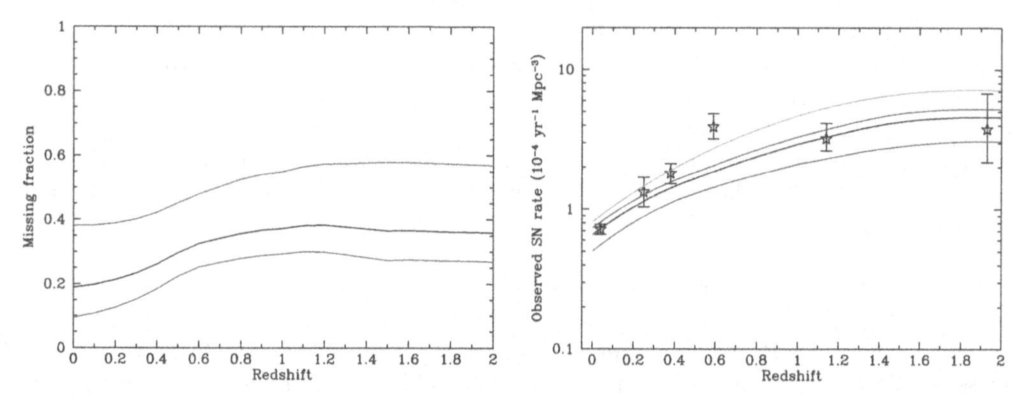

Figure 2. Left: The fraction of SNe missed by rest-frame optical searches as a function of redshift (from Mattila *et al.* 2012). Black line shows the best (nominal) estimate together with low and high missing fraction models as blue and red lines, respectively. Right: Average observed core-collapse SN rates (stars) from Strolger *et al.* (2015) as a function of redshift compared with the expectations from the cosmic SFH (lines) and assuming a Salpeter IMF and stars with initial masses between 8 and 20 solar masses resulting in luminous core-collapse SNe (green line = no missing SNe; black, blue and red lines correspond to the nominal, low and high missing fractions, respectively). This demonstrates that the measured core-collapse SN rates between z = 0 and 2 are consistent with those expected from the cosmic SFH if taking into account the missing SNe in U/LIRGs.

likely classification as type II-P SNe based on the near-IR light curves (Romero-Cañizales *et al.* 2012). Neither of the two objects were detectable in ground-based seeing-limited images highlighting the importance of high spatial resolution for the detection and study of SNe within the LIRG innermost nuclear regions. Also, SN 2010P discovered within the C' component of Arp 299 suffered from a high extinction of $A_V = 7$. An optical spectrum from Gemini-N provided a best match with the type IIb SN 2011dh (Kankare *et al.* 2014) whereas our radio follow-up showed it to be the most distant and most slowly evolving type IIb radio SN detected to date (Romero-Cañizales *et al.* 2014).

More recently, we have expanded our efforts to make use of laser guide star AO with NIRC2 on the Keck telescope, as well as the Gemini Multi-Conjugate Adaptive Optics System (GeMS) on the Gemini-South telescope. The combination of improved angular resolution and increased field of view yielded by GeMS has already allowed the detection of four likely SNe (AT 2013if, 2015ca, 2015cb and 2015cf) within the nuclear regions of the LIRGs IRAS 18293-3413, NGC 3110 and IRAS 17138-1017. Near-IR and radio follow-up observations were obtained for all the four objects confirming their nature as core-collapse SNe (Kool *et al.* 2017, in prep.). In addition, our work has already resulted in some of the sharpest and deepest images of LIRGs ever obtained from the ground allowing also detailed studies of their star formation properties and super star cluster populations (Randriamanakoto *et al.* 2013a,b).

4. Implications of missing core-collapse SNe to cosmic SFH

Our combined use of high angular resolution near-IR and radio observations has allowed an empirical determination of the fraction of core-collapse SNe missed within the nuclear regions of U/LIRGs in the local Universe. With a number of assumptions we extrapolated these estimates up to cosmological redshifts (Mattila *et al.* 2012). For a volume-limited rest-frame optical SN survey we find the missing SN fraction to increase from its average local value of ∼19% to ∼38% at z∼1.2 and then stay roughly constant up to z = 2 (see Fig. 2). The uncertainties in the correction factors for the missing SNe are still substantial due to the limited statistics and our imperfect understanding of the nature of the nuclear SNe and the evolution of U/LIRGs as a function of redshift. A more accurate evaluation of the number of SNe lost in local U/LIRGs will eventually allow a more reliable estimate of the evolution of core-collapse SN rate as a function of redshift. With knowledge of the mass range of the stars producing luminous core-collapse SNe (e.g. Eldridge *et al.* 2013, Smartt 2015) this will allow a detailed comparison with the cosmic star formation rates (e.g. Botticella *et al.* 2012; Mattila *et al.* 2012; Dahlén *et al.* 2012; Cappellaro *et al.* 2015; Strolger *et al.* 2015).

In Figure 2, we compare the measured evolution of the core-collapse SN rate as a function of redshift with the expectations from the latest consensus cosmic SFH from Madau & Dickinson (2014) assuming a Salpeter initial mass function (IMF) with initial mass cut offs for core-collapse SN progenitors of 8 and 20 solar masses. The resulting core-collapse SN rates are shown for (1) no SNe missed within the nuclear regions of U/LIRGs and (2) for the missing fractions evaluated in Mattila *et al.* (2012). This demonstrates that the measured core-collapse SN rates between z = 0 and 2 are consistent with those expected from the cosmic SFRs after accounting for the missing SNe in U/LIRGs. We note that the uncertainties due to the missing SN fraction correction are currently at a similar level as the statistical uncertainties in the measured core-collapse SN rates. However, the SN statistics will be improved significantly in the near-future thanks to the Large Synoptic Survey Telescope (LSST) as well as deep IR surveys (e.g. by the James Webb Space Telescope) also in the highest redshift bins and thus the evaluation of more precise missing SN fractions will become crucial.

References

Botticella, M. T., Smartt, S. J., Kennicutt, R. C., *et al.* 2012, *A&A*, 537, A132
Cappellaro, E., Botticella, M. T., Pignata, G., *et al.* 2015, *A&A*, 584, A62
Dahlén, T. & Fransson, C. 1999, *A&A*, 350, 349
Dahlén, T., Strolger, L.-G., Riess, A. G., *et al.* 2012, *ApJ*, 757, 70
Eldridge, J. J., Fraser, M., Smartt, S. J., Maund, J. R., & Crockett, R. M. 2013, *MNRAS*, 436, 774
Herrero-Illana, R., Pérez-Torres, M. Á., & Alberdi, A. 2012, *A&A*, 540, L5
Horiuchi, S., Beacom, J. F., Kochanek, C. S., *et al.* 2011, *ApJ*, 738, 154
Kangas, T., Mattila, S., Kankare, E., *et al.* 2016, *MNRAS*, 456, 323
Kankare, E., Mattila, S., Ryder, S., *et al.* 2008, *ApJ*, 689, L97
Kankare, E., Mattila, S., Ryder, S., *et al.* 2012, *ApJ*, 744, L19
Kankare, E., Mattila, S., Ryder, S., *et al.* 2014, *MNRAS*, 440, 1052
Madau, P. & Dickinson, M. 2014, *ARA&A*, 52, 415
Magnelli, B., Elbaz, D., Chary, R. R., *et al.* 2011, *A&A*, 528, A35
Mattila, S., Väisänen, P., Farrah, D., *et al.* 2007, *ApJ*, 659, L9
Mattila, S., Dahlen, T., Efstathiou, A., *et al.* 2012, *ApJ*, 756, 111
Mattila, S., Fraser, M., Smartt, S. J., *et al.* 2013, *MNRAS*, 431, 2050
Pérez-Torres, M. A., Romero-Cañizales, C., Alberdi, A., & Polatidis, A. 2009, *A&A*, 507, L17
Pérez-Torres, M. A., Mattila, S., Alberdi, A., *et al.* 2007, *ApJ*, 671, L21
Randriamanakoto, Z., Escala, A., Väisänen, P., *et al.* 2013a, *ApJ*, 775, L38
Randriamanakoto, Z., Väisänen, P., Ryder, S., *et al.* 2013b, *MNRAS*, 431, 554
Romero-Cañizales, C., Mattila, S., Alberdi, A., *et al.* 2011, *MNRAS*, 415, 2688
Romero-Cañizales, C., Pérez-Torres, M. A., Alberdi, A., *et al.* 2012, *A&A*, 543, A72
Romero-Cañizales, C., Herrero-Illana, R., Pérez-Torres, M. A., *et al.* 2014, *MNRAS*, 440, 1067
Ryder, S. D., Mattila, S., Kankare, E., & Väisänen, P. 2014, *Proc. SPIE*, 9148, 91480D
Smartt, S. J. 2015, *PASA*, 32, e016
Strolger, L.-G., Dahlen, T., Rodney, S. A., *et al.* 2015, *ApJ*, 813, 93

Session 6: Conference Summary

The Lives and Death-Throes of Massive Stars
Proceedings IAU Symposium No. 329, 2016
J.J. Eldridge, J.C. Bray, L.A.S. McClelland & L. Xiao, eds.

doi:10.1017/S174392131700285X

Symposium Summary

Emily M. Levesque

Department of Astronomy, University of Washington
Box 351580, Seattle, WA, 98195
email: emsque@uw.edu

Abstract. This proceeding summarizes the highlights of IAU 329, "The Lives and Death-Throes of Massive Stars", held in Auckland, NZ from 28 Nov - 2 Dec. I consider the progress that has been made in the field over the course of these "beach symposia", outline the overall content of the conference, and discuss how the current subfields in massive stellar astrophysics have evolved in recent years. I summarize some of the new results and innovative approaches that were presented during the symposium, and conclude with a discussion of how current and future resources in astronomy can serve as valuable tools for studying massive stars in the coming years.

Keywords. stars: atmospheres, binaries: general, stars: evolution, stars: fundamental parameters (classification, colors, luminosities, masses, radii, temperatures, etc.), stars: interiors, stars: magnetic fields, stars: mass loss; supernovae: general, galaxies: stellar content, sociology of astronomy

---

1. Introduction

IAU 329 was the eleventh meeting in the "beach symposium" series focused on massive stars. As noted by Claus Leitherer in his conference summary of IAU 250, the topics covered at these symposia have shifted noticeably since IAU 49, held in Boulder, CO in 1968. The "median distance" of the objects at each meeting has increased, and the power and scope of computational tools has evolved drastically, allowing our picture of massive stars and the questions we are able to pose about their physical properties and evolution to grow remarkably in complexity.

When preparing the closing summary for this conference, I began by looking over the conference program to try and identify a simple outline or a straightforward progression of topics. At first glance the conference talks could be split into four basic categories: supernovae, stellar evolution, stellar properties, and stellar populations. The organization of the topics was sensible and is typical of how work on massive stars has traditionally been classified.

That said, these topics are related cyclically rather that linearly, and as the field of massive stellar astrophysics has progressed research spanning one of more of these broad areas has become the rule rather than exception. Much of the current work at the forefront of this field is taking place "in the gaps", considering, for example, how detections of core-collapse supernovae (CCSNe) sample star-forming galaxies or how computationally-rigorous treatments of key stellar properties impact evolutionary predictions.

During the opening summary of the conference Georges Meynet laid out the key challenges facing massive stellar astrophysics along with some of the most powerful new observational and computational tools we have at our disposal for addressing these challenges. Following his comments and the conference program, this proceedings summary of my closing talk will offer a brief overview of the new results presented in these four "main" areas of stellar astrophysics while also highlighting cross-disciplinary work and

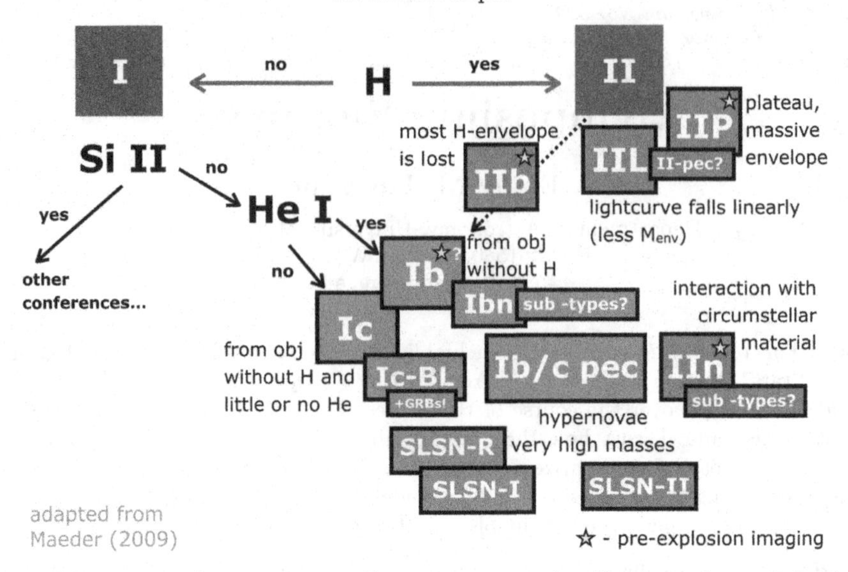

Figure 1. Illustration of current core-collapse supernova classification scheme and subtypes.

emphasizing ways in which we can continue the remarkable progress that has been made in the massive star field in the coming years.

2. Supernovae

The classification of CCSNe and the association of these various subclasses with progenitors has become increasingly complex in recent years (Fig. 1). Despite the increasing diversity of supernova sub-types, based largely on lightcurve evolution and spectroscopic properties, only a handful of definitive subtype+progenitor associations exist, and reproducing the lightcurves and spectra of supernovae from existing massive star models is still a challenge.

That said, much progress has been made in recently years on bringing theoretical models of CCSNe into agreement with their observed properties. Recently, 3D simulations have demonstrated the viability of the neutrino-driven mechanism of supernova explosions, and further work on these simulations should improve our ability to produce actual explosions from core-collapse models (*Muller*). These improvements have also extended to work on the relatively new subclass of superluminous supernovae (SLSNe), with models capable of reproducing lightcurves for Type I (hydrogen-free) SLSNe (*Nomoto*). Work is also underway to better reproduce the spectra of CCSNe based on models of their ejecta (*Hillier*). The ultimate goal of this work is the pursuit of a complete initial mass - metallicity parameter space that associates massive star progenitors with their counterpart CCSN subtypes; unfortunately, at the moment much of this landscape is still theoretical with only a tiny region of the plot ($\sim$11-17$M_{\odot}$ and $\sim Z_{\odot}$) supported by robust observations (*Fraser*).

It is now clear that achieving this goal will demand a much better understanding of not only how massive stars explode, but *which* phases of massive stellar evolution directly precede core-collapse and how their prior evolution affects their terminal explosion. Theoretical work has explored this question in detail, trying to directly reproduce a path from different classes of evolved massive stars to their eventual supernovae based on initial mass (*Groh*) and modeling the predicted lightcurves of different progenitor models to compare with observations (*Bersten*). One good observation-based approach was effectively summarized with the following question: "If [Star] X exploded tomorrow, how would we interpret it?" (*Beasor*). Given that, when studying the supernovae themselves, we often find ourselves working only with pre-explosion imaging (if we're lucky), this represents a practical and rigorous means of approaching this question from the stellar side: considering how known evolved massive stars might appear as supernovae and combining different treatments of available pre-explosion observations to better understand the limits of our current interpretations. Such an approach also leaves room for the growing sample of ambiguous events produced by evolved massive stars, such as luminous transients that cannot always immediately be classified as terminal supernovae or eruptive mass loss events; right now our best means of interpreting these is through studying their host environments and potential parent populations (*Drout, Thöne*). Very massive stars, and their role in producing pair-instability supernovae (particularly at low metallicity), also represent an important region of this relation between initial mass, metallicity, progenitor, and supernova type (*Yusof*).

3. Stellar Evolution

The massive star community now has access to multiple complete grids of stellar evolutionary models with varying treatments of key parameters such as metalliicty, rotation, and binary interaction. In recent years it has become abundantly clear that stellar evolution must be considered within the context of key stellar properties such as multiplicity, mass loss, magnetic fields, and rotation (see also Section 4). As a result, much of the current body of theoretical work on massive stars is focused on how stellar properties impact stellar evolution, and how observations and models can be combined to start placing massive stars into an evolutionary context.

The location of evolved massive stars such as luminous blue variables on the Hertzsprung-Russell diagram can be used to test both single-star and binary evolution models, and to potentially backsolve in the hopes of determining these enigmatic stars' evolutionary histories (*Smith*). Invoking massive binary evolution also opens up a broader range of potential terminal core-collapse products (e.g., the role that mass transfer plays in produced stripped-envelope supernovae produced by mass transfer, direct collapse of very massive stars to black holes, and even the eventual coalescence and gravitational wave signal of binary compact remnants; *Hamann, Podsiadlowski*). Strong surface magnetic fields also appear to decrease the mass loss rates of massive stars during their main sequence evolution, thus impacting the CCSN subtype and compact object that is ultimately produced (*Keszthelyi*). These critical evolutionary effects are not restricted to "external" physical properties such as mass loss and binary interactions; wave energy transport in the interior of a massive star can have a dramatic impact on its evolutionary track during the final years of its life, strongly influencing the question of how we can effectively backtrack from pre-explosion imaging to a progenitor's evolutionary history and initial mass (*Fuller*).

4. Stellar Properties

Much of the research presented at this symposium touched on at least one of the five phenomena in massive stars that currently pose the greatest challenge to models of stellar interiors and evolution: rotation, winds and mass loss, multiplicity, convection, and magnetic fields (Fig. 2).

The evolution of massive stars in binary or multiple systems offers a rich variety of possibilities for observable physical properties and core-collapse outcomes. Models that simulate phenomena such as post-main-sequence mergers and common envelope evolution

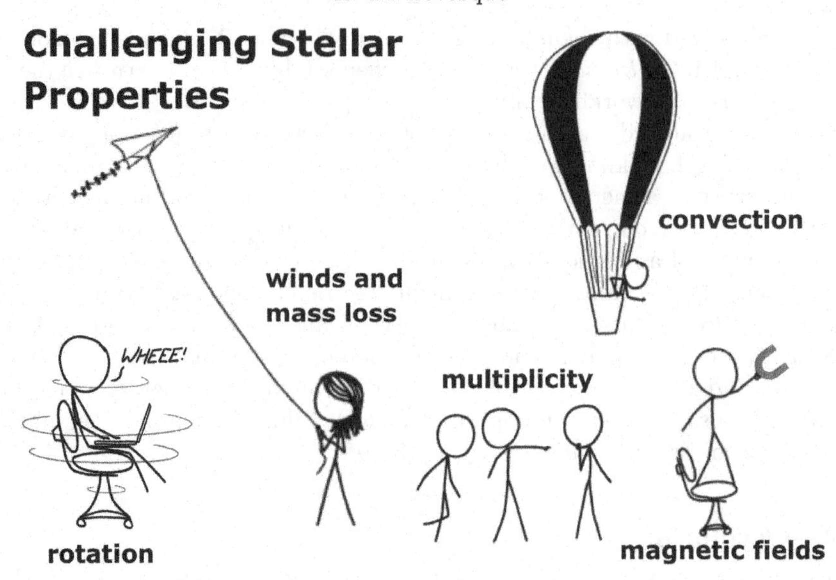

Figure 2. Pictorial summary of the physical properties and phenomena that currently pose the greatest challenge to stellar models. Adapted from illustrations by Randall Munroe available at xkcd.com.

are capable of producing stars such as yellow supergiant SN progenitors, close black hole binaries, and collapsar models for long-duration gamma-ray burst progenitors that agree with observed host environment metallicity effects (*Justham, Ivanova*). Binaries consisting of two massive stars were a particularly compelling topic at this conference given the announcement in February of 2016 of the discovery of gravitational waves from massive black hole binaries: the masses, merger rates, and spins determined from this tremendous discovery all offer new constraints on massive binary evolution (*Postnov*).

Stellar winds and mass loss have long been identified as key physical phenomena in the evolution of massive stars, impacting their observational signatures, angular momentum evolution, and the eventual properties of their core-collapse products. Work on this area in recent years has included high angular resolution imaging observations, using long-baseline interferometry or coronagraphic imaging, that are capable of spatially resolving circumstellar environments and offer us a new means of studying mass loss and dust production in large nearby massive stars (*Ohnaka*). As has been seen at previous massive star symposia, new treatments of phenomena such as clumping and porosity also continue to impact the theoretically-inferred rates of wind-driven mass loss in massive stars, highlighting the importance of incorporating these effects into existing codes (*Gräfener*).

Great progress has also been made in recent years on the effects of magnetic fields, with treatments of both interior and surface magnetic fields now included in models of massive stars. While magnetic field effects can be extremely complex (including orientation-dependent effects on the stars' observed spectra and the effect of stellar properties such as luminosity and changing rotation rates on the dynamo), they are a critical phenomenon to consider given that magnetic massive stars comprise $\sim$10% of the total population (*Erba, Augustson*). It is also worth noting that, while discussions of magnetic massive stars are traditionally focused on main-sequence O- and B-type stars, magnetic fields in massive stars continue to impact their observed spectra and evolution well beyond the main sequence; this highlights the importance of understanding magnetic activity in these stars from formation through death (*Oksala*).

Finally, while each of these phenomena is complex and compelling on its own, a common theme that arose at this symposium was work that has started to combine their effects. Unsurprisingly, a complete stellar model must be able to reproduce how all of these phenomena interact. To take just one example, modeling a dipole magnetic field in a rotating star leads to a decrease in the effective stellar mass loss rate (*Bard*). Interactions (and, from an observational perspective, degeneracies) between effects arising from mass loss, interacting multiple systems, changing rotation rates, and the impact of magnetic fields and convection treatments on interior evolution must all eventually be taken into account when trying to model the realistic evolution of a massive star.

5. Stellar Populations

While individual massive stars still serve as excellent test cases for our models of stellar properties and evolution, it is becoming increasingly common to study these objects on a population-wide scale. Large surveys of massive stars have already proven invaluable for characterizing the typical physical properties of various evolutionary stages.

Spectroscopy of hundreds of O stars in our own galaxy (*Maiz Apellaniz*) and 30 Doradus (*Vink, Sabin-Sanjulian*) has made it possible to determine the positions, parallaxes, proper motions, radial velocities, effective temperature, rotation rate, and evolutionary state. Hundreds of red supergiants in and beyond the Local Group have also been studied both photometrically and spectroscopically. While identifying these populations can be challenging due to contamination from foreground stars or misclassification, determining their physical properties and placing them on the H-R diagram allows them to be used as an extremely effective "magnifying glass" for studying and testing stellar evolutionary theory in the post-main-sequence regime and across a broad range of metallicities (*Dorn-Wallenstein, Massey, Georgy*). Wolf-Rayet stars represent a critical late-time evolutionary stage for massive stars due to their strong mass loss and their presumed (and potentially observationally confirmed) role as the direct progenitors of core-collapse supernovae. Studies of Wolf-Rayet populations in other galaxies have made it possible to calculate an empirical $\dot{M}$-Z relation for these stars (*Hainich*); such a relation is critical in understanding, for example, the role that single or binary Wolf-Rayet stars can play as the progenitors of long-duration gamma-ray bursts and stripped-envelope supernovae. A new population of Wolf-Rayet stars, WN3/O3 stars, has also been discovered in the Magellanic Clouds and appears to be specific to metal-poor environments, highlighting the many questions that still remain regarding post-main-sequence massive stellar evolution and the effects of metallicity (*Neugent*). Finally, population-scale studies of massive stars have also proven invaluable for measuring the binary (and, with increasing frequency, the multiple) fraction of both main-sequence and post-main-sequence massive stars, and for determining the typical angular separations and interactions of these systems (*Barba, Sana*). In all of these cases, new data from Gaia in the next few years will prove invaluable for this work, identifying young clusters of massive stars and making it easier to separate foreground and background populations using exceptionally precise astrometry.

Considering massive stars as a population in their own right also offers an extremely effective means of testing stellar evolutionary models. Massive stars in nearby regions like the Galactic center or nearby clusters in the Milky Way and Magellanic Clouds can be studied in detail and treated as young single-age systems (*Najarro*). Comparing the properties of these clusters with the predictions of simulated massive star samples from stellar population synthesis models serves as a powerful test of evolutionary theory on a large scale (*Vink, Crowther*). At greater distances, individual massive stars can no longer be resolved and we instead observe individual HII regions or starburst galaxies with very young coeval stellar populations such as II Zw 40; however, massive stars still dominate the radiative signature of these regions and galaxies, serving as the main source of ionizing photons and dominating the continuum signature in the blue and ultraviolet regimes. We can therefore use these sites of young active star formation as excellent tests of massive star atmosphere and evolutionary models, comparing observations to synthetic stellar population and examining how different treatments of key properties such as multiplicity and rotation agree with observations of the composite massive star sample (*Stanway, Leitherer*).

This large-scale work - using stellar population synthesis models to simulate large samples of stars in young galaxies and HII regions - brings the study of massive stars, and the summary of this symposium, full circle. Stellar population synthesis and photoionization models, which depend on accurate models of stellar atmospheres and evolution, are the key theoretical tools used for interpreting observations of star-forming galaxies. They are often used as the foundation for developing observational diagnostics used to determine fundamental galaxy properties such as metallicity and star formation rate. Those same properties are in turn the key (indeed, sometimes the only) data points available to observers studying the core-collapse deaths of massive stars in distant galaxies, determining whether supernova and gamma-ray burst subtypes occur preferentially in host environments with a specific metallicity or age profile (*Xiao*).

6. The Next Beach Symposium?

One overarching theme throughout the week of the symposium was the impressive diversity of scientific tools and approaches currently being applied to the study of massive stars. The research presented spanned the entire electromagnetic spectrum, ranging from radio mapping of massive stars' circumstellar environments and winds to observations of high-mass X-ray binaries and the detection of high-energy X-rays and gamma rays from eta Carinae. Imaging studies in stellar astrophysics now operate on massive scales, including a number of different transient surveys and immense missions such as Gaia. The reach of spectroscopic work on massive stars has continued to increase, with observations reaching out to tremendous distances (*Davies*). Challenging observational techniques such as interferometry (*Gies*) and spectropolarimetry (*Hoffmann, Agliozzo*) are offering a new perspective on nearby massive stars, allowing us to spatially resolve our nearest massive neighbors and probe the circumstellar environments of Galactic and even extragalactic massive stars in unprecedented detail. We are also pursuing asteroseismological studies of massive stars and should be able to conduct these on a larger scale in the coming years (*Buysschaert, Fuller*). The study of massive stars has also extended fully into the realm of three-dimensional models and simulations, sometimes literally - practical demonstrations during this symposium included an immersive simulation of the Galactic center based on Chandra data that could be viewed with 3D goggles (*Russell*), and a 3D-printed model of the Homunculus Nebula and interacting winds in eta Carinae (*Daminelli*). Finally, IAU 329 was also held at the dawn of a new era in astrophysics, one particularly pertinent to the study of massive stars: detections of gravitational waves offers us the chance to move beyond the electromagnetic spectrum and conduct multi-messenger explorations of massive stars' evolution, death, and the compact objects they leave behind.

The coming decade of observational facilities is going to continue this incredible rate of progress in the field of massive stellar astrophysics. The James Webb Space Telescope (JWST) is due to launch in October of 2018 and will lead the way in shifting the majority of space-based observational astronomy in the nearby universe into the infrared regime. JWST will also be capable of acquiring spectra of massive star-forming galaxies at z ~9-10, sampling the rest-frame ultraviolet and capturing the composite spectrum of young massive stars in the early universe. The Large Synoptic Survey Telescope (LSST) will achieve first science light in 2021; with a 9.6 square degree field of view and a rapid cadence, LSST will usher in a new era of transient astronomy, demanding new and rigorously-tested models of stellar evolution and supernova properties in order to match the influx of newly-detected massive star transients with their potential progenitors.

Further down the road, the Extremely Large Telescope (ELTs) will be coming online in the next 5-10 years. Multi-object spectrographs on these facilities, supported by multi-conjugate or laser tomography adaptive optics, will open up an astonishing new volume of observable massive stars, capable of acquiring spectrophotometry of individual stars out to tens of Mpc. Future plans for ground- and space-based gravitational wave detectors will significantly expand the detectable strain and frequency parameter space (allowing us to detect gravitational waves from new phenomena such as pre-merger binaries) and improve our ability to localize gravitational wave sources (making it faster and easier to confidently associate these sources with their electromagnetic counterparts). Finally, the Wide-Field Infrared Space Telescope (WFIRST) is scheduled to launch in 2025, with a 0.281 square degree field of view and a 0.11 arcsec/pixel resolution. While the specifications of the WFIRST instrument suite are still being discussed, its incredible resolution will be a valuable tool for expanding the local volume of star-forming galaxies in which stellar populations can be resolved and studied in unprecedented detail. Looking at the potential discoveries that could come from combining these new facilities with the impressive theoretical and observational work that was presented in Auckland, I look forward to learning about more exciting results in massive stellar astrophysics at future "beach symposia" in the coming years.

The Lives and Death-Throes of Massive Stars
Proceedings IAU Symposium No. 329, 2016
J.J. Eldridge, J.C. Bray, L.A.S. McClelland & L. Xiao, eds.

doi:10.1017/S1743921317004446

Talks also presented at the Symposium

Difference between Iib/Ib/Ic/Ic-BL

Fed Bianco
New York University

The Stellar Ultrasound

Matteo Cantiello
KITP - UCSB

Internal rotation and magnetism are key ingredients that largely affect explosive stellar deaths (Supernovae and Gamma Ray Bursts) and the properties of stellar remnants (White Dwarfs, Neutron Stars and Black Holes). However, the study of these subtle internal stellar properties has been limited to very indirect proxies. In the last couple of years, exciting asteroseismic results have been obtained by the Kepler satellite. Among these results are 1) The direct measure of the degree of radial differential rotation in many evolved low-mass stars and in a few massive stars, and 2) The detection of strong ($> 10^5$ G) internal magnetic fields in thousands of red giant stars that had convective cores during their main sequence. I will discuss the impact of these important findings for our understanding of massive star evolution.

Red Supergiants as Cosmic Abundance Probes

Benjamin Davies
Liverpool JMU

Over the past 6 years we have been establishing a novel method for determining chemical abundances from only moderate resolution near-IR spectroscopy of Red Supergiants (RSGs), the brightest stars at infrared wavelengths. We show that we can now routinely perform stellar abundance analysis at distances of 4Mpc, and around 10x this distance for young RSG-dominated clusters. This is a substantial volume of the local Universe, containing hundreds of star-forming galaxies, and we are now determining accurate metallicities and abundance gradients for a substantial sample of galaxies. Ultimately we will provide a robust measurement of the mass-metallicity relation at low redshift, free of the systematic errors that plague HII-region-based work.

Probing the Extremes of Pre-SN Mass Loss with the PanSTARRS1 Medium-Deep Survey

Maria R. Drout
Carnegie Observatories

A non-negligible fraction of massive stars undergo enhanced (possibly violent/eruptive) mass-loss in the final decades before core collapse. Theoretical models of such mass-loss are challenging and observational probes are necessary to help constrain the full diversity of pre-SN mass-loss (e.g. density profile, physical extent), the mechanism by which this mass is ejected, and the progenitors of various sub-classes of events. In this talk I will present new results from PS1 observations of super-luminous Type IIn SN (SLSN-II). In particular, I will highlight new constraints on the progenitors of SLSN-II based on a joint analysis of their explosion/CSM properties and host galaxy environments.

Asteroseismology of massive stars

Jim Fuller
California Institute of Technology

The basic principles of asteroseismic techniques applied to massive stars will first be explained for a non-expert audience and it will be addressed which stellar interior physics can be tested from studying their stellar oscillations. Afterwards, an overview of the main achievements over the last decade will be presented by means of several relevant case studies, especially in the framework of past and ongoing space missions. It will also be illustrated how asteroseismology and spectropolarimetry can be combined to probe magnetic hot stars. The review will end with a discussion on the current challenges and future prospects.

The Elusive Population of Massive Binary Star Products: the far UV Spectra of Stripped Stars

Ylva Louise Linsdotter Goetberg
Anton Pannekoek Institute for Astronomy, University of Amsterdam

Young massive stars are very frequently found in close binary systems, implying that the majority of massive stars interact with a companion during their lives. This raises two questions: How do we identify binary products and why do they appear to be so rare?

Models predict that Roche lobe overflow strips the hydrogen rich envelope of the primary star, exposing its hot helium core. These long-lasting, post-interaction systems are of wide astrophysical interest as (1) direct progenitors of Ib/c supernovae, as (2) unconventional sources of UV radiation and for (3) their capability of calibrating models and thus the theory of binary interaction. Surprisingly, very few stars stripped through binary interaction have been identified observationally - a clear contradiction with theoretical predictions.

We have conducted an extensive computational study using the binary evolutionary code MESA and the radiative transfer code CMFGEN to provide grids of tailor-made atmosphere models of stripped stars. We compute the contribution of stripped stars to the emitted radiation of stellar populations with focus in particular on their high UV-flux. We also propose several concrete observing strategies to systematically search for this elusive population.

Spectroscopic evolution of supernova progenitors

Jose Groh
Trinity College Dublin

Quantitative spectroscopy of supernovae

D. J. Hillier
University of Denver

Current deep, high cadence, untargeted surveys of the sky are revealing the great diversity of core-collapse supernovae of all types. These massive star explosions, understood to follow the gravitational collapse ouillierf the progenitor iron core, produce two distinct types of supernova with a comparable rate. Type II supernovae stem from the explosion of H-rich supergiant stars, while Type Ib/c supernovae stem from the explosion of H-poor and more compact progenitor stars.

For the most nearby SNe II, progenitor identification is sometimes possible. However, in general, all our inferrences on such explosions rely on the analysis of the supernova radiation. This will be even more true for the forthcoming deep surveys of the transient sky (e.g., with the Large Synoptic Survey Telescope). Understanding massive star explosions therefore requires detailed radiative transfer tools to connect progenitor/explosion and supernova observables.

In this talk, I will review the basic properties of core-collapse supernova ejecta and their observed properties. I will then describe the various methods used to model supernova radiation, and focus in particular on two approaches. The first is a local approach that treats the photospheric regions and assumes steady state. The second is a global approach that treats the entire ejecta in a time dependent way. This allows for the accurate computation of the evolution of the escaping radiation, providing multi-band light curves and spectra.

I will show the results from recent studies obtained with the code CMFGEN to emphasize the importance of line blanketing, non-LTE effects, time dependence, and non-thermal processes. I will also summarize the inferrences based on such studies concerning the explosion mechanism and progenitors of core-collapse supernovae.

3D Radiation Magnetohydrodynamic Simulations of Massive Star Envelopes at the Iron Opacity Peak

Yan-Fei Jiang

Smithsonian Astrophysical Observatory

I will describe a set of three-dimensional radiation (magneto-) hydrodynamic simulations of the structure and dynamics of the radiation dominated envelopes of massive stars at the location of the iron opacity peak. One-dimensional hydrostatic calculations predict an unstable density inversion at this location, whereas our simulations reveal a complex interplay of convective and radiative transport whose behavior depends on the ratio of the photon diffusion time to the dynamical time. Our simulations provide the first numerical calibration of mixing length theory in the radiation dominated regime. When diffusion time is shorter than the dynamic time scale, the envelopes show large amplitude oscillations and density fluctuations that allow photons to preferentially diffuse out through low-density regions. Magnetic field enhances the advective energy transport through magnetic buoyancy and increases the density fluctuation as well as the porosity effect. The simulations show that the turbulent velocity field may affect the broadening of spectral lines and therefore stellar rotation measurements in massive stars, while the time variable outer atmosphere could lead to variations in their mass loss and stellar radius.

Massive binary stars

S. Justham
University of the Chinese Academy of Sciences & NAOC

The Spatial Distribution of Massive Stars and Stellar Evolution

Jeremiah Murphy
Physics, Florida State University

We propose that the spatial distribution of O stars encodes information about their evolution, in particular their binary evolution. Smith & Tombleson 2015 noted that Luminous Blue Variables are very far away from other massive O stars, and they suggested that LBVs, as a class, are highly associated with binary evolution, kicks, and mass gainers. We have attempted to model the spatial distribution of O stars and LBVs hoping to provide theoretical constraints on the Smith & Tombleson 2015 observation. In just modeling the distribution of O stars, we are able to reproduce the average separation among O stars, but it is much more difficult to model the variance in separations with simple models. This implies that something is missing in these simple models and that we can learn about stellar evolution from the spatial distribution of O stars. In addition, we crudely model the spatial distribution of LBVs, and we find two models that are consistent with the very large separations between LBVs and other O stars. In model one, LBVs are mass gainers in binary evolution and receive, on average, 200 km/s kick when the primary star explodes, and in model two, LBVs are the product of mergers, in which the merger is triggered by the post main sequence evolution of at least a 17 solar mass star.

Binary Evolution and the Final Fate of Massive Stars

Philipp Podsiadlowski
Oxford University

Binary interactions do not only affect the envelope structure of massive supernova progenitors, thereby determining the appearance of the resulting supernova, but also the final fate of the core, specifically whether the core collapses to a neutron star or black hole or produces a gamma-ray burst or other exotic event. In this talk I will summarize how various binary interactions (mass loss, accretion, mergers, tidal interactions) affect the final fate of stars and its potential implications for a variety of "normal" and exotic supernova events, including supernovae with a circumstellar medium ("LBV supernovae"), superluminous supernovae, gamma-ray burst sources, pair-instability supernovae and aLIGO gravitational-waves sources.

Multi Epoch views of massive stars

Sergio Simon-Diaz
Instituto de Astrofísica de Canarias, Spain

The beginning of the 21st century has witnessed the compilation of several high quality spectroscopic surveys of massive OB stars. The scientific exploitation of this unique observational material, in combination with the imminent information provided by the Gaia mission will, without any doubt, quantitatively change our view of the properties and evolution of massive stars. Some of these surveys include multi-epoch observations scheduled with different time-resolution. In this talk I will benefit from the spectroscopic observations gathered by the IACOB project (Simón-Díaz et al. 2015, http://www.iac.es/proyecto/iacob/) to present an illustrative summary of the spectroscopic variability phenomena which are commonly detected in the O and B star domain. I will also highlight the importance of complementing the empirical information provided by the spectroscopic analysis of single-epoch observations of large samples of O- and B- type stars with the compilation and analysis of specifically designed time-resolved observations.

Challenges to stellar evolution from LBVs, SN Impostors, and Supernovae with Dense CSM

Nathan Smith
Steward Observatory, University of Arizona

I will discuss luminous blue variables (LBVs) and their connections to extragalactic transients and supernova (SN) explosions, as well as the challenges they pose for our understanding of massive star evolution in general. Their role in single-star evolution as transition objects between H and He burning appears to be invalid, and some massive LBVs appear to be exploding as SNe. This makes it challenging to understand what LBVs are, and also poses fundamental challenges to models of single-star evolution. I will report new results from spectroscopic and photometric monitoring of the light echoes of Eta Carinae for the past several years, and will discuss critical clues they provide for understanding the physics of this classic LBV giant eruption and connections to recent spectroscopic studies SN impostors. All proposed physical models, including stellar merger events, still have severe shortcomings and open questions in trying to accound for LBV eruptions. I will also discuss the basic nature of LBVs and the viability of LBVs as progenitors of Type IIn supernovae, considering the environments of LBVs and SNe IIn, combined with the properties of their CSM, explosion parmeters, and direct progenitor detections.

Magnetic fields in massive stars

Gregg Wade
Department of Physics, Royal Military College of Canada

Magnetic fields are directly observable only at the surfaces of stars, and while surface magnetic fields have important consequences for the evolution of OB stars (see Z. Keszthelyi's presentation), interior fields are in principle even more significant. For example, modern models of massive star evolution including interior magnetic field prescriptions find that fields dominate the angular momentum transport. Such models differ fundamentally from non-magnetic models, and are characterized by rigidly-rotating envelopes and strong core-envelope coupling. These affects have fundamental consequences for predicted surface rotation rates and chemical abundances, and ultimately HR diagram positions and evolutionary pathways.

In this presentation we report first results of an effort to employ new and existing observational results to constrain the influence of interior magnetic fields on the internal structure and evolution of hot stars. We adopt two approaches. First, we examine the physical properties of stars with radiative envelopes and detected surface magnetism (in part a legacy of the MiMeS and BoB large surveys), comparing with the properties of their non-magnetic peers, searching for differences attributable to their (interior, fossil) magnetic fields. Second, we examine the larger population of (apparently) non-magnetic hot stars, searching for mass- and rotation-dependent behaviour as predicted by radiative envelope shear and turbulent dynamo models.

Very Massive Stars at Different Metallicities

Norhasliza Yusof
Department of Physics, University of Malaya

In the work, we will present the evolution of Very Massive Stars (M>100 Mo) at different metallicities (Z=0.001 to Z=0.02). This includes the general properties, impact on the chemical abundances due to the rotational impact, dependence on metallicities and mass loss of very massive stars. Very massive stars has very large convective core during the main sequence thus their evolution are not affected strongly by rotational mixing but more by the mass loss due to strong stellar winds. In this presentation, we will also present pair instability supernovae modelled with our VMS progenitor models and compare them to super-luminous supernovae.

Splinter Session: X-ray observations of massive stars

The Lives and Death-Throes of Massive Stars
Proceedings IAU Symposium No. 329, 2016
J.J. Eldridge, J.C. Bray, L.A.S. McClelland & L. Xiao, eds.

doi:10.1017/S1743921317002411

Stellar Winds in Massive X-ray Binaries

Peter Kretschmar[1], Silvia Martínez-Núñez[2], Enrico Bozzo[3], Lidia M. Oskinova[4], Joachim Puls[5], Lara Sidoli[6], Jon Olof Sundqvist[7], Pere Blay[8], Maurizio Falanga[9], Felix Fürst[1], Angel Gímenez-García[10], Ingo Kreykenbohm[11], Matthias Kühnel[11], Andreas Sander[4], José Miguel Torrejón[10], Jörn Wilms[11], Philipp Podsiadlowski[12] and Antonios Manousakis[13,14]

[1]European Space Astronomy Centre (ESA/ESAC), Science Operations Department
P.O. Box 78, E-28691, Villanueva de la Cañada, Madrid, Spain
email: peter.kretschmar@esa.int

[2]Instituto de Física de Cantabria (CSIC-Universidad de Cantabria) E-39005, Santander, Spain

[3]ISDC, University of Geneva, Chemin d'Ecogia 16, Versoix, 1290, Switzerland

[4]Institut für Physik und Astronomie, Universität Potsdam,
Karl-Liebknecht-Str. 24/25, D-14476 Potsdam, Germany

[5]Universitätssternwarte der Ludwig-Maximilians-Universität München,
Scheinerstrasse 1, 81679, München, Germany

[6]INAF, Istituto di Astrofisica Spaziale e Fisica Cosmica - Milano,
via E. Bassini 15, I-20133 Milano, Italy

[7]Instituut voor Sterrenkunde, KU Leuven, Celestijnenlaan 200D, 3001 Leuven, Belgium

[8]Nordic Optical Telescope - IAC, P.O. Box 474, E-38700, Santa Cruz de La Palma
Santa Cruz de Tenerife, Spain

[9]International Space Science Institute (ISSI), Hallerstrasse 6, CH-3012 Bern, Switzerland

[10]Instituto Universitario de Física Aplicada a las Ciencias y las Tecnologías,
University of Alicante, P.O. Box 99, E03080 Alicante, Spain

[11]Dr. Karl Remeis-Observatory & ECAP, Universität Erlangen-Nürnberg,
Sternwartstr. 7, D-96049 Bamberg, Germany

[12]Department of Astronomy, Oxford University, Oxford OX1 3RH, United Kingdom

[13]Centrum Astronomiczne im. M. Kopernika, Bartycka 18, 00-716, Warszawa, Poland

[14]Department of Physics, Sultan Qaboos University, 123 Muscat, Oman

Abstract. Strong winds from massive stars are a topic of interest to a wide range of astrophysical fields. In High-Mass X-ray Binaries the presence of an accreting compact object on the one side allows to infer wind parameters from studies of the varying properties of the emitted X-rays; but on the other side the accretor's gravity and ionizing radiation can strongly influence the wind flow. Based on a collaborative effort of astronomers both from the stellar wind and the X-ray community, this presentation attempts to review our current state of knowledge and indicate avenues for future progress.

Keywords. stars: winds, outflows; supergiants; X-rays: binaries; accretion

---

1. Structures in winds from massive stars

Winds from massive stars are attributed to radiative line-driving, see, e.g., Puls, Vink & Najarro (2008) for a review. Although the standard theory of line-driven winds assumes a stable, time-independent and homogeneous wind, both theoretical considerations and observational features at different wavelengths clearly indicate that the winds of massive stars are not smooth and isotropic, but structured.

Small-scale structures are explained by reverse shocks in the wind, which are caused by a very strong, intrinsic instability in line-drive winds (LDI), already noted by Lucy & Solomon (1970). Numerical hydrodynamical modelling, e.g., by Feldmeier (1995) or Sundqvist & Owocki (2013) finds that the wind plasma becomes compressed into spatially narrow "clumps" separated by large regions of rarefied gas. The characteristic length scale for these structures is the Sobolev length; for typical hot supergiants this leads to an order of magnitude estimate of 10^{18} g for typical clump masses and a few $R_\odot$ for their extent. See, e.g., Oskinova, Feldmeier & Kretschmar (2012) for specific predictions.

Large-scale structures in winds from massive stars are mainly inferred from the so-called Discrete Absorption Components (DACs), observed in most O- and early B-star winds (Howarth & Prinja 1989) and in late B-supergiants (Bates & Gilheany 1990). A widely held candidate mechanism for these structures are Co-rotating Interaction Regions (Mullan 1984,1986), well studied in the solar wind. Another candidate are Rotational Modulations (RMs), as reported, e.g., by Massa *et al.* (1995). The density contrasts for these larger structures are rather low (factors of at most a few), but they may contain large overall masses, e.g., 10^{21-22} g for similar assumptions as above.

2. Wind-accreting High-Mass X-ray Binaries

In High-Mass X-ray Binaries (HMXB) a compact object, mostly a neutron star, sometimes a black hole or a white dwarf accretes, matter from its companion and produces copious X-ray radiation. For a typical neutron star $L_X \approx 0.1\dot{M}c^2$ for a mass accretion rate $\dot{M}$. There are several mechanisms to fuel the X-ray source, e.g., Roche-Lobe overflow, or from the disk around a Be star – neither discussed further here – but also accretion from the massive star wind. This last mechanism is present in two sub-groups: Classical Supergiant X-ray Binaries (SGXB) tend to be mostly persistent sources with erratic variations in flux. The more recently identified sub-group of Supergiant Fast X-ray Transients (SFXTs) has similar system parameters (where known), but remains mostly in a low luminosity state with brief outbursts and much larger flux variations. For a recent overview of different HXMB in our Galaxy see Walter *et al.* (2015).

3. X-ray absorption and fluorescence

A conceptually straightforward method to infer clumps or larger structures in stellar winds is to measure the attenuation of the X-ray flux, i.e., the variations in the measured absorbing column which in HMXB usually is in the range $N_{\rm H} \sim 10^{21-24}$ cm^{-2}. The main caveat is that this requires a good knowledge of the unabsorbed spectral continuum in order to minimise the degeneracy between spectral slope and absorption. An implicit issue is also that accreting X-ray sources are intrinsically variable and thus care has to be taken when comparing different observations. To obtain detailed observational results on wind structures, very extensive campaigns are required, like that reported in Grinberg *et al.* (2015) and previous publications for Cyg X-1. Large scale structures can also be traced in some cases with the lower time resolution of X-ray monitor data as, e.g., Malacaria *et al.* (2016) have demonstrated.

Another diagnostic is from X-ray fluorescence lines which will stem mostly from emission nearby to the compact object at most a few $R_\odot$ from the X-ray source. The line parameters can yield information on distribution, velocities and ionisation of the reprocessing material as described, e.g., in Giménez-García *et al.* (2015) and references therein.

4. Tracing accreted mass

As explained above, the X-ray luminosity of an accreting compact object is a direct measure of the current mass accretion rate. Assuming direct infall of matter, the X-ray source would then be a "local probe" of structures in the wind traversed by it. This approach has been used by various authors to explain flares and low-flux or "off" states in HMXB, e.g., by Ducci *et al.* (2009) or Fürst *et al.* (2010). But the estimates for clump masses from such studies have sometimes been 2–3 order of magnitudes larger than those from hydrodynamical simulations of stellar winds.

A closer look at accretion physics also shows that direct infall of captured matter is not necessarily taking place. According to Oskinova, Feldmeier & Kretschmar (2012), this would also imply orders of magnitude higher variability in many systems than observed. Different studies in recent years discuss, e.g., the possibility of settling envelopes around the compact objects, depending on conditions (Shakura *et al.* 2012). Another possibility is Chaotic Cold Accretion with complex accretion flows and condensation to filaments and cool clumps, as Gaspari, Temi & Brighenti (2017) have put forward as a model for AGNs, noting that the findings may also apply to X-ray binaries. In the common case of an accreting neutron star with a strong magnetic field, the interaction with the magnetosphere will lead to additional complications, including possible inhibition of accretion as detailed, e.g., in Bozzo *et al.* (2008).

5. Feedback on wind flow

The presence of the X-ray emitting compact object evidently also influences the wind flow, sometimes quite dramatically so. The gravitational pull focusses the stellar wind in the orbital plane. The bow shock of the compact object moving through the dense wind can create an "accretion wake" following the compact object in its orbit. Also, the intense X-ray emission of bright sources creates a large Strömgren sphere in which the wind is photoionised and the wind acceleration can be slowed or even cut off. These effects have been discussed in quite some detail already by Blondin *et al.* (e.g., 1990); for recent simulations of these effects see, e.g., Manousakis & Walter (2015) or Čechura, Vrtilek & Hadrava (2015). But so far, these feedback models have been based on smooth winds, while models including LDI and clumpy winds have usually not included an accretor and X-ray feedback.

6. Ongoing efforts and Outlook

The authors of this contribution and other colleagues have met at the International Space Science Institute (ISSI) Bern for meetings in 2013 & 2014 and a differently structured follow-up group is meeting again in 2016 & 2017 in order to discuss the open questions and possible avenues forward.

Among the findings of the first series of meetings are: (1) serious discrepancies in clump sizes and density contrasts used in the literature; (2) systematically lower wind velocities (factor 2–5) in HMXB than those derived for single stars; (3) CIRs should be stable over several orbits, but this is not reflected in HMXB studies of orbital variation; (4) the different behaviour of classical SGXBs and SFXTs remains an open question with no simple explanation. These findings and other results have been published in a detailed review by Martínez-Núñez *et al.* (2017). The ongoing meetings aim to reduce some of the uncertainties recognised in the first set, as well as include more modelling efforts for wind structure and accretion, and also discuss the impact of these findings for population synthesis studies.

For the future, we hope to arrive at models combining intrinsically clumpy winds with the effects from X-ray feedback, including a realistic picture of time varying accretion and X-ray emission. Systematic multi-wavelength observations via coordinated campaigns with space and ground instruments are required to follow variations on time scales of days or faster. The arrival of fast, sensitive optical spectrographs on ground allows to study some wind variations on time scales of seconds. In space, the advent of X-ray calorimeters will open a new era of X-ray line diagnostics. Until that time, further deep, dedicated observations with the existing grating instruments could still shed light on many questions.

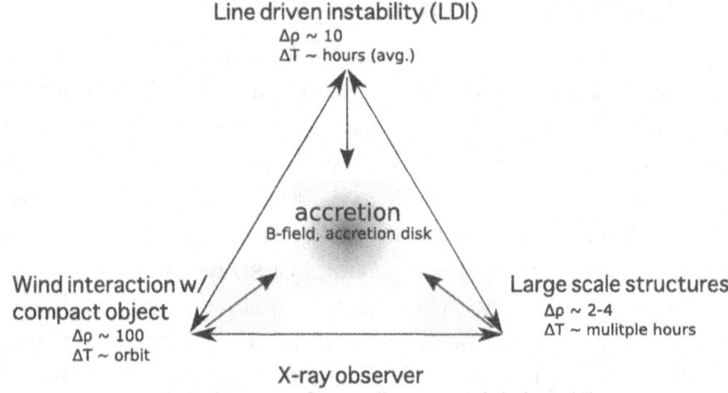

Figure 1. Scheme of interactions in a HMXB, $\Delta\rho$ indicates typical density variations and ΔT typical time scales.

References

Bozzo, E., Falanga, M., & Stella, L., 2008, *ApJ*, 683, 1031
Bates, B. & Gilheany, S., 1990, *MNRAS*, 243,320
Blondin, J. M., Kallman, T. R., Fryxell, B. A., & Taam, R. E., 1990, *ApJ*, 356, 591
Čechura, J., Vrtilek, S. D., & Hadrava, P., 2015, *MNRAS*, 450, 2410
Ducci, L., Sidoli, L., Mereghetti, S., Paizis, A., & Romano, P. 2009, *MNRAS*, 398, 2152
Feldmeier, A., 1995, *A&A*, 299, 523
Fürst, F., Kreykenbohm, I., Pottschmidt, K., Wilms, J., Hanke, M., Rothschild, R. E., Kretschmar, P., Schulz, N. S., Huenemoerder, D. P., Klochkov, D., & Staubert, R. 2010, *A&A*, 519, A37
Gaspari, M., Temi, P., & Brighenti, F., 2017, *MNRAS*, 466, 677
Giménez-García, A., Torrejón, J. M., Eikmann, W., Martínez-Núñez, S., Oskinova, L. M., Rodes-Roca, J. J., & Bernabéu, G., 2015, *A&A*, 576, A108
Grinberg, V., Leutenegger, M. A., Hell, N., Pottschmidt, K., Böck, M., García, J. A., Hanke, M., Nowak, M. A., Sundqvist, J. O., Townsend, R. H. D., & Wilms, J., 2015, *A&A*, 576, A117
Howarth, I. D. & Prinja, R. K., 1989, *ApJS*, 69, 527
Lucy, L. B. & Solomon, P. M., 1970, *ApJ*, 159, 879
Malacaria, C., Mihara, T., Santangelo, A., Makishima, K., Matsuoka, M., Morii, M., & Sugizaki, M., 2016, *A&A*, 588, A100
Manousakis, A. & Walter, R., 2015, *A&A*, 575, A58
Martínez-Núñez, S., Kretschmar, P., Bozzo, E., Oskinova, L. M., Puls, J., Sidoli, L., Sundqvist, J. O., Blay, P., Falanga, M., Fürst, F., Gímenez-García, A., Kreykenbohm, I., Kühnel, M., Sander, A., Torrejón, J. M., & Wilms, J., 2017, *Space Sci. Rev.*, doi:10.1007/s11214-017-0340-1
Massa, D., Fullerton, A. W., Nichols, J. S., Owocki, S. P., Prinja, R. K., & 28 co-authors, 1995, *ApJL*, 452, L53
Mullan, D. J., 1984, *ApJ*, 283, 303
Mullan, D. J., 1986, *A&A*, 165, 157
Oskinova, L. M., Feldmeier, A., & Kretschmar, P., 2012, *MNRAS*, 421, 2820
Puls, J., Vink, J. S., & Najarro, F., 2008, *A&AR*, 16, 209
Shakura, N., Postnov, K., Kochetkova, A., & Hjalmarsdotter, L., 2012, *MNRAS*, 420, 216
Sundqvist, J. O. & Owocki, S. P., *MNRAS*, 428, 1837
Walter, R., Lutovinov, A. A., Bozzo, E., & Tsygankov, S. S., 2015, *A&AR*, 23, 2

The Lives and Death-Throes of Massive Stars
Proceedings IAU Symposium No. 329, 2016
J.J. Eldridge, J.C. Bray, L.A.S. McClelland & L. Xiao, eds.

doi:10.1017/S1743921317001028

X-rays from colliding winds in massive binaries

Yaël Nazé and Gregor Rauw

FNRS/Université de Liège, Belgium, email: naze@astro.ulg.ac.be

Abstract. In a massive binary, the strong shock between the stellar winds may lead to the generation of bright X-ray emission. While this phenomenon was detected decades ago, the detailed study of this emission was only made possible by the current generation of X-ray observatories. Through dedicated monitoring and observations at high resolution, unprecedented information was revealed, putting strong constraints on the amount and structure of stellar mass-loss.

Keywords. X-rays: stars – stars: early-type – stars: winds – stars: binaries

---

Stellar winds are expected to collide in a strong shock if their sources, massive stars, are sufficiently close to each other, as occurs in multiple systems. This shocked plasma should not escape detection as its signature could appear anywhere throughout the electromagnetic spectrum, e.g., hard X-ray emission, non-thermal radio or gamma-ray emissions, or optical Hα emissions. In the X-ray domain in particular, massive stars display intrinsic X-ray luminosities following $L_X \sim 10^{-7} L_{BOL}$: any additional phenomenon, such as colliding winds, should produce a departure from this relation. The first X-ray observations of massive stars with Einstein or ROSAT seemed to support this scenario, but larger surveys with the more sensivitive XMM and Chandra observatories demonstrated that only few O+OB systems are overluminous and harder - the models were thus predicting too large X-ray luminosities. The situation is clearly different for systems comprising WRs or LBVs, where binaries truly are more luminous than single objects. Here we summarize the results of the recent sensitive studies of X-ray bright colliding-wind binaries - for a full review, see Rauw & Nazé (2016).

Since other phenomena may also produce overluminosities and hard X-rays (e.g. magnetically confined winds), the smoking gun for identifying colliding wind emission rather relies on the presence of recurrent variations linked to the orbital period. Two broad categories can be defined. The first one concerns the detection of changing absorption. Indeed, when the two winds have different densities, a modulation in the soft band will be detected as the line-of-sight towards the collision alternatively crosses each wind. This effect is particularly strong in asymmetrical cases, i.e., WR+OB systems. For example, large increases in the observed emission of γ^2 Vel or V444 Cyg are detected when the collision is seen through the O-star wind. Such changes can be used to constrain the opening angle of the collision cone or the wind densities. Another possibility to get an absorption modulation occurs in eccentric systems: as the collision zone plunges into the densest regions of the wind, the absorption increases, lowering the soft X-ray flux observed at Earth (see e.g. the cases of WR22, WR25, or WR140).

The second category only concerns eccentric systems: the changing separation is then the source of additional variability, as it directly impacts on the collision strength (which can be directly probed using the hard X-ray emission). For adiabatic collisions, one expects $L_X \propto 1/D$ where D is the orbital separation (i.e. emission should be maximum at periastron). This is observed for several systems (Cyg OB2 #9, 9 Sgr, WR25) but strong deviations from this relation are also seen and some remain unexplained: the emission in WR140 varies less than from a $1/D$ scaling while the emission even appears constant in the case of WR22 and γ^2 Vel despite their large eccentricities. For radiative collisions, one expects $L_X \propto v^2$ (i.e. minimum at periastron if the winds are still accelerating), but again it is not always clearly detected. Note that some collision may change their nature along the orbit. Such transitions between adiabatic and radiative types are especially prone to occur near periastron, and it can be easily detected by a deviation from the $1/D$ relation at that phase (as seen e.g. for Cyg OB2 #9 or 9 Sgr). The influence of the radiative braking (a slowing of the stellar wind of one star by the UV emission of its companion) plays a crucial role in this context.

The sensitivity of current X-ray observatories, coupled to dedicated, dense monitorings, revealed additional things. For example, since colliding wind emission should mostly arise close to the stagnation point, i.e. the location along the line-of-centers where wind momenta equilibrate, eclipses of the emitting zone by the stellar bodies are expected to occur when the inclination is high - and this was observed for V444 Cyg. Also, for close binaries, the orbital velocities are non-negligible compared to the wind velocities, so that Coriolis deflections of the collision zone are expected - and this leads to lightcurve asymmetries as detected in V444 Cyg. Moreover, when the secondary wind is very weak, it may not be able to maintain a stable collision zone against the primary wind, leading to a crash or collapse of the collision at (or close to) the secondary photosphere. In close and circular systems, this may occur all the time (e.g. the case of CPD $-41°7742$); in eccentric systems, this may occur only at periastron (as e.g. for WR140 or WR21a). Finally, detailed hydrodynamic simulations revealed that the emission from the shocked plasma in eccentric systems does not react instantly to the changing conditions, i.e. it has a "memory" hence the variations recorded on the way towards periastron will be different from those after periastron, even at similar separations - and this hysteresis effect has now been detected in several systems.

High resolution X-ray spectroscopy brings the most stringent constraints, and it has become available in the last two decades but only for the brightest objects: only a handful colliding-wind binaries could be observed (WR48a, WR140, WR147, θ Mus, γ^2 Vel, HD166734). These data first confirmed the origin of X-rays in a collision distant from the photosphere, thanks to the detection of strong f lines in He-like triplets. They also revealed the presence of cool gas through the detection of radiative recombination continua and brought some information on the shock cone geometry thanks to the analysis of line shifts and widths. Obviously, these pioneering datasets have demonstrated the potential of such observations.

The future of colliding wind studies appears twofold. On the one hand, it is crucial to continue observations with the aim of filling the parameter space: studying other binary configurations to probe other collision regimes appears important, as surprises (i.e. deviations from expectations) have been numerous in this field; probing systems with different metallicities are also required, to check our understanding with totally different wind strengths. On the other hand, monitoring the line profiles at high resolution and high sensivity for many systems is the logical next step, as such observations will provide the most stringent constraints on the geometry and properties of the interaction zone - but it will have to await the advent of a new generation of spectrometers, like the forthcoming XIFU onboard Athena.

Reference

Rauw, G. & Nazé, Y. 2016, *Ad. Sp. Research*, 58, 761

The Lives and Death-Throes of Massive Stars
Proceedings IAU Symposium No. 329, 2016
J.J. Eldridge, J.C. Bray, L.A.S. McClelland & L. Xiao, eds.

doi:10.1017/S1743921317001612

The Origin of X-ray Emission from the Enigmatic Be Star γ Cassiopeiae

K. Hamaguchi[1,2], L. Oskinova[3], C. M. P. Russell[4], R. Petre[4], T. Enoto[5,6], K. Morihana[7] and M. Ishida[8]

[1]CRESST II and X-ray Astrophysics Laboratory NASA/GSFC, Greenbelt, MD 20771, USA: Kenji.Hamaguchi@nasa.gov, [2]Department of Physics, University of Maryland, Baltimore County, 1000 Hilltop Circle, Baltimore, MD 21250, USA, [3]Institute of Physics and Astronomy, University of Potsdam, 14476 Potsdam, Germany, [4]X-ray Astrophysics Laboratory NASA/GSFC, Greenbelt, MD 20771, USA, [5]The Hakubi Center for Advanced Research, Kyoto University, Kyoto 606-8302, Japan, [6]Department of Astronomy, Kyoto University, Kitashirakawa- Oiwake-cho, Sakyo-ku, Kyoto 606-8502, Japan, [7]Nishi-Harima Astronomical Observatory, Center for Astronomy, University of Hyogo, 407-2, Nichigaichi, Sayo-cho, Sayo, Hyogo, 670-5313, Japan, [8]The Institute of Space and Astronautical Science, Japan Aerospace Exploration Agency, 3-1-1 Yoshinodai, Chuo-ku, Sagamihara, 252-5210, Japan

Abstract. Gamma Cassiopeiae is an enigmatic Be star with unusually hard, strong X-ray emission compared with normal main-sequence B stars. The origin has been debated for decades between two theories: mass accretion onto a hidden compact companion and a magnetic dynamo driven by the star-Be disk differential rotation. There has been no decisive signature found that supports either theory, such as a pulse in X-ray emission or the presence of large-scale magnetic field. In a ~100 ksec duration observation of the star with the *Suzaku* X-ray observatory in 2011, we detected six rapid X-ray spectral hardening events called "softness dips". All the softness dip events show symmetric softness ratio variations, and some of them have flat bottoms apparently due to saturation. The softness dip spectra are best described by either ~40% or ~70% partial covering absorption to kT ~12 keV plasma emission by matter with a neutral hydrogen column density of ~$2{-}8\times10^{21}\,\mathrm{cm}^{-2}$, while the spectrum outside of these dips is almost free of absorption. This result suggests that two distinct X-ray emitting spots in the γ Cas system, perhaps on a white dwarf companion with dipole mass accretion, are occulted by blobs in the Be stellar wind, the Be disk, or rotating around the white dwarf companion. The formation of a Be star and white dwarf binary system requires mass transfer between two stars; γ Cas may have experienced such activity in the past.

Keywords. stars: emission-line, Be — stars: individual (γ Cassiopeiae) — stars: winds, outflows — X-rays: stars — white dwarfs — blue stragglers

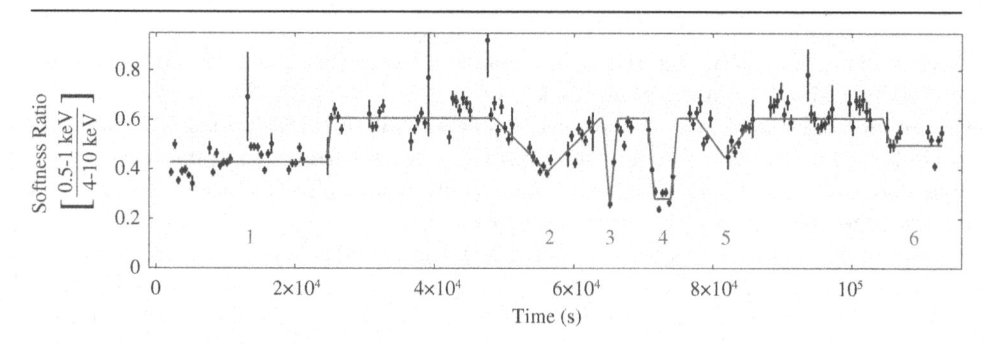

Figure 1. Time series of a softness ratio — the 0.5−1 keV count rate over the 4−10 keV count rate — measured with the *Suzaku* X-ray observatory in 2011. The solid line and numbers in red show the best-fit of an empirical softness dip model and the detected six softness dips. See Hamaguchi *et al.* 2016, Astrophysical Journal, 832, 140 for details.

The Lives and Death-Throes of Massive Stars
Proceedings IAU Symposium No. 329, 2016
J.J. Eldridge, J.C. Bray, L.A.S. McClelland & L. Xiao, eds.

doi:10.1017/S1743921317003532

X-ray Emission from Massive Stars at the Core of Very Young Clusters

Norbert S. Schulz

Kavli Institute of Astrophysics and Space Research, Massachusetts Institute of Technology, NE83-565, One Kendall Square, Bldg. 300, Cambridge 02139, USA
email: nss@space.mit.edu

Abstract. Most cores of very young stellar clusters contain one or more massive stars at various evolutionary stages. Observations of the Orion Nebula Cluster, Trumpler 37, NGC 2362, RCW38, NGC 3603 and many others provide the most comprehensive database to study stellar wind properties of these massive cluster stars in X-rays. In this presentation we review some of these observations and results and discuss them in the context of stellar winds and possible evolutionary implications. We argue that in very young clusters such as RCW38 and M17, shock heated remnants of a natal shell could serve as an alternate explanation to the colliding wind paradigm for the hot plasma components in the X-ray spectra.

Keywords. techniques: spectroscopic, stars: early-type, stars: winds, outflows, stars: formation

---

1. Introduction

The advent of *Chandra* and *XMM-Newton* allowed detailed X-ray diagnostics of X-ray stellar wind properties (Schulz *et al.* (2000), Kahn *et al.* (2001), Waldron & Cassinelli (2001)). A decade of *Chandra* observations of high resolution X-ray spectra produced a convincing correlation between X-ray line ionizations and optical spectral types of OB stars (Walborn, Nichols & Waldron (2009)) asserting that X-rays in stellar winds are indeed connected to fundamental stellar wind parameters. However these observations also produced a zoo of other X-ray production mechanisms ranging from magnetically confined winds, colliding wind activites, as well as massive dense winds as was most recently reported on (Schulz *et al.* (2003), Gagne *et al.* (2005), Pollock & Corcoran (2006), Huenemoerder *et al.* (2015)).

2. Young Massive Cluster Stars

Observations of clusters like the Orion Nebula Cluster (Schulz *et al.* (2001), Feigelson *et al.* (2005)), M17 (Townsley *et al.* (2003)) or the Carina region (Townsley *et al.* (2011)) have produced plenty of evidence for either magnetically confined winds in Orion, colliding winds systems in M17 and Carina. Observing X-rays from wind phenomena in very young dense clusters require high spatial resolution and effective areas leaving only a very few dozen clusters within the reach of *Chandra*.

Observations. In order to diagnose X-ray wind phenomena we need high spectral resolution to resolve X-ray line emissions. This reduces available clusters with reasonable *Chandra* exposures to less than a dozen objects. Table 1 shows about half a dozen of massive stars we have observed with sufficient exposures with the high energy transmission gratings (HETG) onboard *Chandra*. Key here is the fact that we associate the age of the massive star with the age of the cluster. There are three groups to consider. Very young massive stars with ages less the 1 Myr include the Orion Trapezium (not shown),

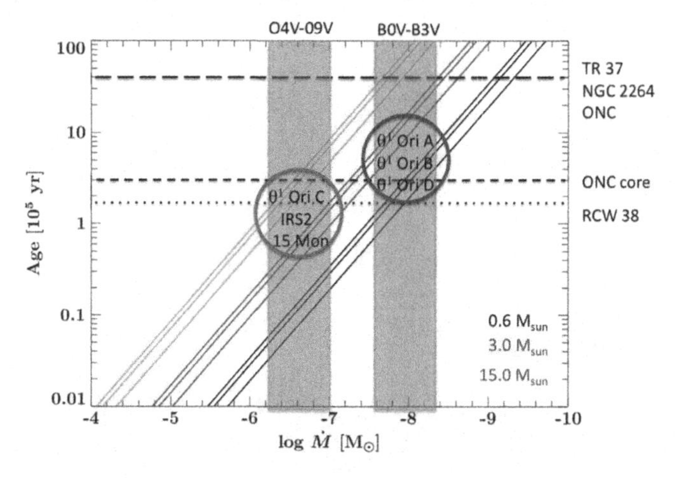

Figure 1. Photoevaporation rates of ultracompact H II envelopes in very young massive stellar winds. The plot shows rate curves for several natal cloud masses, the curves show the range of rate for various stellar mass loss rates, final wind velocities, and UV radiation fields. The black curves refer to 0.6 $M_\odot$ envelope mass, the red curves to 3 $M_\odot$ envelope mass, the green curves to 15 $M_\odot$ envelope mass.

Table 1. Properties of massive stars in very young clusters, young clusters, as well as evolved field stars

Cluster	Age [Myr]	Star	Type	N_H [(1)]	A_V	L_x [erg s^{-1}]	L_x/L_{bol}	kT_1 [MK]	kT_2 [MK]	EM_1 [(2)]	EM_2 [(2)]
Tr37	4	HD206067	O6.5V+09.5V	4.9	1.56	2.7×10^{33}	-5.8	1.6	7.0	21.6	0.7
NGC 2362	4-5	τCMA	O9III	2.1	0.31	2.9×10^{33}	-7.0	2.2	9.1	7.8	0.6
NGC 2264	<2	15 Mon	O7V+B2V	1.6	0.22	2.0×10^{32}	-6.6	1.9	7.4	3.3	0.3
RCW38	<0.2	IRS2	2 O4-5.5V	22.3	15	3.7×10^{33}	-5.5	2.1	15.5	0.7	10.4
M17	<1	KL Main	O4V	10,5	13.2	1.8×10^{33}	-6.0	9.0	87.5	4.4	12.2
		KL Comp	O4V	12.9	10.2	1.2×10^{33}	-6.2	8.8	40.8	5.2	4.0
	>5	ζPup	O4.5I	5.3	0.19	1.0×10^{33}	-6.6	3.7	8.0	3.9	0.4
NGC 3603	<2.5	HD97950C	WN6h	10.0		5.6×10^{33}		5.5	48.3	12.4	13.6
	>5	WR6	WN4	2.1		7.9×10^{32}	-6.3	8.0	50.1	24	11

Notes: (1) 10^{21} ; (2) 10^{55} cm^{-3}

IRS2 in RCW38, and the two Kleinmann stars in M17. Young stars with ages of up to about 5 Myr include HD 206267 in Trumpler 37, τ CMa in NGC 2362, and 15 Mon in NGC 2264. Finally there are evolved field stars arguably with ages above 5 Myr such as ζ Pup and WR6. Except for HD 97950C in the starburst cluster NGC 3603 these are stars where we have high signal to noise HETG spectra available.

Photoevaporation rates. Very young massive cluster stars at these young ages, even though they have already been on the main sequence for quite some time, still have residual circumstellar mass stemming from the star formation process in some form. Massive stars entering the main sequence are still embedded in their ultra-compact H II clouds (Wood & Churchwell (1989)). Depending on the type of the star and its wind properties in terms of mass loss rates and terminal wind velocity it takes between 10^5 yr up to over 10^6 yr to full ridden the system of the H II cloud material. And even though star is optically visible, residual cloud mass can provide various levels of absorption in X-rays. Hollenbach, Johnstone & Shu (1993) descibe the photoevaporation process in some detail and for massive winds project:

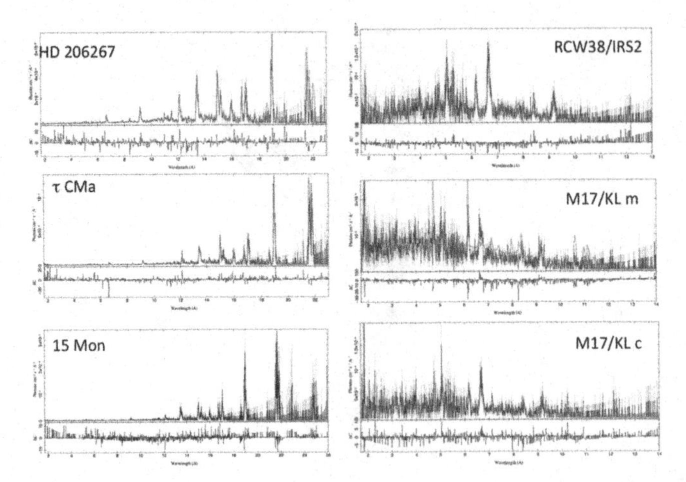

Figure 2. X-ray spectra of very young cluster stars (right panels) versus older cluster stars (left panels). The very young cluster stars are very hard and hot, the older ones conform with the standard wind paradigm independent of binary nature.

$$\dot{M}_{ph} \simeq 5 \times 10^{-6} \frac{\dot{M}_w}{[10^{-7} M_\odot]} \frac{v_w}{[1000 \text{ km s}^{-1}]} \left(\frac{\mathcal{N}_{UV}}{[10^{49} \text{ s}^{-1}]} \right)^{-1/2} M_\odot \text{ yr}^{-1},$$

where $\dot{M}_{ph}$ is the photoevaporaten rate, $\dot{M}_w$ the mass loss rate if the stellar wind, v_w the wind velocity, and $\mathcal{N}_{UV}$ the radiation field of the star. In Fig. 1 we plot the resulting photoevporation time versus the stellar massloss rate of the wind for various wind velocities and radiation fields for H II clouds of various masses.

Predictions for young clusters. Marked in Fig. 1 are mass loss ranges for various O and B-star regimes as well as ages for the clusters in Tab. 1. Massive stars in the red circle indicate cases for very young O-stars. While the cases for 15 Mon, τ CMa, and HD 206267 with a cluster age of 2-4 Myr is quite clear, i.e. all residual material has long been removed, the cases of θ^2 Ori A, and IRS2, and the Kleinmann stars are ambiguous, i.e. residual material may still be in effect to some extent. In contrast, for the later type stars in the Orion Trapezium (see blue circle) it may be the case that these have not yet fully broken free and may not yet be visible at all in X-rays. Schulz *et al.* (2001) and Schulz *et al.* (2003) report that X-rays from these stars are unusually faint and suggest that their spectra resemble more coronal emissions from unseen low-mass companions. More careful extinction studies are necessary to make such an argument as the high A_Vs we observe in RCW38 and M17 appear not so obvious in the Orion Trapezium stars.

3. On the Origins of Hot X-ray Plasma

One of the surprises in early *Chandra* observations of massive stars was the fact that some X-ray spectra show very hot (> 20 MK) plasma components, some do not. One peculiar case is θ^1 Ori C (Schulz *et al.* (2000)) in which a sizeable magnetic field compresses the wind into a dense circumstellar disk producing temperatures in excess of 50 MK (Gagne *et al.* (2005), Schulz *et al.* (2003)). However, claims for magnetically confined winds overall have so far not been substantiated in clusters other than Orion.

Colliding wind emission is another option since many of these cluster stars are multiple systems. This may also true for the two components of Kleinmann's star in M17, which in Tab. 1 are each listed as single O4V. Traditional colliding wind sources such as the

η Carina system (Corcoran *et al.* (2001)) or W140 (Pollock *et al.* (2005)) do show such hot plasma components but attached are also higher luminsities, a pattern of variability and specific X-ray line properties. Fig. 2 shows the HETG spectra of the young cluster stars from Tab. 1. The two temperature collisional plasma fit (red line in Fig. 2) to the spectra finds hot plasma components in the very young clusters stars, but not in the older cluster stars. The sample is certainly not statistically significant, but it is quite peculiar that binarity is a factor in both, very young and older cluster stars, but we do not detect hot plasmas in the latter. We also do not see excessive X-ray luminosities in these sources and within the data available no significant variability. The spectra also indicate very broad and skewed line widths of up to 3000 km s^{-1}. Though all this does not rule out colliding wind activity in these systems, it raises the question whether the hot component in young stars might be of a different origin.

As we have shown above, because of the youth of these systems we have to anticipate the existence of remnants of natal shell material from the star formation process. Here we may invoke the wind model proposed by Owocki, Castor & Rybicki (1988) which features reverse shocks that decelerate material as it rams into dense clumps. Located close to the terminal velocity of the wind, a shocked H II plasma at wind velocities of 2000 km s^{-1} would produce between 40 and 70 MK depending on the degree of ionization of the shocked material. Emission measures as well as column densities can produce consistent plasma volumes and densities but require a high degree of clumpyness, i.e. a low fraction of sufficiently dense clumps are suspended in the shell volume. Residual shell material from the natal cloud can contribute to the hot plasma in very young massive stars but there are many details that need to be worked out to make a valid case.

References

Corcoran, M. F., Swank, J. H., Petre, R., Ishibashi, K., Davidson, K., Townsley, L., Smith, R., White, S., Viotti, R., & Damelli, A. 2001, *ApJ*, 562, 1031

Feigelson, E. D., Getman, K., Townsley, L., Garmire, G., Preibisch, T., Grosso, N., Montmerle, T., Muench, A., & McCaughrean, M. 2005, *ApJS*, 160, 379

Gagne, M., Oksala, M. E., Cohen, D. H., Tonnesen, S. K., ud-Doula, A., Owocki, S. P., Townsend, R. H. D., & MacFarlane, J. J. 2005, *ApJ*, 628, 986

Hollenbach, D., Johnstone, D., & Shu, F. 1993, in: J.P. Cassinelli & E. Churchwell (eds.), *Astronomy of the Pacific Conference Series*, 35, 26

Huenemoerder, D. P., Gayley, K. G., Hamann, W.-R., Ignace, R., Nichols, J. S., Oskinova, L. Pollock, A. M. T., Schulz, N. S., & Shenar, T. 2015, *ApJ*, 815, 29

Kahn, S. M., Leutenegger, M., Cottam, J. Rauw, G., Vreux, J. M., den Boggende, T., Mewe, R & Guedel, M. 2001, *A&A*, 365, L312

Owocki, S. P., Castor, J. L., & Rybicki, G. B. 1988, *ApJ*, 335, 914

Pollock, A.M.T., Corcoran, M.F., Stevens, I.R. & Williams, P.M. 2005 *ApJ*, 629, 482

Pollock, A. M. T. & Corcoran, M. F. 2006, *A&A* 445, 1093

Waldron, W. L. & Cassinelli, J. P. 2001, *ApJ*, 548, L45

Walborn, N. R., Nichols, J. S., & Waldron, W. L. 2009, *ApJ*, 703, 633

Schulz, N. S., Canizares, C. R., Huenemoerder, D. P., & Lee, J. C. 2000, *ApJ*, 545, L135

Schulz, N. S., Canizares, C. R., Huenemoerder, D. P., Kastner, J. H., Taylor, S. C., & Bergstrom, E. J. 2001, *ApJ*, 549, 441

Schulz, N. S., Canizares, C. R., Huenemoerder, D. P., & Tibbets, K. 2003, *ApJ*, 595, 365

Townsley, L. K., Broos, P. S., Feigelson, E. D., Garmire, G., & Getman, K. 2003, *ApJ* 593, 874

Townsley, L. K., Broos, P. S., Chu, Y.-H, Gruendl. R. A., Oey, M. S., & Pittard, J. M. 2011, *ApJS*, 194, 16

Waldron, W. L. & Cassinelli, J. P. 2001, *ApJ*, 548, L45

Walborn, N. R., Nichols, J. S., & Waldron, W. L. 2009, *ApJ*, 703, 633

Wood, D. O. S. & Churchwell, E. 1989, *ApJS*, 69, 831

The Lives and Death-Throes of Massive Stars
Proceedings IAU Symposium No. 329, 2016
J.J. Eldridge, J.C. Bray, L.A.S. McClelland & L. Xiao, eds.

doi:10.1017/S1743921317003180

360-degree videos: a new visualization technique for astrophysical simulations

Christopher M. P. Russell

X-ray Astrophysics Laboratory, NASA/Goddard Space Flight Center,
Greenbelt, MD 20771, USA (NASA Postdoctoral Program Fellow, administered by USRA)
email: crussell@udel.edu

Abstract. 360-degree videos are a new type of movie that renders over all 4π steradian. Video sharing sites such as YouTube now allow this unique content to be shared via virtual reality (VR) goggles, hand-held smartphones/tablets, and computers. Creating 360° videos from astrophysical simulations is not only a new way to view these simulations as you are immersed in them, but is also a way to create engaging content for outreach to the public. We present what we believe is the first 360° video of an astrophysical simulation: a hydrodynamics calculation of the central parsec of the Galactic centre. We also describe how to create such movies, and briefly comment on what new science can be extracted from astrophysical simulations using 360° videos.

Keywords. hydrodynamics, Galaxy: centre

---

1. Introduction

360° videos are a new type of video that displays over all 4π steradian. Spurred by the development of 360° cameras to capture 360° content, it is now possible to share these videos via sites like YouTube and Facebook, thus increasing their reach.

A natural application of this technology to astrophysics is to create 360° videos of simulations. As opposed to traditional movies, which typically view the simulation domain from outside the simulation volume, or which move through the simulation domain but only render each frame in a predetermined direction, 360° videos immerse the viewers in the simulation and allow the viewers to choose where to look. The three methods for viewing 360° videos are on a computer, with a hand-held smartphone or tablet, or in VR goggles. For each method, you click and drag the video on the screen, pan the phone around, or simply look in different directions, respectively, to change the viewing orientation. The VR goggles are by far the most immersive experience, but using a smartphone or computer makes these movies accessible to a much larger audience.

The 360° videos described here are not specifically VR content, in which the user puts on a VR headset plugged into a computer and can walk around to move through the simulation volume; the image in the goggles is then rendered in real time based on the user's position and viewing orientation. On the other hand, sharable 360° videos have a predefined camera location chosen by the creator (just like with traditional movies), but the viewers choose where they look.

In this paper, we first describe how to create 360° videos from astrophysical simulations, then present what we believe is the first such video shared online, which shows the distribution of material expelled by Wolf-Rayet stars in the Galactic centre, and finally discuss potential scientific applications of 360° videos.

2. Constructing a 360-degree video

The first step is to create a movie where the x- and y-axes are the azimuthal and polar axes, respectively. Naturally, the x-axis should range from 0° to 360°, and the y-axis from 0° to 180°, so the aspect ratio is 2-to-1 for the recommended square pixels. Fig. 1 shows a frame from the video presented here.

This is the most challenging step since most visualization software generates images where the x- and y-axes correspond to linear dimensions, not angular dimensions. For our video, we modified the smoothed particle hydrodynamics (SPH) visualization program `Splash` (Price 2007) to render pixels across the full azimuthal and polar ranges at the desired angular width. Once the images are created, the next step is to create a movie with a 2-to-1 aspect ratio using video creation software; we use `ffmpeg`.

The second step is to add metadata to the movie file that states that it is a 360° video file. As we put our videos on YouTube, we followed the instructions here `https://support.google.com/youtube/answer/6178631`, which involves downloading a program called "Spatial Media Metadata Injector." Importing the video, which must be in *.mov or *.mp4 format, to this program adds the necessary metadata.

The final step is to upload the video to YouTube. Besides the aforementioned metadata, nothing special needs to be done to get the video into 360° format. However, the usual YouTube tools to modify videos will overwrite this information, so they can not be used. Therefore, the initial file into which the metadata is injected must be the final form. Once uploaded, the 360° video is available for the world to see.

At present, the maximum resolution for a 360° video shared via YouTube is 2160s through the app and 4320s via a computer. The 's' designation stands for 'spherical,' and the number preceding it is the number of polar pixels; e.g., 2160s is 4320 azimuthal pixels by 2160 polar pixels. The human eye with 20/20 vision can resolve $\sim 1'$, or $\sim$10800s, so the maximum resolution for 360° videos is $\sim$20-40% of what humans can see.

3. 360-degree video of the Galactic centre

Cuadra *et al.* (2008, 2015) constructed numerical simulations of the 30 Wolf-Rayet (WR) stars and their winds orbiting Sgr A* within the central parsec of our galaxy. Starting from the stellar locations 1100 yr ago, the WRs orbit Sgr A* while ejecting their stellar wind material. The central parsec quickly fills up with an ambient medium, into which the newly ejected wind material plows, causing wind-blown bubbles from the slow-moving stars and bow shocks around the fast-moving WRs. The intent was to study the time-dependent accretion history of material onto Sgr A* (the WR winds are the dominant mass-injection source in the region), but was also successful in explaining the thermal X-ray emission resolved with *Chandra* (Russell *et al.* 2017).

We created a 360° video from an updated hydrodynamic simulation of these 30 WR stars and their winds by rendering column density from the position of the centre of the simulation, i.e. at the location of Sgr A*. The link to the video is `https://youtu.be/pK59iu4cNRM` †. Fig. 2 shows the snapshot of Fig. 1 viewed in 360° format. Note that the distortion evident in Fig. 1, which is due to the image being polar vs. azimuthal angle and the viewing region being near the bottom of the plot, is gone when viewed in the 360° format of Fig. 2.

† Alternatively, it might be better, particularly if viewing on a smartphone or VR goggles, to go to CMPR's YouTube channel – either `http://tinyurl.com/cmpr360video`, or search for "Christopher Russell astronomy" in the app – and select the video "Galactic Center Column Density."

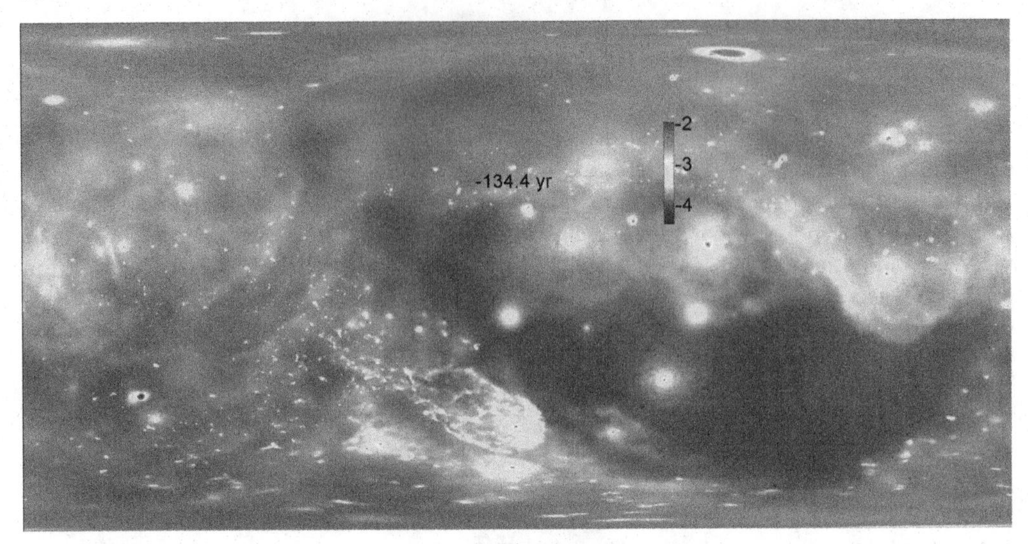

Figure 1. A single frame used to make the 360° video of the column density viewed from the exact centre of our galaxy. The x-axis goes from 0-360° in azimuthal, while the y-axis is 0-180° in polar. Adding the metadata tells YouTube to warp this into a sphere when viewing the movie.

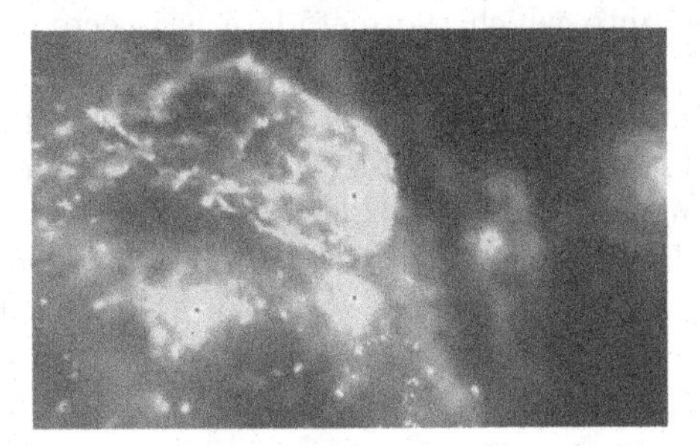

Figure 2. A screenshot of the movie when viewed from inside VR goggles.

To our knowledge, a shorter version of this video, published to YouTube on 15 Nov. 2016, is the first 360° video of an astrophysical simulation ever created and shared. (This video was published 10 days later.)

4. Future work

We are currently studying which science insights can be obtained from viewing simulations in this 360° manner. For example, we have produced a new movie with 10× higher time sampling (`https://youtu.be/BWiBIol7gzQ`). This affords a clear view of the inspiraling and stretching of clumps as they plummet towards Sgr A*. Additionally, a 360° movie of the colliding wind binary η Carinae shows when the primary wind completely engulfs the secondary star around periastron passage (`https://youtu.be/RzF6u0on_tw`). More applications will certainly be developed in the future.

References

Cuadra, J., Nayakshin, S., & Martins, F. 2008, *MNRAS*, 383, 458
Cuadra, J., Nayakshin, S., & Wang, Q. D. 2015, *MNRAS*, 450, 277
Price, D. J. 2007, *PASA*, 24, 159
Russell, C. M. P., Wang, Q. D., & Cuadra, J. 2017, *MNRAS*, 464, 4958

The Lives and Death-Throes of Massive Stars
Proceedings IAU Symposium No. 329, 2016
J.J. Eldridge, J.C. Bray, L.A.S. McClelland & L. Xiao, eds.

doi:10.1017/S1743921317002812

Investigating the Magnetospheres of Rapidly Rotating B-type Stars

C. L. Fletcher[1], **V. Petit**[2], **Y. Nazé**[3], **G. A. Wade**[4], **R. H. Townsend**[5], **S. P. Owocki**[2], **D. H. Cohen**[6], **A. David-Uraz**[2] **and M. Shultz**[7]

[1]Department of Physics and Space Sciences, Florida Institute of Technology, Melbourne, FL, 32904, USA, `cfletcher2013@my.fit.edu`

[2]Department of Physics and Astronomy, Bartol Research Institute, University of Delaware, Newark, DE 19716, USA

[3]FNRS GAPHE - STAR - Institut d'Astrophysique et de Géophysique (B5C), Université de Liège, Allée du 6 Août 19c, 4000-Liége, Belgium

[4]Department of Physics, Royal Military College of Canada, PO Box 17000 Station Forces, Kingston, ON, Canada K7K 0C6

[5]Department of Astronomy, University of Wisconsin-Madison, 5534 Sterling Hall, 475 N Charter Street, Madison, WI 53706, USA

[6]Department of Physics and Astronomy, Swarthmore College, 500 College Ave., Swarthmore, PA 19081, USA

[7]Department of Physics and Astronomy, Uppsala University, Box 516, Uppsala 75120, Sweden

Abstract. Recent spectropolarimetric surveys of bright, hot stars have found that ~10% of OB-type stars contain strong (mostly dipolar) surface magnetic fields (~kG). The prominent paradigm describing the interaction between the stellar winds and the surface magnetic field is the magnetically confined wind shock (MCWS) model. In this model, the stellar wind plasma is forced to move along the closed field loops of the magnetic field, colliding at the magnetic equator, and creating a shock. As the shocked material cools radiatively it will emit X-rays. Therefore, X-ray spectroscopy is a key tool in detecting and characterizing the hot wind material confined by the magnetic fields of these stars. Some B-type stars are found to have very short rotational periods. The effects of the rapid rotation on the X-ray production within the magnetosphere have yet to be explored in detail. The added centrifugal force due to rapid rotation is predicted to cause faster wind outflows along the field lines, leading to higher shock temperatures and harder X-rays. However, this is not observed in all rapidly rotating magnetic B-type stars. In order to address this from a theoretical point of view, we use the X-ray Analytical Dynamical Magnetosphere (XADM) model, originally developed for slow rotators, with an implementation of new rapid rotational physics. Using X-ray spectroscopy from ESA's XMM-Newton space telescope, we observed 5 rapidly rotating B-types stars to add to the previous list of observations. Comparing the observed X-ray luminosity and hardness ratio to that predicted by the XADM allows us to determine the role the added centrifugal force plays in the magnetospheric X-ray emission of these stars.

Keywords. massive stars, magnetic fields, stars, rotation, x-rays

---

1. Characterizing the Magnetic Fields of Massive Stars

Recent surveys have leveraged the development of improved spectropolarimeters to detect and characterize magnetic fields in a large sample of OB-type stars (Wade *et al.* 2016; Fossati *et al.* 2015). These studies have shown that ~10% of massive stars host surface magnetic fields that are strong (~1kG) and mostly dipolar. The stellar winds of massive stars are on the order of the stellar effective temperature and photoionized causing them to interact with the magnetic field.

The Magnetically Confined Wind Shock (MCWS) model presented by Babel & Montemerle (1997) has become the central idea explaining the X-ray emission of magnetic OB stars. In this scenario, the stellar wind plasma is forced to move along the magnetic field lines. In regions near the star, where the magnetic field lines are closed loops, the stellar wind channeled from both footpoints (located in the northern and southern hemispheres) will collide at the magnetic equator. Since the stellar wind is supersonic, the collision creates a shock with high post shock temperatures ($\sim$1 – 10 MK). The shocked material then cools radiatively, emitting X-ray photons (ud-Doula & Owocki 2002). Therefore, X-ray observations are a key tool to probe these magnetospheres.

An analytical approach to modeling the confined material in stars with slow rotation, called the Analytic Dynamical Magnetosphere (ADM) model, was developed by Owocki *et al.* (2016). This model simplifies the complex results from previous MHD simulations (ud-Doula & Owocki 2002; ud-Doula *et al.* 2008, 2009) by assuming a time averaged view of the processes in the confined material. ud-Doula *et al.* (2014) derived predictions for X-ray emission of DMs based on the stellar luminosity and the mass loss rate from the radiative cooling of the magnetically confined material. A main result of the XADM model is an increasing trend of X-ray emission with the spectral type and the size of the last closed field loop (R_A; Alfvén radius) for B-type stars (Nazé *et al.* 2014).

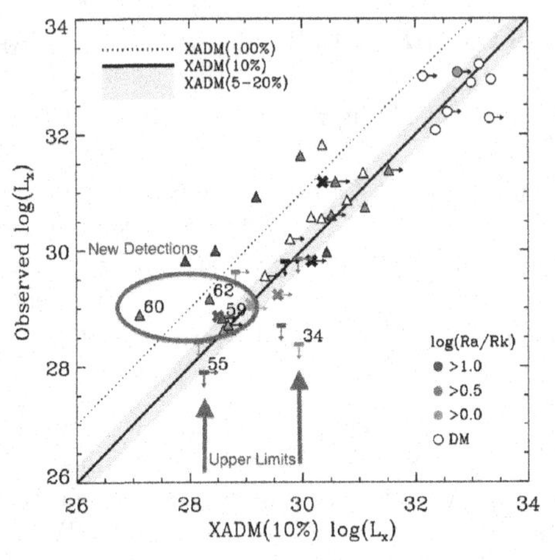

Figure 1. The predicted versus observed X-ray luminosity plot from Nazé *et al.* (2014) updated with new X-ray observations of 5 rapidly rotating B-type stars observed with XMM-Newton. The three new sources that have detections are shown with the numbered triangles and the two upper limits are the numbered rectangles with arrows. The numbers are the identification number taken from Petit *et al.* (2013) with the B–type stars in triangles, the O-type stars in circles and the undetected sources in rectangles. The color scheme corresponds to the size of the CM with the darker the color having a larger CM and the white only having a DM.

The rotational periods of some magnetic B-type stars are short enough for rotation to be dynamically significant, causing an added centrifugal component that affects the magnetically confined material. Townsend & Owocki (2005) developed a rigidly rotating magnetosphere (RRM) model relying on the assumption that the magnetic field lines will be forced to co-rotate with the stellar surface. For closed field lines larger than the Kepler co-rotation radius (R_K; the radius at which the centrifugal acceleration is equal to the gravitational acceleration) the confined material will not fall back to the stellar surface. Therefore, the material will accumulate creating a centrifugal magnetosphere (CM) with a disc-like structure. Perhaps the added centrifugal acceleration on the material trapped in the CM could provide a faster velocity causing larger shock temperatures and, consequently, harder X-ray energies or higher X-ray luminosity. Another possibility to explain this over-luminosity could be an increased X-ray efficiency factor in the CM regions through lack of material fall-back as compared to the DM regions.

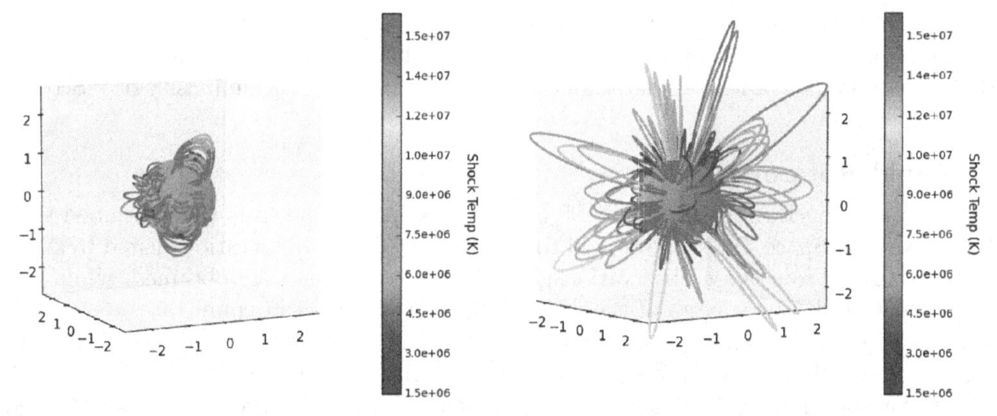

Figure 2. A 3-D representation of the field lines for the source radius $2R_*$ (left) and $5R_*$ (right) with the post shock temperatures plotted along the shock region of the individual loops. In order to reproduce the observed spectral features, there would need to be shock temperatures that are represented in red, however, for the $2R_*$ there no loops producing high enough shock temperatures. The $5R_*$ has larger closed field loops that result in higher shock temperatures.

Nazé *et al.* (2014) performed a study of all the known magnetic OB-type stars for which modern X-ray observations exist. In this study the observed X-ray luminosity from the Chandra X-ray Observatory and/or the XMM-Newton space telescope was compared to the predicted X-ray luminosity from the XADM model (Fig. 1). Although the agreement is generally good, we can see that a number of stars have observed X-ray luminosity higher than that predicted by the XADM model. Most of the overluminous stars in Fig. 1 are fast rotators. However, the one overluminous star that is not a fast rotator is τ Sco; this star has a complex surface field configuration, which is not common for fossil fields and could be the reason for its overluminosity.

2. τ Sco: a Slow Rotator with a Complex Field

One of the overluminous stars identified by Nazé *et al.* (2014) and shown in Fig. 1 is a slow rotator observed to have a complex surface magnetic field (Donati *et al.* 2006). The surface magnetic field can be extrapolated in order to predict the three dimensional structure of the field loops above the surface. A key parameter needed for the extrapolation is the source radius, the radius corresponding to the apex of the largest closed field loop for an arbitrary field configuration, which is determined by the ratio of the magnetic field energy density to the stellar wind kinetic energy density. A smaller mass loss rate will produce a larger source radius. As the mass-loss rate is relatively unconstrained, Donati *et al.* (2006) chose a mid range mass loss rate and extrapolated the surface field out to the corresponding source radius ($2R_*$), resulting in small loops.

τ Sco is an ideal case for comparing the shock temperatures of the field loops to the observed spectral features because of the high resolution grating spectra obtained by Cohen *et al.* (2003) with the Chandra X-ray Observatory. In order to do this, the ADM formalism was adapted for an arbitrary loop configuration, shown in Fig. 2, to determine the shock temperatures for the source radius assumed by Donati *et al.* (2006). The X-ray spectra suggests shock temperatures $\sim$15MK (shown in red in Fig. 2) are needed. However, the shock temperatures from the $2R_*$ source radius in Fig. 2 (left) are insufficient, suggesting that a larger source radius is needed. To find the shock temperatures for larger source radii, a field extrapolation was performed based on the surface field map of Donati *et al.* (2006) but with source radii of $3R_*$, $4R_*$, and $5R_*$. Determining the shock temperatures the same way as before, Fig. 2 (right) shows that the larger source radii produces higher shock temperatures that are ∼15MK. Using these results to constrain the source radius will help to determine what is causing the overluminosity of τ Sco.

3. Rapid Rotators

New X-ray spectral observations of 5 rapidly rotating B-type stars were obtained with XMM-Newton Space Telescope to add to the previous X-ray observations listed by Nazé *et al.* (2014). From these observations, three new detections were obtained, while two were undetected. The spectra of the detected stars with sufficient count rates were modeled using APEC thermal plasma models to determine the total X-ray flux and temperatures. For the two non-detections the upper limit X-ray luminosity was determined. The X-ray luminosities were added to the plot comparing the predicted and observed X-ray luminosity shown in Fig. 1 with the detected stars shown in numbered triangles and the upper limits in the numbered rectangles. Of the three detections, one falls in the overluminous outlier group (#60) with the other two being reasonably close to their predicted luminosity.

Further efforts will help determine whether the added centrifugal force provides sufficient acceleration to the confined material to lead to the high post-shock temperatures inferred observationally. A comparison of the X-ray production efficiency could also lead to an understanding of the discrepancies in the observed X-ray luminosity.

References

Babel, J. & Montmerle, T. 1997 *A&A*, 323, 121

Cohen, D. H., de Messières, G. E., MacFarlane, J. J., Miller, N. A., Cassinell, J. P., Owocki, S. P., & Liedahl, D. A. 2003, *ApJ*, 586, 495

Donati, J. F., Howarth, I. D., Jardine, M. M., Petit, P., Catala, C., Landstreet, J. D., Bouret, J.-C., Alecian, E., Barnes, J. R., Forveille, T., Paletou, F., & Manset, N. 2006, *MNRAS*, 370, 629

Fossati, L., Castro, N., Schöler, M., Hubrig, S., Langer, N., Morel, T., Briquet, M., Herrero, A., Przybilla, N., Sana, H., Schneider, F. R. N., de Koter, A., & BOB Collaboration 2015, *A&A*, 582, A45

Nazé, Y., Petit, V., Rinbrand, M., Cohen, D., Owocki, S., ud-Doula, A., & Wade, G. A., 2014, *ApJS*, 215, 10

Owocki, S. P., ud-Doula, A., Sundqvist, J. O., Petit, V., Cohen, D. H., & Townsend, R. H. D. 2016, *MNRAS*, 462, 3830

Petit, V., Owocki, S. P., Wade, G. A., Cohen, D. H., Sundqvist, J. O., Gagné, M., Maíz Apellániz, J. Oksala, M. E., Bohlender, D. A. Rivinius, T. Henrichs, H. F., Alecian, E., Townsend, R. H. D., ud-Doula, A., & MiMeS Collaboration 2013, *MNRAS*, 429, 398

Townsend, R. H. D. & Owocki, S. P. 2005, *MNRAS*, 357, 251

ud-Doula, A., Owocki, S., Townsend, R., Petit, V., & Cohen, D., 2014, *MNRAS*, 441, 3600

ud-Doula, A., Owocki, S. P., & Townsend, R. H. D. 2009, *MNRAS*, 392, 1022

ud-Doula, A., Owocki, S. P., & Townsend, R. H. D. 2008, *MNRAS* 385, 97

ud-Doula, A. & Owocki, S. P. 2002, *ApJ*, 576, 413

Wade, G. A. *et al.* 2016, *MNRAS*, 456, 2

The Lives and Death-Throes of Massive Stars
Proceedings IAU Symposium No. 329, 2016
J.J. Eldridge, J.C. Bray, L.A.S. McClelland & L. Xiao, eds.

doi:10.1017/S1743921317002757

Hα imaging for BeXRBs in the Small Magellanic Cloud

G. Maravelias[1], A. Zezas[2,3,4], V. Antoniou[3], D. Hatzidimitriou[5] and F. Haberl[6]

[1]Astronomický ústav AVČR, v.v.i., Ondřejov, Czechia, email: maravelias@asu.cas.cz
[2]Department of Physics, University of Crete, Heraklion, Greece
[3]Harvard-Smithsonian Center for Astrophysics, Cambridge, USA
[4]Foundation for Research and Technology-Hellas (FORTH), Heraklion, Greece
[5]Department of Physics, University of Athens, Greece
[6]Max-Planck-Institut für extraterrestrische Physik, Garching, Germany

Abstract. The Small Magellanic Cloud (SMC) hosts a large number of high-mass X-ray binaries, and in particular of Be/X-ray Binaries (BeXRBs; neutron stars orbiting OBe-type stars), offering a unique laboratory to address the effect of metalicity. One key property of their optical companion is Hα in emission, which makes them bright sources when observed through a narrow-band Hα filter. We performed a survey of the SMC Bar and Wing regions using wide-field cameras (WFI@MPG/ESO and MOSAIC@CTIO/Blanco) in order to identify the counterparts of the sources detected in our *XMM-Newton* survey of the same area. We obtained broad-band R and narrow-band Hα photometry, and identified ~10000 Hα emission sources down to a sensitivity limit of 18.7 mag (equivalent to ~B8 type Main Sequence stars). We find the fraction of OBe/OB stars to be 13% down to this limit, and by investigating this fraction as a function of the brightness of the stars we deduce that Hα excess peaks at the O9-B2 spectral range. Using the most up-to-date numbers of SMC BeXRBs we find their fraction over their parent population to be $\sim 0.002 - 0.025$ BeXRBs/OBe, a direct measurement of their formation rate.

Keywords. Magellanic Clouds, stars: early-type, stars: emission-line, Be, X-rays: binaries

1. Introduction

The Small Magellanic Cloud (SMC) has been a major target for X-ray surveys due to our ability to detect sources down to non-outbursting X-ray luminosities ($L_X \sim 10^{33}$ erg s^{-1}) and its impressive large number of High-Mass X-ray Binaries (HMXBs; Haberl & Sturm 2016). However, the X-ray properties alone cannot fully characterize the nature of each source. HMXBs consist of an early-type (OB) massive star and a compact object (neutron star or black hole), which accretes matter from the massive star either through strong stellar winds and/or Roche-lobe overflow in supergiant systems or through an equatorial decretion disk in, non-supergiant, OBe stars (Be/X-ray Binaries; BeXRBs). The compact object dominates the X-ray spectrum while the companion dominates the optical spectrum. Thus, to understand the nature of BeXRBs we need to study their optical counterparts, which should be consistent with OBe stars. These are massive stars that show Balmer lines in emission, of which Hα is typically the most prominent. Although the SMC is close enough to resolve its stellar population, we still lack the identification of the optical counterparts or their optical spectral classification for a large fraction (~ 40% of the candidate HMXBs) of the most recent census (121 candidates in total; Haberl & Sturm 2016). To address this issue we take advantage of the fact that OBe stars display Hα in emission, making them easily discernible from other stars in Hα narrow-band images, and we performed a wide Hα imaging survey of the SMC to reveal prime candidates for BeXRB optical counterparts.

2. Observations and Data Reduction

We used the Wide Field Imager (WFI@MPG/ESO 2.2m, La Silla, on 16/17 November, 2011) and the MOSAIC camera (@CTIO/Blanco 4m, Cerro Tololo, on 15/16 December, 2011) to observe 6 and 7 fields in the SMC, respectively. Given their large field-of-views ($\sim 33' \times 33'$) we covered almost the whole galaxy. Each field was observed in the R broad-band (the continuum) and Hα narrow-band filters. A dithering approach was needed to cover camera chip gaps, and the exposure time was selected to achieve a similar depth ($R \sim 23$ mag) in both campaigns to allow for coverage of late B-type stars at the distance of the SMC. Additionally, a set of spectrophotometric standards was observed to flux calibrate the results. THELI† was used to reduce and produce the final mosaics from the WFI data. For the MOSAIC data we retrieved the reduced data products from the NOAO online pipeline‡ and then combined them using IRAF's `mscred` package. We finally re-sampled (with SWARP¶) the mosaic images for each field, using a common center and frame size (for details see Maravelias 2014).

Because of the high source density we performed PSF photometry with IRAF's `daophot` by properly selecting its parameters for each field. However, we ran the source detection on the broad-band R image only, as the same process in Hα would result to many spurious sources due to the HII regions of the SMC. This (R-selected) source list was used to perform photometry on Hα. We first screened the (flux-calibrated) `daophot` results to select stellar sources, according to their χ^2 (~ 1) and `sharpness` ($|\text{sharp}| < 0.5$) values. Since we were interested in OB stars we kept sources brighter than $R = 18.7$ mag, which corresponds to B8 spectral-type stars at the distance of the SMC. We cross-correlated the two filters (R and Hα) to identify the common sources and then with the MCPS catalog (Zarithsly *et al.* 2002) to obtain their V, B photometry. Using the locus of OB stars (Antoniou *et al.* 2009) we selected the best OB candidate sources, for which we calculated their (H$\alpha - R$) index, its error, and SNR (following Zhao *et al.* 2005).

Since the R filter includes the Hα region the corresponding baseline for stars without any Hα excess would be equal to (H$\alpha - R$)= 0 mag. However, due to the differences between the two filters and the range of spectral types considered, this is not 0 (see Maravelias 2014). To overcome this we define the baseline (H$\alpha - R$) value for non-Hα excess stars individually for each field based on the mode ($\langle$H$\alpha - R\rangle$) and standard deviation (σ) of the (H$\alpha - R$) distribution of all OB stars in each field. We consider as best Hα emitting candidates the sources with: (H$\alpha - R$) $< \langle$H$\alpha - R\rangle - 5 \times \sigma$, and SNR$> 5$.

3. Results and Discussion

Our survey reveals 9808 Hα emitting sources in the SMC. This is 2 to 4 times more sources from other previous surveys (1844 sources; Meyssonnier & Azzopardi 1993), mostly due to our deeper coverage down to $V = 18.5$mag instead of $R \sim 16.5$ mag.

From our analysis we know the number of OB stars and the corresponding number of emission-line stars (i.e. OBe). This allows us to derive the OBe/OB fraction for each field. We find an average value of $\sim$13.3% across the SMC, consistent with previous studies (e.g. $\sim 5 - 11$% from Iqbal & Keller 2013). This fraction is only a lower limit of the actual population of the OBe stars since their activity is a transient phenomenon

† `http://www.astro.uni-bonn.de/~theli/`
‡ `http://portal-nvo.noao.edu/search/query`
¶ `http://www.astromatic.net/software/swarp`

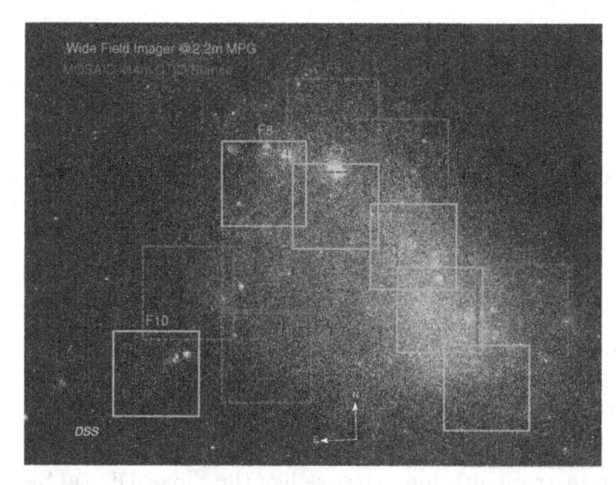

Field-ID	OBs	OBe/OBs	BeXRBs/OBe
1	6916	0.157	0.0184
2	9217	0.128	0.0246
4	9799	0.121	0.0218
5	5396	0.136	0.0136
6	6716	0.101	0.0088
7	8428	0.142	0.0133
8	7305	0.118	0.0081
9	3672	0.132	0.0145
10	1687	0.144	0.0165
11	5542	0.096	0.0019
12	3905	0.183	0.0
13	2693	0.129	0.0
16	3236	0.142	0.0022

Figure 1. *Left:* The fields observed with the WFI (green solid boxes) and the MOSAIC (red dashed boxes) wide-field cameras, overplotted on a DSS image of the SMC. BeXRBs (taken from Haberl & Sturm 2016) are shown as smaller purple circles. *Right:* For each field (col. 1) we show the number of OB stars identified (col. 2), the fraction of $H\alpha$ emitting stars of OBe/OBs (col. 3), and the formation rate BeXRBs/OBe (col. 4). (Not including fields 3 and 14/15 due to reduction issues and shallower exposures, respectively.)

and only a fraction of them is active in a certain epoch. Furthermore, if we examine the relation of this fraction with R magnitude (1 mag $\sim$ 3 spectral sub-types at the distance of the SMC), we notice a peak at $\sim$15 mag (corresponding to O9-B2) and a fast drop with magnitude (equal to later spectral types). This trend is consistent with previous observations, but we extend it to later B-type stars (from Martayan *et al.* 2010: peak at B2 but limited to $\sim$B3). Moreover, it confirms theoretical models that predict a peak of that ratio at B3, as a result of the critical rotational velocity (Maeder & Meynet 2000).

Given the numbers of OBe stars and BeXRBs we derive the BeXRBs/OBe fraction in the range $\sim 0.002 - 0.025$, which provides us with the formation efficiency of these systems with respect to their parent population. This is a direct measurement of their formation rate, which can place constraints on stellar population synthesis models (e.g. Belczynski *et al.* 2008). Currently, we are working on the cross-correlation of this catalog with the most recent list of candidate BeXRBs in the SMC (Haberl & Sturm 2016) in order to identify more optical counterparts.

Acknowledgements

GA ČR (14-21373S); RVO:67985815; NASA Grant NNX10AH47G; The State Scholarships Foundation of Greece (IKY); IAU travel grant.

References

Antoniou, V., Zezas, A., Hatzidimitriou, D., & McDowell, J. C. 2009, *ApJ*, 697, 1695
Belczynski, K., Kalogera, V., Rasio, F. A., *et al.* 2008, *ApJS*, 174, 223
Haberl, F. & Sturm, R. 2016, *A&A*, 586, 81H
Iqbal, S. & Keller, S. C. 2013, *MNRAS*, 435, 3103
Maeder, A. & Meynet, G. 2000, *ARA&A*, 38, 143M
Maravelias, G. 2014, *PhD Thesis*, University of Crete, Heraklion, Greece
Martayan, C., Baade, D., & Fabregat, J. 2010, *A&A*, 509, A11
Meyssonnier, N. & Azzopardi, M. 1993, *A&AS*, 102, 451
Zaritsky, D., Harris, J., Thompson, I. B., Grebel, E. K., & Massey, P. 2002, *AJ*, 123, 855
Zhao, P., Grindlay, J. E., Hong, J. S., Laycock, S., *et al.* 2005, *ApJS*, 161, 429

The Lives and Death-Throes of Massive Stars
Proceedings IAU Symposium No. 329, 2016
J.J. Eldridge, J.C. Bray, L.A.S. McClelland & L. Xiao, eds.

doi:10.1017/S174392131700240X

There Are (super)Giants in the Sky: Searching for Misidentified Massive Stars in Algorithmically-Selected Quasar Catalogs

Trevor Z. Dorn-Wallenstein and Emily Levesque
Astronomy Department, University of Washington,
Physics and Astronomy Building, 3910 15th Ave NE, Seattle, WA 98105, USA
email: tzdw@uw.edu, emsque@uw.edu

Abstract. Thanks to incredible advances in instrumentation, surveys like the Sloan Digital Sky Survey have been able to find and catalog billions of objects, ranging from local M dwarfs to distant quasars. Machine learning algorithms have greatly aided in the effort to classify these objects; however, there are regimes where these algorithms fail, where interesting oddities may be found. We present here an X-ray bright quasar misidentified as a red supergiant/X-ray binary, and a subsequent search of the SDSS quasar catalog for X-ray bright stars misidentified as quasars.

Keywords. astronomical data bases: surveys, stars: supergiants, X-rays: binaries

1. Introduction/Overview

1.1. *Red Supergiant X-ray Binaries*

Over the past decade, many exotic close binary systems with supergiant components have been discovered. Systems like NGC 300 X-1 — a Wolf-Rayet/black hole X-ray binary (Crowther *et al.* 2010) — and SN2010da — a sgB[e]/neutron star X-ray binary (Villar *et al.* 2016) — are two such examples of a coupling between a massive star in a short-lived evolutionary phase and a compact stellar remnant. Interestingly, no X-ray binaries with confirmed red supergiant (RSG) counterparts have been discovered (RSGs have been proposed as the candidate donor star for a few Ultraluminous X-ray Sources, see Heida *et al.* 2016). This may be partially explained by the rarity of RSGs; however, though rare, RSGs are both longer-lived and more common (due to the smaller — 10 - 25 $M_{\odot}$ — initial masses of their zero age main sequence progenitors) than most other evolved massive stars.

RSG X-ray binaries, if they exist, offer a view into an interesting edge case of accretion; their extended envelopes and strong winds ($\dot{M}$ ~10^{-4} $M_{\odot}$ yr^{-1}, van Loon *et al.* 2005) could allow for accretion from both the wind and Roche-Lobe Overflow in an environment continually enriched with dust produced by the RSG. RSG X-ray binaries are also the immediate progenitors of Thorne-Żytkow Objects — stars with embedded neutron star cores (Thorne & Żytkow 1975) — assuming the neutron star plunges into the RSG as it expands (Taam *et al.* 1978).

1.2. *J0045+41*

To search for RSG X-ray binaries, we used the photometry of the Local Group Galaxy Survey (LGGS, Massey & Olsen 2003, Massey *et al.* 2006, 2007), which covers M31, M33, the Magellanic Clouds and 7 dwarf galaxies in the Local Group. Following Massey (1998) to find RSGs among the nearly-identical foreground dwarfs, we cross-referenced

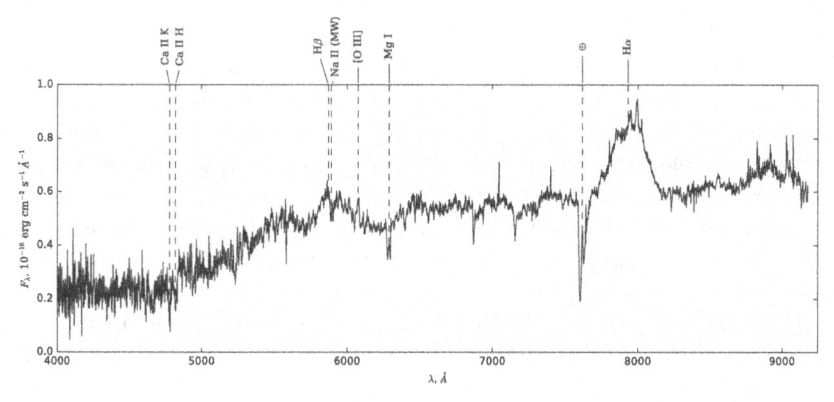

Figure 1. GMOS spectrum of J0045+41 with all identified lines labeled. The Na II feature is intrinsic to the Milky Way/M31, and the ⊕ line is telluric.

the positions of the LGGS RSGs with the *Chandra* Source Catalog (CSC, Evans *et al.* 2010), and found one RSG coincident with an X-ray source.

LGGS J004527.30+413254.3 (J0045+41 hereafter) is a bright ($V \approx 19.9$) object of previously-unknown nature in the disk of M31. Vilardell *et al.* (2006) classify J0045+41 as an eclipsing binary with a period of ~76 days. J0045+41 was also observed with the Palomar Transient Factory (PTF); the *g*-band lightcurve shows evidence for a ~650 day period. On the other hand Kim *et al.* (2007) identify J0045+41 as a globular cluster, and it has been included in catalogs of M31 globular clusters as recently as 2014 (Wang *et al.* 2014). The LGGS photometry is consistent with the color and brightness of a RSG. Indeed, following Levesque *et al.* (2006), we found that, as an RSG, J0045+41 would have an effective temperature of ~3500 K and bolometric magnitude of -6.67, consistent with a 12-15 $M_\odot$ RSG. However, a complete SED fit to photometry from the Panchromatic Hubble Andromeda Treasury (PHAT, Dalcanton *et al.* 2012) using the Bayesian Extinction and Stellar Tool (BEAST, Gordon *et al.* 2016) yields an unphysical result of 300 $M_\odot$, 10^5 K star, extincted by A_V ~4 magnitudes. Furthermore, the object appears extended in the PHAT images (though its radial profile appears similar to that of other nearby stars).

J0045+41 is separated by ~1.18″ from an X-ray source. The source, CXO J004527.3 +413255, is bright ($F_X = 1.98 \times 10^{-13}$ erg s^{-1} cm^{-2}) and hard; fitting a spectrum obtained by Williams *et al.* (in prep) yields a power law with Γ ~1.5. The best-fit neutral hydrogen column density is 1.7×10^{21} cm^{-2}, which corresponds to A_V ~1.

1.2.1. *Observations and Data Reduction*

The apparent periodicity of J0045+41 and its apparent association with a hard and unabsorbed X-ray source prompted us to obtain follow-up spectroscopic observations to determine the true nature of this object. We obtained a longslit spectrum of J0045+41 using the Gemini Multi-Object Spectrograph (GMOS) on Gemini North. Four 875 second exposures were taken 2016 July 5 using the `B600` grating centered on 5000 Å, and four 600 second exposures were taken 2016 July 9 using the `R400` grating centered on 7000 Å, with a blocking filter to remove 2nd-order diffraction. Due to the gaps between GMOS's three CCDs, two of each set of exposures were offset by +50 Å. The data were reduced using the standard `gemini IRAF` package. The final reduced spectrum has continuous signal from ~4000 to ~9100 Åat a resolution of R ~1688 (blue)/1918 (red).

1.2.2. *Spectrum and Redshift Determination*

The spectrum (Figure 1) shows that J0045+41 is a quasar at $z \approx 0.21$ (measured with Hα, Hβ, [OIII]λ5007 and Ca II H and K). While a false positive, this quasar is quite interesting on its own. The (low-significance) detections of periodicity on short timescales by multiple sources are difficult to explain. Furthermore, mistaking a blue quasar for a red star would imply a high reddening along the line of sight, which is consistent with a sightline through M31; however, the low H column density implied by the X-ray counterpart's spectrum and our inability to achieve a satisfactory fit to the optical spectrum by reddening the (redshifted) quasar template spectrum from Vanden Berk *et al.* (2001) indicate that J0045+41 belongs to a small and intriguing class of intrinsically red quasars, observed through a low-extinction region of the ISM in M31 (Richards *et al.* 2003).

2. Stars in the SDSS Quasar Catalog

If a quasar can be misidentified as a RSG, are there red stars — especially X-ray bright stars — in already-existing quasar catalogs? To answer this question, we turned to the quasar catalog of the Sloan Digital Sky Survey (SDSS, York *et al.* 2000).

2.1. *Sample Selection*

We selected all SDSS objects automatically tagged as quasars that were within 0.2 magnitudes of J0045+41 in $g-r$ vs. $r-i$ vs. $i-z$ color-space — J0045+41 is too faint in u to utilize $u-g$ — and ignored any warning flags to avoid throwing out interesting objects that were not easily identified by the SDSS algorithm. 1098 objects in this sample had associated spectroscopic observations. Interestingly, many of the spectroscopically-determined redshifts were unbelievably small or even negative, implying that these objects are in a regime of color-space where classification algorithms may fail. Indeed, on visual inspection of these spectra, many of them are stellar.

2.2. *Stars*

We used `emcee`, a Python implementation of Markov Chain Monte Carlo (MCMC) by Foreman-Mackey *et al.* (2013), to fit Gaussian profiles to the Ca II triplet ($\lambda =$ 8498, 8542, and 8662 Å), which we use to identify stars. The posterior distributions of the parameters allow us to determine if the triplet is well fit, and estimate errors for each parameter. Because the relative centroids and strengths of the lines are fixed, a good fit guarantees the lines are actually detected, while simultaneously measuring — with accurate errors — the radial velocity and equivalent width (W_λ) of the triplet. After a follow-up inspection by eye of spectra that were noisy or missing data, we find 344 confirmed cool stars, representing ~31% of the total sample. Figure 2 shows the distribution of W_λ for $W_\lambda/\sigma_{W_\lambda} > 1$. We follow Jennings & Levesque (2016) to estimate luminosity from W_λ, and find that most stars are dwarfs ($W_\lambda \lesssim 6.5$ Å) but ~40 stars have larger equivalent widths indicating they are likely giants or supergiants (the relationship is dependent on effective temperature and metallicity, so these labels are approximate).

3. Discussion and Future Work

This result demonstrates that, when looking for rare objects like RSG X-ray binaries, it is important to look in unlikely places; e.g., a red and X-ray bright star may be confused for a quasar if the classification algorithm mistakes the continuum between the TiO bands in a RSG spectrum for an emission line. The fact that some of the color-space containing

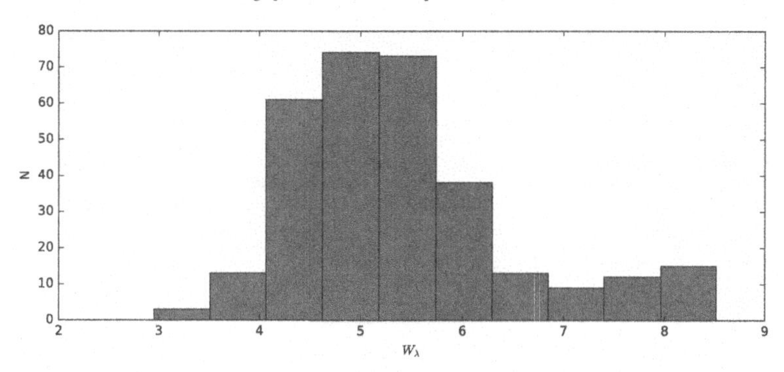

Figure 2. Distribution of measured equivalent widths for confirmed stars.

M-dwarfs — by far the most common type of star — is a regime where classification algorithms fail underlines the importance of improving on these algorithms until they perform as well as the human eye. Indeed, many of these stars were previously identified (see West *et al.* 2011), but are still listed as quasars on the SDSS online data portal.

Future work will focus on improving our star-finding algorithm to use alternate spectral features when the Ca II triplet is missing or obscured by noise, and on finding which areas of color-space contain significant numbers of these misidentified stars, with the goal of finding RSG X-ray binaries as well as improving our knowledge of where classification algorithms fail.

References

Crowther, P. A., Barnard, R., Carpano, S., *et al.* 2010, *MNRAS*, 403, L41
Dalcanton, J. J., Williams, B. F., Lang, D., *et al.* 2012, *ApJS*, 200, 18
Davidsen, A., Malina, R., & Bowyer, S. 1977, *ApJ*, 211, 866
Evans, I. N., Primini, F. A., Glotfelty, K. J., *et al.* 2010, *ApJS*, 189, 37-82
Foreman-Mackey, D., Hogg, D. W., Lang, D., & Goodman, J. 2013, *PASP*, 125, 306
Gordon, K. D., Fouesneau, M., Arab, H., *et al.* 2016, *ApJ*, 826, 104
Heida, M., Jonker, P. G., Torres, M. A. P., *et al.* 2016, *MNRAS*, 459, 771
Jennings, J. & Levesque, E. M. 2016, *ApJ*, 821, 131
Kim, S. C., Lee, M. G., Geisler, D., *et al.* 2007, *AJ*, 134, 706
Levesque, E. M., Massey, P., Olsen, K. A. G., *et al.* 2006, *ApJ*, 645, 1102
Massey, P. 1998, *ApJ*, 501, 153
Massey, P. & Olsen, K. A. G. 2003, *AJ*, 126, 2867
Massey, P., Olsen, K. A. G., Hodge, P. W., *et al.* 2006, *AJ*, 131, 2478
Massey, P., Olsen, K. A. G., Hodge, P. W., *et al.* 2007, *AJ*, 133, 2393
Richards, G. T., Hall, P. B., Vanden Berk, D. E., *et al.* 2003, *AJ*, 126, 1131
Taam, R. E., Bodenheimer, P., & Ostriker, J. P. 1978, *ApJ*, 222, 269
&Thorne, K. S., Żytkow, A. N. 1975, *ApJL*, 199, L19
Vanden Berk, D. E., Richards, G. T., Bauer, A., *et al.* 2001, *AJ*, 122, 549
van Loon, J. T., Cioni, M.-R. L., Zijlstra, A. A., & Loup, C. 2005, *A&A*, 438, 273
Vilardell, F., Ribas, I., & Jordi, C. 2006, *A&A*, 459, 321
Villar, V. A., Berger, E., Chornock, R., *et al.* 2016, *ApJ*, 830, 11
Wang, S., Ma, J., Wu, Z., & Zhou, X. 2014, *AJ*, 148, 4
West, A. A., Morgan, D. P., Bochanski, J. J., *et al.* 2011, *AJ*, 141, 97
York, D. G., Adelman, J., Anderson, J. E., Jr., *et al.* 2000, *AJ*, 120, 1579

Posters

The Lives and Death-Throes of Massive Stars
Proceedings IAU Symposium No. 329, 2016
J.J. Eldridge, J.C. Bray, L.A.S. McClelland & L. Xiao, eds.

doi:10.1017/S1743921317001417

Stellar and wind parameters of massive stars from spectral analysis

Ignacio Araya and Michel Curé

Instituto de Física y Astronomía, Universidad de Valparaíso, Chile
email: ignacio.araya@uv.cl

Abstract. The only way to deduce information from stars is to decode the radiation it emits in an appropriate way. Spectroscopy can solve this and derive many properties of stars. In this work we seek to derive simultaneously the stellar and wind characteristics of a wide range of massive stars. Our stellar properties encompass the effective temperature, the surface gravity, the stellar radius, the micro-turbulence velocity, the rotational velocity and the Si abundance. For wind properties we consider the mass-loss rate, the terminal velocity and the line–force parameters α, k and δ (from the line–driven wind theory). To model the data we use the radiative transport code FASTWIND considering the newest hydrodynamical solutions derived with HYDWIND code, which needs stellar and line–force parameters to obtain a wind solution. A grid of spectral models of massive stars is created and together with the observed spectra their physical properties are determined through spectral line fittings. These fittings provide an estimation about the line–force parameters, whose theoretical calculations are extremely complex. Furthermore, we expect to confirm that the hydrodynamical solutions obtained with a value of δ slightly larger than ~ 0.25, called δ-slow solutions, describe quite reliable the radiation line-driven winds of A and late B supergiant stars and at the same time explain disagreements between observational data and theoretical models for the Wind–Momentum Luminosity Relationship (WLR).

Keywords. stars: early-type, stars: atmospheres, stars: fundamental parameters, stars: winds, outflows

---

Grid of Models

To produce the grid of synthetic line profiles with the code FASTWIND (Puls *et al.* 2005), we first compute the grid of hydrodynamic wind solutions with the stationary code HYDWIND (Curé 2004). These hydrodynamic solutions are calculated with the purpose to obtain the mass loss rate and the terminal velocity from the wind. Only the 30% of our initial grid of hydrodynamic models obtained a physical wind solution. In the case of the FASTWIND model grid, we obtained about 400 000 models. From these models the line profiles of the H, He, and Si elements are obtained in the optical range. Currently, we are developing the tool for the automatic analysis of an observed spectrum in order to derive their stellar and wind parameters. One of the expected applications of our tool has focus on the winds of A and late B supergiant stars. Our purpose try to explain disagreements between observational data and theoretical models for the Wind–Momentum Luminosity Relationship (WLR) utilizing the δ-slow solution (Curé *et al.* 2011).

References

Curé, M. 2004, *ApJ*, 614, 929
Curé, M., Cidale, L., & Granada, A. 2011, *ApJ*, 737, 18
Puls, J., Urbaneja, M. A., Venero, R., Repolust, T., Springmann, U., Jokuthy, A., & Mokiem, M. R. 2005, *A&A*, 435, 669

The Lives and Death-Throes of Massive Stars
Proceedings IAU Symposium No. 329, 2016
J.J. Eldridge, J.C. Bray, L.A.S. McClelland & L. Xiao, eds.

doi:10.1017/S1743921317001600

Mass-ejection events in Be stars triggered by coupled nonradial pulsation modes

D. Baade[1], Th. Rivinius[2], A. Pigulski[3], A. Carciofi[4] and BEST[5]

[1]ESO, Karl-Schwarzschild-Str. 2, 85748 Garching, Germany (email: dbaade@eso.org)
[2]ESO, Casilla 19001, Santiago 19, Chile
[3]Instytut Astronomiczny, Uniwersytet Wrocławski, Wrocław, Poland)
[4]Instituto de Astronomia, Geofísica e Ciências Atmosféricas, Universidade de São Paulo, Brazil
[5]BRITE Executive Science Team

Abstract. Be stars (for an in-depth review see Rivinius, Carciofi & Martayan 2013) rotate at $\geqslant 80\%$ of the critical velocity and are multi-mode nonradial pulsators. Magnetic dipole fields are not detected, and binaries with periods less than 30 days are rare. The name-giving emission lines form in a Keplerian decretion disk, which is viscously re-accreted and also radiatively ablated unless replenished by outburts of unknown origin.

Months-long, high-cadence space photometry with the BRITE-Constellation nanosatellites (Pablo *et al.* 2016) of about 10 early-type Be stars reveals the following (cf. Baade *et al.* 2016a, Baade *et al.* 2016b):

○ Many Be stars exhibit 1 or 2 so-called Δ frequencies, which are differences between two nonradial-pulsation (NRP) frequencies and much lower (mostly less than 0.1 c/d) than the parent frequencies. The associated light curves are roughly sinusoidal. The amplitudes can exceed that of the sum of the parent amplitudes.
○ Conventional beat patterns also occur.
○ Amplitudes of both Δ and beat frequencies can temporarily be enhanced. Around phases of maximal amplitude the mean brightness is in- or decreased, and the scatter can be enhanced.
○ During high-activity phases (outbursts), broad and dense groups of numerous spikes arise in the power spectra. The two strongest groups often have a frequency ratio near 2. The phase coherence seems to be low.
○ Time coverage (less than half a year) is not yet sufficient to infer whether two Δ or beat frequencies can combine to cause long-lasting (years) superoutbursts (cf. Carciofi *et al.* 2012).

From these observations it is concluded:

- The variable mean brightness and the increased Δ-frequency amplitude and scatter trace the amount of near-circumstellar matter.
- Increase or decrease of mean brightness is aspect-angle dependent (pole-on vs. equator-on).
- Increased amounts of near-circumstellar matter are due to rotation-assisted mass ejections caused by coupled NRP modes.
- Observations do not constrain the location of the coupling (atmosphere or stellar interior).
- Broad frequency groups do not represent stellar pulsation modes but circumstellar variability.
- Be stars later than B5 are less active and may in some cases even behave differently.

Keywords. stars: emission-line, Be; stars: mass loss; stars: oscillations

---

References

Baade, D., Rivinius, Th., Pigulski, A., Carciofi, A., & BEST 2016a, *arXiv*1610.02200
Baade, D., Rivinius, Th., Pigulski, A., Carciofi, A., Handler, G., *et al.* 2016b, *arXiv*1611.01113
Carciofi, A., Bjorkman, J. E., Otero, S., Okazaki, A., Štefl, S., *et al.* *ApJ*, 744, L15
Pablo, H., Whittaker, G. N., Popowicz, A., Mochnacki, S. M., *et al.* 2016, *PASP*, 128, 125001
Rivinius, Th., Carciofi, A. & Martayan, Ch. 2013, *A&ARv*, 21, 69

The Lives and Death-Throes of Massive Stars
Proceedings IAU Symposium No. 329, 2016
J.J. Eldridge, J.C. Bray, L.A.S. McClelland & L. Xiao, eds.

doi:10.1017/S1743921317002526

Observations of Bright Massive Stars Using Small Size Telescopes

Sopia Beradze and Nino Kochiashvili
E. Kharadze Abastumani Astrophysical Observatory, Ilia State University
email: sopia.beradze.1@iliauni.edu.ge

Abstract. The size of a telescope determines goals and objects of observations. During the latest decades it becomes more and more difficult to get photometric data of bright stars because most of telescopes of small sizes do not operate already. But there are rather interesting questions connected to the properties and evolution ties between different types of massive stars. Multi-wavelength photometric data are needed for solution of some of them. We are presenting our observational plans of bright Massive X-ray binaries, WR and LBV stars using a small size telescope.

All these stars, which are presented in the poster are observational targets of Sopia Beradze's future PhD thesis. We already have got very interesting results on the reddening and possible future eruption of the massive hypergiant star P Cygni. Therefore, we decided to choose some additional interesting massive stars of different type for future observations. All Massive stars play an important role in the chemical evolution of galaxies because of they have very high mass loss - up to $10^{-4}M_{\odot}$/a year. Our targets are on different evolutionary stages and three of them are the members of massive binaries. We plan to do UBVRI photometric observations of these stars using the 48 cm Cassegrain telescope of the Abastumani Astrophisical Observatory.

Keywords. X-ray Binaries, WR Stars, LBV, Multi-Wavelength Photometry, Small Size Telescopes

---

1. Acknowledgemets

The authors are grateful to E.Kharadze Abastumani Astrophysical Observatory, Ilia State University, Georgia, for funding our travel to New Zealand. N.K. thanks conference organizers for covering part of her local expences at the conference.

The Lives and Death-Throes of Massive Stars
Proceedings IAU Symposium No. 329, 2016
J.J. Eldridge, J.C. Bray, L.A.S. McClelland & L. Xiao, eds.

doi:10.1017/S1743921317001880

Searching for self-enrichment in Cygnus OB2

Sara R. Berlanas[1,2], Artemio Herrero[1,2], Fernando Comerón[3], Anna Pasquali[4] and Sergio Simón-Díaz[1,2]

[1]Instituto de Astrofísica de Canarias, 38205 La Laguna, Tenerife, Spain

[2]Departamento de Astrofísica, Universidad de La Laguna, 38206 La Laguna, Tenerife, Spain

[3]ESO, Karl-Schwarzschild-Strasse 2, 85748 Garching bei München, Germany

[4] Astronomisches Rechen-Institut, Zentrum fr Astronomie der Universität Heidelberg, Mönchhofstr 1214, 69120 Heidelberg, Germany

Cygnus OB2 is a rich and relatively close (d$\sim$1.4 kpc) OB association in our Galaxy. It represents an ideal testbed for our theories about self-enrichment processes produced by pollution of the interstellar medium by successive generations of massive stars. Comerón & Pasquali (2012, A&A, 543, A101) found a correlation between the age of young stellar groups in Cygnus OB2 and their Galactic longitude. If is associated with a chemical composition gradient, it could support these self-enrichment processes.

In this work we have checked whether there is a correlation between abundances (as derived from OB-type stars) and star location in Cygnus OB2. For this purpose, we have performed a spectroscopic analysis of 7 OB stars with low *vsini* in the association that a) are suitable for the spectroscopic analysis and b) cover the positions of all the age groups in the area. We have used FASTWIND stellar atmosphere models to determine their fundamental parameters as well as the Si and O surface abundances.

We have carried out an analysis based on equivalent widths (EW) of metal lines, similar to the method used in Orion by Simón-Díaz (2010, A&A, 510, A22). The stellar parameters (Teff and logg) have been obtained from observed H Balmer lines and the EW ratios of Si II-III-IV. Then, we have applied the curve-of-growth method to derive the associated Si and O abundances.

Our abundance analysis indicates a potential correlation between O abundances and Galactic longitudes (see Fig.1), which we do not find for Si. We derive a total variation of the O abundance of about 0.3-0.5 dex, larger than intrinsic uncertainties. This correlation, together with the one found by Comerón & Pasquali (2012), implies a possible evidence of self-enrichment in the region, taking into account that Cyg OB2 has an extended star-formation history sufficiently long to have included the explosion of its oldest massive members as core-collapse supernovae. Therefore, the young generations of stars could have been enriched with the products from supernovae ejecta from older subgroups.

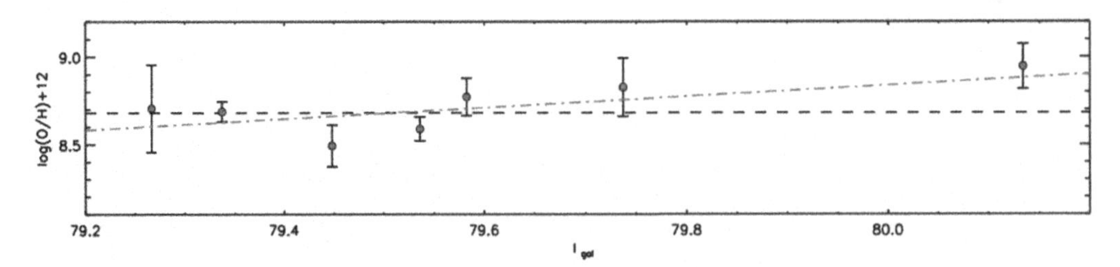

Figure 1. Derived O abundances vs Galactic longitude of our sample of OB stars with low *vsini*. Black dashed line represents the mean abundance value, which is used as a reference to correlate the abundances. Dash-dotted green line represents the fit to the data.

The Lives and Death-Throes of Massive Stars
Proceedings IAU Symposium No. 329, 2016
J.J. Eldridge, J.C. Bray, L.A.S. McClelland & L. Xiao, eds.

doi:10.1017/S1743921317002095

Is the Link Between the Observed Velocities of Neutron Stars and their Progenitors a Simple Mass Relationship?

J. C. Bray

Dept. of Physics, University of Auckland,
Private Bag 92019, Auckland, New Zealand
email: john.bray@auckland.ac.nz

While the imparting of velocity 'kicks' to compact remnants from supernovae is widely accepted, the relationship of the 'kick' to the progenitor is not. We propose the 'kick' is predominantly a result of conservation of momentum between the ejected and compact remnant masses. We propose the 'kick' velocity is given by $v_{\rm kick} = \alpha\,(M_{\rm ejecta}/M_{\rm remnant}) + \beta$, where α and β are constants we wish to determine. To test this we use the BPASS v2 (Binary Population and Spectral Synthesis) code to create stellar populations from both single star and binary star evolutionary pathways. We then use our Remnant Ejecta and Progenitor Explosion Relationship (REAPER) code to apply 'kicks' to neutron stars from supernovae in these models using a grid of α and β values, (from 0 to 200 km s^{-1} in steps of 10 km s^{-1}), in three different 'kick' orientations, (isotropic, spin-axis aligned and orthogonal to spin-axis) and weighted by three different Salpeter initial mass functions (IMF's), with slopes of -2.0, -2.35 and -2.70. We compare our synthetic 2D and 3D velocity probability distributions to the distributions provided by Hobbs *et al.* (1995).

Neutron stars from supernovae in single stars are expected to produce a uni-modal velocity distribution while in binaries a bi-modal distribution is expected. This is because merged systems will have higher ejecta masses then non-merged systems which can experience mass stripping by the companion. We test the fit of the uni- and bi-modal forms of the Gaussian, Maxwell-Boltzmann and lognormal distributions using the maximum likelihood estimator (MLE), and a grid of each of the distribution variables. Our results show a clear preference for a bi-modal distribution suggesting binary evolution for the majority of neutron stars.

While single star models provide a poor fit to both the 2D and 3D distributions, binary models with values of $\alpha = 70\,{\rm km\,s^{-1}}$ and $\beta = 110\,{\rm km\,s^{-1}}$ reproduce the observed 2D velocities and $\alpha = 70\,{\rm km\,s^{-1}}$ and $\beta = 120\,{\rm km\,s^{-1}}$ reproduce their inferred 3D velocity distribution. We find no statistically significant preference for any of the 'kick' orientations, or for any of the three different IMF's.

We use the α and β values identified above, and create synthetic runaway star velocity and NS-NS binary system period probability distributions. These too we find to agree well with the observational data of Tetzlaff *et al.* (2014) and Andrews *et al.* (2015) respectively.

References

Andrews, J., Farr, W., Kalogera, V., & Willems, B. 2015, *ApJ*, 801, 32
Hobbs, G., Lorimer, D., Lyne, A., & Kramer, M. 2005, *MNRAS*, 360, 974
Tetzlaff, N., Torres, G., Bieryla, A., & Neuhauser, R. 2014, *Astronomische Nachrichten*, 335, 981

The Lives and Death-Throes of Massive Stars
Proceedings IAU Symposium No. 329, 2016
J.J. Eldridge, J.C. Bray, L.A.S. McClelland & L. Xiao, eds.

doi:10.1017/S1743921317001570

Detectability of Wolf-Rayet stars in M33 and Beyond the Local Group

Aaron J. Brocklebank, J. L. Pledger and A. E. Sansom
Jeremiah Horrocks Institute, University of Central Lancashire, Preston PR1 2HE, UK
email: ajbrocklebank1@uclan.ac.uk

Abstract. To understand how complete our surveys of Wolf-Rayet (WR) stars can be with the current generation of telescopes, we study images of M33, a galaxy with a nearly complete WR catalogue, and degrade them to investigate the detectability of WRs out to 30Mpc. We lose almost half of our sample at 4.2Mpc, and at 30Mpc we detect only those WRs in bright regions.

Keywords. stars: Wolf-Rayet, stars: statistics, galaxies: stellar content, galaxies: individual: M33

Wolf-Rayet stars (WRs) are expected to be the progenitors of H and H+He deficient Type Ib and Ic supernovae (SNe). With the aim of directly detecting a SNe Ib/c progenitor, as done with the red supergiant progenitors of Type II-P SNe, surveys of WR stars in nearby galaxies are crucial, some of which have been done already. However, there is an uncertainty as to how complete our surveys beyond the Local Group can be with the current generation of telescopes.

In this study we mimicked observations of WR stars in face-on spiral galaxies at different distances based on a deep WR survey of M33 (Neugent & Massey 2011). We investigated how many WR stars were able to be detected at distances out to 30 Mpc. We also looked specifically at the fraction of WC and WN stars recovered to compare their detectability.

The images were blurred then binned (using IRAF) in order to resemble both the seeing effects and CCD sampling of observations at each mimicked distance (4.2, 8.4, 16.8 and 30.2Mpc). Standard photometric techniques were then used to see whether the known WR stars were found to have a helium excess (see Neugent & Massey 2011 for a description of the filters used).

At 4.2Mpc, only 56% of WRs are found. At 16.8Mpc, only 23% of WRs are recovered and 17% at 30.2Mpc, most of which are in dense, bright stellar regions. Beyond $\sim$10Mpc, the WC stars become notably more difficult to detect than the WNs (19% and 11% recovery respectively at 30.2Mpc). Using the Fruchter *et al.* (2006) method to compare the brightness at the location of the WRs with that of the whole galaxy, the distribution shifts to lower brightness and appears to almost follow the galaxy light. At 16.8Mpc the distribution differs significantly (3.9σ) to that of the original WRs.

References

Fruchter, A. S., Levan, A. J., Strolger, L., *et al.* 2006, *Nature*, 441, 463
Neugent, K. & Massey, P. 2011, *ApJ*, 733, 123

The Lives and Death-Throes of Massive Stars
Proceedings IAU Symposium No. 329, 2016
J.J. Eldridge, J.C. Bray, L.A.S. McClelland & L. Xiao, eds.

doi:10.1017/S1743921317002472

B-supergiants in IC1613: testing low-Z massive star physics and evolution

Inés Camacho[1,2], Miriam García[3], Miguel A. Urbaneja[4] and Artemio Herrero[1,2]

[1]Instituto de Astrofísica de Canarias, La Laguna, Tenerife, Spain
[2]Departamento de Astrofísica, Universidad de La Laguna, La Laguna, Tenerife, Spain
[3]Centro de Astrobiología CAB (CSIC/INTA), Madrid, Spain
[4]Institute of Astro and Particle Physics, Universität Innsbruck, Innsbruck, Austria

Abstract. The physical processes taking place in massive stars during their life and death are highly dependent on the metallicity (Z) of their parent cloud. Observations of these stars in low-Z nearby galaxies are crucial to understand these processes. IC1613 is the nearest Local Group galaxy with ongoing star formation and O-abundance lower than the SMC, although UV spectroscopy suggests it is not so metal poor. We performed a spectral analysis of early B-type stars in the galaxy, obtaining physical parameters and abundances. Our results confirm the low O-abundance of IC1613.

Keywords. IC1613, massive stars, galaxy, abundances, metallicity

---

We have studied 11 early B-type stars in IC1613. We performed high quality spectral analysis of the data, using a non-classical automatic analysis method (GPMC, Urbaneja in prep.) and an optimized grid of 800 FASTWIND models. We obtained 11 physical parameters and abundances simultaneously (T_{eff}, $logg$, He, ξ, β, $\dot{M}$, O, N, C, Mg, and Si). From this analysis, we draw three main **conclusions**: **a)** We recover an average O-abundance of $12 + log(O/H) = 7.86 \pm 0.12$, in agreement with other stellar and nebular studies in the galaxy (as shown in Fig. 1-left), and confirm the low O-abundance of IC1613. **b)** The derived T_{eff} of our supergiant sample are consistent with the Markova & Puls (2008, A&A, 478, 823) T_{eff}-scale for Galactic B-type supergiants (See Fig.1-right). This rather suggests there is not a significant effect of metallicity on the effective temperature for B-supergiants. **c)** Comparing the resulting abundances with the predictions of evolutionary models (Brott *et al.*, 2011, A&A, 530, A115), the low metallicity seems to favor high initial rotational velocities, although present Vsini values are lower than predicted.

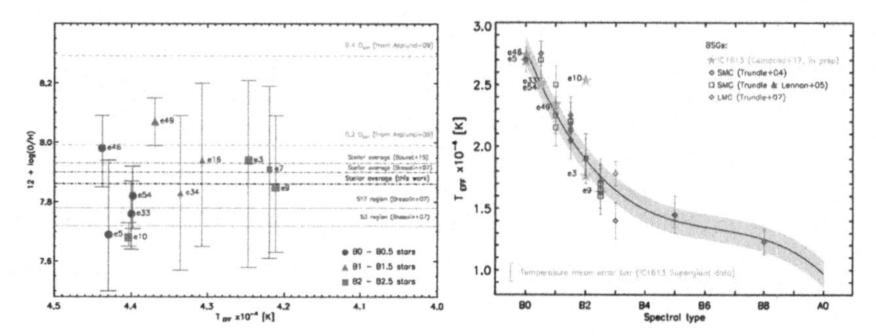

Figure 1. LEFT: Stellar O-abundances in IC1613 derived in this work. Other stellar and nebular studies are also included. **RIGHT:** Temperature scale for Galactic B-supergiants (B-SG) derived by Markova & Puls, together with the B-SG T_{eff}'s derived in this work (orange stars).

The Lives and Death-Throes of Massive Stars
Proceedings IAU Symposium No. 329, 2016
J.J. Eldridge, J.C. Bray, L.A.S. McClelland
& L. Xiao, eds.

doi:10.1017/S1743921317002836

HDUST3 — A chemically realistic, 3-D, NLTE radiative transfer code

Alex C. Carciofi[1], Jon E. Bjorkman[2] and Janos Zsargó[3]

[1]Instituto de Astronomia, Geofísica e Ciências Atmosféricas, Universidade de São Paulo, Rua do Matão 1226, Cidade Universitária, 05508-900 São Paulo, SP, Brazil
email: carciofi@usp.br

[2]Ritter Observatory, Dept. of Phys. & Astron., MS 113, Univ. of Toledo, Toledo, OH 43606

[3]Escuela Superior de Física y Matemática. Instituto Politécnico Nacional. Av. Instituto Politécnico Nacional, Edificio 9, C.P. 07738, DF. México

Abstract. HDUST is a 3D, NLTE radiative transfer code based on the Monte Carlo method. We report on recent advancements on the code, which is now capable of handling He and other elements in the NLTE regime and in 3D configurations. In this contribution we show initial comparisons with CMFGEN, made with spherical wind models composed of H + He.

Keywords. methods: numerical, radiative transfer

The HDUST code (Carciofi & Bjorkman 2006; Carciofi *et al.* 2004) has been used already in over 30 refereed publications, mostly for modeling Be star disks, but also other systems such as hot star winds and dusty envelopes around hot and cool stars. The current stable version includes H in full NLTE for 3D geometries, as well as dust and free electrons.

We present the new version of the code (HDUST3), modified to include a general atomic module, that allows for the inclusion of other chemical elements. An inicial comparison is shown in Fig. 1 for a model with $R_\star = 6.6R_{\rm sun}$, $T_{\rm eff} = 22\,000\,{\rm K}$, $\log g = 4$, $\dot{M} = 2 \times 10^{-10}\,M_{\rm sun}{\rm yr}^{-1}$, and $v_\infty = 1600\,{\rm kms}^{-1}$. The match for H I is quite good, but some discrepancies are found for He I, which are likely due to the fact that the inner boundary condition (namely the photospheric spectrum) is not the same in the two simulations.

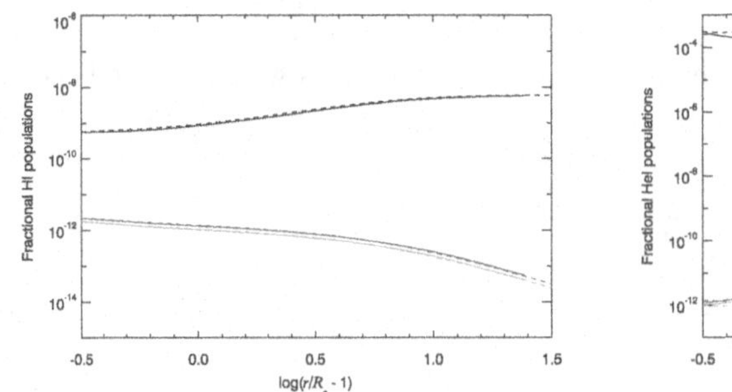

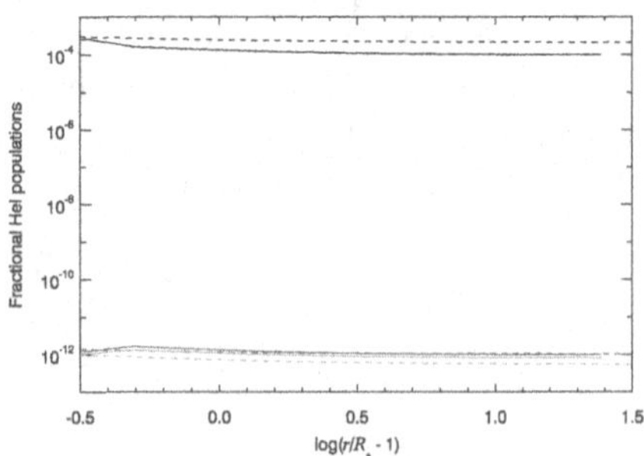

Figure 1. Fractional level populations vs. distance from the star ($n_{\rm level}/n_{\rm element}$). Black: n_1. Red: n_2. Green: n_3. Solid line: HDUST. Dashed line: CMFGEN. *Left:* H I. *Right:* He I.

References

Carciofi, A. C., Bjorkman, J. E., & Magalhães, A. M. 2004, *ApJ*, 604, 238
Carciofi, A. C., & Bjorkman, J. E. 2006, *ApJ*, 639, 1081

The Lives and Death-Throes of Massive Stars
Proceedings IAU Symposium No. 329, 2016
J.J. Eldridge, J.C. Bray, L.A.S. McClelland & L. Xiao, eds.

doi:10.1017/S174392131700120X

Wind-embedded shocks in FASTWIND: X-ray emission and K-shell absorption

L. P. Carneiro[1], J. Puls[1], J. O. Sundqvist[2] and T. L. Hoffmann[1]

[1]LMU Munich, Universitätssternwarte, Scheinerstr. 1, 81679 München, Germany
email: luiz@usm.uni-muenchen.de
[2]Instituut voor Sterrenkunde, KU Leuven, Celestijnenlaan 200D, 3001 Leuven, Belgium

Abstract. EUV and X-ray radiation emitted from wind-embedded shocks can affect the ionization balance in the outer atmospheres of massive stars, and can also be the mechanism responsible for producing highly ionized atoms detected in the wind UV spectra. To investigate these processes, we implemented the emission from wind-embedded shocks and related physics into our atmosphere/spectrum synthesis code FASTWIND. We also account for the high energy absorption of the cool wind, by adding important K-shell opacities. Various tests justfying our approach have been described by Carneiro+(2016, A&A 590, A88).

In particular, we studied the impact of X-ray emission on the ionization balance of important elements. In almost all the cases, the lower ionization stages (O IV, N IV, P V) are depleted and the higher stages (N V, O V, O VI) become enhanced. Moreover, also He lines (in particular He II 1640 and He II 4686) can be affected as well.

Finally, we carried out an extensive discussion of the high-energy mass absorption coefficient, κ_ν, regarding its spatial variation and dependence on $T_{\rm eff}$. We found that (i) the approximation of a radially constant κ_ν can be justified for $r \geqslant 1.2R_*$ and $\lambda \leqslant 18$ Å, and also for many models at longer wavelengths. (ii) In order to estimate the actual value of this quantity, however, the He II background needs to be considered from detailed modeling.

Keywords. methods: numerical, stars: atmospheres, stars: early-type, stars: mass loss

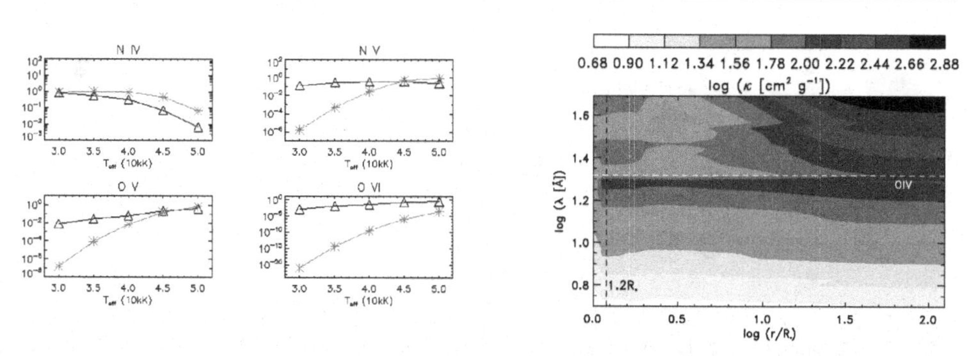

Figure 1. Left: Ionization fractions of important ions at $v(r) = 0.5\ v_\infty$, as a function of $T_{\rm eff}$, for supergiant models with typical X-ray emission (black triangles), and without X-rays (pink asterisks). The figure demonstrates the importance of including the shock emition to obtain a proper description of ions like N V, O V, O VI, particularly in 'cooler' objects.
Right: Contour plot of the radial dependence of the mass absorption coefficient in a supergiant model at $T_{\rm eff}$=40kK, as a function of wavelength. We note that κ_ν(r) becomes constant when r $\geqslant 1.2R_*$ and $\lambda \leqslant 20$Å (log $\lambda \leqslant 1.3$ ~O IV edge), to be on the safe side. In most cases, the quoted radial limit, arising from fluctuations in the background opacity, is even lower. *L.P.C. gratefully acknowledges a grant by the International Astronomical Union. This work has been supported by the Brazilian Coordination for the improvement of Higher Education Personnel* (CAPES).

The Lives and Death-Throes of Massive Stars
Proceedings IAU Symposium No. 329, 2016
J.J. Eldridge, J.C. Bray, L.A.S. McClelland & L. Xiao, eds.

doi:10.1017/S1743921317002538

Red supergiant stars in NGC 4449, NGC 5055, and NGC 5457

Sang-Hyun Chun[1,2], Yong-Jong Sohn[3], Martin Asplund[2] and Luca Casagrande[2]

[1]Astronomy Program, Department of Physics and Astronomy, Seoul National University
email: shyunc.m@gmail.com

[2]Research School of Astronomy & Astrophysics, Australian National University
[3]Department of Astronomy, Yonsei University

Abstract. Nearby galaxies are ideal objects for the study of the mechanisms of galaxy formation and evolution, and massive stars in nearby galaxies are useful sources to investigate the structures and formation of the galaxies. It is important to gather the contents of massive stars for a number of galaxies spanning various metallicities. We focus on the red supergiants (RSGs) in nearby galaxies NGC 4449, NGC 5055, and NGC 5457, and the photometric properties of RSGs of three galaxies were investigated using near-infrared (JHK) imaging data obtained from WFCAM UKIRT. The $(J-K, K)_0$ CMDs are investigated and compared with theoretical isochrones (Figure 1). The majority of RSGs in three galaxies have common age ranges from $\log(t_{yr}) = 6.9$ to $\log(t_{yr}) = 7.3$, and this indicates that these galaxies have experienced recent star formation within 20 Myr. Spatial correlation of RSGs with H II regions and their colour distribution were also investigated. For NGC 4449 and NGC 5457, the RSGs are spatially correlated with the H II regions, which however is not the case for NGC 5055. We found a similar colour distribution and a constant peak magnitude of $M_K = -11.9$ for the RSGs in the three galaxies.

Keywords. (stars:) supergiants, Galaxy: stellar content, infrared: stars.

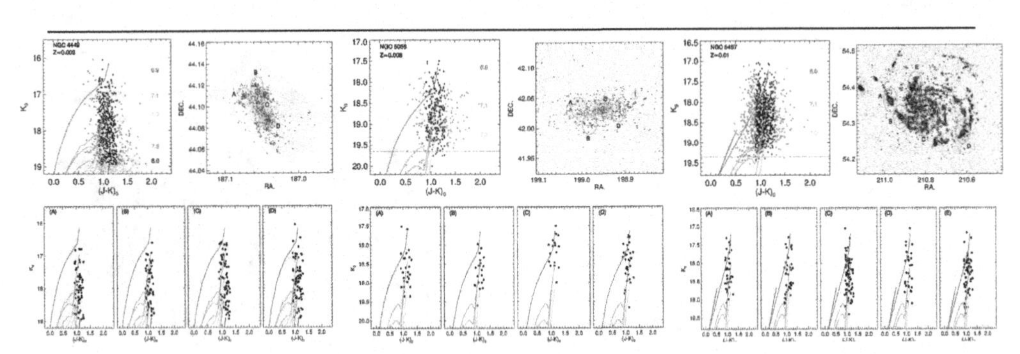

Figure 1. Top-left panels : $(J-K, K)_0$ CMDs of RSGs in the galaxies. The isochrones of PARSEC (Bressan *et al.* 2012) with several ages are plotted with different colours. Black and blue dots are the genuine and possible RSGs, respectively. The ages of $\log(t_{yr})$ are indicated in the panels. Top-right panels: sky distributions on $H\alpha$ images. The dominant HII regions were selected as circles. Bottom panels: $(J-K, K)_0$ CMDs of RSGs in selected HII regions for three galaxies.The label in each panel corresponds to the selected H II region. The theoretical isochrones are the same as those of top-left panel.

Reference

Bressan, A., Marigo, P., Girardi, L., Salasnich, B., Dal Cero, C., Rubele, S., & Nanni A. 2012, *MNRAS*, 427, 127

The Lives and Death-Throes of Massive Stars
Proceedings IAU Symposium No. 329, 2016
J.J. Eldridge, J.C. Bray, L.A.S. McClelland
& L. Xiao, eds.

doi:10.1017/S1743921317001132

Disentangling rotational velocity distribution of stars

Michel Curé[1], Diego F. Rial[2], Julia Cassetti[3] and Alejandra Christen[4]

[1]Universidad de Valparaíso, Chile
[2]Universidad de Buenos Aires, Argentina
[3]Universidad Nacional de General Sarmiento, Argentina
[4]Pontificia Universidad Católica de Valparaíso, Chile

Abstract. Rotational speed is an important physical parameter of stars: knowing the distribution of stellar rotational velocities is essential for understanding stellar evolution. However, rotational speed cannot be measured directly and is instead the convolution between the rotational speed and the sine of the inclination angle $v \sin(i)$. The problem itself can be described via a Fredhoml integral of the first kind. A new method (Curé *et al.* 2014) to deconvolve this inverse problem and obtain the cumulative distribution function for stellar rotational velocities is based on the work of Chandrasekhar & Münch (1950). Another method to obtain the probability distribution function is Tikhonov regularization method (Christen *et al.* 2016). The proposed methods can be also applied to the mass ratio distribution of extrasolar planets and brown dwarfs (in binary systems, Curé *et al.* 2015).

For stars in a cluster, where all members are gravitationally bounded, the standard assumption that rotational axes are uniform distributed over the sphere is questionable. On the basis of the proposed techniques a simple approach to model this anisotropy of rotational axes has been developed with the possibility to "disentangling" simultaneously both the rotational speed distribution and the orientation of rotational axes.

Keywords. methods: analytical - methods: data analysis - methods: numerical - methods: statistical - stars: fundamental- parameters - stars: rotation

Results

We have assumend tha the rotational velocity distribution is given by a Maxwellian distribution and integrate analytically the Fredhoml integral. The rotational data have been selected for twelve clusters from the survey of radial and rotational velocities performed by Mermilliod *et al.* (2009). We fitted (1000 bootstraps) to the distribution of projected velocity with our formula (from Fedhom integral) for a single Maxwellian distribution, but also for a double Maxwellian distribution.

In all cases we find that the distribution of rotational axes is not randomly oriented. The results also show that a mixture of two Maxwellian distribution fits better the observed projected velocity sample. This might indicate the existence of two stellar populations.

References

Chandrasekhar, S. & Münch, G. 1950, *ApJ*, 111, 142
Christen, A, Escarate, P., Curé, M., Rial, D. F., & Cassetti, J. 2016, *A&A* , 595, 50
Curé, M., Rial, D. F., Christen, A., & Cassetti, J. 2014, *A&A* , 565, 85
Curé, M., Rial, D. F., Cassetti, J., Christen, A., & Boffin, H. M. J. 2015, *A&A* , 573, 86
Mermilliod, J.-C., Mayor, M., & Udry, S., 2009, *A&A*, 498, 949

The Lives and Death-Throes of Massive Stars
Proceedings IAU Symposium No. 329, 2016
J.J. Eldridge, J.C. Bray, L.A.S. McClelland & L. Xiao, eds.

doi:10.1017/S1743921317002903

A multi-wavelength view of NGC 1624-2

A. David-Uraz[1,2], V. Petit[1,2], R. MacInnis[1], C. Erba[1], S. P. Owocki[1], A. W. Fullerton[3], N. R. Walborn[3] and D. H. Cohen[4]

[1]Department of Physics and Space Sciences, Florida Institute of Technology, Melbourne, FL 32904, USA

[2]Department of Physics and Astronomy, Bartol Research Institute, University of Delaware, Newark, DE 19716, USA

[3]Space Telescope Science Institute, 3700 San Martin Drive, Baltimore, MD 21218, USA

[4]Department of Physics and Astronomy, Swarthmore College, 500 College Ave., Swarthmore, PA 19081, USA

Abstract. Large magnetometric surveys have contributed to the detection of an increasing number of magnetic massive stars, and to the recognition of a population of magnetic massive stellar objects with distinct properties. Among these, NGC 1624-2 possesses the largest magnetic field of any O-type star; such a field confines the stellar wind into a circumstellar magnetosphere, which can be probed using observations at different wavelength regimes. Recent optical and X-ray observations suggest that NGC 1624-2's magnetosphere is much larger than that of any other magnetic O star. By modeling the variations of UV resonance lines, we can constrain its velocity structure. Furthermore, recent spectropolarimetric observations raise the possibility of a more complex field topology than previously expected. Putting all of these multi-wavelength constraints together will allow us to paint a consistent picture of NGC 1624-2 and its surprising behavior, giving us valuable insight into the very nature of massive star magnetospheres.

Keywords. stars: mass loss, stars: magnetic fields, ultraviolet: stars, X-rays: stars

---

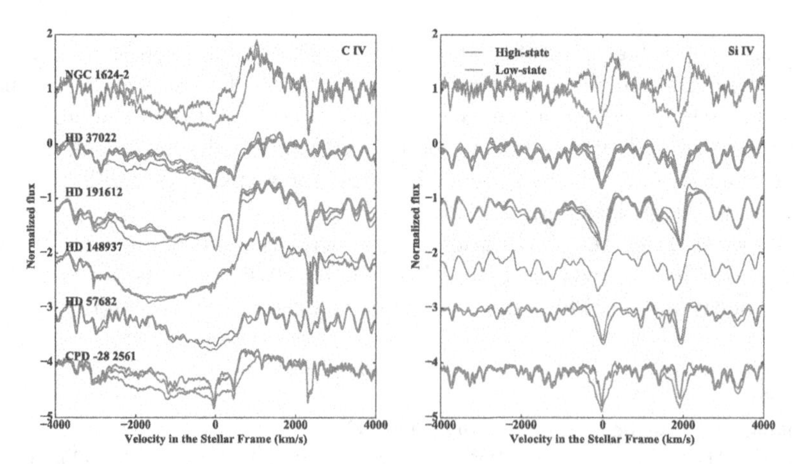

Figure 1. HST/STIS observations of NGC 1624-2 show a remarkable variation of the UV resonance lines between the pole-on and equator-on views (top line); as a means of comparison, high/low state spectra of other magnetic O stars are shown below (David-Uraz *et al.*, in prep.). Ongoing efforts (Erba *et al.*, these proceedings) aim to account for these changes using the Analytical Dynamical Magnetosphere model (ADM; Owocki *et al.*, 2016, *MNRAS*, 462, 3830) rather than full-scale 3D MHD simulations.

The Lives and Death-Throes of Massive Stars
Proceedings IAU Symposium No. 329, 2016
J.J. Eldridge, J.C. Bray, L.A.S. McClelland & L. Xiao, eds.

doi:10.1017/S174392131700206X

Testing the Wind-shock Paradigm for B-Type Star X-Ray Production with θ Car

T. F. Doyle[1], V. Petit[1], D. Cohen[2] and M. Leutenegger[3,4]

[1]Dept. of Physics & Space Sciences, Florida Institute of Technology
Melbourne, FL 32901
email: trisha.mizusawa@gmail.com

[2]Dept. of Physics & Astronomy, Swarthmore College
Swarthmore, PA 19081

[3]NASA/Goddard Space Flight Center
Baltimore, MD 20771

[4]University of Maryland, Baltimore County/CRESST
Baltimore, MD 21250

Abstract. We present Chandra X-ray grating spectroscopy of the B0.2V star, θ Carina. θ Car is in a critical transition region between the latest O-type and earliest B-type stars, where some stars are observed to have UV-determined wind densities much lower than theoretically expected (e.g., Marcolino *et al.* 2009). In general, X-ray emission in this low-density wind regime should be less prominent than for O-stars (e.g., Martins *et al.* 2005), but observations suggest a higher than expected X-ray emission filling factor (Lucy 2012; Huenemoerder *et al.* 2012); if a larger fraction of the wind is shock-heated, it could explain the weak UV wind signature seen in weak wind stars, but this might severely challenge predictions of radiatively-driven wind theory.

We measured the line widths of several He-, H-like and Fe ions and the f/i ratio of He-like ions in the X-ray spectrum, which improves upon the results from Nazé *et al.* (2008) (*XMM-Newton* RGS) with additional measurements (*Chandra* HETG) of MgXI and SiXIII by further constraining the X-ray emission location. The f/i ratio is modified by the proximity to the UV-emitting stellar photosphere, and is therefore a diagnostic of the radial location of the X-ray emitting plasma. The measured widths of X-ray lines are narrow, <300 km s^{-1} and the f/i ratios place the X-rays relatively close to the surface, both implying θ Car is a weak wind star. The measured widths are also consistent with other later-type stars in the weak wind regime, β Cru (Cohen *et al.* 2008), for example, and are smaller on average than earlier weak wind stars such as μ Col (Huenemoerder *et al.* 2012). This could point to a spectral type divide, where one hypothesis, low density, works for early-B type stars and the other hypothesis, a larger fraction of shock-heated gas, explains weak winds in late-O type stars. Archival IUE data still needs to be analyzed to determine the mass loss rate and hydrodynamical simulations will be compared with observations to determine which hypothesis works for θ Car.

Keywords. line: profiles, techniques: spectroscopic, stars: winds, outflows, X-rays: stars

References

Cohen, D. *et al.* 2008, *MNRAS*, 386, 1855
Huenemoerder, D. *et al.* 2012, *ApJL*, 756, L34
Lucy, L. B. 2012, *A&A*, 544, A120
Marcolino, W. L. F. *et al.* 2009, *A&A*, 498, 837
Martins, F. *et al.* 2005, *A&A*, 441, 735
Nazé, Y. *et al.* 2008, *A&A*, 490, 801

The Lives and Death-Throes of Massive Stars
Proceedings IAU Symposium No. 329, 2016
J.J. Eldridge, J.C. Bray, L.A.S. McClelland
& L. Xiao, eds.

doi:10.1017/S1743921317003192

Binary Population and Spectral Synthesis

J. J. Eldridge[1]†, E. R. Stanway[2], L. Xiao[1], L. A. S. McClelland[1], J. C. Bray[1], G. Taylor[1] and M. Ng[1]

[1]Department of Physics, University of Auckland, Private Bag 92019, Auckland, New Zealand
[2]Department of Physics, University of Warwick, Gibbet Hill Road, Coventry, CV4 7AL, UK

Abstract. We have recently released version 2.0 of the Binary Population and Spectral Synthesis (BPASS) population synthesis code. This is designed to construct the spectra and related properties of stellar populations built from ~200,000 detailed, individual stellar models of known age and metallicity. The output products enable a broad range of theoretical predictions for individual stars, binaries, resolved and unresolved stellar populations, supernovae and their progenitors, and compact remnant mergers. Here we summarise key applications that demonstrate that binary populations typically reproduce observations better than single star models.

Keywords. methods: numerical, stars: evolution, stars: statistics, galaxies: stellar content

---

BPASS stellar models were first introduced in 2009 and publically released in 2012. Our v2.0 release in mid-2015 was the most complete and widely-usable to date. This has been reflected in refereed publications using the models in a variety of contexts, including:

Stellar and Transient:

(i) Rates and lightcurves of electron-capture supernovae (Moriya & Eldridge 2016)
(ii) The progenitors of Type Ib SNe (Eldridge & Maund 2016)
(iii) Rates of mergers of compact stellar remnants and their role as gravitational wave transient progenitors (Eldridge & Stanway 2016)
(iv) Evolutionary channels for WR stars in the SMC (Shenar *et al.* 2016)

Extragalactic and Cosmological:

(i) Ionizing photon production at low metallicities and implications for the epoch of reionization (e.g. Stanway, Eldridge & Becker 2016)
(ii) Fitting the integrated light and photometry of young stellar clusters in nearby galaxies (e.g. Wofford *et al.* 2016)
(iii) Stellar populations in galaxies in the distant Universe (e.g. Strom *et al.* 2017)

Other results are discussed in the articles by the co-authors elsewhere this proceedings. We are currently preparing a methodology and verification paper for BPASS (Eldridge *et al.*, in prep), which will be released together with v2.1 of the models. BPASS and its outputs can be found at http://bpass.auckland.ac.nz.

References

Eldridge J. J., Stanway E. R., 2016, MNRAS, 462, 3302
Eldridge J. J., Maund J. R., 2016, MNRAS, 461, L117
Moriya T. J., Eldridge J. J., 2016, MNRAS, 461, 2155
Shenar T., *et al.*, 2016, A&A, 591, A22
Stanway, E. R., Eldridge, J. J., & Becker, G. D. 2016, MNRAS, 456, 485
Strom A. L., *et al.*, ApJ, 836, 164
Wofford A., *et al.*, 2016, MNRAS, 457, 4296

† email: j.eldridge@auckland.ac.nz

The Lives and Death-Throes of Massive Stars
Proceedings IAU Symposium No. 329, 2016
J.J. Eldridge, J.C. Bray, L.A.S. McClelland & L. Xiao, eds.

doi:10.1017/S1743921317003106

Analysis and fit of stellar spectra using a mega-database of CMFGEN models

Celia Fierro-Santillán[1], Janos Zsargó[2], Jaime Klapp[1,3], Santiago Alfredo Díaz-Azuara[4], Anabel Arrieta[5] and Lorena Arias[5]

[1]Centro de Matemática Aplicada y Cómputo de Alto Rendimiento ABACUS-CINVESTAV, Carretera México-Toluca km 38.5, 52740 Estado de México, México.
email: celia.fierro.estrellas@gmail.com

[2]Escuela Superior de Física y Matemáticas, Instituto Politécnico Nacional, México.
[3]Instituto Nacional de Investigaciones Nucleares, México.
[4]Instituto de Astronomía, Universidad Nacional Autónoma de México, México.
[5]Universidad Iberoamericana, México.

Abstract. We present a tool for analysis and fit of stellar spectra using a mega database of 15,000 atmosphere models for OB stars. We have developed software tools, which allow us to find the model that best fits to an observed spectrum, comparing equivalent widths and line ratios in the observed spectrum with all models of the database. We use the H_α, H_β, H_γ, and H_δ lines as criterion of stellar gravity and ratios of He II λ4541/He I λ4471, He II λ4200/(He I+He II λ4026), He II λ4541/He I λ4387, and He II λ4200/He I λ4144 as criterion of $T_{\rm eff}$.

Keywords. stars: atmospheres, stars: abundances, stars: rotation, stars: winds

1. Database and results

The models were calculated with the CMFGEN code (Hillier, D. J. & Miller, D., 1998); database is a matrix arrangement in a 6-dimensional space: surface temperature ($T_{\rm eff}$), luminosity (L), terminal velocity of the wind (V_∞), beta exponent of the wind velocity law (β), clumping filling factor (F_{cl}), and metallicity (Z). The synthetic spectra in the UV (900-2000Å), optical (3500-7000Å) and near IR (10000-30000Å) were rotationally broaden using ROTIN3 (Hubeny & Lanz, 1995) by covering the range between 10 and 350, km/s with steps of 10 km/s. Models were calculated using the ABACUS I super- computer of Centro de Matemáticas Aplicadas y Cómputo de Alto Rendimiento ABACUS-CINVESTAV, México. Software tools allow us to find in the database the model with the best fit to an observed spectrum in a only minutes.

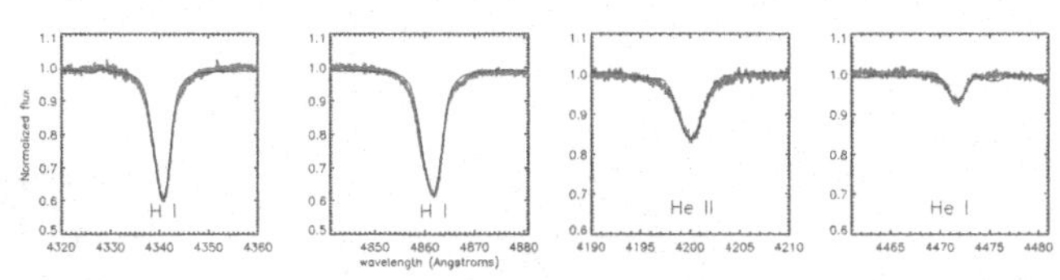

Figure 1. Adjustment example: HD 46223, spectral type O4 V (f) (Sota, A. *et al.* 2011) observed spectrum from public data of the IACOB project, 2016. Parameters of model: $T_{\rm eff}$=40 000K, $logL/L_\odot$=6.05, V_∞=2 370km/s, β=1.1, F_{cl}=0.05, Z=$Z_\odot$

References

Hillier, D. J. & Miller, D. 1998, *ApJ*, 496, 407
Hubeny, I, Lanz, T. 1995, *ApJ*, 439, 875
Sota, A. *et al.* 2011, *ApJS*, 193(2), 24

The Lives and Death-Throes of Massive Stars
Proceedings IAU Symposium No. 329, 2016
J.J. Eldridge, J.C. Bray, L.A.S. McClelland
& L. Xiao, eds.

doi:10.1017/S1743921317001314

Equilibrium structures of rapidly rotating stars with shellular rotation

Kotaro Fujisawa and Yu Yamamoto

Advanced Research Institute for Science and Engineering, Waseda University, 3-4-1 Okubo, Shinjuku-ku, Tokyo 169-8555, Japan
email: `fujisawa@heap.phys.waseda.ac.jp`

Abstract. We have developed a new numerical method for obtaining self-consistent structures of rapidly rotating stars. We obtained self-consistent equilibrium structures of rapidly rotating massive stars with shellular rotation by using the method. These equilibrium structures might be useful for both evolution of rapidly rotating massive stars and progenitor models of core-collapse supernovae simulations.

Keywords. stars: evolution, stars: rotation

1. Introduction

One of the most fascinating challenge in the stellar astrophysics is understanding the structures of rapidly rotating stars and their evolution. Some massive stars have rapid rotations and their outer layers cannot be well described by 1D stellar evolution models. Multi-dimensional models are required to describe evolution of rapidly rotating stars.

2. Method and result

We have developed a new numerical method for obtaining self-consistent structures of rapidly rotating baroclinic stars (Fujisawa 2015) to investigate multi-dimensional stellar evolution models. We calculated rapidly rotating progenitor models with shellular rotation by using the method. A parametric progenitor model developed by Yamamoto & Yamada (2016) is adopted in our calculations. The progenitor has three layers and its mass is 22 $M_{\odot}$ in the case of the non-rotating model. The density distributions of our numerical results are displayed in Fig.1. The outer layers are deformed by their rapid rotations.

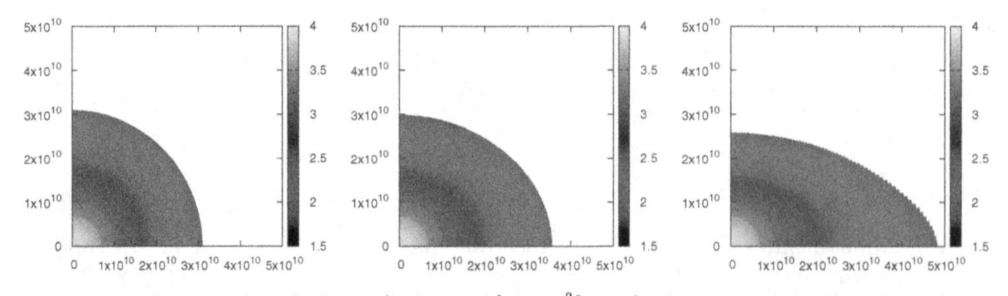

Figure 1. Density distributions ($\log_{10} \rho$ in [g cm^{-3}] unit) of the progenitor models. Left: Non-rotating model. Center: Rapid rotation model. Right: Extreme rotation model.

References

Fujisawa, K. 2015, *MNRAS*, 40, 1635
Yamamoto, Y. & Yamada, S. 2016, *ApJ*, 818, 165

The Lives and Death-Throes of Massive Stars
Proceedings IAU Symposium No. 329, 2016
J.J. Eldridge, J.C. Bray, L.A.S. McClelland & L. Xiao, eds.

doi:10.1017/S1743921317002848

The Ages of High-Mass X-ray Binaries in M33

Kristen Garofali and Benjamin F. Williams
University of Washington Astronomy Department, Box 351580, Seattle, WA, 98195
email: garofali@uw.edu, ben@astro.washington.edu

Abstract. We present initial results on a survey of high-mass X-ray binaries (HMXBs) in the nearby, star-forming spiral galaxy M33. The HMXB population in M33 is identified and characterized using a combination of deep *Chandra* X-ray imaging and archival *Hubble Space Telescope* (HST) observations. We determine ages for the HMXBs to ~5 Myr precision from fits to the color-magnitude diagrams (CMDs) of the surrounding stars and the resultant star formation histories (SFHs). The HMXBs in our M33 sample have measured ages, as well as candidate optical counterparts identified from HST photometry.

Keywords. X-rays: binaries, stars: evolution, surveys

We demonstrate a technique for identifying HMXB candidates and their potential optical counterparts using a combination of deep *Chandra* imaging and archival *HST* data. Ages for each candidate are determined from fitting the CMD of the stars surrounding the HMXB candidate. An example of this technique for a very young (10-20 Myr) candidate HMXB in M33 is shown in the Figure 1.

Previous studies of HMXB populations in the Small Magellenic Cloud, NGC300, and NGC2403 (Antoniou *et al.* 2010; Williams *et al.* 2013) have revealed HMXBs to have a preferred age of ~30-60 Myr, implying large populations HXMBs containing Be donor stars. This survey in M33 will consist of a population of well-characterized HMXBs with measured ages in a large, star-forming spiral galaxy, and will place past HMXB studies in broader context by examining the interplay between host galaxy metallicity and SFH and their combined effects on massive star evolution in binaries.

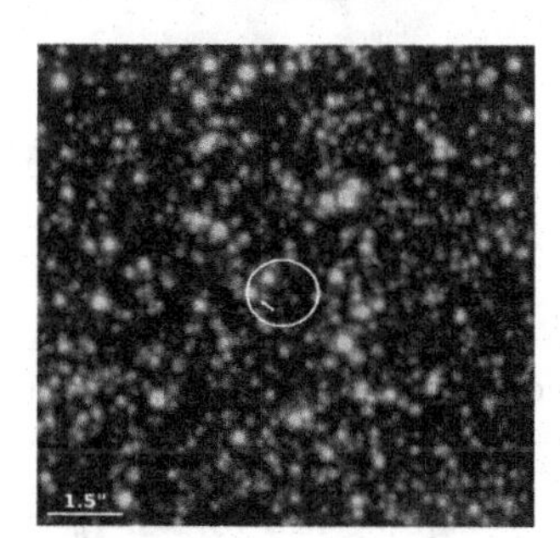

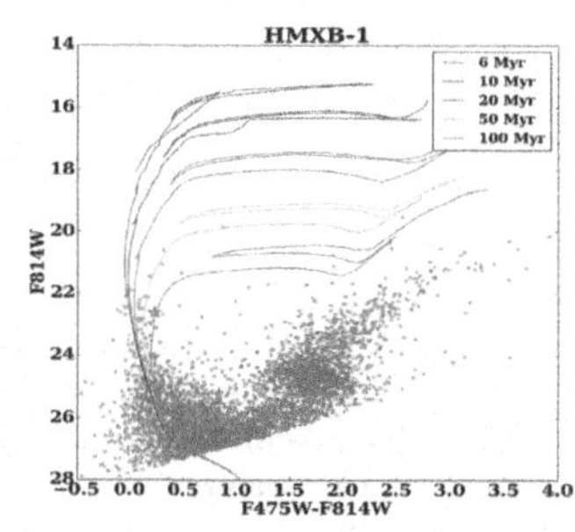

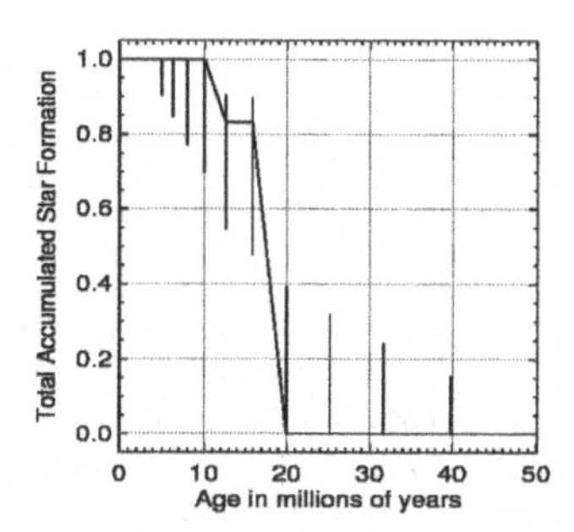

Figure 1. *Left*: The X-ray error circle (0.7″) in white overlaid on a color *HST*/ACS image of a region in the disk of M33. The blue star, denoted by a white arrow, makes this a candidate HMXB. *Center*: The *HST* CMD of the stars within 50 pc of the HXMB candidate (cyan star). Padova group isochrones are overlaid for reference. *Right*: The cumulative age distribution based on CMD fitting for stars within 100 pc of the HMXB candidate.

References

Antoniou, V., Zezas, A., Hatzidimitriou, D., & Kalogera, V. 2010, *ApJL*, 716, 140
Williams, B. F., Binder, B. A., Dalcanton, J. J., *et al.* 2013, *ApJ*, 772, 12

The Lives and Death-Throes of Massive Stars
Proceedings IAU Symposium No. 329, 2016
J.J. Eldridge, J.C. Bray, L.A.S. McClelland & L. Xiao, eds.

doi:10.1017/S1743921317003015

Asymmetric Core-collapse of a Rapidly-rotating Massive Star

Avishai Gilkis

Department of Physics, Technion – Israel Institute of Technology, Haifa 3200003, Israel
email: agilkis@technion.ac.il

Abstract. I find high turbulent shearing in a neutron-rich accretion disk surrounding a proto-neutron star formed at the collapse of a rapidly-rotating $M_{\rm ZAMS} = 54 M_\odot$ star. These might be features of superluminous supernovae powered by jets and/or magnetar spin-down.

Keywords. supernovae: general, stars: rotation

A $54 M_\odot$ star is evolved up to point of core-collapse with the MESA code (Paxton *et al.* 2011), with and without the Spruit dyanmo (Spruit 2002). The dynamo couples the core and envelope, so the loss of mass and angular momentum in winds affects the core and slows it down. At the onset of core-collapse the stellar models are mapped into the hydrodynamic solver FLASH (Fryxell *et al.* 2000), which includes core deleptonization and approximate neutrino transport (Couch & O'Connor 2014). The three-dimensional flow is followed for 225 ms after core bounce. The rotation profile at stellar death significantly influences the flow dynamics and results of the collapse (Fig. 1). These results are also relevant for stellar evolution with magnetic fields where core spin-up occurs due to binary interaction, possibly resulting in a bipolar outflow.

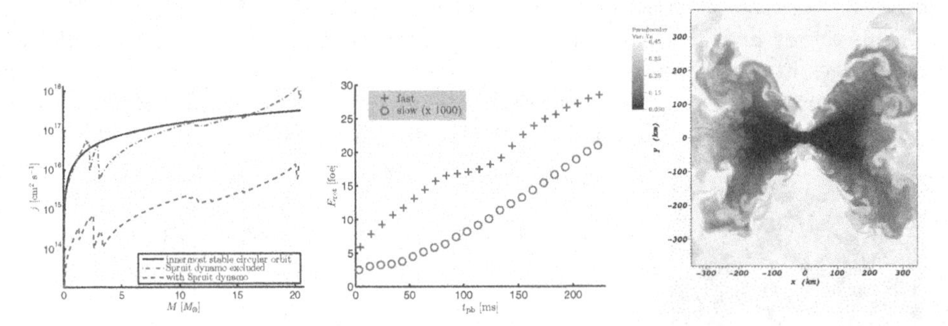

Figure 1. *Left:* Specific angular momentum as function of mass coordinate at the transition from MESA to FLASH. When the Spruit dynamo is excluded, the specific angular momentum is comparable to that of the innermost stable circular orbit around a Schwarzschild black hole. *Center:* Rotational kinetic energy of the proto-neutron star as function of time from core bounce. *Right:* Two-dimensional slice of the three-dimensional flow of the collapsing fast rotator (y is the rotation axis) showing a disk of low electron fraction ($Y_{\rm e}$) matter 200 ms after core bounce.

References

Couch, S. M. & O'Connor, E. P. 2014, *ApJ*, 785, 123
Fryxell, B., *et al.* 2000, *ApJS*, 131, 273
Paxton, B., *et al.* 2011, *ApJS*, 192, 3
Spruit, H. C. 2002, *A&A*, 381, 923

The Lives and Death-Throes of Massive Stars
Proceedings IAU Symposium No. 329, 2016
J.J. Eldridge, J.C. Bray, L.A.S. McClelland & L. Xiao, eds.

doi:10.1017/S1743921317002101

New Solutions to Line-Driven Winds of Hot Massive Stars

Alex C. Gormaz-Matamala[1], Michel Curé[1], Lydia Cidale[2] and Roberto Venero[2]

[1]Instituto de Física y Astronomía, Universidad de Valparaíso, Casilla 5030, Valparaíso, Chile.
email: `alex.gormaz@postgrado.uv.cl`

[2]Departamento de Espectroscopía, Facultad de Ciencias Astronómicas y Geofísicas, UNLP.

Abstract. In the frame of radiation driven wind theory (Castor *et al.*1975), we present self-consistent hydrodynamical solutions to the line-force parameters (k, α, δ) under LTE conditions. Hydrodynamic models are provided by HydWind (Curé 2004). We evaluate these results with those ones previously found in literature, focusing in different regions of the optical depth to be used to perform the calculations. The values for mass-loss rate and terminal velocity obtained from our calculations are also presented.

We also examine the line-force parameters for the case when large changes in ionization throughout the wind occurs (δ-slow solutions, Curé *et al.*2011).

Keywords. Radiative transfer – Hydrodynamics – Stars: winds, outflows – Stars: mass-loss

---

We use the CAK formalism (Castor *et al.*, 1975; Abbott, 1982) to calculate the force multiplier $\mathcal{M}(t)$ by means of the sum of all the spectral lines ($\sim$ 900 thousand) contributing to the radiative-driven processes. Parameters (k, α, δ) are obtained fitting the force multiplier, i.e., $\mathcal{M}(t) = k\,t^{-\alpha}(N_{e,11}/W)^{\delta}$. These calculations are complemented with the solution of the equation of motion by HydWind (Curé, 2004). With the new velocity profile we calculate new line-force parameters and iterate with the hydrodynamic solution until convergence, obtaining therefore a self-consistent wind solution.

Results for fast and δ-slow regimes (Curé *et al.*, 2011) are presented in the Table. Bottom rows corresponds to δ-slow solutions ($\delta > 0.25$) Optical depth t ranges where solutions have been calculated are also displayed.

$T_{\rm eff}$	$\log g$	k	α	δ	$\dot{M}$ $[M_\odot/{\rm yr}]$	v_∞ [km/s]	$\log t_{\rm min}$	$\log t_{\rm max}$
50 000	4.5	0.895	0.572	0.049	1.1×10^{-5}	4 486	−5.305	−0.593
40 000	4.0	0.312	0.585	0.084	2.8×10^{-6}	2 780	−5.732	−1.305
20 000	2.5	0.069	0.601	0.163	4.2×10^{-7}	683	−6.097	−2.392
19 500	2.1	0.07	0.551	0.272	6.18×10^{-6}	204.1	−2.194	−3.259
16 000	2.7	0.05	0.61	0.28	1.15×10^{-9}	477.3	−2.086	−3.974

Present results have been able not only to reproduce previous calculations (Abbott, 1982; Noebauer & Sim, 2015), but also have given values for mass-loss rate in the expected range. A more exhaustive work to properly determine how accurate are our new $\dot{M}$ and v_∞ compared with those measured observationally is ongoing.

References

Abbott, D. C., 1982 *ApJ*, 259, 282 A
Castor, J. I., Abbott, D. C. & Klein, R. I., 1975 *ApJ*, 195, 157C (CAK)
Curé, M., 2004, *ApJ*, 614, 929
Curé, Cidale, L. & Granada, A. 2011, *ApJ*, 737, 18
Noebauer, U. M. & Sim, S. A. 2015, *MNRAS*, 453, 3120

The Lives and Death-Throes of Massive Stars
Proceedings IAU Symposium No. 329, 2016
J.J. Eldridge, J.C. Bray, L.A.S. McClelland
& L. Xiao, eds.

doi:10.1017/S1743921317003313

The supergiant O + O binary system HD 166734: a new study

E. Gosset[1], L. Mahy[1], Y. Damerdji[2,1], C. Nitschelm[3], H. Sana[4], P. Eenens[5] and Y. Nazé[1]

[1]STAR Institute, Liège University, Allée du 6 août, 19c, B4000 Liège, Belgium

[2]CRAAG, route de l'Observatoire, BP63, Bouzareah, 16340 Algiers, Algeria

[3]Unidad de Astronomía, Univ. de Antofagasta, Avenida Angamos 601, Antofagasta, Chile

[4]Inst. voor Sterrenkunde, KU Leuven, Celestijnenlaan 200D, Bus 2401, B3001 Leuven, Belgium

[5]Dept. de Astronomía, Univ. de Guanajuato, Apartado 144, 36000 Guanajuato, GTO, Mexico

Abstract. We present here a modern study of the radial velocity curve and of the photometric light curve of the very interesting supergiant O7.5If + O9I(f) binary system HD 166734. The physical parameters of the stars and the orbital parameters are carefully determined. We also perform the analysis of the observed X-ray light curve of this colliding-wind binary.

Keywords. stars: individual (HD 166734), binaries: spectroscopic, binaries: eclipsing, stars: supergiants, X-ray: binaries

---

1. Introduction

Massive stars are crucial actors in the life of their host galaxy. Despite this importance, our understanding is still poor and the determination of their physical parameters remains an essential task. Comparisons between evolutionary models and observations are essential to improve the former and better understand the evolution of these massive stars. In particular, binary systems give access to accurate values for some of the physical parameters. In this respect, HD 166734, a binary system made of two O supergiants which has been studied in the seventies, is interesting and deserves a modern study.

2. Results

We confirm the 34.54 d period derived by Conti *et al.* (1980) in the sole detailed study of this O7.5If + O9I(f) binary system. In our higher S/N study, the most massive star is now also the brightest and earliest one which is a more standard situation. The new eccentricity (slightly over 0.6) is larger than the previously determined one (0.46). The minimum masses are 28.2 $M_\odot$ for the primary and 24.5 $M_\odot$ for the secondary. The presence of an eclipse was reported by Otero & Wils (2005); we acquired a new light curve that clearly confirms the presence of a secondary eclipse with a depth of 0.2 mag and the lack of primary eclipse. From the CMFGEN analysis of the disentangled spectra, we derive effective temperatures of 32000 K and 30500 K, and log g values of 3.15 and 3.10 for the primary and the secondary, respectively. Our analysis indicates a well evolved object. We also report the first X-ray light curve observed for this star, showing variations in flux by an order of magnitude, further underlining the interest of this colliding-wind binary.

References

Conti, P. S., Massey, P., Ebbets, D., & Niemela, V. S. 1980, *ApJ*, 238, 184
Otero, S. A. & Wils, P. 2005, *Inf. Bull. Var. Stars*, 5644, 1

The Lives and Death-Throes of Massive Stars
Proceedings IAU Symposium No. 329, 2016
J.J. Eldridge, J.C. Bray, L.A.S. McClelland & L. Xiao, eds.

doi:10.1017/S1743921317002861

ALMA Observation of Mass Loss from Massive Stars

D. Y. A. Setia Gunawan[1], M. Curé[1], S. Kanaan[1], J. Puls[2], F. Najarro[3], J. Sundqvist[3], M. M. Rubio Diez[3] and N. Whyborn[4]

[1]IFA, Universidad de Valparaíso, Av. Gran Bretaña 1111, Valparaíso, Chile
[2]LMU Munich, Universitäts-Sternwarte, Scheinerstr. 1, D-81679 Mñchen, Germany
[3]Centro de Astrobiología (CSIC/INTA), 28850 Torrejón de Ardoz, Madrid, Spain
[4]Joint ALMA Observatory, Alonso de Cordova 3107, Vitacura, Santiago, Chile

Abstract. We present a pilot study of using the Atacama Large Millimeter/sub-millimeter Array (ALMA) continuum observations to constrain the density structure in the intermediate wind zone of massive stars, in which the wind is extremely sensitive to clumping.

Keywords. Stars, stars: mass loss, stars: early-type, radio continuum: stars, submillimeter

1. Using ALMA to determine mass loss rates from massive stars

The fast, dense outflows from massive OB type stars are driven by radiative line acceleration. Multiple evidence points to an over-estimation of the currently accepted mass-loss rates from these stars as shown by inconsistencies in results derived using different diagnostics, attributed to wind clumping. Potential downward revisions of mass-loss rates of massive stars have a profound effect on the stars' evolution, the feedback from them, and in turn affects the evolution of the host galaxy. We urgently need to establish the true mass-loss rates of OB stars, by constraining the clumping structure. Models predict that the clumping is radially stratified. If the radial stratification of clumping is known from consistent analyses of different diagnostic methods, realistic mass-loss rates can be derived. (Sub-)mm observations of stellar winds will provide the critical diagnostics for the clumping structure in the currently unconstrained intermediate wind regions of massive OB stars, in which the wind is extremely sensitive to clumping. Our pilot study showed that ALMA can efficiently provide the necessary (sub-)millimeter data.

(Sub-)millimeter fluxes of selected bright OB type stars with available multi-wavelength observation result, HD HD37043, HD37128, HD37742, HD38771, HD66811, HD149757, HD151804, and HD152236, were obtained by ALMA between April 2013 – April 2014, using continuum observing mode in the ALMA bands B3, B6 and B7 (respectively at 100, 230, 345 GHz). The data was reduced and calibrated using the standard ALMA reduction scripts using CASA. Subsequently obtained CLEANed maps have S/N > 8. The (sub-)millimeter SEDs show mostly spectral indices consistent with emission from thermal winds. One target may show a non-thermal emission. Our next step is to combine analysis of all diagnostics from the Far-UV to the radio domain with consistency, to derive the clumping properties throughout the entire wind and help to constrain the physical origin of wind clumping. Expanding the sample and using more ALMA band coverage is necessary to get the complete true mass-loss rate of massive stars.

Acknowledgement

This paper makes use of the following ALMA data: ADS/JAO.ALMA# 2012.1.00941.S (PI Curé) and 2012.1.00955.S (PI Kanaan). ALMA is a partnership of ESO (representing its member states), NSF (USA) and NINS (Japan), together with NRC (Canada), NSC and ASIAA (Taiwan), and KASI (Republic of Korea), in cooperation with the Republic of Chile. The Joint ALMA Observatory is operated by ESO, AUI/NRAO and NAOJ. Este trabajo conta con el apoyo de CONICYT Programa de Astronomia Fondo ALMA-CONICYT cargo proyectos numeros 31AS002 y 31140024.

The Lives and Death-Throes of Massive Stars
Proceedings IAU Symposium No. 329, 2016
J.J. Eldridge, J.C. Bray, L.A.S. McClelland & L. Xiao, eds.

doi:10.1017/S1743921317001594

The origin of the dusty envelope around Betelgeuse

Xavier Haubois[1], Barnaby Norris[2], Peter G. Tuthill[2], Christophe Pinte[3], Pierre Kervella[4], Julien Girard[1], Guy Perrin[4], Sylvestre Lacour[4], Andrea Chiavassa[5] and S. T. Ridgway[6]

[1]European Organisation for Astronomical Research in the Southern Hemisphere, Casilla 19001, Santiago 19, Chile; email: xhaubois@eso.org

[2]Sydney Institute for Astronomy, School of Physics, University of Sydney, NSW 2006, Australia

[3]UJF-Grenoble 1 / CNRS-INSU, Institut de Planétologie et d'Astrophysique de Grenoble, UMR 5274, 38041 Grenoble, France

[4]LESIA, (UMR 8109), Observatoire de Paris, PSL, CNRS, UPMC, Univ. Paris-Diderot, 5 place Jules Janssen, 92195 Meudon, France

[5]Laboratoire Lagrange, Université Côte d'Azur, Observatoire de la Côte d'Azur, CNRS, Boulevard de l'Observatoire, CS 34229, 06304 Nice Cedex 4, France

[6]National Optical Astronomy Observatory, P.O. Box 26732, Tucson, AZ 85726-6732, USA

Abstract. The origin of red supergiant mass loss still remains to be unveiled. Characterising the formation loci and the dust distribution in the first stellar radii above the surface is key to understand the initiation of the mass loss phenomenon. Polarimetric interferometry observations in the near-infrared allowed us to detect an inner dust atmosphere located only 0.5 stellar radius above the photosphere of Betelgeuse. We modelled these observations and compare them with visible polarimetric measurements to discuss the dust distribution properties.

Keywords. Convection-techniques: interferometric- stars: fundamental parameters- infrared: stars- stars: individual: Betelgeuse

---

Summary

Depending on their characteristics and composition, dust grains could interact with the stellar radiation from red supergiants and trigger mass loss via a dust wind. We observed Betelgeuse using the NACO/SAMPol instrument and detected a polarizing structure at 1.5 stellar radius that we modelled as a thin dust shell. The dust-shell-to-stellar-disk flux ratio can be modelled as the fraction of the stellar light that is scattered on the grains of the dust shell. We fit the wavelength variation of this fraction using a Mie scattering routine. Parameters of this model are the grain radius and density, which is translated in a dust shell mass. We derived similar grain sizes of about 300 nm and masses of about 10^{-10} $M_{\odot}$ for three dust species (enstatite $MgSiO_3$, forsterite Mg_2SiO_4 and alumina Al_2O_3). Extrapolating to the visible wavelengths, we can compare our radiative transfer modelling to the SPHERE/ZIMPol data reported in Kervella *et al.* (2016). A preliminary analysis shows that the forsterite and enstatite models match the data better than the alumina dust except in the V-band filter. More spectral filters in the visible should be obtained to infirm/confirm this hint on the dust composition.

Reference

Kervella, P., Lagadec, E., Montargès, M., *et al.* 2016, *A&A*, 585, A28

The Lives and Death-Throes of Massive Stars
Proceedings IAU Symposium No. 329, 2016
J.J. Eldridge, J.C. Bray, L.A.S. McClelland & L. Xiao, eds.

doi:10.1017/S1743921317002770

MASGOMAS project: building a bona-fide catalog of massive star cluster candidates

Artemio Herrero[1,2], Klaus Rübke[1,2], Sebastián Ramírez Alegría[3,4], Miriam Garcia[5] and Antonio Marín-Franch[6]

[1]Instituto de Astrofísica de Canarias, 38205 La Laguna, Tenerife, Spain
[2]Departamento de Astrofísica, Universidad de La Laguna, 38206 La Laguna, Tenerife, Spain
[3]Millenium Institute of Astrophysics, MAS, Chile
[4]Universidad Católica del Norte, Antofagasta, Chile
[5]Centro de Astrobiología, CSIC-INTA, E-28850 Torrejón de Ardoz, Spain
[6]Centro de Estudios de Física del Cosmos de Aragón (CEFCA), E-44001 Teruel, Spain

MASGOMAS (MAssive Stars in Galactic Obscured MAssive clusterS) is a project aiming at discovering OB stars in Galactic, dust enshrouded, star-forming massive clusters (Marín-Franch *et al.* 2009, A&A 502, 559). The project has gone through different phases of increasing automatization, that have allowed us to discover massive clusters like MASGOMAS-1 (Ramírez Alegría *et al.* 2012, A&A 541, A75) (with M≈20,000 $M_\odot$).

Currently the search is carried out through MASA: MASGOMAS Automatic Search Algorithm. MASA consists of three IDL modules: (a) a module that carries out photometric cuts to select massive OB star candidates; (b) a friends-of-friends algorithm based on AUTOPOP (Garcia *et al.*, 2009, A&A 502, 1015) that requires two free parameters: the maximum distance D_s to consider two stars as belonging to the same cluster and N_{min}, the minimum number of members for a cluster; and (c) an output module giving the list of candidates and color-color and color-magnitude diagrams for inspection.

However, automatic methods may introduce many contaminants due both to the high stellar densities in the Galactic Plane and to our inability to set a priori the search parameters in the friends-of-friends algorithm, for which we have no physical constraints. For this reason we have run a series of simulations on synthetic fields of various surface densities with randomly distributed targets. This way we have been able to determine an optimum search distance D_s for a given N_s and stellar density. For these optimum D_s values, no cluster candidates were found in the simulations. Thus, candidates found in real stellar fields have a high chance to be a real cluster or association of massive stars.

We have tested our method in a 6°×6° region around $l = 33°$ and $b = 0°$. This region contains MASGOMAS-1, used as a test. A new promising candidate was found, that we call MASGOMAS-10. Follow-up spectroscopy at $R = 2500$ with LIRIS@WHT of MASGOMAS-10 resulted in 5 OB stars plus one Wolf-Rayet out of 7 observed stars.

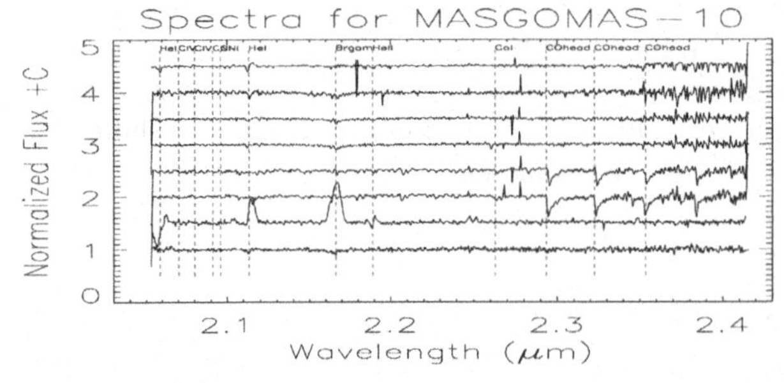

Figure 1. K-band spectra of MASGOMAS-10 with LIRIS@WHT. The top four stars are OB stars, as well as the bottom one. Together with the WR, they clearly characterize it as a massive cluster candidate.The two late-type stars were marked as low probability candidates because their (J-Ks) color was bluer than that of the other stars

The Lives and Death-Throes of Massive Stars
Proceedings IAU Symposium No. 329, 2016
J.J. Eldridge, J.C. Bray, L.A.S. McClelland & L. Xiao, eds.

doi:10.1017/S1743921317001120

Quantitative spectroscopic analyses in the IACOB+OWN project: Massive O-type stars in the Galaxy with the current Gaia information

Gonzalo Holgado,[1,2], Sergio Simón-Díaz[1,2] and Rodolfo Barbá[3]

[1]Instituto de Astrofísica de Canarias, E-38200 La Laguna, Tenerife, Spain.

[2]Departamento de Astrofísica, Universidad de La Laguna, E-38205 La Laguna, Tenerife, Spain.

[3]Departamento de Física y Astronomía, Univ. de la Serena, Av. Juan Cisternas 1200 Norte, La Serena, Chile

We present the results from the quantitative spectroscopic analysis of ~280 likely single O stars targeted by the IACOB and OWN surveys. This implies the largest sample of Galactic O-type stars analyzed homogeneously to date. We used the IACOB-BROAD and IACOB-GBAT tools (see Simón-Díaz *et al.* 2011,2015) to obtain the complete set of spectroscopic parameters which can be determined from the optical spectrum of O-type stars: projected rotational velocity ($v \sin i$), macroturbulence velocity ($v_{\rm mac}$), effective temperature ($T_{\rm eff}$), gravity ($\log g$), wind-strength ($\log Q$), helium abundance ($Y_{\rm He}$), microturbulence ($\xi_{\rm t}$), and the exponent of the wind-law (β).

The results presented in this poster contribution can be also found in Holgado *et al.*(2016). These include a summary of the present completitude in magnitude of our studied sample, the distribution of projected rotational velocities and an overview of the distribution of the stellar properties in the spectroscopic HR diagram. We also evaluate the present situation regarding available information about distances, as provided by the Hipparcos and Gaia missions. We find that the parallaxes in TGAS-DR1 are still not good enough to derive accurate radii, luminosities and masses, as only 2% of our sample have parallaxes with an uncertainty lower than 20%.

For the next steps we plan to :

- Investigate the distribution of projected rotational velocities in Galactic O stars.
- Revisit calibrations of stellar parameters with spectral type and luminosity class.
- Study the origin of the absence of stars near the ZAMS in the range 20-60 M⊙.
- Perform the abundance analysis of the sample.

And, once Gaia will achieve more reliable values of distances (as already expected for DR2):

- Provide a homogeneous and statistically significant empirical overview of the physical properties of Galactic O-type stars.
- Use these results to constraint theoretical predictions about the initial phases of massive stars evolution and evaluate the theory of radiatively-driven winds in the Galactic O Stars.

References

Simón-Díaz, S., Castro, N., Garcia, M., Herrero, A., & Markova, N. 2011d, *Bulletin de la Societe Royale des Sciences de Liege*, 80, 514

Simón-Díaz, S., Negueruela, I., Maíz Apellániz, J., *et al.* 2015, *Highlights of Spanish Astrophysics VIII*, 576

Holgado, G., Simón-Díaz, S., & Barbá, R. H. 2016, arXiv:1611.02634

The Lives and Death-Throes of Massive Stars
Proceedings IAU Symposium No. 329, 2016
J.J. Eldridge, J.C. Bray, L.A.S. McClelland & L. Xiao, eds.

doi:10.1017/S1743921317002939

Time Lapse Spectropolarimetry: Constraining the Nature and Progenitors of Interacting CCSNe

Leah Huk[1]

[1]Dept. of Physics & Astronomy, University of Denver,
2112 E. Wesley Ave., Denver, CO 80210, USA
email: leah.huk@gmail.com

Abstract. SNe of Type IIn are among the brightest supernova explosions due to strong circumstellar interaction. Examining the geometric and optical properties of the circumstellar material (CSM) can help to identify the progenitors of individual IIn SNe. Polarimetry is the optimal method for constraining CSM characteristics, as polarimetric signals both depend upon and preserve geometric information from unresolved sources. I present the results of fitting an ensemble of simulated polarized Hα emission-line profiles of interacting SNe, created using a three-dimensional Monte Carlo radiative transfer code called *SLIP*, to the multi-epoch observed polarized spectra of the Type IIn SN 1997eg. Further study of this model ensemble will allow us to investigate relationships among SNe IIn based on viewing angle and consider how the category should be subdivided based on physical properties of the CSM and/or progenitor.

Inclination is the most influential parameter in our fits. The models best fitting the observed continuum polarization magnitude and depolarization at Hα line center are constrained to inclinations near 90°, implying a nearly edge-on orientation for SN 1997eg.

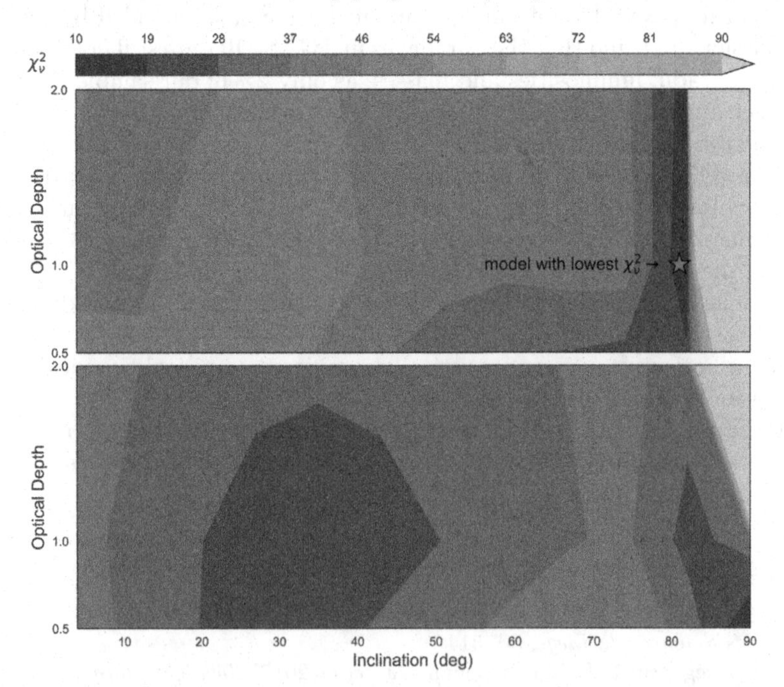

Figure 1. Reduced χ^2 values for fits to the polarized Hα line of SN 1997eg at day 16, with light arising from a central source (*top*) and from a shock region interior to the CSM (*bottom*).

The Lives and Death-Throes of Massive Stars
Proceedings IAU Symposium No. 329, 2016
J.J. Eldridge, J.C. Bray, L.A.S. McClelland & L. Xiao, eds.

doi:10.1017/S1743921317002940

The 2D dynamics of the differentially rotating envelope of massive stars

Delphine Hypolite[1], Stéphane Mathis[1] and Michel Rieutord[2]

[1]Laboratoire AIM Paris-Saclay, CEA/DRF - CNRS - Université Paris Diderot, IRFU/SAp
Centre de Saclay, F-91191 Gif-sur-Yvette Cedex, France
email: delphine.hypolite@cea.fr

[2]Institut de Recherche en Astrophysique et Planétologie, Observatoire Midi-Pyrénées,
Université de Toulouse, 14 avenue Edouard Belin, 31400 Toulouse, France

We build a 2D model of the radiative envelope of main sequence massive stars. We set a dynamical boundary condition at the bottom of the radiative envelope at $\eta = r_C/R$ (where r_C is the core size and R the radius of the star) to account for the differential rotation of the convective core as computed in 3D simulations (e.g. Browning *et al.* (2004, IAUS, 224, 149). We seek the differential rotation and associated meridional circulation induced by such a shear competing with the baroclinic flow of the stably stratified radiative envelope using the Boussinesq approximation. Our study shows that the resulting dynamics depends on the Rossby number $\mathcal{R}o = \Delta\Omega/2\Omega_0$, (e.g. Brown *et al.* (2008, ApJ, 689, 1354), Augustson *et al.* (2012, ApJ, 756, 169), Varela *et al.* (2016, ASR, 58, 1507)) which is the ratio of the differential rotation in the convective core over the global stellar rotation rate. When the Rossby number is higher than one, the differential rotation has a columnar structure as dictated by the Taylor-Proudman theorem, as illustrated in Fig. 1, coming from the shear applied by the boundary condition. The associated circulation does not develop much in the envelope leading to a quasi solid rotation outside the tangent to the core cylinder. When it is lower than one the dynamics is driven by the baroclinic torque. The meridional circulation redistributes angular momentum in the entire envelope through viscosity and leads to a differential rotation with a rapid pole and a slow equator (e.g. Rieutord, 2006, A&A, 451, 1025).

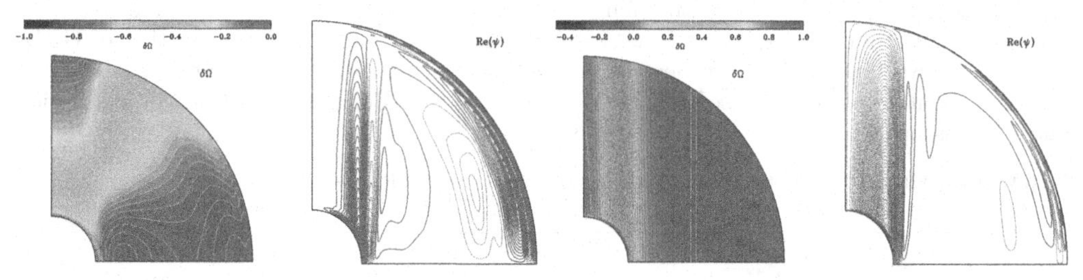

Figure 1. *Left:* Differential rotation $\Delta\Omega$ (relative to the pole rotation) and associated meridional circulation stream function Ψ (red: direct sens) in the radiative envelope, when $E = 10^{-5}$ and $\mathcal{R}o = 10^{-1}$. The rotation axis is vertical. *Right:* Same for $E = 10^{-5}$ and $\mathcal{R}o = 10$.

The Lives and Death-Throes of Massive Stars
Proceedings IAU Symposium No. 329, 2016
J.J. Eldridge, J.C. Bray, L.A.S. McClelland
& L. Xiao, eds.

doi:10.1017/S1743921317003428

The VLT/X-shooter GRB afterglow legacy survey

Lex Kaper[1], Johan P.U. Fynbo[2], Vanna Pugliese[1], Daan van Rest[1], on behalf of the X-shooter GRB collaboration

[1]Anton Pannekoek Institute for Astronomy, University of Amsterdam,
Science Park 904, 1098 XH Amsterdam, the Netherlands
email: L.Kaper@uva.nl

[2]Dark Cosmology Centre, Niels Bohr Institute, University of Copenhagen,
Juliane Maries Vej 30, 2100 Copenhagen 0, Denmark

Abstract. The *Swift* satellite allows us to use gamma-ray bursts (GRBs) to peer through the hearts of star forming galaxies through cosmic time. Our open collaboration, representing most of the active European researchers in this field, builds a public legacy sample of GRB X-shooter spectroscopy while *Swift* continues to fly. To date, our spectroscopy of more than 100 GRB afterglows covers a redshift range from 0.059 to about 8 (Tanvir *et al.* 2009, Nature 461, 1254), with more than 20 robust afterglow-based metallicity measurements (over a redshift range from 1.7 to 5.9). With afterglow spectroscopy (throughout the electromagnetic spectrum from X-rays to the sub-mm) we can hence characterize the properties of star-forming galaxies over cosmic history in terms of redshift, metallicity, molecular content, ISM temperature, UV-flux density, etc.. These observations provide key information on the final evolution of the most massive stars collapsing into black holes, with the potential of probing the epoch of the formation of the first (very massive) stars.

VLT/X-shooter (Vernet *et al.* 2011, A&A 536, A105) is in many ways the ideal GRB follow-up instrument and indeed GRB follow-up was one of the primary science cases behind the instrument design and implementation. Due to the wide wavelength coverage of X-shooter, in the same observation one can detect molecular H_2 absorption near the atmospheric cut-off and many strong emission lines from the host galaxy in the near-infrared (e.g., Friis *et al.* 2015, MNRAS 451, 167). For example, we have measured a metallicity of 0.1 $Z_\odot$ for GRB 100219A at $z = 4.67$ (Thöne *et al.* 2013, MNRAS 428, 3590), 0.02 $Z_\odot$ for GRB 111008A at $z = 4.99$ (Sparre *et al.* 2014, ApJ 785, 150) and 0.05 $Z_\odot$ for GRB 130606A at $z = 5.91$ (Hartoog *et al.* 2015, A&A 580, 139). In the latter, the very high value of [Al/Fe]$= 2.40 \pm 0.78$ might be due to a proton capture process and may be a signature of a previous generation of massive (perhaps even the first) stars. Reconciling the abundance patterns of GRB absorbers, other types of absorbers (in particular QSO DLAs), and old stars in the Local Group is an important long-term goal of this program (see Sparre *et al.* 2014, ApJ 785, 150). Metallicities are also measured from host emission lines (Krühler *et al.* 2015, A&A 581, A125). GRB spectroscopy also allows us to determine the dust content of their environments, both through analysis of the depletion pattern and the measurement of the associated extinction (Japelj *et al.* 2015, A&A 451, 2050). This way one can quantify the dust-to-metals ratio and its evolution with redshift. The detection of GRBs at $z > 6$ shows that GRBs have become competitive as a tool to identifying galaxies at the highest redshifts and unsurpassed in providing detailed abundance information via absorption line spectroscopy.

Keywords. Gamma-ray bursts, massive stars, GRB host galaxies, etc.

The Lives and Death-Throes of Massive Stars
Proceedings IAU Symposium No. 329, 2016
J.J. Eldridge, J.C. Bray, L.A.S. McClelland & L. Xiao, eds.

doi:10.1017/S1743921317002502

Pre-SN neutrino emissions from ONe cores in the progenitors of CCSNe

Chinami Kato[1], S. Yamada[1,2], H. Nagakura[3], S. Furusawa[4], K. Takahashi[5], H. Umeda[5], T. Yoshida[5] and K. Ishidoshiro[6]

[1]School of Advanced Science and Engineering, Waseda University, Japan
[2]Advanced Research Institute for Science and Engineering, Waseda University, Japan
[3]TAPIR, Walter Burke Institute for Theoretical Physics, California Institute of Technology, USA
[4]Frankfurt Institute for Advanced Studies, J.W. Goethe University, Germany
[5]Department of Astronomy, The University of Tokyo, Japan
[6]Research Center for Neutrino Science, Tohoku University, Japan

Abstract. In order to investigate the distinguishability about the progenitors of FeCCSNe and ECSNe, we calculate the luminosities and spectra of their pre-SN neutrinos and estimate the number of events at neutrino detectors.

Keywords. Supernovae:general, neutrinos

1. Overview

According to the stellar evolution theory, the progenitors with relatively light masses ($\lesssim 9M_\odot$) explode as ECSNe, while more massive ones explode as FeCCSNe. Then pre-SN neutrinos have important roles to distinguish two types of progenitors because neutrinos can escape freely from the core and deliver its information directly. Because of the recent development of detectional techniques, there is a possibility to detect them with low energy comparable to detectional threshold. Therefore, we focus on the distinguishability of SN-progenitors by the observation of pre-SN neutrinos.

We employ the realistic progenitor models until core bounce: a $9M_\odot$ model for ECSN, 12 and $15M_\odot$ models for FeCCSNe (Takahashi *et al.* (2013), Nagakura *et al.* (2014)). Based on their results, we calculate the luminosities and spectra of all-flavor neutrinos emitted via thermal and nuclear weak processes. Finally, we estimate the expected number at neutrino detectors including neutrino oscillation, assuming the distance to the progenitors to be 200 pc. It is demonstrated that $\bar{\nu}_e$'s from the ECSNe-progenitors can hardly be detected at almost all detectors, whereas we will be able to detect $\sim$2500 ν_e's at DUNE. From the FeCCSN-progenitors, both ν_e's and $\bar{\nu}_e$'s will be detected: the number of $\bar{\nu}_e$ events will be largest for JUNO, 134-725 $\bar{\nu}_e$'s, depending on the mass hierarchy whereas the number of ν_e events at DUNE is almost the same as that for the ECSN-progenitor. These results imply that the detection of $\bar{\nu}_e$'s is useful to distinghish FeCCSN- from ECSN-progenitors, while ν_e's will provide us with detailed information on core evolutions regardless of progenitor types (Kato *et al.* (2017)).

References

Kato, C., Yamada, S., Nagakura, H., Furusawa, S., Takahashi, K., Umeda, H., Yoshida, T., & Ishidoshiro, K., 2017, in preparation
Takahashi, K., Yoshida, T., & Umeda, H., 2013, *ApJ*, 771, 28
Nagakura, H., Sumiyoshi, K., & Yamada, S., 2014, *ApJS*, 214, 16

The Lives and Death-Throes of Massive Stars
Proceedings IAU Symposium No. 329, 2016
J.J. Eldridge, J.C. Bray, L.A.S. McClelland & L. Xiao, eds.

doi:10.1017/S1743921317002435

Line-Driven Ablation of Circumstellar Disks

Nathaniel Dylan Kee[1], Stan Owocki[2], Rolf Kuiper[1] and Jon Sundqvist[3]

[1]Institute of Astronomy and Astrophysics, University of Tübingen, Germany
email: nathaniel-dylan.kee@uni-tuebingen.de

[2]Department of Physics and Astronomy, University of Delaware, USA
[3]Instituut voor Sterrenkunde, KU Leuven, Belgium

Abstract. Mass is a key parameter in understanding the evolution and eventual fate of hot, luminous stars. Mass loss through a wind driven by UV-scattering forces is already known to reduce the mass of such stars by $10^{-10} - 10^{-4}$ $M_{\odot}$/yr over the course of their lifetimes. However, high-mass stars already drive such strong winds while they are still in their accretion epoch. Therefore, stellar UV-scattering forces will efficiently ablate material off the surface of their circumstellar disks, perhaps even shutting off the final accretion through the last several stellar radii and onto a massive protostar. By using a three-dimensional UV-scattering prescription, we here quantify the role of radiative ablation in controlling the disk's accretion rate onto forming high-mass stars. Particular emphasis is given to the potential impact of this process on the stellar upper mass limit.

Keywords. accretion, accretion disks; circumstellar matter; stars: formation; stars: mass loss

---

Main sequence massive stars are known to drive strong mass loss through the interaction of their UV-photon flux with spectral lines from metal ions. Due to the extreme optical depth of these spectral lines, this interaction is only possible in the case where the lines are Doppler shifted out of their own shadow, for instance by the spherically expanding outflows of massive star winds. While an orbiting circumstellar disk, such as those found during star formation, does not have these same large-scale velocity gradients in the radial direction, Keplerian shear in the disk creates velocity gradients of similar magnitude and line-of-sight variation along non-radial directions. This similarity in scale and morphology of the non-radial line-of-sight velocity gradients through a disk to the radial line-of-sight velocity gradients through a wind implies that an orbiting disk should feel a force comparable to that which launches the stellar wind.

To test this, we use the hydrodynamics code, PLUTO (Mignone *et al.*, 2007, ApJS, 170, 228; 2012 ApJS 198, 7). We have added a three-dimensional UV line-driving prescription following the Cranmer and Owocki (1995, ApJ, 440, 308) extension to Castor, Abbott, and Klein (1975, ApJ, 195, 157). Results from our first paper in the series (Kee *et al.* 2016, MNRAS, 458, 2323) show that the removal of material from a continuum optically thin circumstellar disk, such as might be found in a Classical Be star system, occurs at approximately the spherically symmetric wind mass loss rate. First results for disk densities more comparable to those found in star forming environments show that this basic scaling holds up, with at most a very weak scaling of ablation rate with disk mass. Further work will continue to investigate this effect including additionally the accretion rate of the star forming disk to directly test whether the removal of mass from a star forming disk could play a role in modifying, or perhaps even halting, the flow of material onto a forming star.

The Lives and Death-Throes of Massive Stars
Proceedings IAU Symposium No. 329, 2016
J.J. Eldridge, J.C. Bray, L.A.S. McClelland & L. Xiao, eds.

doi:10.1017/S1743921317002071

Proper Motions of η Carinae's Outer Ejecta and Its Eruptive History

Megan M. Kiminki[1], Megan Reiter[2] and Nathan Smith[1]

[1]Steward Observatory, University of Arizona, 933 N. Cherry Avenue, Tucson, AZ 85721, USA
email: mbagley@email.arizona.edu

[2]Department of Astronomy, University of Michigan, 311 West Hall, 1085 S. University Avenue, Ann Arbor, MI 48109, USA

Abstract. η Carinae, the most extreme luminous blue variable in our Galaxy, underwent a Great Eruption in the 1800s and ejected significant mass into the well-known bipolar Homunculus. But η Car's outer ejecta, a spread of dense, nitrogen-rich knots outside the Homunculus, have led to suspicion that the Great Eruption was not this star's first major mass-loss event. We have measured proper motions for nearly 800 distinct features in the outer ejecta using 21 years of *HST* WFPC2 and ACS imaging. With motions measured across sixteen baselines, we find that the outer ejecta are expanding ballistically and belong to three age groups: one dating to the mid-1200s, another to the mid-1500s, and a third to the early 1800s, associated with but perhaps predating the peak of the Great Eruption. These three age groups are separated in space and radial velocity. There is no evidence for interaction between the dense ejecta that could be powering η Car's soft X-ray shell, which is instead likely driven by fast, rarefied ejecta from the Great Eruption striking the older dense ejecta. The thirteenth-century event was strikingly asymmetric, ejecting mass almost entirely to one side of the star. The sixteenth-century event displays bipolar symmetry, but along a different axis than the current Homunculus. These observations provide constraints on theoretical models of η Car's behavior, as viable models must explain the repetition, timescale, and asymmetry of these major mass-loss events. For more details, see Kiminki, Reiter, & Smith (2016, *MNRAS*, 463, 845).

Keywords. circumstellar matter, stars: individual: η Carinae, stars: mass loss

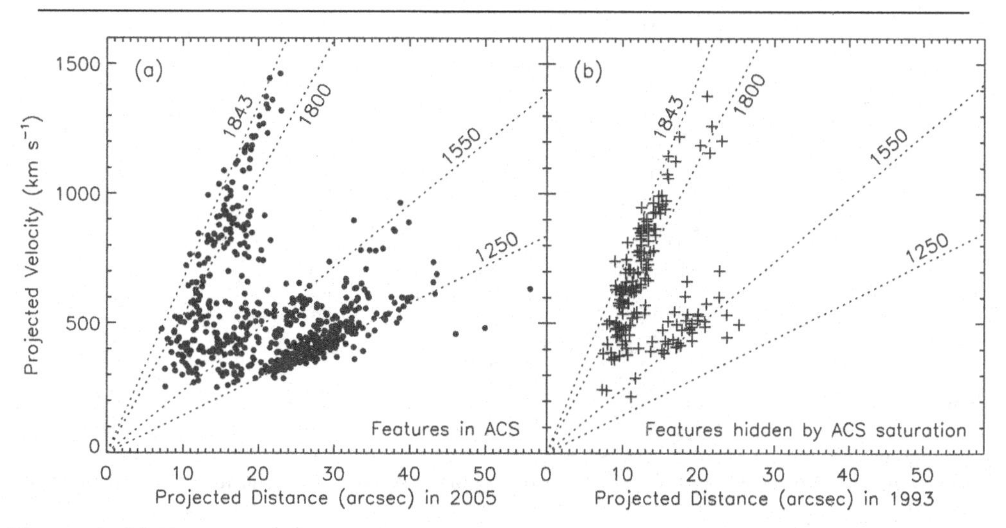

Figure 1. (a) Transverse velocity versus projected distance from η Car for all ejecta detected in ACS imaging. Dotted lines show the expected positions of material ejected at the labeled dates. (b) Same as (a), but for ejecta in a section that was saturated in ACS but visible in WFPC2 images. Note the lack of thirteenth-century features in this section.

The Lives and Death-Throes of Massive Stars
Proceedings IAU Symposium No. 329, 2016
J.J. Eldridge, J.C. Bray, L.A.S. McClelland & L. Xiao, eds.

doi:10.1017/S1743921317003064

The outer disk of the classical Be star ψ Per

Robert Klement[1,2], Alex C. Carciofi[3], Thomas Rivinius[2], Lynn D. Matthews[4], Richard Ignace[5], Jon E. Bjorkman[6], Rodrigo G. Vieira[3], Bruno C. Mota[3], Daniel M. Faes[3], Stanislav Štefl[†]

[1]Astronomical Institute of Charles University, Charles University, V Holešovičkách 2, 180 00, Prague 8, Czech Republic, email: robertklement@gmail.com

[2]European Southern Observatory, Alonso de Córdova 3107, Vitacura, Casilla 19001, Santiago, Chile

[3]Instituto de Astronomia, Geofíisica e Ciências Atmosféricas, Universidade de São Paulo, Rua do Matão 1226, 05508-090, São Paulo, Brazil

[4]MIT Haystack Observatory, off Route 40, Westford MA 01886, USA

[5]Department of Physics & Astronomy, East Tennessee State University, Johnson City, TN 37614, USA

[6]Ritter Observatory, Department of Physics & Astronomy, University of Toledo, Toledo, OH 43606, USA

[†]Deceased

Abstract. To this date ψ Per is the only classical Be star that was angularly resolved in radio (by the VLA at $\lambda = 2\,\mathrm{cm}$). Gaussian fit to the azimuthally averaged visibility data indicates a disk size (FWHM) of ~ 500 stellar radii (Dougherty & Taylor 1992). Recently, we obtained new multi-band cm flux density measurements of ψ Per from the enhanced VLA. We modeled the observed spectral energy distribution (SED) covering the interval from ultraviolet to radio using the Monte Carlo radiative transfer code HDUST (Carciofi & Bjorkman 2006). An SED turndown, that occurs between far-IR and radio wavelengths, is explained by a truncated viscous decretion disk (VDD), although the shallow slope of the radio SED suggests that the disk is not simply cut off, as is assumed in our model. The best-fit size of a truncated disk derived from the modeling of the radio SED is 100^{+5}_{-15} stellar radii, which is in striking contrast with the result of Dougherty & Taylor (1992). The reasons for this discrepancy are under investigation.

Keywords. stars: individual (ψ Per), stars: emission-line, Be, radio continuum: stars

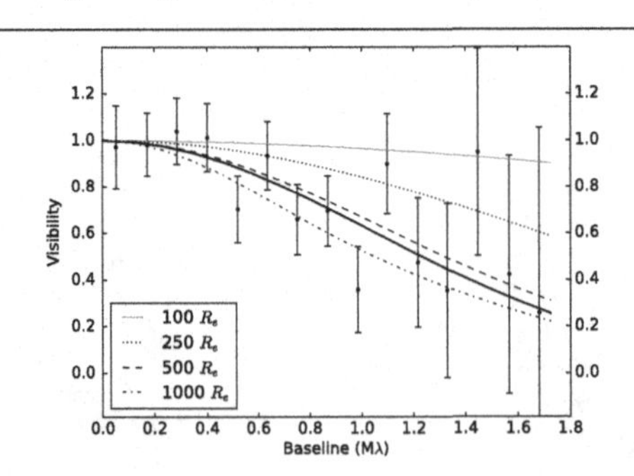

Figure 1. Azimuthally averaged visibility data of Dougherty & Taylor (points) and a Gaussian fit to them (thick line) overplotted with the visibility curves derived from our azimuthally averaged models with different disk sizes (in stellar equatorial radii R_e). The disk size best reproducing the radio SED is $100\,R_e$, which does not agree well with the interferometric data.

References

Carciofi, A.C. & Bjorkman, J.E. 2006, *ApJ*, 639, 1081
Dougherty, S.M. & Taylor A.R. 1992, *Nature*, 359, 808

The Lives and Death-Throes of Massive Stars
Proceedings IAU Symposium No. 329, 2016
J.J. Eldridge, J.C. Bray, L.A.S. McClelland & L. Xiao, eds.

doi:10.1017/S1743921317002514

The Next Possible Outburst of P Cygni

Nino Kochiashvili[1], Sopia Beradze[1], Ia Kochiashvili[1,2], Rezo Natsvlishvili[1], Manana Vardosanidze[1]

[1]E.Kharadze Abastumani Astrophysical Observatory, Ilia State University
email: nino.kochiashvili@iliauni.edu.ge

[2]Dark Cosmology Centre, Niels Bohr Institute, University of Copenhagen

Abstract. On the basis of long-term UBV observations of P Cygni, which were made by Eugene Kharadze and Nino Magalashvili between 1951-1983, is evident that P Cygni undergone reddening during those observations. P cygni is a LBV and a supernova impostor. Corrected on the reddening B-V color has values between about -0.4 (at the beginning of 1950-ies) and -0.1 (for the 1980-ies). It means that the star probably had earlier spectral type at the beginning of 20-th century and accordingly, we are witnesses of its evolutionary changes. It means also that on the HR diagram the star moves gradually to the instability strip of LBVs in Outburst. So, if the rate of the reddening of the P Cygni will the same in near future then the star will have the next eruption (or even supernova explosion) after approximately 80-120 years.

The long (approximately 1500 d, 1160 d, 760 d, 580 d) quasi-periods and the shorter ones (approximatelly 130 d, 68 d and 15-18 days) were revealed using the above observations.

We observed P Cygni on July 23 - October 20, 2014 with the 48 cm Cassegrain telescope and standard B,V,R,I filters. HD 228793 has been used as a comparison star. We revealed that during our observations the star underwent light variations with the mean amplitude of approximately 0.1 magnitudes in all pass-bands and the period of this change was approximately 68 days. There is also a relation between brightness and the Hα EW variability. Therefore, we think that the cause of this behavior may be a variability of rate of the stellar wind that is very strong in this star. Changes in the rate of the stellar wind, on the other hand, maybe due to the pulsation of the star. It seems that quasi-periods of the brightness variability are almost the exact multiples of each other which probably also indicates on pulsation of the star. According to the new photometric observations of 2014 the star continues reddening.

Keywords. P Cyg, UBV Photometry, LBV, supernova impostor, reddening

---

1. Acknowledgements

The authors are grateful to E.Kharadze Abastumani Astrophysical Observatory, Ilia State University, Georgia, for funding our travel to New Zealand. This work was partly supported by Shota Rustaveli National Science Foundation (SRNSF grant No218070: The Next Possible Outburst of P Cygni). N.K. thanks conference organizers for covering part of her local expences at the conference.

The Lives and Death-Throes of Massive Stars
Proceedings IAU Symposium No. 329, 2016
J.J. Eldridge, J.C. Bray, L.A.S. McClelland & L. Xiao, eds.

doi:10.1017/S1743921317002824

First results from Project SUNBIRD: Supernovae UNmasked By Infra-Red Detection

Erik C. Kool[1,2], Stuart D. Ryder[2], Erkki Kankare[3], Seppo Mattila[4]

[1]Dept. of Physics and Astronomy, Macquarie University, Sydney NSW 2109, Australia
email: erik.kool@students.mq.edu.au

[2]Australian Astronomical Observatory, Sydney NSW 1670, Australia

[3]Astrophysics Research Centre, Queen's University Belfast, United Kingdom

[4]Tuorla Observatory, University of Turku, Kaarina, Finland

Abstract. A substantial number of core-collapse supernovae (CCSNe) are expected to be hosted by starbursting luminous infrared galaxies (LIRGs). However, so far very few CCSNe have been discovered in LIRGs, most likely as a result of dust extinction and lack of contrast in their typically luminous and complex nuclear regions. We present the first results of Project SUNBIRD (Supernovae UNmasked By InfraRed Detection), where we aim to uncover dust-obscured nuclear supernovae by monitoring over 30 LIRGs, using near-infrared state-of-the-art Laser Guide Star Adaptive Optics (LGSAO) imaging on the Gemini South and Keck telescopes. Such discoveries are vital for determining the fraction of supernovae which will be missed as a result of dust obscuration by current and future optical surveys.

Keywords. supernovae: general - galaxies: starburst - infrared: galaxies - instrumentation: adaptive optics - instrumentation: high angular resolution

---

1. Summary

The observed rate of CCSNe has been shown to trail the rate predicted from the cosmic star formation rate by as much as a factor of two (Horiuchi *et al.* 2011). One of the main suspects for this discrepancy is dust: the light of supernovae gets obscured and traditional optical supernova surveys fail to detect them, in particular in the nuclear regions of their host galaxies. In the dusty star forming LIRGs this effect is most prominent and earlier work by our collaboration showed that most, if not all of the missing supernovae might be accounted for by dust extinction alone (Mattila *et al.* 2012). However these limits are based on a handful of CCSNe in a single monitored LIRG and thus have large uncertainties.

The aim of Project SUNBIRD is to increase the number of CCSNe found in LIRGs, and in this way improve the constraints on the fraction of CCSNe that are missed in dust obscured nuclear regions. We observe in the near-IR where dust extinction is reduced by a factor of 10 compared to the optical, and make use of telescopes equipped with LGSAO to supply the spatial resolution required to identify nuclear SNe. So far in the first year of SUNBIRD we have discovered three CCSNe, two of which were located within 1 kpc of the nucleus, with extinctions up to 5 magnitudes in V (Kool *et al.* 2017, in prep).

References

Horiuchi S. *et al.* 2011, *ApJ*, 738, 154
Mattila S. *et al.* 2012, *ApJ*, 756, 111

The Lives and Death-Throes of Massive Stars
Proceedings IAU Symposium No. 329, 2016
J.J. Eldridge, J.C. Bray, L.A.S. McClelland & L. Xiao, eds.

doi:10.1017/S1743921317002381

Wind inhibition by X-ray irradiation in high-mass X-ray binaries

Jiří Krtička[1], Jiří Kubát[2] and Iva Krtičková[1]
[1]Masaryk University, Brno, Czech Republic
[2]Astronomical Institute, Ondřejov, Czech Republic

Abstract. Winds of hot massive stars are driven radiatively by light absorption in the lines of heavier elements. Therefore, the radiative force depends on the wind ionization. That is the reason why the accretion powered X-ray emission of high-mass X-ray binaries influences the radiative force and may even lead to wind inhibition. We model the effect of X-ray irradiation on the stellar wind in high-mass X-ray binaries. The influence of X-rays is given by the X-ray luminosity, by the optical depth between a given point and the X-ray source, and by the distance to the X-ray source. The influence of X-rays is stronger for higher X-ray luminosities and in closer proximity of the X-ray source. There is a forbidden area in the diagrams of X-ray luminosity vs. the optical depth parameter. The observations agree with theoretical predictions, because all wind-powered high-mass X-ray binary primaries lie outside the forbidden area. The positions of real binaries in the diagram indicate that their X-ray luminosities are self-regulated.

Keywords. Stars: early-type, stars: mass loss, stars: winds, outflows

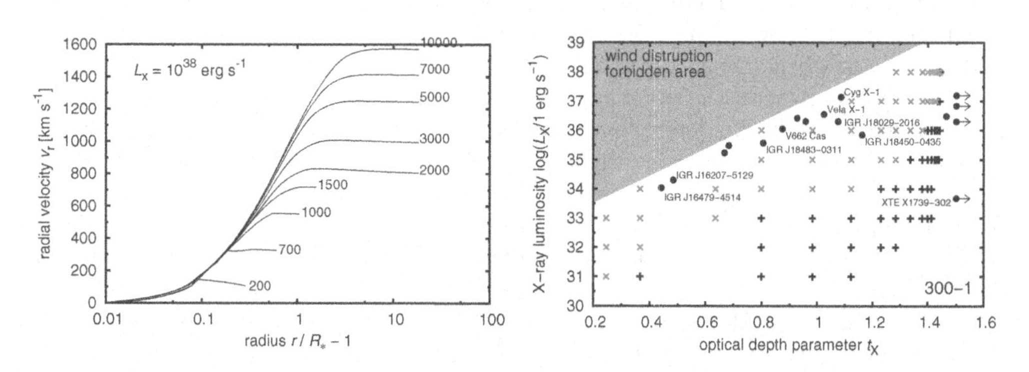

Figure 1. *Left:* Radial dependence of the velocity for different X-ray source distances (in $R_\odot$). The influence of X-rays is stronger for larger X-ray luminosity and lower optical depth parameter (for closer X-ray source). Too strong or too close X-ray source inhibits the flow. *Right:* Diagram of X-ray luminosity vs. the optical depth parameter displaying regions with different effects of the X-ray irradiation. Individual symbols denote positions of: models with negligible influence of X-ray irradiation (black plus, +), models where X-rays lead to the decrease of the wind terminal velocity (blue cross, ×), and non-degenerate components of HMXBs (filled circles, •).

Acknowledgements

This research was supported GA ČR 13-10589S.

Reference

Krtička, J., Kubát, J., & Krtičková, I., 2016, *A&A*, 579, A111

The Lives and Death-Throes of Massive Stars
Proceedings IAU Symposium No. 329, 2016
J.J. Eldridge, J.C. Bray, L.A.S. McClelland & L. Xiao, eds.

doi:10.1017/S1743921317003337

Photometric variability of the Be star population with the KELT survey

Jonathan Labadie-Bartz[1], Joshua Pepper[1], S. Drew Chojnowski[2] and M. Virginia McSwain[1]

[1]Department of Physics, Lehigh University, 16 Memorial Drive East, Bethlehem, PA 18015, USA; email: jml612@lehigh.edu

[2]Apache Point Observatory and New Mexico State University, P.O. Box 59, Sunspot, NM 88349-0059, USA

Abstract. We are using light curves from the KELT exoplanet transit survey (Pepper *et al.* 2007) to study the variability of hundreds of Be stars. Combining these light curves with simultaneous time-series spectra from the APOGEE survey (Majewski *et al.* 2015) provides a glimpse into how changes in the circumstellar environment are correlated to brightness variations.

Keywords. stars: emission-line, Be, stars: variables: other

Be stars are rapidly rotating B-type stars with line emission, which is attributed to a gaseous circumstellar disk. Such systems can be highly variable on timescales ranging from hours to decades. KELT photometry shows numerous cases of periodic variability, including both high and low frequency signals. We attribute high frequency signals to stellar pulsations. Low frequencies may indicate binarity, the coupling of pulsational modes, or other phenomena.

Long-term variability on the order of years is not uncommon, and indicates changes in the circumstellar disk. We detect 'outbursts', where the system brightens suddenly, then fades back towards the baseline brightness, and are understood to be discrete events where material is flung from the star into the circumstellar environment, forming a disk. We find high-frequency signals in 24%, low-frequency signals in 38%, long-term variation in 37%, and outbursts in 36% of our sample. See Labadie-Bartz *et al.* 2017 for more information.

Fig. 1 shows an outburst, as seen in KELT photometry, and two spectra of the Bracket 11 line, as seen by APOGEE. The spectrum taken prior to the outburst shows no sign of a disk, but after the outburst the spectrum shows a clear disk signature. This example supports the idea of photometric outbursts corresponding to the injection of stellar material into the circumstellar environment.

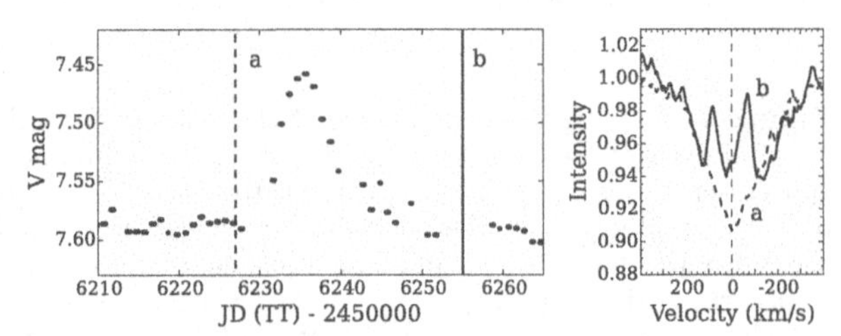

Figure 1. *Left:* KELT light curve showing an outburst. The 2 vertical lines show the epochs of the APOGEE observations. *Right:* APOGEE spectra showing the creation of a disk.

References

Labadie-Bartz, *et al.* 2017, 2017, *AJ*, 153, 252
Majewski, *et al.* 2015, *arXiv*, 1509.05420
Pepper, *et al.* 2007, *PASP*, 119, 923

The Lives and Death-Throes of Massive Stars
Proceedings IAU Symposium No. 329, 2016
J.J. Eldridge, J.C. Bray, L.A.S. McClelland & L. Xiao, eds.

doi:10.1017/S1743921317002599

Massive Stars in M31

Jamie R. Lomax[1], Matthew Peters[2], John Wisniewski[2], Julianne Dalcanton[1], Benjamin Williams[1], Julie Lutz[1], Yumi Choi[3] and Aaron Sigut[4]

[1]Department of Astronomy, University of Washington
Box 351580, Seattle, WA 98195, USA

[2]Homer L. Dodge Department of Physics & Astronomy, University of Oklahoma,
440 W. Brooks St., Norman, OK 73019, USA

[3]Steward Observatory, University of Arizona, Tucson, AZ 85721, USA

[4]Department of Physics and Astronomy, The University of Western Ontario,
London, Ontario Canada N6A 3K7

Abstract. Massive stars are intrinsically rare and therefore present a challenge to understand from a statistical perspective, especially within the Milky Way. We recently conducted follow-up observations to the Panchromatic Hubble Andromeda Treasury (PHAT) survey that were designed to detect more than 10,000 emission line stars, including WRs, by targeting regions in M31 previously known to host large numbers of young, massive clusters and very young stellar populations. Because of the existing PHAT data, we are able to derive an effective temperature, bolarimetric luminosity, and extinction for each of our detected stars. We report on preliminary results of the massive star population of our dataset and discuss how our results compare to previous studies of massive stars in M31.

Keywords. stars: emission-line, Be, stars: Wolf-Rayet

---

1. Emission Line Survey Basics

We carried out a deep, 2 epoch, Hα imaging survey to detect emission line stars in 6 regions of M31 known to host very young stellar populations and large numbers of young massive clusters from the Panchromatic Hubble Andromeda Treasury (PHAT) survey. By leveraging the PHAT dataset, we are able to determine properties such as age, T_{eff}, L_{bol}, and A_V for all our sources. Each of our six targeted regions was observed twice with either WFC3 or ACS. Therefore, each star in our emission line survey has 1 epoch of data from each of the 6 filters used for PHAT plus 2 epochs of Hα and F625W imagery. In total, we detected more than 2 million stars.

2. Preliminary Results and Future Work

We found the B stars in our sample by making cuts on log(T_{eff}) between 4 and 4.5 and log(age) between 7.0 and 8.0. Approximately 6% of the sample is B stars and, of those, 2% are in previously identified clusters. We expect between 10% and 40% of the B star sample are Be stars, which we will determine by comparing the two epochs of data.

Ten previously discovered red supergiants and 12 previously discovered WRs were observed as part of our survey. We are currently trying to identify them in our sample and determine if they show variable Hα emission or absorption that might be indicative of a binary companion.

The Lives and Death-Throes of Massive Stars
Proceedings IAU Symposium No. 329, 2016
J.J. Eldridge, J.C. Bray, L.A.S. McClelland
& L. Xiao, eds.

doi:10.1017/S1743921317000291

4-D Imaging and Modeling of Eta Carinae's Inner Fossil Wind Structures

Thomas I. Madura[1], Theodore Gull[2], Mairan Teodoro[2,3], Nicola Clementel[4], Michael Corcoran[2,3], Augusto Damineli[5], Jose Groh[6], Kenji Hamaguchi[2,7], D. John Hillier[8], Anthony Moffat[9], Noel Richardson[10], Gerd Weigelt[11], Don Lindler[2,12] and Keith Feggans[2,12]

[1]San José State University,
One Washington Square, San José, CA 95192-0106, USA
email: thomas.madura@sjsu.edu

[2]NASA Goddard Space Flight Center, Greenbelt, MD 20771, USA

[3]USRA, 7178 Columbia Gateway Drive, Columbia, MD 20146, USA

[4]South African Astronomical Observatory, P.O. Box 9, Observatory 7935, South Africa

[5]IAG–USP, Rua do Matao 1226, Cidade Universitaria, Sao Paulo 05508-900, Brazil

[6]Trinity College Dublin, University of Dublin, Dublin 2, Ireland

[7]University of Maryland, Baltimore County, 1000 Hilltop Circle, Baltimore, MD 21250, USA

[8]University of Pittsburgh, 3941 OHara Street, Pittsburgh, PA 15260, USA

[9]Universite de Montreal, CP 6128 Succ. A., Centre-Ville, Montreal, Quebec H3C 3J7, Canada

[10]University of Toledo, Toledo, OH 43606-3390, USA

[11]Max-Planck-Institut fur Radioastronomie, Auf dem Hugel 69, D-53121 Bonn, Germany

[12]Sigma Space Corporation, 4600 Forbes Blvd., Lanham, MD 20706, USA

Abstract. Eta Carinae is the most massive active binary within 10,000 light-years and is famous for the largest non-terminal stellar explosion ever recorded. Observations reveal that the supermassive ($\sim$120 $M_\odot$) binary, consisting of an LBV and either a WR or extreme O star, undergoes dramatic changes every 5.54 years due to the stars' very eccentric orbits ($e \approx 0.9$). Many of these changes are caused by a dynamic wind-wind collision region (WWCR) between the stars, plus expanding fossil WWCRs formed one, two, and three 5.54-year cycles ago. The fossil WWCRs can be spatially and spectrally resolved by the *Hubble Space Telescope/Space Telescope Imaging Spectrograph* (*HST/STIS*). Starting in June 2009, we used the *HST/STIS* to spatially map Eta Carinae's fossil WWCRs across one full orbit, following temporal changes in several forbidden emission lines (e.g. [Fe III] 4659 Å, [Fe II] 4815 Å), creating detailed data cubes at multiple epochs. Multiple wind structures were imaged, revealing details about the binary's orbital motion, photoionization properties, and recent ($\sim 5 - 15$ year) mass-loss history. These observations allow us to test 3-D hydrodynamical and radiative-transfer models of the interacting winds. Our observations and models strongly suggest that the wind and photoionization properties of Eta Carinae's binary have not changed substantially over the past several orbital cycles. They also provide a baseline for following future changes in Eta Carinae, essential for understanding the late-stage evolution of this nearby supernova progenitor. For more details, see Gull *et al.* (2016) and references therein.

Keywords. stars: individual (Eta Carinae), stars: mass loss, stars: winds, outflows

---

Reference

Gull, T. R., Madura, T. I., Teodoro, M., *et al.* 2016, *MNRAS*, 462, 3196

The Lives and Death-Throes of Massive Stars
Proceedings IAU Symposium No. 329, 2016
J.J. Eldridge, J.C. Bray, L.A.S. McClelland & L. Xiao, eds.

doi:10.1017/S1743921317002605

The circumstellar environments of B[e] Supergiants

G. Maravelias[1], M. Kraus[1,2], L. Cidale[3,4], M. L. Arias[3,4], A. Aret[1,2] and M. Borges Fernandes[5]

[1]Astronomický ústav AVČR, v.v.i., Ondřejov, Czechia, email: maravelias@asu.cas.cz
[2]Tartu Observatory, Tõravere, Estonia, [3]Facultad de Ciencias Astronómicas y Geofísicas, UNLP, La Plata, Argentina, [4]Instituto de Astrofísica de La Plata, La Plata, Argentina, [5]Observatório Nacional, Rio de Janeiro, Brazil

Abstract. The evolution of massive stars encompasses short-lived transition phases in which mass-loss is more enhanced and usually eruptive. A complex environment, combining atomic, molecular and dust regions, is formed around these stars. In particular, the circumstellar environment of B[e] Supergiants is not well understood. To address that, we have initiated a campaign to investigate their environments for a sample of Galactic and Magellanic Cloud sources. Using high-resolution optical and near-infrared spectra (MPG-ESO/FEROS, GEMINI/Phoenix and VLT/CRIRES, respectively), we examine a set of emission features ([OI], [CaII], CO bandheads) to trace the physical conditions and kinematics in their formation regions. We find that the B[e] Supergiants are surrounded by a series of rings of different temperatures and densities, a probable result of previous mass-loss events. In many cases the CO forms very close to the star, while we notice also an alternate mixing of densities and temperatures (which give rise to the different emission features) along the equatorial plane.

Keywords. stars: circumstellar matter, stars: emission-line, Be, stars: mass loss, line: profiles

1. Analysis and Results

By using a simple kinematical model of a rotating ring as emitting region of each optically thin forbidden emission line and each molecular band, we probe the structure of B[e] Supergiants' disks (for details see [1]). For all studied cases we find a series of rings, and each object displays a unique distribution of these rings in terms of density and temperature. We identify two groups based on: i. the presence of [CaII]/[OI] emitting regions closer to the star than CO, consisting of CPD-529243, HD62623, Hen 3-298 (the only system displaying the [OI]5577 emission line), CPD-572874 (multiple rings, most probably a disk), HD87643, and LHA120-S73 (from [2]), and ii. those without, consisting of HD 327083 and GG Car (circumbinary structures), MWC137 (without any [CaII] emission), and LHA115-S6. Such series of equatorial rings may be the result of a common formation mechanism from mass loss triggered by (non-)radial pulsations and/or other instabilities. A close companion, as seen in two objects, is suitable to clear the innermost atomic gas regions. Minor bodies might be present in the gaps around the single stars as well that help stabilize the ring systems [2].

Acknowledgements: GA ČR (14-21373S); RVO:67985815; IUT40-1; IAU travel grant.

References

Maravelias, *et al.* 2016, arXiv:1610.00607
Kraus, *et al.* 2016, *A&A*, 593, A112

The Lives and Death-Throes of Massive Stars
Proceedings IAU Symposium No. 329, 2016
J.J. Eldridge, J.C. Bray, L.A.S. McClelland & L. Xiao, eds.

doi:10.1017/S1743921317001302

Machine-learning approaches to select Wolf-Rayet candidates

A. P. Marston[1], G. Morello[2,3], P. Morris[2], S. Van Dyk[2] and J. Mauerhan[4]

[1]ESA, STScI, 3700 San Martin Drive, Baltimore, MD 21218, USA
email: tmarston@sciops.esa.int
[2] IPAC, Caltech, 1200 E. California Blvd, Pasadena, CA 91125, USA
[3] Dept. of Physics and Astronomy, UCL, Gower Street, WC1E 6BT, UK
[4] Dept. of Astronomy, University of California, Berkeley, CA, USA

Abstract. The WR stellar population can be distinguished, at least partially, from other stellar populations by broad-band IR colour selection. We present the use of a machine learning classifier to quantitatively improve the selection of Galactic Wolf-Rayet (WR) candidates. These methods are used to separate the other stellar populations which have similar IR colours. We show the results of the classifications obtained by using the 2MASS J, H and K photometric bands, and the Spitzer/IRAC bands at 3.6, 4.5, 5.8 and 8.0μm. The k-Nearest Neighbour method has been used to select Galactic WR candidates for observational follow-up. A few candidates have been spectroscopically observed. Preliminary observations suggest that a detection rate of 50% can easily be achieved.

Keywords. Infrared surveys, Wolf-Rayet stars

We present a machine-learning method for refining our infrared broadband colour selection for determining Galactic Wolf-Rayet (WR) star candidates suitable for follow-up infrared spectroscopy (Mauerhan *et al.* 2011; Hadfield *et al.* 2007). The overall goal is to determine the total numbers of WR stars of different subtypes in our Galaxy.

Our technique uses the k-nearest neighbour machine-learning algorithm (Altman 1992) together with training sets of infrared colours from objects of known spectral type (Morello *et al.* 2017). In preliminary checks, the technique has enabled the discovery of four new WN4-5 stars using the SPEX IR spectrograph on the IRTF (see Fig. 1).

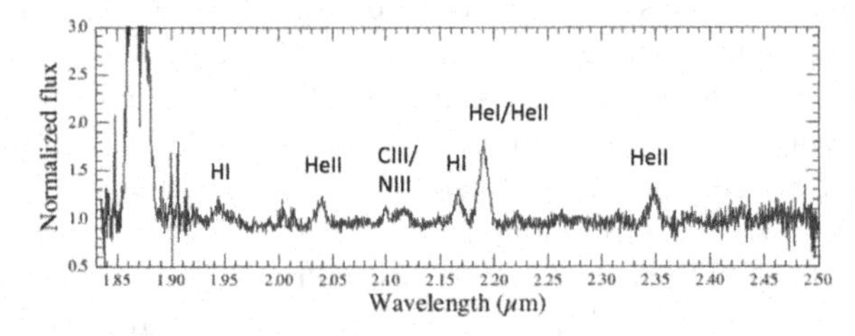

Figure 1. Continuum normalized K band spectrum of a newly found Galactic WR star (2MASSJ18442065-0236510) discovered via the list of machine learning candidates.

References

Altman, N. S. 1992, *The American Statistician*, 46, 175

Hadfield, L. J., Van Dyk, S. D., Morris, P. W., Smith, J. D., Marston, A. P., & Peterson, D. E. 2007, *MNRAS*, 376, 248

Mauerhan, J. C. Van Dyk, S. D., & Morris, P. W. 2011, *Astronomical Journal*, 142, 40

Morello, G., Morris, P. W., Van Dyk, S. D., Marston, A. P., & Mauerhan, J. C. 2017, *MNRAS*, in review

The Lives and Death-Throes of Massive Stars
Proceedings IAU Symposium No. 329, 2016
J.J. Eldridge, J.C. Bray, L.A.S. McClelland & L. Xiao, eds.

doi:10.1017/S1743921317000096

MUSEing about the SHAPE of eta Car's outer ejecta

A. Mehner[1], W. Steffen[2], J. Groh[3], F. P. A. Vogt[1], D. Baade[4], H. M. J. Boffin[4], W. J. de Wit[1], R. D. Oudmaijer[5], T. Rivinius[1] and F. Selman[1]

[1]ESO, Alonso de Cordova 3107, Vitacura, Santiago, Chile
email: amehner@eso.org

[2]Instituto de Astronomía, UNAM, Apdo Postal 106, Ensenada 22800, México

[3]School of Physics, Trinity College Dublin, Dublin 2, Ireland

[4]ESO, Karl-Schwarzschild-Straße 2, 85748 Garching, Germany

[5]School of Physics and Astronomy, The University of Leeds, Leeds, LS2 9JT, UK

Abstract. The role of episodic mass loss in evolved massive stars is one of the outstanding questions in stellar evolution theory. Integral field spectroscopy of nebulae around massive stars provide information on their recent mass-loss history. η Car is one of the most massive evolved stars and is surrounded by a complex circumstellar environment. We have conducted a three-dimensional morpho-kinematic analysis of η Car's ejecta outside its famous Homunculus nebula. SHAPE modelling of VLT MUSE data establish unequivocally the spatial cohesion of the outer ejecta and the correlation of ejecta with the soft X-ray emission.

Keywords. circumstellar matter, mass loss, winds, outflows

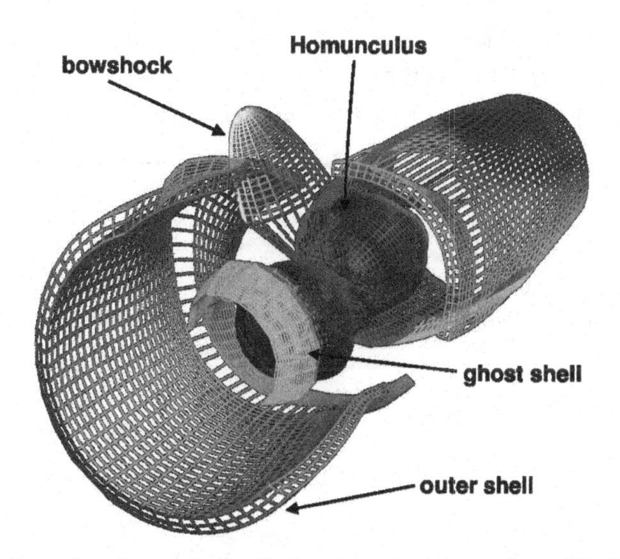

Figure 1. Three-dimensional model of η Car's outer ejecta seen from Earth with the bipolar Homunculus nebula in the center. η Car is surrounded by a cohesive large shell-like emission structure. The outer shell is oriented along a similar direction as the Homunculus nebula. Geometrical considerations point to a origin of this shell during η Car's Great Eruption in the 1840/50s.

The Lives and Death-Throes of Massive Stars
Proceedings IAU Symposium No. 329, 2016
J.J. Eldridge, J.C. Bray, L.A.S. McClelland & L. Xiao, eds.

doi:10.1017/S1743921317001181

Exploring the explosion mechanism of core-collapse supernovae in three dimensions

Tobias Melson[1], Hans-Thomas Janka[1], Alexander Summa[1], Robert Bollig[1], Andreas Marek[2] and Bernhard Müller[3,4]

[1]Max-Planck-Institut für Astrophysik,
Karl-Schwarzschild-Str. 1, 85748 Garching, Germany

[2]Max Planck Computing and Data Facility (MPCDF),
Gießenbachstr. 2, 85748 Garching, Germany

[3]Astrophysics Research Center, School of Mathematics and Physics, Queen's University,
Belfast BT7 1NN, United Kingdom

[4]Monash Center for Astrophysics, School of Physics and Astronomy,
Building 79P, Monash University, Victoria 3800, Australia

Abstract. We present the first successful simulations of neutrino-driven supernova explosions in three dimensions (3D) using the VERTEX-PROMETHEUS code including sophisticated energy-dependent neutrino transport. The simulated models of 9.6 and 20 solar-mass iron-core stars demonstrate that successful explosions can be obtained in self-consistent 3D simulations, where previous models have failed. New insights into the supernova mechanism can be gained from these explosions. The first 3D model (Melson *et al.* 2015a) explodes at the same time but more energetically than its axially symmetric (2D) counterpart. Turbulent energy cascading reduces the kinetic energy dissipation in the cooling layer and therefore suppresses neutrino cooling. The consequent inward shift of the gain radius increases the gain layer mass, whose recombination energy provides the surplus for the explosion energy.

The second explosion (Melson *et al.* 2015b) is obtained through a moderate reduction of the neutral-current neutrino opacity motivated by strange-quark contributions to the nucleon spin. A corresponding reference model without these corrections failed, which demonstrates how close current 3D models are to explosion. The strangeness adjustment is meant as a prototype for remaining neutrino opacity uncertainties.

Keywords. supernovae: general, neutrinos, hydrodynamics, turbulence

---

References

Melson, T., Janka, H.-T., & Marek, A. 1995, *ApJ* (Letters), 801, L24

Melson, T., Janka, H.-T., Bollig, R., Hanke, F., Marek, A., & Müller, B. 2015, *ApJ* (Letters), 808, L42

The Lives and Death-Throes of Massive Stars
Proceedings IAU Symposium No. 329, 2016
J.J. Eldridge, J.C. Bray, L.A.S. McClelland & L. Xiao, eds.

doi:10.1017/S1743921317001934

A new sample of red supergiants in the Inner Galaxy

M. Messineo[1], Q. Zhu[1], D. F. Figer[2], K. M. Menten[3], V. D. Ivanov[4], R.-P. Kudritzki[5] and C.-H. Rosie Chen[3]

[1] Key Laboratory for Researches in Galaxies and Cosmology, University of Science and Technology of China, Chinese Academy of Sciences, Hefei, Anhui, 230026, China
[2] Center for Detectors, Rochester Institute of Technology, 54 Memorial Drive, Rochester, NY 14623, USA
[3] Max-Planck-Institut für Radioastronomie, Auf dem Hügel 69, D-53121 Bonn, Germany
[4] European Southern Observatory, Karl Schwarzschild-Strasse 2, D-85748 Garching bei Munchen, Germany
[5] Institute for Astronomy, University of Hawaii, 2680 Woodlawn Drive, Honolulu, HI 96822

Abstract. We carried out a pivot experiment to select distant luminous late-type stars on the basis on their 2MASS and GLIMPSE photometry. Low-resolution infrared spectra enabled us to measure the equivalent widths (EWs) of their CO band-heads at 2.293 μm, and to confirm an extraordinarily high detection rate of red supergiants (RSGs), .i.e. 61% (Messineo *et al.* (2016)).

Keywords. Galaxy: disk, stars: late-type, supergiants

---

1. Overview of our survey

RSGs are an important probe of Galaxy formation and evolution. Star formation is coupled with the Galactic potential and occurs in preferential locations, such as the two end sides of the Bar where a large number of RSGs have been detected. These locations contain an extraordinary number of RSGs that are easily detectable. Inspired by these findings, we tried to detect individual RSGs, independently of clusters.

Approximately one hundred targets were selected from the 2MASS and GLIMPSE North I surveys, by following the prescriptions of Messineo *et al.* (2012) with $Q1$ and $Q2$ extinction-free colors. We selected stars with $0.1 < Q1 < 0.5$ mag and $0.5 < Q2 < 1.5$ mag. This range includes about 46% of known RSGs (Messineo *et al.* (2012)). Low-resolution HK spectra were collected with the SofI spectrograph on the ESO-NTT 4m-telescope. The EW of the CO band-head at 2.293 μm is a good indicator of temperature. Giants and supergiants follow two distinct relations, and late-type RSGs have large EWs. Contaminating AGB Miras can be classified by the shape of their continuum that is affected by broad H_2O absorption. The spectroscopic analysis has resulted in an extraordinarily large number of new RSGs, obtaining a detection rate of > 60%.

Distances vary from 3.6 to 8.6 kpc; they were estimated with surrounding clump stars (primary indicators of distance) by deriving a relation between reddening and distance for a given line-of-sight. Luminosities confirm that the sample is dominated by RSGs with initial masses from 12 to 20 $M_\odot$. In conclusion, we successfully searched for RSGs and we increased by about 25% the number of previously known RSGs with $|b| < 1°$ and $10° < l < 65°$. Only about 1.5% of these RSGs are found in clusters.

References

Messineo, M., Menten, K. M., Churchwell, E., & Habing, H. 2012, *A&A*, 537, A10
Messineo, M., Zhu, Q., Menten, K. M., *et al.* 2016, *ApJL*, 822, L5

The Lives and Death-Throes of Massive Stars
Proceedings IAU Symposium No. 329, 2016
J.J. Eldridge, J.C. Bray, L.A.S. McClelland & L. Xiao, eds.

doi:10.1017/S1743921317000187

Light-curve and spectral properties of ultra-stripped core-collapse supernovae

Takashi J. Moriya

Division of Theoretical Astronomy, National Astronomical Observatory of Japan, National Institutes of Natural Sciences, 2-21-1 Osawa, Mitaka, Tokyo 181-8588, Japan
email: takashi.moriya@nao.ac.jp

Abstract. We discuss light-curve and spectral properties of ultra-stripped core-collapse supernovae. Ultra-stripped supernovae are supernovae with ejecta masses of only $\sim 0.1 M_\odot$ whose progenitors lose their envelopes due to binary interactions with their compact companion stars. We follow the evolution of an ultra-stripped supernova progenitor until core collapse and perform explosive nucleosynthesis calculations. We then synthesize light curves and spectra of ultra-stripped supernovae based on the nucleosynthesis results. We show that ultra-stripped supernovae synthesize $\sim 0.01 M_\odot$ of the radioactive ^{56}Ni, and their typical peak luminosity is around 10^{42} erg s^{-1} or -16 mag. Their typical rise time is $5-10$ days. By comparing synthesized and observed spectra, we find that SN 2005ek and some of so-called calcium-rich gap transients like PTF10iuv may be related to ultra-stripped supernovae.

Keywords. supernovae: general, gravitational waves

Ultra-stripped supernovae (SNe) are SNe with ejecta masses of only $\sim 0.1\ M_\odot$. When a SN progenitor has a compact companion, this kind of SNe with extreme stripping can occur (e.g., Tauris *et al.* 2013). We show light-curve and spectral properties of ultra-stripped SNe in Fig. 1. See Moriya *et al.* (2016) for more details.

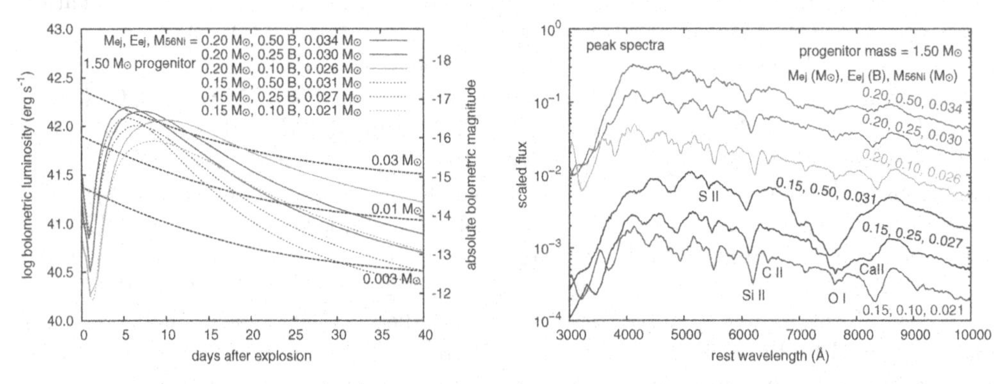

Figure 1. Light-curve (left) and spectral (right) properties of ultra-stripped SNe.

References

Moriya, T. J., Mazzali, P. A., Tominaga, N., Hachinger, S., Blinnikov, S. I., Tauris, T. M., Takahashi, K., Tanaka, M., Langer, N., & Podsiadlowski, Ph. 2016, *Mon. Not. R. Astron. Soc.*, accepted (arXiv:1612.02882)

Tauris, T. M., Langer, N., Moriya, T. J., Podsiadlowski, Ph., Yoon, S.-C., & Blinnikov, S. I. 2013, *Astrophys. J. Letters*, 778, L23

The Lives and Death-Throes of Massive Stars
Proceedings IAU Symposium No. 329, 2016
J.J. Eldridge, J.C. Bray, L.A.S. McClelland & L. Xiao, eds.

doi:10.1017/S174392131700357X

WR 148 and the not so compact companion

Melissa Munoz[1], Anthony J. Moffat[1], Grant M. Hill[2], Tomer Shenar[3], Noel D. Richardson[4], Herbert Pablo1[3], Nicole St-Louis[1] and Tahina Ramiaramanantsoa[1]

[1]Université de Montréal, [2]W. M. Keck Observatory, [3]Institut fürr Physik und Astronomie, [4]University of Toledo

Abstract. The objective is to determine the nature of the unseen companion of the single-lined spectroscopic binary, WR 148 (= WN7h+?). The absence of companion lines supports a compact companion (cc) scenario. The lack of hard X-rays favours a non-compact companion scenario. Is WR 148 a commonplace WR+OB binary or a rare WR+cc binary?

Keywords. binaries: spectroscopic, stars: Wolf-Rayet , stars: individual (WR 148).

WR 148 (WN7h) is a single-lined spectroscopic binary with an established period of 4.3174 d suspected to harbour either a low mass B star or a black hole companion (Marchenko *et al.* 1996). We obtained two nights of spectra from the Keck Observatory at both quadratures complemented by additional other spectra from l'Observatoire du Mont-Mgantic (OMM) in the summers of 2014 and 2015.

The high resolution and high signal-to-noise ratio Keck spectra reveal absorption lines moving in anti-phase to the WR emission lines with similar velocity amplitude (see Table 1 for the orbital elements). Considering an orbital inclination of $\sim 67°$, derived from previous polarimetry observations (Drissen *et al.* 1986), the systems total mass would be a mere 2-3 $M_\odot$; an unprecedented result for suspected massive binary.

Table 1. Orbital solution for the WR star and companion.

Object	P [days]	E [HJD - 2,440,000]	γ [kms s$^-$1]	K [kms s$^-$1]
WR star	4.317336 ± 0.000026	4825.04 ± 0.03	88.1 ± 3.8	-131.4 ± 2.7
Companion			79.2 ± 3.1	-120.1 ± 1.2

We apply the shift-and-add technique to disentangle the spectra and obtain a companion spectrum compatible with an O5 spectral type. Assuming a typical mass for a O5 V type from Martins *et al.* 2003, we obtain a new orbital inclination of $\sim 20°$. This discrepancy in inclination angle can be reconciled with a revised error assessment on the polarisation data. In fact, Wolinski & Dolan (1994) demonstrate that polarimetrically-derived inclination angles have a statistical bias towards higher angles.

To summarise, WR 148 is found to be a normal, massive, close WN7+O5 binary system. Though not discussed here, evidence of colliding winds has been discovered as well as thermal X-rays compatible with other WR+O colliding wind binaries.

References

Drissen, L., Lamontagne, R., Moffat, A. F. J., Bastien, P., & Seguin, M., 1986, *ApJ*, 304, 188
Marchenko, S. V., Moffat, A. F. J., Lamontagne, R., & Tovmassian, G. H. 1996, *ApJ*, 461, 386
Martins, F., Schaerer, D., & Hillier, D. J. 2003, *A&A*, 436, 1049
Wolinski, K. G. & Dolan, J. F. 1994, *MNRAS*, 267, 5

The Lives and Death-Throes of Massive Stars
Proceedings IAU Symposium No. 329, 2016
J.J. Eldridge, J.C. Bray, L.A.S. McClelland
& L. Xiao, eds.

doi:10.1017/S1743921317001193

Multi-messenger signals from core-collapse supernovae

Ko Nakamura[1], Shunsaku Horiuchi[2], Masaomi Tanaka[3], Kazuhiro Hayama[4], Tomoya Takiwaki[3] and Kei Kotake[1]

[1]Department of applied physics, Fukuoka University,
Nanakuma 8-19-1, Johnan, Fukuoka 814-0180, Japan
email: nakamurako@fukuoka-u.ac.jp

[2]Center for Neutrino Physics, Department of Physics, Virginia Tech, Virginia 24061, USA

[3]National Astronomical Observatory of Japan, Osawa 2-21-1, Mitaka, Tokyo 181-8588, Japan

[4]Institute for Cosmic Ray Research, University of Tokyo, Kamioka, Gifu 506-1205, Japan

Abstract. The next Galactic supernova is expected to bring great opportunities for the direct detection of gravitational waves, full flavor neutrinos, and multi-wavelength photons. To prepare for appropriate observations of these multi-messenger signals, we use a long-term numerical simulation of the core-collapse supernova and discuss detectability of the signals in different situations. By exploring the sequential multi-messenger signals of a nearby CCSN, we discuss preparations for maximizing successful studies of such an unprecedented stirring event.

Keywords. Galaxy: general, gravitational waves, neutrinos, supernovae: general

---

Our supernova model is based on a long-term numerical simulation of the core-collapse supernova (CCSN) of a 17 $M_{\odot}$ red supergiant progenitor, which self-consistently models the multi-messenger signals expected in gravitational wave, neutrino, and electromagnetic messengers. In this article, we briefly summarize our findings because of page limitation. Refer to Nakamura *et al.* (2016) for details. The numerical code we employ for the CCSN simulation is essentially the same as found in Nakamura *et al.* (2015).

Galactic Center Supernovae. The neutrino signal from a CCSN occurring at the Galactic Center determines the time of core bounce to within several milliseconds. This high accuracy estimation of the bounce time allows the time window of GW analysis to be reduced, which greatly reduces the background noise. The neutrino signal also provides pointing information, which will facilitate optical followup.

Extremely Nearby Supernovae. This rare event would provide unique information. The pre-CCSN neutrino signals enable us to diagnose the core structure of the progenitor. A high-precision gravitational waveform reconstruction will tell us what happens deeply inside the core. Ceaseless monitoring of nearby massive stars listed in Nakamura *et al.* (2016) is essential not to miss the possible opportunities of such a rare event.

Extragalactic Supernovae. Next-generation neutrino detectors will have sensitivity to the neutrino burst from CCSN in nearby galaxies within a few Mpc. Within the horizon of next-generation detectors, the top 10 galaxies host more than 60% of the total CCSN rate, so the number of galaxies which need followup is rather limited. A list of nearby galaxies in Nakamura *et al.* (2016) will assist early followup by optical telescopes.

References

Nakamura, K., Takiwaki, T., Kuroda, T., & Kotake, K. 2015, *PASJ*, 67, 107
Nakamura, K., Horiuchi, S., Tanaka, M., *et al.* 2016, *MNRAS*, 461, 3296

The Lives and Death-Throes of Massive Stars
Proceedings IAU Symposium No. 329, 2016
J.J. Eldridge, J.C. Bray, L.A.S. McClelland & L. Xiao, eds.

doi:10.1017/S1743921317001119

Up and downs of a magnetic oblique rotator viewed at high resolution

Y. Nazé[1], S. A. Zhekov[2] and A. ud-Doula[3]

[1]FNRS/Université de Liège, Belgium
email: naze@astro.ulg.ac.be
[2]National Astronomical Obsservatory, Bulgaria; [3]Penn State Worthington Scranton, USA

Abstract. In 2006, the Of?p star HD191612 became the second O-star where a magnetic field was discovered. It provided a benchmark to understand the Of?p phenomenon as a whole. Ten years later, an X-ray monitoring performed at high-resolution reveals the behaviour of the hottest magnetospheric plasma: it is located at $\sim 2R_\odot$, hot but not extreme ($\log(T) \sim 7$), producing unshifted lines, and displaying a very repetitive variability. A direct comparison with simulations yields an overall good agreement, with only a few further improvements needed.

Keywords. X-rays: stars – stars: individual (HD191612) – stars: winds – stars: magnetic fields

---

The category of Of?p stars gather stars with spectral peculiarities (e.g. narrow emission component in Balmer lines, strong X-ray emission). They are all magnetic and their periodic variations are explained in the context of magnetic oblique rotators (i.e. they are linked to our changing angle-of-view on the magnetically confined winds).

HD191612 (Of?p, $P = 537.2d$) was the second magnetic O-star identified. Its X-ray variations occur in phase with UV/visible changes. To further study them, we gathered Chandra-HETG data at the phases of minimum and maximum fluxes for 196 ks and 142 ks, respectively (Nazé *et al.* 2016). These high-resolution spectra reveal similarities with other magnetic stars (θ^1 Ori C, HD148937): strong lines from H-like ions, no significant line shift and FWHMs$\sim$400-1200 km s^{-1}. The lines appear slightly broader at minimum flux and may be slightly more redshifted at maximum flux but that requires confirmation. The line ratios constrain the temperature to $\log(T) \sim 7$, with a value 25% smaller at minimum flux. The line ratios also indicate a formation radius $R_{form} = 1.7 - 2.4R_\odot$, which agree well with the expected Alfven radius. Note that the observed flux modulation (by $\sim$40%) shows a great stability over a decade, but cannot be explained by a simple occultation of an axisymmetric equatorial structure at or near this radius, requiring the presence of asymmetries in the confined winds.

A 3D MHD model with the known stellar parameters was computed. It indicates that the wind material should be trapped within 1.6-2.8 $R_\odot$, in agreement with observations. The confined winds should however have a temperature $\log(T) \sim 7.5$, larger than observed. Finally, the model predict very narrow line profiles but with broad wings (which are slightly broader at maximum flux). The average DEM from this model was entered into Xspec and directly fitted to the data. A scaling factor of 0.15-0.28 was needed (probably due to a lower $\dot{M}$), but an overall good fit was achieved: the broad-band fit is adequate up to 3keV (but there is too much harder flux, as could be expected from the higher plasma temperature), and the lines are well fitted by both minimum flux and maximum flux profile models - though an even better fit can be achieved with simple Gaussian broadening (i.e. observed lines are slightly broader than predicted).

Reference

Nazé, Y., ud-Doula, A., & Zhekov, S. A. 2016, *ApJ*, 831, 138

The Lives and Death-Throes of Massive Stars
Proceedings IAU Symposium No. 329, 2016
J.J. Eldridge, J.C. Bray, L.A.S. McClelland
& L. Xiao, eds.

doi:10.1017/S1743921317000990

The quest for magnetic massive stars in the Magellanic Clouds

Y. Nazé[1], S. Bagnulo[2], N. R. Walborn[3], N. Morrell[4], G. A. Wade[5], M. K. Szymanski[6] and R. H. D. Townsend[7]

[1]FNRS/Univ. Liège, Belgium; [2]Armagh Obs., Ireland; [3]STScI, USA; [4]LCO, Chile; [5]RMC, Canada; [6]Warsaw Univ., Poland; [7]Univ. Wisconsin-Madison, USA

Abstract. The Of?p category was introduced more than 40 years ago to gather several Galactic stars with some odd properties. Since 2000, spectropolarimetry, high-resolution spectroscopy, long-term photometry, and X-ray observations have revealed their nature: magnetic oblique rotators - they all have magnetic fields that confine their winds. Several Of?p stars have now been detected in the Magellanic Clouds, likely the prototypes of magnetic massive stars at low metallicity. This contribution will present the most recent photometric, spectroscopic, and spectropolarimetric data, along with the first modeling of these objects.

Keywords. stars: early-type – stars: winds – stars: magnetic fields

---

Of?p are the sole class of (strongly) magnetic O-stars in the Galaxy. Five stars presenting the same spectral peculiarities were discovered in the Magellanic Clouds, and we investigated them in more detail to assess the impact of metallicity on the magnetism of massive stars. Photometric data from ASAS, EROS-2, and OGLE II-III-IV were analyzed using a modified Fourier algorithm, conditional entropy, binned analysis of variances and string length methods (Nazé *et al.* 2015). Spectroscopic data were obtained at LCO, AAT, ESO, and Siding Spr. Observ. (Walborn *et al.* 2015). Spectropolarimetric data were gathered with FORS2 and underwent several checks (Bagnulo *et al.*, A&A, in press): observation of standard magnetic stars and of non-magnetic field stars, use of null diagnostics. Individual results for each star are:

- SMC159-2 (P=14.91d) has very strong emissions (e.g. EW(Hα)$\sim$19Å), only rivalled by the most magnetic O-star NGC1624-2; B_z >5kG is ruled out at maximum phase.
- 2dFS936 (P=1370d + 2d because of line-of-sight object?) recently shows a large increase in emission strength and a small increase in brightness, an unexplained departure from an otherwise repetitive behaviour; B_z <2.5kG at maximum emission phase.
- AV220 is varying (and currently declining) but no periodicity could be identified - is the period long or are irregular variations possible for Of?p stars?; B_z <3kG.
- BI57 (P=787d) shows a single maximum in optical line strength while having two photometric maxima, i.e. it is the first Of?p where photometric and spectroscopic behaviour are (partially) decoupled; B_z <3kG (at a photometric max/spectroscopic min).
- LMC164-2 (P=7.96d) has B_z <1.7kG close to maximum phase.

Finally, we performed Monte-Carlo radiative transfer simulating light scattering in a circumstellar envelope following the confined wind structure simulated for HD191612, with initial photon emission launched from star at a random angle (Wade *et al.*, in prep.). The derived lightcurves strongly depend on the inclination i of the star's rotation axis and obliquity β of the magnetic axis, helping to constrain the systems' geometries.

References

Nazé, Y., Walborn, N. R., Morrell, N., Wade, G. A. & Szymanski, M. K. 2015, *A&A*, 577, A107
Walborn, N. R., Morrell, N. I., Nazé, Y., *et al.* 2015, *AJ*, 150, 99

The Lives and Death-Throes of Massive Stars
Proceedings IAU Symposium No. 329, 2016
J.J. Eldridge, J.C. Bray, L.A.S. McClelland & L. Xiao, eds.

doi:10.1017/S1743921317001582

Mass of dust in core-collapse supernovae as viewed from energy balance in the ejecta

Takaya Nozawa

National Astronomical Observatory of Japan, Osawa 2-21-1, Mitaka, Tokyo 181-8588, Japan
email: takaya.nozawa@nao.ac.jp

Abstract. Recent far-infrared (FIR) observations have revealed the presence of freshly formed dust with the masses exceeding 0.1 $M_\odot$ in young remnants of core-collapse supernovae (CCSNe) such as SN 1987A and Cassiopeia A. Meanwhile, dust masses derived from near- to mid-infrared (N/MIR) observations of CCSNe a few years after explosions are on the order of 10^{-5}–10^{-3} $M_\odot$. Here, we demonstrate that such small dust masses as seen from N/MIR observations would not necessarily reflect the formation history of dust but could be just limited by the luminosity of the SN that can heat up dust formed in the ejecta.

Keywords. dust, extinction, infrared: stars, supernovae: general

---

1. Mass, temperature, and luminosity of dust formed in the SN ejecta

Suppose that the heating source of dust grains formed in the ejecta is the radiation of the SN. The luminosity of the SN, $L_{\rm SN}$, at late phases ($\gtrsim$~100 days) is estimated, given that it is powered by decay energy of ^{56}Co (e.g., Woosley *et al.* 1989). With ^{56}Co of 0.075 $M_\odot$, $L_{\rm SN}$ ranges from 6×10^{40} erg s^{-1} down to 8×10^{38} erg s^{-1} at 300 days to 700 days. On the other hand, the N/MIR observations show that the dust temperature is $\simeq$300–700 K at 300–700 days (e.g., Wooden *et al.* 1993). If all of the newly formed dust grains are heated up to the identical temperature, $T_{\rm dust}$, by absorbing some fraction of the SN radiation, $f_{\rm abs}$ (i.e., IR luminosity of dust emission is given by $L_{\rm IR} = f_{\rm abs} L_{\rm SN}$), the dust mass, $M_{\rm dust}$, is determined from the energy (luminosity) balance as follows:

$$\begin{aligned} M_{\rm dust} &= \frac{f_{\rm abs} L_{\rm SN}}{4\sigma T_{\rm dust}^4 \langle \kappa(T_{\rm dust}) \rangle} \\ &\simeq 0.01 \left(\frac{f_{\rm abs}}{0.5}\right) \left(\frac{L_{\rm SN}}{6\times10^{40}\ {\rm erg\ s^{-1}}}\right) \left(\frac{T_{\rm dust}}{300\ {\rm K}}\right)^{-4} \left(\frac{\langle \kappa(T_{\rm dust}) \rangle}{660\ {\rm cm^2\ g^{-1}}}\right)^{-1} M_\odot, \end{aligned}$$

where $\langle \kappa(T_{\rm dust}) \rangle = 660$ cm^2 g^{-1} is the lowest value of the Planck-averaged mass absorption coefficients in the range of $T_{\rm dust}$ = 300–700 K for both silicate and carbon grains.

The above equation points out that the mass of dust which can be heated to $T_{\rm dust} \geqslant 300$ K should be less than 10^{-2} $M_\odot$. In other words, even if dust grains in excess of 0.1 $M_\odot$ are formed at 1–2 years post-explosion, the SN luminosity is not high enough to heat all of them to temperatures higher than $\simeq$150 K. Hence, it is likely that N/MIR observations, which is sensitive to dust emission with >200 K, have detected only a tiny fraction of newly formed dust. This may suggest that the majority of dust grains form inside dense gas clumps so that their temperatures are too low to be detected at N/MIR wavelengths.

References

Wooden, D. H., *et al.* 1993, *ApJS*, 88, 477
Woosley, S. E., Pinto, P. A., & Hartmann, D. 1989, *ApJ*, 346, 395

The Lives and Death-Throes of Massive Stars
Proceedings IAU Symposium No. 329, 2016
J.J. Eldridge, J.C. Bray, L.A.S. McClelland & L. Xiao, eds.

doi:10.1017/S1743921317003143

Tidal tearing of circumstellar disks in Be/X-ray and gamma-ray binaries

Atsuo T. Okazaki

Faculty of Engineering, Hokkai-Gakuen University, Toyohira-ku, Sapporo 062-8605, Japan
email: okazaki@lst.hokkai-s-u.ac.jp

Abstract. About one half of high-mass X-ray binaries host a Be star [an OB star with a viscous *decretion* (slowly outflowing) disk]. These Be/X-ray binaries exhibit two types of X-ray outbursts (Stella *et al.* 1986), normal X-ray outbursts ($L_X \sim 10^{36-37}$ erg s^{-1}) and occasional giant X-ray outbursts ($L_X > 10^{37}$ erg s^{-1}). The origin of giant X-ray outbursts is unknown. On the other hand, a half of gamma-ray binaries have a Be star as the optical counterpart. One of these systems [LS I +61 303 ($P_{\rm orb} = 26.5$ d)] shows the *superorbital* (1,667 d) modulation in radio through X-ray bands. No consensus has been obtained for its origin. In this paper, we study a possibility that both phenomena are caused by a long-term, cyclic evolution of a highly misaligned Be disk under the influence of a compact object, by performing 3D hydrodynamic simulations. We find that the Be disk cyclically evolves in mildly eccentric, short-period systems. Each cycle consists of the following stages:

1) As the Be disk grows with time, the initially circular disk becomes eccentric by the Kozai-Lidov mechanism.
2) At some point, the disk is tidally torn off near the base and starts precession.
3) Due to precession, a gap opens between the disk base and mass ejection region, which allows the formation of a new disk in the stellar equatorial plane (see Figure 1).
4) The newly formed disk finally replaces the precessing old disk.

Such a cyclic disk evolution has interesting implications for the long-term behavior of high energy emission in Be/X-ray and gamma-ray binaries.

Keywords. hydrodynamics; stars: emission-line, Be; X-rays: binaries

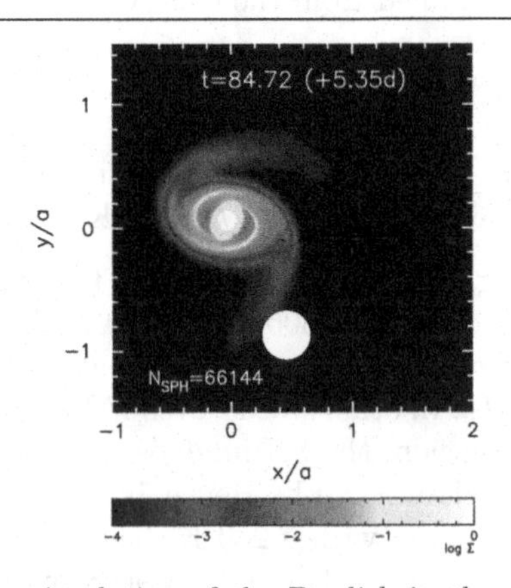

Figure 1. A snapshot from a simulation of the Be disk in the mildly-eccentric, short-period Be/X-ray binary 4U 0115+634. The disk consists of a precessing outer part and a newly formed inner part. The white filled circle denotes the Roche lobe radius of the neutron star.

Reference

Stella L., White N. E., Rosner R. 1986, ApJ, 208, 669

The Lives and Death-Throes of Massive Stars
Proceedings IAU Symposium No. 329, 2016
J.J. Eldridge, J.C. Bray, L.A.S. McClelland & L. Xiao, eds.

doi:10.1017/S1743921317002460

Observational signatures of hot-star magnetospheres

Mary E. Oksala[1,2]

[1]Department of Physics, California Lutheran University
email: moksala@callutheran.edu
[2]LESIA, Observatoire de Paris

Abstract. Magnetic fields play an important role in shaping the circumstellar environment of hot, massive stars. Observational diagnostics give clues to the presence of magnetism across the entire electromagnetic spectrum. Infrared features can show more complex structure, indicating they may probe deeper opacities than optical features. Optical and infrared features mimic each other, with identical blue and red peak variations and identical peak velocity of material. These comparisons indicate the location of the infrared and optical emitting material is similar. Longer wavelength diagnostics are currently being developed and tested. IR spectroscopy is a viable tool to detect magnetic candidates in the Galactic center and star forming regions.

Keywords. techniques: spectroscopic, infrared: stars, stars: early-type, stars: magnetic fields

---

Optical and infrared spectra identify material trapped in the star's magnetosphere, detected via hydrogen recombination lines; the emission pattern depends on magnetic and stellar properties (i.e. field strength, mass-loss rate, stellar geometry).

1. Motivation

- Identify tools to understand dynamics and structure of massive star magnetospheres
- Explore the possibility of infrared to probe deeper through optically thick regions
- Confront differences between observation and theory. Can infrared reveal structures theoretically predicted, but undetected in optical?

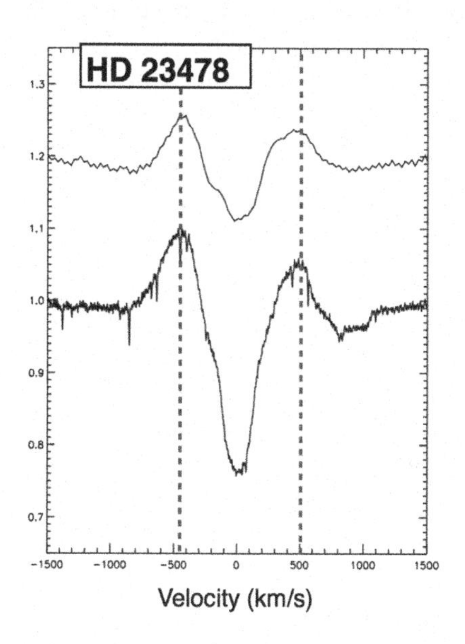

Figure 1. Spectral features for the magnetic, rapidly-rotating B-type star HD 23478. The star was identified as a magnetic candidate due to its emission features in the infrared, prior to confirmation of the presence of a magnetic field. The Brγ line (top, Gemini/GNIRS, R $\sim$ 6000) and the Hα line (bottom, TBL/Narval, R $\sim$ 65000) exhibit similar line shape and maximum peak velocity. The blue dashed lines indicate the peak velocity in the optical. These comparisons indicate the location of the infrared and optical emitting material is similar. The decreased stellar flux in the infrared could make the emission features more easy to observe, and allowing for detection of lower density magnetospheres, which remain invisible in the optical.

The Lives and Death-Throes of Massive Stars
Proceedings IAU Symposium No. 329, 2016
J.J. Eldridge, J.C. Bray, L.A.S. McClelland & L. Xiao, eds.

doi:10.1017/S1743921317002551

Shear-driven transport and mixing in the interiors of massive stars

Vincent Prat and Stéphane Mathis

Laboratoire AIM Paris-Saclay, CEA/DRF - CNRS - Université Paris Diderot, IRFU/SAp
Centre de Saclay, F-91191 Gif-sur-Yvette, France

Abstract. Turbulent transport and mixing generated by hydrodynamic instabilities triggered by rotation gradients are key mechanisms in the evolution of massive stars. We present here a summary of the progresses on shear-induced mixing obtained with numerical simulations, along with a new prescription for horizontal turbulence.

Keywords. Hydrodynamics, instabilities, turbulence, diffusion, stars: interiors, stars: rotation

---

Since Zahn (1992) proposed the formalism for shear mixing in stellar radiative zones, several prescriptions based on phenomenological arguments have been proposed. Spectroscopic and asteroseismic data show that transport in stars is much stronger than what is predicted by these models (e.g. Aerts 2015). Numerical simulations appear as a crucial tool to reliably estimate the transport due to magneto-hydrodynamical processes.

Stellar radiative zones are characterised by a very high thermal diffusivity. To reduce the prohibitive computational cost that this induces, Prat & Lignières (2013) used the small-Péclet-number approximation (Lignières 1999) and showed that it is consistent with the model of Zahn (1992). Later, Prat & Lignières (2014) and Prat *et al.* (2016) investigated the effect of chemical stratification and viscosity on vertical shear mixing. They showed that these two ingredients are able to inhibit the transport and proposed new prescriptions, which generally predict even less transport than older models.

In Mathis *et al.* (2017b), we also proposed a new description of horizontal mixing which accounts for the strong anisotropy resulting from the interplay between stable stratification in the radial direction and the Coriolis acceleration in the horizontal directions. Our new prescription allows us to recover a weak differential rotation in the radiative core of the Sun (Mathis *et al.* 2017a). The validity of this new model needs to be confirmed by numerical simulations, and the implications for massive stars have to be investigated.

Acknowledgements

The authors acknowledge funding by ERC grant SPIRE 647383.

References

Aerts, C. 2015, *AN* **336**, 477
Lignières, F. 1999, *A&A* **348**, 933
Mathis, S., Amard, L., Charbonnel, C., Palacios, A., & Prat, V. 2017a, *A&A*, submitted
Mathis, S., Prat, V., Amard, L., Charbonnel, C., & Palacios, A. 2017b, *A&A*, submitted
Prat, V., Guilet, J., Viallet, M., & Müller, E. 2016, *A&A* **592**, A59
Prat, V. & Lignières, F. 2013, *A&A* **551**, L3
Prat, V. & Lignières, F. 2014, *A&A* **566**, A110
Zahn, J.-P. 1992, *A&A* **265**, 115

The Lives and Death-Throes of Massive Stars
Proceedings IAU Symposium No. 329, 2016
J.J. Eldridge, J.C. Bray, L.A.S. McClelland & L. Xiao, eds.

doi:10.1017/S1743921317000229

FASTWIND reloaded: Complete comoving frame transfer for "all" contributing lines

Joachim Puls

LMU Munich, Universitätssternwarte, Scheinerstr. 1, D-81679 München, Germany
email: uh101aw@usm.uni-muenchen.de

Abstract. FASTWIND is a unified NLTE atmosphere/spectrum synthesis code originally designed (and frequently used) for the optical/IR spectroscopic analysis of massive stars with winds. Until the previous version (v10), the line transfer for background elements (mostly from the iron-group) was performed in an approximate way, by calculating the individual line-transitions in a single-line Sobolev or comoving frame approach, and by adding up the individual opacities and source functions to quasi-continuum quantities that are used to determine the radiation field for the complete spectrum (see Puls *et al.* 2005, A&A 435, 669, and updates).

We have now updated this approach (v11) and calculate, for *all* contributing lines (from elements H to Zn), the radiative transfer in the comoving frame, thus also accounting for line-overlap effects in an "exact" way. Related quantities such as temperature, radiative acceleration and formal integral have been improved in parallel. For a typical massive star atmospheric model, the computation times (from scratch, and for a modern desktop computer) are 1.5 h for the atmosphere/NLTE part, and 30 to 45 minutes (when not parallelized) for the formal integral (i.e., SED and normalized flux) in the ranges 900 to 2000 and 3800 to 7000 Å($\Delta\lambda = 0.03$ Å).

We compare our new with analogous results from the alternative code CMFGEN (Hillier & Miller 1998, ApJ 496, 407, and updates), for a grid consisting of 5 O-dwarf and 5 O-supergiant models of different spectral subtype. In most cases, the agreement is very good or even excellent (i.e., for the radiative acceleration), though also certain differences can be spotted. A comparison with results from the previous, approximate method shows equally good agreement, though also here some differences become obvious. Besides the possibility to calculate the (total) radiative acceleration, the new FASTWIND version will allow us to investigate the UV-part of the spectrum in parallel with the optical/IR domain.

Keywords. methods: numerical, stars: atmospheres, stars: early-type, stars: mass loss

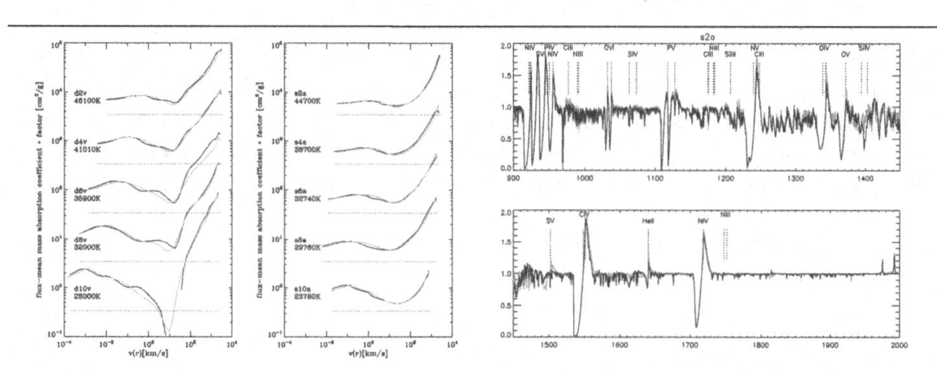

Figure 1. Left: Flux-mean mass absorption coefficient, a quantity directly proportional to the total radiative acceleration, for dwarf (left) and supergiant models (right) from FW v11 (black) and CMFGEN (green). The red lines indicate the corresponding "acceleration" by pure Thomson scattering. All curves (but the lowest ones) have been shifted by multiples of 1 dex. **Right:** Predicted UV spectrum ($v \sin i = 80 \mathrm{km\,s^{-1}}$) in the range between 900 and 2000 Å, for a supergiant model at $T_{\rm eff} = 44{,}700$ K, as synthesized by FW v11 (black) and CMFGEN (green).
JP gratefully acknowledges a travel grant by the German DFG under grant Pu 117/9-1.

The Lives and Death-Throes of Massive Stars
Proceedings IAU Symposium No. 329, 2016
J.J. Eldridge, J.C. Bray, L.A.S. McClelland & L. Xiao, eds.

doi:10.1017/S1743921317001855

CHIPS: The Carina High-contrast Imaging Project of massive Stars

Alan Rainot[1], Hugues Sana[1], Carlos A. Gomez-Gonzalez[2], Olivier Absil[2], Philippe DeLorme[3], Leonardo Almeida[4], Saida Caballero-Nieves[5], Kaitlin Kratter[6], Sylvestre Lacour[7], Jean-Baptiste Le Bouquin[3], Laurent Pueyo[8] and Hans Zinnecker[9]

[1]KU Leuven, Belgium; [2]Université de Liège, Belgium; [3]IPAG Grenoble, France; [4]Sao Paulo, Brazil; [5]FIT Florida, USA; [6]Arizona, USA; [7]Observatoire de Paris, France; [8]STScI, Baltimore, USA; [9]USRA Sofia, USA
Email: alan.rainot@kuleuven.be

Abstract. The formation of massive stars remains one of the most intriguing questions in astrophysics today. The main limitations result from the difficulty to obtain direct observational constraints on the formation process itself. In this context, the Carina High-contrast Imaging Project of massive Stars (CHIPS) aims to observe all 80+ O stars in the Carina nebula using the new VLT 2nd-generation extreme-AO instrument SPHERE. This instrument offers unprecedented imaging contrast allowing us to detect the faintest companions around massive stars. These novel observational constraints will help to discriminate between the different formation scenarios by comparing their predictions for companion statistics and properties.

Keywords. binaries: general, binaries (including multiple): close, binaries: general, stars: imaging, techniques: interferometric, methods: data analysis

---

Description of the Project

CHIPS builds on the foundations of the Southern Massive Stars at High angular resolutions survey (SMaSH+, Sana *et al.* 2014), which found a larger number of faint companions at separations roughly corresponding to expectations of the outer edge of the accretion disk. CHIPS will investigate whether low-mass companions exist at closer separations or whether there is a characteristic length at which the flux versus separation distribution changes. In this project we use the second VLT generation extreme-AO instrument SPHERE which allows us to reach objects 100 times fainter than achieved by SMaSH+ and other high contrast imaging campaigns (Turner *et al.* 2008, Maíz Apellániz 2010; see comparison in Sana & Evans 2010). CHIPS will observe 84 O and WR stars reaching a flux contrast of $\Delta mag = 10$ at separations as close as $0.15''$. The data analysis is performed using a state-of-the-art code named VIP, the Vortex Image Processing package (Gomez Gonzalez *et al.* 2016).

References

Gomez-Gonzalez, C. A., *et al.* 2016, *A&A*, 589, A54
Maíz Apellániz, J. 2010, *A&A*, 518, A1
Sana, H., *et al.* 2014, *AJ*, 215, 1
Sana, H. & Evans, C. J. 2011, *IAU Symp.* 272 in Active OB Stars: Structure, Evolution, Mass Loss, and Critical Limits, ed. C. Neiner, G. Wade, G. Meynet, & G. Peters (Cambridge: Cambridge Univ. Press), 474
Turner, N. H., ten Brummelaar, T. A., Roberts, L. C., *et al.* 2008, *AJ*, 136, 554

The Lives and Death-Throes of Massive Stars
Proceedings IAU Symposium No. 329, 2016
J.J. Eldridge, J.C. Bray, L.A.S. McClelland & L. Xiao, eds.

doi:10.1017/S1743921317003027

Properties of the O-type giants and supergiants in 30 Doradus

O. H. Ramírez-Agudelo and VFTS consortium
UK ATC, Royal Observatory Edinburgh, Blackford Hill, Edinburgh, EH9 3HJ, UK
email: oscar.ramirez@stfc.ac.uk

Abstract. We discuss the stellar and wind properties of 72 presumably single O-type giants, bright giants, and supergiants in the 30 Doradus region. This sample constitutes the largest and most homogeneous sample of such stars ever analyzed and offers the opportunity to test models describing their main-sequence evolution.

Keywords. stars: early-type – stars: evolution – stars: fundamental parameters – Magellanic Clouds – Galaxies: star clusters: individual: 30 Doradus

---

The 30 Doradus (30 Dor) region in the Large Magellanic Cloud (LMC) is one of the best possible laboratory to investigate aspects of the formation and evolution of massive stars. In the framework of the VLT-FLAMES Tarantula Survey (VFTS; Evans *et al.* 2011), we determine stellar and wind properties of 72 presumably single O-type stars in 30 Dor with luminosity class (LC) III (giants), II (bright giants), and I (supergiants).

We apply an automated fitting method for quantitative spectroscopic analysis of O-stars to determine the following key stellar properties of our sample of stars: the effective temperature, the surface gravity, the mass-loss rate, the helium abundance, the micro-turbulent velocity, and the projected rotational velocity. Details on the method are given in Ramírez-Agudelo *et al.* 2017.

Our main findings can be summarized as follows:

- We present an empirical effective temperature versus spectral subtype calibration for the stars with LC III and LC I. The calibration for the LC III shows a +1 kK offset compared to similar Galactic calibrations.
- According to the spectroscopic and classical Hertzsprung-Russell diagrams our sample of O stars occupy the region predicted for the core hydrogen-burning phase. Late LC III and LC II stars occupy the region where O9.5-9.7 V stars are expected, but where few morphologically-classified dwarfs are seen. This behavior may reflect an intricacy in the luminosity classification at late O spectral subtype.
- The surface helium abundances of our sample stars are generally in agreement with the initial composition of the LMC, but five stars display enriched helium abundances. These He-enriched stars have moderate rotational velocities ($< 200\,\mathrm{km\,s^{-1}}$), and hence do not agree with predictions of rotational mixing in main-sequence stars. They may reveal other physics not including in the models, such as binary-interaction effects.
- Adopting theoretical results for the wind velocity law, for stars brighter than $10^5\,L_\odot$ the measured (unclumped) mass-loss rates can be brought in agreement with predictions if the clump filling factor in the Hα and He II λ4686-forming region were $\sim 1/8$ to $1/6$.

References

Evans *et al.*, 2011, A&A, 530, A108
Ramírez-Agudelo *et al.*, 2017, A&A, 600, A81

The Lives and Death-Throes of Massive Stars
Proceedings IAU Symposium No. 329, 2016
J.J. Eldridge, J.C. Bray, L.A.S. McClelland
& L. Xiao, eds.

doi:10.1017/S1743921317002873

Massive stars in young VVV clusters

S. Ramírez Alegría[1,2], J. Borissova[3,2] and A.-N. Chené[4]

[1]Instituto de Astronomía, UCN, Chile [2]Millennium Institute of Astrophysics, MAS [3]Instituto de Física y Astronomía, U. de Valparaíso, Chile [4]Gemini North Observatory, USA

The role of massive stars in the Galactic evolution is crucial. During their lifetime these stars change the kinematics around them through stellar winds, affect the formation of new stars, ionise and chemically enrich the media with the final supernova explosion. But the census of both massive stars and their host clusters is still poor. We expect that still $\sim$100 of galactic massive stellar clusters remains unknown (Hanson & Popescu, 2008).

Trying to improve this census, we built an homogeneous sample of physically characterized clusters. This long-term effort ($\sim$5 years), combines near-infrared spectrophotometric data acquisition, reduction, and analysis. Currently we have a representative sample of 65 clusters, allowing the study of relations between clusters, the stellar content and our Galaxy. Some types of clusters included in our database are:

• **Wolf-Rayet clusters (Chené *et al.* 2013, 2015; Borissova *et al.* 2014)**: A total of 7 clusters with 9 WR (plus 3 WR/O If, 2 RSG & 1 BSG) objects have been discovered by our group. The clusters are young (2-7 Myr), moderately massive (800-3.000 $M_\odot$), highly obscured and compact (1-2 pc). We observe a (M_{CL}-m_{max}) relation, except for VVV CL041, which has a total mass of $3 \cdot 10^3 M_\odot$ and it hosts a WN8h object, with an initial mass of $\sim 100 M_\odot$. This object apparently would not follow a Kroupa's IMF with optimal sampling (Kroupa *et al.* 2013). A review of this object can be found in the article by Chené *et al.*, in this proceeding.

• **OB-stars clusters (Ramírez Alegría *et al.* 2014, 2016)**: Six very young ($<$ 20 Myr, via main -Lejeune & Schaerer, 2001- and PMS -Siess *et al.* 2000- isochrone fitting), without signs of an evolved population clusters, characterized using VVV PSF-photometry & ISAAC-VLT spectroscopy. Their total masses follow a (M_{CL}-m_{max}) relation, similar to the presented by Weidner *et al.* (2013).

• **Clusters with MYSOs (Borissova *et al.* 2016)**: Eight clusters with confirmed MYSO, selected using VVV (NIR) and GLIMPSE (MIR) photometry, and observed variability for some YSO. The variability was observed using the VVV multi-epoch survey.

References

Borissova, J., Chené, A.-N., Ramírez Alegría, S., *et al.* 2014, *A&A* 569, A24
Borissova, J., Ramírez Alegría, S., Alonso, J., *et al.* 2016, *AJ* 152, 74
Chené, A.-N., Borissova, J., Bonatto, C., *et al.* 2013, *A&A* 549, A98
Chené, A.-N., Ramírez Alegría, S., Borissova, J., *et al.* 2015, *A&A* 584, A31
Hanson, M. M. & Popescu, B. 2008, in: F. Bresolin, P. A. Crowther & J. Puls (eds.), *Massive stars as cosmic engines*, Proc. IAU Symposium No. 250, p. 307-312
Kroupa, P., Weidner, C., Pflamm-Altenburg, J., *et al.* 2013, in: Oswald T. D & Gilmore, G. (eds.), *Planets, Stars and Stellar Systems*, Volume 5: Galactic Structure and Stellar Populations (Springer, New York), p. 115
Lejeune, T. & Schaerer, D. 2001, *A&A* 366, 538
Ramírez Alegría, S, Borissova, J. Chené, A.-N., *et al.* 2014, *A&A* (Letters) 564, L9
Ramírez Alegría, S, Borissova, J. Chené, A.-N., *et al.* 2016, *A&A* 588, 40
Siess, L., Dufour, E., & Forestini, M. 2000, *A&A* 358, 593
Weidner, C., Kroupa, P., & Pflamm-Altenburg, J. 2013, *MNRAS* 434, 84

The Lives and Death-Throes of Massive Stars
Proceedings IAU Symposium No. 329, 2016
J.J. Eldridge, J.C. Bray, L.A.S. McClelland & L. Xiao, eds.

doi:10.1017/S1743921317002897

Massive pre-main-sequence stars in M17

M. C. Ramírez-Tannus[1], L. Kaper, A. de Koter, F. Tramper, H. Sana, O. H. Ramírez-Agudelo, A. Bik, L. E. Ellerbroek and B. B. Ochsendorf

[1]API, University of Amsterdam, The Netherlands; email: m.c.ramireztannus@uva.nl

We obtained VLT/X-shooter spectra of twelve candidate young massive stars previously selected by Hanson *et al.* (1997) in the giant HII region M17. An analysis of their spectra using FASTWIND models (Puls *et al.* 2005) shows that they span a mass range of 6 - 20 $M_\odot$. We identify the presence of gaseous and dusty disks around six sources based on emission lines in the spectrum and infrared continuum excess. By comparing their position in the HRD with theoretical PMS tracks we conclude that these are genuine pre-main-sequence (PMS) stars contracting towards the main sequence after having experienced high mass-accretion ($M_{\rm acc} \sim 10^{-4} - 10^{-3}$ $M_\odot$ yr^{-1}). Three sources are close to the ZAMS, for two objects we did not detect a gaseous disk, and one object might be a PMS star without a disk. Though the sample is small, it does allow for a first analysis of statistical properties of a massive PMS star population (Ramírez-Tannus *et al.* 2017, A&A, in press). We find a strikingly small radial velocity dispersion, which is at odds with similar distributions of few-Myr-old OB star populations. This may either point to M17 having a low binary fraction ($f_{\rm bin}$ = 0.12) or to a lack of short period binaries (in our sample $P_{\rm cutoff}$ > 9 months) relative to these older populations (Sana *et al.* 2017), characterised by $f_{\rm bin}$ = 0.7 and $P_{\rm cutoff}$ > 1.4 days (Sana *et al.* 2012).

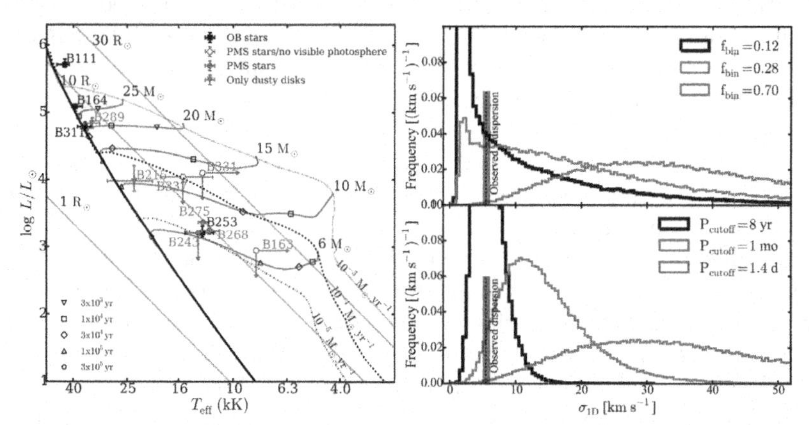

Figure 1. *Left:* HRD for our sources in M17. The light-blue dots show the PMS stars surrounded by gaseous and dusty disks, the black squares the OB stars, the brown triangles objects surrounded by dusty disks. The filled symbols represent the stars that we modelled with FASTWIND and the open symbols represent the objects for which we derived $T_{\rm eff}$ and $L/L_\odot$ from their spectral type (from Ramírez-Tannus *et al.* 2017). *Right:* Simulated $\sigma_{\rm 1D}$: Top: For different binary fractions, we discard $f_{\rm bin}$ > 0.34 and > 42 at the 10 and 5% level, respectively. Bottom: For different cutoff periods, we discard $P_{\rm cutoff}$ < 131 and < 47 days at the 10 and 5% level, respectively. The vertical line represents the observed $\sigma_{\rm 1D}$ and the associated error for our M17 stars (from Sana *et al.* 2017).

References

Hanson, M. M., Howarth, I. D., & Conti, P. S. 1997, *ApJ*, 489, 698
Hosokawa, T. & Omukai, K. 2009, *ApJ*, 691, 823
Puls, J., Urbaneja, M. A., Venero, R., *et al.* 2005, *A&A*, 435, 669
Sana, H., de Mink, S. E., de Koter, A., *et al.* 2012, *Science*, 337, 444
Sana, H., Ramirez-Tannus, M., de Koter, A., & Kaper, L., *et al.* 2017, ArXiv:1702.02153

The Lives and Death-Throes of Massive Stars
Proceedings IAU Symposium No. 329, 2016
J.J. Eldridge, J.C. Bray, L.A.S. McClelland & L. Xiao, eds.

doi:10.1017/S1743921317001673

Not so lonely: The LBV binary HR Car

T. Rivinius[1], H. M. J. Boffin[1,2], A. Mérand[1,2], A. Mehner[1], J.-B. LeBouquin[3], D. Pourbaix[4], W.-J. de Wit[1] and C. Martayan[1]

[1] ESO - European Organisation for Astronomical Research in the Southern Hemisphere, Chile
[2] ESO, Karl-Schwarzschild-Str. 2, 85748 Garching, Germany
[3] Institut de Planétologie et d'Astrophysique de Grenoble, 38041 Grenoble, France
[4] Institut d'Astronomie et d'Astrophysique, Université Libre de Bruxelles (ULB), Belgium

Abstract. Luminous Blue Variables (LBVs) are a brief phase in the evolution of massive stars, but a very important one. The giant eruptions remain enigmatic, but the discovery of the flagship LBV η Car to be a five-year highly eccentric binary put focus on possible binarity induced mechanism for these outbursts, and prompted binarity searches among LBVs. While several wide LBV binaries were identified, HR Car is the first system found to be similar to η Car, i.e., relatively close & eccentric.

Keywords. binaries: general, stars: individual: HR Car, stars: variables: other

---

After an interferometric snapshot campaign with AMBER at the VLTI in 2013/14, HR Car was suspected to be a close binary. A follow-up campaign with PIONIER, a 4-beam interferometric instrument working in the H-band continuum confirmed the binarity and allowed initial constraints on the orbit. Analysis of the obtained PIONIER data firmly establishes the presence of two different sources with a contrast ratio of about 1:6 to 1:9. The components are marginally resolved (P: 0.38 ± 0.08 mas, S: 0.85 ± 0.2 mas). The best orbital fit parameters from current data are shown in Table 1, but since the first orbit is not yet complete, the parameters are preliminary: While shorter orbits are excluded, much longer ones remain possible. See Boffin *et al.*(2016) for a complete discussion of the observations and fitting process.

At the distance of 5.4 ± 0.4 kpc, the primary's size would be $220 \pm 60\,R_{\odot}$, too large for the LBV in minimum, but acceptable if wind/nebulosity is taken into account. The secondary, much larger than the primary, could be either a red supergiant, or disturbed from equilibrium by a previous orbital passage: HR Car was in a maximum until 2001.

During LBV minimum, the periastron distance is about three times larger than the sum of radii. The last periastron was about 2014.5, around the times of AMBER observations, which show some sign of wind-wind interaction. During LBV maximum, though, the secondary may even pass through the extended outer layers of the primary, a scenario that may lead to clearly observable effects and new insights into η Car type binarity.

Table 1. Parameters of the best orbit that fits the PIONIER observations.

Ω (deg.)	46.9±0.6	P (days)	4557.5±21.0	T_0 (MJD)	56990.6±16.0
a (mas)	3.324±0.026	e	0.4±0.2	i (deg.)	119.2±0.7
ω (deg.)	201.9±2.1	Flux ratio H-band	1:6 to 1:9		

Reference

Boffin, H. M. J., Rivinius, T., Mérand, A., *et al.* 2016, A&A, 593, A90

The Lives and Death-Throes of Massive Stars
Proceedings IAU Symposium No. 329, 2016
J.J. Eldridge, J.C. Bray, L.A.S. McClelland & L. Xiao, eds.

doi:10.1017/S1743921317003039

Kinematics of SNRs CTB 109 and G206.9+2.3

Margarita Rosado[1], **Mónica Sánchez-Cruces**[2] **and Patricia Ambrocio-Cruz**[3]

[1]Instituto de Astronomía, UNAM,
Apartado Postal 70-264, Ciudad Universitaria 04510,
Ciudad de México, México
email: margarit@astro.unam.mx

[2]ESFM-IPN, México,
email:

[3]ITSOEH, México,
email: silviap@uaeh.edu.mx

Abstract. We present results of optical observations in the lines of Hα and [SII] (λ 6717 and 6731 Å) obtained with the UNAM Scanning Fabry-Perot Interferometer PUMA (Rosado *et al.* 1995,RMxAASC, 3, 263) aimed at obtaining the kinematical distance, shock velocity and other important parameters of two supernova remnants (SNRs) with optical counterparts. We discuss on how kinematical distances thus obtained fit with other distance determinations. The studied SNRs are CTB 109 (SNR G109.1 − 1.0) hosting a magnetar (Sánchez-Cruces *et al.* 2017, in preparation) and the SNR G206.9 + 2.3 (Ambrocio-Cruz *et al.* 2014,RMxAA, 50, 323), a typical supernova remnant, to have a comparison. In Fig. 1 is depicted the [SII] line emission of two filaments of the optical counterpart of SNR CTB 109. We find complex radial velocity profiles obtained with the Fabry-Perot interferometer, revealing the presence of different velocity components. From these velocity profiles we obtain the kinematical distance, an expansion velocity of 188 km/s and an initial energy of 8.1 x 10^{50} ergs. These values are rather typical of other SNRs regardless that SNR CTB 109 hosts a magnetar. Thus, the mechanical energy delivered in the supernova explosion forming the magnetar does not seem to impact more than other SNe explosions the interstellar medium. This work has been funded by grants IN103116 and 253085 from DGAPA-UNAM and CONACYT, respectively.

Keywords. ISM: supernova remnants, ISM: kinematics and dynamics, shock waves.

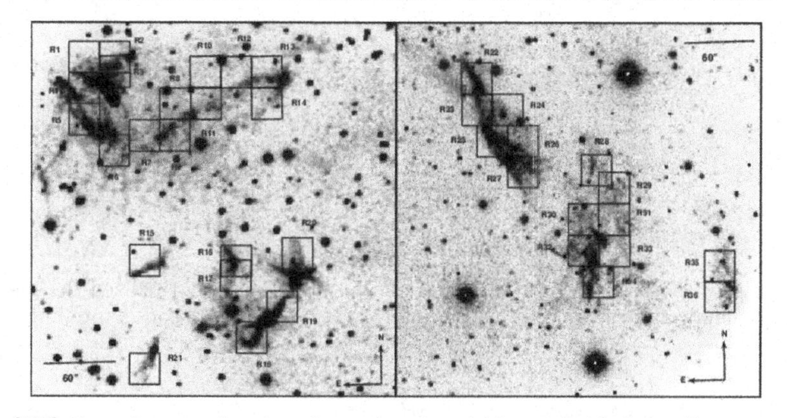

Figure 1. [SII] direct images showing the regions in which we divided the filaments in order to carry out our analysis (R1 to R36). Left and right panels correspond to the Northeastern and Southeastern filaments, respectively.

The Lives and Death-Throes of Massive Stars
Proceedings IAU Symposium No. 329, 2016
J.J. Eldridge, J.C. Bray, L.A.S. McClelland & L. Xiao, eds.

doi:10.1017/S1743921317003477

HD151018: strong magnetic field in a giant O-type star?

M. S. Rubinho, A. Damineli, A. Carciofi and D. Moser
Instituto de Astronomia, Ciências Atmosféricas e Geofísica, Universidade de São Paulo

Abstract. HD151018 is a normal O-type giant with periodic variability of the wind seen in both spectroscopy and linear polarimetry. The characteristics of the wind emission strongly indicate a magnetically confined wind. This work presents the observational results and an initial modelling of the polarimetric modulation of this star.

Keywords. magnetic fields, polarimetric, spectroscopic.

While there are many **B-type stars** with known magnetic fields, the number drops to less than a dozen when it comes to **O stars** (Wade & MiMeS, 2015). HD 151018 is a O-type giant star with wind emission, which strongly indicates a wind modulated by the magnetic field (confining here has a very specific meaning eg., Ud-Doula, 2008). The aim of this work is to report our observations of the polarimetric modulation of HD 151018. Also, to get insights of the magnetic wind confinement configuration in this star. HD 151018 was monitored both spectroscopically and polarimetrically over the last few years at the OPD/LNA observatory in Brazil. The H_α emission profiles resembles others similar O stars with magnetic field. The Fig.1a shows a phase-locked line profile variations in H_α with a period of about 4.3d. The Fig.1b presents the results of linear polarization which shows a data variation with amplitude of $\geqslant 0.02\%$ and a period of $\approx 4.5d$. We also present our first findings on the application of physically motivated model (PMM), for polarimetry, which consists of two blobs of gas connected by a disk, as created by Carciofi *et al.* (2013). Although the PMM model is preliminary, the model reproduces qualitatively well the polarimetric curve.

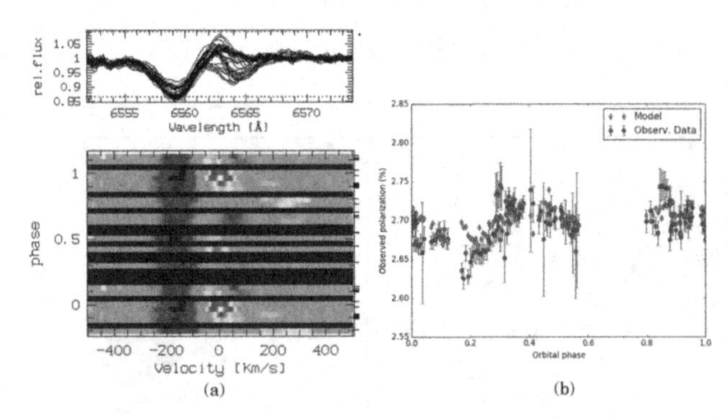

Figure 1. Observational data of HD151018 *Right top:* superposed normalized Hα spectra. *Right bottom:* Hα dynamical spectrum. *Left:* Polarimetric data and the proposed preliminar model.

References

Wade, G. A., MiMeS Collaboration 2015, *ASPCS*, 494, 30
Carciofi, A., Faes, D., Townsend, R., & Bjorkman, J. 2013, *ApJ*, 766, L9
Ud-Doula, A. 2008, *Clumping in Hot-Stars Winds*,125

The Lives and Death-Throes of Massive Stars
Proceedings IAU Symposium No. 329, 2016
J.J. Eldridge, J.C. Bray, L.A.S. McClelland
& L. Xiao, eds.

doi:10.1017/S1743921317003131

Modelling the thermal X-ray emission around the Galactic centre from colliding Wolf-Rayet winds

Christopher M. P. Russell[1], Q. Daniel Wang[2] and Jorge Cuadra[3]

[1]X-ray Astrophysics Laboratory, NASA/Goddard Space Flight Center,
Greenbelt, MD 20771, USA (NASA Postdoctoral Program Fellow, administered by USRA)
email: crussell@udel.edu

[2]Department of Astronomy, University of Massachusetts, Amherst, MA 01003, USA

[3]Instituto de Astrofísica, Pontificia Universidad Católica de Chile, 782-0436 Santiago, Chile

Abstract. We compute the thermal X-ray emission from hydrodynamic simulations of the 30 Wolf-Rayet (WR) stars orbiting within a parsec of Sgr A*, with the aim of interpreting the *Chandra* X-ray observations of this region. The model well reproduces the spectral shape of the observations, indicating that the shocked WR winds are the dominant source of this thermal emission. The model X-ray flux is tied to the strength of the Sgr A* outflow, which clears out hot gas from the vicinity of Sgr A*. A moderate outflow best fits the present-day observations, even though this supermassive black hole (SMBH) outflow ended ~100 yr ago.

Keywords. Galaxy: centre, X-rays: stars, stars: Wolf-Rayet, stars: winds, outflows

---

Fig. 1 shows the main results. See Russell *et al.* (2017) for further details of this work.

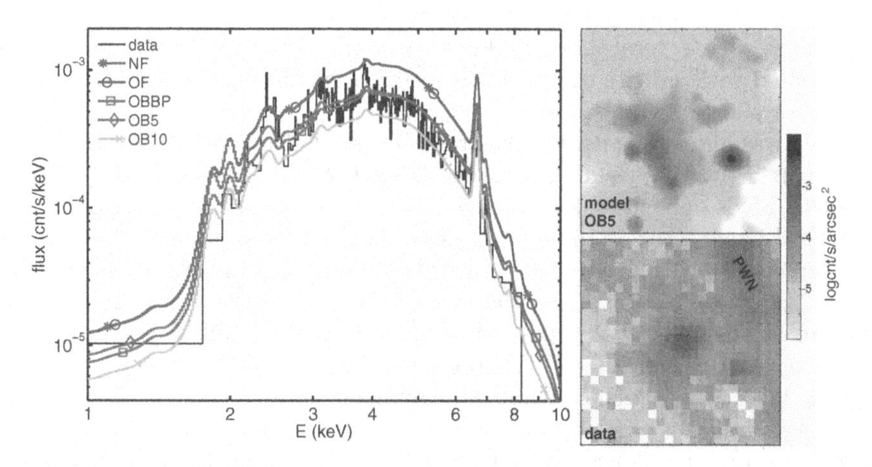

Figure 1. *Left*: ACIS-S/HETG zeroth-order spectra from 2″–5″ ring around Sgr A*. The SMBH outflow increases from NF (no feedback) to OB10 (strongest feedback). *Right*: ACIS-S/HETG zeroth-order 4-9 keV images of 12″×12″ centered on Sgr A*, showing the best-fit model (OB5) and the data. The non-thermal emission from the SMBH and pulsar wind nebula is not modeled.

Reference

Russell, C. M. P., Wang, Q. D., & Cuadra, J. 2017, *MNRAS*, 464, 4958

The Lives and Death-Throes of Massive Stars
Proceedings IAU Symposium No. 329, 2016
J.J. Eldridge, J.C. Bray, L.A.S. McClelland & L. Xiao, eds.

doi:10.1017/S1743921317001351

New radio observations of the Type IIn Supernova 1978K

Stuart D. Ryder[1], Erik C. Kool[2,1] and Rubina Kotak[3]

[1]Australian Astronomical Observatory, Sydney NSW 1670, Australia
email: sdr@aao.gov.au
[2]Dept. of Physics & Astronomy, Macquarie University, Sydney NSW 2109, Australia
[3]Astrophysics Research Centre, Queens University Belfast, United Kingdom

Abstract. SN 1978K is the oldest-known Type IIn supernova, and one of the closest. We report new radio observations at high frequency and spatial resolution. SN 1978K has been detected at 34 and 94 GHz with the Australia Telescope Compact Array, while Very Long Baseline Interferometry at 8.4 GHz has allowed us to derive the past average expansion velocity, which indicates significant deceleration as the blast wave interacts with the dense circumstellar medium.

Keywords. supernovae: general – supernovae: individual: SN 1978K – galaxies: individual: NGC 1313

Summary

At just 4.6 Mpc away, SN 1978K in the late-type barred spiral galaxy NGC 1313 is the second-closest example (after SN 1996cr in the Circinus Galaxy) of a Type IIn supernova, in which the Balmer optical emission lines show a narrow (few hundred km s^{-1}) emission peak atop a broader (several thousand km s^{-1}) profile. These characteristics are thought to be associated with a dense circumstellar medium arising from significant mass-loss by the progenitor star prior to explosion.

SN 1978K was observed with the Australia Telescope Compact Array at 34 GHz and at 94 GHz in Sep 2014, yielding fluxes of 2.9 ± 0.2 mJy and 1.2 ± 0.3 mJy, respectively. SN 1978K is only the third evolved extragalactic supernova (after SN 1987A and SN 1996cr) to be detected at these frequencies, but is >400× brighter than SN 1987A.

We observed SN 1978K with the Australian Long Baseline Array at 8.4 GHz on 29 March 2015. After deconvolving the resultant image by the 3 milli-arcsec beam, we find the current diameter of the remnant of SN 1978K is <5 milli-arcsec (0.1 pc), giving a past average expansion velocity <1 500 km s^{-1}.

Our radio light curve fitting constrains the remnant radial expansion to have $R \sim t^{0.78}$, which implies significant deceleration of the ejecta by a substantial circumstellar medium, and predicts a current expansion velocity ~1 000 km s^{-1}. Recent optical spectroscopy of SN 1978K by Kuncarayakti *et al.* (2016) shows remarkable similarities with SN 1987A, and line-widths indicating a current expansion velocity ~500–600 km s^{-1}, within a factor of 2 of that predicted by this model for SN 1978K.

Full details of the results outlined here can be found in Ryder *et al.* (2016).

References

Kuncarayakti, H., Maeda, K., Anderson, J., Hamuy, M., Nomoto, K., Galbany, L., & Doi, M. 2016, *MNRAS*, 458, 2063

Ryder, S., Kotak, R., Smith, I., Tingay, S., Kool, E., & Polshaw, J. 2016, *A&A*, 595, L9

The Lives and Death-Throes of Massive Stars
Proceedings IAU Symposium No. 329, 2016
J.J. Eldridge, J.C. Bray, L.A.S. McClelland & L. Xiao, eds.

doi:10.1017/S1743921317001995

The formation of Wolf-Rayet stars in the SMC is not dominated by mass transfer

Tomer Shenar, R. Hainich, H. Todt, A. Sander, W.-R. Hamann, A. Moffat, J. J. Eldridge, H. Pablo, L. Oskinova and N. Richardson

Institut für Physik & Astronomie, Universität Potsdam,
Karl-Liebknecht-Str. 24-25, 14476 Potsdam, Germany
email: shtomer@astro.physik.uni-potsdam.de

Abstract. We analyzed spectra of all Wolf-Rayet stars in the Small Magellanic Cloud (SMC). We find that, unlike predicted, mass-transfer in binaries is not needed to explain their formation.

Massive Wolf-Rayet (WR) stars, characterized by strong mass-loss and hydrogen depletion, can form either as single stars (e.g. Cassinelli 1979) or as mass donors in binaries (e.g. Paczynski 1973). Population synthesis studies (e.g. Meynet & Maeder 2005, Georgy *et al.* 2015) suggest that the majority or all of the WR stars in the SMC ($Z \approx 0.1-0.2\,Z_{\odot}$, Larsen *et al.* 2000) formed via the binary formation channel (see also Foellmi *et al.* 2003).

The Potsdam Wolf-Rayet (PoWR) model atmosphere code (Gräfener *et al.* 2002, Sander *et al.* 2015) is a state-of-the art tool for the analysis of hot stars (e.g. Shenar *et al.* 2015). Using the PoWR code, we performed a spectral analysis of all known single (Hainich *et al.* 2015) and binary (Shenar *et al.* 2016) WR stars in the SMC. By comparing the observed properties of each binary system to evolution tracks which account for mass-transfer (Eldridge *et al.* 2008), we could derive the complete set of initial masses and periods for the WR binaries in the SMC. We find that, while mass-transfer has likely occurred in most WR binaries, the WR primaries were initially massive enough to enter the WR phase regardless of mass-transfer ($M_{\rm i} \gtrsim 60\,M_{\odot}$). Combined with our results for the single WR stars, we conclude that the binary formation channel is not needed to account for the WR population in the SMC, in contradiction to previous predictions.

Why are there more massive WR stars in the SMC than expected? Where are the WR stars forming via mass-transfer in the SMC? These are key questions that should drive future studies and observational campaigns in the coming years.

References

Cassinelli, J. P. 1979, *ARA&A* , 17, 275
Eldridge, J. J., Izzard, R. G., & Tout, C. A. 2008, *MNRAS*, 384, 1109
Foellmi, C., Moffat, A. F. J., & Guerrero, M. A. 2003a, *MNRAS*, 338, 360
Georgy, C., Ekström, S., Hirschi, R., *et al.* 2015, in Wolf-Rayet stars proceedings, ed. W.-R. Hamann, A. Sander, H. Todt
Gräfener, G., Koesterke, L., & Hamann, W.-R. 2002, *A&A*, 387, 244
Hainich, R., Pasemann, D., Todt, H., *et al.* 2015, *A&A*, 581, A21
Larsen, S. S., Clausen, J. V., & Storm, J. 2000, *A&A*, 364, 455
Meynet, G. & Maeder, A. 2005, *A&A*, 429, 581
Paczynski, B. 1973, in IAU Symposium, Vol. 49, Wolf-Rayet and High-Temperature Stars, ed. M. K. V. Bappu & J. Sahade, 143
Sander, A., Shenar, T., Hainich, R., *et al.* 2015, *A&A*, 577, A13
Shenar, T., Hainich, R., Todt, H., *et al.* 2016, *A&A*, 591, A22
Shenar, T., Oskinova, L., Hamann, W.-R., *et al.* 2015, *ApJ*, 809, 135

The Lives and Death-Throes of Massive Stars
Proceedings IAU Symposium No. 329, 2016
J.J. Eldridge, J.C. Bray, L.A.S. McClelland & L. Xiao, eds.

doi:10.1017/S1743921317003155

CFHT/SITELLE Observations of the Ejected WR Nebula M1-67

N. St-Louis[1], M. Sévigny[2], L. Drissen[2] and T. Martin[2]

[1]Département de physique, Université de Montréal
C.P. 6128, Succ. Centre Ville, Montréal (Qc), H3C 3J7, Canada

[2]Département de physique, de génie physique et d'optique,
Université Laval, Québec (QC) G1V 0A6, Canada

Abstract. We present preliminary results of **Sitelle** observations of the M1-67 nebula surrounding the Wolf-Rayed stars WR 124.

Keywords. stars: Wolf-Rayet, stars: winds, outflows, ISM: individual (M1-67), ISM: bubbles

1. The M1-67 Nebula

M1-67, which surrounds the WN8h star WR 124, consist of N-enhanced and O-deficient gas (Esteban *et al.* 1991) with mixed-in dust (Vamvatira-Nakou *et al.* 2016). Sirianni *et al.* (1998) found that it consist of an expanding shell (46 km/s) and a bipolar outflow (88 km/s). The nebula was interpreted by Van der Sluys & Lamers (2003) as an LBV outburst interacting with the bow shock caused by the high velocity of this runaway star.

2. Sitelle Observations

In May 2016, we obtained a data cube of M1-67 using the optical imaging Fourier transform spectrograph **Sitelle**. This instrument, installed on the CFHT, covers a wavelength range from 350 to 900 nm with a field of view is $11'' \times 11''$ (with $0.32''$ pixels). In Figure 1, the Hα flux is shown in red and the [NII]λ6548,6584 flux in blue. Two different structures can clearly be seen: a NE-SW oval shape region where hydrogen is dominant (bottom) and a NW-SE filamentary ring in which nitrogen dominates (top).

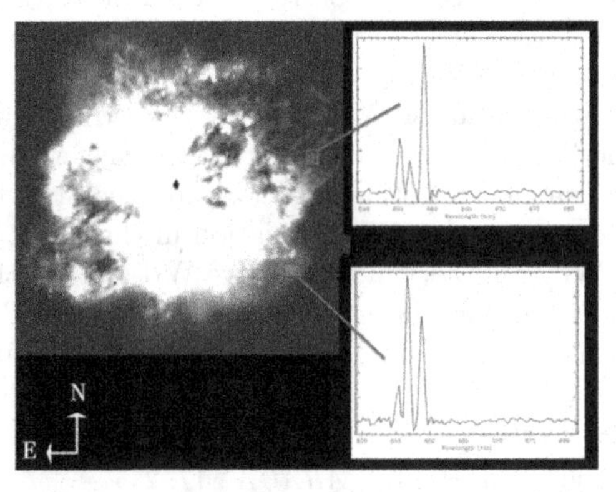

Figure 1. Sitelle line-map images of M1-67. Red is Hα, Blue is [N II].

This data will provide density, temperature and abundances for the entire nebula and allow us to kinematically separate the different components of the expanding gas. This will allow us to shed new light on the past evolutionary phases undergone by the star.

References

Esteban, C., Vilchez, J. M., Manchado, A., & Smith, L. J. 1991, *A& A*, 205
Sirianni, M., Nota, A., Pasquali, A., & Clampin, M. 1998, *A& A*, 335, 1089
Vamvatira-Nakou, *et al.* 2016, *A& A*, 588, A92
Van der Sluys, M. V. & Lamers, H. J. G.. L. M.. 2003, *A& A*, 398, 181

The Lives and Death-Throes of Massive Stars
Proceedings IAU Symposium No. 329, 2016
J.J. Eldridge, J.C. Bray, L.A.S. McClelland & L. Xiao, eds.

doi:10.1017/S174392131700103X

Spectropolarimetry of the BL-Ic SN 2014ad

H. F. Stevance and J. R. Maund

University of Sheffield, Department of Physics and Astronomy, Hounsfield Rd, Sheffield S3 7RH, UK. Contact
email: fstevance1@sheffield.ac.uk

Abstract. Over the past two decades a clear relation between Gamma-Ray Bursts (GRBs) and some broad-lined Type Ic Supernovae (SNe) has been observed. The use of spectropolarimetry allows us to directly probe the 3D geometry of the unresolved ejecta of SNe, which can help us better understand the physics of their explosion and the SN-GRB connection. We present 7 epochs of spectropolarimetry of the broad-lined Type Ic SN 2014ad and highlight its similarities with SN 1998bw.

1. Introduction

Massive stars that have been stripped of their outer hydrogen and helium layers end their lives as Type Ic SNe. Some of these SNe exhibit very broad spectral features caused by high ejecta velocities (as high as 30,000 km/s): they are called broad-lined Type Ic (BL-Ic). A number of BL-Ic have been connected to LGRBs and X-ray Flashes: e.g SN 1998bw, SN 2003dh and SN 2006aj (Patat *et al.* 2001, Stanek *et al.* 2003, Sollerman *et al.* 2006). The question of the differences in the explosion mechanism between normal Type Ic SNe and BL-Ic SNe has not yet been fully answered.

2. Overview

The spectropolarimetric data obtained for SN 2014ad consists of 7 epochs spanning 68 days, starting 2 days before V-band maximum, which is the best spectropolarimetric data set ever obtained for a BL-Ic SN. Additionally we also acquired 8 epochs of spectroscopy, ranging from 2 days before to 107 after V-band maximum. The spectroscopic data revealed that SN 2014ad closely resembles SN 1998bw. The similar levels of broadening suggest similar photospheric velocities. The spectra of SN 2014ad are dominated by blended Fe II features in the blue and the Ca II IR P-Cygni profile in the red. Additionally a strong Si II feature appears by 5 days after V-band maximum. The degree of polarisation of SN 2014ad shows the most structure at the first epoch, with polarised Si II, O I and Ca II features. By +5 days there are only 3 visible peaks, associated with O I, Si II and a blend of Ca II and O I. The $q-u$ plots of the first 3 epochs show a strong dominant axis suggesting an axial symmetry, as well as O I and Ca II IR loops indicating departures from axial symmetry. The evolution of the O I and Ca II loops is also reminiscent of the other BL-Ic SN 2002ap (Patat *et al.* 2003).

Further results and figures will be published in Stevance *et al.* (2017).

References

Patat *et al.* 2001, *ApJ*, 55:900-917
Patat *et al.* 2003, *ApJ*, 592:457-466
Sollerman *et al.* 2006, *A&A*, 454:503-509
Stanek *et al.* 2003, *ApJ*, 591:L17-L20

The Lives and Death-Throes of Massive Stars
Proceedings IAU Symposium No. 329, 2016
J.J. Eldridge, J.C. Bray, L.A.S. McClelland & L. Xiao, eds.

doi:10.1017/S1743921317003325

Stellar wind measurements for Colliding Wind Binaries using X-ray observations

Yasuharu Sugawara[1], **Yoshitomo Maeda**[1] **and Yohko Tsuboi**[2]

[1]Institute of Space and Astronautical Science, Japan Aerospace Exploration Agency, 3-1-1 Yoshinodai, Chuo-ku, Sagamihara, Kanagawa 229-8510, Japan
email: sugawara.yasuharu@jaxa.jp

[2]Department of Physics, Faculty of Science & Engineering, Chuo University, 1-13-27 Kasuga, Bunkyo, Tokyo 112-8551, Japan

Abstract. We report the results of the stellar wind measurement for two colliding wind binaries. The X-ray spectrum is the best measurement tool for the hot postshock gas. By monitoring the changing of the the X-ray luminosity and column density along with the orbital phases, we derive the mass-loss rates of these stars.

Keywords. binaries: spectroscopic, stars: Wolf-Rayet, X-rays: stars, stars: mass loss

1. Introduction

Mass-loss is one of the most important and uncertain parameters in the evolution of a massive star. Colliding wind binary is the best testing ground for plasma shock physics, because plasma properties vary with binary separations. We have reported an approach to determine a mass-loss rate of a binary system with a highly eccentric orbit with X-ray multi-phase spectroscopy (Sugawara *et al.* 2015). In this paper. we select WR21a (O3/WN5ha+O3Vz((f*))) and WR25 (WN6h+O4f) as samples.

2. Wind measurements

In the case of WR21a, we fitted the X-ray spectra from XMM-Newton with two absorbed thin thermal plasma model. As periastron approached, the local column density increased, which can be explained as self-absorption by the W-R wind. Therefore, the local column density (mass-loss rate) can be expressed as the integral from the X-ray emitting region (a compact region around periastron) to infinite distance along the line of sight. Using the column densities from spectral fitting, our estimated mass-loss rate for primary star of WR21a is $(5.5\pm1.0)\times10^{-6}$ $M_\odot$ yr^{-1}. This is consistent the results using the similar method (Gosset & Nazé 2016).

In the case of WR25, we used the reported local column density data measured by XMM-Newton (Pandy *et al.* 2014). The column density data could be reproduced by the model using low eccentricity value (e =0.35) and low inclination angle (i <10°). Our estimate for W-R (WN) star is 5.1×10^{-5} $M_\odot$ yr^{-1}. By using this method for WR25, we could limit not only mass-loss rate but also orbital inclination angle.

This work was supported by JSPS KAKENHI Grant Number JP16K17667.

References

Gosset, E. & Nazé, Y. 2016, *A&A*, 590, A113
Pandey, J. C., Pandey, S. B., & Karmakar, S. 2014, *ApJ*, 788, 84
Sugawara, Y., Maeda, Y., Tsuboi, Y., *et al.* 2015, *PASJ*, 67, 121

The Lives and Death-Throes of Massive Stars
Proceedings IAU Symposium No. 329, 2016
J.J. Eldridge, J.C. Bray, L.A.S. McClelland
& L. Xiao, eds.

doi:10.1017/S174392131700117X

Exploring the physics of core-collapse supernovae with multidimensional simulations: from axisymmetry to three dimensions

Alexander Summa[1], Hans-Thomas Janka[1], Florian Hanke[1,2], Tobias Melson[1,2], Andreas Marek[3] and Bernhard Müller[4,5]

[1]Max-Planck-Institut für Astrophysik,
Karl-Schwarzschild-Str. 1, 85748 Garching, Germany
email: asumma@mpa-garching.mpg.de

[2]Physik Department, Technische Universität München,
James-Franck-Straße 1, 85748 Garching, Germany

[3]Max Planck Computing and Data Facility (MPCDF),
Gießenbachstr. 2, 85748 Garching, Germany

[4]Astrophysics Research Centre, School of Mathematics and Physics,
Queen's University Belfast, Belfast, BT7 1NN, United Kingdom

[5]Monash Centre for Astrophysics, School of Physics and Astronomy,
Monash University, Victoria 3800, Australia

Abstract. Multidimensional effects are essential for the success of the neutrino-driven explosion mechanism of core-collapse supernovae. Although astrophysical phenomena in nature involve three spatial dimensions, the huge computational demands still allow only for a few self-consistent, three-dimensional (3D) simulations focusing on specific aspects of the explosion physics, whereas systematic studies of larger sets of progenitor models or detailed investigations of different explosion parameters are restricted to the axisymmetric (2D) modeling approach at the moment. Employing state-of-the-art neutrino physics, we present the results of self-consistent core-collapse supernova simulations performed with the Prometheus-Vertex code in 2D and 3D. The 2D study of 18 successfully exploding pre-supernova models in the range of 11 to 28 solar masses shows the progenitor dependence of the explosion dynamics: if the progenitor exhibits a pronounced decline of the density at the Si/Si-O composition shell interface, the rapid drop of the mass-accretion rate at the time the interface arrives at the shock induces a steep reduction of the accretion ram pressure. This causes a strong shock expansion supported by neutrino heating and thus favors an early explosion. In case of a more gradually decreasing accretion rate, it takes longer for the neutrino heating to overcome the accretion ram pressure and explosions set in later. By considering the effects of turbulent pressure in the gain layer, we derive a generalized condition for the critical neutrino luminosity that captures the explosion behavior of all models very well. We show that this concept can also be extended to describe the effects of rotation as well as the behavior of recent 3D simulations and that the conditions necessary for the onset of explosion can be defined in a similar way.

Keywords. supernovae: general, hydrodynamics, turbulence, neutrinos

---

The Lives and Death-Throes of Massive Stars
Proceedings IAU Symposium No. 329, 2016
J.J. Eldridge, J.C. Bray, L.A.S. McClelland
& L. Xiao, eds.

doi:10.1017/S1743921317001867

The Search for Wolf-Rayet Stars in IC10

Katie Tehrani, Paul Crowther and Isabelle Archer

Department of Physics & Astronomy, University of Sheffield,
Hounsfield Road, S3 7RH, Sheffield, UK
email: k.tehrani@sheffield.ac.uk

Abstract. We present a deep imaging and spectroscopic survey of the Local Group starburst galaxy IC10 using Gemini North/GMOS to unveil the global Wolf-Rayet population. It has previously been suggested that for IC10 to follow the WC/WN versus metallicity dependence seen in other Local Group galaxies, a large WN population must remain undiscovered. Our search revealed 3 new WN stars, and 5 candidates awaiting confirmation, providing little evidence to support this claim. We also compute an updated nebular derived metallicity of log(O/H)+12=8.40 ± 0.04 for the galaxy using the direct method. Inspection of IC10 WR average line luminosities show these stars are more similar to their LMC, rather than SMC counterparts.

Keywords. stars: Wolf-Rayet, galaxies: individual (IC10), ISM: abundances

---

The Local Group dwarf galaxy IC10, located at a distance of 740kpc, is host to an unusual population of Wolf-Rayet (WR) stars, which are massive stars progressing through the end stages of evolution. Previous WR surveys have identified 26 WR stars, corresponding to an unusually high spectral type ratio of WC/WN=1.3, in view of the low metallicity environment of IC10 (Crowther *et al.* 2003). One explanation suggested these prior surveys are incomplete for WN stars (Massey & Holmes 2002), therefore prompting the deep imaging and spectroscopy presented here.

Using narrow HeII and continuum filters with Gemini North/GMOS, image subtraction techniques were used to reveal HeII excess sources, a robust marker for WR stars. This search yielded 9 WR candidates, and follow up spectroscopy on 4 of these targets successfully confirmed 3 new WN stars. One of these stars, associated with the HII region HL45, provided a clear measurement of the [OIII] λ4363 nebular emission line, allowing an updated oxygen abundance to be obtained. Prior metallicity measurements recorded a value of log(O/H)+12$\approx$8.26 (Garnett 1990), however our analysis suggests a higher value of log(O/H)+12=8.40 ± 0.04.

The WR star total in IC10 has now been raised to 29, with an updated WC/WN ratio of 1. With deep imaging revealing a further 5 WR candidates, we are confident a huge hidden population of WR stars does not exist. The sensitivity limit of this survey is $M_{\mathrm{HeIIC}}=-2.4$ mag (for $A_{\mathrm{HeIIC}}=3.4$ mag) which should be sufficient to identify these stars, even those which may fall into the recently proposed new WR class of characteristically faint WR stars (Massey *et al.* 2014). Average HeII λ4686 and CIV λ5808 line luminosities for WN and WC stars respectively are comparable to those of LMC WR stars and therefore consistent with the new higher metallicity measurement obtained.

References

Crowther, P. A., Drissen, L., Abbott, J. B., Royer, P., & Smartt, S. J. 2003, *A&A*, 404, 483
Garnett, D. R. 1990, *ApJ*, 363, 142
Massey, P. & Holmes, S. 2002, *ApJ*, 580, L35
Massey, P., Neugent, K. F., Morrell, N., & Hillier, D. J. 2014, *ApJ*, 788, 83

The Lives and Death-Throes of Massive Stars
Proceedings IAU Symposium No. 329, 2016
J.J. Eldridge, J.C. Bray, L.A.S. McClelland & L. Xiao, eds.

doi:10.1017/S174392131700254X

Core-Collapse Supernovae in the Early Universe: Radiation Hydrodynamics Simulations of Multicolor Light Curves

Alexey Tolstov[1], Ken'ichi Nomoto[1,6], Nozomu Tominaga[2,1], Miho Ishigaki[1], Sergei Blinnikov[3,4,1] and Tomoharu Suzuki[5]

[1]Kavli Institute for the Physics and Mathematics of the Universe (WPI), The University of Tokyo Institutes for Advanced Study, The University of Tokyo, 5-1-5 Kashiwanoha, Kashiwa, Chiba 277-8583, Japan
email: alexey.tolstov@ipmu.jp

[2] Department of Physics, Faculty of Science and Engineering, Konan University, 8-9-1 Okamoto, Kobe, Hyogo 658-8501, Japan

[3] Institute for Theoretical and Experimental Physics (ITEP), 117218 Moscow, Russia

[4] All-Russia Research Institute of Automatics (VNIIA), 127055 Moscow, Russia

[5] College of Engineering, Chubu University, 1200 Matsumoto-cho, Kasugai, Aichi 487-8501, Japan

[6] Hamamatsu Professor

Abstract. The properties of the first generation of stars and their supernova (SN) explosions remain unknown due to the lack of their actual observations. Pop III stars may have been very massive and predicted to be exploded as pair-instability SNe, but the observed metal-poor stars show the abundance patterns which are more consistent with yields of core-collapse SNe. We study the multicolor light curves for a metal-free core-collapse SN models (massive stars of 25-100 solar mass range) to determine the indicators for the detection and identification of first generation SNe. We use mixing-fallback supernova explosion models which explain the observed abundance patterns of metal poor stars. Numerical calculations of the multicolor light curves are performed using the multigroup radiation hydrodynamic code STELLA. The calculated light curves of metal-free SNe are compared with our calculations of non-zero metallicity models and observed SNe.

Keywords. radiative transfer, shock waves, stars: abundances, supernovae: general, stars: Population III

1. Overview

Blue supergiants are typical presupernovae for Pop III core-collapse SNe with $M_{\rm MS} \lesssim$ 40-60 $M_\odot$ and their structure determines the properties of the light curves. We found that first SNe are bluer, shorter and fainter than ordinary SNe. The most important result is the flat color evolution curve with typical values B-V from 0.0 to 0.6 during the plateau phase can be used as an indicator of Pop III and low-metallicity SNe. In contrast, all solar metallicity models show gradual reddening during the plateau. The low amount of ^{56}Ni that is used to explain carbon-enhanced metal-poor stars with mixing-fallback and sharp luminosity decline after the plateau phase can also be used as indicators of a low-metallicity progenitor.

Our results can be used to identify first-generation SNe in the current (Subaru/HSC) and future transient surveys (LSST, JWST). They are also suitable for identifying of the low-metallicity SNe in the nearby Universe (PTF, Pan-STARRS, Gaia).

The Lives and Death-Throes of Massive Stars
Proceedings IAU Symposium No. 329, 2016
J.J. Eldridge, J.C. Bray, L.A.S. McClelland
& L. Xiao, eds.

doi:10.1017/S1743921317001168

A new prescription for the mass-loss rates of hydrogen-free WR stars

Frank Tramper[1], Hugues Sana[2] and Alex de Koter[2,3]

[1]European Space Astronomy Centre (ESA/ESAC), Operations Department, Villanueva de la Cañada (Madrid), Spain
email: ftramper@sciops.esa.int

[2]Institute of Astrophysics, KU Leuven, Celestijnenlaan 200 D, B-3001, Leuven, Belgium

[3]Anton Pannekoek Institute for Astronomy, University of Amsterdam, P.O. Box 94249, 1090 GE Amsterdam, The Netherlands

Abstract. We present a new empirical prescription for the mass-loss rates of hydrogen-free Wolf-Rayet stars based on results of detailed spectral analyses of WC and WO stars. Compared to the prescription of Nugis & Lamers (2000), $\dot{M}$ is less sensitive to the surface helium abundance, implying a stronger mass loss at the late stages of Wolf-Rayet evolution. The winds of hydrogen-free WN stars have a strong metallicity dependence, while those of WC and WO stars have a very weak metallicity dependence.

Keywords. stars: Wolf-Rayet, stars: mass loss, stars: winds, outflows

---

Wolf-Rayet (WR) stars are characterised by emission-line spectra as a result of their very strong stellar winds. The amount of mass lost in the WR phase strongly impacts the final surface abundances (and hence the supernova type), as well as the nature of the compact remnant left behind. The current standard prescription from Nugis & Lamers (2000) works well for WN and WC stars, but cannot reproduce the mass-loss rates of WO stars (Tramper *et al.* 2015).

Here we present an improved prescription for the mass-loss rates of hydrogen-free WN, WC, and WO stars. Details of the derivation of the prescription and a discussion on its dependences are given in Tramper *et al.* (2016). The mass-loss prescription with 1σ uncertainties on the coefficients is

$$\log \dot{M} = -9.20(\pm 0.35) + 0.85(\pm 0.06) \log \frac{L}{L_\odot} + 0.44(\pm 0.08) \log Y + \alpha \log \frac{Z_{\rm ini}}{Z_\odot},$$

where $\alpha = 0.25(\pm 0.08)$ for WC and WO stars and $\alpha = 1.3(\pm 0.2)$ for hydrogen-free WN stars. The relation provides mass-loss rates with an uncertainty of $\sigma = 0.06$ dex for WC and WO stars, and $\sigma = 0.2$ dex for hydrogen-free WN stars.

Compared to Nugis & Lamers (2000), WR stars have a higher mass loss towards the end of their evolution, when the surface helium abundance is low, and at low metallicities *if* the WC stage can be reached. There is a strong change in metallicity dependence from the WN to the WC states, implying that carbon driving may be relatively important. Implications of the new prescription include lower final masses, higher final surface abundances of carbon and oxygen, and a longer duration of the WC and WO stages.

References

T. Nugis & H. J. G. L. M. Lamers 2000, *A&A*, 360, 227
F. Tramper, S. M. Straal, D. Sanyal, *et al.* 2015, *A&A*, 581, 110
F. Tramper, H. Sana, A. de Koter 2016, *ApJ*, 833, 133

The Lives and Death-Throes of Massive Stars
Proceedings IAU Symposium No. 329, 2016
J.J. Eldridge, J.C. Bray, L.A.S. McClelland & L. Xiao, eds.

doi:10.1017/S1743921317000199

Destruction of Be star disk by large scale magnetic fields

Asif ud-Doula[1], Stanley Owocki [2], Nathaniel (Dylan) Kee [3] and Michael Vanyo[1]

[1]Penn State Worthington Scranton, Dunmore, PA, USA,
email: asif@psu.edu

[2] University of Delaware, Newark, DE, USA,

[3]University of Tubingen, Tubingen, Germany

Abstract. Classical Be stars are rapidly rotating stars with circumstellar disks that come and go on time scale of years. Recent observational data strongly suggests that these stars lack the 10% incidence of global magnetic fields observed in other main-sequence B stars. Such an apparent lack of magnetic fields may indicate that Be disks are fundamentally incompatible with a significant large scale magnetic field. In this work, using numerical magnetohydrodynamics (MHD) simulations, we show that a dipole field of only 100G can lead to the quick disruption of a Be disk. Such a limit is in line with the observational upper limits for these objects.

Keywords. stars: emission-line, Be, stars: magnetic fields, stars: mass loss, (magnetohydrodynamics:) MHD

There is no unique definition of a Be star. The most common one, popularized by Collins (1987) defines a Be star as 'a non-supergiant B star whose spectrum has, or had at some time, one or more Balmer lines in emission'. Despite some problems with this broad definition, implication is that a Be star at some point of its life hosted a disk of circumstellar dense material. Material in such disks have been shown to be in Keplerian orbit and the host stars themselves have been found to be rapidly rotating. Recent extensive search of globally organized magnetic field in Be stars, especially by teams from MiMeS project, yielded null results, putting an upper limit of few tens G to such possible fields. In this work, we demonstrate that strong global magnetic field (dipolar) in excess of 50-100G are fundamentally incompatible with Keplerian disk in Be stars, at least partially explaining the lack of detectable magnetic fields in such objects.

We set up a numerical parameter study of disk-field interaction (along with weak radiatively driven wind) of a typical Be star. We assume, initially, a pure global dipole field anchored in the interior of the star, and a pre-existing Keplerian disk surrounding the star. Weak radiatively driven wind has very little dynamical effects on the disk, but nonetheless is kept for the sake consistency. For various models, we change only the strength of the magnetic field. We find that any field in access of 10G has significant dynamical effects on the disks. Our highest field strength model (100G) completely disrupts the Keplerian disk.

Our models do not account for rapid stellar rotation, and associated with it gravity darkening. All these could affect the threshold of our results, although we do not believe they would alter our basic conclusion that strong magnetic fields are incompatible with Keplerian disks in Be stars.

The Lives and Death-Throes of Massive Stars
Proceedings IAU Symposium No. 329, 2016
J.J. Eldridge, J.C. Bray, L.A.S. McClelland & L. Xiao, eds.

doi:10.1017/S1743921317001326

The potential of using KMOS for multi-object massive star spectroscopy

Michael Wegner[1], Ralf Bender[1,3], Ray Sharples[2] and the KMOS Team[1,2,3,4,5,6]

[1]University Observatory Munich, Scheinerstr. 1, 81679 München, Germany,
email: wegner@usm.lmu.de
[2]Durham University, Departement of Physics, South Rd, Durham DH1 3LE, UK,
[3]Max Planck Institute for Extraterrestrial Physics, Gießenbachstr. 1, 85748 Garching, Germany
[4]UK Astronomy Technology Centre Edinburgh, Blackford Hill, EH9 3HJ, Edinburgh, UK
[5]Department of Physics, University of Oxford, Parks Road, Oxford OX1 3PU, UK
[6]European Southern Observatory, Karl-Schwarzschild-Str. 2, 85748 Garching, Germany

Abstract. KMOS, the "K-Band Multi-Object Spectrometer", was built by a British-German consortium as a second generation instrument for the ESO Paranal Observatory. It is available to the user community since its successful commissioning in 2013 (Sharples *et al.* 2013). As a multi-object integral field spectrometer for the near infrared, KMOS offers 24 deployable IFUs of 2.8x2.8 arcsec and 14x14 spatial pixels each, which can either be placed individually within a 7.2 arcmin field of view or combined in a *Mosaic* mode in order to map contiguous fields on sky. The instrument covers the whole range of NIR atmospheric windows (0.8...2.5μm) with 5 spectral bands and a resolution of R $\approx$ 3000...4000.

Although the main science driver for KMOS was to enable the study of galaxy formation and evolution through multiplexed observations of high-redshift galaxies, KMOS also already exhibited its tremendous potential for the spectroscopy of massive stars: A quantitative study of 27 RSGs in NGC 300 (Gazak *et al.* 2015) proves its applicability for the spectroscopy of individual stars even beyond the Local Group. A *Mosaic* observation of the Galactic centre (Feldmeier-Krause *et al.* 2015) demonstrates how spectra of early-type stars can be extracted from a contiguous field. Other applications include (but need not be limited to) velocity determinations of globular cluster stars, observations of jets/outflows of high mass protostars, or contiguous mapping of star-forming regions.

We therefore aim at presenting the excellent capabilities of KMOS to a wider community and indicate potential applications.

Keywords. instrumentation: spectrographs, infrared: stars, stars: early-type, techniques: spectroscopic

---

References

Sharples, R., Bender, R., Agudo Berbel, A., Bezawada, N., Castillo, R., Cirasuolo, M., Davidson, G., Davies, R., Dubbeldam, M., Fairley, A., Finger, G., Förster Schreiber, N., Gonte, F., Hess, H.-J., Jung, I., Lewis, I., Lizon, J.-L., Muschielok, B., Pasquini, L., Pirard, J., Popovic, D., Ramsay, S., Rees, P., Richter, J., Riquelme, M., Rodrigues, M., Saviane, I., Schlichter, J., Schmidtobreick, L., Segovia, A., Smette, A., Szeifert, T., van Kesteren, A., Wegner, M., & Wiezorrek, E. 2013, *The Messenger*, 151, 21

Gazak, J. Z., Kudritzki, R., Evans, C., Patrick, L., Davies, B., Bergemann, M., Plez, B., Bresolin, F., Bender, R., Wegner, M., Bonanos, A. Z., & Williams, S. J. 2015, *ApJ*, 805, 182

Feldmeier-Krause, A., Neumayer, N., Schödel, R., Seth, A., Hilker, M., de Zeeuw, P. T., Kuntschner, H., Walcher, C. J., Lützgendorf, N., & Kissler-Patig, M. 2015, *A&A*, 584, A2

The Lives and Death-Throes of Massive Stars
Proceedings IAU Symposium No. 329, 2016
J.J. Eldridge, J.C. Bray, L.A.S. McClelland & L. Xiao, eds.
doi:10.1017/S1743921317004458

Posters also presented at the Symposium

3D Hydrodynamic Simulations of O-Shell Convection

Robert Andrassy
Department of Physics and Astronomy, University of Victoria

I am reporting on our team's progress in investigating fundamental properties of convective shells in the deep stellar interior during advanced stages of stellar evolution. We have performed a series of 3D hydrodynamic simulations of convection in conditions similar to those in the O-shell burning phase of massive stars. We focus on characterizing the convective boundary and the mixing of material across this boundary. Results from 768^3 and 1536^3 grids are encouragingly similar (typically within 20%). Several global quantities, including the rate of mass entrainment at the convective boundary and the driving luminosity, are related by scaling laws. We investigate the effect of several of our assumptions, including the treatment of the nuclear burning driving the convection or that of neutrino cooling. The burning of the entrained material from above the convection zone could have important implications for pre-supernova nucleosynthesis.

The Upper Initial Mass Function in Nearby Dwarf Galaxies

J. Andrews, D. Calzetti, D. Cook, D. Dale, and M. Krumholz
University of Arizona

Dwarf galaxies are likely the building blocks of all galaxies observed today, but star formation is still poorly-understood in these unevolved systems. We have obtained ground-based optical spectroscopy with the MMT and Palomar-200 inch for a sample of young, H-alpha bright clusters in ~ 20 dwarfs within ~ 3.5 Mpc; our targets all have extensive HST broad-band coverage as well. By comparing these extinction corrected spectra to stochastic stellar synthesis models we can obtain ages and masses of the clusters and associations and determine whether the dearth of ionizing photons in dwarfs is an effect of deficiency in the production of massive stars. Here we will present the results of those observations, and show that stellar clusters with massive stars are not as rare as generally thought, and are commonly found in smaller clusters in the low metallicity dwarfs.

Constraining Disk Properties of a Survey of Southern Be Stars

Arcos, C. ; Kanaan, S. ; Curé, M.;
Universidad de Valparaíso

Be stars represent a challenge in the study of the physical mechanism required to form and support disks around hot stars. Using different techniques, such as photometry, interferometry and spectroscopy and thanks to the development of powerful radiative codes, we can study the conditions for which disk exist as well as disk features such as the size of the emission region, inclination angle, mass, angular momentum, etc. Using `BEDISK` a non-LTE radiative transfer code and `BERAY` that solves the transfer equation throughout the disk, we studied 63 Be southern stars by modeling H_α emission line profiles. We found for early stellar *emission* types that the density structure in the surface of the disk can be modeled by a power law with base density between $\rho_0 \sim$ (4.00 - 6.30)$\times 10^{-11}$ g cm^{-3} with a power-law fall of between $n \sim$ 2.0 - 2.5. We also found that outer disk radii of the H_α emitting regions are between $R_{disk} \sim$ 20 - 30$R_\star$, the angular momentum between $J_d \sim$ (1.0 - 3.1)$\times 10^{-7} J_\star$ and the disk mass between $M_d \sim$ (1.0 - 3.1)$\times 10^{-9} M_\star$.

RADFLAH: A New, Open Numerical Framework to Model Supernova Light Curves with the FLASH Code

Chatzopoulos, E., Wheeler, J. C., Vinko, J.
Louisiana State University

We present the newly-incorporated gray radiation hydrodynamics capabilities of the FLASH code based on a radiation flux-limiter aware hydrodynamics numerical implementation designed specifically for applications in astrophysical problems. The newly incorporated numerical methods consist of changes in the pristine unsplit hydrodynamics solver using operator splitting and adjustments in the flux-limited radiation diffusion unit. Our method can treat problems in both the strong and weak radiation-matter coupling limits as well as transitions between the two regimes. Appropriate extensions in the "Helmholtz" equation of state are implemented to treat two-temperature astrophysical plasmas involving the interaction between radiation and matter. A set of radiation-hydrodynamics test problems is presented aiming to showcase the new capabilities of FLASH and to provide direct comparison to similar codes like CASTRO. To illustrate the capacity of FLASH to simulate phenomena occurring in stellar explosions, such as shock break-out, radiative precursors and ejecta heating due to the decays of radioactive nickel and cobalt, we also present an 1D supernova simulation yielding a model gray light curve. The latest public release of FLASH with these enhanced capabilities is freely available for download and use by the broader astrophysics community.

The X-Ray Origin of the Be Star Gamma Cassiopeiae and the Implication to Its Stellar Evolution

Michael F. Corcoran, Neetika Sharma, Ted Gull, Hiromitsu Takahashi, Christopher M. Russell, Tom Madura, Anthony Moffat, Takayuki Yuasa, Julian Pittard, Manabu ISHIDA, Jose Groh, Stan Owocki, Noel Richardson
CRESST NASA/GSFC & UMBC

Gamma Cassiopeiae is an enigmatic Be star with unusually hard, strong X-ray emission compared with normal main-sequence B stars. The origin is controversial for decades between two theories: mass accretion onto a hidden compact companion and magnetic dynamo driven by the star-Be disk differential rotation. There has been found no decisive signature that support either theory, such as a pulse in X-ray emission or the presence of large-scale magnetic field. In a $\sim$ 100 ksec duration observation of the star with the Suzaku X-ray observatory in 2011, we detected six rapid X-ray spectral hardening events called "softness dips". All the softness dip events show symmetric softness ratio variations, and some of them have flat bottoms apparently due to saturation. The softness dip spectra are best described by either $\sim$ 40% or $\sim$ 70% partial covering absorption to kT $\sim$ 12 keV plasma emission by matter with a neutral hydrogen column density of

$\sim 2 - 8 \times 10^{21}\,\mathrm{cm}^{-2}$, while the spectrum outside of these dips is almost free of absorption. This result suggests that two distinct X-ray emitting spots in the gamma Cas system, perhaps on a white dwarf companion with dipole mass accretion, are occulted by blobs in the Be stellar wind, the Be disk, or rotating around the white dwarf companion. The formation of a Be star and white dwarf binary system requires mass transfer between two stars; gamma Cas may have experienced such activity in the past. We discuss the gamma Cas type Be stellar evolution, based on this result.

Connecting Nuclear Physics and Massive Star Models to Galactic Chemical Evolution

Benoit Cote, Christian Ritter, Falk Herwig, Brian W. O'Shea, Chris L. Fryer

University of Victoria / Michigan State University

Modeling the chemical evolution of the Milky Way and Local galaxies represents a significant challenge, as it contains several sources of uncertainties and ideally requires multidisciplinary collaborations. Nuclear physics provides nuclear reaction rates, stellar models provide the composition of stellar ejecta, galaxy models and simulations follow the evolution of chemical species driven by multiple stellar populations, and observations provide constraints to test and improve numerical recipes. Continuous communication and feedback between all these fields is a key component in improving our understanding of how, where, and when elements have been created and recycled across cosmic time. We built a numerical pipeline that connects NuGrid stellar yields to galactic chemical evolution models, going from a classical one-zone model to a multi-zone model able to post-process cosmological hydrodynamical and dark-matter-only simulations. During this talk, I will present how we used this pipeline, in a circular way, to probe the impact of nuclear astrophysics and massive star evolution assumptions on the chemical signatures predicted by chemical evolution models. I will highlight the importance of core-collapse nucleosynthesis and how the uncertainties associated with the stellar remnant mass and black hole formation prescriptions in massive star models affect our ability to provide reliable numerical predictions in the Milky Way and in dwarf spheroidal galaxies.

3D MHD Simulations of Radiatively Driven Winds with Inclined Magnetic Fields

Simon Daley-Yates

Astrophysics and Space Research Group, University of Birmingham

We present results of 2D and 3D numerical simulations of magnetically confined, radiatively driven stellar winds of massive stars, conducted using the astrophysical MHD code Pluto, with a focus on understanding the rotational variability of radio and sub-mm emission. Radiative driving is implemented according to the Castor, Abbott and Klein theory of radiatively driven winds. Many magnetic massive stars posses a magnetic axis which is inclined with respect to the rotational axis. This misalignment leads to a complex wind structure as magnetic confinement, centrifugal acceleration and radiative driving act to channel the circumstellar plasma into a warped disk whose properties should be apparent in multiple wavelengths. Building upon work carried out by ud-Doula, Owocki and Townsend, simulating magnetically channeled radiatively outflow. Light curves in multiple wavelengths via Monte Carlo radiative transfer are presented. Parameters such as the mass-loss rate, magnetic field strength, rotational period, magnetic and rotational axis misalignment are investigated.

Photometric Variability of Luminous Blue Variables in M33 on Short Timescales

Gantcho Gantchev, Petko Nedialkov, Valentin Ivanov, Evgeni Ovcharov, Antoniya Valcheva, Milen Minev
Department of Astronomy, Physics Faculty, Sofia

We used SDSS r-band aperture photometry and astrometry of ~500 000 stellar-like objects in the M33 galaxy performed by the CASU (Cambridge Astronomy Survey Unit) Astronomical Data Centre in the Institute of Astronomy, University of Cambridge. The observations were carried out with the 2.6m VISTA telescope at the Cerro Paranal, Chile. More than 500 images in that passband were obtained with the OmegaCAM, a large format (16k×16k pixels) CCD camera, and each of them covers a field of view of $1° \times 1°$. The current time span of the data is 2.1 yrs until the end of 2014. The structure function analysis (Hughes *et al.* 1992) was applied in order to study the variability of ~ 30 known or suspected LBVs in the M33 galaxy (Massey *et al.* 2007) on different time scales. In some cases like Var C the time resolution of the data allows us to confirm an enhanced weekly variations Dm ~ 0.3m which is somehow shorter than the previously know typical monthly variations with the same maximum amplitude thought to be caused by non-radial pulsations.

Diversified Core-Collapse Supernovae Through A Jet-Feedback Mechanism

Avishai Gilkis, Noam Soker
Physics Department, Technion - Israel Institute of Technology

I propose that the energy output of explosions following massive stellar core-collapse is regulated through a jet-feedback mechanism. This scenario advocates for jet-powered central engines producing different outcomes depending on the efficiency of the interaction between the jets and their surrounding. An efficient coupling of the jets with the infalling material will result in the halt of material accretion and the central engine shutting down early-on, resulting in a regular energy supernova. A lower efficiency will allow a longer accretion period and a higher energy output from the central engine, possibly resulting in black hole formation and superluminous supernovae or gamma-ray bursts. I demonstrate these concepts on models of massive stars with the aid of analytic approximations and numerical codes, and discuss potential consequences of this model.

Wolf-Rayet Stars in M81: Detection and Characterization Using GTC/OSIRIS Spectra and HST/ACS Images

Víctor Mauricio Alfonso Gómez-Gonzólez, Yalia Divakara Mayya, Daniel Rosa Gonzólez
Instituto Nacional de Astrofísica, Óptica y Electrónica

We here report the properties of Wolf-Rayet (W-R) stars in 14 locations in the nearby spiral galaxy M81. These locations were found serendipitously while analysing the slit spectra of a sample of $\sim$ 150 star-forming complexes, taken using the long-slit and Multi-Object spectroscopic modes of the OSIRIS instrument at the 10.4m Gran Telescopio Canarias. Colours and magnitudes of the identified point sources in the Hubble Space Telescope images compare well with those of individual W-R stars in the Milky Way. Using templates of individual W-R stars, we infer that the objects responsible for the observed W-R features are single stars in 12 locations, comprising of 3 WNLs, 3 WNEs, 2 WCEs and 4 transitional WN/C types. In diagrams involving bump luminosities and the width of the bumps, the W-R stars of the same sub-class group together, with the transitional stars occupying locations intermediate between the WNE and WCE groups, as expected from the evolutionary models. However, the observed number of 4 transitional stars out of our sample of 14 is statistically high as compared to the 4% expected in stellar evolutionary models.

The Yellow and Red Supergiants of M31/M33 and Post-Red Supergiant Evolution

Michael S. Gordon, Roberta Humphreys, and Terry J. Jones
Minnesota Institute for Astrophysics

From surveys of luminous star populations in nearby galaxies, we have selected a sample of yellow and red supergiant candidates in M31 and M33 for study of their spectral characteristics and spectral energy distributions to place them on an HR Diagram. As the position of intermediate and late-type supergiants on the color-magnitude diagram can be heavily contaminated by foreground dwarfs, we used spectral classification and multi-band photometry from optical and near-infrared surveys to confirm membership, in addition to careful measurements of the extinction along the line-of-sight to each source to determine bolometric luminosities. Based on spectroscopic evidence for stellar winds and mass loss and the presence of circumstellar dust in their SEDs, we find that 30-40% of the yellow supergiants in M31 and M33 are likely in a post-red supergiant state. Comparison with evolutionary tracks suggests that these mass-losing, post-RSGs have initial masses between $20 - 40M_{\odot}$. More than half of the observed red supergiants are producing dusty circumstellar ejecta, and thus may be more likely to evolve back to warmer temperatures due to their high mass loss.

Feedback by Massive Stars and the Self-Regulation of Star Formation

Gerhard Hensler
Dept. of Astrophysics, Univ. of Vienna

Although star formation (SF) in action becomes directly noticeable in galaxies by various observational signatures, its process, its empirical relations to gas properties, and the reason for a much lower SF rate than expected are still far from being understood in detail. Most importantly, massive stars act as the main driver of galaxy evolution by various modes of stellar feedback, energy, mass and element releases, the first regulating the SF and energizing the interstellar medium (ISM), the last one enriching the ISM with heavier elements. For cosmological and galaxy evolutionary models the "star-formation efficiency" is an essential parameter to describe which fraction of star-forming gas is really converted into stars. Instead as being assumed temporally and empirically constant, this parameter depends not only on the stellar energy release, but also on the conditions of the formed star cluster and its surrounding gas. The formation of extremes like super star clusters as those in starburst galaxies and required previously for Globular Clusters as well as SF in low-surface brightness galaxies prove this conclusion. From our former numerical models of radiation-driven and wind-blown HII regions around single massive stars (Freyer *et al.*) we evaluate the energy transfer efficiency and found much lower values than those generally assumed in galaxy evolution modeling. Moreover, ongoing numerical studies of lacking massive stars at very low SF rates has also strong consequences for our understanding of feedback and SF self-regulation. Both will be discussed in our presentation.

Stellar parameters of the reference O-type stars

Anthony Hervé
Stellar Department ASU

Massive stars play a key role in various fields of astrophysics. In the 70's, the two dimensional spectral classification of O stars was developed by Walborn. Standard stars have been selected to be the reference object of each stellar type / luminosity class. These standard stars are still used today for the classificationof the any newly discovered O stars. However, the stellar properties of these reference objects have never been determined in a homogeneous way. Using high-resolution, high signal-to-noise ratio from OHP and ESO and the modeling code CMFGEN, we are currently determining the properties of the standard O-type stars. During this presentation we will present the preliminary results of our study.

Rising from the Ashes: Mid-Infrared Re-Brightening of the Impostor SN 2010da in NGC 300

Ryan M. Lau, Mansi M. Kasliwal, Howard E. Bond, et al.
Caltech/JPL

We present multi-epoch mid-infrared (IR) photometry and the optical discovery observations of the "impostor" supernova (SN) 2010da in NGC 300 using new and archival Spitzer Space Telescope images and ground-based observatories. The mid-IR counterpart of SN 2010da was detected as SPIRITS 14bme in the SPitzer InfraRed Intensive Transient Survey (SPIRITS), an ongoing systematic search for IR transients. A sharp increase in the 3.6 μm flux followed by a rapid decrease measured $\sim$150 d before and $\sim$80 d after the initial outburst, respectively, reveal a mid-IR counterpart to the coincident optical and high luminosity X-ray outbursts. At late times after the outburst ($\sim$ 2000 d), the 3.6 and 4.5 μm emission increased to over a factor of 2 times the progenitor flux. We attribute the re-brightening mid-IR emission to continued dust production and increasing luminosity of the surviving system associated with SN 2010da. We analyze the evolution of the dust temperature, mass, luminosity, and equilibrium temperature radius in order to resolve the nature of SN 2010da. We address the leading interpretation of SN 2010da as an eruption from a luminous blue variable (LBV) high-mass X-ray binary (HMXB) system. We propose that SN 2010da is instead a supergiant (sg)B[e]-HMXB based on similar luminosities and dust masses exhibited by two other known sgB[e]-HMXB systems. Additionally, the SN 2010da progenitor occupies a similar region on a mid-IR color-magnitude diagram (CMD) with known sgB[e] stars in the Large Magellanic Cloud. The lower limit estimated for the orbital eccentricity of the sgB[e]-HMXB ($e > 0.82$) from X-ray luminosity measurements is high compared to known sgHMXBs and supports the claim that SN 2010da may be associated with a newly formed HMXB system.

Chandra X-Ray Grating Spectroscopic Diagnostics of the X-Ray Emitting Plasma in the Magnetic O+O Binary Plaskett's Star

M. Leutenegger, J. Grunhut, G. Wade, D. Cohen, T. Doyle, M. Gagne, R. Ignace, J. C. Leyder, L. Mahy, T. Moffat, Y. Naze, S. Owocki, V. Petit, J. Sundqvist, A. ud-Doula
NASA/Goddard Space Flight Center Code 662

Recent spectropolarimetric observations have revealed that the rapidly rotating secondary in the massive O+O binary system Plaskett's star harbors a strong magnetic field with a surface dipole field strength of ~ 3 kG (Grunhut *et al.* 2013), suggesting the likely presence of a centrifugally supported magnetosphere, unique among O stars. Previous X-ray observations interpreted the X-ray emission in terms of a collision between the primary and secondary wind (Linder *et al.* 2006),but the presence of a strong magnetic field on the secondary suggests that X-ray emission might originate from the magnetosphere or even in a wind magnetosphere collision. To distinguish between these competing hypotheses, we have obtained high resolution Chandra grating spectra of Plaskett's star, with good orbital and rotational phase coverage. The forbidden-to-intercombination line ratios of He-like Si XIII and Mg XI rule out formation of X-ray emitting plasma inside the magnetosphere of the secondary, and are consistent with formation in the region between the two stars. The lines are broad, with a Gaussian sigma of ~ 700 km/s. We surprisingly find no evidence for variability of the line shape or shift with either secondary rotational phase or orbital phase. This is contrary to expectations for either a magnetospheric or a colliding wind origin for the X-rays.

The Progenitor Masses of ~ 100 Core-Collapse SNe

Jeremiah Murphy, Mariangelly Díaz Rodríguez
Physics, Florida State University

By age-dating the stellar populations in the vicinity of supernova remnants (SNRs), we derive the progenitor masses for more than 200 core-collapse SNe. With this large statistical sample, we are able to characterize the distribution of progenitor masses. Using Bayesian statistical inference, we find that the minimum mass of SNR progenitors is 7.2 ± 0.3 solar masses, the maximum mass is 33^{+17}_{-6} solar masses, and the power law slope in between is 2.8 ± 0.5, consistent with the Saltpeter IMF. The accuracy of the minimum mass may provide interesting constraints on stellar evolution. With regard to the maximum mass, either the most massive of massive stars are not exploding, or there is severe bias against forming SNRs by the explosions of the most massive stars.

Analytic Conditions for Core-Collapse Supernova Explosions

Jeremiah Murphy, Joshua Dolence, and Quintin Mabanta
Department of Physics, Florida State University

We derive an integral condition for core-collapse supernova explosions and use it to construct a new diagnostic of explodability. The fundamental challenge in core-collapse supernova theory is to explain how a stalled accretion shock revives to explode a star. In this talk, we assume that shock revivalis initiated by the delayed-neutrino mechanism and derive an integral condition for shock expansion, $v_s > 0$. In general, there are five parameters in the core-collapse problem: the neutrino luminosity, the temperature of neutrinos, the neutron star radius, the neutron star mass, and the accretion rate onto the stalled shock. The integral condition represents an explosion condition that incorporates these five parameters, and this integral condition is represented by a single equation, which is characterized by a single dimensionless parameter, $\Psi > 0$. $\Psi = 0$ defines a critical five-dimensional hypersurface, below this surface, stalled-shock solutions exists, and above this hypersurface only $v_s > 0$ solutions exist. Therefore, $\psi = 0$ defines a critical hypersurface forexplosion, and we show that the critical neutrino luminosity curve proposed by Burrows & Goshy 1993 is a projection of this more generalcritical condition. Finally, we propose and verify with 1D simulations that $\Psi > 0$ is a reliable and accurate explosion diagnostic.

The Spectral Temperature of Light Echoes From Eta Carinae's Giant Eruption

Stan Owocki & Nir Shaviv
Dept. of Physics and Astronomy, U. of Delaware

The 1840's era giant eruption of eta Carinae remains a challenge for understanding the lives and death throes of the most massive stars. In recent years the detection by Rest *et al.* (2012) of light echoes from this era has provided important new observational constraints on the nature of the eruption. In particular, spectra of the echoes suggest a relative cool spectral temperature of about 5500K, below the lower limit of about 7000K derived from the optically thick wind outflow analysis of Davidson (1987). This has lead to a debate about the viability of this steady wind model relative to alternative, explosive scenarios. Here we present an updated analysis of the wind outflow model using newer low-T opacity tabulations and accounting for the stronger mass loss associated with the $> 10 M_{\odot}$ mass now inferred for the Homunculus. A major conclusion is that, because of the sharp drop in opacity due to free electron recombination for T<6500K, a low spectral temperature of about 5000K is compatible with, and indeed expected from, a wind with the extreme mass loss inferred for the eruption. Within a spherical gray model in radiative equilibrium, we derive spectral energy distributions for various assumptions for the opacity variation of the wind, and make initial comparisons with observed light echo spectra.

A Very Luminous Rosetta Stone to Decipher Massive Stellar Evolution: Linking LBVs, SN Impostors, Bright Radio Emitters and ULXs

Manfred W. Pakull
Observatoire Astronomique de Strasbourg

The subject of this talk is a previously unknown, very luminous ($M_V \sim -10$) evolved star embedded in a supergiant HII region in the nearby dwarf galaxy NGC 5408. Hα has a HWZI of $\sim$ 5000km/s and the equivalent width is presently 2000Å. It displays non-thermal radio emission which is 8 times more powerful than the Cas A SNR, and long-slit spectra reveal a large surrounding HeIII region (300 pc diameter) with intense HeII 4686 and [NeV]3426 emission (ionisation potential 54 and 97 eV, respectively). This points to the presence of an ultraluminous X-ray source ($L_X \sim$ several 10^40 erg/s, hereafter referred to as "NGC 5408 X-2"), whereas Chandra only detected a 3 orders of magnitude lower flux level. Curiously, this object is very close to, but not identical with the famous ULX NGC 5408 X-1, which is currently being studied in much detail in the literature. I will present ESO/VLT medium resolution spectra of "X-2" revealing variable P Cygni profiles of H, HeI and various ionisation stages of (forbidden) metal lines, suggesting a hot LBV nature. However, its optical continuum level appears to be constant whereas the Hα emission had increased tenfold over the last 30 years. The system appears to be in a very rapid stage of stellar evolution with currently ongoing huge mass-loss/mass-transfer between the binary components. LBV/ULX NGC 5408 X-2 appears to be a Rosetta stone worthwhile to be deciphered for a better understanding of the latest phases of massive (binary) stars.

A revised magnetic topology for the magnetic O-type star θ^1 Ori C

V. Petit, P. Mohanty, G. Wade
Florida Institute of TEchnology

θ^1 Orionis C was the first O-type star to be measured to host a magnetic field at its surface, through circular polarization induced by the Zeeman effect. Since then, the discovery and systematic characterization of magnetic fields in OB stars by the new generation of powerful spectropolarimeters has enabled a new understanding of the influence of magnetism on their radiation-driven stellar winds, and on their observed spectral characteristics at all wavelengths. However θ^1 Orionis C, the usual benchmark of magnetic O-type stars, has yet to be characterized with modern spectropolarimetric observations. We present a new ESPaDOnS monitoring of this poster-child object and a new determination of its surface field topology.

The Berkeley Sample of Stripped-Envelope Supernovae: 20+ Years of Spectroscopy and Photometry

Isaac Shivvers, Alexei Filippenko
Astronomy Department, University of Berkeley at California

Stripped-envelope supernovae (SNe of types IIb/Ib/Ic, as well as some rarer subtypes) are associated with the core collapse of some of the most massive stars in the universe. These SNe exhibit heterogenous properties in their spectral and photometric evolution, the study of which is confounded by the scarcity of well-observed examples (these subtypes, on average, are both less luminous and less volumetrically common than SNe of types II and Ia). I will present the ongoing effort at Berkeley to analyze, understand, and make public the large sample of stripped-envelope SNe observations accumulated by the Filippenko group - data both previously published and as-yet unpublished, including 700+ spectra of 250+ SNe.

The Impacts of Pre-Collapse Structures In CCSNe Theory

Yu Yamamoto, Kotaro Fujisawa
Department of Physics and Applied Physics, Waseda University

The main topic of CCSNe theory is now starting to pay attention to what is exact condition of powerful explosion and particularly what kind of pre-collapse structure expedite to obtain the canonical explosion, i.e. "initial problem". So far the practical stellar evolution calculations of massive stars are, however, still in progress due to the complicated convection theory, the effect of rotation, the presence of mass loss events and the resolution dependence. In Yamamoto *et al.* (2016), we studied the influence of the progenitor structure on the dynamics systematically by constructing progenitor models parametrically instead of employing realistic models provided by stellar evolution calculations. The mass of the iron core and that of silicon plus surfer (Si+S) layer are chosen as the parameters and dynamical evolutions after collapse are calculated in one and two dimensions. We found that the explosion energy is tightly correlated with mass accretion rate at shock revival irrespective of dimension and the lighter iron cores but with rather high entropies, which are yet to be produced by realistic stellar evolution calculations, may produce the more energetic explosions and the larger amounts of nickel masses. Our simulation confirmed necessity of early time explosion to reproduce 1051 erg.The authors further studied the parametric pre-collapse structures by altering the entropy distributions in core so as to see the difference in the time evolution of mass accretion rates. Furthermore, the models are also extended to shellular rotation stars by employing an approach in Fujisawa *et al.* (2015). These additional effects will be demonstrated and discussed in the conference.

Author index